PRIMATE ORIGINS: ADAPTATIONS AND EVOLUTION

DEVELOPMENTS IN PRIMATOLOGY: PROGRESS AND PROSPECTS

Series Editor: Russell H. Tuttle, University of Chicago, Chicago, Illinois

---

This peer-reviewed book series melds the facts of organic diversity with the continuity of the evolutionary process. The volumes in this series exemplify the diversity of theoretical perspectives and methodological approaches currently employed by primatologists and physical anthropologists. Specific coverage includes: primate behavior in natural habitats and captive settings; primate ecology and conservation; functional morphology and developmental biology of primates; primate systematics; genetic and phenotypic differences among living primates; and paleoprimatology.

ANTHROPOID ORIGINS: NEW VISIONS
Edited by Callum F. Ross and Richard F. Kay

MODERN MORPHOMETRICS IN PHYSICAL ANTHROPLOGY
Edited by Dennis E. Slice

BEHAVIORAL FLEXIBILITY IN PRIMATES: CAUSES AND CONSEQUENCES
By Clara B. Jones

NURSERY REARING OF NONHUMAN PRIMATES IN THE 21ST CENTURY
Edited by Gene P. Sackett, Gerald C. Ruppenthal and Kate Elias

NEW PERSPECTIVES IN THE STUDY OF MESOAMERICAN PRIMATES: DISTRIBUTION, ECOLOGY, BEHAVIOR, AND CONSERVATION
Edited by Paul Garber, Alejandro Estrada, Mary Pavelka and LeAndra Luecke

HUMAN ORIGINS AND ENVIRONMENTAL BACKGROUNDS
Edited by Hidemi Ishida, Russel H. Tuttle, Martin Pickford, Naomichi Ogihara and Masato Nakatsukasa

PRIMATE BIOGEOGRAPHY
Edited by Shawn M. Lehman and John Fleagle

REPRODUCTION AND FITNESS IN BABOONS: BEHAVIORAL, ECOLOGICAL, AND LIFE HISTORY PERSPECTIVES
Edited By Larissa Swedell and Steven R. Leigh

RINGAILED LEMUR BIOLOGY: *LEMUR CATTA* IN MADAGASCAR
Edited by Alison Jolly, Robert W. Sussman, Naoki Koyama and Hantanirina Rasamimanana

PRIMATE ORIGINS: ADAPTATIONS AND EVOLUTION
Edited by Matthew J. Ravosa and Marian Dagosto

LEMURS: ECOLOGY AND ADAPTATION
Edited by Lisa Gould and Michelle L. Sauther

Books are to be returned on or before
the last date below.

PRIMATE ORIGINS: ADAPTATIONS AND EVOLUTION

Edited by

Matthew J. Ravosa and Marian Dagosto

Department of Cell and Molecular Biology
Feinberg School of Medicine, Northwestern University
Chicago, Illinois

 Springer

Matthew J. Ravosa
Department of Cell and
Molecular Biology
Feinberg School of Medicine,
    Northwestern University
Chicago, Illinois

Marian Dagosto
Department of Cell and
Molecular Biology
Feinberg School of Medicine,
    Northwestern University
Chicago, Illinois

Library of Congress Control Number: 2005937517

ISBN 10: 0-387-30335-9
ISBN 13: 978-0-387-30335-2

Printed on acid-free paper.

9 8 7 6 5 4 3 2 1

springer.com

CONTENTS

v

PART II: ADAPTATIONS AND EVOLUTION OF THE CRANIUM

CONTRIBUTORS

Jonathan I. Bloch, Florida Museum of Natural History, University of Florida, Gainesville, FL 32611-7800

Douglas M. Boyer, Department of Anatomical Sciences, Stony Brook University, Stony Brook, NY 11794-8081

Matt Cartmill, Department of Biological Anthropology and Anatomy, Duke University Medical Center, Durham, NC 27710

Marian Dagosto, Department of Cell and Molecular Biology, Northwestern University Feinberg School of Medicine, Chicago, IL 60611-3008, and Department of Zoology/Mammals Division, Field Museum of Natural History, Chicago, IL 60605-2496

Wilfried W. deJong, Department of Biochemistry, University of Nijmegen, Netherlands

Brigitte Demes, Department of Anatomical Sciences, Health Sciences Center, Stony Brook University, Stony Brook, NY 11794-8081

Nathaniel J. Dominy, Department of Anthropology, University of California, Santa Cruz, CA 95064

Christophe J. Douady, Department of Biology, University of California, Riverside, CA 92521, and Department of Biochemistry and Molecular Biology, Dalhousie University, Halifax B3H 4H7, Canada

Eduardo Eizirik, Laboratory of Genomic Diversity, National Cancer Institute, Frederick, MD 21702

Marc Godinot, Ecole Pratique des Hautes Etudes, and Paléontologie, Muséum National d'Histoire Naturelle, Paris, France 61801

Margaret I. Hall, Department of Anatomical Sciences, Stony Brook University, Stony Brook, NY 11794-8081

Mark W. Hamrick, Department of Cellular Biology and Anatomy, Medical College of Georgia, Augusta, GA 30912

Christopher P. Heesy, Department of Anatomy, New York College of Osteopathic Medicine, Old Westbury, NY 11568

William L. Hylander, Department of Biological Anthropology and Anatomy, Duke University Medical Center, and Duke Primate Center, Durham, NC 27710

Kirk R. Johnson, Department of Biological Anthropology and Anatomy, Duke University Medical Center, Durham, NC 27710

Susan G. Larson, Department of Anatomical Sciences, Health Sciences Center, Stony Brook University, Stony Brook, NY 11794-8081

Pierre Lemelin, Division of Anatomy, Faculty of Medicine and Dentistry, University of Alberta, Edmonton T6G 2H7, Canada

William R. Leonard, Department of Anthropology, Northwestern University, Evanston, IL 60208

Wen-Hsiung Li, Department of Ecology and Evolution, University of Chicago, Chicago, IL 60637

Peter W. Lucas, Department of Anthropology, George Washington University, Washington DC 20037

Ole Madsen, Department of Biochemistry, University of Nijmegen, Netherlands

Charles R. Marshall, Department of Earth and Planetary Sciences, Harvard University, Cambridge, MA 02138

Robert D. Martin, Department of Anthropology and Office of Academic Affairs, Field Museum of Natural History, Chicago, IL 60605-2496

Alexandra E. Mueller, Anthropologisches Institüt und Museum, University of Zurich, 8057 Zurich, Switzerland

William J. Murphy, Laboratory of Genomic Diversity, National Cancer Institute, Frederick, MD 21702

Stephen J. O'Brien, Laboratory of Genomic Diversity, National Cancer Institute, Frederick, MD 21702

Daniel Osorio, School of Biological Sciences, University of Sussex, Brighton BN1 9QG, UK

Wanda Peterson-Pereira, Escuela de Biologia, Universidad de Costa Rica, San Pedro, San José, Costa Rica

Todd M. Preuss, Yerkes Primate Research Center, Emory University, Atlanta, GA 30329

D. Tab Rasmussen, Department of Anthropology, Washington University, Saint Louis, MO 63130

Matthew J. Ravosa, Department of Cell and Molecular Biology, Northwestern University Feinberg School of Medicine, Chicago, IL 60611-3008, and Department of Zoology/Mammals Division, Field Museum of Natural History, Chicago, IL 60605-2496

Pablo Riba-Hernandez, Escuela de Biologia, Universidad de Costa Rica, San Pedro, San José, Costa Rica

Marcia L. Robertson, Department of Anthropology, Northwestern University, Evanston, IL 60208

Callum F. Ross, Department of Organismal Biology and Anatomy, University of Chicago, Chicago, IL 60637

Eric J. Sargis, Department of Anthropology, Yale University, New Haven, CT 06520

Denitsa G. Savakova, Department of Cell and Molecular Biology, Northwestern University Feinberg School of Medicine, Chicago, IL 60611-3008

Mark Scally, Department of Biology, University of California, Riverside, CA 92521, and Queen's University of Belfast, Department of Biology and Biochemistry, Belfast, UK

Daniel Schmitt, Department of Biological Anthropology and Anatomy, Duke University Medical Center, Durham, NC 27710

Brian T. Shea, Department of Cell and Molecular Biology, Northwestern University Feinberg School of Medicine, Chicago, IL 60611-3008

Mary T. Silcox, Department of Anthropology, University of Winnipeg, Winnipeg, MB R3B 2E9, Canada

J. Josh Snodgrass, Department of Anthropology, University of Oregon, Eugene, OR 97403

Christophe Soligo, Human Origins Programme, Department of Palaeontology, Natural History Museum, London SW7 5BD, UK

Silvia Solis-Madrigal, Escuela de Biologia, Universidad de Costa Rica, San Pedro, San José, Costa Rica

Mark S. Springer, Department of Biology, University of California, Riverside, CA 92521

Michael J. Stanhope, Queen's University of Belfast, Department of Biology and Biochemistry, Belfast, UK, and Bioinformatics, GlaxoSmithKline, Collegeville, PA 19426

Kathryn E. Stoner, Centro de Investigaciones en Ecosistemas, Universidad Nacional Autónoma de México, Morelia, Michoacán 58089, México

Robert W. Sussman, Department of Anthropology, Washington University, Saint Louis, MO 63130

Frederick S. Szalay, Departments of Anthropology, and Ecology and Evolutionary Biology, City University of New York, New York, NY 10021

Simon Tavaré, Program in Molecular and Computational Biology, University of Southern California, Los Angeles, CA 90089-1340

Emma C. Teeling, Department of Biology, University of California, Riverside, CA 92521, and Laboratory of Genomic Diversity, National Cancer Institute, Frederick, MD 21702

Urs Thalmann, Anthropologisches Institüt und Museum, University of Zurich, 8057 Zurich, Switzerland

Christopher J. Vinyard, Department of Anatomy, Northeastern Ohio Universities College of Medicine, Rootstown, OH 44272

Christine E. Wall, Department of Biological Anthropology and Anatomy, Duke University Medical Center, Durham, NC 27710

Oliver A. Will, Department of Statistics, University of Washington, Seattle, WA 98195-4322

Susan H. Williams, Department of Biomedical Sciences, Ohio University, Athens, OH 45701

Nayuta Yamashita, Department of Cell and Neurobiology, Keck School of Medicine, University of Southern California, Los Angeles, CA 90089-9112

Soojin Yi, Department of Ecology and Evolution, University of Chicago, Chicago, IL 60637

PREFACE

Primates of modern aspect are characterized by several traits of the skull and postcranium, most notably increased encephalization, olfactory reduction, postorbital bars, larger and more convergent orbits, an opposable hallux, and nails instead of claws on the digits. When, where, how, and why a group of mammals with this distinctive morphology emerged continues to capture the interest of biologists. The past 15 years have witnessed the discovery of numerous well-preserved basal forms and sister taxa from the Paleocene and Eocene of Asia, Africa, North America, and Europe. These new findings are particularly fascinating because they extend the antiquity of several higher-level clades, greatly increase our understanding of the taxonomic diversity of the first Primates, and document a far greater spectrum of variation in skeletal form and body size than noted previously. Not surprisingly, the past decade also has witnessed molecular and paleontological attempts to resolve primate supraordinal relationships. Many of our current notions about the *adaptations* of the first primates, however, are based on research performed 20–35 years ago—a period when the fossil record was much less complete. For instance, there remains considerable debate over the leaping versus quadrupedal component of early primate locomotion, as well as differing views regarding the function of certain mandibular and circumorbital features in basal primates. Indeed, the absence of a forum for the integration of past, recent, *and* ongoing research on the origin of primates has greatly hindered a better understanding of the significance of marked anatomical and behavioral transformations during this important and interesting stage of primate evolution. Accordingly, our December 2001 conference and this accompanying edited volume on Primate Origins and Adaptations capitalize on an increasing amount of independent museum, field and laboratory-based research on many important outstanding problems surrounding the adaptive synapomorphies of the earliest primates. Moreover, it couples the emerging views of junior researchers with those who have made significant contributions to the study of early primate phylogeny over the past three decades.

Due to the evident need for a reassessment of primate origins and adaptations, there were two principal goals of our conference and volume. First, we aim to provide a broad focus on adaptive explanations for locomotor and postural patterns, craniofacial form, neuro-visual specializations, life history patterns, socioecology, metabolism, and biogeography in basal primates. Second, to offer an explicit evolutionary context for the analysis of major adaptive transformations, we aim to provide a detailed morphological and molecular review of the phylogenetic affinities of basal primates relative to later primate clades, as well as other mammalian orders. As Plesiadapiformes have figured so heavily in discussions of primate origins, this, and the focus of our volume on adaptive scenarios, helps to explain the overt emphasis on the evolution of anatomical features. Therefore, in addition to strictly systematic or paleontological questions regarding primate origins, we concentrate primarily on the adaptive importance of unique primate characters via a comprehensive consideration of anatomical, behavioral, experimental, and ecological investigations of primate and nonprimate mammals. In this regard, a phylogenetic framework is critical for detailing the functional and evolutionary significance of specific character states and morphological complexes. Given ongoing debate regarding the appropriate content of the taxon Primates, we have decided to let authors use the terms Primates and Euprimates as they see fit. The meaning is usually obvious from the context. Likewise, for the tooth-combed lemurs, we have let authors use the spelling Strepsirrhini or Strepsirhini as they choose.

Since an increasingly evident fact about the earliest primates is their very diminutive body size, another important related goal is to characterize those adaptive trends, morphological features, and behaviors which vary and covary allometrically. Obviously, this has figured heavily in certain explanations for the evolution of grasping appendages in small-bodied basal primates. In addition, the negative allometry of neural and orbital size, coupled with relatively larger convergent orbits, has important structural consequences for explaining increased orbital frontation and the correlated evolution of a postorbital bar at small skull sizes. Perhaps the most important contribution of our volume to bioanthropology and paleontology is that it develops a forum for evaluating past *and* current research on primate origins. In doing so, we directly address a series of competing long-standing scenarios regarding the adaptive significance of important primate synapomorphies. By examining hypotheses that have dominated our notions regarding early primate evolution and coupling this with an emergent body of novel evidence due to fossil discoveries, as well as technological and methodological advances, our edited volume will provide a long overdue multidisciplinary reanalysis of a suite of derived life history, socioecological, neural, visual, circumorbital, locomotor, postural, and masticatory specializations of the first primates. This integrative neontological and paleontological perspective is critical for understanding major behavioral and morphological transformations during the later evolution of higher primate clades.

This volume collects a wide-ranging series of contributions by experts actively performing novel research relevant to the adaptive synapomorphies of the Order Primates. The authors and original conference participants are identical due to the enthusiastic response of each. For this reason, we gather together virtually every researcher, or one of their former graduate students, currently performing important research relevant to primate origins and adaptations. The series of chapters are divided into the following sections: The Supraordinal Relationships of Primates and Their Time of Origin; Adaptations and Evolution of the Cranium; Adaptations and Evolution of the Postcranium; Adaptations and Evolution of the Brain, Behavior, Physiology, and Ecology. The contents of each chapter are briefly as follows:

Springer et al. address the molecular data regarding primate supra- and infraordinal affinities. Soligo et al. reassess the antiquity and biogeography of primates and related mammals. Sargis considers the implications of tree shrew postcranial morphology for understanding early primate phylogeny. Godinot similarly stresses the importance of tree shrews for understanding primate origins. Silcox reexamines the fossil evidence regarding primate–plesiadapiform affinities. Ross et al. examine the evidence for activity patterns of early Primates. Ravosa et al. and Heesy et al. discuss comparative and experimental data regarding circumorbital form and function in primates and other vertebrates. Vinyard et al. integrate novel *in vivo* and morphological evidence regarding masticatory form and function in archontans and primates. Lemelin and Schmitt provide novel information about cheiridial morphology and performance in a series of primate and nonprimate mammals. Hamrick discusses the basis of evolvability of the mammalian autopod with special reference to the evolution of digital proportions in primates. Cartmill et al. consider the novelty and significance of primate diagonal gaits among mammals. Larson examines kinematic and skeletal evidence regarding forelimb adaptations unique to primates. Bloch and Boyer discuss the important implications of previously unknown North American plesiadapiform postcrania for understanding the evolution of basal primate locomotor adaptations. Szalay reviews the philosophy of model construction in primate locomotor evolution. Dagosto considers the evidence regarding locomotor adaptations of ancestral primates. Shea investigates the evolution of encephalization in archontans vis-à-vis life history, ecological, and allometric factors. Preuss employs neuroanatomical data to provide insight into neural specializations of the primate visual system. Mueller et al. employ a systematic analysis of extant primates to consider the evolution of basal primate social systems. Snodgrass et al. review the evolutionary and adaptive significance of variation in metabolic rate in the evolution of brain size. Yi and Li evaluate examples of protein evolution during primate evolution. Sussman and Rasmussen review the ecological underpinnings of early primate adaptations in marsupial analogs. Lucas et al. analyze the relation between dietary evolution and color vision. Apart from a consideration of new fossil discoveries and their direct relevance to outstanding issues regarding the evolution of the locomotor apparatus in early primates, these presentations represent a significant increase in the wealth of kinematic and developmental data aimed at the question of primate origins.

Numerous individuals and institutions have contributed greatly to the success of our conference and this accompanying edited volume. On the publishing end, the following at Springer/Kluwer are thanked for their support, diligence, and patience—Andrea Macaluso, Krista Zimmer, Joanne Tracey, as well as the series editor Russ Tuttle (University of Chicago). Our international conference benefited significantly from the financial support of the Wenner-Gren Foundation for Anthropological Research, Physical Anthropology Program of the National Science Foundation, Field Museum of Natural History (especially the Mammals Division), and Department of Cell and Molecular Biology at Northwestern University Feinberg School of Medicine. The following individuals are singled out for providing unique assistance with the organization and implementation of our conference—Bill Stanley, Bruce Patterson, Larry Heaney, Bob Martin, Bob Goldman, and Gail Rosenbloom. The following graduate students offered technical and logistical help that ensured the symposium went off without a hitch—Aaron Hogue, Kristin Wright, Barth Wright, and Kellie Heckman. Lastly, and most importantly, we thank our spouses—Sharon Stack and Dan Gebo—for their continued support and our respective children—Nico and Luca, and Anne Marie—for inspiration.

Matthew J. Ravosa

Marian Dagosto

INTRODUCTIONS FOR SECTIONS I–IV

Section I: Supraordinal Relationships of Primates and Their Time of Origin

Despite new fossil discoveries, new sources of data, and new methods of analysis, several important issues concerning the origin and phylogeny of Primates remain unresolved. One currently controversial issue is the time of origin of the Order Primates. Both the analysis of molecular data (**Springer et al.**) and mathematical modeling (**Soligo et al.**) suggest a time of origin in the middle of the Cretaceous period (80–90 MYA), while the earliest fossil record of primates is only 55 MYA. The fossil record can only provide a minimum age for the origin of any taxon, while these other approaches may be measuring the initial divergence between a taxon and its sister group—an event that may be not marked by any morphological differentiation. **Soligo et al.** discount this latter possibility, since the molecular estimate for the Strepsirhine–Haplorhine split is 80 MYA. They calculate, therefore, that there is a 25-MY gap between the origin of identifiable primates and their first appearance in the fossil record.

The supraordinal relationships among mammals have been an area of intense interest among paleontologists, and primates are no exception. Although primatologists have reached some consensus about the content of the Order, there is still little agreement as to which living or fossil group is the sister taxon of Primates. The molecular analysis of nuclear and mitochondrial genes by **Springer et al.** provides support for the clade Euarchontoglires, consisting of Primates, Dermopterans, Scandentia, Rodentia, and Lagomorpha. Within this clade, Primates are most closely related to Dermoptera and Scandentia (=clade Euarchonta). Neither tree shrews nor flying lemurs are the exclusive sister group of Primates, but form a clade with each other. Unfortunately, this analysis does not include *Ptilocercus*, a tree shrew that may be the most primitive of its clade and thus may have particular relevance to primate origins (**Sargis, Godinot**). Nor can the relationships of fossil taxa be addressed. The morphological analysis of the postcranium by **Sargis,** which does include *Ptilocercus,* finds, like the molecular analyses, that Scandentia and Dermoptera form a group (but only if Chiroptera is excluded). On the other hand, **Godinot's** analysis, which includes cranial, dental, and postcranial characters, makes a strong case for a special relationship between tree shrews, particularly *Ptilocercus,* and primates. Plesiadapiformes, the Paleogene fossil group that has been traditionally most closely linked to Primates, were too incomplete to be analyzed effectively in these analyses. **Silcox,** by using the more ubiquitous dental characters (as well as cranial and postcranial features) supports a sister-group relationship between Plesiadapiformes and Primates to the exclusion of tree shrews or flying lemurs. Therefore she, like Bloch and Boyer (Section III), supports the assignment of this fossil group within the Order Primates following the conventional paleontological interpretation.

Section II: Adaptations and Evolution of the Cranium

Vinyard et al. inquire if primates have unique aspects of jaw mechanics or morphology that might indicate a role for dietary change in primate origins. Although certain jaw-adductor muscles (e.g., temporalis) show similar patterns of firing during chewing, others (e.g., deep masseter) are quite variable, thus primates are not homogeneous in jaw-muscle activity patterns. There are few differences between tree shrews and strepsirhine primates in jaw morphology or the timing and relative activity levels of the jaw adductors, suggesting that the Origin of Primates may not have been accompanied by any major dietary shift.

Compared to any likely sister group, primates evidence a reorganization of the skull characterized by relatively large convergent orbits and a postorbital bar. The adaptational significance of these features is still debated today. One question concerns the activity pattern of ancestral primate. Based on an analysis of eye and orbit shape in mammals and birds **Ross et al.** are able to show that nocturnality is the best explanation for the increased orbital convergence, large eyes, and large corneas that were likely present in primitive primates, although only the first two features are primate apomorphies. These features improve image brightness and visual acuity.

The postorbital bar, one of the hallmark features of Primates, also has been hypothesized to play a role in visual acuity by functioning as a barrier between the eyeball and the masticatory muscles to prevent distortion of the visual image during chewing. **Heesy et al.**, however, provide experimental evidence that the anterior temporalis and medial pterygoid can cause deformation of the eye even in animals with postorbital bars (*Otolemur, Felis*), and thus speculate that other compensatory mechanisms for maintaining visual acuity must be present. The action of the extraocular muscles is one such mechanism, and a postorbital bar would help maintain the integrity of the lateral orbit, giving a stable substrate from which these muscles could act.

Ravosa et al. marshal comparative and experimental comparative evidence to support a modified version of the NVP's "rigidity" argument where increased orbital convergence *and* orbital frontation (due to increased encephalization) both play a role in postorbital bar formation. They stress the independent and interactive roles of asymmetrical jaw-adductor recruitment patterns (characteristic of insectivores and frugivores), nocturnality, encephalization and small body size on the evolution and function of the circumorbital region and skull in basal primates.

Section III: Adaptations and Evolution of the Postcranium

Definitions of the Order Primates always have made reference to traits of the postcranial skeleton, most notably the opposable hallux and the presence of nails instead of claws on the digits, and it always has been the received wisdom that something about an arboreal lifestyle has influenced the diagnostic limb features of primates. Several workers, especially Matt Cartmill, refined these early, vague ideas. As part of the Nocturnal Visual Predation (NVP) model he proposed a more specific relationship between hindlimb opposition and one aspect of primate-style arborealism, namely the need to balance and move slowly on small supports when stalking and capturing prey. By comparing primates with marsupials of similar habitus, **Lemelin and Schmitt** are able to demonstrate that additional prehensility enhancing features (phalangeal proportions, metapoodial/phalangeal proportions) are correlated with superior ability to deal with a fine-branch substrate. **Hamrick** discusses the experimental and morphological evidence for the evolvability of the distal limb, explaining why the origin and adaptive radiation of primates is accompanied by high diversity in digit proportions.

There are also many behavioral aspects of primate locomotor behavior that distinguish them from most other arboreal mammals. The use of the diagonal sequence/diagonal couplets gait, emphasis on the hindlimb for support and propulsion, low stride frequencies but longer stride lengths, large angular excursions of the limbs, and the compliant gait are among them. **Cartmill et al.** and **Lemelin and Schmitt** address these behavioral differences concluding that the diagonal sequence gait is a solution to moving on small branches with a prehensile extremity. They stress that this would only be a successful strategy for an animal with relatively deliberate locomotor habits. **Lemelin and Schmitt** also show that *Caluromys* (a marsupial small-branch specialist) differs from terrestrial marsupials in sharing many of these behavioral traits, strongly suggesting that all of them are related to moving quadrupedally on small branches. **Larson** examines the morphological correlates of the highly protracted forelimb that results in the large angular excursion of the forelimb during quadrupedal walking. These are: a more obtuse spinoglenoid angle, a reduction in the anterior and superior projection of the greater tubercle possibly produced by an anterior shift of the humeral head. *Smilodectes*, the only early prosimian included in the study, appears more primitive in the humeral features than any extant primate.

In contrast to the slow moving ancestor envisioned by Cartmill and colleagues, **Szalay** and **Dagosto**, in their grasp-leaping model, propose a much more agile creature. In their view, leaping is a component of locomotor behavior equal in importance to grasping in defining the postcranial morphotype of Primates. **Szalay** offers a sharply critical account of previous reconstructions of the locomotor abilities of early primates and the philosophies underlying the logic of these reconstructions. **Dagosto** echoes these ideas, citing a number of derived leaping related features of the limb skeleton shared by all primates, including the presumably paraphyletic adapids and omomyids. She also points out the difficulty in attempting to explain all of the derived characters of a higher-level clade with a single adaptive hypothesis. A staged model for the acquisition of key locomotor related adaptations is proposed.

Bloch and Boyer describe the variation in postcranial bones and inferred locomotor adaptations present among plesiadapiform primates, identifying a variety of arboreal behavioral adaptations in this group. They find no evidence of gliding or phylogenetic links to Dermoptera among micromomyids, refuting Beard's Eudermoptera hypothesis. Believing the similarities of the hallucal-grasping complex in *Carpolestes* to be a synapomorphy with true primates, they posit that grasping was in place before anatomical adaptations for visual predation or leaping, thus contesting both the NVP and grasp-leaping hypotheses.

Section IV: Adaptations and Evolution
of the Brain, Behavior, Physiology, and Ecology

As evidenced by the first three sections, features of the skull and skeleton have figured prominently in evaluations of primate origins. But primates differ from other mammals in many other features including size and structure of the brain, social organization, life history, physiology, and biochemistry. The papers in this section discuss some of these attributes and the relationships among them.

Primates are among the "brainiest" mammals. **Shea** argues that the unusual combination of a precocial life history strategy at small size by early primates, possibly allowed by a stable resource base, set the stage for a grade-shift in encephalization that was preserved as they diversified into larger sizes. In addition to greater size, the structure and organization of the brain distinguish primates from other mammals. In an extensive review, **Preuss** demonstrates how primates have developed new cortical areas, reorganized existing structures, changed the way existing structures connect, and established new kinds of connections. Many of these changes reflect the integration of information from the eyes and forelimb, contributing to a distinct kind of "looking and reaching" in primates, which fits nicely with models of primate origins that stress the role of visual foraging.

Mueller et al. review aspects of the social organization of primates and mammals in order to reconstruct the ancestral pattern of primate social organization. They argue that a "dispersed" system (solitary foraging with social networks formed by a core of related females) was present in early primates. The presence of social networks and contacts that are maintained throughout the year, rather than being restricted to the breeding season, are derived features compared to primitive mammals. Factors that may explain the development of sociality in primates are frugivory, prolonged mother-infant relationships, and large body size.

Snodgrass et al. show that ecological factors (diet quality, habitus, activity pattern) are only partially successful in explaining the difference in basal metabolic rates between lower and higher primates. The shared hypometabolism of strepsirhines, tarsiers, and tree shrews indicates that the ancestral primate inherited this physiology, small body size, and dependence on insects as a food source from an archontan ancestor. Hypometabolism is possibly an adaptation to environments with low productivity and/or marked seasonality.

Thus, the **Shea**, **Mueller et al.**, and **Snodgrass et al.** models sometimes contrast with each other in their reconstruction of ancestral diet (insectivory versus frugivory), environment (stable versus unstable resource base), and body size (small versus large). In addition to metabolic rates, other aspects of physiology and biochemistry distinguish lower and higher primates. **Yi and Li** discuss the evidence for adaptive evolution of the physiologically important proteins growth hormone (GH), growth hormone receptor (GHR), and chorionic gonadotropin (CR). Each shows evidence of rapid change sometimes associated with gene duplication and changes in site of expression.

We close the volume with two papers that address ecological aspects of primate evolution and tie in themes from several of the previous sections. Most contributors use the comparative method to elucidate the adaptational significance of primate apomorphies. As shown by the extensive review of phalangeroid marsupial biology and ecology provided by **Rasmussen and Sussman**, this radiation provides many interesting opportunities for understanding the origin of primate diet, locomotion, foraging strategies, orbital convergence, life history, and physiology. While noting how much still needs to be learned about these mammals, they use what is known to critique current hypotheses of primate origins.

Lucas et al. manage to tie in almost every primate apomorphy in a model of foraging evolution based, like Sussman's, on the coevolution of primates and angiosperms. In this model, early Primates were small, nocturnal, and dichromatic, living in angiosperms, but subsisting primarily on insects. As angiosperms developed spines and thorns to protect themselves against dinosaurs, primates developed nails and pads to protect themselves against spines and thorns. The trichromatic color vision typical of catarrhines, and independently developed in some platyrrhines and lemurs, thus is a more recent evolutionary development.

CHAPTER ONE

A Molecular Classification for the Living Orders of Placental Mammals and the Phylogenetic Placement of Primates

Mark S. Springer, William J. Murphy,
Eduardo Eizirik, Ole Madsen, Mark Scally,
Christophe J. Douady, Emma C. Teeling,
Michael J. Stanhope, Wilfried W. de Jong,
and Stephen J. O'Brien

INTRODUCTION

For more than a century, systematists have debated higher-level relationships among the orders of placental mammals. The order Primates is no

Mark S. Springer • Department of Biology, University of California, Riverside, CA 92521 William J. Murphy, Eduardo Eizirik, and Stephen J. O'Brien • Laboratory of Genomic Diversity, National Cancer Institute, Frederick, MD 21702 Ole Madsen and Wilfried W. de Jong • Department of Biochemistry, University of Nijmegen, Netherlands Mark Scally • Department of Biology, University of California, Riverside, CA 92521; Queen's University of Belfast, Biology and Biochemistry, Belfast, UK Christophe J. Douady • Department of Biology, University of California, Riverside, CA 92521; Department of Biochemistry and Molecular Biology, Dalhousie University, Halifax, Nova Scotia, B3H 4H7 Canada Emma C. Teeling • Department of Biology, University of California, Riverside, CA 92521; Laboratory of Genomic Diversity, National Cancer Institute, Frederick, MD 21702 Michael J. Stanhope • Department of Population Medicine and Diagnostic Sciences, College of Veterinary Medicine, Cornell University, Ithaca, NY 14853, USA.

exception. One prominent hypothesis is Archonta that was originally proposed as a superorder by Gregory (1910) to include primates, bats, flying lemurs, and menotyphlan insectivores (i.e., tree shrews, elephant shrews). Minus elephant shrews, the Archonta hypothesis has survived for nearly a century. The bulk of support for this hypothesis derives from modifications of the tarsus (Novacek and Wyss, 1986; Shoshani and McKenna, 1998; Szalay, 1977; Szalay and Drawhorn, 1980; Szalay and Lucas, 1993). Archonta is a recurrent theme in higher-level mammalian classifications (McKenna, 1975; McKenna and Bell, 1997; Szalay, 1977). There are also morphological studies (Cartmill and MacPhee, 1980; Luckett, 1980; Novacek, 1980; Simpson, 1945) that question the monophyly of Archonta. Simpson (1945) suggested that Archonta is "almost surely an unnatural group." Even among studies that advocate Archonta, the possibility that bats are an independently arboreal group from the Archonta has been noted (Szalay and Drawhorn, 1980). In part, this reservation was expressed because bats lack shared, derived tarsal specializations that unite other archontans (Szalay, 1977; Szalay and Drawhorn, 1980). Szalay and Drawhorn (1980) attribute this to the major functional transformation that the chiropteran ankle has undergone in association with the "extreme reorientation of the femoral–acetabular articulation." Given the absence of tarsal modifications that unite bats with other archontans, the primary rationale for including bats in Archonta is the suite of novel features that bats share with flying lemurs (Gregory, 1910; Simmons, 1995; Simmons and Quinn, 1994; Szalay and Drawhorn, 1980).

Aside from whether or not primates belong to a monophyletic Archonta, there are questions pertaining to the sister-group of primates. Several studies resolve archontans into a trichotomy between primates, tree shrews, and Volitantia (i.e., flying lemurs + bats) (Novacek, 1990; Novacek et al., 1988; Novacek and Wyss, 1986; Szalay, 1977). Other studies, some of which support Archonta and others of which do not, support a sister-group relationship between primates (or euprimates) and tree shrews (Martin, 1990; Shoshani and McKenna, 1998; Simpson, 1945; Wible and Covert, 1987; Wible and Novacek, 1988). Beard (1993) has argued for the Primatomorpha hypothesis that postulates a sister-group relationship between flying lemurs and primates. Another alternative is a sister-group relationship between tree shrews and flying lemurs, with this collective group as the sister-group to primates (Sargis, 2001).

Over the last three decades, molecular data have become increasingly important for testing and proposing hypotheses of interordinal relationships. In the mid-1970s, Goodman (1975) summarized immunological and amino acid data bearing on higher-level primate affinities. The latter included parsimony analyses of amino acid sequences for proteins such as myoglobin, and α- and β-hemoglobins. Based on a consideration of the available molecular evidence, which was not entirely congruent, Goodman (1975) concluded that "the tentative solution adopted from immunodiffusion evidence of grouping Primates and Tupaioidea (also Dermoptera) in the superorder Archonta would seem a valid compromise." Almost two decades later, Stanhope et al. (1993) evaluated higher-level affinities of primates based on nucleotide and amino acid sequences. Addressing the Archonta hypothesis, Stanhope et al. (1993) concluded that there is "marked divergence of Chiroptera from Primates, Scandentia, and Dermoptera" based on analyses of IRBP and ε-globin data sets. Instead, Stanhope et al. (1993) suggested that "a more likely primate supraordinal clade consists of Primates, Dermoptera, Lagomorpha, Rodentia, and Scandentia." Adkins and Honeycutt (1993) also concluded that Archonta is not monophyletic based on mitochondrial DNA sequences.

Additional support for the "supraordinal clade" suggested by Stanhope et al. (1993) comes from two analyses of molecular supermatrices (Madsen et al., 2001; Murphy et al., 2001a). In both studies, maximum likelihood analyses resolved placental mammals into the same four major groups: Xenarthra, Afrotheria, Laurasiatheria, and Euarchonta + Glires (= Euarchontoglires of Murphy et al., 2001b,c). The latter group includes Primates, Scandentia, Dermoptera, Lagomorpha, and Rodentia. However, Madsen et al. (2001) and Murphy et al. (2001a) did not provide convincing support for the placement of primates, tree shrews, and flying lemurs relative to each other and to Glires (Lagomorpha, Rodentia). To achieve additional resolution among the orders of placental mammals, including the phylogenetic position of primates, we combined and expanded the molecular data sets of Madsen et al. (2001) and Murphy et al. (2001a). The resulting supermatrix is 16.4 kb in length (comprising 10,059 variable and 7,785 informative characters) and includes segments of 19 nuclear genes and three mitochondrial genes for 44 taxa. Primary analyses of this data set are provided in Murphy et al. (2001b). Here, we provide an expanded set of analyses and suggest a higher-level classification for the living orders of placental mammals based on our molecular results.

MATERIALS AND METHODS

Murphy et al. (2001b) concatenated and expanded the data sets of Madsen et al. (2001) and Murphy et al. (2001a) to generate a data set that included 19 nuclear segments and three mitochondrial genes (12S rRNA, tRNA valine, 16S rRNA) for 42 placental taxa and two marsupial outgroups. Some taxa were chimeric, being composed of sequences from species belonging to the same well-supported (noncontroversial) monophyletic group (see Murphy et al., 2001b). After excluding regions of the data set that were judged alignment-ambiguous, the data set was 16,397 bp nucleotides in length. Of these, 14,750 nucleotides were from nuclear genes and 1,647 nucleotides were from mitochondrial genes.

Data were analyzed using likelihood-based analyses, including Bayesian phylogenetic analyses (Huelsenbeck et al., 2001; Huelsenbeck and Ronquist, 2001; Larget and Simon, 1999; Mau et al., 1999). Likelihood methods are statistically consistent given a correct model of sequence evolution and have the potential to resolve complex phylogenetic problems (Whelan et al., 2001). In both maximum likelihood and Bayesian analyses, we used the general time reversible (GTR) model of sequence evolution with a gamma (Γ) distribution of rates and an allowance for a proportion of invariant sites (I) based on the results of Modeltest (Posada and Crandall, 1998). Additional details on model parameters are given in Murphy et al. (2001b). PAUP 4.0 (Swofford, 1998) was used to perform maximum likelihood (ML) analyses, including nonparametric bootstrapping. However, it was necessary to employ phylogenetic constraints (see asterisks in Figure 1) and limit searching to nearest neighbor interchanges in ML bootstrap analyses because of computational demands. Whereas ML analyses search for tree(s) having the highest likelihood score, Bayesian methods sample trees according to their posterior probability properties (Huelsenbeck et al., 2001). An advantage of the Bayesian approach is that complex models of sequence evolution, including GTR + Γ + I, can be employed with large data sets and without the need for phylogenetic constraints.

Even though the Bayesian approach is feasible for large data sets, analytical calculation of Bayesian posterior probabilities requires summation over all topologies and integration over all possible combinations of branch length and substitution model parameter values (Huelsenbeck et al., 2001). These calculations become analytically intractable for even small phylogeny problems (Huelsenbeck et al., 2001), and posterior probabilities must be estimated using other methods. One method is the Markov chain Monte Carlo (MCMC) approach with Metropolis–Hastings sampling (Huelsenbeck et al., 2001; Huelsenbeck and Ronquist, 2001). New states for the Markov chain are proposed using a stochastic mechanism, acceptance probabilities for the new state are calculated, and the new state is accepted if the acceptance probability is higher than a uniform random variable between 0 and 1. Using this approach, a large set of trees can be evaluated from the universe of potential phylogenetic trees (Huelsenbeck et al., 2001; Huelsenbeck and Ronquist, 2001). This provides a powerful alternative to searching for a single maximum likelihood tree and evaluating the reliability of this tree using the nonparametric bootstrap. We used MrBayes 2.01 (Huelsenbeck and Ronquist, 2001), which performs Bayesian analyses using Metropolis-coupled Markov chain Monte Carlo (MCMCMC) sampling, to approximate posterior probabilities distributions for the topology and parameters of the model of sequence evolution. MCMCMC runs n chains simultaneously and allows for state swaps between chains. Relative to approaches that employ a single Markov chain, MCMCMC is less susceptible to local entrapment and is more efficient at crossing deep valleys in a landscape of phylogenetic trees (Huelsenbeck and Ronquist, 2001).

Bayesian analyses employed four independent chains (three heated, one cold; see Huelsenbeck and Ronquist, 2001), all starting from random trees, and were run for 300,000 or 600,000 generations. Chains were sampled every 20 generations and burnin values were set at 75,000 generations based on empirical evaluation. Additional details are given in Murphy et al. (2001b). Bayesian analyses were also performed with single-taxon outgroup jackknifing and with subsets of nucleotide sequences. The latter included nuclear genes only and mt rRNA genes only. We also partitioned the nuclear data set in two different ways. First, protein-coding genes (12,988 bp) versus UTRs (untranslated regions) (1762 bp). Second, 1st+2nd codon positions (8658 bp) versus 3rd codon positions + UTRs (6092 bp). All supplementary analyses were run for 300,000 generations with burnin set at 60,000 or 75,000 generations based on empirical evaluation.

RESULTS

Likelihood and Bayesian Analyses with the Full Data Set

Figure 1 shows a maximum likelihood cladogram (-ln likelihood = 211,110.54) for the 16.4 kb data set under the GTR + Γ + I model of sequence evolution.

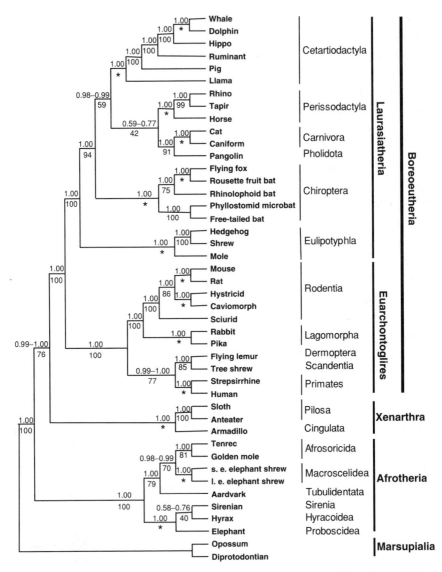

Figure 1. Maximum likelihood cladogram, with bootstrap values shown below branches, for the 16.4 kb data set. Asterisks indicate clades that were constrained in maximum likelihood analyses. The range of Bayesian posterior probabilities from four independent MCMCMC runs with the 16.4 kb data set is given above each branch (also see Tables 1–4). In cases where only a single value is shown, this value was obtained in all four runs. Abbreviations: s.e.=short-eared; i. e=long-eared.

Bootstrap support percentages are shown below each branch. Four independent Bayesian (MCMCMC) analyses, two that were run for 600,000 generations and two for 300,000 generations, resulted in the same topology. The range of posterior probabilities (all four runs) is shown above each branch, with posterior probabilities expressed as percentages. Tables 1–4 show posterior probabilities for individual MCMCMC runs. Although the maximum likelihood tree and Bayesian trees (from independent runs) were topologically identical, bootstrap proportions were generally lower than posterior probabilities.

Consistent with the maximum likelihood analyses presented in Madsen et al. (2001) and Murphy et al. (2001a), our analyses recovered four major clades of placental mammals: Xenarthra (Cingulata, Pilosa); Afrotheria (Afrosoricida, Macroscelidea, Tubulidentata, Hyracoidea, Proboscidea, Sirenia); Euarchontoglires (Rodentia, Lagomorpha, Scandentia, Dermoptera, Primates); and Laurasiatheria (Eulipotyphla, Chiroptera, Carnivora, Pholidota, Perissodactyla, Cetartiodactyla). There is also moderate to strong support for relationships between these groups. Euarchontoglires and Laurasiatheria are sister-taxa that together constitute a clade named Boreoeutheria (Springer and de Jong, 2001). The basal split among living placental mammals is between Afrotheria and Boreoeutheria + Xenarthra (herein named Notolegia, see below for definition). Bootstrap support for Notolegia was 76% and posterior probabilities were 0.99–1.00 in independent MCMCMC analyses with the complete data set. Two competing hypotheses for the position of the root that had the next highest likelihood scores are: (i) at the base of Xenarthra (i.e., a basal split between Xenarthra and Epitheria; -ln likelihood = 211,119.66) and (ii) between Atlantogenata (i.e., Afrotheria + Xenarthra) and Boreoeutheria (-ln likelihood = 211,115.95). SOWH (Swofford-Olsen-Waddell-Hillis) tests (Swofford et al., 1996; Goldman et al., 2000) rejected these locations for the root (Murphy et al., 2001b). There were no other positions for the root that had nonzero bootstrap percentages or nonzero probabilities in likelihood, and Bayesian analyses, respectively.

Most relationships within the major clades were also resolved. In Xenarthra, pilosans (sloth and anteater) cluster to the exclusion of the armadillo (Cingulata). In Afrotheria, there is a basal separation of paenungulates (Hyracoidea, Proboscidea, Sirenia) and a clade containing Tubulidentata, Macroscelidea, and Afrosoricida, with the latter two orders as sister-taxa. The basal split in Laurasiatheria is between Eulipotyphla and other taxa (i.e., Scrotifera of Waddell et al., 1999; hereafter called Variamana for reasons that are discussed

Mark S. Springer et al.

Table 1. Likelihood bootstrap support and Bayesian posterior probabilities for the major clades of placental mammals*

Data sets and analysis		Clade							
		Afrotheria	Xenarthra + Boreoeutheria	Epitheria	Atlantogeneta	Xenarthra	Boreoeutheria	Laurasiatheria	Euarchontoglires
1. Full data set (16,397 bp)	ML-Boot	100	76	4	18	NA	100	100	100
	600,000 – 1st	1.00	0.99	0.00	0.01	1.00	1.00	1.00	1.00
	600,000 – 2nd	1.00	1.00	0.00	0.00	1.00	1.00	1.00	1.00
	300,000 – 1st	1.00	1.00	0.00	0.00	1.00	1.00	1.00	1.00
	300,000 – 2nd	1.00	1.00	0.00	0.00	1.00	1.00	1.00	1.00
2. Outgroup jackknifing (16,397 bp)	-Opossum	1.00	0.96	0.00	0.04	1.00	1.00	1.00	1.00
	-Diprotodontian	1.00	0.99	0.00	0.01	1.00	1.00	1.00	1.00
3. Subsets of genes	Nuclear (14,750 bp)	1.00	1.00	0.00	0.00	1.00	1.00	1.00	1.00
	Protein-coding	1.00	0.98	0.00	0.02	1.00	1.00	1.00	1.00
	UTRs	1.00	0.75	0.20	0.05	1.00	1.00	1.00	1.00
	Protein-coding 1+2	1.00	0.26	0.07	0.67	1.00	1.00	1.00	1.00
	UTRs+protein-coding 3	1.00	0.98	0.02	0.00	1.00	1.00	1.00	1.00
	Mitochondrial rRNA	1.00	0.01	0.00	0.99	1.00	0.01	0.00	0.00

*With the exception of the maximum likelihood bootstrap analysis that was run for the full data set, all other analyses reported in this table are MCMCMC runs with MrBayes 2.01 (Huelsenbeck and Ronquist, 2001). All Bayesian analyses with outgroup jackknifing and subsets of genes were run for 300,000 generations, with burnins set at 60,000 or 75,000 generations based on empirical evaluation. NA, not applicable because clade was constrained in ML bootstrap analysis.

Table 2. Likelihood bootstrap support and Bayesian posterior probabilities for relationships within Afrotheria*

Data sets and analysis		Clade						
		Paen-ungulata	Tethytheria	Proboscidea + Hyracoidea	Sirenia + Hyracoidea	Tubulidentata + Macroscelidea + Afrosoricida	Macroscelidea + Afrosoricida	Afrosoricida
1. Full data set (16,397 bp)	ML-Boot	NA	38	22	40	79	70	81
	600,000 – 1st	1.00	0.02	0.22	0.76	1.00	0.99	1.00
	600,000 – 2nd	1.00	0.03	0.40	0.58	1.00	0.99	1.00
	300,000 – 1st	1.00	0.04	0.23	0.73	1.00	0.98	1.00
	300,000 – 2nd	1.00	0.01	0.27	0.72	1.00	0.99	1.00
2. Outgroup jackknifing (16,397 bp)	-Opossum	1.00	0.04	0.48	0.48	1.00	1.00	1.00
	-Diprotodontian	1.00	0.02	0.21	0.77	1.00	1.00	1.00
3. Subsets of genes	Nuclear (14,750 bp)	1.00	0.01	0.88	0.11	1.00	1.00	1.00
	Protein-coding	1.00	0.00	0.83	0.18	1.00	0.98	1.00
	UTRs	1.00	0.28	0.49	0.23	0.95	0.97	0.09
	Protein-coding 1+2	1.00	0.01	0.99	0.00	1.00	0.83	1.00
	UTRs+protein-coding 3	1.00	0.08	0.17	0.75	0.54	0.88	0.99
	Mitochondrial rRNA	1.00	0.95	0.01	0.04	0.98	0.02	1.00

*With the exception of the maximum likelihood bootstrap analysis that was run for the full data set, all other analyses reported in this table are MCMCMC runs with MrBayes 2.01 (Huelsenbeck and Ronquist, 2001). All Bayesian analyses with outgroup jackknifing and subsets of genes were run for 300,000 generations, with burnins set at 60,000 or 75,000 generations based on empirical evaluation.

Table 3. Likelihood bootstrap support and Bayesian posterior probabilities for relationships within Euarchontoglires*

Data sets and analysis		Clade					
		Glires	Rodentia	Euarchonta	Primatomorpha	Scandentia+Primates	Scandentia+Dermoptera
1. Full data set (16,397 bp)	ML-Boot	100	100	77	15	0	85
	600,000 – 1st	1.00	1.00	1.00	0.00	0.00	1.00
	600,000 – 2nd	1.00	1.00	1.00	0.00	0.00	1.00
	300,000 – 1st	1.00	1.00	1.00	0.00	0.00	1.00
	300,000 – 2nd	1.00	1.00	0.99	0.00	0.00	1.00
2. Outgroup jackknifing (16,397 bp)	-Opossum	1.00	1.00	0.99	0.00	0.00	1.00
	-Diprotodontian	1.00	1.00	0.99	0.00	0.00	1.00
3. Subsets of genes	Nuclear (14,750 bp)	1.00	1.00	0.94	0.03	0.00	0.97
	Protein-coding	1.00	1.00	0.56	0.58	0.01	0.41
	UTRs		1.00	0.02	0.25	0.01	0.03
	Protein-coding 1+2	1.00	1.00	0.61	0.77	0.04	0.13
	UTRs+protein-coding 3	1.00	1.00	0.88	0.01	0.00	0.99
	Mitochondrial rRNA	0.00	0.00	0.00	0.00	0.00	0.98

*With the exception of the maximum likelihood bootstrap analysis that was run for the full data set, all other analyses reported in this table are MCMCMC runs with MrBayes 2.01 (Huelsenbeck and Ronquist, 2001). All Bayesian analyses with outgroup jackknifing and subsets of genes were run for 300,000 generations, with burnins set at 60,000 or 75,000 generations based on empirical evaluation.

Table 4. Likelihood bootstrap support and Bayesian posterior probabilities for relationships within Laurasiatheria*

Data sets and analysis		Clade							
		Vari-amana	Chiroptera+ Eulipotyphla	Fereu-ungulata	Carnivora+ Pholidota	Hippo+ Cetacea	Eulipo-typhla	Erinaceidae+ Soricida	Yinptero-chiroptera
1. Full data set (16,397 bp)	ML-Boot	94	<5	59	91	100	100	100	75
	600,000 – 1st	1.00	0.00	0.98	1.00	1.00	1.00	1.00	1.00
	600,000 – 2nd	1.00	0.00	0.98	1.00	1.00	1.00	1.00	1.00
	300,000 – 1st	1.00	0.00	0.99	1.00	1.00	1.00	1.00	1.00
	300,000 – 2nd	1.00	0.00	0.99	1.00	1.00	1.00	1.00	1.00
2. Outgroup jackknifing (16,397 bp)	-Opossum	1.00	0.00	0.95	1.00	1.00	1.00	1.00	1.00
	-Diprotodontian	1.00	0.00	0.92	1.00	1.00	1.00	1.00	1.00
3. Subsets of genes	Nuclear (14,750 bp)	1.00	0.00	0.65	1.00	1.00	1.00	1.00	1.00
	Protein-coding	1.00	0.00	0.98	1.00	1.00	1.00	1.00	1.00
	UTRs	0.06	0.00	0.00	1.00	0.97	1.00	0.99	0.69
	Protein-coding 1+2	1.00	0.00	0.06	0.91	0.53	1.00	0.84	0.69
	UTRs+protein-coding 3	0.48	0.01	0.38	1.00	1.00	1.00	1.00	1.00
	Mitochondrial rRNA	0.95	0.00	0.00	0.02	0.11	0.03	0.03	0.92

* With the exception of the maximum likelihood bootstrap analysis that was run for the full data set, all other analyses reported in this table are MCMCMC runs with MrBayes 2.01 (Huelsenbeck and Ronquist, 2001). All Bayesian analyses with outgroup jackknifing and subsets of genes were run for 300,000 generations, with burnins set at 60,000 or 75,000 generations based on empirical evaluation.

below). Within Variamana, the next split is between Chiroptera and Fereuungulata (carnivores, pholidotans, perissodactyls, cetartiodactyls). There is strong support for a carnivore–pholidotan clade within Fereuungulata. Euarchontoglires is divided into Glires (Lagomorpha + Rodentia) and Euarchonta (Dermoptera + Primates + Scandentia), both of which receive high bootstrap support percentages and posterior probabilities. Within Euarchonta, Scandentia and Dermoptera are sister-taxa (bootstrap support = 85%; posterior probabilities = 1.00 in four independent MCMCMC runs).

Analyses with Outgroup Jackknifing

In Bayesian analyses that deleted either the opossum or the diprotodontian, the ingroup topology remained unchanged. Posterior probabilities remained ≥ 0.95 for all but two clades (Sirenia + Hyracoidea = 0.48; Carnivora + Pholidota + Perissodactyla = 0.76) with the diprotodontian outgroup. With opossum outgroup, all clades were supported with posterior probabilities > 0.95 except for Sirenia + Hyracoidea (0.77), Carnivora + Pholidota + Perissodactyla (0.70), and Fereuungulata (0.92). Within Euarchontoglires, posterior probabilities for Euarchonta were 0.99 in both analyses that deleted one of the marsupial outgroups. Similarly, tree shrew + flying lemur support remained high (posterior probability = 1.00 for both analyses).

Analyses with the Nuclear Data Set

Bayesian analyses with nuclear genes only resulted in a tree that was identical to that shown in Figure 1, except that Proboscidea and Hyracoidea are sister-taxa within Paenungulata (posterior probability = 0.88). In addition to the Hyracoidea + Proboscidea clade, other groups that were supported with posterior probabilities less than 0.95 included Fereuungulata (0.65), Carnivora + Pholidota + Perissodactyla (0.53), and Euarchonta (0.94). Support for tree shrew + flying lemur remained high (posterior probability = 0.97).

Analyses with Subsets of Nuclear Genes

Among trees based on subsets of nuclear genes, the tree based on protein-coding genes only was most similar to the tree for the complete nuclear data set. All branching relationships were identical except that Primatomorpha received higher support (posterior probability = 0.58) than Scandentia + Dermoptera (posterior probability = 0.42). Euarchonta was still supported, but the posterior probability was only 0.56.

Analyses with the remaining partitions of the nuclear data (UTRs; 1st and 2nd codon positions; 3rd codon positions + UTRs) resulted in trees with more topological differences. Posterior probabilities remained high for some clades, but were lower in other cases. Posterior probabilities were 1.00 for the four major groups (Xenarthra, Afrotheria, Laurasiatheria, Euarchontoglires) and for Boreoeutheria. At the base of the tree, Xenarthra + Boreoeutheria was strongly supported by 3rd codon positions + UTRs (posterior probability = 0.98); other partitions did not resolve the root of the placental tree with high probabilities (i.e., >0.95). Within Afrotheria, Paenungulata, Tubulidentata + Macroscelidea + Afrosoricida, Macroscelidea + Afrosoricida, and Afrosoricida were generally supported. Within Euarchontoglires, posterior probabilities for Glires ranged from 0.02 to 1.00 with different partitions. Posterior probabilities for Euarchonta were lower than for Glires and ranged from 0.02 (UTRs only) to 0.88 (UTRs + 3rd codon positions). Within Euarchonta, 3rd codon positions + UTRs supported Scandentia + Dermoptera (posterior probability = 0.99) whereas, 1st + 2nd codon positions favored Primatomorpha (posterior probability = 0.77). Within Laurasiatheria (Table 4), Variamana was supported by 1st + 2nd codon positions of nuclear genes (posterior probability = 1.00). Chiroptera+ Eulipotyphla was not supported by any of the partitions of the nuclear genes. Support for Fereuungulata was not evident in analyses with UTRs, 1st + 2nd codon positions, and UTRs + 3rd codon positions. Carnivora + Pholidota was consistently supported in analyses with different nuclear partitions.

Analyses with Mt rRNA Genes

In contrast to the nuclear partitions, all of which provided robust support for the four major clades of placentals and for Boreoeutheria, analyses with the mt rRNA partition only provided robust support for Xenarthra and Afrotheria. Within Afrotheria, there was support for Paenungulata, Tubulidentata+ Macroscelidea + Afrosoricida, and Afrosoricida. Within Paenungulata, mt rRNA was the only partition that favored Tethytheria (i.e., Sirenia + Proboscidea) over competing hypotheses. In addition to not supporting Euarchontoglires, the mt rRNA partition failed to support Glires, rodent monophyly, and Euarchonta. Scandentia + Dermoptera was supported (posterior probability = 0.98). Within

Laurasiatheria, Variamana was strongly supported (posterior probability = 0.95), but relationships within this group were not well resolved.

DISCUSSION

Likelihood Versus Bayesian Results

Given that Bayesian methods are relatively new in phylogenetics, it is reassuring that independent Bayesian runs with the full data set resulted in trees that are topologically identical to each other and to the maximum likelihood tree when analyses were performed under the GTR + Γ + I model of sequence evolution. Maximum likelihood bootstrap percentages were generally lower than Bayesian posterior probabilities. The observation that bootstrap support proportions are lower than Bayesian posterior probabilities has now become common (Douady et al., 2003; Huelsenbeck et al., 2002). As noted by Murphy et al. (2001b), this result is consistent with the suggestion of Hillis and Bull (1993) that nonparametric bootstrap support may be too conservative. Specifically, Hillis and Bull (1993) found that bootstrap proportions $\geq 70\%$ almost always defined a true clade in their study of a known bacteriophage T7 phylogeny. In our maximum likelihood bootstrap analysis, only four clades had support percentages below 70%. Efron et al. (1996) showed that bootstrap proportions (to a first approximation) are unbiased, but also that properties of the bootstrap underlie results like those of Hillis and Bull (1993). In an analysis using computer simulations, Wilcox et al. (2002) concluded that posterior probabilities are more reliable indicators of statistical confidence than bootstrap proportions. Huelsenbeck et al. (2002) suggested that the discrepancy between bootstrap proportions and posterior probabilities may reflect a statistical bias in uncorrected bootstrap proportions.

Whether or not bootstrap proportions are too conservative (for the best supported clades), other authors have concluded that Bayesian posterior probabilities may be too high (Suzuki et al., 2002; Waddell et al., 2001). Waddell et al. (2001) cautioned that Bayesian results are less robust than nonparametric bootstrap results in the face of model-violations. Elevated posterior probabilities may also result if Markov chain Monte Carlo runs fail to incorporate adequate mixing. In this context, it may be important to distinguish between posterior probabilities that are calculated analytically versus posterior probabilities that are estimated using MCMC (with or without Metropolis-coupling). Given that Bayesian phylogenetics is in its infancy, we can almost certainly expect improvements in methods that estimate posterior probabilities. Even if we ignore posterior probabilities, maximum likelihood bootstrap percentages above 90% occur over most of the tree and allow for only localized rearrangements. Maximum likelihood bootstrap support values for the four major groups (Xenarthra, Afrotheria, Laurasiatheria, Euarchontoglires), as well as for Boreoeutheria, were all 100%.

Early History of Placentalia

In likelihood and Bayesian analyses with the complete data set, as well as in Bayesian analyses with subsets of nuclear genes, only three positions for the root of the placental tree had nonzero probabilities. These were between Afrotheria and Notolegia (Xenarthra + Boreoeutheria), between Atlantogenata (Afrotheria + Xenarthra) and Boreoeutheria, and between Xenarthra and Epitheria (Afrotheria + Boreoeutheria). Of these, a basal split between Afrotheria and Notolegia received the highest support. Waddell et al. (2001) also favored rooting at the base of Afrotheria based on extensive analyses with amino acid sequences. SOWH tests reported by Murphy et al. (2001b) rejected rooting at the base of Xenarthra, and between Atlantogenata and Boreoeutheria. However, likelihood scores for these root locations are only slightly lower than for the Afrotheria root. This raises the possibility that the sensitivity of the SOWH test may be too high. Buckley (2002) has shown that SOWH tests can give overconfidence in a topology when the assumptions of a model of sequence evolution are violated. Accordingly, we regard Afrotheria versus Notolegia as the best-supported hypothesis, but also recognize Boreoeutheria versus Atlantogenata, and Xenarthra versus Epitheria as valid alternatives for the placental root.

Given Gondwanan origins for Xenarthra and Afrotheria (Madsen et al., 2001; Murphy et al., 2001b; Scally et al., 2001), two of the three viable locations for the placental root (Afrotheria versus other placentals; Xenarthra versus other placentals) allow for the possibility of a paraphyletic southern hemisphere group at the base of Placentalia (i.e., crown-group Eutheria). Southern hemisphere paraphyly, in turn, suggests that crown-group eutherians may have their common ancestor in Gondwana, with subsequent dispersal to Laurasia. An alternate hypothesis that is consistent with these root locations is that crown-group placentals have their most recent common ancestor in Laurasia, and that there were independent dispersal events from Laurasia to

Africa (ancestor of Afrotheria) and South America (ancestor of Xenarthra), respectively (Archibald, 2003). This hypothesis is more in keeping with the conventional view that crown-group eutherians originated in the northern hemisphere (Matthew, 1915). The third alternative for the placental root, Boreoeutheria versus Atlantogenata, suggests reciprocal monophyly of Laurasian and Gondwanan moieties at the base of Placentalia, and is potentially compatible with a placental root in either Gondwana or Laurasia. As discussed below, stem eutherian fossils are also important for evaluating the geographic provenance of the last common ancestor of Placentalia.

Under any of the three rooting scenarios, the basal or near-basal separation of Afrotheria and Xenarthra may be accounted for by the vicariant event that sundered South America and Africa approximately 100–120 million years ago. If this plate tectonic event is causally related to placental cladogenesis, we should expect a divergence date at approximately 100–120 million years (MY) for the split between Afrotheria and Xenarthra. In agreement with this prediction, molecular divergence dates for the divergence of Afrotheria and Notolegia are approximately 103 MY based on linearized trees and quartet dating (Murphy et al., 2001b). Springer et al. (2003) used relaxed molecular clock methods and estimated this split at 97–112 MY. Archibald (2003) regards the agreement between plate tectonic and molecular dates as coincidental and unrelated. However, there are at least hints from the fossil record that stem eutherians have a deeper history in the southern hemisphere than previously recognized. Rich et al. (1997) described ausktribosphenids from the Early Cretaceous of Australia and suggested placental affinities. This hypothesis remains controversial (Helgen, 2003; Kielan-Jaworowska et al., 1998; Luo et al., 2001, 2002; Rich et al., 1998, 1999, 2001, 2002; Woodburne et al., 2003). Perhaps more significant is *Ambondro*, which is the oldest tribosphenic mammal from the Middle Jurassic of Gondwana (i.e., Madagascar; Flynn et al., 1999). Luo et al. (2001, 2002) argue that *Ambondro* belongs in the clade Australosphenida, along with ausktribosphenids and monotremes, and that tribospheny in this clade evolved independently from its origin in Theria. In contrast, Sigogneau-Russell et al. (2001) suggest that *Ambondro* is antecedent to Laurasian Cretaceous Tribosphenida and further state (p. 146) that "the tribosphenic molar may thus have evolved in Gondwana in the late Early Jurassic and from there have spread to (and diversified in) the two hemispheres..." In addition, cladistic analyses offer the possibility that *Ambondro* is a stem eutherian (Woodburne et al., 2003). This hypothesis demands a marsupial-placental split no later than 167 MY, which is in good agreement with molecular dates for the marsupial-placental split at 173–190 MY (Kumar and Hedges, 1998; Penny et al., 1999; Woodburne et al., 2003). If stem eutherians were in Gondwana in the Jurassic, a Gondwanan origin for crown-group placentals becomes more plausible.

Major Clades of Placental Mammals

Our analyses extend studies with the Murphy et al. (2001b) data set and provide robust support for four major clades of placental mammals (Afrotheria, Xenarthra, Euarchontoglires, Laurasiatheria). We also find robust support for a Euarchontoglires + Laurasiatheria clade (i.e., Boreoeutheria; Springer and de Jong, 2001). Sequences for a 1.3 kb segment of the apolipoprotein B gene also recover these clades (Amrine-Madsen et al., 2003). In addition, Waddell et al. (2001) argued for these clades based on analyses of amino acid sequences. In contrast, Arnason et al. (2002) failed to recover monophyletic Laurasiatheria and Euarchontoglires clades in their analysis of mitogenomic sequences. Instead, both Laurasiatheria and Eulipotyphla were diphyletic with Erinaceomorpha (hedgehog, moon rat) as the first placental branch followed by a paraphyletic Euarchontoglires (and Rodentia). In contrast, Lin et al. (2002) found that mitogenomic sequences recover Xenarthra, Afrotheria, Laurasiatheria, and Euarchontoglires in unrooted analyses. This arrangement is in fundamental agreement with nuclear data presented here and elsewhere (Delsuc et al., 2002; Madsen et al., 2001; Murphy et al., 2001a,b; Scally et al., 2001; Waddell et al., 2001). In rooted analyses, however, results similar to Arnason et al. (2002) were recovered with eulipotyphlan diphyly, rodent paraphyly, and Euarchontoglires paraphyly. As noted by Lin et al. (2002), adding an outgroup should not result in changes within the ingroup if the model of sequence evolution is correct. Lin et al. (2002) concluded that the unrooted mitochondrial tree is correct and that peculiar features of the rooted mitochondrial tree, such as eulipotyphlan diphyly and rodent paraphyly, are the result of inadequate models that do not take into account changes in mutational mechanisms in murid rodents, erinaceomorphs, and some marsupials. Further, Hudelot et al. (2003) performed rooted analyses with mitochondrial RNA (tRNA and rRNA) gene sequences that incorporated secondary structure information and found support for Xenarthra, Afrotheria, Laurasiatheria, and Euarchontoglires. The monophyly

of Euarchontoglires is especially compelling in view of two different deletions in nuclear genes that support this clade (de Jong et al., 2003; Poux et
al., 2002).

Morphology agrees with our molecular results in supporting Xenarthra.
However, Afrotheria, Euarchontoglires, Laurasiatheria, and Boreoeutheria
are all without morphological support. Instead, analyses of morphological
characters have placed taxa with similar morphotypes together even though
constituent taxa belong to different clades, e.g., paenungulates
(Afrotheria) and perissodactyls (Laurasiatheria) are sometimes united
together in the superordinal group Altungulata. From a molecular perspective, the occurrence of similar morphotypes in different clades (e.g.,
ungulates in Afrotheria versus Laurasiatheria) must be regarded as paral-
lel/convergent evolution (Helgen, 2003; Madsen et al., 2001; Scally et al.,
2001).

Relationships in Afrotheria

In agreement with most molecular studies and some morphological studies, our
analyses support Paenungulata. Resolution of relationships within
Paenungulata remains one of the major challenges for future studies of interor-
dinal relationships. Novel hypotheses that emerge from our analyses are
Tubulidentata + Macroscelidea + Afrosoricida and Macroscelidea Afrosoricida.
The latter clade also receives support from fetal membrane characters.
Specifically, afrosoricidans and elephant shrews are the only afrotherians with
haemochorial placentas (Carter, 2001).

Relationships Within the Euarchontoglires Clade

Our results agree with morphology in supporting Glires. In contrast, our results
disagree with morphological studies that support the Archonta and Volitantia
hypotheses. Rather, our analyses support an emended archontan clade that
Waddell et al. (1999) dubbed Euarchonta. This hypothesis requires that characters associated with volancy are convergent in bats and flying lemurs. Indeed,
advocates of the Archonta hypothesis have long recognized that bats lack tarsal
features that occur in other archontan orders (Szalay and Drawhorn, 1980).
Within Euarchonta, our analyses favor Scandentia + Dermoptera over competing hypotheses. In his analysis of morphological data that forced Chiroptera

outside of Archonta, Sargis (2001) also found support for a sister-group relationship between Scandentia and Dermoptera.

Relationships in Laurasiatheria

Within Laurasiatheria, our analyses suggest a basal split between Eulipotyphla (moles, shrews, hedgehogs) and Variamana (i.e., Chiroptera + Perissodactyla + Cetartiodactyla + Pholidota + Carnivora). Mitochondrial studies are divided between those that are consistent with Variamana (Lin and Penny, 2001) and those that favor a sister-group relationship between bats and Eulipotyphla (represented by a mole) (Nikaido et al., 2001). Within Variamana, there is strong support for a sister-group relationship between carnivores and pangolins. Interestingly, these taxa are unique among living placentals in possessing an osseous tentorium (Shoshani and McKenna, 1998).

MOLECULAR CLASSIFICATION FOR THE LIVING ORDERS OF PLACENTAL MAMMALS

The well-resolved molecular tree that we present provides a basis for classifying the living orders of placental mammals (Table 5). Following McKenna and Bell (1997), we used mirorder, grandorder, superorder, magnorder, cohort, supercohort, and infralegion as successively more inclusive taxonomic ranks above the rank of order. Our classification includes clades that have been recognized in previous classifications, although in many cases these have not been recognized with formal Linnaean ranks. Our classification also includes newly named clades. All names above the ordinal level are suggested for crown-clades with node-based definitions.

We recognize Placentalia as a clade of crown-group eutherians at the taxonomic rank of infralegion. Placentalia is divided into the supercohorts Afrotheria and Notolegia. The name Notolegia is new and is suggested for a clade with southern (Gondwanan) origins that subsequently gave rise to legions of placental taxa. Our basal split between Afrotheria and Notolegia contrasts with two competing hypotheses for the base of the placental tree, both of which are associated with likelihood scores that are only slightly lower than when the root is placed between Afrotheria and Notolegia. First, Atlantogenata (Waddell et al., 1999) versus Boreoeutheria (Springer and de Jong, 2001), which places the root between Gondwanan- and Laurasian-origin clades. Second, between Xenarthra and Epitheria; Epitheria is compatible with some morphological analyses (McKenna and Bell, 1997).

Within Afrotheria, we recognize a fundamental split between the grandorders Fossoromorpha and Paenungulata. Fossoromorpha is a newly named clade that includes aardvarks, elephant shrews, and afrosoricidans. The name Fossoromorpha is suggested based on the occurrence of fossorial adaptations in many taxa within this clade (e.g., golden moles, aardvarks). Among fossoromorphs, as well as other afrotherians that have been investigated, Macroscelidea and Afrosoricida share a haemochorial placenta (Carter, 2001). In recognition of this feature, we suggest the new name Haemochorialia for this clade. Paenungulata (= Uranotheria of McKenna and Bell, 1997), although not previously recognized at the rank of grandorder, is a feature of some morphological classifications (e.g., Simpson, 1945). Because relationships within Paenungulata have proved difficult to resolve with molecular data, we do not recognize additional classificatory structure within this group.

Notolegia includes Xenarthra and Boreoeutheria as its constituent cohorts. Xenarthra includes the orders Cingulata (armadillos) and Pilosa (sloths, anteaters). Boreoeutheria is divided into the magnorders Laurasiatheria and Euarchontoglires. Within Laurasiatheria, the superorder Variamana includes all taxa excepting the order Eulipotyphla. We suggest the name Variamana for the clade that includes Chiroptera, Perissodactyla, Cetartiodactyla, Pholidota, and Carnivora in recognition of the variable hand that occurs in different members of this group, e.g., flippers in cetaceans, wings in bats, hooves in

Table 5. Classification of living orders of placental mammals[1]

Infralegion Placentalia Owen, 1837, new rank
  Supercohort Afrotheria Stanhope, Waddell, Madsen, de Jong, Hedges, Cleven, Kao &
    Springer, 1998, new rank
      Grandorder Fossoromorpha, new[2]
        Order Tubulidentata Huxley, 1872
      Mirorder Haemochorialia, new
        Order Macroscelidea Butler, 1956
        Order Afrosoricida Stanhope, Waddell, Madsen, de Jong, Hedges, Cleven, Kao
          & Springer, 1998
      Grandorder Paenungulata Simpson 1945
        Order Hyracoidea, Huxley, 1869
        Order Sirenia Illiger, 1811
        Order Proboscidea Illiger 1811

Table 5. (*Continued*)

Supercohort Notolegia, new [3]
 Cohort Xenarthra Cope, 1889, new rank
   Order Cingulata Illiger, 1811
   Order Pilosa Flower, 1883
 Cohort Boreoeutheria Springer & de Jong 2001, new rank
  Magnorder Laurasiatheria Waddell, Okada & Hasegawa, 1999, new rank
    Order Eulipotyphla Waddell, Okada & Hasegawa, 1999
   Superorder Variamana, new [4]
    Order Chiroptera Blumenbach, 1779
    Grandorder Fereuungulata Waddell, Okada & Hasegawa, 1999, new rank
     Order Cetartiodactyla Montgelard, Catzeflis & Douzery, 1997
     Order Perissodactyla Owen, 1848
    Mirorder Ostentoria, new [5]
     Order Carnivora Bowdich, 1821
     Order Pholidota Weber, 1904
  Magnorder Euarchontoglires Murphy, Stanyon & O'Brien, 2001
   Grandorder Glires Linnaeus, 1758, new rank
    Order Lagomorpha Brandt, 1855
    Order Rodentia Bowdich, 1821
   Grandorder Euarchonta Waddell, Okada & Hasegawa, 1999, new rank
    Order Primates Linnaeus, 1758
    Mirorder Paraprimates, new [6]
     Order Dermoptera Illiger, 1811
     Order Scandentia Wagner, 1855

---

[1] To maintain stability, and subject to the requirement of monophyly, traditional mammalian orders have been retained at the Linnaean rank of order. All taxa above the rank of order are intended as crown clades with node-based definitions (i.e., the most recent common ancestor of all living members of a group, plus all of the descendants, living or extinct, of this common ancestor; see examples below for newly defined groups). Our classification explicitly avoids redundant taxonomic names that fail to convey additional phylogenetic information. For example, the grandorder Fossoromorpha includes the orders Tubulidentata, Afrosoricida, and Macroscelidea. Of these, Afrosoricida and Macroscelidea are hypothesized as sister-taxa in the mirorder Haemochorialia. Tubulidentata is the sister-taxon to Haemochorialia, but we have not erected a redundant mirorder for Tubulidentata. This does not imply that Tubulidentata is *incertae sedis* in Fossoromorpha. We appreciate that categorical subordination is a convenient approach to tag sister-groups in Linnaean classifications, but agree with Wiley (1981) that redundant or empty categories should not be introduced unnecessarily. Also, categorical subordination is less relevant for sister-groups with node-based definitions than for sister-groups with stem-based definitions. Only in the latter case can we expect sister-groups that have precisely equivalent origination times that result from the same cladogenic event. Our classification retains flexibility for adding names for taxa with stem-based definitions.

[2] Definition: for the most recent common ancestor of Fossoromorpha and Tubulidentata and all of its descendants. The name Fossoromorpha was suggested by Kris Helgen.

[3] Definition: for the most recent common ancestor of Xenarthra, Eulipotyphla, Chiroptera, Cetartiodactyla, Perissodactyla, Carnivora, Pholidota, Rodentia, Lagomorpha, Primates, Scandentia, and Dermoptera and all of its descendants.

[4] Definition: for the most recent common ancestor of Chiroptera, Cetartiodactyla, Perissodactyla, Carnivora, and Pholidota and all of its descendants. The name Variamana was suggested by Kris Helgen.

[5] Definition: for the most recent common ancestor of Carnivora and Pholidota and all of its descendants.

[6] Definition: for the most recent common ancestor of Dermoptera and Scandentia and all of its descendants.

ungulates. Variamana is suggested as an alternative to Scrotifera (Waddell et al., 1999), which we find less appropriate for this clade. Within Variamana, our analyses support Fereuungulata (Waddell et al., 1999). Interordinal relationships within Fereuungulata are not resolved except for Carnivora + Pholidota. Numerous morphological studies unite pholidotans with xenarthrans, but there is also support for a carnivore-pholidotan alliance (Shoshani and McKenna, 1998). We suggest that Ostentoria is an appropriate name for this hypothesis given the osseous tentorium that occurs in carnivores and pangolins (Shoshani and McKenna, 1998). Waddell et al. (1999) suggested the name Ferae for this clade, but Gregory's (1910) monograph reveals that this name has a long and complicated history that begins in 1758 with Linnaeus, who included carnivores in Ferae and pangolins in Bruta. Simpson (1945) included carnivores and creodonts in the superorder Ferae and pangolins in the cohort Unguiculata.

Within Euarchontoglires, we recognize the grandorders Glires and Euarchonta. Glires includes lagomorphs and rodents and is recognized in most morphological classifications. The name Euarchonta was suggested by Waddell et al. (1999) for an emended archontan clade that includes primates, scandentians, and dermopterans, but not chiropterans. Peters (1864) included tree shrews, flying lemurs, and elephant shrews in one of his two great groups of insectivores (reviewed in Gregory, 1910). Although Haeckel (1866) removed flying lemurs from this group when he erected Menotyphla, it is now evident that Scandentia and Dermoptera are more closely related to each other than either is to elephant shrews. We suggest the new name Paraprimates for the Scandentia + Dermoptera hypothesis in recognition of the phylogenetic proximity of this clade to Primates.

CONCLUSIONS

Likelihood and Bayesian analyses of the Murphy et al. (2001b) data set provide a well-resolved phylogeny for the orders of placental mammals that includes four major clades (Afrotheria, Xenarthra, Laurasiatheria, Euarchontoglires). In addition, Laurasiatheria and Euarchontoglires are sister-taxa. Among these clades, only Xenarthra was previously hypothesized based on morphology. This reorganization of the placental tree has implications for early placental biogeography and the deployment of morphological character evolution. Remaining uncertainties in the placental tree are confined to local rearrangements, i.e., unresolved trifurcations.

In contrast to morphological studies that group Primates in Archonta, our molecular results suggest archontan diphyly with bats in Laurasiatheria and the remaining archontan orders (primates, tree shrews, flying lemurs) in Euarchontoglires. The latter clade is divided into Glires (rodents + lagomorphs) and Euarchonta (primates + tree shrews + flying lemurs). Future morphological studies that examine relationships among primates and their relatives should not blithely assume archontan monophyly as a basis for outgroup choice.

REFERENCES

Adkins, R. M., and Honeycutt, R. L., 1993, A molecular examination of archontan and chiropteran monophyly, in: *Primates and Their Relatives in Phylogenetic Perspective*, R. D. E. MacPhee, ed., Plenum, New York, pp. 227–249.

Amrine-Madsen, H., Koepfli, K.-P., Wayne, R. K., and Springer, M. S., 2003, A new phylogenetic marker, apolipoprotein B, provides compelling evidence for eutherian relationships, *Mol. Phylogenet. Evol.* 28:186–196.

Archibald, J. D., 2003, Timing and biogeography of the eutherian radiation: Fossils and molecular compared, *Mol. Phylogenet. Evol.* 28:350–359.

Arnason, U., Adegoke, J. A., Bodin, K., Born, E. W., Esa, Y. B., Gullberg, A. et al., 2002, Mammalian mitogenomic relationships and the root of the eutherian tree, *Proc. Natl. Acad. Sci. USA* **99**: 8151–8156.

Beard, K. C., 1993, Phylogenetic systematics of the Primatomorpha, with special reference to Dermoptera, in: Mammal Phylogeny: Placentals, F. S. Szalay, M. J. Novacek, and M. C. McKenna, eds., Springer-Verlag, Berlin, pp. 129–150.

Buckley, T. R., 2002, Model misspecification and probabilistic tests of topology: Evidence from empirical data sets, *Syst. Biol.* **51**: 509–523.

Carter, A. M., 2001, Evolution of the placenta and fetal membranes seen in the light of molecular phylogenetics, *Placenta* **22**: 800–807.

Cartmill, M., and MacPhee, R. D. E., 1980, Tupaiid affinities: The evidence of the carotid arteries and cranial skeleton, in: *Comparative Biology and Evolutionary Relationships of Tree Shrews*, W. P. Luckett, ed., Plenum, New York, pp. 95–132.

de Jong, W. W., van Dijk, M. A. M., Poux, C., Kappe, G., van Rheede, T., and Madsen, O., 2003, Indels in protein-coding sequences of Euarchontoglires constrain the rooting of the eutherian tree, *Mol. Phylogenet. Evol.* 28:328–340.

Delsuc, F., Scally, M., Madsen, O., Stanhope, M. J., de Jong, W. W., Catzeflis, F. M. et al., 2002, Molecular phylogeny of living xenarthrans and the impact of character and taxon sampling on the placental tree rooting, *Mol. Biol. Evol.* **19**: 1656–1671.

Douady, C. J., Delsuc, F., Boucher, Y., Doolittle, W. F., and Douzery, E. J. P., 2003, Comparison of Bayesian and maximum likelihood bootstrap measures of phylogenetic reliability, *Mol. Biol. Evol.* **20**: 248–254.

Efron, B., Halloran, E., and Holmes, S., 1996, Bootstrap confidence levels for phylogenetic trees, *Proc. Natl. Acad. Sci. USA* **93**: 13429–13434.

Flynn, J. J., Parrish, J. M., Rakotosamimanana, B., Simpson, W. F., and Wyss, A. E., 1999, A Middle Jurassic mammal from Madagascar, *Nature* **401**: 57–60.

Goldman, N., Anderson, J. P., and Rodrigo, A. G., 2000, Likelihood-based tests of topologies in phylogenetics, *Syst. Biol.* **49**: 652–670.

Goodman, M., 1975, Protein sequence and immunological specificity: Their role in phylogenetic studies of Primates, in: *Phylogeny of the Primates*, W. P. Luckett and F. S. Szalay, eds., Plenum, New York, pp. 219–248.

Gregory, W. K., 1910, The orders of mammals, *Bull. Amer. Mus. Nat. Hist.* **27**: 1–524.

Haeckel, E., 1866, *Generelle Morphologie der Organismen*, Berlin. (Secondary citation from Gregory, 1910).

Helgen, K. M., 2003, Major mammalian clades: A review under consideration of molecular and palaeontological evidence, *Mammal. Biol.* **68**: 1–15.

Hillis, D. M., and Bull, J. J., 1993, An empirical test of bootstrapping as a measure of assessing confidence in phylogenetic analysis, *Syst. Biol.* **42**: 182–192.

Hudelot, C., Gowri-Shankar, V., Jow, H., Rattray, M., and Higgs, P. G., 2003, RNA-based phylogenetic methods: Application to mammalian mitochondrial RNA sequences, *Mol. Phylogenet. Evol.* 28:241–252.

Huelsenbeck, J. P., and Ronquist, F., 2001, MRBAYES: Bayesian inference of phylogenetic trees, *Bioinformatics* 17: 754–755.

Huelsenbeck, J. P., Ronquist, F., Nielsen, R., and Bollback, J. P., 2001, Bayesian inference of phylogeny and its impact on evolutionary biology, *Science* **294**: 2310–2314.

Huelsenbeck, J. P., Larget, B., Miller, R. E., and Ronquist, F., 2002, Potential applications and pitfalls of Bayesian inference of phylogeny, *Syst. Biol.* **5**: 673–688.

Kielan-Jaworowska, Z.-K., Cifelli, R. L., and Luo, Z., 1998, Alleged Cretaceous placental from down under, *Lethaia* **31**: 267–268.

Kumar, S., and Hedges, S. B., 1998, A molecular timescale for vertebrate evolution, *Nature* **392**: 917–920.

Larget, B., and Simon, D., 1999, Markov chain Monte Carlo algorithms for the Bayesian analysis of phylogenetic trees, *Mol. Biol. Evol.* **16**: 750–759.

Lin, Y.-H., and Penny, D., 2001, Implications for bat evolution from two complete mitochondrial genomes, *Mol. Biol. Evol.* **18**: 684–688.

Lin, Y.-H., McLenachan, P. A., Gore, A. R., Phillips, M. J., Ota, R., Hendy, M. D. et al., 2002, Four new mitochondrial genomes and the increased stability of evolutionary trees of mammals from improved taxon sampling, *Mol. Biol. Evol.* **19**: 2060–2070.

Luckett, W. P., 1980, The use of reproductive and developmental features in assessing tupaiid affinities, in: *Comparative Biology and Evolutionary Relationships of Tree Shrews*, W. P. Luckett, ed., Plenum, New York, pp. 245–266.

Luo, Z.-X., Cifelli, R. L., and Kielan-Jaworowska, Z., 2001, Dual origin of tribosphenic mammals, *Nature* **409**: 53–57.

Luo, Z.-X., Kielan-Jaworowska, Z., and Cifelli, R. L., 2002, In quest for a phylogeny of Mesozoic mammals, *Acta Palaeontologica Polonica* **47**: 1–78.

Madsen, O., Scally, M., Douady, C. J., Kao, D. J., DeBry, R. W., Adkins, R. et al., 2001, Parallel adaptive radiations in two major clades of placental mammals, *Nature* **409**: 610–614.

Martin, R. D., 1990, *Primate Origins and Evolution: A Phylogenetic Reconstruction*, Princeton University Press, Princeton, NJ.

Matthew W. D., 1915, Climate and evolution. *Annals of the New York Academy of Sciences* 24:171–318.

Mau, B., Newton, M., and Larget, B., 1999, Bayesian phylogenetic inference via Markov chain Monte carlo methods, *Biometrics* **55**: 1–12.

McKenna, M. C., 1975, Toward a phylogenetic classification of the Mammalia, in: *Phylogeny of the Primates*, W. P. Luckett and F. S. Szalay, eds., Plenum, New York, pp. 21–46.

McKenna, M. C., and Bell, S. K., 1997, *Classification of Mammals Above the Species Level*, Columbia University Press, New York.

Murphy, W. J., Eizirik, E., Johnson, W. E., Zhang, Y. P., Ryder, O. A., and O'Brien, S. J., 2001a, Molecular phylogenetics and the origins of placental mammals, *Nature* **409**: 614–618.

Murphy, W. J., Eizirik, W., O'Brien, S. J., Madsen, O., Scally, M., Douady, C. J. et al., 2001b, Resolution of the early placental mammal radiation using Bayesian phylogenetics, *Science* **294**: 2348–2351.

Murphy, W. J., Stanyon, R., and O'Brien, S. J., 2001c, Evolution of mammalian genome organization inferred from comparative gene mapping, *Genome Biology* **2**(6): reviews0005.1–0005.8.

Nikaido, M., Kawai, K., Harada, M., Tomita, S., Okada, N., and Hasegawa, M., 2001, Maximum likelihood analysis of the complete mitochondrial genomes of eutherians and a reevaluation of the phylogeny of bats and insectivores, *J. Mol. Evol.* **53**: 508–516.

Novacek, M. J., 1980, Cranioskeletal features in tupaiids and selected eutherians as phylogenetic evidence, in: *Comparative Biology and Evolutionary Relationships of Tree Shrews*, W. P. Luckett, ed., Plenum, New York, pp. 35–93.

Novacek, M. J., 1990, Morphology, paleontology, and the higher clades of mammals, in: *Current Mammalogy*, H. H. Genoways, ed., vol. 2, pp. 507–543, Plenum, New York.

Novacek, M. J., and Wyss, A. R., 1986, Higher-level relationships of the Recent eutherian orders: Morphological evidence, *Cladistics* **2**: 257–287.

Novacek, M. J., Wyss, A. R., and McKenna, M. C., 1988, The major groups of eutherian mammals, in: *The Phylogeny and Classification of the Tetrapods*, M. J. Benton, ed., vol. 2, pp. 31–71, Oxford University Press, London.

Penny, D., Hasegawa, M., Waddell, P. J., and Hendy, M. D., 1999, Mammalian evolution: Timing and implications from using the logdeterminant transform for proteins of differing amino acid composition, *Syst. Biol.* **48**: 76–93.

Peters, W. C. H., 1864, Uber die Saugethiere Gattung Solenodon, *Abhandl. Akad. Wissensch (Berlin)* **1864**: 1–22. (Secondary citation from Gregory, 1910.)

Posada D., and Crandall, K. A., 1998, MODELTEST: testing the model of DNA substitution. *Bioinformatics* **14**: 817–8.

Poux, C., van Rheede, T., Madsen, O., and de Jong, W.W., 2002, Sequence gaps join mice and men: Phylogenetic evidence from deletions in two proteins, *Mol. Biol. Evol.* **19**: 2035–2037.

Rich, T. H., Vickers-Rich, P., Constantine, A., Flannery, T. F., Kool, L., and van Klaveren, N., 1997, A tribosphenic mammal from the Mesozoic of Australia, *Science* **278**: 1438–1442.

Rich, T. H., Flannery, T. F., and Vickers-Rich, P., 1998, Alleged Cretaceous placental from down under: Reply, *Lethaia* **31**: 346–348.

Rich, T. H., Vickers-Rich, P., Constantine, A., Flannery, T. F., Kool, L., and van Klaveren, N., 1999, Early Cretaceous mammals from Flat Rocks, Victoria, Australia, *Rec. Queen Victoria Mus.* **106**: 1–29.

Rich, T., H., Flannery, T. F., Trusler, P., Kool, L., van Klaveren, N., and Vickers-Rich, P., 2001, A second tribosphenic mammal from the Mesozoic of Australia, *Rec. Queen Victoria Mus.* **110**: 1–10.

Rich, T. H., Flannery, T. F., Trusler, P., Kool, L., van Klaveren, N., and Vickers-Rich, P., 2002, Evidence that monotremes and ausktribosphenids are not sistergroups, *J. Vert. Paleont.* **22**: 466–469.

Sargis, E., 2001, The phylogenetic relationships of archontan mammals: Postcranial evidence, *J. Vert. Paleont.* **21**: 97A.

Scally, M., Madsen, O., Douady, C. J., de Jong, W. W., Stanhope, M. J., and Springer, M. S., 2001, Molecular evidence for the major claddes of placental mammals. *Journal of Mammalian Evolution.* **8**: 239–278.

Shoshani, J., and McKenna, M. C., 1998, Higher taxonomic relationships among extant mammals based on morphology, with selected comparisons of results from molecular data, *Mol. Phylogenet. Evol.* **9**: 572–584.

Sigogneau-Russell, D., Hooker, J. J., and Ensom, P. C., 2001, The oldest tribosphenic mammal from Laurasia (Purbeck Limestone Group, Berriasian, Cretaceous, UK) and its bearing on the 'dual origin' of Tribosphenida. *C. R. Acad. Sci. Paris de la Terre et des planets* **333**: 141–147.

Simmons, N. B., 1995, Bat relationships and the origin of flight, in: *Ecology, Evolution and Behavior of Bats*, P. A. Racey and S. M. Swift, eds., *Symp. Zool. Soc. London*, **67**: 27–43.

Simmons, N. B., and Quinn, T. H., 1994, Evolution of the digital tendon locking mechanism in bats and dermopterans: A phylogenetic perspective, *J. Mammal. Evol.* **2**: 231–254.

Simpson, G. G., 1945, The principles of classification and a classification of mammals, *Bull. Amer. Mus. Nat. Hist.* **85**: 1–350.

Springer, M. S., and de Jong, W. W., 2001, Phylogenetics: Which mammalian supertree to bark up? *Science* **291**: 1709–1711.

Springer, M. S., Murphy, W. J., Eizirik, E., and O'Brien, S. J., 2003, Placental mammal diversification and the Cretaceous-Tertiary boundary. *Proc Natl Acad Sci U S A* **100**: 1056–61.

Stanhope, M. J., Bailey, W. J., Czelusniak, J., Goodman, M., Si, J.-S., Nickerson, J. et al., 1993, A molecular view of primate supraordinal relationships from the analysis of both nucleotide and amino acid sequences, in: *Primates and Their Relatives in Phylogenetic Perspective*, R. D. E. MacPhee, ed., Plenum, New York, pp. 251–292.

Suzuki, Y., Glazko, G. V., and Nei, M., 2002, Overcredibility of molecular phylogenies obtained by Bayesian phylogenetics, *Proc. Natl. Acad. Sci. USA* **99**: 16138–16143.

Swofford, D. L., 1998, *PAUP*. Phylogenetic Analysis Using Parsimony (* and Other Methods)*, Version 4, Sinauer Associates, Sunderland, Massachusetts.

Szalay, F. S., 1977, Phylogenetic relationships and a classification of the eutherian Mammalia, in: *Major Patterns in Vertebrate Evolution*, M. K. Hecht, P. C. Goody, and B. M. Hecht, eds., Plenum, New York, pp. 315–374.

Szalay, F. S., and Drawhorn, G., 1980, Evolution and diversification of the Archonta in an arboreal milieu, in: *Comparative Biology and Evolutionary Relationships of Tree Shrews*, W. P. Luckett, ed., Plenum, New York, pp. 133–169.

Szalay, F. S., and Lucas, S. G., 1993, Cranioskeletal morphology of archontans, and diagnoses of Chiroptera, Volitantia, and Archonta, in: *Primates and Their Relatives in Phylogenetic Perspective*, R. D. E. MacPhee, ed., Advances in primatology series, Plenum, New York, pp. 187–226.

Swofford, D. L., 1998, PAUP*: *Phylogenetic Analysis Using Parsimony (* and Other Methods)*, Version 4. Sinauer, Sunderland, MA.

Waddell, P. J., Kishino, H., and Ota, R., 2001, A phylogenetic foundation for comparative mammalian genomics, *Genome Inform.* **12**: 141–154.

Waddell, P. J., Okada, N., and Hasegawa, M., 1999, Toward resolving the interordinal relationships of placental mammals, *Syst. Biol.* **48**: 1–5.

Whelan, S., Lio, P., and Goldman, N., 2001, Molecular phylogenetics: State-of-the-art methods for looking into the past, *Trends Genet.* **17**: 262–272.

Wible, J. R., and Covert, H. H., 1987, Primates: Cladistic diagnosis and relationships, *J. Hum. Evol.* **16**: 1–22.

Wible, J. R., and Novacek, M. J., 1988, Cranial evidence for the monophyletic origin of bats, *Amer. Mus. Novit.* **2911**: 1–19.

Wilcox, T. P., Zwickl, D. J., Heath, T. A., and Hillis, D. M., 2002, Phylogenetic relationships of the dwarf boas and a comparison of Bayesian and bootstrap measures of phylogenetic support, *Mol. Phylogenet. Evol.* **25**: 361–371.

Wiley, E. O., 1981, *Phylogenetics: The Theory and Practice of Phylogenetic Systematics*, Wiley, New York.

Woodburne, M. O., Rich, T. A., and Springer, M. S., 2003, The evolution of tribospheny and the antiquity of mammalian clades, *Mol. Phylogenet. Evol.* **28**:360–385.

CHAPTER TWO

New Light on the Dates of Primate Origins and Divergence

Christophe Soligo, Oliver A. Will, Simon Tavaré,
Charles R. Marshall, and Robert D. Martin

INTRODUCTION

The known fossil record for undoubted primates of modern aspect (i.e., confined to Euprimates and excluding Plesiadapiformes) dates back to the beginning of the Eocene epoch, about 55 million years ago (MYA), and it is widely accepted among primate paleontologists that primates originated during the preceding Paleocene epoch, some 60–65 MYA. A parallel conclusion has been reached for most orders of placental mammals, and it is generally assumed that the origin and radiation of most if not all placental orders with extant representatives took place after the extinction of dinosaurs at the end of the Cretaceous. In common parlance, the Age of Mammals followed on from the Age of Dinosaurs. A comparable explanation has been given for the adaptive radiation of modern birds. All such interpretations depend on the

Christophe Soligo • Human Origins Programme, Department of Palaeontology, The Natural History Museum, London, UK **Oliver Will** • Statistics Department, University of Washington, Seattle, WA **Simon Tavaré** • Program in Molecular and Computational Biology, University of Southern California, Los Angeles, CA **Charles R. Marshall** • Department of Invertebrate Paleontology, Museum of Comparative Zoology, Harvard University, Oxford Street, Cambridge, MA **Robert D. Martin** • The Field Museum, Chicago, IL

common procedure of dating the origin of a group by the earliest known fossil representative, perhaps adding a safety margin of a few million years in tacit but conservative recognition of the fact that the earliest known fossil is unlikely to coincide exactly with the time of origin. Such direct dating from the fossil record faces two problems: (1) if the fossil record represents a very poor sample, the first known fossil representative of a given group is likely to be considerably more recent than the actual origin of that group, and (2) various kinds of bias in the fossil record may introduce further error. In this light, it has been suggested that a relatively low sampling level of the fossil record for primates has led to substantial underestimation of their time of origin (Martin 1986, 1990, 1993; Tavaré et al., 2002).

Correct timing of the initial emergence of a group such as the primates is of great importance if the mechanisms that led to its evolution are to be understood, as both biotic and abiotic environmental conditions can be taken into account only if the origin of the group and the prevailing environmental conditions can be accurately correlated chronologically.

In this chapter, we review available paleontological and molecular evidence pertinent to the timing of the origin of the primates. We also present new analyses using a recently developed statistical method that estimates times of origin of clades based on their modern diversity, their known fossil record, diversification models, and estimates of relative sampling intensities.

THE FOSSIL RECORD

Before proceeding any further, it is necessary to draw a crucial distinction between the time of initial divergence of a given group, such as the primates, and the age of the last common ancestor of all known, diagnosable members of that group (Figure 1). In a phylogenetic tree, the initial time of origin of any given taxon is indicated by the point of divergence between that taxon and its most closely related sister taxon (node 1 in Figure 1). Initially, the taxon of interest might diverge from its closest relatives as a lineage lacking the characteristic morphological features of its later descendants and then exist for some time before developing recognizable diagnostic characters. A considerable temporal gap may, therefore, occur between the initial divergence of a taxon and the emergence of diagnostic morphological characteristics as recognized by paleontologists (i.e., between nodes 1 and 2 in Figure 1). With respect to the evolution of placental mammals, this point has been succinctly expressed by Madsen et al. (2001): "Easteal (1999) suggested that primitive placentals from the Cretaceous may have diversified phylogenetically before they diverged morphologically and acquired the diagnostic features of ordinal level crown-group clades." The upper limit for the temporal gap between the initial divergence of a taxon and the emergence of diagnostic morphological characteristics is set by the estimated age of the last common ancestor of modern lineages within the taxon (node 3 in Figure 1), or by the age of the oldest known clearly recognizable fossil representative of the taxon, whichever is older.

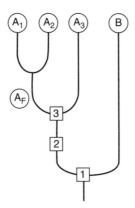

Figure 1. In a molecular phylogeny, the time of origin of taxon A (with living representatives A_1, A_2, and A_3) is indicated by node 1, the point of inferred divergence from the most closely related sister taxon with living representatives (B). The time of initial divergence of living representatives of taxon A from their last common ancestor may be considerably younger, as indicated by node 3. Molecular estimates can also be used to infer the date of node 3, in this case the time of divergence between A_1 and (A_2 + A_3). Derived morphological features shared by the living representatives of taxon A may have developed at any time between nodes 1 and 3. The earliest morphologically recognizable member of taxon A exhibiting derived diagnostic features shared with the living representatives is indicated by node 2. The first known fossil representative allocated to taxon A (A_F), on the basis of derived features shared with living representatives, yields a minimum date for the origin of the taxon. It should be noted that A_F may be nested within the adaptive radiation leading to living representatives (as is widely presumed to be the case for Eocene adapiforms and omomyiforms), but it is also possible that A_F diverged at some time prior to the common ancestor of living representatives (i.e., prior to node 3).

It should be noted that inferred phylogenetic relationships, in conjunction with the fossil record, may be used to extend minimum estimates of divergence times in some cases (Norell, 1992; Smith, 1994). Under the assumption that sister groups had the same time of origin, the later-appearing sister group is assumed to have existed at least by the time of first appearance of the earlier-appearing sister group. The range extension for the later-appearing sister group is referred to as a ghost lineage (Norell, 1992). In the case of the primates, the uncertainties that prevail regarding both the composition of and the relationships within Archonta—the supraordinal grouping to which primates are often allocated—make it difficult to apply the concept of the ghost lineage. It can be noted, however, that none of the modern orders of Archonta extends back much beyond the time of the earliest known primate fossils. The oldest known fossils belonging to Scandentia are from Eocene deposits (Tong, 1988), while the oldest fossils tentatively attributed to Volitantia (Dermoptera + Chiroptera) are late Paleocene (Stucky and McKenna, 1993), which would extend the expected range of the primates back by no more than a few million years. Among extinct groups of archontans, the Plesiadapiformes and the Mixodectidae (as possible members of Dermoptera) are potentially relevant (Hooker, 2001). If confirmed to represent the sister group of primates, either of these would extend the expected range of primates back to the early Paleocene.

Undoubted primates (equated here with Euprimates) first appeared in the fossil record at the beginning of the Eocene period in Western Europe, Asia, and North America. A reported primate from the late Paleocene of Morocco (Sigé et al., 1990), *Altiatlasius*, has recently been reassigned to the Plesiadapiformes (Hooker et al., 1999) and is, therefore, not considered here. The absence from the known fossil record of any pre-Eocene primates of modern aspect is usually interpreted as evidence that the order originated not long before that period, around 60 MYA and no earlier than 65 MYA.

However, the ages of the first known fossil representatives of certain other mammalian groups are in themselves incompatible with the interpretation that the placental lineage leading to primates diverged only 60–65 MYA. The best illustration of this is provided by studies of artiodactyl relationships. It has long been accepted that cetaceans and artiodactyls are sister-groups, but recent molecular evidence has uniformly indicated that cetaceans are actually nested within the artiodactyls and that their closest relatives are hippopotamuses. This conclusion, initially suggested by immunological data (Sarich,

1993), is now supported by nuclear gene sequences (Gatesy, 1997; Gatesy et al., 1996, 1999; Graur and Higgins, 1994; Madsen et al., 2001; Murphy et al., 2001a, b), by insertions of interspersed elements (retroposons) in the nuclear genome (Nikaido et al., 1999), and by complete mitochondrial genomes (Ursing and Arnason, 1998). In fact, evidence from two early terrestrial relatives of cetaceans: *Ichthyolestes* and *Pakicetus* (Thewissen et al., 2001), has confirmed that they share the unique tarsal morphology of artiodactyls and are, therefore, more closely related to them than to mesony-chians, which were long thought to be the direct sister group of cetaceans. Although a cladistic analysis of the morphological data did not confirm a specific link between cetaceans and hippopotamuses, there is undoubtedly a closer link between cetaceans and artiodactyls than hitherto believed by paleontologists. The molecular evidence now uniformly indicates that the fol-lowing sequence of divergences occurred during the evolution of the hoofed mammals (ungulates): (1) between odd-toed perissodactyls and even-toed artiodactyls; (2) within artiodactyls between camels + pigs and ruminants + hippos + cetaceans; (3) between ruminants and hippos + cetaceans; (4) between hippos and cetaceans. Given that the first known fossil representative of the cetaceans is dated to about 53.5 MYA (Bajpai and Gingerich, 1998), it follows that the initial divergence in this well-supported sequence of 4 splits in ungulate evolution must have occurred at a relatively early date and that the separation between ungulates and the lineage leading to primates must have taken place even earlier. A date of only 60–65 MYA for the divergence of the primate lineage from other lineages of placental mammals hence seems inher-ently improbable. It seems likely, instead, that the early evolution of primates has simply remained undocumented in the known fossil record.

Early placental mammals seem to be generally poorly documented in the known fossil record. This is strikingly illustrated by the case of bats (order Chiroptera). Modern bats constitute a widespread and diverse group con-taining around a thousand species, including at least 165 megachiropterans (Old World fruit bats) and at least 815 microchiropterans (Corbet and Hill, 1991). As with primates of modern aspect, the earliest known clearly identifiable bat fossils date back to the beginning of the Eocene (about 55 MYA) in North America, Europe, Africa, and Australia, although one report extends this back into the latest Paleocene, to 56 MYA. The first relatively complete bat skeletons are known from early Eocene deposits in North America (*Icaronycteris*) and from Early/Middle Eocene deposits in Europe

(*Archaeonycteris, Hassianycteris,* and *Palaeochiropteryx*). By this time, all of the major defining morphological features of bats can be identified, notably including the development of a wing membrane (patagium) between digits II and V of the hand and extreme backward rotation of hindlimbs for suspension, involving extensive remodeling of the pelvis and ankle joint. Furthermore, all four Eocene bat genera documented by relatively complete skeletons show weak to moderate enlargement of the cochlea, indicating the development of some degree of echolocation capacity. For this and other reasons, a recent review of morphological evidence (Simmons and Geisler, 1998) concludes that these 4 genera are more closely related to microchiropterans than to megachiropterans and branched off successively from the lineage leading to the common ancestor of microchiropterans, such that they are an integral part of the adaptive radiation that led to modern bats. Yet there are no known fossils documenting the transition from a generalized early placental ancestor to the highly specialized, immediately recognizable condition of the earliest known bat skeletons. Furthermore, there is an obvious and extreme bias in the geographical occurrence of well-preserved bat fossils. Whereas at least 4 skeletons of *Icaronycteris* have been reported from a single site in North America (Green River, Wyoming, approx. 53 MYA), all the others (some 100 skeletons of *Archaeonycteris, Hassianycteris,* and *Palaeochiropteryx*) have been discovered at the European site of Messel, southern Germany (approx. 49 MYA). With some of the exquisitely preserved bat skeletons from Messel, remains of the stomach contents are also present. Analysis of these has revealed moth wing scales indicating dietary habits comparable to those of modern microchiropteran bats.

The fossil record for Old World fruit-bats (megachiropterans) is even less informative. The earliest known remnant is a single tooth identified as that of a megachiropteran found in upper Eocene deposits of Thailand (Ducrocq et al., 1993). Given that microchiropterans are reliably documented from the earliest Eocene, this could indicate a ghost lineage of some 15 MY prior to the earliest known megachiropteran.

Furthermore, a recent cladistic analysis of archontan relationships using both cranial and postcranial characters has provided evidence for a Cretaceous origin of bats (Hooker, 2001). In the cladogram issued from that study, bats branch off at a lower node than both the extinct genus *Deccanolestes*—a possible primitive Archontan—and the extinct family Nyctitheriidae. Therefore, the early Paleocene age of the oldest known nyctithere and the latest

Cretaceous age of *Deccanolestes* imply that the divergence of bats from other known mammals occurred at least as long ago as the latest Cretaceous (Hooker, 2001).

Overall, it is obvious that there are very large gaps in the fossil record for bats. In particular, the transition to the shared morphology of all known bats is not documented at all.

THE MOLECULAR EVIDENCE

Since 1994, evidence concerning the time of divergence between primates and other orders of placental mammals, which conflicts with a direct reading of the known fossil record, has been steadily accumulating from several independent studies of DNA sequence data. In a comparative analysis of the marsupial *Didelphis virginiana* and several placentals, taking sequence data for 8 mitochondrial genes with rates of evolution not significantly differing from a molecular clock model, a calibration date of 130 MYA for the marsupial/placental divergence yielded a date of 93 ± 2 MYA for the divergence between human (representing primates) and a group representing carnivores, artiodactyls, and cetaceans (ferungulates) (Janke et al., 1994). Subsequently, using sequence information for a large sample of nuclear genes showing relatively constant rates of change in mammals and birds, and taking a calibration date of 310 MYA for the separation between diapsid and synapsid reptiles, divergence times between primates and artiodactyls and between primates and rodents were both estimated to be around 90 MYA or older (Hedges et al., 1996). In a follow-up study based on a larger sample of species and nuclear gene sequences, it was found that inferred molecular dates calibrated in this way agree with most early (Paleozoic) and late (Cenozoic) paleontological dates, but that major gaps are apparent in the Mesozoic fossil record. It was inferred that at least five lineages of placental mammals arose more than 100 MYA and that most modern orders diverged before the end of the Cretaceous (Kumar and Hedges, 1998). On a separate tack, combined analysis of DNA sequences from three mitochondrial genes and two nuclear genes indicated that adaptive radiation from a specific common ancestor gave rise to a group of African mammals containing golden moles, hyraxes, manatees, elephants, elephant shrews, and aardvarks ("Afrotheria"). Using nine different calibration points within the mammalian tree (including a date of 130 MY for the marsupial/eutherian split and a date of 60 MY for the ruminant/cetacean split), the mean divergence time between Afrotheria and other orders of mammals (including primates) was estimated at about 90 MYA (Springer et al., 1997). In yet another approach, sequence data for the complete cytochrome b gene were used to generate a tree showing divergences between various mammal species, including 10 primates, and the tree was calibrated by taking a date of 60 MYA for the split between artiodactyls and cetaceans. This calibration indicated that primates diverged from other orders of mammals at about 90 MYA and that the split between haplorhine and strepsirrhine primates took place about 80 MYA (Arnason et al., 1996). The data set was subsequently expanded to include new sequence data for the baboon, and a double calibration based on the fossil record for ungulates was applied: 60 MY for the divergence between artiodactyls and cetaceans and 50 MY for the divergence between equids and rhinocerotids among perissodactyls. The time of divergence between ungulates and primates was estimated at 95 MYA, while the split between strepsirrhines and haplorhines was confirmed to be in the region of 80 MYA (Arnason et al., 1998). These studies consistently indicate that primates diverged from other placental mammals about 90 MYA.

A date of 90 MYA for the divergence between primates and other placentals has received further consistent support from several very recent studies. A new statistical technique for handling the variation of the molecular clock between lineages was applied to complete mitochondrial genome sequences for 23 mammalian species. Using a calibration of 56.5 MYA for the split between hippos and cetaceans, the method found a divergence time of 97.6 MYA for primates from a sister clade containing Artiodactyla, Perissodactyla, and Carnivora (Huelsenbeck et al., 2000). Another group of investigators constructed a phylogenetic tree for 26 placental taxa using up to 8665 bp of nuclear DNA. In supplementary information for their paper, they report only the time of the basal split of placental mammals at 111–118 MYA using two calibration points: elephants and hyraxes splitting at 60 MYA and hippos and cetaceans splitting at 55 MYA. However, we can interpolate their figure and conclude that their tree supports a primate divergence of approximately 90 MYA (Madsen et al., 2001). Subsequently, this data set was combined with that used in a parallel study (Murphy et al., 2001a) to yield an overall sequence set of 16,397 bp and to generate a consensus phylogeny for placental mammals (Murphy et al., 2001b). This combined study provided further confirmation for the existence of 4 superordinal groupings (Afrotheria,

Xenarthra, Laurasiatheria, and Euarchontoglires, the latter including Primates). Afrotheria was the first of these groupings to diverge, at an estimated date of 103 MYA, while the divergence between Laurasiatheria and Euarchontoglires was estimated at 79–88 MYA.

It should be noted that all of the molecular trees cited were calibrated using the ages of various known fossil representatives of lineages external to the order Primates. Given that first recorded fossil representatives must in all cases indicate *minimum* dates for times of divergence, it is striking that a relatively consistent result emerges with respect to inference of the time of divergence of primates. (This is perhaps because comparatively well-documented parts of the mammalian tree were selected as sources of calibration dates). It should also be emphasized that the primary concern in calibration of molecular trees to date has been the time of divergence of primates from other orders of placental mammals (node 1 in Figure 1). There has been relatively little interest in dating the last common ancestor of living primates (node 3 in Figure 1), although genetic distances uniformly indicate that the temporal gap between the initial divergence of primates and their common ancestor must have been relatively small.

QUANTIFYING THE INCOMPLETENESS OF THE FOSSIL RECORD

As already noted above, the earliest known unequivocal fossil primates are of basal Eocene age (about 55 MYA), and the standard view is that primates originated no earlier than about 65 MYA, close to and probably above the K/T-boundary, with their initial radiation following the extinction of the dinosaurs at the end of the Cretaceous.

Although the molecular evidence, when calibrated with various fossil dates outside the primate tree, consistently indicates that the lineage leading to living primates diverged from other placental mammal lineages about 90 MY ago (node 1 in Figure 1), it is conceivable that the diagnostic features of known living and fossil primates did not emerge until some time after this divergence (node 2 in Figure 1), and that the last common ancestor of living primates (node 3 in Figure 1) may be even more recent. It might, therefore, be imagined that a species-poor lineage with barely differentiated morphological features did indeed diverge from other placental mammals some 90 MYA, but did not lead to morphologically recognizable primates until 60–65 MYA. This could potentially explain the disparity between the known fossil record and molecular-based estimates of the time of divergence between primates and other mammals. However, available molecular evidence concerning the first divergence among living primates, between strepsirrhines and haplorhines, indicates that it took place relatively soon after the primates diverged from other placental mammals. In what appear to be the only published calibrations of the first divergence among living primates, a date of about 80 MYA is indicated (Arnason et al., 1996, 1998). Hence, if the diagnostic morphological features shared by all living primates and their known fossil relatives can be attributed to common ancestry rather than to convergent evolution (as is generally assumed), these features must have been present at an early stage. If primates diverged from other placental mammals about 90 MYA, the diagnostic features of the group must accordingly have been developed by about 80 MYA, well before the end of the Cretaceous, and a major gap must, therefore, exist preceding the known fossil record. The extent of that gap may in part be due to the K/T mass extinction. A loss of taxa at the K/T boundary and the possibility that some taxa were slow to recover from that event might to some extent explain the difficulty encountered in finding primates of modern aspect in the Paleocene. It should be noted that there is evidence indicating that biological recovery from major extinctions may take as long as 10 MY (Kirchner and Weil, 2000). However, in order to adequately interpret apparent discrepancies between molecular and fossil data it is necessary to develop methods that can quantitatively estimate degrees of incompleteness within the fossil record.

A simple calculation by Martin (1993) indicated that only 3% of extinct primate species have so far been documented. Rough correction for underestimation of the time of origin led to the inference that ancestral primates existed about 80 MYA. This preliminary inference has now been confirmed by our newly developed statistical approach (Tavaré et al., 2002), which is based on an estimate of species preservation derived from a model of the diversification pattern of the analyzed group. The method takes into account the number of extant species, the mean species lifetime, the ages of the bases of the relevant stratigraphic intervals, the numbers of fossil species found in those intervals, and the relative sizes of the sampling intensities in each interval. It can be used to estimate either: (1) the age of the last common ancestor of living primates or (2) the age of the first morphologically recognizable primate. A logistic diversification model was chosen in which logistic growth is parametrized by the time at which diversity reached 90% of its present value. Various diversification models can be explored with our method, but logistic growth is the most biologically realistic model (Raup et al., 1973), as it matches the general expectation of an equilibrium diversity level. The great diversity of Holarctic primates during the Eocene indicates that at least 90% of modern diversity would already have been reached by the middle Eocene. Consequently logistic growth was parametrized at 49 MYA. We used a mean species lifetime of 2.5 MY, but our results were relatively insensitive to changes in this value.

Our approach is based on modeling the speciation process as a nonhomogeneous Markov branching process with a specified diversification curve. This is a process in which species live for a random amount of time, go extinct and are replaced by a random number of species. The lifetime of the species and the number of descendant species are not affected by any of the other species alive at that time. This is a commonly accepted model for the diversification of a clade (Kubo and Iwasa, 1995; MacArthur and Wilson, 1963; Nee et al., 1994). The branching process allows us to compute the expected number of species alive in a given stratigraphic interval. Assuming that any species alive in such an interval can be fossilized and found with the same probability, we may calculate the expected number of species found as fossils in each stratigraphic interval. Our statistical method is based on matching the observed and expected number of fossil finds in each interval as closely as possible, and a parametric bootstrap approach is used to assess bias in the estimates and to find approximate confidence intervals.

Using this approach, we first determined an estimate of the age of the last common ancestor of living primates, (i.e., the time of divergence between strepsirrhines and haplorhines) as 81.5 MYA, with an approximate 95% confidence interval of (72.0, 89.6) MYA (Tavaré et al., 2002). This closely agrees with the only available molecular estimates of the strepsirrhine–haplorhine divergence (Arnason et al., 1996, 1998).

The age of the last common ancestor of living primates thus determined corresponds to node 3 in Figure 1. It gives the minimum age for the presence of morphological characteristics considered to be shared-derived features (autapomorphies) of primates of modern aspect, assuming that all known fossil primates of modern aspect belong within the phylogenetic tree for extant primates. The present consensus view is that the earliest known primates of modern aspect (early Eocene adapiforms and omomyiforms) are sister groups of modern strepsirrhines and haplorhines, respectively. However, it is conceivable that the adapiforms and/or the omomyiforms diverged prior to the last common ancestor of modern primates. To allow for this possibility, we here extend our previously published analyses (Tavaré et al., 2002) to estimate the time of the initial diversification of the primate clade (node 2 in Figure 1), which can be taken as the age of the first morphologically diagnosable primates of modern aspect. In our initial estimations of the time of divergence between living strepsirrhines and haplorhines, we considered only simulated trees in which an initial bifurcation led to living representatives on both sides. This implicitly incorporated the assumption that all known fossil primates of modern aspect are nested within the tree including all living primates. To allow for the possibilities that defining features of living primates might have emerged prior to their last common ancestor and that some fossil primates might have diverged prior to that ancestor, the analysis was repeated without the constraint of an initial bifurcation with surviving representatives on both sides of the tree.

In order to estimate the age of node 3 in Figure 1, we start the speciation process from 2 initial species, both leading to living descendants. However, to estimate the age of node 2 in Figure 1, the speciation process starts from a single species. We are then assuming that this first species and all its descendants would be identifiable as primates of modern aspect by a paleontologist. It is important to recognize that the combination of features distinguishing primates from their mammalian relatives—and probably distinguishing the first primates from earlier ancestors in the lineage leading to them—are unlikely to have evolved simultaneously. As a result, designation of the first morphologically recognizable primates on a temporal scale can only be hypothetical, and the estimate of their age is an approximate indication of when the acquisition of primate characteristics took place.

Repeating the model specifications that were used to estimate the age of the strepsirrhine-haplorhine divergence, the age of the first morphologically recognizable primates (node 2 in Figure 1) is estimated at 85.9 MYA, with a 95% confidence interval of (73.3, 95.7) MYA. Note that the estimate for the strepsirrhine–haplorhine divergence is only 4.5 MY younger than the best estimate for age of the first morphologically recognizable primate. If the notion of a first morphologically recognizable primate provokes discomfort, it is reassuring to know that the relatively short time span between this construct and the last common ancestor of living primates allows one to use the age of the first morphologically recognizable primate as a proxy for the age of the last common ancestor of living primates.

Therefore, for the most realistic model settings (i.e., assuming a logistic growth model with 90% of modern diversity reached by the base of the middle Eocene), our estimates of the emergence and subsequent diversification of primates of modern aspect are in broad agreement with molecular estimates of divergence times (Table 1). Other diversification models such as linear or exponential growth, as well as parametrization of the logistic growth with more recent dates, result in age estimates for the presence of the first morphologically recognizable primates that are even older.

In stark contrast with our results, Gingerich and Uhen (1994) argued, on the basis of a formalization of Martin's (1993) heuristic approach, that there is only a 5 in a billion chance (5×10^{-9}) that primates originated 80 MYA, and that, at a 95% confidence level, the origin of primates was located somewhere between 55 and 63 MYA. Using our updated data on the number of fossil primate species, the probability that primates originated 80 MYA calculated in this way in fact declines even further to a mere 2×10^{-18}. However, although modern species diversity is initially entered into the model by Gingerich and Uhen (1994), it eventually falls out of the equation

Table 1. Molecular and paleontological estimates of divergence and diversification times during early primate evolution

Estimated Node	Molecular estimates	Paleontological estimate, mean and 95% confidence limits, MYA
Node 1: divergence of the primate lineage from other modern mammals	~90 mya	NA
Node 2: initial diversification of primates/first morphologically recognizable primates	NA	85.9 (73.3, 95.7)
Node 3: Divergence of strepsirrhines and haplorhines/last common ancestor of living primates	~80 mya	81.5 (72.0, 89.6)

Paleontological estimates are derived from a statistical approach developed by Tavaré et al. (2002). Estimated nodes refer to the nodes in Figure 1. Molecular estimates are from Arnason et al. (1996, 1998), Hedges et al. (1996), Janke et al. (1994), Kumar and Hedges (1998), Springer et al. (1997).

that is applied. The results of the calculation are the same regardless of the number of modern primate species or estimated preservation rates and are as such based solely on the existing fossil record. As a consequence, their model is set to return the highest probability for the scenario in which the time of origin of a group is equal to the age of the oldest known fossil of that group. It, therefore, simply states that the more a scenario differs from a direct reading of the existing fossil record, the less likely it is to be real, thus entering precisely the kind of circularity which we have aimed to eliminate (Tavaré et al., 2002). The problem with such an approach can be illustrated by applying the method of Gingerich and Uhen (1994) to the complete gap that exists in the primate fossil record during the middle Oligocene. That gap, between the Fayum primates of the early Oligocene and the earliest occurrence of platyrrhines in the fossil record of South America in the late Oligocene, is likely to cover around 6 MY. Application of the method of Gingerich and Uhen (1994) yields a vanishingly small probability of 2×10^{-19} that primates existed during that gap.

In an analysis of evolutionary and preservational constraints on the times of divergence of eutherian mammals, Foote et al. (1999) concluded that molecular estimates of the times of origin of the living eutherian orders could be correct only if the preservation potential per lineage per million years was at least an order of magnitude smaller than it appeared to be. They consequently argued that it was unlikely for these ordinal divergences to have occurred as deep in the Cretaceous as the molecular clock data suggest. This conclusion, however, is not matched by our analyses of the fossil record of primates (Tavaré et al., 2002). The reason for this discrepancy seems to lie in the estimated preservation potential of mammalian lineages. Foote et al. estimate the preservation potential for Cenozoic mammals to be between 0.25 and 0.37/lineage/MY (Foote, 1997, Foote et al., 1999), and that of Cretaceous mammals to be 0.03/lineage/MY. Significantly, the average values for the preservation potential based on our approach are 0.023/lineage/MY for the known fossil record of primates, and 0.003/lineage/MY for the time prior to the first known fossils. These values are, in fact, an order of magnitude smaller than those determined by Foote et al. (1999). It thus seems that our two very different methods of analyses of the fossil record are not in conflict; where we differ is in the estimated preservability of taxa.

PRESERVATIONAL BIAS IN THE FOSSIL RECORD

There are several reasons why the preservation rates calculated by Foote et al. (1999) are likely to be overestimated. These all relate to the problem of circularity when interpreting the completeness of the fossil record through analysis of the fossil record alone. First, methods for assessing the completeness of the fossil record based exclusively on the fossil record can only account for gaps that occur within known lineages. They are insensitive to the existence of larger gaps, both chronological and geographical, and will overestimate completeness where such gaps occur. Foote (1997) demonstrated that the method used by Foote et al. (1999) will overestimate preservation potential where chronological gaps occur, with larger gaps within a given chronological range resulting in a larger overestimation. Even simple temporal variation in preservation probability will in most cases cause a slight to moderate overestimation of completeness (Foote, 1997). The primate fossil record as a whole has two large gaps. One, already noted above, extends over a period of about 6 MY during the middle Oligocene. The other is the gap between the origin of the order and its first fossil appearance, a gap which most would agree to be at least 5–10 MY and which we estimate to be over 25 MY. More gaps become apparent when individual lineages are considered. In the most dramatic primate example, documentation of Malagasy lemurs was, until very recently, strictly limited to subfossils just a few thousand years old. Yet it was known that lemurs must have existed much earlier, as the sister-group (lorisiforms) is documented by fossils that are at least 20 MY old (Szalay and Delson, 1979), and possibly over 30 MY old (Simons, 1995), thus documenting a ghost lineage (Norell, 1992) for lemurs extending at least that far back in time. Recent discoveries of fossil lorisiforms in the Fayum have now increased the minimum age of the lemur ghost lineage to about 40 MY (Seiffert et al. 2003). Very recently, a strepsirrhine primate (*Bugtilemur*) interpreted as a possible relative of the lemur family Cheirogaleidae has been recovered from Early Oligocene deposits of Pakistan (Marivaux et al., 2001). Rather than closing a gap, however, this new find illustrates just how little may be known about key aspects of primate evolution. The lemurs are a diverse group of modern primates known, until now, exclusively from the island of Madagascar. To explain the presence of a lemur in the Oligocene of Pakistan combined with the, as yet, total absence of fossil lemurs from anywhere else in the world, requires the contemplation of some fairly elaborate biogeographical scenarios.

Substantial geographical gaps are, in fact, likely to be the rule during the earlier phases of primate evolution. Living primates are essentially confined to tropical and subtropical climates (Martin 1990; Figure 2A). Support for the inference that this was also true in the past comes from the fact that primates only ever populated substantial parts of the northern continents when these areas supported subtropical climates at times of markedly increased global temperatures, during the Eocene and the Miocene. Yet, 47% of all known fossil primate species come from restricted areas of North America and Europe and, for the first half of paleontologically documented primate

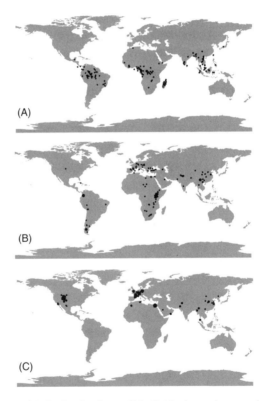

Figure 2. Geographical distribution of individual modern and fossil euprimate species, taking the mid-range point in each case and plotting in relation to present-day continental positions (updated from Tavaré et al., 2002): (A) modern and subfossil primates; (B) fossil species between the Late Pleistocene and the Late Oligocene; (C) fossil species between the early Oligocene and the early Eocene. (B) and (C) are separated by a fossil-free gap of 6 MY. Note the progressive southward shift from (C) to (A).

evolution, sites yielding fossil primates are largely restricted to these two regions (Figure 2C). A direct reading of the known fossil record would suggest that primates originated some time during the Paleocene in the northern continents and subsequently migrated southwards. An alternative interpretation is that primates originated earlier in the relatively poorly documented southern continents and expanded northwards when climatic conditions permitted during the Eocene and, to a lesser degree, during the Miocene. The preservation rates proposed by Foote et al. (1999) for modern eutherian mammals as a whole are based either entirely (for the Cenozoic rates) or to more than three quarters (for the Cretaceous rates) on North American faunas. North America is the best-sampled region in the world, and estimates based on that region will necessarily overestimate the preservation rates of groups with an almost worldwide distribution.

Our method also implies that approximately 5% and no more than 7% of all primate species that have ever existed are known from the fossil record. This low value does not seem unrealistic, as only 6–7% of all living primate species are known from the fossil record, a record that is expected to be better than the average, given that it is dominated by easily collected and relatively common Pleistocene sediments. In addition, the belief underlying any direct reading of the fossil record—namely that most of primate evolution has by now been unearthed and described—is easily refuted by the ongoing rate of publication of new species of fossil primates (Figure 3).

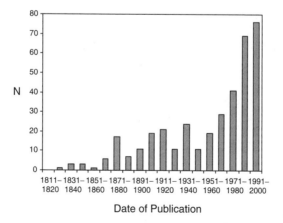

Figure 3. Histogram showing the numbers of new species of fossil primates by year of publication, grouped into decades.

To produce a precise estimate for the time of origin of primates (or any clade) using our method requires knowledge of the true diversity curve of the clade, of the relative sampling intensities of each stratigraphic interval, and of the mean species longevity (although the first is the most important in influencing the resulting estimates). As the values of these parameters are not known in detail, our estimates of the time of origin of primates must remain provisional. It is significant, however, that a number of our models produce dates concordant with various molecular estimates using calibration with fossil dates outside the primate tree (Arnason et al., 1996, 1998; Eizirik et al., 2001; Hedges et al., 1996; Huelsenbeck, 2000; Janke et al., 1994; Kumar and Hedges, 1998; Murphy et al., 2001b; Springer et al., 1997).

The poor sampling that we have inferred for the primate fossil record is unlikely to be restricted to that group. Cretaceous divergence times for primates and other modern orders of mammals should now be considered the more likely scenario, in which case the influence of continental drift has probably been considerable (Easteal et al., 1995; Hedges et al., 1996; Martin, 1990; Murphy et al., 2001b). Clearly, fossil evidence from appropriate regions is needed to test this proposition. In the case of primates, it can be predicted that early members of the order showing characteristic morphological features lived somewhere in the southern continents (i.e., on parts of the former Gondwanaland) approximately 85 MYA.

ACKNOWLEDGMENTS

The authors thank Marian Dagosto and Matt Ravosa for inviting them to contribute to this volume and for organizing the conference on primate origins. ST and OW were supported in part by NSF grant DBI 95–04393. CRM was supported in part by NSF grant, EAR–9258045, by NASA grant, NCC2–1053 and by American Chemical Society Petroleum Research Fund Grant, 31316–AC8. CS was supported by Janggen-Poehn-Stiftung, by the G. & A. Claraz Schenkung and by a grant to RDM from the A. H. Schultz-Stiftung.

REFERENCES

Arnason, U., Gullberg, A., Janke, A., and Xu, X.-F., 1996, Pattern and timing of evolutionary divergences between hominoids based on analyses of complete mtDNAs, *J. Mol. Evol.* **43**: 650–661.

Arnason, U., Gullberg, A., and Janke, A., 1998, Molecular timing of primate divergences as estimated by two nonprimate calibration points, *J. Mol. Evol.* **47**: 718–727.

Bajpai, S., and Gingerich, P. D., 1998, A new Eocene archaeocete (Mammalia, Cetacea) from India and the time of origin of whales, *Proc. Natl. Acad. Sci. USA* **95**: 15464–15468.

Corbet, G. E., and Hill, J. E., 1991, *A World List of Mammalian Species*, 3rd Ed., Oxford University Press, Oxford.

Ducrocq, S., Jaeger, J.-J., and Sigé, B., 1993, Un mégachiroptère dans l'Eocène supérieur de Thaïlande: Incidence dans la discussion phylogénique du groupe. *Neues Jahrb. Geol. Paläontol, Monatschr.* **9**: 561–575.

Easteal, S., Collett, C., and Betty, D., 1995, *The Mammalian Molecular Clock*, R.G. Landes, Austin, Texas.

Eizirik, E., Murphy, W. J., and O'Brien, S. J., 2001, Molecular dating and biogeography of the early placental mammal radiation, *J. Hered.* **92**: 212–219.

Foote, M., 1997, Estimating taxonomic durations and preservation probability, *Paleobiology* **23**: 278–300.

Foote, M., Hunter, J. P., Janis, C. M., and Sepkoski, J. J. Jr., 1999, Evolutionary and preservational constraints on origins of biological groups: Divergence times of eutherian mammals, *Science* **283**: 1310–1314.

Gatesy, J., 1997, More DNA support for a Cetacea/Hippopotamidae clade: The blood-clotting protein gene γ-fibrinogen, *Mol. Biol. Evol.* **14**: 537–543.

Gatesy, J., Hayashi, C., Cronin, M. A., and Arctander, P., 1996, Evidence from milk casein genes that cetaceans are close relatives of hippopotamid artiodactyls, *Mol. Biol. Evol.* **13**: 954–963.

Gatesy, J., Milinkovitch, M., Waddell, V., and Stanhope, M., 1999, Stability of cladistic relationships between Cetacea and higher-level artiodactyl taxa, *Syst. Biol.* **48**: 6–20.

Gingerich, P. D., and Uhen, M. D., 1994, Time of origin of primates, *J. Hum. Evol.* **27**: 443–445.

Graur, D., and Higgins, D. G., 1994, Molecular evidence for the inclusion of cetaceans within the order Artiodactyla, *Mol. Biol. Evol.* **11**: 357–364.

Hedges, S. B., Parker, P. H., Sibley, C. G., and Kumar, S., 1996, Continental breakup and the ordinal diversification of birds and mammals, *Nature (London)*, **381**: 226–229.

Hooker, J. J., 2001, Tarsals of the extinct insectivoran family Nyctitheriidae (Mammalia): Evidence for archontan relationships, *Zool. J. Linn. Soc.* **132**: 501–529.

Hooker, J. J., Russell, D. E., and Phélizon, A., 1999, A new family of Plesiadapiformes (Mammalia) from the Old World Lower Paleogene, *Palaeontology* **42**: 377–407.

Huelsenbeck, J. P., Larget, B., and Swofford, D., 2000, A compound Poisson process for relaxing the molecular clock, *Genetics* **154**: 1879–1892.

Janke, A., Feldmaier-Fuchs, G., Thomas, W. K., von Haeseler, A., and Pääbo, S., 1994, The marsupial mitochondrial genome and the evolution of placental mammals, *Genetics* **137**: 243–256.

Kirchner, J. W., and Weil, A., 2000, Delayed biological recovery from extinctions throughout the fossil record, *Nature (London)*, **404**: 177–180.

Kubo, T., and Iwasa, Y., 1995, Inferring the rates of branching and extinction from molecular phylogenies, *Evolution* **49**: 695–704.

Kumar, S., and Hedges, S. B., 1998, A molecular timescale for vertebrate evolution, *Nature (London)*, **392**: 917–920.

MacArthur, R. H., and Wilson, E. O., 1963, An equilibrium theory of insular zoogeography, *Evolution* **17**: 373–387.

Madsen, O., Scally, M., Doudy, C. J., Kao, D., DeBry, R.W., Adkins, R. et al., 2001, Parallel adaptive radiations in two major clades of placental mammals, *Nature (London)*, **409**: 610–614.

Marivaux, L., Welcomme, J.-L., Antoine, P.-O., Métais, G., Baloch, I. M., Benammi, M., et al., 2001, A fossil lemur from the Oligocene of Pakistan, *Science* **294**: 587–591.

Martin, R. D., 1986, Primates: A definition, in: *Major Topics in Primate and Human Evolution*, B. A. Wood, L. B. Martin, and P. Andrews, eds., Cambridge University Press, Cambridge; pp 1–31.

Martin, R. D., 1990, *Primate Origins and Evolution: A Phylogenetic Reconstruction*, Princeton University Press, New Jersey.

Martin, R. D., 1993, Primate origins: Plugging the gaps, *Nature (London)*, **363**: 223–234.

Murphy, W. J., Eizirik, E., Johnson, W. E., Zhang, Y. P., Ryder, O. A., and O'Brien, S. J., 2001a, Molecular phylogenetics and the origins of placental mammals, *Nature (London)*, **409**: 614–618.

Murphy, W. J., Eizirik, E., O'Brien, S. J., Madsen, O., Scally, M., Douady, C. J., et al., 2001b, Resolution of the early placental mammal radiation using Bayesian phylogenetics, *Science* **294**: 2348–2351.

Nee, S., May, R. M., and Harvey, P. H., 1994, The reconstructed evolutionary process. *Phil. Trans. R. Soc. Lond. B* **344**: 305–311.

Nikaido, M., Rooney, A. P., and Okada, N., 1999, Phylogenetic relationships among certartiodactyls based on insertions of short and long interspersed elements: Hippopotamuses are the closest extant relatives of whales. *Proc. Natl. Acad. Sci. USA* **96**: 10261–10266.

Norell, M. A., 1992, Taxic origin and temporal diversity: The effect of phylogeny, in: *Extinction and Phylogeny*, M. J. Novacek and Q. D. Wheller, eds., Columbia University Press, New York, pp. 89–118.

Raup, D. M., Gould, S. J., Schopf, T. M., and Simberloff, D. S., 1973, Stochastic models of phylogeny and the evolution of diversity. *J. Geol.* **81**: 525–542.

Sarich, V. M., 1993, Mammalian systematics: twenty-five years among their albumins and transferrins, in: *Mammal Phylogeny, vol. 1: Placentals*, F. S. Szalay, M. J. Novacek, and M. C. McKenna, eds., Berlin Heidelberg New York, pp. 103–114.

Seiffert, E. R., Simons, E. L., and Attia, Y., 2003, Fossil evidence for an ancient divergence of lorises and galagos. *Nature (London)*, **422**: 421–424.

Sigé, B., Jaeger, J.-J., Sudre, J., and Vianey-Liaud, M., 1990, *Altiatlasius koulchii* n.gen. et sp., primate omomyidé du Paléocène supérieur du Maroc, et les origines des Euprimates. *Palaeontographica. Abt. A* **214**: 31–56.

Simmons, N. B., and Geisler, J. H., 1998, Phylogenetic relationships of *Icaronycteris, Archaeonycteris, Hassianycteris*, and *Palaeochiropteryx* to extant bat lineages, with comments on the evolution of echolocation and foraging strategies in Microchiroptera. *Bull. Amer. Mus. Nat. Hist.* **235**: 4–182.

Simons, E. L., 1995, Egyptian Oligocene primates: A review. *Yb. Phys. Anthropol.* **38**: 199–238.

Smith, A. B., 1994, *Systematics and the Fossil Record*, Blackwell Scientific, Oxford.

Springer, M. S., Cleven, G. C., Madsen, O., de Jong, W. W., Waddell, V. G., Amrine, H. M., and Stanhope, M. J., 1997, Endemic African mammals shake the phylogenetic tree. *Nature (London)*, **388**: 61–64.

Stucky, R. K., and McKenna, M. C., 1993, Mammalia in: *The Fossil Record II*, M. J. Benton, ed., Chapman & Hall, London, pp. 739–771.

Szalay, F. S., and Delson, E., 1979, *Evolutionary History of the Primates*, Academic Press, New York.

Tavaré, S., Marshall, C. R., Will, O., Soligo, C., and Martin, R. D., 2002, Estimating the age of the last common ancestor of extant primates using the fossil record. *Nature (London)*, **416**: 726–729.

Thewissen, J. G. M., Williams, E. M., Roe, L. J., and Hussain, S. T., 2001, Skeletons of terrestrial cetaceans and the relationship of whales to artiodactyls. *Nature (London)*, **413**: 277–281.

Tong, Y-S., 1988, Fossil tree shrews from the Eocene Hetaoyuan Formation of Xichuan, Henan. *Vert. PalAsiat.* **26**: 214–220.

Ursing, B. M., and Arnason, U., 1998, Analyses of mitochondrial genomes strongly support a hippopotamus-whale clade. *Proc. Roy. Soc. Lond. B* **265**: 2251–2255.

CHAPTER THREE

The Postcranial Morphology of *Ptilocercus lowii* (Scandentia, Tupaiidae) and Its Implications for Primate Supraordinal Relationships

Eric J. Sargis

INTRODUCTION

"[I]t is certain that the *tree shrews represent a highly important group of mammals*, and, for this reason, *they demand an intensive study from all aspects*."

—Le Gros Clark (1927, p. 255, italics added)

"Among living non-primates the tupaiids are apparently the closest primate relatives, and *these conclusions in no way lessen the value of tupaiids to primatology*."

—McKenna (1966, p. 9, italics added)

Eric J. Sargis • Department of Anthropology, Yale University, New Haven, CT 06520

Tree shrews (Scandentia, Tupaiidae) are small-bodied mammals from South and Southeast Asia. They have long been considered to have close affinities with primates and are often used as an outgroup in analyses of relationships among primate taxa (e.g., Shoshani et al., 1996). Despite volumes of recent debate concerning their relationships to primates and other mammals (see Luckett, 1980; MacPhee, 1993), the supraordinal relationships of tree shrews remain poorly understood. A better understanding of tupaiid evolutionary relationships has been hindered by their poor fossil record, which consists of teeth and skull fragments from the Miocene of India, Pakistan (Chopra and Vasishat, 1979; Chopra et al., 1979; Jacobs, 1980), China (Ni and Qiu, 2002; Qiu, 1986), and Thailand (Mein and Ginsburg, 1997), as well as the Eocene of China (Tong, 1988). Only one postcranial specimen, a ribcage from the Pliocene of India reported by Dutta (1975), has been suggested to represent a tupaiid (see Sargis, 1999, for a review of tupaiid fossils). Furthermore, while tupaiid craniodental morphology has been relatively well studied (see Butler, 1980; Cartmill and MacPhee, 1980; MacPhee, 1981; Steele, 1973; Wible and Martin, 1993; Wible and Zeller, 1994; Wöhrmann-Repenning, 1979; Zeller, 1986a,b, 1987), tupaiid postcranial morphology was poorly known and had not been studied from a functional morphological perspective prior to Sargis (2000). In order to gain a better understanding of the character states found in the Tupaiidae, which, in turn, should provide primate systematists with a better understanding of this often-used outgroup, I conducted a functional morphological study of the tupaiid postcranium (Sargis, 2000, 2001, 2002a,b,c).

TAXONOMY AND PHYLOGENY OF THE FAMILY TUPAIIDAE

Taxonomy

The order Scandentia is represented by the single family Tupaiidae, which includes the subfamilies Ptilocercinae and Tupaiinae (Table 1). Ptilocercinae is represented only by *Ptilocercus lowii*, while Tupaiinae consists of *Tupaia* (14 species), *Dendrogale* (2 species), *Urogale everetti*, and *Anathana ellioti* (see Table 1; Wilson, 1993). Differences in postcranial morphology are often split down the subfamilial line, as *Ptilocercus'* postcranium is adapted for arboreal locomotion, while that of the tupaiines is adapted for terrestrial locomotion (Sargis, 2000, 2001, 2002a,b,c).

Table 1.　Classification of tree shrews (Wilson, 1993)

Order Scandentia
　Family Tupaiidae
　　Subfamily Tupaiinae
　　　Tupaia (14 species)
　　　Dendrogale (2 species)
　　　Anathana ellioti
　　　Urogale everetti
　　Subfamily Ptilocercinae
　　　Ptilocercus lowii

Supraordinal Relationships of Tupaiids

Tree shrews were originally included in the order Insectivora by Wagner (1855), and Haeckel (1866) later grouped them in the insectivoran suborder Menotyphla with the elephant shrews (see Table 2; Butler, 1972). A close relationship between tupaiids and primates was first suggested in 1910 when Gregory proposed the superorder Archonta, which included Chiroptera, Dermoptera, Primates, and Menotyphla (which he recognized as an order). Carlsson (1922) moved tupaiids to the order Primates, and Le Gros Clark (1924a,b, 1925, 1926) strongly supported this grouping with his studies of tupaiid anatomy (Table 2). Simpson (1945) considered Archonta to be an unnatural group, but at the same time he supported the inclusion of tupaiids in the order Primates (Table 2). Tupaiids continued to be classified as Primates until the 1960s (see Napier and Napier, 1967), when they were removed from

Table 2.　History of tree shrew ordinal designations

Order Insectivora	Wagner (1855)
	Haeckel (1866)
Order Menotyphla	Gregory (1910)
Order Primates	Carlsson (1922)
	Le Gros Clark (1924a,b, 1925, 1926)
	Simpson (1945)
	Napier and Napier (1967)
Removed from order Primates	Van Valen (1965)
	Jane et al. (1965)
	McKenna (1966)
	Campbell (1966a,b)
	Martin (1966, 1968a,b)
	Szalay (1968, 1969)
Order Scandentia	Butler (1972)

the order by Campbell (1966a,b), Jane et al. (1965), Martin (1966, 1968a,b), McKenna (1966), Szalay (1968, 1969), and Van Valen (1965) (Table 2). Most similarities between the two groups were deemed to be erroneous observations, shared primitive characters, or convergences found only in derived representatives of the groups rather than ancestral morphotypes. Once tupaiids were removed from the order Primates, Butler (1972) classified them in their own order Scandentia (Table 2)—a name used (at the family level) by Wagner (1855) for tupaiids. In 1975, McKenna accepted Butler's (1972) classification of tupaiids as an independent order Scandentia, and he revised Gregory's (1910) superorder Archonta by including the orders Scandentia, Primates, Chiroptera, and Dermoptera and excluding the Macroscelidea (elephant shrews). Szalay (1977) supported this grouping with evidence from the tarsus, and within the Archonta, he recognized a Primate–Scandentia clade and a Chiroptera–Dermoptera clade (called Volitantia; see Figure 1A; Table 3).

Szalay and Drawhorn (1980) found further support for the archontan hypothesis in the tarsus of fossil plesiadapiforms and other archontan mammals. Novacek and Wyss (1986) came to support the archontan hypothesis (Table 3) based on one tarsal character (from Szalay and Drawhorn, 1980) and one penial character (from Smith and Madkour, 1980), despite Novacek's (1980, 1982, 1986) previous agreement with Cartmill and MacPhee (1980) that this hypothesis was not supported by cranial or postcranial evidence. Since 1986, Novacek has become one of the major proponents of both the Archonta and Volitantia hypotheses (Novacek, 1989, 1990, 1992, 1993, 1994; Novacek et al., 1988). Wible (Wible and Covert, 1987; Wible and Novacek, 1988) has also supported these hypotheses, and has provided evidence for a Scandentia–Euprimates clade (Table 3) that does not include the "archaic primates" (Plesiadapiformes) (also supported by Kay et al., 1992). Some subsequent studies have continued to bolster support for the Archonta hypothesis (Johnson and Kirsch, 1993; Shoshani and McKenna, 1998; Szalay and Lucas, 1993, 1996), while others have rejected it (e.g., Kay et al., 1990, 1992). Recently, McKenna and Bell (1997) have reconfirmed support of the Archonta in their classification of mammals.

Within the Archonta, however, interordinal relationships have not been agreed upon. For instance, Beard's (1989) detailed functional morphological analysis of archontan postcranial morphology led him to reject the Volitantia hypothesis. Beard argued that plesiadapiforms should be included in the order Dermoptera (a hypothesis also supported by Kay et al., 1990, 1992)

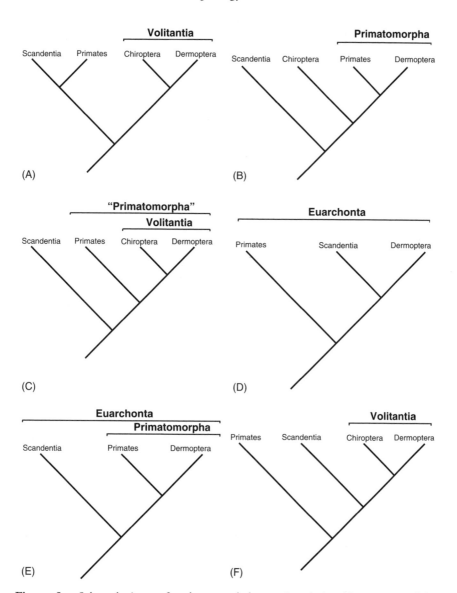

Figure 1. Selected views of archontan phylogenetic relationships supported by: (A) Novacek (1992), Szalay (1977), Wible and Covert (1987), Wible and Novacek (1988); (B) Beard (1993b); (C) Shoshani and McKenna (1998); (D) Liu and Miyamoto (1999), Liu et al. (2001), Murphy et al. (2001a,b), this study (Analysis 3); (E) Waddell et al. (1999); and (F) Silcox (2001a,b, 2002), this study (Analysis 2).

Table 3. Clades supported by various phylogenetic analyses discussed in the text

	Archonta or Euarchonta	Within Archonta
Szalay, 1977	Archonta	Volitantia, Primates + Scandentia
Novacek and Wyss, 1986	Archonta	Volitantia
Wible and Novacek, 1988	Archonta	Volitantia, Euprimates + Scandentia
Beard, 1993b	Archonta	Primatomorpha
Shoshani and McKenna, 1998	Archonta	Volitantia
Waddell et al., 1999	Euarchonta	Primatomorpha
Liu and Miyamoto, 1999	Euarchonta	Dermoptera + Scandentia
Liu et al., 2001	Euarchonta	Dermoptera + Scandentia
Murphy et al., 2001a,b	Euarchonta	Dermoptera + Scandentia
Silcox, 2001a,b, 2002	Archonta	Volitantia, Primates (*sensu lato*)

and that it is Dermoptera and Primates, rather than Dermoptera and Chiroptera that form a natural group, which he named Primatomorpha (Figure 1B; Table 3). The evidence for this grouping includes similarities between the morphology of the intermediate manual phalanges of dermopterans and paromomyids that may be related to gliding (Beard, 1990, 1993a), but other characters have been used to support this hypothesis as well (Beard, 1993b). This grouping of Dermoptera with Primates contrasts sharply with the grouping of Dermoptera with Chiroptera in Volitantia, which has been strongly supported in numerous morphological studies using extremely varied databases and phylogenetic methods (see Bloch et al., 2002; Johnson and Kirsch, 1993; Kriz and Hamrick, 2001; Novacek, 1982, 1986, 1989, 1990, 1992, 1993, 1994; Novacek and Wyss, 1986; Novacek et al., 1988; Sargis, 2002d; Shoshani and McKenna, 1998; Silcox, 2001a,b, 2002; Szalay, 1977; Szalay and Lucas, 1993, 1996; Thewissen and Babcock, 1991, 1993; Wible, 1993; Wible and Covert, 1987; Wible and Novacek, 1988). Simmons (1995), Simmons and Quinn (1994), and Thewissen and Babcock (1992) have also supported Volitantia rather than Primatomorpha, but they incorporated Beard's (1990, 1993a) results by including the paromomyids in the order Dermoptera (but see Bloch and Silcox, 2001). McKenna and Bell (1997), while not recognizing Beard's Primatomorpha, also included the Paromomyidae in Dermoptera and further recognized Beard's (1993a,b) results by including Dermoptera as a suborder of the order Primates (but see Szalay, 1999, for criticisms). Most of the other plesiadapiform families were placed by McKenna and Bell (1997) in the order Primates, but not specifically in the suborder Dermoptera. A curious exception was the placement of the Carpolestidae within the Tarsiiformes in the suborder Euprimates (see Szalay, 1999). Shoshani and McKenna (1998) recognized both Volitantia and Primatomorpha, but their "Primatomorpha" was a grouping of Volitantia and Primates (Figure 1C; Table 3). This concept of Primatomorpha has surely lost the meaning that Beard (1993a,b) intended because Shoshani and McKenna (1998) argued that Chiroptera was the closest relative of Dermoptera, while Beard (1993a,b) supported a Dermoptera–Primates clade. It is significant that when Beard's (1993b) data set was incorporated into the much larger data set of Shoshani and McKenna (1998), the signal for a Dermoptera–Primates clade was lost, whereas a Dermoptera–Chiroptera clade was supported. Recently, Bloch et al. (2002) and Silcox (2001a,b, 2002) supported both a Volitantia–Scandentia clade (Figure 1F) and a Plesiadapiformes–Euprimates clade (i.e., Primates, *sensu lato*; Table 3). Their studies rejected Beard's (1993a,b) Primatomorpha, and Silcox's (2001a,b, 2002) classification included plesiadapiforms in Primates rather than in Dermoptera (contra Beard, 1989, 1993a,b).

Beard (1989, 1990, 1991, 1993a,b) has clearly advanced the debate about archontan phylogenetics, but his studies have been criticized on numerous grounds. Krause (1991) questioned the identifications and associations of the paromomyid specimens that Beard (1989, 1990) analyzed. Szalay and Lucas (1993, 1996) questioned and reevaluated many of the postcranial characters that Beard used to support both his concept of Primatomorpha and his hypothesis concerning the "mitten" or finger-gliding capabilities of paromomyids (based on intermediate phalangeal proportions). Simmons (1994) showed that 2 of Beard's (1993b) 29 characters included erroneous observations, while Stafford and Thorington (1998) showed that 2 additional characters included erroneous observations and another character was misinterpreted (see Sargis, 2002d). Shoshani and McKenna (1998) used only 12 of Beard's (1993b) 29 characters in their phylogenetic analysis because they said that Beard himself stated that the others were questionable. Hamrick et al. (1999) rejected the "mitten" or finger-gliding capabilities of paromomyids based on their analysis of phalangeal proportions. They did recognize several similarities between the phalanges of dermopterans and paromomyids; however, they interpreted these features not as gliding adaptations but as adaptations for vertical clinging and climbing on large arboreal supports. They also showed that a phalangeal feature previously believed to be unique to paromomyids and dermopterans (Beard, 1993b) is also found in chiropterans (see below; Thewissen and Babcock, 1992), and they identified an additional derived phalangeal character shared by chiropterans, dermopterans, and paromomyids (Hamrick et al., 1999). These characters support the amended volitantian concept (where paromomyids are included in Dermoptera; but see Bloch and Silcox, 2001) of Simmons (1995), Simmons and Quinn (1994), and Thewissen and Babcock (1992) rather than Beard's (1993a,b) Primatomorpha. Lemelin (2000) further supported Volitantia rather than Primatomorpha with a unique feature of the volar skin that is shared by dermopterans and chiropterans. Finally, Sargis (2002d) demonstrated that 12 of Beard's (1993b) 22 postcranial characters should be interpreted differently when *Ptilocercus*, rather than *Tupaia*, is used to represent Scandentia, which greatly reduces the evidence for Primatomorpha.

In contrast to many of these morphological studies, molecular studies have consistently supported a group that includes Dermoptera, Scandentia, and Primates to the exclusion of Chiroptera (Adkins and Honeycutt, 1991, 1993; Allard et al., 1996; Cronin and Sarich, 1980; Honeycutt and Adkins, 1993; Liu and Miyamoto, 1999; Liu et al., 2001; Murphy et al., 2001a,b; Porter et al., 1996; Waddell et al., 1999). This group has been called Euarchonta by Waddell et al. (1999). Despite this apparent consensus regarding the exclusion of bats from the Archonta based on molecular evidence, there has been little agreement concerning which order represents the closest relative of the remaining archontan orders (see Allard et al., 1996). Several orders, including Macroscelidea, Lagomorpha, Rodentia, and occasionally both Lagomorpha and Rodentia (placed in the supraordinal grouping Glires), have been proposed to be more closely related to the remaining members of the Archonta than are bats (Allard et al., 1996; Bailey et al., 1992; Goodman et al., 1994; Honeycutt and Adkins, 1993; Madsen et al., 2001; Miyamoto, 1996; Murphy et al., 2001a,b; Porter et al., 1996; Stanhope et al., 1993, 1996; Waddell et al., 1999). Recently, a Euarchonta–Glires clade has received the most support (Murphy et al., 2001a,b), and this clade has been named Euarchontoglires by Murphy et al. (2001b).

Some molecular studies have specifically supported a Scandentia–Lagomorpha clade. Graur et al. (1996) argued that lagomorphs are very closely related to primates and tree shrews, and they tentatively concluded that Lagomorpha represents the sister taxon of Scandentia. This hypothesis was also supported by Schmitz et al. (2000). However, Liu and Miyamoto

(1999) recently found the most support for a Dermoptera–Scandentia clade (also supported by the molecular data of Liu et al., 2001; Madsen et al., 2001; Murphy et al., 2001a,b; see Figure 1D; Table 3), while Waddell et al. (1999), in the same volume, grouped Dermoptera with Primates in the Primatomorpha (also supported by the molecular analyses of Teeling et al., 2000; Killian et al., 2001; see Figure 1E; Table 3). The latter conclusion is particularly significant with regard to the analysis conducted by Graur et al. (1996), who stated that the "phylogenetic position of Dermoptera relative to Primates and Lagomorpha could not be resolved with the available data" (p. 335). Liu and Miyamoto's (1999) conclusion is also noteworthy because Graur et al. (1996) never tested a Scandentia–Dermoptera relationship and Schmitz et al. (2000) did not include Dermoptera in their analysis. Perhaps Graur et al. (1996) and Schmitz et al. (2000) would not have supported a Scandentia–Lagomorpha relationship if they had included a test of a Scandentia–Dermoptera clade in their studies (see Liu and Miyamoto, 1999; Liu et al., 2001; Madsen et al., 2001; Murphy et al., 2001a,b). Similarly, it is possible that Primatomorpha would not have been supported by Teeling et al. (2000) if they had included Scandentia in their analysis, thereby testing a Scandentia–Dermoptera relationship.

The exclusion of chiropterans from Archonta based on molecular data is not only a revision of the morphological concept of Archonta, but it is also a rejection of the Volitantia hypothesis. Hence, the molecular concept of Euarchonta is incompatible with the morphological concepts of Archonta and Volitantia. It seems, therefore, that there is little congruence between morphological and molecular data concerning these alternative phylogenetic hypotheses. Beard's Primatomorpha hypothesis (1989, 1993a,b), however, is based on morphological evidence and is concordant with the molecular concept of Euarchonta. It is the competing hypotheses of Volitantia and Primatomorpha that will be considered here in a reexamination of some of the postcranial evidence.

Significance of *Ptilocercus*

A study of the postcranium of the least well-known order in the Archonta, Scandentia, was undertaken by Sargis (2000) in order to provide additional information to contribute to an understanding of the relationships among archontan mammals. The inclusion of *Ptilocercus* in this study was critical because *Ptilocercus* has long been considered to be the living taxon most closely resembling the ancestral tupaiid in both its ecology and morphological attributes (Butler, 1980; Campbell, 1974; Emmons, 2000; Gould, 1978; Le Gros Clark, 1926; Martin, 1990; Sargis, 2000, 2001, 2002a,b,c,d; Szalay, 1969; Szalay and Drawhorn, 1980; Szalay and Lucas, 1993, 1996) and thus must play a paramount role in any supraordinal phylogenetic analysis that includes the Tupaiidae. A better understanding of tupaiid supraordinal relationships is likely confounded by the common use of *Tupaia*, a relatively derived tupaiid (see Martin, 1990), to represent Scandentia in studies of mammalian supraordinal relationships (e.g., Beard, 1989, 1993b), and these relationships would likely be better understood if *Ptilocercus* was included in the analysis (Sargis, 2002d). Most previous studies have also used *Tupaia*, rather than *Ptilocercus*, as an outgroup when the relationships among various groups of primates were being examined (e.g., Shoshani et al., 1996).

In this chapter, I reanalyzed Beard's (1993b) data specifically in light of the fact that Beard (1993b) did not use *Ptilocercus* to represent Scandentia in his analysis. I also recoded characters based on an *a priori* character analysis (Sargis, 2002d), and added additional postcranial characters to see how they affect the results.

MATERIALS AND METHODS

I examined tupaiid skeletal specimens at the following institutions: American Museum of Natural History (AMNH), New York; Field Museum of Natural History (FMNH), Chicago; United States National Museum of Natural History (USNM), Washington, DC; Museum of Comparative Zoology (MCZ) at Harvard University, Cambridge; Natural History Museum (NHM), London; Muséum national d'Histoire naturelle (MNHN), Paris; Nationaal Natuurhistorisch Museum (NNM), Leiden; Muséum d'Histoire Naturelle (MHN), Geneva; Forschungsinstitut Senckenberg (FS), Frankfurt; Zoologische Staatssammlung (ZS), Munich; and the Swedish Museum of Natural History (SMNH), Stockholm. In addition, I studied plesiadapiform postcranial specimens at the AMNH, the MNHN, and the Carnegie Museum of Natural History (CMNH) in Pittsburgh and examined postcranial specimens of other archontans at the AMNH, MCZ, and in the research collection of F. S. Szalay (FSS). All of the specimens examined in this study are listed in Sargis (2000, Table 2.2).

In this chapter, I tested Beard's (1993b) hypotheses using his own methods in order to determine the effects of changing some of the variables (i.e., taxa and characters) and therefore, followed Beard (1993b) in conducting a phylogenetic analysis using PAUP (Swofford, 1993, Version 3.1.1). Like Beard (1993b), I performed an exhaustive search. No characters were ordered or weighted, and trees were rooted using a hypothetical outgroup (all characters coded as zeros). Three separate analyses were conducted using Beard's (1993b) methods, each with some change to the variables included in the analysis.

Analysis 1: In this analysis, two variables are different from those in Beard (1993b). A) While Beard's 29-character data set was used, several characters were recoded because some of Beard's character codings included erroneous observations (e.g., see Hamrick et al., 1999; Simmons, 1994; Stafford and Thorington, 1998). The relevant references for character states that were subsequently corrected are included with all of the characters and character states in Appendix A. B) While I included the same taxa as Beard, I used a different genus to represent the order Scandentia. Beard used *Tupaia* to represent Scandentia in his analysis, but *Ptilocercus* was used to represent Scandentia in this analysis because it has been proposed to be the most primitive living tupaiid (see earlier section). The purpose of this analysis was to determine if the use of *Ptilocercus* to represent Scandentia had any effect on the results. Characters #1–22 in Appendix A represent Beard's (1993b) postcranial characters (#8–29). Beard's (1993b) craniodental characters (#1–7) do not vary between *Tupaia* and *Ptilocercus*.

Analysis 2: In this analysis, only postcranial data were used, but additional postcranial characters, some of which have been used to support the Volitantia hypothesis (e.g., see Simmons, 1995), were added to the data set. The combined postcranial data set included Beard's (1993b) 22 postcranial characters and 20 other postcranial characters from additional sources (see Appendix A). Appendix B represents the character–taxon matrix for this analysis. In addition to the exhaustive search (following Beard, 1993b), a bootstrap analysis with 1000 replicates was also performed. The purpose of this analysis was to determine if adding additional postcranial characters to the data set changes the results.

Analysis 3: In this analysis, Chiroptera was excluded because molecular analyses have consistently supported the exclusion of bats from Archonta (e.g., see Murphy et al., 2001a,b). The purpose of this analysis was to determine if Primatomorpha is supported when it is impossible for Volitantia to be supported. In other words, if the possibility of a Dermoptera–Chiroptera clade (Volitantia) is eliminated, then is a Dermoptera–Primates clade (Primatomorpha) supported?

Functional analyses of tupaiid postcranial characters, as well as an *a priori* analysis of Beard's (1993b) characters, have been reported elsewhere (Sargis, 2001, 2002a,b,c,d). In this chapter, I report the results of an *a posteriori* character analysis conducted in MacClade (Maddison and Maddison, 2001, Version 4.03) in which I mapped the characters in Appendix A onto the tree produced in recent molecular analyses (Figure 1D).

RESULTS

Analysis 1: The single most parsimonious tree from Beard's (1993b) analysis is shown in Figure 2. When *Ptilocercus* is used to represent Scandentia, the only change in topology is that Scandentia, rather than Chiroptera, is the sister to Primatomorpha (Figure 3). Primatomorpha, however, is still supported (Figure 3). The tree is 52 steps long, the consistency index (CI) is 0.85, and the retention index (RI) is 0.83.

Analysis 2: A single most parsimonious tree with a length of 64 steps (CI: 0.77; RI: 0.67) was recovered (Figure 4). Ten trees were 65 steps long and 10 trees 66 steps long. These 21 trees varied in terms of the placement of the 3

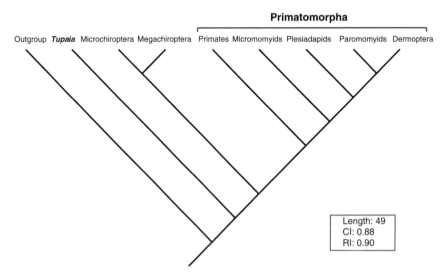

Figure 2.   Phylogeny from Beard (1993b). CI = consistency index; RI = retention index. Note that *Tupaia* was used to represent Scandentia.

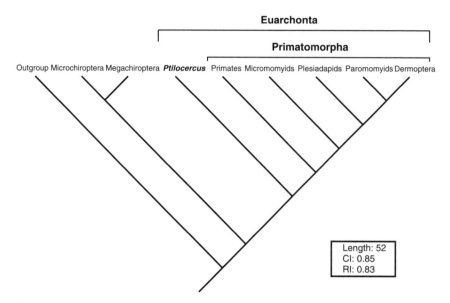

Figure 3. Phylogeny from Analysis 1. CI = consistency index; RI = retention index. Note that *Ptilocercus* was used to represent Scandentia and that Primatomorpha was still supported. However, the molecular concept of Euarchonta is also supported when *Ptilocercus* is used to represent Scandentia.

plesiadapiform families due to missing data in these fossil groups. Volitantia was supported in all 21 of the most parsimonious trees (Figure 4), while Primatomorpha was not supported in any of them. Bootstrap support for Volitantia (79%) was also relatively strong (Figure 5). Volitantians, scandentians, and plesiadapiforms consistently formed a clade (with primates as the sister taxon) that had relatively strong bootstrap support (78%). When Beard's (1993b) tree was reproduced using this data set, the resulting tree was 82 steps long. In other words, forcing Beard's (1993b) tree topology to be recovered requires 18 additional steps. In fact, 82 steps is closer to the longest tree at 94 steps than it is to the shortest tree at 64 steps, and there were 2647 trees more parsimonious than Beard's (1993b) topology (i.e., those that were 64–81 steps long).

Analysis 3: A single most parsimonious tree with a length of 51 steps (CI: 0.88; RI: 0.74) was recovered (Figure 6). When Chiroptera was removed from the analysis (as suggested by molecular evidence), Primatomorpha was still not supported. Instead, a Dermoptera–Scandentia clade was supported (Figure 6).

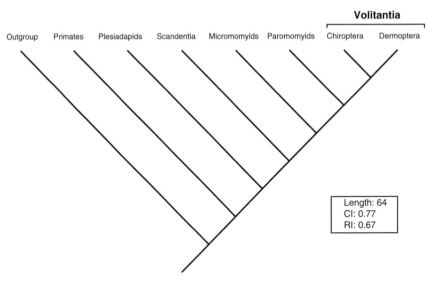

Figure 4. Single most parsimonious tree from Analysis 2. CI = consistency index; RI = retention index. Note that Volitantia is supported and that Scandentia is the extant sister to Volitantia.

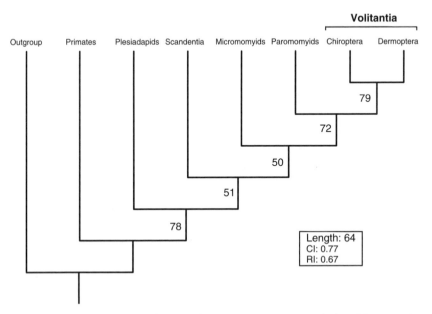

Figure 5. Bootstrap analysis from Analysis 2. CI = consistency index; RI = retention index. Note that Volitantia is well supported at 79%.

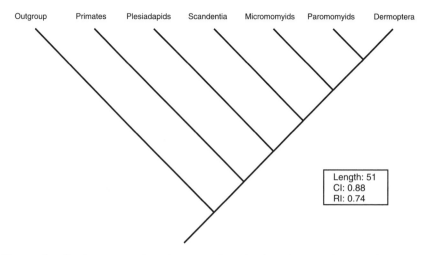

Outgroup   Primates   Plesiadapids   Scandentia   Micromomyids   Paromomyids   Dermoptera

Length: 51
CI: 0.88
RI: 0.74

Figure 6. Single most parsimonious tree from Analysis 3. Note that Scandentia, not Primates, is the extant sister taxon to Dermoptera. Hence, Primatomorpha is not supported, but a Dermoptera–Scandentia clade is supported. The latter conclusion is congruent with the results of molecular analyses (see text).

DISCUSSION

Analysis 1: Replacing *Tupaia* with *Ptilocercus* in Beard's (1993b) analysis produces only a minor change in tree topology. Whereas Beard's (1993b) analysis supported a Primatomorpha–Chiroptera clade (Figure 2), this analysis instead supports a Primatomorpha–Scandentia clade (Figure 3). This change in tree topology is not particularly surprising because Beard (1993b) stated that support for a Primatomorpha–Chiroptera clade was weak. The change is significant, however, in light of recent molecular results. Beard's (1993b) tree topology did not support the molecular concept of Euarchonta (see Figure 1D and E), but the inclusion of *Ptilocercus* in the analysis changes the tree topology so that Euarchonta is supported (Figure 3). Hence, when *Ptilocercus* is included in the analysis, the results are more congruent with those from molecular studies (see Adkins and Honeycutt, 1991, 1993; Allard et al., 1996; Cronin and Sarich, 1980; Honeycutt and Adkins, 1993; Liu and Miyamoto, 1999; Liu et al., 2001; Murphy et al., 2001a,b; Porter et al., 1996; Waddell et al., 1999).

Analysis 2: When additional characters were added to the analysis, Primatomorpha was no longer supported. In fact, in all 21 most parsimonious trees, Volitantia was supported (Figure 4) and bootstrap support for this clade (79%) was relatively strong (Figure 5). That Scandentia was the extant sister taxon to Volitantia is congruent with other recent phylogenetic analyses of cranial, dental, and postcranial data (Bloch et al., 2002; Silcox, 2001a,b, 2002).

It is not surprising that the placement of the three plesiadapiform families was not consistent in the 21 most parsimonious trees, as many of their character states could not be coded due to missing data in these extinct groups (Appendix B). Hence, little can be said about the relationships of plesiadapiforms based on this analysis. In studies of larger data sets that include cranial and dental data in addition to postcranial data, however, plesiadapiforms are consistently found to be the sister taxon to Euprimates, and should therefore be included with Euprimates in the order Primates (Bloch et al., 2002; Silcox, 2001a,b, 2002).

Analysis 3: Support of Volitantia in *Analysis 2* is not congruent with molecular results, which exclude bats from Archonta altogether. Chiroptera was, therefore, removed from the analysis as suggested by molecular data, yet Primatomorpha was still not supported (Figure 6). Hence, even when a Dermoptera–Chiroptera clade could not be supported, a Dermoptera–Primates clade was still not supported. In other words, although it was impossible for Volitantia to be supported, there was still no support for Primatomorpha. Alternatively, a Dermoptera–Scandentia clade was supported (Figure 6), and this is congruent with molecular results (Liu and Miyamoto, 1999; Liu et al., 2001; Madsen et al., 2001; Murphy et al., 2001a,b). This is a particularly interesting phylogenetic hypothesis from a biogeographic perspective because both of these taxa are endemic to South and Southeast Asia. If the Dermoptera–Scandentia clade represents a natural grouping, then Volitantia represents an unnatural grouping based on convergences rather than homologies, and dermopterans and chiropterans must have evolved their similarities independently in relation to gliding and flying, respectively (Sargis, 2002d). Insofar as Primatomorpha is concerned, it is not supported whether bats are included in the analysis or not, so there is more evidence for either Volitantia or the Dermoptera–Scandentia clade than there is for Primatomorpha. Hence, Primatomorpha likely represents an unnatural grouping (Sargis, 2002d), and it must be rejected based on these analyses.

In an attempt to further examine potential morphological evidence for a Dermoptera–Scandentia clade, the characters in Appendix A were mapped onto the molecular phylogeny shown in Figure 1D. Possible synapomorphies

Table 4. Possible synapomorphies of Euarchonta and a Dermoptera–Scandentia clade

Euarchonta

1. Robust humeral lesser tuberosity with strong medial protrusion (character #2)
2. Spherical capitulum (character #3)
3. Circular and deeply excavated radial central fossa (character #4)
4. Elliptical acetabulum that is elongated craniocaudally (character #13)
5. Cranial expansion of bony buttressing on acetabulum (character #14)
6. Enlarged, flattened, triangular area between greater and lesser trochanters for insertion of quadratus femoris (character #16)
7. Synovial distal tibiofibular joint (character #18)
8. Concave cuboid facet of calcaneus (calcaneocuboid pivot) (character #21)
9. Wide distal facet on the entocuneiform (character #22)

Dermoptera–Scandentia Clade

1. Cuneiform contacts two bones radially (character #9)
2. Deep ungual phalanges that are highly compressed mediolaterally and tall dorsoventrally (character #12)
3. Short, wide, shallow patellar groove (character #17)
4. Craniocaudally wide atlas vertebra (character #23)
5. Short, wide thoracic spinous processes (character #24)
6. Short lumbar spinous processes (character #25)
7. Lumbar transverse processes short and face laterally (character #26)
8. Short, wide scapula (character #27)
9. Small greater trochanter (character #28)
10. Anteroposteriorly shallow femoral condyles (character #29)
11. Short cervical (C3–C7) spinous processes (character #32)
12. Craniocaudally expanded ribs (character #33)

of both Euarchonta and the Dermoptera–Scandentia clade are listed in Table 4. Nine synapomorphies of Euarchonta were found when bats were excluded from the analysis (Table 4). Several of these were considered by Beard (1993b) to be synapomorphies of Primatomorpha, but they are also found in *Ptilocercus* (Sargis, 2002a,b,d). The Dermoptera–Scandentia clade, on the other hand, is supported by 12 synapomorphies (Table 4). Hence, there appears to be some morphological support for this clade, which previously has been supported only by molecular evidence. Here again, the inclusion of *Ptilocercus* in the analysis is critical because this morphological support is based on similarities between *Cynocephalus* and *Ptilocercus*, not *Tupaia*.

CONCLUSIONS

In summary, while Beard's (1993b) methods were followed as closely as possible, the results of this exercise do not support the Primatomorpha hypothesis. This is true even when bats are removed from the analysis so that the competing Volitantia hypothesis cannot possibly be supported (Analysis 3). When the results of these phylogenetic analyses are considered together with the results of the character analyses conducted by Hamrick et al. (1999), Sargis (2002d), Stafford and Thorington (1998), and Szalay and Lucas (1993, 1996), as well as the phylogenetic analyses of Bloch et al. (2002), Kriz and Hamrick (2001), Shoshani and McKenna (1998), and Silcox (2001a,b, 2002), the Primatomorpha hypothesis must be rejected. Analysis 2 does, however, support Volitantia (Figures 4 and 5), and this hypothesis has been further corroborated by the character analyses of Hamrick et al. (1999), Stafford and Thorington (1998), Szalay and Lucas (1993, 1996), and Thewissen and Babcock (1992), as well as the phylogenetic analyses of Bloch et al. (2002), Kriz and Hamrick (2001), Shoshani and McKenna (1998), and Silcox (2001a,b, 2002). Based on morphological evidence, therefore, it would appear that the sister taxon of Dermoptera is Chiroptera, not Primates (Bloch et al., 2002; Johnson and Kirsch, 1993; Kriz and Hamrick, 2001; Novacek, 1982, 1986, 1989, 1990, 1992, 1993, 1994; Novacek and Wyss, 1986; Novacek et al., 1988; Sargis, 2002d; Shoshani and McKenna, 1998; Silcox, 2001a,b, 2002; Simmons, 1995; Simmons and Quinn, 1994; Szalay, 1977; Szalay and Lucas, 1993, 1996; Thewissen and Babcock, 1991, 1992, 1993; Wible, 1993; Wible and Covert, 1987; Wible and Novacek, 1988; contra Beard, 1989, 1993a,b; McKenna and Bell, 1997). Again, it seems that Primatomorpha represents an unnatural grouping (see previous section; Sargis, 2002d).

If Volitantia is better supported by morphological studies than Primatomorpha, then what is the sister taxon to Volitantia? Shoshani and McKenna's (1998) analysis supported a Volitantia–Euprimates clade, but Analysis 2 of this study supported a Volitantia–Scandentia clade (Figures 4 and 5). The latter clade has also been supported by the analyses of Silcox (2001a,b, 2002) and Bloch et al. (2002).

The most significant problem with the morphological support of Volitantia is that molecular studies have continually rejected this clade (Adkins and Honeycutt, 1991, 1993; Allard et al., 1996; Cronin and Sarich, 1980; Honeycutt and Adkins, 1993; Liu and Miyamoto, 1999; Liu et al., 2001; Murphy et al., 2001a,b; Porter et al., 1996; Waddell et al., 1999). It is, of course, possible that dermopterans and chiropterans evolved their similarities independently (see above), so Volitantia may represent an unnatural grouping based on convergent, rather than homologous, characters (Sargis, 2002d).

While a Scandentia–Euprimates clade has been supported by cranial evidence (Kay et al., 1992; Wible and Covert, 1987; Wible and Novacek, 1988), such a clade is not supported by this analysis of postcranial data. Alternatively, molecular evidence has repeatedly supported a Scandentia–Dermoptera clade (Liu and Miyamoto, 1999; Liu et al., 2001; Madsen et al., 2001; Murphy et al., 2001a,b) with Primates as the sister taxon to this clade (Figure 1D; Liu and Miyamoto, 1999; Liu et al., 2001; Murphy et al., 2001a,b). A Scandentia–Dermoptera clade is also supported by postcranial evidence when Chiroptera is removed from the analysis (Analysis 3; Figure 6). This clade is supported by craniodental evidence as well, but only when Chiroptera is removed from the phylogenetic analysis (Bloch et al., 2002; Silcox, personal communication). The most probable sister taxon of Dermoptera, therefore, may be Scandentia rather than either Primates or Chiroptera.

The relationship of plesiadapiforms to Euprimates could not be fully assessed here because of missing data in fossil plesiadapiforms, as well as the fact that only postcranial data were analyzed. In recent studies that included craniodental evidence, however, a Plesiadapiformes–Euprimates clade (i.e., Primates, *sensu lato*) was supported (Bloch et al., 2002; Silcox, 2001a,b, 2002). Plesiadapiforms should, therefore, be included in Primates rather than Dermoptera (Bloch et al., 2002; Silcox, 2001a,b, 2002).

Finally, the inclusion of *Ptilocercus* in these analyses had significant effects on the results. For instance, when *Ptilocercus* is used to represent Scandentia in Beard's (1993b) analysis (Analysis 1), the molecular concept of Euarchonta is supported (Figure 3; Table 4), whereas this clade was not supported when Beard (1993b) used *Tupaia* to represent Scandentia (Figure 2). The inclusion of *Ptilocercus* also resulted in the identification of morphological synapomorphies for a Dermoptera–Scandentia clade (Table 4), a grouping previously supported only by molecular evidence. Hence, in studies of primate supraordinal relationships that include comparisons of postcranial characters to those in tupaiids, *Ptilocercus* should certainly be included in the analysis (Sargis, 2000, 2002a,b,d). Similarly, if tupaiids are chosen as an outgroup in primate phylogenetic analyses that include postcranial evidence, then *Ptilocercus* should be used as the outgroup because its attributes are more conservative for Scandentia (Sargis, 2000, 2002a,b,d).

ACKNOWLEDGMENTS

For access to specimens, I am grateful to the following people and institutions: Ross MacPhee and Darrin Lunde, Department of Mammalogy; Mark Norell, Malcolm McKenna, and John Alexander, Department of Vertebrate

Paleontology, American Museum of Natural History, New York; Larry
Heaney and Bill Stanley, Field Museum of Natural History, Chicago; Richard
Thorington and Linda Gordon, National Museum of Natural History,
Washington, DC; Maria Rutzmoser, Museum of Comparative Zoology at
Harvard University, Cambridge; Chris Beard, Carnegie Museum of Natural
History, Pittsburgh; Marc Godinot, Christian de Muizon, Pascal Tassy;
Brigitte Senut, Department of Paleontology, and Michel Tranier and Jacques
Cuisin, Department of Mammalogy, Muséum national d'Histoire naturelle,
Paris; Paula Jenkins, Natural History Museum, London; Chris Smeenk,
Nationaal Natuurhistorisch Museum, Leiden; Louis de Roguin and Albert
Keller, Muséum d'Histoire Naturelle, Geneva; Gerhard Storch,
Forschungsinstitut Senckenberg, Frankfurt; Richard Kraft and Michael
Hiermeier, Zoologische Staatssammlung, Munich; and Olavi Grönwall,
Swedish Museum of Natural History, Stockholm. I thank Matt Ravosa and
Marian Dagosto for inviting me to contribute to this volume, as well as for
inviting me to present these data at the International Conference on Primate
Origins and Adaptations: A Multidisciplinary Perspective. Thanks also to
Larissa Swedell for her many helpful comments on previous drafts of this
manuscript and thanks to Katie Binetti and Michael Muehlenbein for check-
ing the character mapping in my Topics and Issues in Systematics graduate
seminar. This work was funded by a National Science Foundation Doctoral
Dissertation Improvement Grant (SBR-9616194), a Field Museum of
Natural History Visiting Scholarship, a Sigma Xi Scientific Research Society
Grant-in-Aid of Research, and a New York Consortium in Evolutionary
Primatology graduate fellowship.

APPENDIX A

Descriptions of Characters and Character States

1. Position of deltopectoral crest: anterior (0); lateral (1) (character #8 in Beard, 1993b).
2. Robusticity of lesser tuberosity: gracile, no strong medial protrusion (0); robust, strong medial protrusion (1) (character #9 in Beard, 1993b; chiropteran condition corrected by Simmons, 1994).
3. Shape of capitulum: spindle-shaped (0); spherical (1) (character #10 in Beard, 1993b).

4. Shape and degree of excavation of radial central fossa: ovoid and shallow (0); circular and deep (1) (character #11 in Beard, 1993b).

5. Extent of lateral lip around perimeter of proximal radius: broad, limited to lateral side (0); narrow, extends approximately halfway around (1) (character #12 in Beard, 1993b).

6. Form of ulnocarpal articulation: mediolaterally and dorsopalmarly extensive, lies in transverse plane (0); limited to radial and palmar aspects of distal ulna, lies in proximodistal plane (1) (character #13 in Beard, 1993b; chiropteran condition corrected by Stafford and Thorington, 1998).

7. Shape of cuneiform in dorsal view: quadrate (0); triangular (1) (character #14 in Beard, 1993b).

8. Spatial relationships of lunate and scaphoid: lunate ulnar to scaphoid (0); lunate distal to scaphoid (1) (character #15 in Beard, 1993b; dermopteran condition corrected by Stafford and Thorington, 1998).

9. Radial articular contacts of cuneiform: contact with single bone (lunate) (0); contact with two bones (1) (character #16 in Beard, 1993b; chiropteran, dermopteran, and scandentian conditions corrected by Stafford and Thorington, 1998).

10. Size of pisiform: moderately robust (0); reduced (1) (character #17 in Beard, 1993b; chiropteran condition corrected by Simmons, 1994; and Stafford and Thorington, 1998).

11. Phalangeal proportions: proximal longer than intermediate (0); intermediate longer than proximal (1) (character #18 in Beard, 1993b; chiropteran condition corrected by Thewissen and Babcock, 1992; and Hamrick et al., 1999).

12. Shape of distal phalanges: moderately laterally compressed and moderately high dorsoventrally (0); highly compressed mediolaterally and tall dorsoventrally (1); mediolaterally wide and dorsoventrally flattened (2) (character #19 in Beard, 1993b; chiropteran condition corrected by Szalay and Lucas, 1993, 1996; Simmons, 1995; Lemelin, 2000).

13. Acetabular shape: circular in lateral view (0); elliptical, elongated craniocaudally (1) (character #20 in Beard, 1993b).

14. Pattern of bony buttressing around acetabulum: evenly developed around circumference (0); emphasized on cranial side (1) (character #21 in Beard, 1993b).

15. Position of fovea capitis femoris: centrally placed on femoral head (0); posterior to midline (1) (character #22 in Beard, 1993b).

16. Area of insertion of quadratus femoris: limited area on posterior side of proximal femoral shaft (0); enlarged, flattened, triangular area between greater and lesser trochanters (1) (character #23 in Beard, 1993b).

17. Shape of patellar groove: long, narrow, and moderately excavated (0); short, wide, and shallow (1); deeply excavated (2) (character #24 in Beard, 1993b).

18. Nature of distal tibiofibular joint: syndesmosis (0); synovial (1) (character #25 in Beard, 1993b).

19. Position of flexor fibularis groove on posterior side of astragalus: midline (0); lateral (1); groove absent (2) (character #26 in Beard, 1993b).

20. Secondary articulation between posterior side of sustentaculum and astragalus: absent (0); articulation between medial malleolus of tibia and posterior side of sustentaculum (1); present (2); sustentaculum reduced or absent (3) (character #27 in Beard, 1993b; recoded to reflect autapomorphous condition of sustentaculum in Scandentia).

21. Nature of calcaneocuboid articulation: cuboid facet on calcaneus moderately concave or flat, which articulates with calcaneal facet on cuboid that is evenly convex, oval, and elongated mediolaterally (0); plantar pit or concavity on distal calcaneus that articulates with proximally projecting process on cuboid (calcaneocuboid pivot) (1) (character #28 in Beard, 1993b).

22. Form of distal facet on entocuneiform: narrow distally (0); wide distally (1); entocuneiform proximodistally short with flat and triangular distal facet for first metatarsal (2) (character #29 in Beard, 1993b; the condition of the plantodistal process was excluded from these character states because its size does not correspond to the width of the distal entocuneiform facet as Beard, 1993b, originally proposed).

23. Size of atlas vertebra: craniocaudally narrow (0); craniocaudally wide (1) (see Sargis, 2001).

24. Shape of thoracic spinous processes: long and narrow (0); short and wide (1) (see Sargis, 2001).

25. Size of lumbar spinous processes: long (0); short (1) (see Sargis, 2001).

26. Size and orientation of lumbar transverse processes: long and face ventrally (0); short and face laterally (1) (see Sargis, 2001).

27. Shape of scapula: long and narrow (0); short and wide (1) (see Sargis, 2002a).

28. Size of greater trochanter: large (0); small (1) (see Sargis, 2002b).

29. Size of femoral condyles: anteroposteriorly deep (0); anteroposteriorly shallow (1) (see Sargis, 2002b).

30. Nature of volar skin: papillary ridges present (0); papillary ridges absent (1) (see Lemelin, 2000).

31. Shape of proximal articular surface on pedal intermediate phalanges: mediolaterally wide and dorsoplantarly compressed (0); dorsoplantarly high and mediolaterally compressed (1) (see Hamrick et al., 1999).

32. Size of cervical (C3–C7) spinous processes: long (0); short or absent (1) (see Simmons, 1995; Wible and Novacek, 1988).

33. Size of ribs: craniocaudally narrow (0); craniocaudally wide (1) (see Simmons, 1995; Szalay and Lucas, 1993, 1996; Wible and Novacek, 1988).

34. Size of forelimb: no elongation (0); markedly elongated (1) (see Simmons, 1995; Wible and Novacek, 1988).

35. Form of deltopectoral crest: shelf that extends distally (0); single torus displaced proximally (1) (see Simmons, 1995; Szalay and Lucas, 1993, 1996).

36. Humeropatagialis muscle: absent (0); present (1) (see Simmons, 1995; Wible and Novacek, 1988).

37. Form of proximal ulna: not reduced, contacts anterior humerus (0); reduced, anterior humeral contact reduced (disengagement) (1) (see Simmons, 1994, 1995; Szalay and Lucas, 1993, 1996; Wible and Novacek, 1988).

38. Form of distal radius and ulna: radius and ulna unfused, distal radius narrow, and no deep grooves for carpal extensors on dorsal surface of radius (0); radius and ulna fused, distal radius transversely widened, and deep grooves for carpal extensors on dorsal surface of radius (1) (see Simmons, 1995; Szalay and Lucas, 1993, 1996).

39. Fusion of carpals: unfused (0); fusion of scaphoid and lunate into scapholunate, centrale free (1); fusion of scaphoid, lunate, and centrale into scaphocentralunate (2) (see Simmons, 1995; Szalay and Lucas, 1993, 1996).

40. Patagium between manual digits: absent (0); present (1) (see Simmons, 1995; Szalay and Lucas, 1993, 1996; Wible and Novacek, 1988).

41. Size of fourth and fifth pedal rays: no elongation (0); elongated (1) (see Simmons, 1995; Szalay and Lucas, 1993, 1996).

42. Pedal digital tendon locking mechanism: absent (0); present (1) (see Simmons, 1995; Simmons and Quinn, 1994).

APPENDIX B

Character–Taxon Matrix*

	1	2	3	4	5	6	7	8	9	10	11	12	13	14	15	16	17	18	19	20	21	22	23	24	25	26	27	28	29	30	31	32	33	34	35	36	37	38	39	40	41	42
Scandentia	0	1	1	0	0	0	0	1	0	0	1	1	1	1	0	0	1	1	0	1	1	1	1	1	1	1	1	1	0	1	0	1	1	0	0	0	0	0	1	0	0	0
Chiroptera	0	1	0	0	0	1	0	0	1	0	1	1	1	1	2	1	0	0	0	2	3	0	1	1	1	1	1	1	1	0	1	1	1	1	1	1	0	1	2	1	1	1
Primates	1	1	1	1	1	0	0	0	0	0	2	1	1	1	1	1	2	1	1	0	1	1	0	1	0	0	0	0	1	1	0	0	0	0	0	0	0	0	0	0	1	0
Micromomyidae	1	1	1	1	1	0	?	?	?	?	?	1	1	1	1	1	1	1	1	0	1	?	1	1	?	?	?	?	1	1	?	?	?	?	?	?	0	0	?	1	?	?
Plesiadapidae	1	1	1	1	1	0	0	?	?	0	1	1	1	1	0	1	0	1	1	1	1	1	0	?	1	?	?	?	1	1	0	1	0	?	0	?	0	0	?	1	1	1
Paromomyidae	1	1	1	1	1	1	?	?	1	0	1	1	1	1	1	1	1	1	1	2	1	1	1	1	1	1	1	1	1	1	1	1	1	1	1	1	1	1	0	1	1	1
Dermoptera	0	1	1	1	1	1	1	0	1	0	1	1	1	1	1	1	1	1	1	2	1	1	1	1	1	1	1	1	1	1	1	1	1	1	1	1	1	1	2	1	1	0
Outgroup	0	0	0	0	0	0	0	0	0	0	0	0	0	0	0	0	0	0	0	0	0	0	0	0	0	0	0	0	0	0	0	0	0	0	0	0	0	0	0	0	0	0

*Characters 1 through 22 are from Beard (1993b). Character states in bold were coded differently in this analysis than in Beard's (1993b) study (see text and Appendix A).

REFERENCES

Adkins, R. M., and Honeycutt, R. L., 1991, Molecular phylogeny of the superorder Archonta, *Proc. Nat. Acad. Sci.* **88**: 10317–10321.

Adkins, R. M., and Honeycutt, R. L., 1993,. A molecular examination of archontan and chiropteran monophyly, in: *Primates and Their Relatives in Phylogenetic Perspective*, R. D. E. MacPhee, ed., Plenum Press, New York, pp. 227–249.

Allard, M. W., McNiff, B. E., and Miyamoto, M. M., 1996, Support for interordinal eutherian relationships with an emphasis on primates and their archontan relatives, *Mol. Phylo. Evol.* **5**: 78–88.

Bailey, W. J., Slightom, J. L., and Goodman, M., 1992, Rejection of the "flying primate" hypothesis by phylogenetic evidence from the ε-globin gene, *Science* **256**: 86–89.

Beard, K. C., 1989, *Postcranial Anatomy, Locomotor Adaptations, and Paleoecology of Early Cenozoic Plesiadapidae, Paromomyidae, and Micromomyidae (Eutheria, Dermoptera)*, Ph.D. Dissertation, Johns Hopkins University.

Beard, K. C., 1990, Gliding behavior and paleoecology of the alleged primate family Paromomyidae (Mammalia, Dermoptera), *Nature* **345**: 340–341.

Beard, K. C., 1991, Vertical postures and climbing in the morphotype of Primatomorpha: Implications for locomotor evolution in primate history, in: *Origine(s) de la Bipédie chez les Hominidés*, Y. Coppens and B. Senut, eds., CNRS, Paris, pp. 79–87.

Beard, K. C., 1993a, Origin and evolution of gliding in Early Cenozoic Dermoptera (Mammalia, Primatomorpha), in: *Primates and Their Relatives in Phylogenetic Perspective*, R. D. E. MacPhee, ed., Plenum Press, New York, pp. 63–90.

Beard, K. C., 1993b, Phylogenetic systematics of the Primatomorpha, with special reference to Dermoptera, in: *Mammal Phylogeny: Placentals*, F. S. Szalay, M. J. Novacek, and M. C. McKenna, eds., Springer-Verlag, New York, pp. 129–150.

Bloch, J. I., and Silcox, M. T., 2001, New basicrania of Paleocene-Eocene *Ignacius*: Re-evaluation of the plesiadapiform-dermopteran link, *Am. J. Phys. Anthro.* **116**: 184–198.

Bloch, J. I., Silcox, M. T., and Sargis, E. J., 2002, Origin and relationships of Archonta (Mammalia, Eutheria): Re-evaluation of Eudermoptera and Primatomorpha, *J. Vert. Paleo.* **22** (Supp. to No. 3): 37A.

Butler, P. M., 1972, The problem of insectivore classification, in: *Studies in Vertebrate Evolution*, K. A. Joysey and T. S. Kemp, eds., Oliver and Boyd, Edinburgh, pp. 253–265.

Butler, P. M., 1980, The tupaiid dentition, in: *Comparative Biology and Evolutionary Relationships of Tree Shrews*, W. P. Luckett, ed., Plenum Press, New York, pp. 171–204.

Campbell, C. B. G., 1966a, Taxonomic status of tree shrews, *Science* **153**: 436.

Campbell, C. B. G., 1966b, The relationships of the tree shrews: The evidence of the nervous system, *Evolution* **20**: 276–281.

Campbell, C. B. G., 1974, On the phyletic relationships of the tree shrews, *Mamm. Rev.* **4**: 125–143.

Carlsson, A., 1922, Über die Tupaiidae und ihre Beziehungen zu den Insectivora und den Prosimiae, *Acta Zool., Stockholm* **3**: 227–270.

Cartmill, M., and MacPhee, R. D. E., 1980, Tupaiid affinities: The evidence of the carotid arteries and cranial skeleton, in: *Comparative Biology and Evolutionary Relationships of Tree Shrews*, W. P. Luckett, ed., Plenum Press, New York, pp. 95–132.

Chopra, S. R. K., and Vasishat, R. N., 1979, Sivalik fossil tree shrew from Haritalyangar, India, *Nature* **281**: 214–215.

Chopra, S. R. K., Kaul, S., and Vasishat, R. N., 1979 Miocene tree shrews from the Indian Sivaliks, *Nature* **281**: 213–214.

Cronin, J. E., and Sarich, V. M., 1980, Tupaiid and Archonta phylogeny: The macromolecular evidence, in: *Comparative Biology and Evolutionary Relationships of Tree Shrews*, W. P. Luckett, ed., Plenum Press, New York, pp. 293–312.

Dutta, A. K., 1975, Micromammals from Siwaliks, *Indian Minerals* **29**: 76–77.

Emmons, L. H., 2000, *Tupai: A Field Study of Bornean Treeshrews*, Berkeley: University of California Press.

Goodman, M., Bailey, W. J., Hayasaka, K., Stanhope, M. J., Slightom, J., and Czelusniak, J., 1994, Molecular evidence on primate phylogeny from DNA sequences, *Am. J. Phys. Anthro.* **94**: 3–24.

Gould, E., 1978, The behavior of the moonrat, *Echinosorex gymnurus* (Erinaceidae) and the pentail tree shrew, *Ptilocercus lowii* (Tupaiidae) with comments on the behavior of other Insectivora, *Zeit. Tierpsychologie* **48**: 1–27.

Graur, D., Duret, L., and Gouy, M., 1996, Phylogenetic position of the order Lagomorpha (rabbits, hares and allies), *Nature* **379**: 333–335.

Gregory, W. K., 1910, The orders of mammals, *Bull. Am. Mus. Nat. Hist.* **27**: 1–524.

Haeckel, E., 1866, *Generelle Morphologie der Organismen*, Berlin: Georg Reimer.

Hamrick, M. W., Rosenman, B. A., and Brush, J. A., 1999, Phalangeal morphology of the Paromomyidae (?Primates, Plesiadapiformes): The evidence for gliding behavior reconsidered, *Am. J. Phys. Anthro.* **109**: 397–413.

Honeycutt, R. L., and Adkins, R. M., 1993, Higher level systematics of eutherian mammals: An assessment of molecular characters and phylogenetic hypotheses, *Ann. Rev. Ecol. Syst.* **24**: 279–305.

Jacobs, L. L., 1980, Siwalik fossil tree shrews, in: *Comparative Biology and Evolutionary Relationships of Tree Shrews*, W. P. Luckett, ed., Plenum Press, New York, pp. 205–216.

Jane, J. A., Campbell, C. B. G., and Yashon, D., 1965, Pyramidal tract: A comparison of two prosimian primates, *Science* **147**: 153–155.

Johnson, J. I., and Kirsch, J. A. W., 1993, Phylogeny through brain traits: Interordinal relationships among mammals including Primates and Chiroptera, in: *Primates and*

Their Relatives in Phylogenetic Perspective, R. D. E. MacPhee, ed., Plenum Press, New York, pp. 293–331.

Kay, R. F., Thorington, R. W., and Houde, P., 1990, Eocene plesiadapiform shows affinities with flying lemurs not primates, *Nature* **345**: 342–344.

Kay, R. F., Thewissen, J. G. M., and Yoder, A. D., 1992, Cranial anatomy of *Ignacius graybullianus* and the affinities of the Plesiadapiformes, *Am. J. Phys. Anthro.* **89**: 477–498.

Killian, J. K., Buckley, T. R., Stewart, N., Munday, B. L., and Jirtle, R. L., 2001, Marsupials and eutherians reunited: Genetic evidence for the Theria hypothesis of mammalian evolution, *Mamm. Genome* **12**: 513–517.

Krause, D. W., 1991, Were paromomyids gliders? Maybe, maybe not, *J. Hum. Evol.* **21**: 177–188.

Kriz, M., and Hamrick, M. W., 2001, The postcranial evidence for primate superordinal relationships, *Am. J. Phys. Anthro.* Supp. **32**: 93.

Le Gros Clark, W. E., 1924a, The myology of the tree shrew (*Tupaia minor*), *Proc. Zool. Soc. London* **1924**: 461–497.

Le Gros Clark, W. E., 1924b, On the brain of the tree shrew (*Tupaia minor*), *Proc. Zool. Soc. London* **1924**: 1053–1074.

Le Gros Clark, W. E., 1925, On the skull of *Tupaia*, *Proc. Zool. Soc. London* **1925**: 559–567.

Le Gros Clark, W. E., 1926, On the anatomy of the pen-tailed tree shrew (*Ptilocercus lowii*), *Proc. Zool. Soc. London* **1926**: 1179–1309.

Le Gros Clark, W. E., 1927, Exhibition of photographs of the tree shrew (*Tupaia minor*). Remarks on the tree shrew, *Tupaia minor*, with photographs, *Proc. Zool. Soc. London* **1927**: 254–256.

Lemelin, P., 2000, Micro-anatomy of the volar skin and interordinal relationships of primates, *J. Hum. Evol.* **38**: 257–267.

Liu, F.-G. R., and Miyamoto, M. M., 1999, Phylogenetic assessment of molecular and morphological data for eutherian mammals, *Syst. Biol.* **48**: 54–64.

Liu, F.-G. R., Miyamoto, M. M., Freire, N. P., Ong, P. Q., Tennant, M. R., Young, T. S., and Gugel K. F., 2001, Molecular and morphological supertrees for eutherian (placental) mammals, *Science* **291**: 1786–1789.

Luckett, W. P., ed., 1980, *Comparative Biology and Evolutionary Relationships of Tree Shrews*, Plenum Press, New York.

MacPhee, R. D. E., 1981, *Auditory Regions of Primates and Eutherian Insectivores: Morphology, Ontogeny, and Character Analysis*, Karger, Basel.

MacPhee, R. D. E., ed., 1993, *Primates and Their Relatives in Phylogenetic Perspective*, Plenum Press, New York.

Maddison, D. R., and Maddison, W. P., 2001, *MacClade 4: Analysis of Phylogeny and Character Evolution*, Version 4.03. Sinauer Associates, Sunderland, Massachusetts.

Madsen, O., Scally, M., Douady, C. J., Kao, D. J., DeBry, R. W., Adkins, R. M., Amrine, H. M., Stanhope, M. J., de Jong, W. W., and Springer, M. S., 2001, Parallel adaptive radiations in two major clades of placental mammals, *Nature* **409**: 610–614.

Martin, R. D., 1966, Tree shrews: Unique reproductive mechanism of systematic importance, *Science* **152**: 1402–1404.

Martin, R. D., 1968a, Towards a new definition of primates, *Man* **3**: 377–401.

Martin, R. D., 1968b, Reproduction and ontogeny in tree shrews (*Tupaia belangeri*), with reference to their general behavior and taxonomic relationships, *Zeit. Tierpsychologie* **25**: 409–532.

Martin, R. D., 1990, *Primate Origins and Evolution*, Princeton University Press, Princeton.

McKenna, M. C., 1966, Paleontology and the origin of the primates, *Folia Primatol.* **4**: 1–25.

McKenna, M. C., 1975, Toward a phylogenetic classification of the Mammalia, in: *Phylogeny of the Primates: A Multidisciplinary Approach*, W. P. Luckett and F. S. Szalay, eds., Plenum Press, New York, pp. 21–46.

McKenna, M. C., and Bell, S. K., 1997, *Classification of Mammals Above the Species Level*. Columbia University Press, New York.

Mein, P., and Ginsburg, L., 1997, Les mammifères du gisement miocène inférieur de Li Mae Long, Thailande: Systématique, biostratigraphie et paléoenvironnement, *Geodiversitas* **19**: 783–844.

Miyamoto, M. M., 1996, A congruence study of molecular and morphological data for eutherian mammals, *Mol. Phylo. Evol.* **6**: 373–390.

Murphy, W. J., Eizirik, E., Johnson, W. E., Zhang, Y. P., Ryder, O. A., and O'Brien, S. J., 2001a, Molecular phylogenetics and the origins of placental mammals, *Nature* **409**: 614–618.

Murphy, W. J., Eizirik, E., O'Brien, S. J., Madsen, O., Scally, M., Douady, C. J., Teeling, E. C., Ryder, O. A., Stanhope, M. J., de Jong, W. W., and Springer, M. S., 2001b, Resolution of the early placental mammal radiation using Bayesian phylogenetics, *Science* **294**: 2348–2351.

Napier, J. R., and Napier, P. H., 1967, *A Handbook of Living Primates*, Academic Press, London.

Ni, X., and Qiu, Z., 2002, The micromammalian fauna from the Leilao, Yuanmou hominoid locality: Implications for biochronology and paleoecology, *J. Hum. Evol.* **42**: 535–546.

Novacek, M. J., 1980, Cranioskeletal features in tupaiids and selected Eutheria as phylogenetic evidence, in: *Comparative Biology and Evolutionary Relationships of Tree Shrews*, W. P. Luckett, ed., Plenum Press, New York, pp. 35–93.

Novacek, M. J., 1982, Information for molecular studies from anatomical and fossil evidence on higher eutherian phylogeny, in: *Macromolecular Sequences in Systematic and Evolutionary Biology*, M. Goodman, ed., Plenum Press, New York, pp. 3–41.

Novacek, M. J., 1986, The skull of leptictid insectivorans and the higher-level classification of eutherian mammals, *Bull. Am. Mus. Nat. Hist.* **183:** 1–112.

Novacek, M. J., 1989, Higher mammal phylogeny: The morphological-molecular synthesis, in: *The Hierarchy of Life*, B. Fernholm, K. Bremer, and H. Jornvall, eds., Elsevier, Amsterdam, pp. 421–435.

Novacek, M. J., 1990, Morphology, paleontology, and the higher clades of mammals, in: *Current Mammalogy*, H. H. Genoways, ed., Plenum Press, New York, pp. 507–543.

Novacek, M. J., 1992, Mammalian phylogeny: Shaking the tree, *Nature* **356:** 121–125.

Novacek, M. J., 1993, Reflections on higher mammalian phylogenetics, *J. Mamm. Evol.* **1:** 3–30.

Novacek, M. J., 1994, Morphological and molecular inroads to phylogeny, in: *Interpreting the Hierarchy of Nature*, L. Grande and O. Rieppel, eds., Academic Press, New York, pp. 85–131.

Novacek, M. J., and Wyss, A. R., 1986, Higher-level relationships of the recent eutherian orders: Morphological evidence, *Cladistics* **2:** 257–287.

Novacek, M. J., Wyss, A. R., and McKenna, M. C., 1988, The major groups of eutherian mammals, in: *The Phylogeny and Classification of the Tetrapods, vol. 2: Mammals*, M. J. Benton, ed., Clarendon Press, Oxford, pp. 31–71.

Porter, C. A., Goodman, M., and Stanhope, M. J., 1996, Evidence on mammalian phylogeny from sequences of exon 28 of the von Willebrand factor gene, *Mol. Phylo. Evol.* **5:** 89–101.

Qiu, Z., 1986, Fossil tupaiid from the hominoid locality of Lufeng, Yunnan, *Vertebrata PalAsiatica* **24:** 308–319.

Sargis, E. J., 1999, Tree shrews, in: *Encyclopedia of Paleontology*, R. Singer, ed., Fitzroy Dearborn, Chicago, pp. 1286–1287.

Sargis, E. J., 2000, *The Functional Morphology of the Postcranium of* Ptilocercus *and* Tupaiines (Scandentia, Tupaiidae): Implications for the Relationships of Primates *and other Archontan Mammals*, Ph.D. Dissertation, City University of New York.

Sargis, E. J., 2001, A preliminary qualitative analysis of the axial skeleton of tupaiids (Mammalia, Scandentia): Functional morphology and phylogenetic implications, *J. Zool. London* **253:** 473–483.

Sargis, E. J., 2002a, Functional morphology of the forelimb of tupaiids (Mammalia, Scandentia) and its phylogenetic implications, *J. Morph.* **253:** 10–42.

Sargis, E. J., 2002b, Functional morphology of the hindlimb of tupaiids (Mammalia, Scandentia) and its phylogenetic implications, *J. Morph.* **254:** 149–185.

Sargis, E. J., 2002c, A multivariate analysis of the postcranium of tree shrews (Scandentia, Tupaiidae) and its taxonomic implications, *Mammalia* **66:** 579–598.

Sargis, E. J., 2002d, The postcranial morphology of *Ptilocercus lowii* (Scandentia, Tupaiidae): An analysis of primatomorphan and volitantian characters, *J. Mamm. Evol.* **9:** 137–160.

Schmitz, J., Ohme, M., and Zischler, H., 2000, The complete mitochondrial genome of *Tupaia belangeri* and the phylogenetic affiliation of Scandentia to other eutherian orders, *Mol. Biol. Evol.* **17**: 1334–1343.

Shoshani, J., and McKenna, M. C., 1998, Higher taxonomic relationships among extant mammals based on morphology, with selected comparisons of results from molecular data, *Mol. Phylo. Evol.* **9**: 572–584.

Shoshani, J., Groves, C. P., Simons, E. L., and Gunnell, G. F., 1996, Primate phylogeny: Morphological vs molecular results, *Mol. Phylo. Evol.* **5**: 102–154.

Silcox, M. T., 2001a, A phylogenetic analysis of Plesiadapiformes and their relationship to euprimates and other archontans, *J. Vert. Paleo.* **21** (Supp. to No. 3): 101A.

Silcox, M. T., 2001b, *A Phylogenetic Analysis of Plesiadapiformes and Their Relationship to Euprimates and Other Archontans*, Ph.D. Dissertation, Johns Hopkins University.

Silcox, M. T., 2002, The phylogeny and taxonomy of plesiadapiforms, *Am. J. Phys. Anthro.* (Supp.) **34**: 141–142.

Simmons, N. B., 1994, The case for chiropteran monophyly, *Am. Mus. Novitates* **3103**: 1–54.

Simmons, N. B., 1995, Bat relationships and the origin of flight, *Symp. Zool. Soc. London* **67**: 27–43.

Simmons, N. B., and Quinn, T. H., 1994, Evolution of the digital tendon locking mechanism in bats and dermopterans: A phylogenetic perspective, *J. Mamm. Evol.* **2**: 231–254.

Simpson, G. G., 1945, The principles of classification and a classification of mammals, *Bull. Am. Mus. Nat. Hist.* **85**: 1–350.

Smith, J. D., and Madkour, G., 1980, Penial morphology and the question of chiropteran phylogeny, in: *Proceedings of the Fifth International Bat Research Conference*, D. E. Wilson and A. L. Gardner, eds., Texas Tech Press, Lubbock Texas, pp. 347–365.

Stafford, B. J., and Thorington, R. W., 1998, Carpal development and morphology in archontan mammals, *J. Morph.* **235**: 135–155.

Stanhope, M. J., Bailey, W. J., Czelusniak, J., Goodman, M., Si, J.-S., Nickerson, J., Sgouros, J. G., Singer, G. A. M., and Kleinschmidt, T. K., 1993, A molecular view of primate supraordinal relationships from the analysis of both nucleotide and amino acid sequences, in: *Primates and Their Relatives in Phylogenetic Perspective*, R. D. E. MacPhee, ed., Plenum Press, New York, pp. 251–292.

Stanhope, M. J., Smith, M. R., Waddell, V. G., Porter, C. A., Shivji, M. S., and Goodman, M., 1996, Mammalian evolution and the interphotoreceptor retinoid binding protein (IRBP) gene: Convincing evidence for several superordinal clades, *J. Mol. Evol.* **43**: 83–92.

Steele, D. G., 1973, Dental variability in the tree shrews (Tupaiidae), in: *Craniofacial Biology of Primates: Symposium of the IVth International Congress of Primatology*, vol. 3, M. R. Zingeser, ed., Karger, Basel, pp. 154–179.

Swofford, D. L., 1993, *PAUP: Phylogenetic Analysis Using Parsimony*, Version 3.1.1. Smithsonian Institution, Washington, DC.

Szalay, F. S., 1968, The beginnings of primates, *Evolution* **22**: 19–36.

Szalay, F. S., 1969, Mixodectidae, Microsyopidae, and the insectivore-primate transition, *Bull. Am. Mus. Nat. Hist.* **140**: 193–330.

Szalay, F. S., 1977, Phylogenetic relationships and a classification of the eutherian Mammalia, in: *Major Patterns in Vertebrate Evolution*, M. K. Hecht, P. C. Goody, and B. M. Hecht, eds., Plenum Press, New York, pp. 315–374.

Szalay, F. S., 1999, Review of "Classification of Mammals Above the Species Level" by M. C. McKenna and S. K. Bell, *J. Vert. Paleo.* **19**: 191–195.

Szalay, F. S., and Drawhorn, G., 1980, Evolution and diversification of the Archonta in an arboreal milieu, in: *Comparative Biology and Evolutionary Relationships of Tree Shrews*, W. P. Luckett, ed., Plenum Press, New York, pp. 133–169.

Szalay, F. S., and Lucas, S. G., 1993, Cranioskeletal morphology of archontans, and diagnoses of Chiroptera, Volitantia, and Archonta, in: *Primates and Their Relatives in Phylogenetic Perspective*, R. D. E. MacPhee, ed., Plenum Press, New York, pp. 187–226.

Szalay, F. S., and Lucas, S. G., 1996, The postcranial morphology of Paleocene *Chriacus* and *Mixodectes* and the phylogenetic relationships of archontan mammals, *Bull. New Mexico Mus. Nat. Hist. Sci.* **7**: 1–47.

Teeling, E. C., Scally, M., Kao, D. J., Romagnoli, M. L., Springer, M. S., and Stanhope, M. J., 2000, Molecular evidence regarding the origin of echolocation and flight in bats, *Nature* **403**: 188–192.

Thewissen, J. G. M., and Babcock, S. K., 1991, Distinctive cranial and cervical innervation of wing muscles: New evidence for bat monophyly, *Science* **251**: 934–936.

Thewissen, J. G. M., and Babcock, S. K., 1992, The origin of flight in bats, *Bioscience* **42**: 340–345.

Thewissen, J. G. M., and Babcock, S. K., 1993, The implications of the propatagial muscles of flying and gliding mammals for archontan systematics, in: *Primates and Their Relatives in Phylogenetic Perspective*, R. D. E. MacPhee, ed., Plenum Press, New York, pp. 91–109.

Tong, Y., 1988, Fossil tree shrews from the Eocene Hetaoyuan Formation of Xichuan, Henan, *Vertebrata PalAsiatica* **26**: 214–220.

Van Valen, L. M., 1965, Tree shrews, primates, and fossils, *Evolution* **19**: 137–151.

Waddell, P. J., Okada, N., and Hasegawa, M., 1999, Towards resolving the interordinal relationships of placental mammals, *Syst. Biol.* **48**: 1–5.

Wagner, J. A., 1855, *Die Säugethiere in Abbildungen nach der Natur.*: Weiger, Leipzig, Wagner 1855 Supplementband, Abt. **5**: 1–810.

Wible, J. R., 1993, Cranial circulation and relationships of the colugo *Cynocephalus* (Dermoptera, Mammalia), *Am. Mus. Novitates* **3072**: 1–27.

Wible, J. R., and Covert, H. H., 1987, Primates: Cladistic diagnosis and relationships, *J. Hum. Evol.* **16:** 1–22.

Wible, J. R., and Martin, J. R., 1993, Ontogeny of the tympanic floor and roof in archontans, in: *Primates and Their Relatives in Phylogenetic Perspective*, R. D. E. MacPhee, ed., Plenum Press, New York, pp. 111–148.

Wible, J. R., and Novacek, M. J., 1988, Cranial evidence for the monophyletic origin of bats, *Am. Mus. Novitates* **2911:** 1–19.

Wible, J. R., and Zeller, U. A., 1994, Cranial circulation of the pen-tailed tree shrew *Ptilocercus lowii* and relationships of Scandentia, *J. Mamm. Evol.* **2:** 209–230.

Wilson, D. E., 1993, Order Scandentia, in: *Mammal Species of the World: A Taxonomic and Geographic Reference*, 2nd ed., D. E. Wilson and D. M. Reeder, eds., Smithsonian Institution Press, Washington, pp. 131–133.

Wöhrmann-Repenning, A., 1979, Primate characters in the skull of *Tupaia glis* and *Urogale everetti* (Mammalia, Tupaiiformes), *Senck. Biologica* **60:** 1–6.

Zeller, U. A., 1986a, Ontogeny and cranial morphology of the tympanic region of the Tupaiidae, with special reference to *Ptilocercus, Folia Primatol.* **47:** 61–80.

Zeller, U. A., 1986b, The systematic relations of tree shrews: Evidence from skull morphogenesis, in: *Primate Evolution*, J. G. Else and P. C. Lee, eds., Cambridge University Press, Cambridge, pp. 273–280.

Zeller, U. A., 1987, Morphogenesis of the mammalian skull with special reference to *Tupaia*, in: *Morphogenesis of the Mammalian Skull*, H. J. Kuhn and U. A. Zeller, eds., Verlag Paul Parey, Hamburg, pp. 17–50.

CHAPTER FOUR

Primate Origins: A Reappraisal of Historical Data Favoring Tupaiid Affinities

Marc Godinot

INTRODUCTION

The origin of primates remains a fascinating question. In spite of many anatomical and molecular studies, the identification of the living sister group of primates is not clearly settled. The lack of consensus about primate origins is the result of the great antiquity of the events that marked primate differentiation. Morphological as well as molecular signals are masked by the amount of subsequent evolution in primates and in potential sister groups, and by the extinction of some critical intermediates. In addition, the Paleocene fossil record of mammals is still particularly poor in Africa and in the southern tropical regions of Asia, where some of the important steps presumably took place. What makes the problem especially puzzling is the realization that the increase in and the enhanced quality of the anatomical and molecular data sets extracted from the living forms did not result in any increased consensus. On the contrary, the cladistic treatment of a large morphological data set concerning archontan phylogeny, assembled by Simmons (1993) from previous

Marc Godinot • Ecole Pratique des Hautes Etudes, UMR 5143, Paris, France

studies, did not detect a strong signal concerning the sister group of primates. This constitutes a challenge for phylogeneticists. Molecular data sets, which also did not yield a consensus in 1993, continue to expand. It seems now that a strong molecular signal favors a close relationship between Primates, Scandentians, and Dermopterans (Liu et al., 2001; Madsen et al., 2001; Murphy et al., 2001; see Springer, this volume).

During the first half of the 20th century, tree shrews (family *Tupaiidae*) were often considered as the most primitive representatives of the order Primates. Carlsson (1922) compared many features in *Tupaia*, Lemuriformes, Macroscelididae, and Lipotyphla. She found many similarities between *Tupaia* and lemuriforms, including the presence of a postorbital bar, and she was the first to formally include tupaiids in primates. Her view was strongly supported by the work of Le Gros Clark (1959, 1971), and was adopted by Simpson (1945) and Saban (1963) among others. However, a remarkable book on "Comparative Biology and Evolutionary Relationships of Tree Shrews" (Luckett, 1980a) placed the long-standing belief in a close tupaiid–primate relationship into question, with the rigor of cladistic methodology. Many similarities between the two groups appeared as primitive retentions or convergences, and tree shrews were left in their own order Scandentia without any clear affinity within the mammals. Another attempt at putting "primates and their relatives in phylogenetic perspective" led to the contributions assembled by MacPhee (1993a). No consensus emerged from this attempt, but it did lead to "the rehabilitation of scandentians as being at least reasonably close relatives of primates (and colugos)" (MacPhee, 1993b).

A group of Early Cenozoic fossil mammals, including the Plesiadapidae and Paromomyidae, were described as primates by most early paleontologists. They show some general dental similarities with primates, and there are detailed similarities between some molars of *Plesiadapis* and *Cantius*. However, it has been long recognized that these must be convergences because they do not exist in primitive plesiadapids and primitive adapids such as *Donrussellia*. Furthermore, *Plesiadapis* was believed to have a petrosal bulla, which is a hallmark of primates (Szalay, 1969). Hence, plesiadapids and related families, which are usually assembled in the taxon Plesiadapiformes (Simons, 1972), were considered as representatives of an early radiation of the primates by many (e.g., Gingerich, 1976; Romer, 1966; Simons, 1972; Szalay and Delson, 1979).

At the same time, other scholars were working toward a better definition of primates based on primarily the derived characters shared by the living forms (Cartmill, 1972; Martin, 1968, 1985). This research generated a new understanding of the adaptive significance of primate characteristics (Cartmill, 1972, 1974a). In fact, our understanding of primate characteristics has probably progressed much further during the past decade than our understanding of archontan phylogeny (e.g., Dagosto, 1988; Rasmussen, 1990; Ravosa et al., 2000; see other contributions in this volume). The series of derived characters shared by virtually all living primates—petrosal bulla, a complete postorbital bar encircling large forward-oriented orbits, an opposable hallux, and nails instead of claws (Cartmill, 1972; expanded in Martin, 1986)— is widely accepted. However, the taxonomic consequences of this understanding are not treated in the same way by all authors. Those who maintain a broader view of the order Primates and their close relationships to the plesiadapiforms use the term 'Euprimates' (Hoffstetter, 1977), to include the living ones and their close relatives, or the informal "primates of modern aspect" of Simons (1972). Moreover, they consider the Plesiadapiformes a suborder of the Primates (Fleagle, 1988; Gunnell, 1989; Szalay et al., 1987; Van Valen, 1994). Others restrict the order Primates to the euprimates and consider the Plesiadapiformes as a separate order of mammals—an opinion that is becoming more widely accepted (e.g., Fleagle, 1999). This choice is adopted here because it gives the taxon primates adaptive significance and also because the sister taxon of the modern primates is unresolved.

Between 1990 and 1993, a quite different hypothesis emerged from the study of new fossil material. Postcranial studies of the plesiadapiform family Paromomyidae seemed to favor a close affinity of this family with Dermoptera, leading to the concept of Primatomorpha—a mirorder including primates and dermopterans (the plesiadapiforms being included within dermopterans; Beard, 1993a,b). Similar relationships are accepted by McKenna and Bell (1997). These new hypotheses are accompanied by radical changes in classifications (e.g., the inclusion of the Eudermoptera in the Plesiadapiformes). Much worse, the inclusion of Dermoptera as a suborder of the order Primates (McKenna and Bell, 1997) simply destroys all previous constructions of the taxon primates. Moreover, all these dramatic changes reflect a hypothesis which is very questionable. This is scrutinized in the next section. A more recent treatment of the problem by Silcox (2001) again nests modern primates within plesiadapiform groups.

As the quantity of information has enormously increased and the current hypotheses are so numerous and contradictory, it has become a challenge to reconcile the multiple lines of evidence and to expose the most likely hypothesis. Here the author proposes such a view, that of a paleontologist, inclined to favor that part of the evidence for which there is some historical insight. The primary reliance on cranioskeletal characters is based on the conviction that there are lessons to be learned from traditional systematics and from the paleontological record. Of course, such a view needs to be confronted with insights inferred from the study of living forms (e.g., molecules, neural, and reproductive traits, etc.).

LIMITS OF CLADISTICS CONFRONTED WITH LARGE DATA SETS

The synthetic treatment of very large data sets by Simmons (1993) and Silcox (2001) is very interesting. Simmons searched for all possible morphological characters studied in the literature, established in 10 presumably monophyletic units (Scandentia, Strepsirhini, Tarsiiformes, Anthropoidea, Galeopithecidae, Megachiroptera, Microchiroptera, Plesiadapidae, Paromomyidae, and Micromomyidae). The most striking result of her analysis is that, in partitioning the data set into six different subsets corresponding to different anatomical systems, despite the elimination of three fossil families to avoid too much missing data, six different phylogenies were obtained. Nonauditory cranial (33 characters), auditory (20), anterior axial skeleton and forelimb (31), hindlimb (35), reproductive tract and fetal membranes (12), and neural (23) data sets—all gave different phylogenies that sometimes did not recover primate or chiropteran monophyly (Simmons, 1993). In spite of her will to be as objective as possible, Simmons (1993) made one big decision that affected her results—the exclusion of dental characters. On the one hand, this is understandable. It is common experience that the dentition is "very useful in differentiating species and genera, but at higher taxonomic levels its value is diminished because it is particularly subject to parallel evolution" (Butler, 1980). Moreover, dental characters are also subject to numerous successive transformations, rendering the *a posteriori* deciphering of successive states difficult. However, on the other hand, all characters are susceptible to convergence and successive transformations, and dental characters do not appear more affected by homoplasy than others (Sanchez-Villagra and Williams, 1998). As dentitions constitute a large part of the fossil record, their importance should not be undervalued (Silcox, 2001; Van Valen, 1994).

When the fossil record is sufficiently dense, teeth may decisively point toward evolutionary continuities, which demonstrate phylogenetic relationships. The work of Silcox (2001), including all the plesiadapiforms even if only known dentally, and using a large number of dental characters, is thus very welcome, complementing that of Simmons (1993). Silcox also conducted three different analyses using three different anatomical systems: dental (97 characters), cranial (30), postcranial (54), and all the characters were reevaluated in light of recent studies. Despite this reevaluation, she again found three different phylogenies (Silcox, 2001)! A careful look at both studies reveals the following:

1. There is no strong phylogenetic signal contained in Simmons' summary data set; it is sufficient to replace the treatment of the 154 characters from equal weight given to each change in a character to equal weight given to each character to completely lose primate monophyly and find the Anthropoidea as the sister group of Dermoptera + Chiroptera (Simmons, 1993, Figure 8). If the character transformations that can be ordered are ordered (something the author would consider mandatory), the preceding orders and primates appear monophyletic, but their relationships are completely unresolved, with the exception of scandentians being the primitive sister to all others (*idem*, Figure 10). In the equal transformation weighting, Simmons acknowledges that with only a few additional steps, there are a large number of trees (indicating that the phylogenetic signal is very weak).

2. The fact that different anatomical systems give different results suggests that some of these phylogenetic signals must be wrong. If the corresponding data are kept, they will in any such analysis, consistently yield wrong signals. One problem is that similar functional requirements and other mechanisms may lead to common evolutionary trends or convergences. This is well-known for some locomotor and dental specializations; however, it may occur in other anatomical systems as well. Such homoplasies introduce consistent signals, which drive parsimony analyses to incorrect solutions.

3. The comparison of the trees obtained from the three data sets of Silcox (2001) and that from the summary data set shows that the total evidence tree primarily reflects the largest of the subsets—the dental one. It also reveals that this total evidence solution yields phylogenetic signals that were so weak that they had not appeared on the tree extracted from the subset giving the signal. The sister group relationship between primates

and Toliapinidae, which appears on the total evidence tree, must come from the dental evidence because this is the only data set known for Toliapinidae. However, that relationship was not apparent in the strict or Adams' consensus trees of the dental data set at the family level, which is a subset of the total evidence tree (this relationship appeared in the dental analysis at the species level). This is a clear demonstration that the larger summary data set, as attractive as it may be, in fact produces very weak phylogenetic signals. Silcox (2001) is aware of the fragility of the node uniting primates and Toliapinidae.

4. The characters that have been recognized by all authors as having systematic importance should be weighted heavily. In Plesiadapiformes, the shape of the incisors is recognized as having more systematic significance than the dental reductions that occurred repeatedly. For example, Silcox finds that *Picromomys* is closely related to *Niptomomys*, something "rather surprising, in light of the fact that *Picromomys* lacks the characteristic I/1 morphology of microsyopids" (Silcox, 2001). She then rightly assumes that Microsyopidae is monophyletic excluding *Picromomys*, a choice with which the author agrees, but forces the conclusion that the methodology is unsatisfactory. To avoid having several more commonly derived characters outweigh the unique I/1 morphology, this unique morphology should be more heavily weighted. The deduction that the microsyopid I/1 is characteristic of the family is common knowledge of systematicians, and it should be translated into the cladistic analysis by an appropriate weighting procedure (see e.g., Neff, 1986).

In conclusion, these cladistic analyses are unsatisfactory because in the search for "objectivity", they refuse to weight heavily characters that we know have high systematic significance (Szalay et al., 1987). As such, rejecting previously acquired knowledge is regressive. Secondly, in adding more and more characters and then more and more homoplasies between different anatomical systems of many groups, what is finally privileged is global similarity at the detriment of more specific signals; this procedure leads ironically in part to a return to phenetics (this is especially true when transformations are not oriented). The experience of systematics is that a small number of characters can diagnose many groups, including higher systematic groups. The astragalus of artiodactyls, the toothcomb of lemuriforms, and the petrosal bulla of primates bear testimony that single characters or character complexes can mean a lot, and that such characters can be found in dental, cranial, or postcranial anatomy. Put in historical perspective, the differentiation of a higher taxon could be linked to a small number of characters.

To avoid the trap of excessive data sets effectively driving a return to phenetics, different strategies are possible. One would suggest scrutinizing the data to try to eliminate the characters that are likely to introduce wrong signals. This is more easily said than done; however, during the course of anatomical studies, a number of characters have been shown to be not pertinent. Likewise, paleontological information usually leads to the elimination of derived characters shared by living forms, which are not present in early members of a group, and to the inclusion of more primitive character states that are known to be represented in these early members. Another idea is to try not to lose sight of the characters that have been shown by others to be crucial, in our case, crucial primate characteristics (related to vision, locomotion with grasping, and nails). Furthermore, the fossil record clearly demonstrates that there are features very unlikely to exhibit reversal. For example, teeth once lost are never regained. Strangely, Rose and Bown (1996) allowed such reversals that partly explain the bizarre results of their cladistic analysis of a group of plesiadapiforms. Likewise, Bloch et al. (2001) support a phylogeny in which a P/2 has "reevolved" in *Carpocristes oriens*; this is unlikely, although such a possibility must be addressed. There are morphological characters other than tooth presence that can be lost, without the entire morphology ever reverting to ancestral states. Unlikely reversals can help to evaluate or even refute some phylogenetic hypotheses, as is argued later. In this chapter, the author emphasizes the role of historical data and the importance of understanding character changes in polarity and in function. In doing the author finds himself in agreement with the position repeatedly advocated by (Szalay, 2000; Szalay and Lucas, 1996; Szalay et al., 1987) and with the suggestion that "perhaps it is time to return to scenarios about adaptational history as better devices for understanding primate evolution" (MacPhee, 1991) and adds, especially primate origins.

An evaluation of the Primatomorpha hypothesis is done first because its acceptance or rejection will determine the content and the systematic meaning of Plesiadapiformes. In the course of this chapter, Plesiadapiformes is used as a separate order of mammals—a choice which has become common (e.g., Fleagle, 1999; Rose, 1995) and will be justified later (following Szalay and many others, Microsyopidae is provisionally not included in Plesiadapiformes).

FACING PRIMATOMORPHA

Are Paromomyid Dental Characters Compatible with Dermopteran Origins?

When Beard interpreted postcranial features of paromomyids as reflecting a gliding adaptation of a type similar to that of colugos, he included paromomyids in Dermoptera (Beard, 1990, 1991). He further interpreted the paromomyid dentition as convergent to that of the sugar-glider, *Petaurus*, and inferred a similar gliding and tree-exudate-eating adaptation for paromomyids. It is difficult to accept this interpretation, partly because paromomyid dentitions look extremely divergent from, and hardly ancestral to, those of colugos (see also Rose, 1995; Szalay and Lucas, 1996). They also seem as much adapted to insectivory as to exudate-feeding (Godinot, 1984). More importantly, the skulls of paromomyids are quite similar to those of plesiadapids (Kay et al., 1992), showing very few characters indicating a link with dermopterans. The evidence concerning paromomyid dental adaptation and affinities is accordingly reconsidered.

It is well-known that the teeth of paromomyids are very peculiar in morphology (Simpson, 1955; Szalay and Delson, 1979). The upper molars of *Phenacolemur* and *Ignacius* have a marked posterolingual expansion, delimited by the strong, posteriorly directed postprotocingulum (a crest going down from the tip of the protocone, posteriorly, and added to the major pre- and postprotocrista). A *Nannopithex*-fold as found in primates is different because it usually differentiates through a breaking of the postprotocrista and subsequent increase of the fold, which more or less clearly "replaces" the postprotocrista (Figure 1; one exception is found on the omomyid *Trogolemur*, on which the posterior fold has developed without a decrease of the postprotocrista, resulting in a structure exactly similar to a postprotocingulum; see e.g., Gunnell, 1995). The postprotocingulum is already well differentiated in the primitive paromomyid *Paromomys*, and as it is also present in plesiadapids and carpolestids, it is usually considered a shared derived character of the plesiadapoids. Beard (1993a) also considers this character as shared derived between these groups (plus primates), and subsequently lost in dermopterans. Is such a reversal likely? The author has explained elsewhere why the loss of a *Nannopithex*-fold in early primates was very unlikely (Godinot, 1994; Kay and Williams, 1994; see also Kay et al., 1992). Like the *Nannopithex*-fold, the postprotocingulum also pertains to the common trend of broadening and

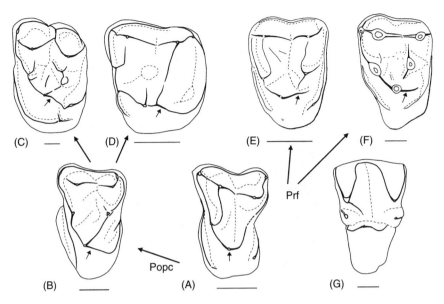

Figure 1. Drawings of left upper second molars of several archontans in occlusal views. A, *Purgatorius*; B, *Palaechthon*; C, *Plesiadapis*; D, *Arcius*; E, *Nannopithex*; F, *Cantius*; and G, *Cynocephalus*. All drawn at the same transverse width, posterior side on the right; scale bars are 1 mm. Small arrows on the teeth indicate the protocone in A; the postprotocingulum in B, C, and D; and the protocone-fold in E and F. The long arrows indicate steps of morphological changes; however, they do not represent true phylogenies. They show on the left the development of a postprotocingulum (Popc), a third posterolingual crest issued from the tip of the protocone, in plesiadapiforms, (C is a plesiadapid, D a paromomyid). On the right, they show the formation of a protocone-fold (Prf) or *Nannopithex*-fold, via the breaking of the postprotocrista and elongation of its posterior part in two primates (E is a microchoerid; F a notharctid); orientation of this crest is posterolabial. Upper molars of living colugos (G) seem primitively narrow in their lingual part.

reinforcing the lingual part of upper molars (initially triangular in early mammals, having later become quadrangular in many groups). Of course, a structure, and even more a detail on a structure, can be lost, however not in any fashion. For example, in this case, one can find large plesiadapid teeth on which the postprotocingulum is weakly expressed (Figure 1C). However, in such cases, this happens to much larger species having quite different proportions and functional adaptations. This weakening and possible loss of the crest does not modify the acquired quadrate lingual shape of the tooth. Even with the possible complete loss of the crest (which would be coded as a "reversion"), there is in fact no reversion of the whole structure, no going back to an ancestral morphology.

The living colugos, in spite of their specialized labial shearing crests, have transversely very elongated upper molars. These molars are primitive in their narrow lingual part (Figure 1G); it is not very likely that this is derived from teeth that would have been lingually broadened by a postprotocingulum (or a hypocone on a lingual cingulum). One should be cautious because there is a very long duration between the Early Eocene fossils and living species. On the other hand, these species remained relatively small, which renders more likely simple preservation of a primitive structure (large changes masking preceding adaptations are much more frequent when associated with a marked change in size—a change in niche). The teeth of the Plagiomenidae, putative early dermopteran relatives, are very specialized and convergent on those of living galeopithecids (e.g., Rose and Simons, 1977). Interestingly, the upper molars of *Plagiomene* and *Elpidophorus*, which are relatively lingually broad, do not show a trace of a postprotocingulum. Those of *Elpidophorus* have a hypocone at the lingual extremity of the posterior cingulum, strongly suggesting that those animals and their ancestors never had a postprotocingulum. Overall, it appears very improbable that the upper molars of galeopithecids, as well as those of plagiomenids, evolved from upper molars with a postprotocingulum. This dental analysis contradicts the cladogram of Beard (1993a), which implies such a morphological transformation. Concerning primates, Beard (1993a) considered them, as many earlier authors did (Hoffstetter, 1986; Szalay et al., 1987), as having primitively possessed a postprotocingulum, equated with a *Nannopithex*-fold. This is probably not true, but it does not greatly affect that part of his cladogram; this character has simply to be moved one node up on the main line. However, this is not entirely trivial because it eliminates the only non-postcranial character supporting the primatomorph node.

Dermopteran Incisors

Colugos are believed to have three lower incisors, two of which are very specialized, pectinate (MacPhee et al., 1989; Vaughan, 1972). It would seem impossible to derive such a dental formula from that of a paromomyid or plesiadapoid because the latter have one enlarged incisor and at most a second one, usually very reduced (Figure 2). [Several years ago, it was commonplace to consider the reduction of the number of incisors from three to two as a shared derived character uniting primates and plesiadapiforms; however, this character was eliminated by the discovery of a third upper incisor in car-polestids (Bloch and Gingerich, 1998; Fox, 1993).] An interesting point about those lower incisors is that some plagiomenid incisors show how a pectinate lower incisor can form (Rose and Simons, 1977). Such pectinate incisors can form from relatively large anterior incisors. However, one would have to go back to a paromomyid endowed with three incisors, including two relatively large ones to evolve the pectinate incisors of colugos. Such an animal would neither be a member of a paromomyoid nor a plesiadapoid clade because these clades are partly defined by the possession of one enlarged lower incisor and with a second lower incisor that is reduced or lost. The new paromomyid genus *Acidomomys*, which retains an I/2, has indeed a very small

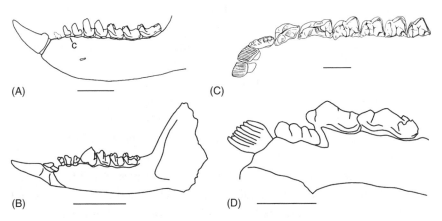

(A)

(C)

(B)

(D)

Figure 2. Drawings of two plesiadapiform left jaws (A and B) compared to two right jaws of living dermopterans (C and D). Drawn to similar lengths; scale bars are 5 mm. Here, (A) is the primtive plesiadapid *Pronothodectes* (from Gingerich, 1976); (B) the *Chronolestes*, a primitive carpolestid or primitive plesiadapoid (from Beard and Wang, 1995); (C) the jaw of living *Cynocephalus volans* (from MacPhee et al., 1989); and (D) a juvenile *C. variegatus* (Anatomie Comparée collection, A-3958, M. N. H. N., Paris). Living dermopterans have three incisors: two of them pectinate (C). (D) shows details of I/2 and the erupting I/3. It seems impossible to derive a dentiton like that of living colugos from the dentitions of the early plesiadapoids, which have an enlarged I/1; and the other teeth between I/1 and P/4, including 1/2 and 1/3, very reduced.

I/2 in an advanced stage of reduction (Bloch et al., 2002). Even the primitive plesiadapid *Pronothodectes*, and *Chronolestes*, which may be a primitive carpolestid or lie at the base of a plesiadapoid clade, have teeth posterior to I/1 much too reduced to be possibly ancestral to dermopteran incisors. Thus, the Galeopithecidae cannot be nested within the plesiadapoid clade (or a paromomyoid clade).

Concerning upper incisors, they are lost in the living colugos, whereas paromomyids have the large multilobate I1/ typical of plesiadapoids (Gingerich, 1976; Godinot, 1984; Rose et al., 1993), and of paromomyids if they are considered as belonging to another clade. The known trends in plesiadapoids seem not to be toward the loss of the upper I1. Overall, concerning both lower and upper anterior incisors, the evolutionary trend(s) largely started in paromomyids and their plesiadapoid relatives do not lead toward a dermopteran-like anterior dentition. In fact, to be possibly ancestral to dermopterans, an animal would have to have kept three lower incisors, and to have increased the second approximately as much as the first—two conditions at odds with known plesiadapiforms. If we add the difficulties from the upper molars, the hypothesis of dermopterans being more closely related to paromomyids than to plesiadapids becomes so intractable that it is quasi impossible. The scenario would imply rooting paromomyids in a form as primitive as, or even more primitive than *Purgatorius*, with dermopterans branching off to one side and paromomyids on the other side converging in many characters with plesiadapids. It is not impossible; however, it would destroy the plesiadapoid synapomorphies recognized by Beard and others, and require a new ad hoc dental scenario.

Known dental trends of paromomyids and other plesiadapiforms make it so unlikely that one of them could be ancestral to dermopterans that this hypothesis should be considered as quasi impossible, and thus abandoned. It is not definitively proven to be impossible, but decisive new evidence would be needed to justify reconsideration of this hypothesis. It is similarly intractable to adjust dental characters to the cladogram of Bloch and Boyer (2002), which proposes a sister group relationship between carpolestids and primates, nested within plesiadapoids. It would seem "easier" to redevelop two vertically implanted lower incisors than to evolve the dermopteran incisal device; however, this is likewise so opposite to dental trends in plesiadapoids that any nesting of primates or dermopterans within plesiadapoid families is dentally quasi impossible.

Paromomyid Postcranials, Gliding, and Apatemyid Adaptations

The core of the primatomorph hypothesis developed by Beard (1990, 1991, 1993a,b) was the interpretation of paromomyid phalanges and carpal characters as reflecting a gliding adaptation similar to the peculiar finger-gliding adaptation of living colugos (dermopterans), and homologous with it. However, several papers since 1990 have raised doubts concerning this interpretation. Krause (1991) questioned the allocation of the isolated middle phalanges to the hand or foot of *Phenacolemur*; however, Beard (1993b) answered that question by quantifying the elongation of isolated middle phalanges. Later, Runestad and Ruff (1995) showed that the humerus of paromomyids does not present the distinct diaphyseal dimensions found in living gliding mammals. Szalay and Lucas (1996) noticed that the associated metapodials were lacking the two articular lobules of living colugos. More recently, Hamrick et al. (1999) found that the intermediate phalanges of paromomyids "are similar in their relative length and midshaft dimensions to those from the hand of vertical clingers (e.g., *Tarsius* and *Glaucomys*) as well as those from the foot of *Cynocephalus*" (p. 408). They conclude that the "existing phalanges of paromomyids, ..., therefore provide no conclusive evidence that paromomyids possesed a colugo-like patagium" (*idem*, p. 409). Hence, arguments have accumulated against the view of paromomyids as finger-gliders.

In 1993, Beard had written, "detractors of the hypothesis ... have yet to offer an alternative explanation for the function of the elongated intermediate phalanges of these animals" (Beard, 1993b), and Szalay and Lucas (1996) mentioned the need to find a functional explanation for them as well. The explanation given by Hamrick et al. (1999) is "that vertical climbing and clinging were frequent locomotor and postural behaviors practiced by these animals," in partial agreement with Beard. A partly similar explanation of extreme phalangeal lengthening was given earlier in the study which during these years had become the most challenging to the dermopteran hypothesis, and which remained unnoticed: the description of the Messel apatemyid *Heterohyus* skeleton by Koenigswald (1990; Koenigswald and Schiernig, 1987). This author described in detail the anatomy of *H. nanus*, which shows extremely elongated hands (Figure 3), with apparently straight and elongated intermediate phalanges (and also very elongated proximal phalanges) and interpreted this animal as an arboreal insectivorous creature

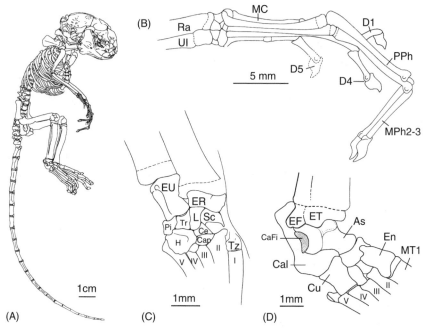

Figure 3. Elements of the skeleton of *Heterohyus nanus* from Messel, Germany, redrawn and slightly modified from Koenigswald (1990). Whole skeleton (A); Right hand (B) with extremities of radius (Ra) and ulna (Ul), carpals, metacarpals (MC), proximal phalanges (PPh), middle phalanges (MPh), and distal phalanges (D); numbers refer to digits 1–5. The proximal part of a right hand (C) shows the extremities of radius (ER) and ulna (EU), the proximal carpal row reinterpreted as scaphoid (Sc), lunate (L), triquetrum (Tr), and pisiform (Pi); the second carpal row with hamate (H), capitate (cap), centrale (ce), and trapezium (Tz); metacarpals are numbered from I to V. On the proximal part of a right foot (D), a piece of bone is reinterpreted as being possibly the extremity of the fibula (EF), which would slide on a calcaneofibular facet (CaFi) on the calcaneum (Cal). The astragalus is "As" and cuboid "Cu." Note the curved calcaneocuboid contact, suggesting the presence of some kind of pivot joint, and the relatively large entocuneiform (En) and first metatarsal (MT1), which suggest some hallucial opposition. The long hand and especially long, straight, middle phalanges of *Heterohyus* are believed to be a good analog for those of paromomyids.

convergent with *Daubentonia* and *Dactylopsila* (two taxa not present in the comparative tables of Beard, 1993b, and Hamrick et al., 1999). Because of the peculiar preservation of Messel skeletons, slightly crushed, there are some details of phalangeal anatomy that cannot be checked. They should become available on the beautiful material recovered by Bloch and Boyer (2001). The possibility that *Heterohyus* might have had a patagium needs to be briefly addressed; however, it seems unlikely given the way the two Messel specimens are preserved. On Messel bats the wings are often visible and a large patagium should be recognizable on a Messel glider (three specimens). Von Koenigswald (1990) rejected a patagium in *Heterohyus* based on the extreme elongation of only two digits, whereas more digits should be elongated to support a gliding membrane. Overall, the convergence put forth by von Keonigswald seems to be quite a convincing one. The author proposes to extend the same convergence to paromomyids, which also have very similar elongated incisors and presumed insectivorous adaptation (with variations; *Heterohyus* might have been partially wood boring and the apatemyid *Sinclairella* much more, its incisors being convergent on those of *Daubentonia*). Because a number of the morphological peculiarities, functionally related to gliding in living colugos, are lacking in paromomyids, vertical climbing and clinging in trees seems a much better explanation for their peculiar extremities. The new material of apatemyids and paromomyids (Bloch and Boyer, 2001) should provide conclusive evidence concerning this.

On Postcranial Characters and Archontan Phylogeny

It seems important to comment on the postcranial characters used by Beard (1993a) in his phylogenetic analysis. The primatomorph node was considered the most robust, supported by eleven characters: one dental and 10 postcranial. Some of these characters may now appear problematic in their definition and distribution. From the first three of them, the humeral ones, two are mentioned as examples. The lesser tuberosity is considered robust and medially protruding in "primatomorphs" by Beard (1993a). This tuberosity is in fact less protruding in many primates than in *Phenacolemur* and *Plesiadapis* (Szalay et al., 1975), and it has been shown to be as robust and as protruding in *Ptilocercus*, giving the more probable state of the character in primitive scandentians (Sargis, 2002a). The subspheroidal to spheroidal shape of the capitulum is also found in *Ptilocercus*, and is considered as a likely primitive archontan character by several authors (Sargis, 2002a,c; Szalay and Lucas, 1993, 1996). These two humeral characters have interesting functional implications, but they do not support a primatomorph node.

A complete reappraisal of all the postcranial characters studied by Beard (1993a), Silcox (2001), and Bloch and Boyer (2002) is far beyond the scope of this chapter. Some general comments are provisionally given. All but one of the characters taken by Beard (1993a) to support Primatomorpha are linked to some kind of arboreal adaptation (these characters would increase to 13 with the inclusion of the coxo-femoral characters which are part of the same functional complex, and which also occur in Chiroptera and were listed as primatomorph in Beard, 1991). A number of them appear to be found in *Ptilocercus* and are considered as probably primitive archontan (Sargis, 2000, 2002a,b,c; Szalay and Lucas, 1993, 1996). Some others, linked to leaping in *Tarsius* and *Hemiacodon*, are questionably part of the primate morphotype. In fact, it would seem hazardous to give a list of all the postcranial attributes of the primate morphotype, because this is a debated notion, which depends on one's preferred interpretation of anthropoid characters (e.g., Dagosto, 1990; Ford, 1988; Godinot, 1992). This choice should be justified. Furthermore, the plesiadapiform radiation is a large radiation of presumably mainly arboreal animals, which leads me to suspect a very complex history of arboreal adaptations in the group. Some of the characters studied by Beard show an evolution within plesiadapiforms. For example, plesiadapids have a limited area for insertion of the *M. quadratus* femoris on the posterior side of the femur. Beard (1993a) interpreted this primitive state as a reversal; however, it could as well indicate that plesiadapids were less specialized than other "primatomorphs" in their arboreal adaptation. This could be confirmed by the variation in the calcaneocuboid articulation alluded to by Beard (1993a). Plesiadapids is taken as having a derived pivot-like calcaneocuboid joint—a putative primatomorph character, whereas Szalay and Decker (1974) carefully described the calcaneocuboid joint as much more primitive in *Plesiadapis* than in primates, without a real pivot [Wible and Covert (1987) also mention that the calcaneum of *Plesiadapis cookei* has no pronounced groove for the tendon of M. flexor fibularis. Does this imply intraspecific variability, or more?]. One would then infer that plesiadapids, and other more primitive plesiadapiforms, were probably less specialized than Beard's reconstructed primatomorph morphotype. A complex history of arboreal adaptations in the different families of plesiadapiforms appears likely, implying a correlative history of some anatomical structures. The calcaneocuboid joint could have undergone several changes in plesiadapiforms, as it has in primates (Gebo, 1988; Gebo et al., 2001). A complex history is already demonstrated by some of the new skeletons, with the remarkable finding of a fully opposable hallux bearing a nail in a carpolestid (Bloch and Boyer, 2001, 2002). In such a context, postcranial characters, which are known to be prone to convergence due to similar functional demands, will probably turn out to be, as the dental characters, very good for intrafamilial phylogeny and locomotor history, but much more difficult to use for the higher level phylogeny. For example, Hamrick et al. (1999) find that a series of phalangeal characters of paromomyids and dermopterans are shared derived. However, including *Dactylopsila, Daubentonia,* and perhaps *Heterohyus* would probably have destroyed the support that they found in favor of paromomyid–dermopteran relationships.

It has been realized that eutherians have a history of arboreal adaptation older than what was previously hypothesized, dating back to the Late Cretaceous, that the broad polarity of some postcranial characters of eutherians has become less secure (Godinot and Prasad, 1994; Prasad and Godinot, 1994), and that many more arboreally adapted groups should be taken into account [e.g., mixodectids (Szalay and Lucas, 1996)], apatemyids, possibly also nyctitheriids (Hooker, 2001). With such a complex background, high-level phylogenetic inferences will require either a more continuous record of morphologies, or enough experience to spot rare characters, unique transformational series. Within the wealth of postcranial characters, there is no reason why some would not turn out to have a high phylogenetic value, as some carpal and tarsal characters have proven to have in primates and beyond. For the time being, the author is doubtful about the carpal characters used by Beard (1993a) because they rely too heavily on the sole triquetrum of *Phenacolemur*, which is on the one hand not too different from that of *Plesiadapis*, and on the other so different from that of *Cynocephalus* that homologies between them are not straightforward (see also Stafford and Thorington, 1998; Szalay and Lucas, 1996). The more complete skeletons found in the Bighorn Basin are very promising (Bloch and Boyer, 2001, 2002). They should help both testing some of Beard's hypotheses and deciphering good phylogenetic signals. On the whole, despite remarkable progress in the knowledge of plesiadapiform postcrania and their functional interpretation, the use of postcranial characters for deciphering early archontan phylogeny (excepting the question of the Volitantia) is considered very conjectural and risky. However, tarsal characters will be mentioned again in later section.

Skull Characters and Conclusion

The study of paromomyid skulls underscores the difficulty of interpreting partial and distorted specimens. Studying a crushed skull of *Phenacolemur jepseni* showing parts of the middle ear, Szalay (1972) inferred the presence of a petrosal bulla, in agreement with his hypothesis of a petrosal bulla in *Plesiadapis* (Szalay, 1969). He also tentatively suggested that the base of a ridge on the promontorium continuous with a longitudinal septum could house a bony canal for a promontory artery (Szalay, 1972). The basicranial morphology of *Phenacolemur* was subsequently reconstructed with a promontory canal (Szalay, 1975; Szalay and Delson, 1979). A petrosal bulla and an osseous promontory canal would have been primate-like characters. However, on another fragmentary skull of the closely related *Ignacius*, it was shown that the "canal" for the promontory artery was in fact imperforate, and it was suggested that *Ignacius* had an ascending pharyngeal artery entering the brain cavity through a middle lacerate foramen (MacPhee et al., 1983).

However, much better preserved specimens later recovered from calcareous nodules and extracted by acid-attack allowed progressively better interpretations (Bloch and Silcox, 2001; Kay et al., 1990, 1992). Study of these new beautiful fossils revealed that the bulla of *Ignacius* is not petrosal, but made by an independently derived entotympanic bone (Kay et al., 1992). This contrasts with dermopterans, which have an ectotympanic bulla (Hunt and Korth, 1980; Wible and Martin, 1993). On their specimen, Kay et al. (1990, 1992) could see only a crest below the ectotympanic, which they interpreted as the crista tympanica. They concluded that *Ignacius* had a tympanic ring fused with the bulla, a morphology that might have been a shared derived similarity between *Ignacius* and dermopterans (absence of annular bridge, Beard and MacPhee, 1994). However, one of the partial skulls described by Bloch and Silcox (2001) clearly shows part of an ectotympanic ring: the crista seen on the other specimen had to be the remnant of an annular bridge (Bloch and Silcox, 2001). With a ringlike ectotympanic suspended by an annular bridge, isolating an epitympanic recess, *Ignacius* appeared similar to *Plesiadapis*, possibly primate-like, and very distinct from dermopterans.

The internal carotid artery (ICA) provided provisional support for a paromomyid–dermopteran link. Szalay (1972) had guessed that a large part of the blood supply to the brain would be carried through the vertebral arteries in *Phenacolemur* and *Plesiadapis* as in other primitive mammals. However, subsequent analyses of arterial circulation in mammals have shown that large promontory and stapedial arteries were probably primitive in eutherian mammals (Wible, 1987). When a carotid foramen and canal were found in *Ignacius*, they were so small that Kay et al. (1992) interpreted them as carrying only nerves, as in lorisids. The complete loss of the ICA could then be a derived similarity shared with the similar loss in dermopterans. This gave some support for a plesiadapiform–dermopteran link (non-microsyopoid plesiadapiforms, Kay et al., 1992). However, study of better-preserved specimens led Bloch and Silcox (2001) to conclude that *Ignacius* still preserved a small promontory artery. More importantly, the lateral course of the internal carotid artery and nerves in *Ignacius* was different from the medial course of these nerves in dermopterans, strongly suggesting that the reduction of the ICA in the two groups was convergent (Bloch and Silcox, 2001). Wible and Martin (1993) had pointed out that a partial involution of the internal carotid system is not unusual in eutherians. Bloch and Silcox conclude the most complete study of paromomyid skulls done until now with: "there remain no unequivocal cranial synapomorphies linking paromomyids and dermopterans to the exclusion of other archontans" (Bloch and Silcox, 2001). Their arguments appear quite convincing.

The absence of finger gliding in paromomyids destroys the most compelling evidence put forward by Beard (1990, 1993a,b) in favor of paromomyids being dermopterans. In view of the quasi-impossible dental morphological transformation implied by this hypothesis, and the absence of any significant cranial character supporting it (Bloch and Silcox, 2001), the primatomorph hypothesis should be abandoned. The primatomorph hypothesis is not definitively refuted, and the problem of dermopteran origins is not solved. An origin within the radiation of archontan claw climbers remains likely. However, the paromomyid connection is eliminated, and we do not know either the living sister group of Dermoptera, nor the fossil sister group of Galeopithecidae, which could be Plagiomenidae (which have a very peculiar basicranium; MacPhee et al., 1989), Mixodectidae, or yet another unknown group. Under these conditions, the concept of Primatomorpha should be abandoned until a better connection suggests dermopterans to be the living sister taxon of primates.

THE PLESIADAPIFORM RADIATION AND PRIMATE ANCESTRY

Plesiadapiformes have grown into an assemblage of 11 families, including more than 40 genera and over a hundred species, if microsyopids are included. With the recognition of *Purgatorius* at the family level as

Purgatoriidae (Gunnell, 1989), there are 10 families listed in Fleagle (1999), and the recently named Toliapinidae (Hooker et al., 1999) has to be added. Surveys of the group can be found in textbooks (e.g., Fleagle, 1999; Szalay and Delson, 1979); however, they are quickly outdated due to the continual description of new genera and species [e.g., *Russellodon* (Sigé and Marandat, 1997), *Toliapina* and *Sarnacius* (Hooker et al., 1999), *Carpomegodon* (Bloch et al., 2001), *Acidomomys* (Bloch et al., 2002)], and the sometimes rapid questioning of some of the new taxa (e.g., *Sarnacius* synonymized with *Berruvius* in Silcox, 2001). The phylogeny of plesiadapiforms is a very active field of research. While a grouping into two major superfamilies, Plesiadapoidea and Microsyopoidea, was advocated by many authors (e.g., Fleagle, 1999; Gingerich, 1976; Gunnell 1989), new accumulated evidence shows that Paromomyidae is possibly less closely related to other plesiadapoids than had been thought (Van Valen, 1994), and a third superfamily, Paromomyoidea, recognized by Silcox (2001), which partly reflects relationships earlier defended by Szalay: a more inclusive family Paromomyidae, with Palaechthonina as part of Paromomyini (Szalay and Delson, 1979), or paromomyids closely related to palaechthonids. The family Carpolestidae received a lot of attention following the recovery of a large quantity of new material (Beard, 2000; Beard and Wang, 1995; Bloch and Gingerich, 1998; Fox, 1984, 1993; Silcox et al., 2001). Its phylogeny has been much scrutinized since Rose (1975), and the family has become a model for stratocladistics (Bloch et al., 2001). The details of plesiadapiform phylogeny are left aside (microsyopids are provisionally excluded). Here, only aspects of the evidence which relate closely to primate origins are considered.

Temporal and Geographical Extension

The radiation of the Plesiadapiformes is in large part Paleocene, predating the well-documented fossil record of primates, which is Eocene (however, *Altiatlasius* is Paleocene, Sigé et al., 1990). This is another reason why they have always been scrutinized in the search of possible primate ancestors or sister groups. Until recently, the plesiadapiforms were considered as a mainly North American radiation, with some families present in Europe having a North American origin. However, several Asiatic plesiadapiforms have been described, which increases the biogeographical complexity of the previous picture. Two carpolestids were reported by Beard and Wang (1995) from the Paleocene or Early Eocene Wutu fauna: a new derived species, *Carpocristes oriens*, close to North American *Carpolestes*, and the new genus *Chronolestes*, interpreted as a primitive carpolestid. This genus was subsequently reinterpreted as a much more primitive plesiadapiform (Silcox et al., 2001). A paromomyid was mentioned in the Wutu fauna (Tong and Wang, 1998), which is probably of North American origin. Another putative carpolestid, *Parvocristes*, was described from Pakistan (Thewissen et al., 2001). However, its P/4 has well-formed cingulids, in contrast with typical carpolestids. It does not resemble *Chronolestes* either. The referred I1/ is also unlike those of North American carpolestids; however, it resembles more closely the I1/ of *Chronolestes*. Such fragmentary material is difficult to identify; however, its attribution to carpolestids is dubious. From the same beds, the same authors describe a presumed plesiadapid, *Jattadectes*, which is also quite difficult to allocate. Because the posterior molar seems much more salient lingually than the preceding one, it would be important to check if the last one could not be a DP4/. If the posterior one were an M1/, it could possibly pertain to *Panobius* found in the same locality and having apparently a comparable size (*Panobius* would then appear similar to some omomyids like *Trogolemur*). If examination of the specimen confirmed that the posterior tooth is an M3/ (e.g., if the metacone is much lower than the paracone), other affinities would be indicated, although probably not plesiadapid. With one paromomyid and one carpolestid pertaining to North American families and having likely dispersed from America to Asia close to the Paleocene–Eocene boundary, *Chronolestes* appears to be the only new genus, which creates an interesting problem. According to the analysis of Silcox et al. (2001), it branches between *Pandemonium* and [(carpolestids, plesiadapids) saxonellids], implying a very early branching. In the broader analysis of Silcox (2001), *Chronolestes* lies at the base of plesiadapoids. Because plesiadapoids have some molar characters that could be primitive (e.g., a centrally placed protocone) in comparison with other plesiadapiforms and *Purgatorius* (Beard and Wang, 1995; Silcox, 2001; Szalay and Delson, 1979), *Chronolestes* raises the possibility that plesiadapoids could have an Asiatic origin. Their North American origin seems to require reversals of important dental characters, which are considered unlikely by this author to occur during a phase of radiation. After all, *Purgatorius* also should be of Asiatic origin.

Plesiadapiform Dental Characters and Primate Origins

Many plesiadapiforms have enlarged anterior incisors and a degree of reduction of the posterior teeth, between I/1 and P/4. It seems clear that the increase of the anterior incisor results in a crowding of the following teeth, which leads in many advanced plesiadapiforms to a diastema between the large incisor and the remaining teeth, or to a series of tiny crowded teeth as in carpolestids. Such specializations have commonly been recognized as preventing any possible ancestral relationship between advanced plesiadapiforms and primates. However, many authors have considered that the most primitive members of the group would be primitive enough to root primates within them (e.g., Simons, 1972; Szalay, 1975; Szalay and Delson, 1979; Van Valen, 1994; Van Valen and Sloan, 1965). Nevertheless, the possibility that *Purgatorius* might be ancestral to primates has also been questioned (Gunnell, 1989; Rose et al., 1994). Having done a very complete survey of plesiadapiform material, Silcox (2001) writes that "every plesiadapiform for which the lower central incisor is known has an enlarged, procumbent I/1." She was able to verify that this is also true for *Purgatorius janisae*. Incisor size increase is common in plesiadapiforms, and it suffers rare exceptions [e.g., the lower incisor decreased in size between *Elphidotarsius* and *Carpolestes* in relation with a shift in function involving the hypertrophied fourth premolars (Biknevicius, 1986)]. However, even in such an exception, the lower incisor stayed procumbent and did not reevolve a more primitive shape. A size decrease of the anterior incisor accompanied by increased orthality seems very unlikely. The reduction of the following teeth also started early in the group; most plesiadapiforms have lost P/1, which is retained only in *Purgatorius, Palenochtha weissae, Anasazia williamsoni,* and one other undetermined specimen (Silcox, 2001; Van Valen, 1994). Because early primates have a P/1, they could be rooted only in one of these earliest plesiadapiforms. The P/4 of *P. janisae* has a relatively large paraconid, which is located quite high on the protoconid. This morphology is very unlikely ancestral to the small incipient paraconid of the earliest primates, which is set at a lower level on the protoconid (Figure 4). *P. unio* has a smaller P/4 paraconid; however, it is tall and the P/4 seems to already display some increase in height in comparison with the P/4 of the earliest primates (see also Rose et al., 1994). The upper P4/ of *P. janisae* has a distinct metacone, whereas early primates have a simple P4/ (Figure 4). The very simple upper and lower P4 of *Donrussellia* and *Teilhardina* are suggestive of an ancestry in a group having simple fourth premolars, and not the higher

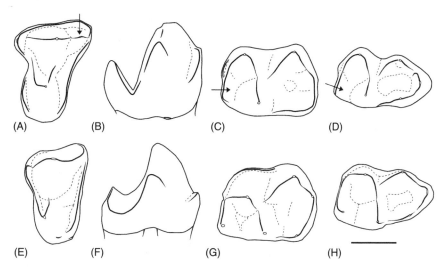

Figure 4. Isolated teeth of *Purgatorius* (A–D) compared with those of early primates (E–H). Drawn to same length or same transverse width (A and E); scale bar in H is 1 mm. All are occlusal views except B and F, which are lingual. On the P4/ (A and E), *Purgatorius* (A) has a metacone (arrow) and a more developed trigon basin than *Teilhardina* (E). On P/4, *Purgatorius* (B) has higher metaconid, paraconid, and talonid summit, than *Donrussellia* (F). On M/1, *Purgatorius* (C) has a reduced trigonid due to its reduced paraconid (arrow) in comparison with *Eosimias* (G). On M/3, the trigonid is also anteroposteriorly narrower in *Purgatorius* (D), with a slightly reduced paraconid (arrow) and moderately elongated posterior lobe, in comparison with *Eosimias* (H), which has a broad trigonid, broad paraconid, and a relatively short posterior lobe. All the character states in *Purgatorius* are interpreted as more advanced in comparison with the states in the various primitive primates. (Redrawn from Buckley, 1997; Kielan-Jaworowska et al., 1979; Rose et al., 1994; Szalay and Delson, 1979; Tong, 1997.)

and more complex ones of *P. janisae* [it is possible that the P4/ metacone secondarily decreased and disappeared in some plesiadapiforms, Silcox (2001); however, this is unlikely to have happened in the primate ancestral group, because some primates later develop a large P4/ metacone. When this happens, it typifies whole groups, such as adapines or sivaladapines, and does not show high plasticity]. The lower molars of *Purgatorius* have a trigonid that is antero-posteriorly compressed, with a relatively reduced paraconid; the paraconid and paracristid are even smaller in *P. unio* than in *P. janisae*. On the contrary, the earliest primates have a trigonid, which is longer antero-posteriorly because they have a large paraconid (Figure 4). This clearly is a character for which both species of *Purgatorius* and all other plesiadapiforms are more derived than the earliest primates are and cannot be ancestral to them (Buckley, 1997, also mentions the incipient postprotocingulum of *Purgatorius* and the "strongly mesiobuccally shifted and mesially inclined molar protocone," valid if it did not revert in plesiadapoids). Considered in isolation, some of these characters could be debated and suspected of some possible degree of reversion. Taken together, however, they show that the typical plesiadapiform specializations were well under way in the earliest known members of the group. These characters, especially the procumbency of the large I1 and the shorter lower molar trigonid, exclude any possible ancestral link between purgatoriids or other plesiadapiforms, and primates.

Despite the absence of an ancestral relationship of plesiadapiforms to primates, the former are still in part dentally primate-like, and a better understanding of the polarity of other dental characters would be important to flesh out the preceding point of view, and to enhance our understanding of early archontan relationships. Silcox (2001) mentions that many plesiadapiforms have the trigonid mesially inclined, excluding *Palenochtha* and plesiadapoids. *Purgatorius* has a small degree of trigonid mesial inclination (Buckley, 1997; Silcox, 2001), and this could be a further indication of divergence from primates, which do not have it. The place of the protocone on the upper molars, relatively mesial in many genera but central in *P. janisae* and plesiadapoids (Beard and Wang, 1995; Szalay and Delson, 1979; Van Valen, 1994) should be scrutinized. Silcox (2001) codes the protocone "skewed mesiobuccally" in tupaiids and *Donrussellia*, whereas it is coded as central in other primates. On the described upper molar of *Eosimias* (Tong, 1997), the protocone seems quite mesial; however, it is not particularly labial (= buccal) in position. An understanding of the polarity of this character would be important. It is likely to be complex; the labial shift and long lingual slope of the protocone is one consideration (discussed by Godinot, 1994, and poorly understood functionally), and the mesiodistal place of the protocone is another, hopefully to be distinguished from the simple posterior extension of the posterior face, which makes the protocone appear "mesially shifted" in relation to the whole mesiodistal diameter; another aspect is the distolingual angle or shift of the lingual part of the upper molars, precisely defined by Van Valen (1994). Hence, at least three characters should be evaluated to better describe the complexity of the upper molar "basic" structure.

For upper molar and other dental characters, it must be stressed that the primate ancestral dental morphotype has not been elucidated. The description of eosimiid dentitions (Beard et al., 1994, 1996; Tong, 1997) has introduced new primitive dental character states that need to be taken into account in deciphering the primitive primate morphotype. As in *Tarsius*, eosimiids have a trigonid on the lower molars, which does not become more compressed antero-posteriorly from M/1 to M/3 (judged from Beard et al., 1994, 1996; the isolated lower molars figured by Tong, 1997, suggest a low degree of trigonid compression and paraconid labial shift on M/2-3, derived in comparison to the other specimens). On the contrary, trigonid compression markedly increases from M/1 to M/3 in other primates and a number of plesiadapiforms, again demonstrating convergence ·for this character. The absence of trigonid compression from M/1 to M/3 must be primitive for primates. *Eosimias* has a relatively broad stylar shelf, primitive; its very large parastyle is also reminiscent of a parastylar lobe, which would make it a primitive character state. On the lower molars, the cristid obliqua does not join the metaconid on M/1, which is primitive in comparison with *Donrussellia* (a cristid obliqua ascending on a posteriorly shifted metaconid, the "stepped postvallid" of Silcox, 2001, is also convergent in many primates and plesiadapiforms). The relatively short M/3 talonid of *Eosimias* is also probably primitive for primates (see suggestion of a possibly short talonid on the M/3 of *Altiatlasius* in Godinot, 1994). There was again convergence in M/3 third lobe elongation between many plesiadapiforms and primates.

In this context, the interpretation of *Altanius*, a genus most often considered as a primitive primate, but repeatedly suspected of having plesiadapiform affinities (Rose and Krause, 1984; Rose et al., 1994), remains intriguing. Silcox (2001) places *Altanius* within the primates, corroborating the view of Gingerich et al. (1991). *Altanius* clearly does not fit in the radiation of North American plesiadapiforms as understood here. However, it shows a series of similarities with some plesiadapiforms, which must be convergent (and not primitive as in the parsimony analysis of Gingerich et al., 1991), and would be very interesting to understand adaptively (exodaenodonty, high M/1 trigonid). Because it is quite autapomorphic within primates, it appears very difficult to decipher its affinities; and its placement in a subfamily Altaniinae *inc. sed.* as proposed by Van Valen (1994) seems warranted. Whether this early genus could reveal some characters of the primate morphotype is unknown (P3/ and P4/ triangular, with small protocone?).

Based on dental characters, Silcox (2001) suggests that Toliapinidae, restricted to *Toliapina* and *Avenius*, could be closely related to primates. However, these genera are known only from isolated teeth from P4 to M3 and the node is weakly supported; she does not give much credence to this cladistic result. *Avenius* seems to have a rather typical plesiadapiform-like P/4—a correlative enlarged incisor would prove such an affinity and rule out close primate ties. These tiny low-crowned forms would in any case be rather distant from the well-known primates, which have higher, more pointed cusps, suggestive of a more insectivorous diet. A more convincing close affinity with primates would necessitate more complete dentitions and dental intermediates, or the confirmation of primate affinity through the non-dental characters cited below.

To sum up, not only do derived plesiadapiform families show specializations of their dentitions very divergent from those of primates, evidently irreconcilable with primate ancestry, but also even the most primitive plesiadapiforms (with or without microsyopids) have derived characters excluding them from possible primate ancestry. A sister group relationship of *Purgatorius* or a primitive plesiadapoid to primates would reflect a dichotomy going back to the Earliest Paleocene or Late Cretaceous. It is quite hazardous to evaluate such a hypothesis without a better knowledge of the primate dental morphotype and some ideas about primitive tupaiid dentitions (hardly compensated for by very distant insectivore outgroups as in Hooker, 2001). However, a better understanding of the functional meaning of molar character evolution in plesiadapiforms would greatly assist the evaluation of possible morphological changes in primate ancestry.

Other Characters and Conclusion

Progress in the interpretation of paromomyid cranial characters engendered by the discovery of new and well-preserved specimens has been discussed earlier. The discovery of a tympanic ring and an annular bridge in *Ignacius* (Bloch and Silcox, 2001) lends considerable support to the link between *Plesiadapis* and *Ignacius* advocated by Kay et al. (1992). It is thus very likely that, like paromomyids, plesiadapiforms (excluding microsyopids) are not closely related to dermopterans. However, the more relevant question is if they have a close relationship to primates. Important non-dental support for this hypothesis came from the initial interpretation of the *Plesiadapis* bulla as petrosal (Russell,

1959; Szalay, 1969, 1975; Szalay et al., 1987)—a view abandoned by Russell (1964) and considered doubtful by others because bullar sutures can fuse early (Gingerich, 1976; MacPhee et al., 1983; MacPhee and Cartmill, 1986). Given these uncertainties for *Plesiadapis*, and given its above-mentioned similarity with *Ignacius*, *Plesiadapis* probably also had an entotympanic bulla (Kay et al., 1992; Wible and Martin, 1993). The case for primate affinities for the plesiadapiforms is thus considerably weakened.

Other potential cranial plesiadapiform-primate synapomorphies are also debatable. The central position of the promontorium within the auditory bulla is linked to the medial expansion of the middle ear cavity in paromomyids, plesiadapids, and adapids (Szalay et al., 1987). However, Kay et al. (1992) suggested that this similarity arose through different developmental pathways: plesiadapiforms [in fact paromomyids and not plesiadapids, Beard and MacPhee, 1994] differ from adapids in having a narrow basisphenoid across which the bullae nearly touch, whereas in adapids the basisphenoid is much broader and the two bullae are widely separated (Kay et al., 1992). Whether this really implies convergence for this character might require further scrutiny. The lateral route of the ICA in plesiadapiforms and primates is unusual and significant (Bloch and Silcox, 2001; Wible, 1993). However, there is some variability within primates: *Shoshonius* has the lemur-like lateral position of the posterior carotid foramen, which leads us to consider this as the primitive pathway in primates (Beard and MacPhee, 1994; Bloch and Silcox, 2001). However, this implies that a reversion to the more primitive location "occurred in omomyids more derived than *Shoshonius*" (Bloch and Silcox, 2001). In fact, this hypothesis would require two independent reversals in the North American *Rooneyia* and *Omomys* (Ross and Covert, 2000) and in European microchoerids. Such a scenario is questionable, and in any case, it shows that this character either did revert or changed convergently in primates, diminishing its phylogenetic value. Several other potential plesiadapiform-primates synapomorphies were mentioned and subsequently dismissed. A maxillary-frontal contact in the orbit occurs in plesiadapiforms and primates; however, it is not a convincing synapomorphy as it is not unusual among eutherians (e.g., lipotyphlans, rodents, lagomorphs, Wible and Covert, 1987). The ventral shielding of the fenestra cochleae was listed (Szalay, 1975; Szalay et al., 1987); however, MacPhee (1981) suggested that shielding by a caudal tympanic process of the petrosal is primitive rather than derived. The presence of an annular bridge linking the tympanic ring to the bulla wall was considered an important plesiadapid-primate similarity (Cartmill, 1975; Gingerich, 1976); however, this character is also present in tupaiids, and it is not present in all primates, rendering its polarity problematic. I would provisionally follow Beard and MacPhee (1994), who consider a complete annular bridge as primitive for primates, due to its presence in tupaiids, plesiadapids (+ paromomyids), and omomyids. It appears that most of the proposed synapomorphies between plesiadapiforms and primates either have been refuted or appeared problematic. Basicranial features have been the subject of much attention, including remarkable developmental studies. They provide characters very important for the study of primate phylogeny; however, they proved relatively deceptive in the search for archontan phylogeny (e.g., MacPhee, 1981; Starck, 1975; Wible and Martin, 1993). Recent authors observe that the basicranium is not a taxonomic touchstone (Bloch and Silcox, 2001; Wible and Martin, 1993).

If we cannot rely on decisive characters from the basicranium, what would the rest of the cranium suggest? Until now, the general shapes of known plesiadapiform skulls are very unlike those of primates. The skulls of *Plesiadapis* and *Ignacius* are primitive in having small laterally directed orbits, a broad interorbital breadth reflecting large olfactory bulbs, a large infraorbital foramen suggestive of important blood supply to the anterior part of the muzzle (probably with well-developed vibrissae), as was inferred in the beautiful study of the skull of *Palaechthon nacimienti* by Kay and Cartmill (1977). They seem to have had a small brain case in comparison with primates. From the size of the optic foramen, Kay et al. (1992) inferred that *Ignacius* had eyes similar in size to those of *Erinaceus*. These characters reflect the absence of any evolutionary step toward the crucial primate visual apomorphies. One can find some isolated apomorphies (e.g., a relative reduction of the infraorbital foramen seen in *Plesiadapis*), or the beginning of a postorbital process in *Palaechthon* (palaechthonid), and in the microsyopid *Megadelphus*; however, these seem to be of limited significance, and probably not homologous with primate states. *Plesiadapis* and *Ignacius* skulls also have cross specializations, such as a tubular ectotympanic, absent in the ancestral primate morphotype. One could guess that some of the cross-specializations seen in later plesiadapiforms could be absent in earlier forms, and they may not therefore disallow a close phylogenetic relationship between plesiadapiforms and primates (Bloch and Silcox, 2001). However, given the absence of any crucial primate-like characters, as those of the orbit, in later forms, one can infer their absence in primitive plesiadapiforms and conclude that there is no strong cranial evidence in favor of a close plesiadapiform-primate phylogenetic relationship. In fact, there are the specializations of the muzzle, narrow and elongated in correlation with the large anterior incisors, which argue strongly against any plesiadapiform being ancestral to primates. If anything, the known skull evidence does not favor plesiadapiform-primate close relationship.

On the other hand, it is possible that a series of derived cranial characters would well support plesiadapiform monophyly, in addition to the dental ones. A series of cranial synapomorphies of plesiadapiforms was provided by Kay et al. (1992), among which several seem to hold: suboptic foramen present, ossified external auditory meatus, and strong mastoid tubercle. There is also a degree of reduction of carotid blood supply to the brain, and probably other common characters to extract from the specializations of *Plesiadapis* and *Ignacius* skulls. When known, plesiadapiforms (with the exception of the hallux of *Carpolestes*) have claws and not nails. Their tarsals do not show a close approximation with those of primates, contrary to those of tupaiids (see below). Perhaps some plesiadapiform postcranial synapomorphies will be found? As stated above, a number of characters formerly believed to be plesiadapiform-primate synapomorphies are now suspected to be archontan. The history of arboreal adaptations is much more complex than previously thought, requiring further analysis before we can delineate which characters retain a high phylogenetic value.

Beyond the restricted plesiadapiforms considered here, several other families should be taken into account in a search to elucidate archontan phylogeny. Microsyopidae, which have a more primitive auditory region (Szalay, 1969) are sometimes mentioned as a primate or dermopteran sister group; Mixodectidae and Plagiomenidae are more commonly suspected to have dermopteran affinities (Szalay and Lucas, 1996); Apatemyidae, with their dental and arboreal specializations, could pertain to the same broader group. How far should such a group be extended in order to include primates and their close relatives? Probably, as far as the living tree shrews, order Scandentia. The broad and not well-delineated Archonta appears then as a group of arboreal eutherians, whose history is certainly very complex and which might even include the Cretaceous *Deccanolestes* (Prasad and Godinot, 1994). They may also be related to nyctitheres (Hooker, 2001).

RETURNING TO TUPAIIDAE

When Luckett's (1980a) book on tree shrews was published, plesiadapiforms were considered by all the contributors primitive primates. This was a relatively broad, though not unanimous, consensus. It is very interesting to examine how the elimination of the Plesiadapiformes from primates affects our interpretation of certain characters. Surprisingly, this elimination renders a whole series of derived characters possible synapomorphies of tupaiids and primates. Figure 5 illustrates how the informative part of a cladogram of Luckett (1980a) can be reinterpreted simply in favor of a close tupaiid–primate affinity. A simple redrawing of the cladogram without imposing the Plesiadapidae in primates leads to a more parsimonious solution (Figure 5B, with 4 convergences instead of 5). The solution chosen by Luckett was based on the assumption that Plesiadapidae had a petrosal bulla, and this character had to be given a high weight in the phylogenetic analysis, a procedure one would be willing to endorse. However, since then it has been shown convincingly that the Paromomyidae do not have a petrosal bulla (Bloch and Silcox, 2001; Kay et al., 1992), and as a consequence, the closely related and extremely similar Plesiadapidae very probably also do not have a petrosal bulla. This new information renders a tupaiid–primate affinity quite well supported (Figure 5C).

Reinterpreting several other chapters of Luckett's book in light of new data gives similar results. In Novacek (1980), several cranial features (one orbital and two auditory) and one postcranial feature appear as characters shared by tupaiids and primates (not plesiadapiforms) and derived in comparison with the eutherian morphotype (there are also five characters: one orbital, one auditory, and four tarsal—derived and shared by tupaiids, primates, and plesiadapiforms). In their contribution on carotid arteries and cranial characters, Cartmill and MacPhee (1980) came up with eight characters shared by tupaiids and primates and derived in comparison with the ancestral eutherian morphotype. Four of these were eliminated from further consideration because they were not present in the Plesiadapoidea: eliminating this group makes them reappear as potential synapomorphies of tupaiids and primates (three more may reappear if they are shown to be derived in relation to the eutherian morphotype, which is unlikely). In their study of tarsal characters, Szalay and Drawhorn (1980) listed seven tupaiid autapomorphies. However, several of them are present in the tarsals of *Teilhardina* and, if plesiadapiforms are removed from primates, they become potential

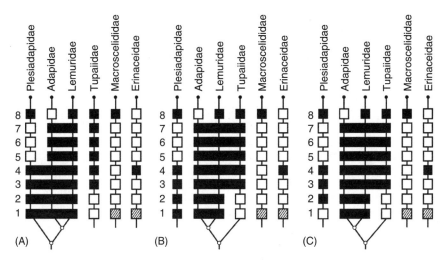

Figure 5. The informative part of the cladogram published by Luckett (1980b, Figure 5) is shown as published by him (A), and reinterpreted in a more parsimonious way favoring tupaiid affinities for primates (B). There are four convergences in B, cf. six in A. However, (A) was chosen because character 1 was correctly believed to be highly significant. In C, the reinterpretation of character 1 (see text) confirms the tupaiid affinities of primates from this simplified data set. Such simple reinterpretation shows that adding or subtracting one or two characters can change a phylogenetic hypothesis, rendering a formerly favored hypothesis weak, whereas establishing good characters is much more important if phylogenetic interpretations are to be well grounded. Characters are from those utilized by Carlsson (1922). They are shown in primitive state (open square), derived state (black square), and intermediately derived (obliquely-lined square). 1: Petrosal wing of bulla minute or absent (Pr) or forming virtually entire bulla (De); 2: Foramen rotundum confluent with (Pr) or distinct from (De) sphenoidal fissure; 3: Fibular facet on calcaneus prominent (Pr) or reduced or absent (De); 4: Sustentacular facet of astragalus separate from (Pr) or continuous with (De) navicular facet; 5: Ectotympanic exposed at least partially at lateral margin of bulla (Pr) or "intrabullar" (De); 6: postorbital processes of frontal and jugal absent (Pr) or postorbital bar complete (De); 7: Jugal (zygomatic) foramen absent (Pr) or present (De); 8: Fibular crista of astragalar trochlea subequal (Pr) or higher (De) than tibial crista.

synapomorphies of tupaiids and primates. A series of potential synapomorphies between tupaiids and primates appear from these studies because of the removal of plesiadapiforms from primates. It is beyond the scope of this chapter to make a complete reappraisal of all those characters. However, after discarding those characters for which the preceding authors disagreed or those which now appear as likely archontan traits, briefly listed in the following are the cranial and soft anatomical characters which seem to be still valid, including those proposed in more recent studies. The tarsal characters are considered in the next section.

1. Postorbital bar complete (consensual);
2. Optic foramen distinctly enlarged (also present in some plesiadapiforms, Novacek, 1980; needs further scrutiny, also enlarged in *Ptilocercus*?);
3. Canals around intratympanic portions of facial nerve and stapedial artery formed from outgrowths of the petrosal (Wible and Covert, 1987; petrosal tube around promontory artery listed as distinct character by Cartmill and MacPhee, 1980; canal around the internal carotid artery petrosal in primates, entotympanic in tree shrews, MacPhee, 1981; Wible and Covert, 1987; homologous or convergent?);
4. Anterior carotid foramen converted into a long tube (Wible and Covert, 1987);
5. Tegmen tympani that is greatly expanded anterolaterally to cover the entire middle-ear ossicular chain (MacPhee, 1981; Wible and Covert, 1987; in tupaiids and strepsirhines, presence of an epitympanic crest, MacPhee, 1981; Zeller, 1986);
6. Maxillary artery pierces the ectopterygoid plate (Kay et al., 1992);
7. Tympanohyal large, isolates stylomastoid foramen from tympanic chamber (Novacek, 1980);
8. Tympanic process of petrosal partial bullar element (intermediate state in Tupaiinae [? Ptilocercinae], primary bullar element in primates, Novacek, 1980);
9. Large jugal or zygomatic foramen (relatively consensual, tupaiine and lemurid states distinguished by Cartmill and MacPhee, 1980);
10. Jugular foramen dual (Kay et al., 1992; *Tupaia* potential intermediate, *Loris* reversion);
11. Olfactory bulbs intermediately reduced (reduced in primates, Luckett, 1980b; less reduced in *Ptilocercus* than in tupaiines, Le Gros Clark, 1926);
12. Uterus, simplex in Anthropoidea, intermediately derived state in Tupaiidae and other primates (Luckett, 1980b);
13. Volar skin with serial papillary ridges ("finger-prints," Lemelin, 2000); low value because present in other mammals, however, absent in Insectivores; could be valid within archontans.

It is clear that some of these characters have a rich history of study, including for example, the bony canals around the arteries and nerves in the middle ear. The latter were sometimes interpreted as a single character (for the presence of bony canals), but subsequently portions of canals and osseous derivation were taken into account (opening up the difficult issue of possible phylogenetic replacement of one bone by another, as for the bulla). The number of characters represented here is not clear. The interpretation of the bulla is difficult; its petrosal composition in primates as opposed to essentially entotympanic in tupaiids has often been regarded as an obstacle to tupaiid–primate ties. However, the existence of a petrosal part in the tupaiid bulla may well signal a common heritage with primates, as hypothesized by Novacek (1980). Other characters on the above list, more recently suggested, may be shown in the future to be convergent for tupaiids and primates, as was the aphaneric ectotympanic (intrabullar, not seen in ventral view, condition unique in tupaiids and primitive strepsirhines among eutherians, Cartmill and MacPhee, 1980; but likely convergent because of the probable absence in the ancestral primate morphotype). Even if some characters were to be discarded in the future, the list contains several important and very consensual synapomorphies. It is noteworthy that, in their reappraisal of basicranial characters based on ontogenetic studies, Wible and Martin (1993) concluded that: "If euprimates share a special relationship with any archontans, it is with scandentians based on the basicranial evidence. Both have an enlarged tegmen tympani that roofs the entire middle-ear ossicular chain, and there are further unique resemblances in the tegmen tympani of lemuriforms and scandentians."

In sum, there is strong evidence in support of tupaiid–primate affinity in basicranial characters, including tegmen tympani and carotid circulation characters. There are characters linked to the crucial primate orbital apomorphies, the complete postorbital bar, enlarged optic foramen suggesting larger eyes, associated with some reduction of the olfactory bulbs (however, *Ptilocercus* relies less on vision and more on olfaction than tupaiines do; Le Gros Clark, 1926). These visual characters should be confirmed or questioned by further studies of the eyes of living forms and the neural characters linked with optic function (see Ross, Preuss, this volume). It has been mentioned that other mammals did evolve a postorbital bar; however, what is remarkable in the case of tree shrews is that they are small "insectivore-like" mammals with a diverse group of primitive forms, which must have an ancient origin. This in turn suggests that their acquisition of a postorbital bar was very ancient, early enough to possibly signal common ancestry with primates. The beginning of a reduction of the number of incisors in tupaiids could also pertain to the global transformation of the face, possibly homologous with primates. Added to these characters is a series of other, more isolated characters, proposed by Novacek (1980), Kay et al. (1992), some of which might well turn out to give a real phylogenetic signal. Despite the differences between the nocturnal or crepuscular *Ptilocercus* and the diurnal, more visually evolved tupaiines, the cranial evidence in favor of a close tupaiid–primate affinity appears impressive.

Important Tarsal Characters

A series of tupaiid tarsal characters were interpreted by Szalay and Drawhorn (1980) as derived (as opposed to primitive archontan retentions): astragalus with the groove for the tendon of M. flexor (digitorum) fibularis aligned parallel to the long axis and located upon a ventrally projecting medial body; length of the astragalus relatively large in contrast to the squat bones in Paleogene archontans; sulcus astragalus not approaching the trochlear groove for the tendon of M. flexor fibularis (this may be related to preceding character); calcaneum having lost the primitively large peroneal process found in early archontans; on calcaneum, the anterior plantar tubercle is greatly reduced and has receded more distally [sic, they meant proximally] than it probably was on the ancestral archontan. In fact, these characters and some others can be reinterpreted as potential synapomorphies of tupaiids and primates.

The greater length of the astragalus in comparison with that of plesiadapiforms makes it primate-like (Figure 6). This neck elongation has been linked by Dagosto (1988) with the elaboration of subtalar motion. The most striking difference between the *Ptilocercus* astragalus and those of all plesiadapiforms (Paleocene "primates") described by Szalay and Drawhorn (1980) is that the latter have a body that is lower medially than laterally; the medial trochlear crest is lower and less salient—a primitive state. In contrast, *Ptilocercus* has a body that is almost as high medially as laterally, and the two rims of its trochlea are more similar (the medial rim is still more rounded than the lateral one). These derived features make the astragalus of *Ptilocercus* much more primate-like than all the plesiadapiform astaragali described so far (from the first drawings of *Carpolestes* published by Bloch and Boyer, 2002, it seems that this fossil too has an astragalus more plesiadapiform-like than primate-like). Why should these characters not be considered shared derived

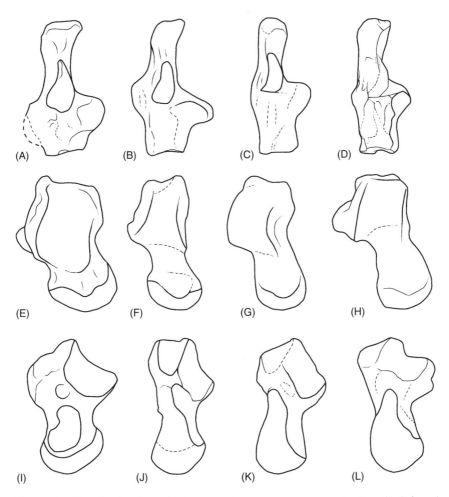

Figure 6. Tarsals of archontans drawn at the same antero-posterior length. Calcanei in dorsal views (A–D) and astragali in dorsal (E–H) and ventral (I–L) views. *Nannodectes gidleyi* (A, E, and I), *Ptilocercus lowii* (B, F, and J); eosimiids (C, G, and K), *Notharctus* (D), and *Teilhardina* (H and L). Redrawn from different authors and not entirely accurate (different orientation in g, some facets partially drawn because not clearly delineated on photographs). Note especially that *Ptilocercus* (B) has a per-oneal tubercle in a proximal position as in primates, a high and wedge-shaped astra-galar trochlea (F), and a long neck (F and J), whereas *Nannodectes* has a short astragalar neck (E), and a sustentacular facet distinct from the navicular (I).

with primates? In *Ptilocercus*, the astragalar sustentacular facet is continuous with the navicular facet, whereas on several plesiadapiforms there is a clear discontinuity between the facets (on others this continuity does occur). Another interesting fact is that in distal view, the astragalar head is mediolaterally elongated in several plesiadapiforms; it is less elongated, more ovoid, in some other plesiadapiforms and *Ptilocercus*. Whereas the astragalar head in *Ptilocercus* is less spherical than in primates, its morphology appears intermediate in form [this would deserve quantification; I note that some of the Shanghuang astragali have heads that are less spherical than in other primates (Gebo et al., 2001; and this could be a primitive state)]. The last intriguing character seen on Szalay and Drawhorn's Figure 9 (1980), and which was observed on one specimen, is a clearly wedge-shaped astragalar trochlea in dorsal view. This character was considered primate-like by Szalay and Lucas (1996), who proposed its interpretation in the framework of the grasp-leaping theory. The distal broadening of the trochlea would be well-suited to the transmission of stress during landing after a leap, with dorsiflexed feet. However, *Ptilocercus* is not a leaper, and there is no reason to believe that it could have inherited the character from a leaping ancestor. Gebo et al. (2001) link a wedge-shaped trochlea with enhanced dorsiflexed foot positions and greater use of vertical supports, something which is more in line with the behavior of *Ptilocercus*, which "spend relatively more time [than other sympatric tree shrews] on large vertical supports" (Emmons, cited in Stafford and Thorington, 1998). The wedge-shaped trochlea could also be related to some foot rotation in conjunction with flexion-extension, as is recognized for monkeys having such a trochlea, and linked with *Ptilocercus'* frequent use of hallucal opposition (Sargis, 2001). In any case, for this character as well, *Ptilocercus* is remarkably primate-like. There are still differences between *Ptilocercus* and primates, including an astragalus with a less spherical head, with a lower medial side of the body, possibly a more elongate ventral groove for the tendon of M. flexor fibularis, and other differences that might be more or less accentuated depending on how one reconstructs the primate morphotype (shallow astragalar facet, small or absent posterior trochlear shelf, etc.; see below).

On the calcaneum of *Ptilocercus*, the peroneal process is reduced in comparison with that of early archontans (Szalay and Drawhorn, 1980). It is also more proximally placed, being at the level of the posterior astragalar facet (Figure 6). This position is remarkably primate-like and unlike the distal position of the peroneal process of plesiadapiforms and many other mammals. Dagosto (1988) noted that "The change in placement of this tubercle has been related to the elongation of the tarsus (Decker and Szalay, 1974), and the shift in the primary function of this muscle from a foot evertor to an hallucal adductor (Gebo, 1986; PhD), but there is as yet no satisfactory explanation for its reduction in size [in primates]." The presence of a reduced proximal process in *Ptilocercus* suggests that the correlation with tarsal elongation is faulty. Given the hallucial grasping behavior of *Ptilocercus* (Sargis, 2001), it appears that M. peroneus longus may be a hallucal adductor in these animals. The reduction in the size of the peroneal process in primates may well be partly inherited from a ptilocercine-like ancestor; however, further reduction was manifested in the subsequent primate ancestral morphotype (here too it is interesting to note that some of the Shanghuang calcanei attributed to a new taxon of protoanthropoids have a peroneal tubercle varying from small and moderate (mostly) to "prominent" in one specimen: a primitive retention?). The calcaneocuboid joint of *Ptilocercus* is a circular pivot (Sargis, 2002b; observed by the author). Because primates typically have a pivot joint, this provides a tempting *Ptilocercus*-primate synapomorphy. However, we must remember that the calcaneocuboid joint has changed a lot during its evolutionary history, and therefore, a more detailed analysis of the relevant morphology is required before homologous stages can be infered (direct historical evidence might be necessary).

The above list of derived similarities of ptilocercine tupaiids with primates is impressive. Several of them either appear related to enhanced subtalar mobility or to enhanced hallucial opposition, which makes *Ptilocercus* functionally intermediate with primates (Sargis, 2001). In view of the value of tarsal characters for phylogeny reconstruction, the above-mentioned derived similarities of *Ptilocercus* and primates make a strong case in favor of their close phylogenetic affinity, to the exclusion of known plesiadapiforms. A similar conclusion was drawn by Hooker (2001) in his analysis of the tarsal characters of the archontans, *Deccanolestes*, and nyctitheriids; however, this author lost this signal by introducing dental characters in the same analysis. The lengthening of the astragalus, associated with a tall medial body and concomitant reduction of the peroneal tubercle signals an interesting tarsal transformation, possibly associated with frequent hallucal opposition (still far from the powerful hallucal-grasping of primates). This set of characters deserves further functional scrutiny.

Conclusion: Tupaiids as Sister-Group of Primates

In summary, it appears that tupaiids share a series of derived skull characters with primates, among which are basicranial characters supported by ontogenetic studies (Wible and Martin, 1993) and mutually agreed upon characters linked with crucial primate visual apomorphies. Several other cranial characters also proposed in recent studies lend support to this hypothesis (Kay et al., 1992; Wible and Covert, 1987). Among those cranial characters, the most intriguing is the complete postorbital bar, which may require an explanation beyond its role in living primates. Some soft anatomical characters may add support to this view (olfactory bulbs, uterus, Luckett, 1980b). Finally, tarsal characters offer strong support in favor of the same view, and evidence against a closer affinity with plesiadapiforms. Thus, tupaiids appear as the best living or fossil sister group of primates, and they remain so even in comparison with Paleocene–Eocene plesiadapiforms (Figure 7). Plesiadapiforms appear as a likely North American monophyletic group; tupaiids are Asiatic and do not show, at least for *Ptilocercus*, specializations (other than dental and carpal), which would exclude them from being the best model for primate ancestry. Their behavior, especially their manual insect seizing (Le Gros Clark, 1926; Sargis, 2001), may be considered particularly well-suited to lead to the acquisition of primate characteristics (see later section).

If the hypothesis of a close tupaiid–primate relationship is true, it should be corroborated in the future by dental characters. Dental characters have proven to be of little use as there is no Paleocene tree shrew, which would show more primitive tupaiid dental characters. From the dentition of living tupaiids, one would easily infer that all of them have derived characters that prevent them of being ancestral to primates, among which specializations in their anterior dentitions or molar characters in tupaiines (see Butler, 1980). This is not astonishing, knowing that teeth have continuously evolved in most mammalian groups. The genus *Ptilocercus* is dentally more primitive than other tupaiids, and it also has a few derived characters that would exclude it as a possible ancestor; however, these characters do not appear important. The absence of conules in tupaiids is sometimes mentioned as an obstacle; however, Butler (1980) mentioned that the preprotocrista sometimes develops a paraconule in *Ptilocercus*, and the postprotocrista, abruptly interrupted on the M2/ figured by Hooker (2001, Figure 33), which would also appear unlike that of primitive primates, is in fact extended past the base of the metacone in some specimens (Butler, 1980). The hypocone is very small. What is

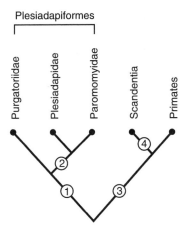

Figure 7. Cladogram summarizing the main characters, which unite primates and scandentians to the exclusion of plesiadapiforms. 1: Derived dental characters of purgatoriids excluding possible ancestry to primates, or cross-specializations: enlarged procumbent I/1, P4/ having a metacone, high P/4 with high metaconid and talonid, lower molars with some anteroposterior trigonid compression, some paraconid reduction. 2: Accentuated plesiadapoid specializations, including reduction of the teeth located between I/1 and P/4, strong postprotocingulum on upper molars, long muzzles, etc. 3: Possible homologous synapomorphies shared by tupaiids and primates, including basicranial characters (enlarged tegmen tympani, bony canals around middle ear arteries, and others), postorbital bar, a series of tarsal characters (see text). 4: There are autapomorphies of living tupaiids, including several dental characters.

attractive in the dentition of *Ptilocercus* is that the upper and lower P4 are simple, not molarized, and close in shape to those of primitive primates. Importantly, the lower molars are relatively low but retain an anteroposteriorly elongated trigonid—a condition that must have been present in the ancestral primate, but that was lost in a number of primitive proteutherians and insectivores. Hence, it appears that a more primitive ptilocercine, which from common evolutionary trends can be predicted to have had upper molars more transversely elongated than those of *Ptilocercus*, without a hypocone, with small conules and more usual protocristae, and with a more primitive anterior dention, could come close to the (problematic) primate morphotype. The reality of these trends is documented in tupaiines by the isolated teeth of the Chinese Eocene *Eodendrogale*, which has transversely elongated upper molars without hypocone (Tong, 1988). Even if this may appear quite speculative, the dentition of *Ptilocercus* in spite of its autapomorphies, presents basic similarities with primates, which probably contain some phylogenetic signal. Due to the temporal gap, it seems difficult to be more precise at this moment. Early tupaiid dentitions should be found in the future, and provide a test of these ideas.

REMARKS ON SCENARIOS OF THE ACQUISITION OF NAILS

In the initial formulation of the visual predation theory, Cartmill (1972, 1974a,b) explained the acquisition of the opposable hallux by the ancestral primate and the subsequent loss of claws in several groups of primates by the invasion of the fine branch milieu. It is usually now accepted that the possession of nails on the extremities is also a part of the primate morphotypic condition. Recent comparative and functional studies of arboreal primates and marsupials went further in the study of the convergences between these two groups and led to an increase in our knowledge of the arboreal adaptation of some of them (Rasmussen, 1990; Lemelin, 1999; Larson et al., 2000; Hamrick, 1998, 2001). These inquiries have shown the importance of long phalanges to grasp fine branches. They revealed that the peculiarities of primate "hindlimb domination" and diagonal sequence in footfall exist also in the highly arboreal marsupials, which suggests that these peculiarities are also adaptations for locomotion on fine branches (Schmitt and Lemelin, 2002). These discoveries are interpreted in favor of the fine branch milieu as the adaptive shift explaining the origin of primate appendicular characteristics (Hamrick, 1998, 2001; Larson et al., 2000; Lemelin, 1999; Rasmussen, 1990; Schmitt and Lemelin, 2002). However, are these findings sufficient to confirm the fine branch milieu hypothesis concerning primate origins? Consideration of both marsupials and primates leads me to embrace a more complex scenario.

Concerning marsupials, why did the arboreal ones fail to evolve nails on their non-hallucal digits, if habitual grasping of fine branches is the key adaptational shift explaining the transformation of claws into nails? A marked reduction of claws has been reported in the Australian *Tarsipes*, *Cercartetus*, and *Burramys* (Cartmill, 1974b). A detailed study of their terminal phalanges would be important. The American *Marmosa* and *Caluromys* retain claws. Are those claws reduced? Are their terminal phalanges broadened? Is their arboreal adaptation too recent to have allowed the replacement of claws by nails? American Marmosidae are known since 12.5 MYA (Flynn et al., 1997;

Goin, 1997). The author has not found information concerning fossil caluromyids (a family in Nowak, 1999). However, *Burramys* is found in the Late Oligocene–Middle Miocene Ngama fauna, and Burramyidae, as well as Pseudocheiridae and Petauridae, are reported since the Late Oligocene–Early Miocene of Australia (Rich, 1991). It is also well known that Didelphidae, in a restricted sense including tarsal characters and opposable hallux (i.e., having arboreal adaptations), are known at least since the Early Eocene (Szalay, 1982). In fact, Ameridelphians, and possibly Metatheria, are thought to have possessed arboreal characters. In this context, it is very strange that there are not many marsupials having reached the "ultimate" fine branch adaptation of nails, if the last indeed are due to fine branch locomotion. Marsupial analogs suggest that claw loss may have been a very long process.

A real effort to understand the replacement of claws by nails was made by Cartmill (1974b). He suggested that, if a mammal, whose first toe has become divergent enough to oppose the other four toes in grasping slender branches, continues to emphasize prolonged cautious locomotion among slender branches, its "first digit will become proportionately more powerful, and claw grip will be proportionately enfeebled"—something he reported as true of marsupials such as *Marmosa*, *Didelphis*, and *Petaurus* (*idem*, p. 71). What makes little sense is that the process was not conducted to complete claw loss in a larger number of marsupials, particularly given how ancient their arboreal adaptations are. One would also guess in such a scenario that a similar opposition of the thumb to the other digits of the manus would occur (or another type such as digits one and two opposed to the three others). However, it seems that opposability in the hand has rarely been developed in arboreal "didelphid-like" marsupials, whereas it did in larger forms such as *Phascolarctos*. A better understanding of hand grasping in small marsupials might help understand why they did not evolve opposable hands and nails on their extremities. Some degree of hallucial opposition is also known in other mammals, including some rodents, tree shrews, and fossil plesiadapiforms. It is suggested that a "didelphid-like" adaptation, with opposable hallux, claws on other digits and non-opposable thumbs represents a successful arboreal adaptation, allowing the possible use of fine branches if the digits become long enough. It would represent a step toward but not a complete convergence with primate appendicular characteristics. It seems that opposition in the hand and the presence of nails are less common; could these two properties be somehow related? Some primates have very little thumb opposition, and among them are the callitrichids, which have reacquired claws. Further study is needed to clarify this relationship. In fact, nails could be acquired on hands and feet as a result of selection on feet only, or hands only, due to the developmental modules that link structures between the limbs (Hallgrimsson et al., 2002). Their role in removing skin parasites and social grooming should also be explored.

Concerning primates, when we first discussed the proportions of the primitive morphotype hand, we proposed that a relatively long hand within the forearm, as found in living claw climbers, vertical clingers and leapers, and *Notharctus*, was probably primitive for the order (Jouffroy et al., 1991). Concerning the intrinsic proportions of the hand, we hypothesized that hands with long digits close to those of galagines and *Tarsius* (i.e., with third digit amounting to 62–65% of total hand length), would make a good hypothetical morphotype. *Notharctus* is close, with similar proportions of its third digit, but with a different position on the diagram due to its very short metacarpals (Jouffroy et al., 1991). Subsequent reconstructions of proportions for other available Middle Eocene primates underscored a group of long-digited fossil species, having proportions similar to those of *Galago* and differing from *Tarsius* only in their longer carpus and shorter metacarpus. This group of long-digited fossils was proposed as giving a good approximation of the primitive primate morphotype for hand proportions (Godinot, 1992). In these studies, cheirogaleids are at some distance from this group, being closer to other lemurs and a group of platyrrhines. If these speculations about the ancestral primate hand were valid, the cheirogaleid hand would not be a good analog for the most primitive primate hands. Since our work, a much more complete *Notharctus* hand was described (Hamrick and Alexander, 1996), and its describers concurred with us that a long hand with short metacarpus and long digits probably was morphotypic for primates. Possibly more significantly, Hamrick (2001; this volume) shows that such digital and metacarpal proportions are manifested early in primate ontogeny and are thus likely to reflect a strong phylogenetic signal. This suggests that the proportions of Middle Eocene primate hands are very significant, and much closer to the primitive primate morphotype than those of most living primates.

What is the adaptive significance of hands with very long digits? The hands of *Tarsius* are especially good at catching insects, and apparently not very good for grasping small branches. Likewise, galagines are well-known for their ability to catch insects flying away from a branch. The working hypothesis is that primitive primate hands, long and with very long digits, were especially well adapted for catching insects. In a preliminary functional study of Eocene primate hands, the author speculated about the function of distal phalanges having retained from ancestral claw-bearing a high proximal part, but having acquired a broadened distal part for nail bearing (Godinot, 1991). The author, in this regard, could come up only with stabilization as the major functional difference between a nail-bearing and a claw-bearing digit. The mediolateral stabilization of the tip of the digits is transmitted proximally through the broad proximal expansion of primate distal phalanges so that the whole hand and foot must have had increased control of items being grasped. This in turn would be beneficial for both stabilizing insect prey in the hand, and more importantly stabilizing the feet on the support and controlling lower limb movements during insect catching. *Galago*, for example, is known to jump and catch an insect by sweeping the air while remaining attached to a branch by its feet, which allows it to retract back on the branch and eat the prey. For such an acrobatic behavior, a firm stabilization of the feet is certainly important. This stabilization is greater than would be provided by a weakly opposable hallux. It demands that the hallux and the opposed digits achieve a powerful grasp so that forces can be transmitted from the digits to the limbs appropriately. Thus, the hypothesis developed in 1991 states that primate morphotypic appendicular characters, nails, the powerful opposable hallux, and correlated postcranial characters are better explained by insect catching in the arboreal milieu than by invasion of the fine branch niche. This modest modification of the visual predation theory in its initial formulation has the advantage of making it simpler, the postcranial characters being an integral part of the behavioral and functional complex implied by the visual predation hypothesis. This view is in complete accord with the fact that the recently studied arboreal marsupials do not show the transition to nails: they do not have the insect pre-dation specialization, despite the fact that they snatch insects (Cartmill, 1974b; Nowak, 1999; Rasmussen, 1990). It appears that primate locomotor charac-teristics were acquired in at least two steps: (1) a "didelphid-like" step with opposable hallux and long phalanges allowing the grasping of small branches, with the correlative gait characteristics found in both groups (a "*Ptilocercus*-like" step with incipient hallucial grasping may illustrate either a preceding step or a different scenario) and (2) a second step reached only by primates, with the insect-catching specialization implying long hands, powerful hallucial grasp-ing, nails, and other correlated tarsal and long bones characters (for examples of more detailed scenarios, see Dagosto, this volume).

The visual predation theory has been challenged by Sussman (1991, 1995), who proposed that angiosperm feeding played a major role in the acquisition of primate characteristics. When Szalay (1968) interpreted the evolution of the earliest primate dentitions by a shift toward a frugivorous diet, he was mainly concerned with early plesiadapiforms. Since then, functional studies of dentitions have been made (Kay, 1975; Strait, 1993). Williams and Covert (1994) found *Teilhardina americana* just at the limit of predominantly insect eaters. The author suspects that the earliest primate dentitions (*Altiatlasius, Teilhardina belgica, Donrussellia*) would appear more insectivorous; however, a real quantification needs to be done. Even if the history of the group is one of mixed-feeding, small mammals need a source of protein, which is often constituted by small arthropods. Thus, during the post-Cretaceous radiation of small mammals, there must have been strong competition among all the small species requiring some insects. It is very possible that this competition was the most stringent, and therefore, provided the driving force for evolutionary change. Plesiadapiform insect-catching was through the incisors, in animals having claws and relying on olfaction (possibly also on audition, as they had big bullae). This is in sharp contrast with primates, which appear to have specialized on insect snatching via hand capture, relying on visual and auditory cues. This adaptive contrast could be the consequence of competition; however, it could also be linked with very different types of forests. The lesser development of high canopy in the Paleocene of North America and Europe may have favored claw-climbing species. On the other hand, we do not know which taxa were competing with primates during the Paleocene in Asia; as the diversity of plesiadapiforms was apparently not great in Asia, we may suspect other groups. To conclude, even if mixed-feeding was their actual dietary adaptation, primates may have nevertheless acquired their characteristics as the result of a specialization for the capture of their insect prey by audiovisually directed predation with hand capture.

CONCLUDING REMARKS

Fossils, Methods, and Primate Origins

In any attempt at deciphering primate origins within archontans, we should never forget the significance of: (1) the divergence of skull form in paromomyids, plesiadapids (?carpolestids) and (2) a divergent, cross specialized, and continuous record of dentitions, which further eliminates plesiadapiforms from any ancestral role in primate origins. The primatomorph hypothesis is dentally impossible, and cranially very unlikely. The carpolestid hypothesis (Bloch and Boyer, 2002) is as dentally impossible as the primatomorph hypothesis, strongly suggesting that carpolestid hallucial grasping with a nail must be convergent with primates (the tarsals and skull should confirm, or refute, this). Even if *Plesiadapis* turns out to have a real petrosal bulla, its skull and dentition would strongly suggest convergence. In fact, any sister-group relationship of one plesiadapoid family with primates which would be nested within other plesiadapoid families is dentally impossible (as it would imply that all the dental characters supporting the notion of plesiadapoids are convergent, destroying the systematic validity of this group). Despite the attractiveness of cladograms, numbers, and computers, a partial data set, or a very large one, will never render likely a hypothesis that one well-known anatomical system renders impossible. For this reason, the dental record of plesiadapiforms, which excludes any ancestral relationship to primates within them, should not be ignored (Figure 7). Whereas a very ancient common ancestor with primates is possible, a nesting of primates within the plesiadapiform radiation cannot be taken seriously.

If the hypothesis of a close tupaiid–primate relationship is valid, it should be corroborated in the future via the study of dental characters. Dental characters have proven to be of little use until now because there is no Paleocene record of a tree shrew. An Eocene species from China shows two primitive tupaiine dental characters (Tong, 1988). *Ptilocercus* is dentally more primitive than other tupaiids; however, it certainly has a number of derived ptilocercine characters that render the reconstruction of the ancestral dentition of Tupaiidae difficult (Butler, 1980), in turn hindering a complete evaluation of the tupaiid–primate hypothesis. At this moment, the primate dental morphotype is also very difficult to reconstruct: *Altiatlasius* is of critical importance, yet a confirmation of its primate status would be welcome (Godinot, 1994; Sigé et al., 1990; Silcox, 2001). The meaning of the dental characters of *Altanius* is still ambiguous. As seen in an earlier section, the eosimiids also introduce new characters in the primitive primate morphotype (e.g., trigonids not decreasing in anteroposterior breadth from M/1 to M/3, absence of well-developed third lobe on M/3). For dental traits too, the removal of plesiadapiforms from the primates considerably modifies some of the ideas we might hold about primitive primate dental characters and their polarities.

For example, relatively large conules on the upper molars could be presumed to be primitive based on their presence in several early plesiadapiforms. Discarding the last group, directly comparing Early Eocene Primates with *Ptilocercus* or *Purgatorius*, it appears that the earliest primates may have possessed only very small conules (which would better fit with the later enlargement of conules in some groups of primates, e.g., microchoerids and parapithecids). Similar changes in our interpretation of the polarity of other dental traits might occur. The primitive primate dental morphotype is at present in a state of great uncertainty. In this context, a better understanding of the polarity of dental traits in plesiadapiforms, already well advanced by Silcox (2001), is still critical. It should help us understand polarities in early "primatelike" dentitions.

It was argued in the introduction that, confronted with very large and contradictory data sets, the best strategy is to progressively delineate those data that are likely to be misleading, and to eliminate them, or consider only their possibly informative part. Our knowledge of postcranial characters in plesiadapiform families and in other early archontans is still not complete enough to extract secure phylogenetic signals. This means that the forelimb axial skeleton and the hindlimb data sets in Simmons (1993), largely generated from Beard (1993a,b), should be completely reassessed (in agreement with Szalay and Lucas, 1993, 1996; Sargis, 2002). The tarsal evidence appears more convincing; however, it needs to be tested through the detailed study of *Carpolestes* tarsals. Despite the carpal autapomorphies found in the living *Ptilocercus* (Stafford and Thorington, 1998), further work on the carpals should also provide phylogenetic signal.

Many of the changes in cranial and dental character interpretations mentioned above resulted from the study of new fossils (e.g., the beautiful skulls of paromomyids and the dental remains of eosimiids). This shows one more time the critical importance of historical information to reach better phylogenetical hypotheses (e.g., Gauthier et al., 1988; Donoghue et al., 1989). Soft anatomical characters are less useful for the deciphering of ancient branching events because we usually do not have enough information on their patterns of evolution (frequency of convergences, reversals, factors involved, etc.). We must try to integrate lessons from the fossil record when searching for those few characters that provide the best phylogenetic signal. We should also examine the adaptive significance of those characters and carefully scrutinize alternative scenarios of their morphofunctional transformation. Such a strategy differs from conducting a parsimony analysis of as many characters as possible. Other anatomical evidence and molecular studies will help confirm or contradict the tupaiid hypothesis; however, ultimately, a denser record of dental and tarsal characters will lead to a richer documentation of the scenario, or to its refutation.

Primate Morphotype Locomotor Mode

The realization that scandentians, as best represented by *Ptilocercus*, are probably the closest sister group of primates, has implications for scenarios of primate origins. The present author has always been reluctant to accept the leaping part of the grasp-leaping theory (Dagosto, 1988, 1990; Szalay and Dagosto, 1980, 1988; Szalay and Delson, 1979), for a variety of reasons: (1) the fact that leaping is a behavior that strongly shapes morphology and would subsequently leave strong anatomical signals (e.g., elongated tarsals in *Otolemur crassicaudatus*); (2) grasp-leaping would presumably require powerful grasping hands, which, as believed, were not present in the primate ancestral morphotype, or a rapid shift to vertical supports and vertical clinging and leaping (leaping on horizontal supports seems mechanically problematic with regard to landing with grasping hands); and (3) grasp-leaping would imply the coordinated reversal of a lot of characters in the early simiiforms, something considered very unlikely during the basal phase of an adaptive radiation. This locomotor reversal was considered likely because it was presumed to be linked to a shift from grasp-leaping toward a more quadrupedal above-branch locomotion, probably also linked with an increase in size (Dagosto, 1990; Gebo, 1986). Taking *Ptilocercus* as a primitive reference would diminish the support for this view because some of the primitive primate characters (e.g., elongated astragalus with a relatively high body) presumed to reflect grasp-leaping are present in *Ptilocercus*, while frequent leaping is not; however, it would be important to better document the locomotion of *Ptilocercus* (expert climber, hopping on the ground, Nowak, 1999; see Le Gros Clark, 1926; Stafford and Thorington, 1998; Sargis, 2001). More importantly, the description of the small tarsals from the Shanghuang fissures provides arguments in favor of the author's view (contra Gebo et al., 2001). One astragalar character in particular, the reduced medial facet, is considered by Gebo et al. (2001) to be primitive for mammals but reversed from a derived prosimian state in eosimiids. This is incredibly unlikely! Now that the character has been documented in

Middle Eocene primates of 50–125 g, it is much more probably primitive for primates, and also for simiiforms. A reappraisal of the tarsal characters of the Shanghuang primates is beyond the scope of this chapter. However, a few other characters need to be mentioned. The more salient sustentaculum of the "protoanthropoid—new taxon" projects medially to a greater degree than in the other taxa, rendering this calcaneum wider than the others. This character is said to be especially "platyrrhine-like" by Gebo et al. (2001). However, it is also *Ptilocercus*-like, and evidently primitive. The short distal part of the calcaneum of the "new protoanthropoids" is also *Ptilocercus*-like and probably primitive. The very circular shape of the calcaneocuboid joint in eosimiids is *Ptilocercus*-like, and thus could be primitive, whereas the removed wedge on the plantar side is derived and appears as one of the rare derived anthropoid-like features of these tarsals. The small Shanghuang tarsals are exciting because they document some new primitive tarsal characters that must be included in the primate morphotype (because primitive characters are not indicative of phylogenetic affinity, their link with simiiforms remains weak). Their comparison with *Ptilocercus* should enlighten our understanding of early primate locomotor evolution and phylogeny. It is the conviction of this author that leaping and climbing specializations are more accentuated in prosimians, which are derived for the related characters (and many parallel acquisitions of these characters are to be expected during the early radiation of primates), whereas the more quadrupedal eosimiids and "new protoanthropoids" are more similar to *Ptilocercus*, and more likely approach the primitive primate morphotype. Simiiforms are likely primitive for many tarsal characters; however, tarsals from early simiiforms (= "telanthropoids") will be critical to test this hypothesis (the posterior astragalar shelf present in *Ptilocercus* and eosimiids needs further study; its absence in simiiforms might be derived). Let us simply add that the simiiform ovoid entocuneiform facet for the first metatarsal is more likely to be derived from a more primitive (? ptilocercine-like) morphology than reverted from the prosimian sellar joint (contra Szalay and Dagosto, 1988). Entocuneiforms might be too small to be found in eosimiids; however, first metatarsals when found should test these ideas.

In sum, the primitive reference offered by *Ptilocercus* and the eosimiid tarsals lead to further question the leaping component of the grasp-leaping theory and to favor a hypothesis of rapid grasp-quadrupedalism for the primitive primate morphotype locomotor mode. This would be associated with insect manual catching. This hypothesis avoids multiple postcranial character reversals in the origin of simiiforms and considers leaping features as apomorphies developing in prosimian groups. This sketchy view should be expanded in much more detailed scenarios (see Dagosto, this volume).

SUMMARY

Paromomyid dental characters appear incompatible with the hypothesis of a sister group relationship between paromomyids and dermopterans (the primatomorph hypothesis). All hypotheses placing dermopterans or primates nested within plesiadapoid families are dentally quasi impossible.

Paromomyids were not gliders but more probably claw climbers having locomotor and insect capture specializations close to those of *Heterohyus* and *Dactylopsila*. Skull characters do not show any paromomyid-dermopteran synapomorphy (Bloch and Silcox, 2001). The Primatomorpha hypothesis, which now lacks postcranial support and cannot be reconciled with dental or cranial evidence, must be abandoned.

The plesiadapiforms are a radiation of forms combining a series of primitive skull characters, small orbits and large olfactory bulbs, and very derived characters of the anterior incisors and muzzle. Despite their dental convergences with primates, early acquired dental specializations exclude them of having primates rooted within them. They are a radiation of clawed arboreal mammals, within which *Carpolestes* hallucial grasping represents a remarkable convergence with primates. Many of them are North American and they are probably monophyletic (Kay et al., 1992; e.g., all those descended from *Purgatorius*). An interesting question remains concerning the possibility that plesiadapoids might have an independent Asiatic origin, as raised by some phylogenetic interpretations of *Chronolestes* dentition; however, this would not alter the broad picture inferred from their cranial and postcranial characters. Hence, Plesiadapiformes are best considered as an order of their own (admittedly having an imprecise content). The evidence favoring a sister group relationship of plesiadapiforms with primates is ambiguous.

The exclusion of plesiadapiforms from primates renders more plausible a series of potential synapomorphies between primates and Tupaiidae. These include characters from the basicranium, the orbit, other parts of the cranium (Kay et al., 1992; Wible and Covert, 1987; Wible and Martin, 1993), and a series of important characters of the astragalo–calcaneum complex. Tupaiidae appear again to be the best available sister group of primates.

Large data sets of morphological characters have proven unsuitable to clearly decipher the sister group of primates. A strategy of pruning the data and adding as much historical information as possible seems more appropriate than simply adding characters to the list. The likelihood of particular reversals should be considered when evaluating alternative hypotheses.

A complex history of arboreal adaptations in the various families of plesi-adapiforms and in early archontans is strongly implicated. More comparative work, taking in account *Ptilocercus*, is needed before phylogenetic signals can be extracted from the postcranial anatomy. However, tarsal characters are sufficiently understood to support the hypothesis of *Ptilocercus*-primate synapo-morphy for a series of tarsal characters. The author suggests that several tarsal characters of the eosimiids and other Shanghuang primates are primitive for the order. These hypotheses in turn challenge the leaping aspect of the theory of grasp-leaping as the primate morphotype locomotor mode, favor-ing a mode closer to grasp-quadrupedalism, and not requiring multiple rever-sals for the origin of simiiform locomotor characteristics.

Small arboreal marsupials such as *Marmosa* and *Caluromys* show similarities with primates linked to locomotion on fine branches; however, they do not have nails on all digits. Insect-catching in the arboreal milieu probably better explains the acquisition of primate postcranial characterisics (powerful hallucial grasping, nails, and the peculiar proportions of early primate hands), and visual predation explains the cranial characters as proposed by Cartmill (1972, 1974a,b).

Soft anatomical and molecular characters should be used to test the tupaiid–primate hypothesis. A denser record of tarsal and dental characters also should give decisive confirmation, or refutation, of this hypothesis. The pri-mate dental morphotype must be reassessed after discarding the plesiadapi-forms and including the eosimiids. It is presently far from being established.

Since this essay was completed, several papers related to the subject were published. Among them, only one would chage my view of the plesiadapiform radiation, that is if *Dralestes* and the Azibiidae were Eocene African plesiadapi-forms (Tabuce et al., 2004). However, the lingual part of the upper molars of *Dralestes* is relatively narrow, suggesting that their posterolingual crest is not a postprotocingulum homologous with that of known plesiadapiforms but more likely represents convergent structure. Tabuce et al. (idem, p 318) recognize that "the phylogenetic position of azibiids needs to be confirmed with more relevant data". The author agrees, doubts their plesiadapiform affinities, and suggests elsewhere that they might be unexpected euprimates.

ACKNOWLEDGMENTS

The author thanks Marian Dagosto and Matt Ravosa for having invited me to participate in the Primate Origins Conference. Mary Silcox very kindly sent a copy of her dissertation and provided helpful criticism on an early version of this paper. Chris Beard helped me to understand some of the characters that he published. François Catzeflis, Eric Sargis, and Carlos Jaramillo shared some interesting information. Aaron Hogue and Jerry Hooker provided casts of *Ptilocercus* dentitions. Access to specimens in Anatomie Comparée, Muséum National d'Histoire Naturelle, Paris, was granted by M. Robineau, and access to specimens in the Field Museum of Natural History was granted by Bruce Patterson and William Stanley. Study of important fossils was kindly allowed by Dan Gebo and Chris Beard. The figures were cleaned and mounted by H. Lavina. Laurie Godfrey edited the English of an almost definitive draft, and Marian Dagosto completed the task; both gave helpful suggestions on the content.

REFERENCES

Beard, K. C., 1990, Gliding behaviour and palaeoecology of the alleged primate family *Paromomyidae* (Mammalia, Dermoptera), *Nature* **345**: 340–341.

Beard, K. C., 1991, Vertical postures and climbing in the morphotype of Primatomorpha: Implications for locomotor evolution in primate history, in: *Origine(s) de la Bipédie chez les Hominidés*, Y. Coppens and B. Senut, eds., Cahiers de Paléoanthropologie, Editions du C. N. R. S., Paris pp. 79–87.

Beard, K. C., 1993a, Phylogenetic systematics of the Primatomorpha, with special reference to Dermoptera, in: *Mammal Phylogeny: Placentals*, F. S. Szalay, M. J. Novacek, and M. C. McKenna, eds., Springer-Verlag, New York, pp. 129–150.

Beard, K. C., 1993b, Origin and evolution of gliding in early Cenozoic Dermoptera (Mammalia, Primatomorpha), in: *Primates and Their Relatives in Phylogenetic Perspective*, R. D. E. MacPhee, ed., Plenum Press, New York, pp. 63–90.

Beard, K. C., 2000, A new species of *Carpocristes* (Mammalia: Primatomorpha) from the middle Tiffanian of the Bison Basin, Wyoming, with notes on carpolestid phylogeny, *Ann. Carnegie Mus.* **69**: 195–208.

Beard, K. C., and MacPhee, R. D. E., 1994, Cranial anatomy of *Shoshonius* and the antiquity of Anthropoidea, in: *Anthropoid Origins*, J. G. Fleagle and R. F. Kay, eds., Plenum Press, New York, pp. 55–97.

Beard, K. C., and Wang, J., 1995, The first Asian plesiadapoids (Mammalia, Primatomorpha), *Ann. Carnegie Mus.* **64**: 1–33.

Beard, K. C., Qi, T., Dawson, M. R., Wang, B., and Li, C., 1994, A diverse new primate fauna from middle Eocene fissure-fillings in southeastern China, *Nature* **368**: 604–609.

Beard, K. C., Tong, Y., Dawson, M. R., Wang, J., and Huang, X., 1996, Earliest complete dentition of an anthropoid primate from the late Middle Eocene of Shanxi Province, China, *Science* **272**: 82–85.

Biknevicius, A., 1986, Dental function and diet in the Carpolestidae, *Am. J. Phys. Anthrop.* **71**: 157–171.

Bloch, J. I., and Boyer, D. M., 2001, Taphonomy of small mammals in freshwater limestones from the Paleocene of the Clarks Fork Basin. *Univ. Michigan Pap. Paleont.* **33**: 185–198.

Bloch, J. I., and Boyer, D. M., 2002, Grasping primate origins, *Science* **298**: 1606–1610.

Bloch, J. I., and Gingerich, P. D., 1998, *Carpolestes simpsoni*, new species (Mammalia, Proprimates) from the late Paleocene of the Clarks Fork Basin, Wyoming, *Contrib. Mus. Paleontol. Univ. Michigan* **30**: 131–162.

Bloch, J. I., and Silcox, M. T., 2001, New basicrania of Paleocene-Eocene *Ignacius*: Re-evaluation of the plesiadapiform-dermopteran link, *Am. J. Phys. Anthro.* **116**: 184–198.

Bloch, J. I., Fisher, D. C., Rose, K. D., and Gingerich, P. G., 2001, Stratocladistic analysis of Paleocene Carpolestidae (Mammalia, Plesiadapiformes) with description of a new late Tiffanian genus, *J. Vert. Pal.* **21**: 19–131.

Bloch, J. I., Boyer, D. M., Gingerich, P. D., and Gunnell, G. F., 2002, New primitive paromomyid from the Clarkforkian of Wyoming and dental eruption in Plesiadapiformes, *J. Vert. Pal.* **22**: 366–379.

Buckley, G. A., 1997, A new species of *Purgatorius* (Mammalia, Primatomorpha) from the lower Paleocene Bear Formation, Crazy Mountains Basin, south-central Montana, *J. Paleont.* **71**: 149–155.

Butler, P. M., 1980, The tupaiid dentition, in: *Comparative Biology and Evolutionary Relationships of Tree Shrews,* W. P. Luckett, ed., Plenum Press, New York, pp. 171–204.

Carlsson, A., 1922, Über die Tupaiidae und ihre Beziehungen zu den Insectivora und den Prosimiae, *Acta Zool.* **3**: 227–270.

Cartmill, M., 1972, Arboreal adaptations and the origin of the Order Primates, in: *The Functional and Evolutionary Biology of Primates,* R. Tuttle, ed., Aldine-Atherton, Chicago, pp. 97–122.

Cartmill, M., 1974a, Rethinking primate origins, *Science* **184**: 436–443.

Cartmill, M., 1974b, Pads and claws in arboreal locomotion, in: *Primate Locomotion,* F. A. Jenkins, ed., Academic Press, New York, pp. 45–83.

Cartmill, M., and MacPhee, R. D. E., 1980, Tupaiid affinities: The evidence from the carotid arteries and cranial skeleton, in: *Comparative Biology and Evolutionary Relationships of Tree Shrews,* W. P. Luckett, ed., Plenum Press, New York, pp. 95–132.

Dagosto, M., 1988, Implications of postcranial evidence for the origin of euprimates, *J. Hum. Evol.* **17**: 35–56.

Dagosto, M., 1990, Models for the origin of the anthropoid postcranium, *J. Hum. Evol.* **19**: 121–139.

Donoghue, M. J., Doyle, J. A., Gauthier, J., Kluge, A. G., and Rowe, T., 1989, The importance of fossils in phylogeny reconstruction, *Annu. Rev. Ecol. Syst.* **20**: 431–460.

Fleagle, J. G., 1988, *Primate Adaptation and Evolution*, Academic Press, San Diego.

Fleagle, J. G., 1999, *Primate Adaptation and Evolution*, 2nd Ed., Academic Press, San Diego.

Flynn, J. J., Guerrero, J., and Swisher, C. C., 1997, Geochronology of the Honda Group, in: *Vertebrate Paleontology in the Neotropics—The Miocene Fauna of La Venta, Columbia*, R. F. Kay, R. H. Madden, R. L. Cifelli, and J. J. Flynn, eds., Smithsonian Institution Press, Washington, pp. 44–59.

Ford, S. F., 1988, Postcranial adaptations of the earliest platyrrhine, *J. Hum. Evol.* **17**: 155–192.

Fox, R. C., 1984, A new species of the Paleocene primate *Elphidotarsius* Gidley: Its stratigraphic position and evolutionary relationships, *Can. J. Earth Sci.* **21**: 1268–1277.

Fox, R. C., 1993, The primitive dental formula of the Carpolestidae (Plesiadapiformes, Mammalia) and its phylogenetic implications, *J. Vert. Pal.* **13**: 516–524.

Gauthier, J. A., Kluge, A. G., and Rowe, T., 1988, Amniote phylogeny and the importance of fossils, *Cladistics* **4**: 105–209.

Gebo, D. L., 1986, Anthropoid origins—The foot evidence, *J. Hum. Evol.* **15**: 421–430.

Gebo, D. L., 1988, Foot morphology and locomotor adaptation in Eocene primates, *Folia Primatol.* **50**: 3–41.

Gebo, D. L., Dagosto, M., Beard, K. C., and Qi, T., 2001, Middle Eocene primate tarsals from China: Implications for haplorhine evolution, *Am. J. Phys. Anthrop.* **116**: 83–107.

Gingerich, P. D., 1976, Cranial anatomy and evolution of early Tertiary Plesiadapidae (Mammalia, Primates), *Univ. Michigan Pap. Paleont.* **15**: 1–141.

Gingerich, P. D., Dashzeveg, D., and Russell, D. E., 1991, Dentition and systematic relationships of *Altanius orlovi* (Mammalia, Primates) from the early Eocene of Mongolia, *Geobios* **24**: 637–646.

Godinot, M., 1984, Un nouveau genre de Paromomyidae (Primates) de l'Eocène Inférieur d'Europe, *Folia Primatol.* **43**: 84–96.

Godinot, M., 1991, Approches fonctionnelles des mains de primates paléogènes, *Geobios, Mém. Spéc.* **13**: 161–173.

Godinot, M., 1992, Early euprimate hands in evolutionary perspective, *J. Hum. Evol.* **22**: 267–283.

Godinot, M., 1994, Early North African primates and their significance for the origin of Simiiformes (= Anthropoidea), in: *Anthropoid Origins*, J. G. Fleagle and R. F. Kay, eds., Plenum Press, New York, pp. 235–295.

Godinot, M., and Prasad, G. V. R., 1994, Discovery of Cretaceous arboreal eutherians, *Naturwiss.* **81:** 79–81.

Goin, F. J., 1997, New clues for understanding Neogene marsupial radiations, in: *Vertebrate Paleontology in the Neotropics—The Miocene Fauna of La Venta, Columbia,* R. F. Kay, R. H. Madden, R. L. Cifelli, and J. J. Flynn, eds., Smithsonian Institution Press, Washington, pp. 187–206.

Gunnell, G. F., 1989, Evolutionary history of Microsyopoidea (Mammalia, Primates) and the relationship between Plesiadapiformes and Primates, *Univ. Michigan Pap. Paleont.* **27:** 1–157.

Gunnell, G. F., 1995, Omomyid primates (Tarsiiformes) from the Bridger Formation, middle Eocene, southern Green River Basin, Wyoming, *J. Hum. Evol.* **28:** 147–187.

Hallgrimsson, B., Willmore, K., and Hall, B. K., 2002, Canalization, developmental stability, and morphological integration in primate limbs, *Yearbk. Phys. Anthrop.* **45:** 131–158.

Hamrick, M. W., 1998, Functional and adaptive significance of primate pads and claws: Evidence from New World anthropoids, *Am. J. Phys. Anthrop.* **106:** 113–127.

Hamrick, M. W., 2001, Primate origins: Evolutionary change in digital ray patterning and segmentation, *J. Hum. Evol.* **40:** 339–351.

Hamrick, M. W., and Alexander, J. P., 1996, The hand skeleton of *Notharctus tenebrosus* (Primates, Notharctidae) and its significance for the origin of the primate hand, *Am. Mus. Novitates* **3182:** 1–20.

Hamrick, M. W., Rosenman, B. A., and Brush, J. A., 1999, Phalangeal morphology of the Paromomyidae (?Primates, Plesiadapiformes): The evidence for gliding behavior reconsidered, *Am. J. Phys. Anthrop.* **109:** 397–413.

Hoffstetter, R., 1977, Phylogénie des Primates—Confrontation des résultats obtenus par les diverses voies d'approche du problème, *Bull. Mém. Soc. Anthrop. Paris* **13**(4): 327–346.

Hoffstetter, R., 1986, Limite entre Primates et non-Primates; position des Plesiadapiformes et des Microsyopidae, *C. R. Acad. Sci. Paris* **302**(2): 43–45.

Hooker, J. J., 2001, Tarsals of the extinct insectivoran family Nyctitheriidae (Mammalia): Evidence for archontan relationships, *Zool. J. Linn. Soc.* **132:** 501–529.

Hooker, J. J., Russell, D. E., and Phélizon, A., 1999, A new family of Plesiadapiformes (Mammalia) from the Old World Lower Paleogene, *Palaeontology* **42:** 377–407.

Hunt, R. M., and Korth, W. K., 1980, The auditory region of Dermoptera: Morphology and function relative to other living mammals, *J. Morph.* **164:** 167–211.

Jouffroy, F. K., Godinot, M., and Nakano, Y., 1991, Biometrical characteristics of primate hands, *Hum. Evol.* **6:** 269–306.

Kay, R. F., 1975, The functional adaptations of primate molar teeth, *Am. J. Phys. Anthrop.* **43:** 195–215.

Kay, R. F., and Cartmill, M., 1977, Cranial morphology and adaptations of *Palaechthon nacimienti* and other Paromomyidae (Plesiadapoidea, ?Primates), with a description of a new genus and species, *J. Hum. Evol.* **6:** 19–53.

Kay, R. F., and Williams, B. A., 1994, Dental evidence for anthropoid origins, in: *Anthropoid Origins,* J. G. Fleagle and R. F. Kay, eds., Plenum Press, New York, pp. 361–445.

Kay, R. F., Thorington, R. W., and Houde, P., 1990, Eocene plesiadapiform shows affinities with flying lemurs not primates, *Nature* **345:** 342–344.

Kay, R. F., Thewissen, J. M. G., and Yoder, A. D., 1992, Cranial anatomy of *Ignacius graybullianus* and the affinities of the Plesiadapiformes, *Am. J. Phys. Anthrop.* **89:** 477–498.

Koenigswald, W. von., 1990, Die Paläobiologie des Apatemyiden (Insectivora s.l.) und die Ausdeutung der Skelettfunde von *Heterohyus nanus* aus dem Mitteleozän von Messel bei Darmstadt, *Palaeontographica A* **210:** 41–77 (4 pl.).

Koenigswald, W. von., and Schiernig, H.-P., 1987, The ecological niche of an extinct group of mammals, the early Tertiary apatemyids, *Nature* **326:** 595–597.

Krause, D. W., 1991, Were paromomyids gliders? Maybe, maybe not, *J. Hum. Evol.* **21:** 177–188.

Larson, S. G., Schmitt, D., Lemelin, P., and Hamrick, M., 2000, Uniqueness of primate forelimb posture during quadrupedal locomotion, *Am. J. Phys. Anthrop.* **112:** 87–101.

Le Gros Clark, W. E., 1926, On the anatomy of the pen-tailed tree-shrew (*Ptilocercus lowii*), *Proc. Zool. Soc. London* **1926:** 1179–1309 (5 pl.).

Le Gros Clark, W. E., 1959, *The Antecedents of Man,* Edinburgh University Press, Edinburgh.

Le Gros Clark, W. E., 1971, *The Antecedents of Man,* 3rd Ed., Edinburgh University Press, Edinburgh.

Lemelin, P., 1999, Morphological correlates of substrate use in didelphid marsupials: Implications for primate origins, *J. Zool. London* **247:** 165–175.

Lemelin, P., 2000, Micro-anatomy of the volar skin and interordinal relationships of primates, *J. Hum. Evol.* **38:** 257–267.

Liu, F.-G. R., Miyamoto, M. M., Freire, N. P., Ong, P. Q., Tennant, M. R., Young, T. S., and Gugel, K. F., 2001, Molecular and morphological supertrees for eutherian (placental) mammals, *Science* **291:** 1786–1789.

Luckett, W. P., ed., 1980a, *Comparative Biology and Evolutionary Relationships of Tree Shrews,* Plenum Press, New York.

Luckett, W. P., 1980b, The suggested evolutionary relationships and classification of tree Shrews, in: *Comparative Biology and Evolutionary Relationships of Tree Shrews,* W. P. Luckett, ed., Plenum Press, New York, pp. 3–31.

MacPhee, R. D. E., 1981, Auditory regions of primates and eutherian insectivores—Morphology, ontogeny, and character analysis, *Contrib. Primatol.* **18:** 1–282.

MacPhee, R. D. E., 1991, The supraordinal relationships of Primates: A prospect, *Am. J. Phys. Anthrop. Supp.* **12:** 121–122.

MacPhee, R. D. E., ed., 1993a, *Primates and Their Relatives in Phylogenetic Perspective,* Plenum Press, New York.

MacPhee, R. D. E., 1993b, Summary, in: *Primates and Their Relatives in Phylogenetic Perspective,* R. D. E. MacPhee, ed., Plenum Press, New York, pp. 363–373.

MacPhee, R. D. E., and Cartmill, M., 1986, Basicranial structures and primate systematics, in: *Comparative Primate Biology,* D. R. Swindler and J. Erwin, eds., vol. 1, pp. 219–275, Liss, New York.

MacPhee, R. D. E., Cartmill, M., and Gingerich, P. D., 1983, New Paleogene primate basicrania and the definition of the order Primates, *Nature* **301:** 509–511.

MacPhee, R. D. E., Cartmill, M., and Rose, K. D., 1989, Craniodental morphology and relationships of the supposed Eocene dermopteran *Plagiomene* (Mammalia), *J. Vert. Pal.* **9:** 329–349.

Madsen, O., Scally, M., Douady, C. J., Kao, D. J., DeBry, R. W., Adkins, R., Amrine, H. M. et al., 2001, Parallel adaptive radiations in two major clades of placental mammals, *Nature* **409:** 610–614.

Martin, R. D., 1968, Towards a new definition of Primates, *Man* **3:** 377–401.

Martin, R. D., 1986, Primates: A definition, in: *Major Topics in Primate and Human Evolution,* B. A. Wood, L. B. Martin, and P. J. Andrews, eds., Cambridge University Press, Cambridge, pp. 1–31.

McKenna, M. C., and Bell, S. K., 1997, *Classification of Mammals Above the Species Level,* Columbia University Press, New York.

Murphy, W. J., Eizirik, E., Johnson, W. E., Zhang, Y. P., Ryder, O. A., and O'Brien, S. J., 2001, Molecular phylogenetics and the origins of placental mammals, *Nature* **409:** 614–618.

Neff, N. A., 1986, A rational basis for a priori character weighting, *Syst. Zool.* **35:** 110–123.

Novacek, M. J., 1980, Cranioskeletal features in tupaiids and selected eutheria as phylogenetic evidence, in: *Comparative Biology and Evolutionary Relationships of Tree Shrews,* W. P. Luckett, ed., Plenum Press, New York, pp. 35–93.

Nowack, R. M., 1999, *Walker's Mammals of the World,* 6th Ed., vol. 1, The Johns Hopkins University Press, Baltimore.

Prasad, G. V. R., and Godinot, M., 1994, Eutherian tarsal bones from the late Cretaceous of India, *J. Paleont.* **68:** 892–902.

Rasmussen, D. T., 1990, Primate origins: Lessons from a Neotropical marsupial, *Am. J. Primatol.* **22:** 263–277.

Ravosa, M. J., Noble, V. E., Hylander, W. L., Johnson, K. R., and Kowalski, E. M., 2000, Masticatory stress, orbital orientation and the evolution of the primate postorbital bar, *J. Hum. Evol.* **38**: 667–693.

Rich, T. H., 1991, Monotremes, placentals, and marsupials: Their record in Australia and its biases, in: *Vertebrate Palaeontology of Australasia*, P. Vickers-Rich, J. M. Monaghan, R. F. Baird, and T. H. Rich, eds., Pioneer Design Studio and Monash Univ. Pub. Com., Melbourne, pp. 893–1004.

Romer, A. S., 1966, *Vertebrate Paleontology*, 3rd Ed., The University of Chicago Press, Chicago.

Rose, K. D., 1975, The Carpolestidae—Early Tertiary primates from North America, *Bull. Mus. Comp. Zool.* **147**: 1–74.

Rose, K. D., 1995, The earliest primates, *Evol. Anthrop.* **3**: 159–173.

Rose, K. D., and Bown, T. M., 1996, A new plesiadapiform (Mammalia: Plesiadapiformes) from the early Eocene of the Bighorn Basin, Wyoming, *Ann. Carnegie Mus.* **65**: 305–321.

Rose, K. D., and Krause, D. W., 1984, Affinities of the primate *Altanius* from the early Tertiary of Mongolia, *J. Mamm.* **65**: 721–726.

Rose, K. D., and Simons, E. L., 1977, Dental function in the Plagiomenidae: Origin and relationships of the mammalian order Dermoptera, *Contr. Mus. Pal. Univ. Michigan* **24**: 221–236.

Rose, K. D., Beard, K. C., and Houde, P., 1993, Exceptional new dentitions of the diminutive plesiadapiforms *Tinimomys* and *Niptomomys* (Mammalia), with comments on the upper incisors of Plesiadapiformes, *Ann. Carnegie Mus.* **62**: 351–361.

Rose, K. D., Godinot, M., and Bown, T. M., 1994, The early radiation of Euprimates and the initial diversification of Omomyidae, in: *Anthropoid Origins*, J. G. Fleagle and R. F. Kay, eds., Plenum Press, New York, pp. 1–28.

Ross, C. F., and Covert, H. H., 2000, The petrosal of *Omomys carteri* and the evolution of the primate basicranium, *J. Hum. Evol.* **39**: 225–251.

Runestad, J. A., and Ruff, C. B., 1995, Structural adaptations for gliding in mammals with implications for locomotor behavior in paromomyids, *Am. J. Phys. Anthrop.* **98**: 101–119.

Russell, D. E., 1959, Le crâne de *Plesiadapis*, note préliminaire, *Bull. Soc. Géol. France* 7(1): 312–315.

Russell, D. E., 1964, Les Mammifères paléocènes d'Europe, *Mém. Mus. Nat. Hist. Nat. C* **13**: 1–324 (16 pl.).

Saban, R., 1963, Contribution à l'étude de l'os temporal des primates, *Mém. Mus. Nat. Hist. Nat. A* **29**: 1–378 (30 pl.).

Sanchez-Villagra, M. R., and Williams, B. A., 1998, Levels of homoplasy in the evolution of the mammalian skeleton, *J. Mamm. Evol.* **5**: 113–126.

Sargis, E. J., 2000, The postcranium of *Ptilocercus lowii* and other archontans: An analysis of primatomorphan and volitantian characters, *J. Vert. Pal. Supp.* **20**: 67A.

Sargis, E. J., 2001, The grasping behaviour, locomotion and substrate use of the tree shrews *Tupaia minor* and *T. tana* (Mammalia, Scandentia), *J. Zool. London* **253**: 485–490.

Sargis, E. J., 2002a, Functional morphology of the forelimb of tupaiids (Mammalia, Scandentia) and its phylogenetic implications, *J. Morph.* **253**: 10–42.

Sargis, E. J., 2002b, Functional morphology of the hindlimb of tupaiids (Mammalia, Scandentia) and its phylogenetic implications, *J. Morph.* **254**: 149–185.

Sargis, E. J., 2002c, The postcranial morphology of *Ptilocercus lowii* (Scandentia, Tupaiidae): An analysis of primatomorphan and volitantian characters, *J. Mamm. Evol.* **9**: 137–160.

Schmitt, D., and Lemelin, P., 2002, Origins of primate locomotion: Gait mechanics of the woolly opossum, *Am. J. Phys. Anthrop.* **118**: 231–238.

Sigé, B., and Marandat, B., 1997, Apport à la faune du Paléocène inférieur d'Europe: un Plésiadapiforme du Montien de Hainin (Belgique), in: *Actes du Congrès BiochroM'97*, J.-P. Aguilar, S. Legendre, and J. Michaux, eds., Mémoires et Travaux de l'E. P. H. E., Institut de Montpellier, N° 21, pp. 679–686.

Sigé, B., Jaeger, J.-J., Sudre, J., and Vianey-Liaud, M., 1990, *Altiatlasius koulchii* n. gen. et sp., Primate omomyidé du Paléocène Supérieur du Maroc, et les origines des Euprimates, *Palaeontographica A* **214**: 31–56.

Silcox, M. T., 2001, *A Phylogenetic Analysis of Plesiadapiformes and Their Relationship to Euprimates and Other Archontans*, Ph.D. Dissertation, Johns Hopkins University, p. 728.

Silcox, M. T., Krause, D. W., Maas, M. C., and Fox, R. C., 2001, New specimens of *Elphidotarsius russelli* (Mammalia, ?Primates, Carpolestidae) and a revision of plesiadapoid relationships, *J. Vert. Pal.* **21**: 132–152.

Simmons, N. B., 1993, The importance of methods: Archontan phylogeny and cladistic analysis of morphological data, in: *Primates and Their Relatives in Phylogenetic Perspective*, R. D. E. MacPhee, ed., Plenum Press, New York, pp. 1–61.

Simons, E. L., 1972, *Primate Evolution: An Introduction to Man's Place in Nature*, MacMillan, New York.

Simpson, G. G., 1945, The principles of classification and a classification of mammals, *Bull. Am. Mus. Nat. Hist.* **85**: 1–350.

Simpson, G. G., 1955, The Phenacolemuridae, new family of early primates, *Bull. Am. Mus. Nat. Hist.* **105**: 411–442.

Stafford, B. J., and Thorington, R. W., 1998, Carpal development and morphology in archontan mammals, *J. Morph.* **235**: 135–155.

Starck, D., 1975, The development of the chondrocranium in Primates, in: *Phylogeny of the Primates—A Multidisciplinary Approach*, W. P. Luckett and F. S. Szalay, eds., Plenum Press, New York, pp. 127–155.

Strait, S. G., 1993, Differences in occlusal morphology and molar size in frugivores and faunivores, *J. Hum. Evol.* **25**: 471–484.

Sussman, R. W., 1991, Primate origins and the evolution of angiosperms, *Am. J. Primatol.* **23**: 209–223.

Sussman, R. W., 1995, How primates invented the rainforest and vice versa, in: *Creatures of the Dark—The Nocturnal Prosimians*, L. Alterman, G. A. Doyle, and M. Kay Izard, eds., Plenum Press, New York, pp. 1–10.

Szalay, F. S., 1968, The beginnings of Primates, *Evolution* **22**: 19–36.

Szalay, F. S., 1969, Mixodectidae, Microsyopidae, and the insectivore-primate transition, *Bull. Am. Mus. Nat. Hist.* **140**: 193–330 (31 pl.).

Szalay, F. S., 1972, Cranial morphology of the early Tertiary *Phenacolemur* and its bearing on primate phylogeny, *Am. J. Phys. Anthrop.* **36**: 59–75.

Szalay, F. S., 1975, Phylogeny of primate higher taxa: The basicranial evidence, in: *Phylogeny of the Primates—A Multidisciplinary Approach*, W. P. Luckett and F. S. Szalay, eds., Plenum Press, New York, pp. 91–125.

Szalay, F. S., 1982, Phylogenetic relationships of the marsupials, *Geobios Mém. Spéc.* **6**: 177–190.

Szalay, F. S., 2000, Function and adaptation in paleontology and phylogenetics: Why do we omit Darwin? *Palaeontologia Electronica* **3**(2): 1–25.

Szalay, F. S., and Dagosto, M., 1980, Locomotor adaptations as reflected on the humerus of Paleogene primates, *Folia Primatol.* **34**: 1–45.

Szalay, F. S., and Dagosto, M., 1988, Evolution of hallucial grasping in the primates, *J. Hum. Evol.* **17**: 1–33.

Szalay, F. S., and Decker, R. L., 1974, Origins, evolution, and function of the tarsus in late Cretaceous Eutheria and Paleocene primates, in: *Primate Locomotion*, F. A. Jenkins, ed., Academic Press, New York, pp. 223–259.

Szalay, F. S., and Delson, E., 1979, *Evolutionary History of the Primates*, Academic Press, New York.

Szalay, F. S., and Drawhorn, G. D., 1980, Evolution and diversification of the Archonta in an arboreal milieu, in: *Comparative Biology and Evolutionary Relationships of Tree Shrews*, W. P. Luckett, ed., Plenum Press, New York, pp. 133–169.

Szalay, F. S., and Lucas, S. G., 1993, Cranioskeletal morphology of archontans, and diagnoses of Chiroptera, Volitantia, and Archonta, in: *Primates and Their Relatives in Phylogenetic Perspective*, R. D. E. MacPhee, ed., Plenum Press, New York, pp. 187–226.

Szalay, F. S., and Lucas, S. G., 1996, The postcranial morphology of Paleocene *Chriacus* and *Mixodectes* and the phylogenetic relationships of archontan mammals, *Bull. New Mexico Mus. Nat. Hist. Sci.* **7**: 1–47.

Szalay, F. S., Tattersall, I., and Decker, R. L., 1975, Phylogenetic relationships of *Plesiadapis*—Postcranial evidence, *Contr. Primatol.* **5**: 136–166.

Szalay, F. S., Rosenberger, A. L., and Dagosto, M., 1987, Diagnosis and differentiation of the Order Primates, *Yearbk. Phys. Anthrop.* **30**: 75–105.

Tabuce, R., Mahboubi, M., Tafforeau, P., and Sudre, J., 2004, Discovery of a highly-specialized plesiadapiform primate in the early-middle Eocene of northwestern Africa. *J. Hum. Evol.* **47**: 305–321.

Thewissen, J. G. M., Williams, E. M., and Hussain, S. T., 2001, Eocene mammal faunas from northern Indo-Pakistan, *J. Vert. Pal.* **21**: 347–366.

Tong, Y., 1988, Fossil tree shrews from the Eocene Hetaoyuan Formation of Xichuan, Henan. *Vertebrata PalAsiatica* **26**: 214–220.

Tong, Y., 1997, Middle Eocene small mammals from Liguanqiao Basin of Henan Province and Yuanqu Basin of Shanxi Province, central China, *Palaeontol. Sinica* **186**(C26): 1–268.

Tong, Y., and Wang, J., 1998, A preliminary report on the early Eocene mammals of the Wutu fauna, Shandong Province, China, *Bull. Carnegie Mus. Nat. Hist.* **34**: 186–193.

Van Valen, L. M., 1994, The origin of the plesiadapid primates and the nature of *Purgatorius*, *Evol. Monographs* **15**: 1–79.

Van Valen, L., and Sloan, R. E., 1965, The earliest primates, *Science* **150**: 743–745.

Vaughan, T. A., 1972, *Mammalogy*, Saunders Company, Philadelphia.

Wible, J. R., 1987, The eutherian stapedial artery: Character analysis and implications for superordinal relationships, *Zool. J. Linn. Soc.* **91**: 107–135.

Wible, J. R., 1993, Cranial circulation and relationships of the colugo *Cynocephalus* (Dermoptera, Mammalia), *Am. Mus. Novitates* **3072**: 1–27.

Wible, J. R., and Covert, H. H., 1987, Primates: Cladistic diagnosis and relationships, *J. Hum. Evol.* **16**: 1–22.

Wible, J. R., and Martin, J. R., 1993, Ontogeny of the tympanic floor and roof in archontans, in: *Primates and Their Relatives in Phylogenetic Perspective*, R. D. E. MacPhee, ed., Plenum Press, New York, pp. 111–148.

Williams, B. A., and Covert, H. H., 1994, New early Eocene anaptomorphine primate (Omomyidae) from the Washakie Basin, Wyoming, with comments on the phylogeny and paleobiology of anaptomorphines, *Am. J. Phys. Anthro.* **93**: 323–340.

Zeller, U., 1986, Ontogeny and cranial morphology of the tympanic region of the Tupaiidae, with special reference to *Ptilocercus. Folia Primatol.* **47**: 61–80.

CHAPTER FIVE

Primate Taxonomy, Plesiadapiforms, and Approaches to Primate Origins

Mary T. Silcox

INTRODUCTION

In biology, there is currently a debate being waged about the basic principles of doing taxonomy (e.g., Benton, 2000; Cantino and De Queiroz, 2000; De Queiroz, 1994, 1997; De Queiroz and Gauthier 1990, 1992, 1994; Lee, 1996; Lidén and Oxelman, 1996; Lidén et al., 1997; Moore, 1998; Nixon and Carpenter, 2000; Pennisi, 1996; Schander and Thollesson, 1995). This debate stems from the common opinion that taxonomy should reflect evolution in some manner, combined with a disagreement about the practical details of how to do this. Although some authors have provided suggestions for making the Linnean system of taxonomy work within the context of a cladistic approach to phylogeny reconstruction (e.g., McKenna and Bell, 1997; Nixon and Carpenter, 2000; Wiley, 1981), others have advocated scrapping the entire Linnean system (De Queiroz, 1994; De Queiroz and Gauthier, 1990, 1992, 1994; Griffiths, 1976), culminating in the dissemination

Mary T. Silcox • Dept. of Anthropology, University of Winnipeg, 515 Portage Avenue, Winnipeg MB R3B 2E9, Canada

way of the Internet of a new code for phylogenetic nomenclature, the Phylocode (Cantino and De Queiroz, 2000). Although this document does not yet include guidelines for species taxa, and in spite of the fact that the Phylocode has not yet been "activated" by its authors (as of May, 2003; but see below), there are nonetheless a growing number of instances of the principles codified by this system being applied in real taxonomic practice (e.g., the redefinitions of Mammalia by Rowe, 1988). As such, even if the Phylocode is never adopted or accepted in full, it can still be considered to represent many current ideas about the practicalities of doing taxonomy.

In light of these debates a reconsideration of the meaning, content, and status of the taxon name "Primates" seems timely. Anthropologists have sometimes been criticized for ignoring taxonomic principles common to other areas of Biology (e.g., Mayr, 1950; Simpson, 1963), and the fact that the intense debates over taxonomic practice that have been waged in the biological literature in recent years are only rarely reflected in the contemporary anthropological literature suggests that this problem is ongoing. One of the central goals in understanding primate origins must be forming an understanding of where the group lies in relation to non-primate groups, since only against that comparative background can the relative uniqueness of primate features be fully understood. Without such an understanding it is impossible to create plausible adaptive scenarios for why changes occurred in the early evolution of the group. In light of this, it is clear that anthropologists cannot work in a vacuum from current evolutionary and taxonomic practice as applied to other groups of mammals.

Thus, it seems prudent to consider how Primates would stand in the context of the new system if the Phylocode were enacted, and how our common conceptions of what this term means could be dealt with in this framework. Even if the Phylocode is never accepted by all, it is worth considering the relative merits of the philosophical position that it represents. This has particular relevance in relation to the inclusion or exclusion of plesiadapiforms from the order Primates, since a determination of whether or not this cluster of extinct forms can be designated as primates depends not only on the supported pattern of relationships but also the taxonomic philosophy being applied. Finally it must be asked whether or not these disagreements over taxonomic approach influence the way in which we do, and should, ask questions about primate origins.

BACKGROUND ON TAXONOMIC DEBATES

Problems with Combining Cladistics and Linnean Taxonomy

The Linnean system of taxonomy is based on hierarchically internested sets of ranks. Common membership in a group indicated by a particular taxonomic name is shown by common usage of that name for the members of the group—therefore, all mammals are included in Mammalia. Although this may seem trivial, the result is that the taxonomy communicates a hypothesis about common group membership and, in an evolutionary context, common descent. The result is that Linnean taxonomic names can and often do provide an indication of the pattern of relationships thought to underlie the taxonomy. Therefore, when a cladist, such as M. C. McKenna (see McKenna and Bell, 1997, and below), includes a particular cluster of forms in Primates, the implication is that these are a monophyletic group or clade (Hennig, 1966) that shares a most recent common ancestor not shared by other forms.

In an evolutionary context, a Linnean approach to taxonomy allows, or even requires, some kind of hypothesis about how animals are related, and uses ranks to indicate how these relationships are internested. This element of the system creates some problems when it comes to a strict cladistic approach. The first is the multiplication of names necessary if all, or even a significant proportion, of nodes from a hypothesis of relationships are to be named. McKenna and Bell (1997), for example, needed to employ numerous unfamiliar ranks (e.g., infralegion, magnorder, mirorder, etc.) between the traditional ranks to accommodate all the groupings that they wished to recognize. Second, any change in the pattern of inferred relationships requires a cascade of changes in the taxonomic designations of not only the taxon in question, but of all taxa surrounding it on the phylogenetic tree. This is necessary to fit the taxonomy to the available ranks, and in some cases to accommodate the required (i.e., by the International Code of Zoological Nomenclature) endings to rank designations for any taxon reclassified at or below the family level. For example, if workers wish to name a node that sits between available ranks, shifting of taxa up- and downstream to fit the taxonomy to the tree is required. If this necessitates changing a previously recognized family into a subfamily, the ending of the group name must change to indicate this (i.e., from "idae" to "inae").

Third, a philosophical objection has been raised to the way in which Linnean names are designated, in using types and in being based on a list of characters (diagnoses) to indicate membership. This approach has been labeled "essential-ist" and "Aristotelian" (De Queiroz, 1994; De Queiroz and Gauthier, 1990), and has been criticized based on the impossibility of defining taxa using lists of characters if they are to be considered individuals in a strict philosophical sense (Ghiselin, 1984). The basis for this view is that characters are mutable, and that some nonmutable aspects of a taxon should be pointed to if they are going to be properly defined. Rowe (1987) indicated that a nonmutable aspect of any evolutionary group is the most recent common ancestor from which it evolved, and that the definition of a taxon should be based on delineating the group aris-ing from a particular common ancestor rather than a list of diagnosing charac-ters. Although in cladistically applied Linnean taxonomy groups are generally monophyletic (that is, they share a common ancestor not shared by other forms—this corresponds to the usage of the term "holophyletic" by some work-ers, e.g., Szalay et al., 1987), the definition or diagnosis of this group does not aim to specifically indicate that common ancestor. In the Linnean system there is nothing inherent in the way in which names are applied, therefore, to prevent the usage of paraphyletic groupings, since it is not mandated that common ancestry be demonstrated for a group to be recognized.

Fourth, the lack of consistent use of taxonomic ranks can lead to problems when data is analyzed at any level higher than the species. For example, De Queiroz and Gauthier (1994) enumerate several instances of nonsensical results that have come from workers tallying data across the family level without accounting for the fact that families can be very different in their composition.

Finally, the binominal system implies that the name of a species (the fun-damental unit of evolution) is dependent on a hypothesis of relationships, indicated by membership in a particular genus, which may change. Combining a rank designation (genus) with a fundamental biological entity (species) is considered inappropriate by some because of the potential insta-bility that it causes at the most basic level of taxonomy (Cantino et al., 1999; De Queiroz and Gauthier, 1992).

Phylogenetic Taxonomy's Solutions to the Problems Posed by Linnean Taxonomy

Phylogenetic taxonomy (De Queiroz 1994; De Queiroz and Gauthier, 1990, 1992, 1994), as codified in the Phylocode (Cantino and De Queiroz, 2000), is a rankless system of taxonomy, implying that it avoids all of the problems (multiplication of names, cascading rank changes, shifting taxon name endings, inappropriate comparisons based on rank) that arise because of the Linnean system being based on a set of internested ranks. What this implies, of course, is a different set of priorities about what taxonomy should be communicating from those who wish to at least partially mirror the pattern of relationships. Workers who object to phylogenetic taxonomy list as one of their key criteria for a nomenclatural system "high hierarchical information content" (Lidén et al., 1997: 735). Phylogenetic taxonomists, on the other hand, prioritize giving clades and species "names that explicitly and unambiguously refer to those entities and do not change with time" (Cantino and De Queiroz, 2000: 3), at the expense of having interrelationships between the names themselves communicate anything about the hierarchical pattern of relationships.

Names in phylogenetic taxonomy must be associated with a particular ancestor, so that membership in a group implies common descent by necessity. In this case the distinction between definition and diagnosis is clear—the definition of the taxon is a statement indicating to which ancestor the name refers, while the diagnosis is a list of characters that can be used to recognize members of the group. Within the context of the Phylocode the name is tied to the definition, which is fixed, while elements of the diagnosis can change. There are three ways that the ancestor associated with a taxonomic name can be designated, depending on the desired composition of the taxon being labeled (Cantino and De Queiroz, 2000; De Queiroz and Gauthier, 1990, 1992, 1994; see Figure 1). The first is a node-based definition (Figure 1A): "the clade stemming from the most recent common ancestor of two other taxa" (De Queiroz and Gauthier, 1990: 310). A special case of the node-based definition is the crown-clade of Jeffries (1979), in which both of the taxa specified are extant. This approach has been applied to Mammalia (De Queiroz, 1994; Rowe, 1988; Wible, 1991), leading to the definition of this taxon as all the members of the clade stemming from the most recent common ancestor of extant Monotremata and Theria (Rowe, 1988). A similar approach for Primates could define the group as all members of the clade originating with the common ancestor of extant Strepsirhini and Haplorhini (assuming that both of these groups are monophyletic). Alternatively, a node (but not crown) clade could be defined based on the clade stemming from the most recent common ancestor of *Purgatorius* and Haplorhini (see discussion given in a later section).

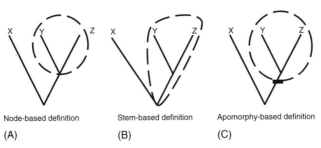

Figure 1. Types of taxonomic definitions advocated by Phylogenetic Taxonomy. The dashed shape surrounds the group being indicated by a taxonomic label. (A) Node-based definition: The taxonomic label refers to the clade deriving from the most recent common ancestor of Y and Z, but not any of the stem taxa from the lineage preceding that taxon. (B) Stem-based definition: The taxonomic label refers to the entire closed descent community (Ax, 1985) including not only the products of the most recent common ancestor of Y and Z, but also all taxa more closely related to this common ancestor than to X. (C) Apomorphy-based definition: In this case an apomorphy (indicated by the solid black bar) is used to delineate a common ancestor to the group indicated by the taxonomic label. The implication is that those members of the stem lineage that possess this apomorphy would be included, while those lacking it would be excluded, making this an intermediate between the node and stem based approaches.

The second type of definition under the Phylocode is the stem-based approach in which taxa are considered to belong to a "closed descent community," (Ax, 1985; Figure 1B) or encompassing "all those entities sharing a more recent common ancestor with one recognized taxon than with another" (De Queiroz and Gauthier, 1990: 310). A useful example of this is Anthropoidea, for which a stem-based definition could be constructed based on the clade including Catarrhini and all organisms that share a more recent common ancestor with Catarrhini than with *Tarsius* (modified from Williams and Kay, 1995; but see Wyss and Flynn, 1995). For Primates it would be possible to formulate the following definition: the clade consisting of Haplorhini and all organisms that share a more recent common ancestor with Haplorhini than with Scandentia. The precise composition of the resulting group will depend, of course, on the pattern of relationships supported in the systematic "gap" between haplorhines and scandentians, as discussed in later section.

The third type of definition is apomorphy-based: "the clade stemming from the first ancestor to possess a particular synapomorphy" (De Queiroz and Gauthier, 1990: 310; Figure 1C). Unlike a more traditional approach, in which the characters themselves are diagnosing the taxon, in this case the derived feature chosen serves merely to identify a particular ancestor. One might define Aves, for example, as the clade stemming from the first animal to possess wings homologous with those in *Archaeopteryx*. This implies that you cannot safely use more than a single character to identify the common ancestor, since it is impossible to be sure that the features did not arise in a step-like fashion. For example, until recently it would have seemed quite reasonable to identify the common ancestor of Aves as the first taxon with wings and with feathers. With the discovery of a number of feathered dinosaurs lacking wings (e.g., *Sinosauropteryx* Ji and Ji, 1996), this became a problematic definition because the two features (feathers and wings) arose at different times, implying that the definition would point to two different ancestors. Using a complex characteristic that may have arisen in a step-like fashion could also lead to problems in attempts to identify a single ancestor. Wings might, therefore, be a poor choice since they are complex structures with many parts that likely did not appear in an instant. In fact, a recent discovery of an apparently non-avian dinosaur with four wing-like structures (Xu et al., 2003) demonstrates how complicated even this simple example could become. The same argument can be made for most characters considered "significant enough" to be the indicator apomorphy for a clade.

For Primates, one possibility would be to define the group as the clade stemming from the first species to possess a petrosal bulla synapomorphic with that in Haplorhini. There are several reasons why this is not a viable approach, however, as detailed in later section.

By associating taxonomic names with the common ancestor of a monophyletic group, names cannot be applied to non-monophyletic taxa under the rules of the Phylocode, making it impossible to accidentally (or intentionally) name paraphyletic or polyphyletic taxa. In addition to doing away with ranks, phylogenetic taxonomy will presumably require uninominal names for species (Cantino et al., 1999; as it stands the Phylocode does not include guidelines for the species level) or, at least, that the first part of the binominal name not be considered to indicate anything about phylogenetic affinity or group membership. In this case, the species could include a "forename" or "praenomen" (following the terminology of Griffiths, 1976) rather than a genus name as the first part of the binominal.

Other Taxonomic Priorities

Two conflicting priorities for taxonomy have been mentioned previously: hier-archical information content, and stability in the entities to which names refer. There are, unquestionably, other priorities that can be considered key in assigning names to groups of taxa. In particular, some nomina have been viewed as bearing an implied significance that extends beyond their correct application according to taxonomic rules. A good example of this is the genus name *Homo*. Workers wanting to apply this name have generally sought some standard of humanity by which to judge the appropriateness of referring a taxon to our own genus. Louis Leakey and coworkers, for example, considered the ability to make tools key to recognition of human status when they named the species *Homo habilis* in 1964. Wood and Collard (1999) have sought to recognize a major adaptive shift with membership in the genus *Homo*, requiring that taxa designated with this nomen show such features as a human-like pattern of development and postcranial features associated with obligate bipedalism. Proponents of this view require that taxa not only be evolutionarily congruous, but also adaptively coherent. Of course, this approach is not terribly helpful to designating the taxonomic position of forms that fall outside of the key taxon in question. Wood and Collard, for example, while arguing for the removal of *Homo habilis* and *Homo rudolfensis* from the genus *Homo*, suggest only that these species be transferred to *Australopithecus*, making that genus even more paraphyletic, and poorly defined adaptively, than it is likely to be in the context of its commonly accepted composition. Such an approach is fine if you are only interested in the very tip of the phylogenetic tree, but is manifestly inappropriate if the goal is to actually place the "end taxa" in an evolutionary context. The relationships among the members of the genus *Australopithecus* are important to understanding the evolutionary background against which features of *Homo* must be assessed, in the same way the relationships of archontans (or euarchontans) that are not definitively primates are vital to an understanding of the adaptive sequence that went into building primates of modern aspect.

Another set of taxonomic priorities must also be recognized: stability and continuity through time. Although phylogenetic taxonomy considers stability a key priority, the type of stability that it fosters is more metaphysical than practical (Nixon and Carpenter, 2000). In fact, De Queiroz and Gauthier (1990: 312) explicitly state: "The use of phylogenetic definitions will effectively initiate a new era in biological taxonomy. In this new era there will be, in one sense, *no existing taxa* (named entities), for the names have not yet been tied explicitly to the entities through phylogenetic definitions" (emphasis added). Some authors, upon recognizing that traditionally applied taxonomic labels may refer to paraphyletic taxa, have nonetheless consciously retained them, prioritizing historical stability above other considerations (e.g., Silcox, 2001; Silcox et al., 2001; Van Valen, 1994). Such paraphyletic taxa have been called "metataxa" by Archibald (1994), and "natural paraphyletic groups" by Van Valen (1994), and often exist as a series of primitive branches off the stem leading to some cohesive monophyletic group. As such "natural paraphyletic groups" can be recognizable and diagnosable, if only by primitive or intermediate traits. The carpolestid genus *Elphidotarsius*, for example, includes a cluster of animals that can be differentiated from non-carpolestid by the presence of a mitten-shaped, blade-like P_4, and from more derived carpolestids by fewer apical cuspules on the P_4 and a less expanded P^3 and P^4 (Rose, 1975; Silcox, 2001; Silcox et al. 2001).

PREVIOUS DEFINITIONS AND DIAGNOSES OF PRIMATES

With this background in place, it is now appropriate to turn to the question of how Primates has been defined and diagnosed in the past, what the status of these past approaches is in relation to current taxonomic considerations, and the relevance of plesiadapiforms to this issue.

Mivart (1873) is generally cited as providing the first coherent definition of the order Primates. Mivart listed a series of characters seen in extant Primates including features unique or unusual to Primates (e.g., hallux with a nail) together with more widely distributed features that are likely to be primitive at a lower level (e.g., scrotal testes). Le Gros Clark (1959; see also Napier and Napier, 1967) translated these features into a series of evolutionary trends, again focusing on characteristics of extant primates, and again including both unusual primate features (e.g., elaboration of visual system) and more primitive eutherian traits (e.g., preservation of pentadactyly). Martin (1968, 1986) followed a similar approach in focusing on modern primates and in using a list of characters. In contrast to previous studies, however, Martin sought to apply Hennigean principles to filter the characters so that the diagnosing list included only synapomorphies.

These authors used a character-based approach focusing on extant forms. The problematic aspect of this method can be seen in the difficulties faced by Martin (1986) in trying to fit fossil adapids and omomyids into a modern conception of Primates. Some adapids and omomyids exhibit primitive features missing in all modern primates, such as the retention of the first premolar. Defining Primates based on a list of characters derived from observations of modern primates does not allow for loss of primitive traits that were present in the common ancestor. Such losses could occur either in the stem lineage leading to modern primates, or in parallel in groups that diverged off that ancestral lineage (i.e., as likely occurred with the loss of P_1). Any taxonomy that makes the implicit or explicit assumption that modern primates exhaust all the forms that the group has taken in its >65 MY history are likely to fail when faced with the real complexities of the extinct species to be found in the fossil record.

The approach taken by these authors is most consistent with the node-based, crown-clade method of defining taxa from phylogenetic taxonomy. Particularly, Martin requires fossil taxa to share a preponderance of features with extant taxa, indicating that they arose from the same common ancestor. Martin excluded forms (e.g., plesiadapiforms) which may lay along the stem leading to that ancestor, and which would therefore lack some or all of the features recognized in extant forms. Martin's justification for this approach is as follows (1968, p. 385): "If the term 'order' is to express anything in concrete terms, one should be able to picture a common ancestor with a distinct 'total functional pattern' at the base of each order, distinguishing this from other orders and providing the specific basis for the evolution of the descendants included in the order." Cartmill (1972, 1974, 1992) took a similar approach to the problem in advocating an adaptational methodology for delimiting major taxa, requiring that taxonomic boundaries mirror the pattern of major adaptive shifts. Under this approach plesiadapiforms could be excluded from Primates, independent of their phylogenetic position (but see Szalay, 1975), because they lack features associated with some novel adaptive complex. How this complex may be characterized so as to exclude all plesiadapiforms is becoming unclear, however, particularly in light of the discovery of some key adaptive primate features in plesiadapforms (e.g., a divergent hallux with a nail in *Carpolestes simpsoni*; Bloch and Boyer, 2002). The presence of arboreal features in *Ptilocercus lowii*, a scandentian that may be our best representative of the ancestral archontan (Bloch et al., 2003; Sargis, 2002a,b, and 2006), also complicates our understanding of the supposed adaptive uniqueness of definitive primates, since it appears that arboreality has a much more ancient origin relative to the group than had previously been assumed. The arboreal features of plesiadapiforms also underline this (Bloch and Boyer, 2002, 2003, and 2006; Bloch et al., 2003; Boyer et al., 2001; Szalay, 1975).

As with the example given above relating to the genus *Homo*, an additional problem with this approach is that it suggests no meaningful way of reclassifying those taxa that are excluded from this privileged primate status. Particularly, Cartmill (1972, 1974, 1992) advocated removing plesiadapiforms to Insectivora, explicitly favoring their transfer to a wastebasket taxon. This approach is further discussed in later section.

An additional reason why Martin (1968, 1986) felt that plesiadapiforms could be safely excluded from Primates was his opinion that they are not, in fact, the sister group of definitive (i.e., undisputed) Primates. If this is true, a good case could be made for excluding the group from the order (see a later section). Martin, however, provided no strong evidence for this, and failed to identify either a more plausible sister-group for Primates, or a more convincing alternative placement for plesiadapiforms.

Hoffstetter (1977) expressed no such doubts about the sister group relationship between plesiadapiforms and definitive primates. Hoffstetter's approach was fundamentally node-based, designating Primates as the group stretching from plesiadapiforms up to anthropoids (his Simiiformes), and introducing the very useful term Euprimates for the group referred to until now as "definitive primates" (i.e., Primates of modern aspect; crown-clade Primates). MacPhee et al. (1983; see also Gingerich, 1986) also included plesiadapiforms in Primates to avoid the information loss that they felt would result from their removal (Gingerich, 1973, 1986; but see Gingerich, 1989). These authors differed radically from Hoffstetter in their basic approach, however, forming boundaries based on three grades separated by morphological novelties supposedly associated with changes in adaptive pattern. Their view of Primates included not only plesiadapiforms, but also potentially mixodectids, apatemyids, tupaiids, and even dermopterans. As such, these authors' scheme would allow for paraphyletic and even polyphyletic groupings, making this approach explicitly opposed to the basic assumptions and goals of phylogenetic taxonomy, so that it would be difficult or impossible to accommodate this view in the context of the Phylocode. Although this does not immediately invalidate this approach, Szalay et al. (1987; see also Beard, 1990b) make some very good points about problems with this classification. Particularly, they note that of the three "grades" recognized by MacPhee et al. (1983), the only one that was adaptively coherent was their "Grade 3": Anthropoidea. It is not clear what adaptive features are supposed to link their "Grade 1" (plesiadapiforms, apatemyids, mixodectids, tupaiids, and dermopterans) in light of the very divergent postcranial and dental characteristics in the included taxa—consider, for example, the contrasting postcranial features of dermopterans and tree shrews, and the differences in dental morphology between picrodontids and carpolestids. As Szalay et al. (1987) note, Grade II (living and fossil "Prosimii"), by including tarsiers, exhibits a patchwork of features present in anthropoids and in more primitive forms. In a modern perspective, with our increased understanding of the primitive nature of early anthropoids (i.e., eosimiids), it is becoming clear that the discontinuities that were thought to separate these various grades were a sampling phenomenon rather than something real, so that even "Grade 3" is no longer really adaptively coherent. This highlights a more general point. The appearance of distinct adaptive groupings in the modern world likely owes as much to extinction or pseudoextinction as it does to evolution *per se*. As the fossil record continues to improve, the distinctness of such gaps continues to degrade as more intermediates are found. This is true in Primates as in non-primates. We now have, for example, a plesiadapiform with a supposed defining feature of definitive Primates (a divergent hallux bearing a nail; Bloch and Boyer, 2002, 2003, and 2006) and non-avian dinosaurs that bear not only feathers, but multiple wings that may have been used for gliding (Xu et al., 2003). Any taxonomy that fails to allow for the discovery and classification of such intermediates between adaptive "grades" is fundamentally unworkable if fossils are to be classified along with living organisms.

Szalay et al.'s (1987) own approach returned to a more phylogenetic methodology, although they explicitly advocated allowing the taxonomic labeling of paraphyletic as well as monophyletic (their "holophyletic") groupings. These authors included plesiadapiforms in Primates and used Hoffstetter's term Euprimates, providing lists of derived characters for the common ancestor of each of the major divisions of the order. In spite of their allowance for paraphyletic groupings these authors' approach, in practice, is most similar to the apomorphy-based technique of definition from phylogenetic taxonomy, in using uniquely derived features to identify the common ancestor that sits at the base of a named group.

Wible and Covert (1987; Figure 2), writing in the same year as Szalay et al. (1987), were unimpressed by similarities that had been noted between plesiadapiforms and euprimates (see later section). In an explicitly cladistic framework, these authors argued that the lack of consensus about the sister group to Euprimates implied that a stem-based approach could not be employed. They advocated a crown-clade approach to the definition of Primates, synonymizing Euprimates with Primates. Kay et al. (1992) and Beard (1993a)

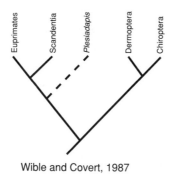

Wible and Covert, 1987

Figure 2.   Wible and Covert's (1987; redrawn from their Fig. 7a) preferred cladogram, allying tree shrews and Euprimates to the exclusion of *Plesiadapis* and supporting the link between dermopterans and chiropterans (Volitantia). Note that their use of "*Plesiadapis*" alone, rather than "Plesiadapiformes," reflects both their almost sole reliance on cranial data (well-known only for *Plesiadapis* and microsyopids at the time they were writing), and their view that "...the affinities of microsyopids remain muddled (p. 18)."

applied a similar usage of Primates. In both of these cases the support that the authors saw for a relationship between Dermoptera and some plesiadapiforms made it unnecessary to include any taxa outside of Euprimates in Primates. Beard (1991, 1993a) defined a taxon, Primatomorpha (note that this is incorrectly spelled "Primatamorpha" in Beard, 1993a: Figure 10.2), based on a list of synapomorphies shared by dermopterans, some plesiadapiforms, and Euprimates. This grouping would be consistent with a node-based type of definition, as indicated by the recognition of a comparable cluster by McKenna and Bell (1997), who favor this approach. In this case, however, McKenna and Bell (1997) call the assemblage of taxa Primates, rather than Primatomorpha. The inclusion of Dermoptera in their conception of Primates is interesting in that it would force a fundamental shift in what most contemporary authors think of as the adaptive limits of the order.

THE PHYLOGENETIC POSITION OF PLESIADAPIFORMES

Background

One of the two factors key to determining whether or not plesiadapiforms can be considered primates is to establish the phylogenetic relationships of the group to Euprimates. If some or all plesiadapiforms are sister taxa to Euprimates, then determining where to draw the primate/non-primate line depends only on taxonomic philosophy about whether to prioritize adaptive cohesiveness, a crown-clade model, or an apomorphy-based approach using some feature lacking in plesiadapiforms over a stem-based approach or a nodal definition that includes plesiadapiforms. If, however, some other group could be shown to be the sister taxon to Euprimates, or to be more closely related to some or all plesiadapiforms than plesiadapiforms are to Euprimates, the decision becomes more complicated and depends on the preferred pattern of relationships as well as the taxonomic philosophy employed.

Wible and Covert (1987) argued that tree shrews are a better sister taxon to Euprimates than any plesiadapiform (Figure 2). As alluded to above, their view of the dental evidence linking plesiadapiforms to Euprimates was that it consists only of vague trends that are not unique in the broader context of Eutheria. Their study lacks, however, any attempt at a detailed analysis of dentitions. Discussion of the postcranium is similarly lacking, although this is more forgivable since the major revision of this material (Beard, 1989) was not completed until two years after their study was published. Wible and Covert's study was strongly biased toward the basicranium. It is rather ironic to compare this study to that by Cartmill and MacPhee (1980), who argued for the exclusion of scandentians from Primates on the basis of a similar comparative sample of basicrania. These earlier authors used plesiadapiform basicrania as the ancestral morphotype for Primates, based on the evidence from other systems. They noted the profound differences between scandentian and plesiadapiform basicrania, and concluded that the similarities between primates and tree shrews must, therefore, be convergent. Wible and Covert, beginning with much of the same data, essentially excluded the importance of non-cranial systems, and therefore came to a different conclusion. Obviously what is needed to form a consensus between these two approaches is an unbiased look at the cranial evidence combined with a detailed analysis of other systems.

The possibility that some group other than Euprimates is most closely related to plesiadapiforms was advocated by Beard (1989, 1990a, 1993a,b) and Kay et al. (1990, 1992), who documented new postcranial and cranial material and interpreted it to indicate a tie between Dermoptera and at least some plesiadapiforms. By demonstrating not only that euprimates and plesiadapiforms are very different, but also using evidence from multiple systems that the latter may have ties elsewhere, these studies seemed to toll a death knell to a euprimate–plesiadapiform relationship. This changed perspective was heralded as a major breakthrough in popular and non-specialist accounts (e.g., Martin, 1993; Shipman, 1990; Zimmer, 1991) and seems to have been broadly accepted by most anthropologists whose expertise lies in different areas.

Careful reading of Kay et al. and Beard's studies reveals, however, a number of inconsistencies between the authors' viewpoints. Beard (1993a,b) considered plesiadapiforms, primates, and modern dermopterans to be linked in a mono-phyletic clade, Primatomorpha, to the exclusion of Scandentia and Chiroptera (Figure 3A). This implies that if one were willing to include Dermoptera in Primates one could include plesiadapiforms in Primates by simply equating Beard's Primatomorpha with Primates. Such a solution was indicated by McKenna and Bell (1997; see an earlier section). Beard also considered some plesiadapiforms (paromomyids and micromomyids) to be more closely related to dermopterans than they are to other plesiadapiforms, implying that Plesiadapiformes is only monophyletic if dermopterans are included.

Kay et al. (1992), on the other hand, while supporting the link between modern dermopterans and paromomyids, presented evidence that paro-momyids and plesiadapids form a monophyletic clade to the exclusion of dermopterans (Figure 3B). Their position on the relationship between

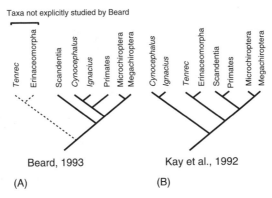

Figure 3. (A) Relationships of the taxa included by Kay et al. (1993) taken from Beard's (1993a) analysis. Beard did not explicitly include any lipotyphlan insectivores, implying that he considered them to lie outside of the ingroup. (B) Maximum parsi-mony cladogram found by Kay et al. (1994; redrawn from their Fig. 11). The only areas of congruence between these cladograms are the monophyly of Chiroptera and the relationship between *Cynocephalus* and *Ignacius*, with profound differences in the relative positions of both Scandentia and Primates (=Euprimates as employed here).

euprimates and plesiadapiforms also differed from Beard's view. They considered Scandentia to be the sister group of Euprimates (their Primates), and placed plesiadapiforms and dermopterans outside of a clade linking Scandentia, Euprimates, Chiroptera, and even lipotyphlan insectivores. In this context there is no way to include plesiadapiforms in Primates without including all of Archonta, lipotyphlan insectivores, and any unsampled taxa that may lie in between the nodes represented on Kay et al.'s tree. This would be nonsensical, since it would likely include a significant proportion of Mammalia including rabbits and rodents at the very least (see a later section). This discussion reveals that although Beard and Kay et al.'s studies are often cited together as supporting the plesiadapiform–dermopteran tie, in virtually all other aspects their results are not congruent.

Although Beard and Kay et al.'s studies looked at a broader range of data than had ever previously been brought to bear in a cladistic analysis considering the question of plesiadapiform relationships, there were nonetheless some holes in the sampling of both taxa and characters. First and foremost is the lack of dental data. Kay et al. included no dental data whatsoever, while Beard (1993a) included only a single dental character, the postprotocingulum, although he invoked unspecified dental evidence to link saxonellids and carpolestids with plesiadapids on his tree. In light of the fact that the traditional association between Euprimates and plesiadapiforms had been based on dental similarities, ruling out this relationship without considering characters from the teeth seems premature. This also substantially limited the taxon sampling with respect to plesiadapiforms, since the vast majority of species are known only from teeth. Although both Beard (1993a) and Kay et al. (1992) included at least some cranial data, they did not include any characters for plesiadapiform taxa outside of Plesiadapidae and Paromomyidae. The exclusion of the scrappy cranial material known at that time for *Palaechthon nacimienti* (Kay and Cartmill, 1977) and *Tinimomys graybulliensis* (Gunnell, 1989; but see MacPhee et al., 1995) is likely a product of the fact that very few cranial characters can be scored for either of these taxa. More surprising, however, is the exclusion of cranial material of Microsyopidae, particularly in light of the suggestion by Szalay et al. (1987) of a special link between microsyopids and dermopterans. At the time of the studies by Beard (1993a) and Kay et al. (1992), excellent cranial material for microsyopids was already well known (Gunnell, 1989; MacPhee et al., 1983, 1988; McKenna, 1966; Szalay, 1969).

Beard's (1989, 1993a) study did incorporate all of the plesiadapiform postcranials known at that time. In this case, however, there were some holes in the character sampling. Particularly, postcranial features that have been used to unite Volitantia (Dermoptera + Chiroptera) were not assessed (Simmons, 1995; Simmons and Geisler, 1998; Simmons and Quinn, 1994).

A More Comprehensive Analysis

Toward filling the gaps in these studies, the author collected data on 181 characters of the dentition (97), cranium (30), and postcranium (54). A detailed description of these characters and a discussion of their distribution is available elsewhere (Silcox, 2001). All the characters that have been used to support Primatomorpha (Beard, 1993a), and Volitantia (Simmons, 1995; Simmons and Geisler, 1998; Simmons and Quinn, 1994), that could be considered for fossils were assessed.

In selecting taxa to study, the author included members of all 11 families of plesiadapiforms known in 2001 (Carpolestidae, Plesiadapidae, Microsyopidae, Paromomyidae, Picromomyidae, Purgatoriidae, Micromomyidae, Toliapinidae, Palaechthonidae, Saxonellidae, and Picrodontidae). Plesiadapiform species to be sampled were selected based on two criteria. First, all the basal-most members of groups for which previous studies are informative were included (i.e., *Elphidotarsius* and *Chronolestes* for Carpolestidae—Simpson, 1928, 1935b; Rose, 1975, 1977, Beard and Wang, 1995; *Pandemonium* and *Pronothodectes* for Plesiadapidae—Gingerich, 1976; Van Valen, 1994; *Navajovius*, *Niptomomys*, and *Arctodontomys* for Microsyopidae—Gunnell, 1985, 1989; *Paromomys* for Paromomyidae—Simpson, 1955; Bown and Rose, 1976; and *Picromomys* for Picromomyidae—Rose and Bown, 1996). For all the other families poor sampling and/or a lack of consensus about the internal relationships of the group mandated the study of all known species (Purgatoriidae, Micromomyidae, Toliapinidae, Palaechthonidae, Saxonellidae, and Picrodontidae). Second, taxa preserving features of the cranium or postcranium were included, even if they are considered to be well nested within their respective families (e.g., *Plesiadapis tricuspidens*). The end result was a list of 62 species of plesiadapiforms to be studied, representing about half the total number currently recognized. Representatives of Chiroptera, Dermoptera, Scandentia, Mixodectidae, Plagiomenidae, and Euprimates were also chosen for analysis using similar criteria. Primitive eutherians (leptictids and *Asioryctes nemegtensis*) were employed as outgroups.

Data was initially collected at the species level. Cladistic analyses were run on the three major data partitions (dental, cranial, and postcranial) at the species level using PAUP* 4.0β (Swofford, 2001), and character distributions were studied using MacClade 3 (Maddison and Maddison, 1992). Families whose monophyly (all but Purgatoriidae, Palaechthonidae, and Toliapinidae *sensu* Hooker et al., 1999) were well supported were then combined. In cases where a family-level grouping could not be used, genera were employed when a genus was supported as monophyletic (i.e., *Toliapina*). The resulting dataset, using higher taxonomic groupings, was analyzed for each of the three data partitions and in a total evidence analysis (following the reasoning of Kluge, 1989). The discussion here will focus on this total evidence analysis, since this approach allows all for conflicting patterns of character distribution to compete directly. It is worth noting that the conclusions of the partitioned analyses did not always coincide with those arising from the total evidence analysis. Particularly, the postcranial analyses showed support for a Paromomyidae + Volitantia clade, while the cranial analyses indicated (very weak) support for a Microsyopidae + Volitantia grouping. The former conclusion is subject to revision in light of recent discoveries of new plesiadapiform postcranials (Boyer et al., 2001; Bloch and Boyer 2002, 2003 and 2006; Bloch et al., 2002, 2003), and new descriptions of the extant scandentian *Ptilocercus lowii* (Sargis, 2002a, b, and 2006), which were not available at the time of data collection. A collaborative project that includes these new data is currently underway (Bloch and Boyer, 2003; Bloch et al., 2002, 2004). The microsyopid/volitantian node is so poorly supported as to be unconvincing in light of the evidence from other parts of the study. The cranial analysis did not uphold Wible and Covert's (1987) claim of basicranial support for a Euprimate-Scandentia clade that excludes plesiadapiforms, or Kay et al.'s (1990, 1992) claim for cranial support of a paromomyid–dermopteran clade that excludes Euprimates (see also Bloch and Silcox, 2001, 2006). In fact, recent discoveries have documented one of the features that Wible and Covert cited as being key to supporting a euprimate–scandentian clade in a plesiadapiform (a bony tube for the internal carotid nerves in *Ignacius*; Silcox, 2003).

The total evidence analysis is open to criticism in that the results may largely reflect patterns of relationships indicated by the dental data. Many workers seem to consider dental data to be less reliable than other types of information, on the grounds that it supposedly shows more homoplasy than other parts of the skeleton (see discussion in Van Valen, 1994; Silcox, 2001). A quantitative study that actually analyzed the amount of homoplasy in different skeletal systems found no significant differences between dental, cranial, and postcranial regions (Sánchez-Villagra and Williams, 1998). What is more, teeth offer an important advantage over other parts of the skeleton. While for plesiadapiforms in general the only cranial and postcranial remains that are available belong to advanced members of different families, dentitions are known for very primitive as well as very derived forms. This allows one to get closer to the actual branching points, minimizing the confounding effects of interfamilial evolution producing convergences to other derived forms. In other words, study of dental features avoids long branch problems that are likely to be marked in more poorly sampled systems.

A series of heuristic searches totaling 3000 replicates, starting from different random trees and swapping on all starting trees, was performed on the total dataset with monophyletic higher taxa (i.e., families or genera) combined. This dataset included 38 taxa scored for all 181 characters. The search found 20 most parsimonious trees of length = 788 steps, CI(consistency index) = 0.490, RI(retention index) = 0.521, and RC(rescaled consistency index) = 0.255. The strict consensus tree resulting from these 20 trees was largely unresolved as a result of a couple of "wildcard" taxa that occupied very different positions on the various trees (i.e., *Mixodectes* and *Eudaemonema*). An effective way of dealing with such wildcard taxa is to calculate an Adams consensus tree. In an Adams consensus tree wildcard taxa appear unresolved at the highest node at which their position can be ascertained, with no loss of resolution "upstream". As such, examining an Adams consensus tree provides a better view of the pattern of relationships suggested by the data, since relationships that are well documented will be retained. The resulting Adams consensus tree is given as Figure 4.

All of the 20 most parsimonious trees include a clade that contains all plesiadapiforms and euprimates, and that excludes Scandentia, Chiroptera, and Dermoptera—this is also reflected in the Adams consensus tree at the node labeled "Primates". Rather than exhibiting a close relationship to Paromomyidae, Dermoptera is part of a monophyletic Volitantia, which may be more closely related to Scandentia than plesiadapiforms and Euprimates. Removal of Chiroptera from the analysis (as suggested by molecular results; Miyamoto et al., 2000; Pumo et al., 1998; Waddell et al., 1999; Springer et al., 2006) does not impact the relationships between the remaining taxa (e.g., Primatomorpha and Eudermoptera are not re-formed).

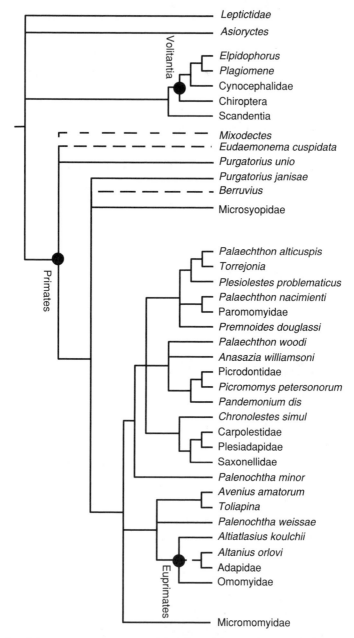

Figure 4. Adams consensus tree resulting from 3000 replicates of a heuristic search of the combined dental, cranial, and postcranial datasets (181 total characters) from Silcox, 2001. Twenty most parsimonious trees were found of length =788 steps, CI = 0.490, RI = 0.521, and RC = 0.255. The dotted lines leading to *Mixodectes* and *Eudaemonema* indicate the lack of certainty surrounding the relationships of these taxa—they were not included in Primates for this reason. Although *Berruvius* appears in an unresolved position, this is purely a product of missing data and the available dental evidence indicates a sister group relationship with Microsyopidae.

Plesiadapiformes itself is not a monophyletic taxon to the exclusion of Euprimates. This is a feature of every result (i.e., species-level, family-level, partitioned, and total evidence) found from the various analyses run in this study. The paraphyletic nature of Plesiadapiformes implies a fairly high level of homoplasy in the evolution of the group, as also indicated by the relatively low CI. The structure of this tree implies that some unusual features, such as an I^1 with an apical division, evolved more than once. Although most of the clearest evidence of homoplasy rests with the dental data partition, this effect turns out to be a product of effective sample size (Silcox, 2001).

Three families were also not found to be monophyletic–Purgatoriidae, Palaechthonidae, and Toliapinidae *sensu* Hooker et al., 1999. The former two taxa appear to be generally primitive, paraphyletic clusters. Toliapinidae can be rendered monophyletic if *Berruvius* is transferred back to the Microsyopidae (where it had generally been considered to reside until Hooker et al., 1999; see Gunnell, 1989; Russell, 1981).

Taxonomic Implications of the Current Analysis

In light of these results the issue of whether or not to include plesiadapiforms becomes one entirely of taxonomic philosophy. If one wishes to emphasize adaptive cohesiveness there is no question that the euprimate clade is more adaptively cohesive than any grouping that includes plesiadapiforms. The common ancestor at the euprimate node can be reconstructed as possessing a long list of newly derived characters which, together, seem to be associated with improvements to the visual system (reduced snout; complete postorbital bar; large optic foramina) and modifications of the postcranial skeleton for leaping (third trochanter at the same level as the lesser trochanter; deep distal femur; long astragalar neck; humerofemoral index less than 70; distal calcaneus elongate; lateral side of the femoral trochlea more anteriorly projecting than the medial side). The plesiadapiform + euprimate node is largely supported by dental features that are less easily interpreted as a functional complex (e.g., postprotocingulum present on P_4, enlarged M_3 hypoconulid)—this is hardly surprising, however, in light of the fact that the basal-most plesiadapiforms are known only from teeth. A crown-clade approach would also exclude plesiadapiforms from Primates, since all plesiadapiforms appear to lie outside of the clade that would include modern Primates (although no modern Primates were actually included in the list of taxa studied). An apomorphy-based approach would include or exclude plesiadapiforms depending on which feature was chosen. As the only character that has been identified as being unique to Primates (Wible and Covert, 1987), the presence of a petrosal bulla is likely the only choice that could hope to garner a consensus as such a key "indicator apomorphy". In spite of this, the petrosal bulla is a manifestly impractical choice. First, we do not yet know exactly how a petrosal bulla was acquired, but it seems unlikely that it appeared in an evolutionary instant. Wible and Covert (1987), for example, highlight similarities in scandentians and modern Primates in the expansion of part of the petrosal, the tegmen tympani, which might indicate part of the pathway leading from a nonpetrosal bulla to one formed exclusively by this bone. Specifying just "presence of a petrosal bulla" as the indicator apomorphy has the potential, therefore, to open up questions about how much of a petrosal bulla is adequate, particularly if fossils representing intermediate stages do become available. Second, and more importantly, it has been pointed out that it is impossible to demonstrate conclusively whether a fossil has a petrosal bulla. The morphology of both *Plesiadapis* (Russell, 1959, 1964; Silcox, 2001; Szalay, 1972; Szalay et al., 1987) and *Carpolestes* (Bloch and Silcox, 2003, 2006; Silcox, 2001) is consistent with the presence of a petrosal bulla. As MacPhee et al. (1983; see also Beard and MacPhee, 1994; MacPhee and Cartmill, 1986) point out, however, this identification cannot be confirmed without developmental evidence, which will likely never be available for these taxa. A character that cannot be confidently identified in the taxa of greatest interest and dispute is singularly inappropriate as a delimiter for a taxonomic group. It is unlikely, however, that any other single character would be supported as a "key apomorphy" for Primates, making this approach generally impractical for delimiting the order.

In any case, none of these options are informative about how plesiadapiforms should be classified if they are not to be considered Primates. One option is to place plesiadapiforms in a wastebasket Insectivora. Cartmill (1972, 1974, 1992) has advocated this approach in the past, in spite of dental (Gidley, 1923; Simpson, 1935a), cranial (Russell, 1959, 1964) and postcranial (Russell, 1964; Szalay et al., 1975) features that were already well known at that time to be shared by plesiadapiforms and Primates, and that are missing in insectivorans. The end result of this re-classification might make Primates easier to define, but makes Insectivora a meaningless, polyphyletic assemblage. This also ignores the fact that Insectivora is not, in fact, "available" to be a wastebasket for taxa that anthropologists do not want to deal with. The makeup and evolutionary relationships of forms traditionally included in Insectivora (more properly Lipotyphla) are a subject of very intense current debate. This debate has been stirred up by molecular discoveries (e.g., Stanhope et al., 1998), which suggest multiple origins of Insectivora, and their apparent conflict with morphological data (e.g., Asher, 1999). If one's only goal is to provide a clear definition of Primates these debates may seem irrelevant—what an insectivore is, and who they are related to, becomes an unimportant question. However, if the order is to be understood in the broader context of mammalian evolution, these debates are vitally important. A common finding of molecular results is that primates are part of a clade with dermopterans and scandentians (Euarchonta; Liu and Miyamoto, 1999; Liu et al., 2001; Pumo et al., 1998; Waddell et al., 1999) that may be closely related to Glires (rodents + rabbits) and only very distantly related to any traditional insectivorans (Waddell et al., 1999; Murphy et al., 2001a,b; Madsen et al., 2001; Springer et al., 2006). As such, Insectivora would be an entirely unsuitable place to put taxa that are closely related to modern primates (or dermopterans).

Also, we cannot simply ignore or dump into a wastebasket taxon those forms that do not already display all the features present in extant groups, rendering them by implication unimportant to questions of primate evolution. It is these forms that will be most crucial, in fact, in helping us understand which unusual euprimate features did actually arise as part of a particular adaptive complex. Recent discoveries highlight this fact. It is now clear from novel discoveries of plesiadapiform postcranial material that most of the features associated with grasping predate the common ancestor of modern Primates (Bloch and Boyer, 2002, 2003 and 2006), and evolved much earlier than characteristics such as convergent orbits or the postorbital bar. As such, any adaptive scenario for the origin of euprimates that links together grasping and visual features in a single pattern of change must be incorrect, because the evolutionary transitions involved occurred at significantly different points in time.

Another option would be to use Plesiadapiformes (or Proprimates; see Gingerich, 1989) itself as a separate order. However, the fact that Plesiadapiformes does not appear to be a monophyletic grouping implies that it is not possible to classify that cluster as a separate order, unless one is willing to provide a formal taxonomic label for a non-monophyletic group. I would argue against this approach for two reasons. First, modern taxonomic practice frowns on the use of non-monophyletic groupings. It is likely that any such taxonomy would be subject to rapid revision, or at least derision, along these lines. Second, and more importantly, dumping these forms into a different, common wastebasket (Plesiadapiformes or Proprimates rather than Insectivora) still obfuscates the central point that some plesiadapiforms are more closely related to euprimates than others. The structure of the tree, in terms of the branches leading up to the group that is of central interest to most (Euprimates), is actually very important, since it documents what steps precede that node and what forms are the best models for the common euprimate ancestor. Failing to recognize this taxonomically by dumping all plesiadapiforms into a group together, therefore suggesting that they can be treated as a unit, would lead to obfuscation of this important point. Although I think continuing to use "plesiadapiforms" informally is useful, I would not advocate applying this term as a formal taxonomic label.

Applying a stem-based definition is also problematic. Primates could be defined as the clade consisting of Euprimates and all organisms that share a more recent common ancestor with Euprimates than with the common ancestor of Volitantia + Scandentia. This definition assumes, however, that the Volitantia + Scandentia clade is well supported, which is really not the case. Particularly, some of the results of this study suggested that Scandentia might be the basal-most group in Archonta. This would imply that a common ancestor of Volitantia and Scandentia not shared with Euprimates never existed. Also, this possibility means that the stem-based definition suggested earlier in this article (i.e., the clade consisting of Haplorhini and all organisms that share a more recent common ancestor with Haplorhini than with Scandentia) could mandate the inclusion of Volitantia in Primates. I believe that this would stretch the adaptive boundaries of Primates in a way unacceptable to most. Finally, the assumption of the monophyly of Archonta made for this analysis does not take into consideration recent molecular results (e.g., Miyamoto et al., 2000; Pumo et al., 1998; Waddell et al., 1999;) that suggest chiropterans belong in a clade with carnivores and ungulates, rather than as part of a monophyletic Volitantia with dermopterans. In light of these uncertainties, and although I am generally an advocate of a stem-based approach to delimiting groups, it appears to be an impractical approach for Primates at the current time.

I suggest that a node-based approach is the most appropriate here. If Primates is defined as the clade stemming from the most recent common ancestor of *Purgatorius* and Euprimates (or Haplorhini, etc.), we have a solution that provides a satisfactory and historically consistent ordinal designation for both plesiadapiforms and Euprimates. What is more, because *Purgatorius* is a very primitive form, most future discoveries of euprimate stem taxa or plesiadapiforms will be easily accommodated in this definition. As such, this is essentially equivalent to a stem-based definition, while avoiding the problems created by uncertainty over the relationships of the rest of Archonta (or Euarchonta). This definition does not rely on any characters that cannot be assessed in fossils (e.g., the petrosal bulla), and avoids the defeatist and short-sighted approach that dumps all difficult to classify fossils into a meaningless wastebasket taxon.

PRIMATE TAXONOMY AND THE STUDY OF EUPRIMATE ORIGINS

How does this debate over where to draw the primate–nonprimate boundary relate to the study of primate origins? For most workers a key issue is the identification of those taxa that are important to elucidating the evolutionary relationships of unquestionable Primates (i.e., Euprimates), independent of taxonomy. That is, we need to understand which forms have features that tell us something about the common ancestor of Euprimates, and which more primitive taxa might give us clues about the order in which euprimate features were added through time. Those key taxa are not adapids, omomyids, or other fossil forms whose inclusion in Primates is not debated, no matter which definition is employed. Adapids and omomyids exhibit many features of the cranium, postcranium, and dentition (Silcox, 2001; Szalay et al., 1987; characters listed at "N" on Figure 5) that clearly demonstrate that they have already markedly diverged from the primitive archontan stock. As such, these taxa already postdate all the evolutionary events of interest—that is, the points at which euprimate features were added to some ancestral non-euprimate stock. It is necessary to go at least a node down to study euprimate origins, to forms that give us information on the order, pattern, and adaptive context in which euprimate characters were acquired—that is, to the "protoeuprimates" indicated on Figure 5. According to my results, these "protoeuprimate" positions are occupied by plesiadapiforms. Therefore, no matter what you call them, plesiadapiforms are the key taxa to this question. Whether you call them primates, proprimates, primatomorphs or something else is irrelevant to that fact.

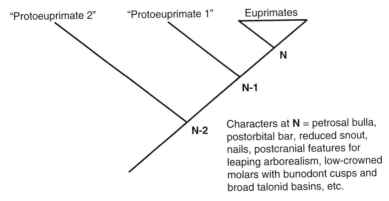

Figure 5. Hypothetical cladogram indicating the stepwise acquisition of distinctive euprimate features (present at N) through intermediates that possess only some of these traits ("protoeuprimates"). The author's analysis would place plesiadapiforms in the "protoeuprimate" positions, implying that it is these animals that will tell us about the order in which euprimate traits were acquired. Figure modified from a slide shown by D. Gebo at the Anthropoid Origins conference (April, 2001).

Having said that, one of the inevitable facts about taxonomy is that it directs the way that we formulate questions about evolution. If plesiadapiforms are excluded from the order Primates, studies aiming to consider primate origins may omit them (e.g., Soligo and Martin, in press). This is particularly true if they are relegated to some meaningless wastebasket taxon, which carries with it the implication that they are fundamentally unimportant. This assertion is based in part on my observation that, with the wide popularity of the view that plesiadapiforms and dermopterans are sister taxa, there is a general feeling that plesiadapiforms can be safely ignored by anthropologists. This is in spite of the fact that one of the major proponents of this view, Beard, published a cladogram that supported a closer relationships between plesiadapiforms + dermopterans and Primates than between Primates and any other taxon (Beard, 1993a).

A decision on the question of whether or not to include plesiadapiforms in Primates has real practical implications in terms of the way in which we ask questions about euprimate origins. Although it is not absolutely necessary to recognize with taxonomy the fact that plesiadapiforms are the key taxa to the study of euprimate origins, from a practical viewpoint their classification as primates ensures the appropriate focus when asking questions about the origins of the group of central interest to most.

CONCLUSIONS

Phylogenetic taxonomy has added an extremely useful refinement to the process of classifying organisms in emphasizing the need to associate taxonomic names with a pattern of relationships, and thus an ancestor, rather than with lists of mutable characters. When dealing with the fossil record it is inevitable that the closer to a branching point we get (and therefore to the answers that we are most interested in), the fewer and more subtle the characters differentiating groups will become. This is why classifications based only on modern forms work poorly when applied to the > 99% of life on Earth that has gone extinct (Raup, 1992; Schopf, 1982). Following this reasoning Primates is defined here according to a node-based approach to include plesiadapiforms, even though all known plesiadapiforms lack some features seen in modern Primates. No other classificatory position available for plesiadapiforms both conforms to modern taxonomic practice (i.e., in rejecting non-monophyletic groups) and emphasizes their importance as the key taxa to the study of the sequence of adaptations leading up to the origination point of Euprimates.

In spite of my enthusiasm for some of the tenets of phylogenetic taxonomy (i.e., its methods for defining groups), I feel that by aiming to replace the ranked Linnean system of taxonomy with a rankless system, the Phylocode has gone too far. For biologists whose central interest is the understanding and interpretation of evolution, using a ranked system offers the benefit of communicating details of the preferred phylogeny that are simply not outweighed by the metaphysical stability offered by the Phylocode. Also, historical stability is a consideration that should not be overlooked. By endeavoring to start from scratch, the adoption of the Phylocode would lead to a long period of flux, in which the central goal of taxonomy, communication, will not be met (Benton, 2000).

The basic premise upon which the Phylocode is based seems to be that we have, or will have very soon, a complete understanding of the phylogeny of all organisms. If this were true it would be a relatively simple matter to apply their guidelines to point at the nodes that we wish to name (Nixon and Carpenter, 2000, call it the "node-pointing" system to recognize this). In light of the fact that to date we have likely uncovered only a tiny fraction of the species that have lived on this planet, and considering the disagreements that still surround details of the branching pattern of known organisms, such a view seems extremely naïve. The Linnean system, in spite of all its faults, offers flexibility in that the taxonomist can make choices about how names should be applied. Particularly, if the Phylocode were enacted, the term Primates would have to be established using one of the allowed definition types, and then registered, before it could be utilized. Once that procedure was executed, the decision made by whoever performed this conversion would stand for the rest of time. If this definition were found to be based on an incorrect cladogram, the meaning of Primates as understood by most workers could easily come into conflict with the technical definition, which seems a problematic situation for a system intended first and foremost for clear communication. Although the end result of the flexibility allowed under the Linnean system is some lack of consistency, and a good deal of arguing, this flexibility seems essential in a world where there is so much variation in what we know for different groups, and in how widely accepted patterns of relationships are.

ACKNOWLEDGMENTS

This chapter began as a section of the author's dissertation mandated by D.B. Weishampel. The author would like to thank him, in addition to her dissertation advisor, K.D. Rose, and the rest of her dissertation committee, K.C. Beard and G.F. Gunnell. This work benefited from conversations with J.I. Bloch, A.E. Chew, and A.C. Walker and the comments of two anonymous reviewers. For access to specimens the author would like to thank J.P. Alexander, R.D.E. MacPhee, M.C. McKenna (American Museum of Natural History), K.C. Beard (Carnegie Museum of Natural History), J.I. Bloch (South Dakota School of Mines), P.D. Gingerich, G.F. Gunnell (University of Michigan Museum of Paleontology), G. Buckley (Field Museum of Natural History), M. Cassiliano, J.A. Lillegraven (University of Wyoming), W.A. Clemens, P. Holroyd (University of California Museum of Paleontology, Berkeley), J. Erfurt, H. Haubold (Geiseltal Museum), R.C. Fox (University of Alberta Laboratory of Vertebrate Paleontology), M. Godinot, P. Tassy (Muséum National d'Histoire Naturelle), L. Gordon (U.S. National Museum, Vertebrate Zoology), R.F. Kay (Duke University), D.W. Krause (SUNY Stony Brook), S.G. Lucas, T.E. Williamson (New Mexico Museum of Natural History), D. Miao (University of Kansas Museum of Natural History), R. Purdy (U.S. National Museum, Paleobiology), K.D. Rose (Johns Hopkins University School of Medicine), B. Sigé (University of

Montpellier II), M.A. Turner (Yale Peabody Museum), and L. Van Valen (University of Chicago). I also wish to thank the following for extending exceptional kindness to me in the course of my research travel: A. Aumont, K.C. Beard, J.I. Bloch, S. Ducrocq, C. Forster, Mme Gallette, J. Gardner, G.F. Gunnell, M. Klinger, E. Kowalski, V.E. Noble, S. Olsen, A. Voss, and the Clemens and Erfurt families. Finally, the author would like to offer her appreciation to M.J. Ravosa and M. Dagosto for their invitation to participate in this symposium. This research was funded by the National Science Foundation (NSF SBR-9815884), NSERC, the Wenner-Gren Foundation for Anthropological Research, Sigma Xi, and the Paleobiological Fund.

REFERENCES

Archibald, J. D., 1994, Metataxon concepts and assessing possible ancestry using phylogenetic systematics, *Syst. Biol.* **43**: 27–40.

Asher, R. J., 1999, A morphological basis for assessing the phylogeny of the "Tenrecoidea" (Mammalia, Lipotyphla), *Cladistics* **15**: 231–252.

Ax, P., 1985, Stem species and the stem lineage concept, *Cladistics* **1**: 279–287.

Beard, K. C., 1989, *Postcranial anatomy, locomotor adaptations, and paleoecology of Early Cenozoic Plesiadapidae, Paromomyidae, and Micromomyidae (Eutheria, Dermoptera).* Ph.D. Dissertation, Johns Hopkins University School of Medicine, Baltimore MD, 661 pp. (unpublished).

Beard, K. C., 1991, Vertical postures and climbing in the morphotype of Primatomorpha: implications for locomotor evolution in primate history. in: *Origines de la Bipédie chez les Hominidés,* Y. Coppens and B. Senut, eds., Editions du CNRS (Cahiers de Paléoanthropologie), Paris. pp. 79–87.

Beard, K. C., 1990a, Gliding behavior and palaeoecology of the alleged primate family Paromomyidae (Mammalia, Dermoptera), *Nature* **345**: 340–341.

Beard, K. C., 1990b, Do we need the newly proposed order Proprimates? *J. hum. Evol.* **19**: 817–820.

Beard, K. C., 1993a, Phylogenetic systematics of the Primatomorpha, with special reference to Dermoptera, in: *Mammal Phylogeny: placentals,* F. S. Szalay, M. J. Novacek, and M. C. McKenna, eds., Springer-Verlag, New York, pp. 129–150.

Beard, K. C., 1993b, Origin and evolution of gliding in Early Cenozoic Dermoptera (Mammalia, Primatomorpha), in: *Primates and their Relatives in Phylogenetic Perspective,* R. D. E. MacPhee, ed., Plenum Press, New York, pp. 63–90.

Beard, K. C., and MacPhee, R. D. E., 1994, Cranial anatomy of *Shoshonius* and the antiquity of Anthropoidea, in: *Anthropoid Origins,* J. G. Fleagle and R. F. Kay, eds., Plenum Press, New York, pp. 55–97.

Beard, K. C., and Wang, J., 1995, The first Asian plesiadapoids (Mammalia: Primatomorpha), *Ann. Carnegie Mus.* **64**: 1–33.

Benton, M. J., 2000, Stems, nodes, crown clades, and rank-free lists: Is Linnaeus dead? *Biol. Rev.* **75**: 633–648.

Bloch, J. I., and Boyer, D. M., 2002, Grasping primate origins, *Science* **298**: 1606–1610.

Bloch, J. I., and Boyer, D. M., 2003, Response to comment on "Grasping primate origins", *Science* **300**: 741c.

Bloch, J. I., and Boyer, D. M., (2006), New skeletons of Paleocene-Eocene Plesiadapiformes: a diversity of arboreal positional behaviors in early primates, in: *Primate Origins and Adaptations: A Multidisciplinary Perspective*, M. J. Ravosa and M. Dagosto, eds., Plenum Press, New York.

Bloch, J. I., and Silcox, M. T., 2001, New basicrania of Paleocene-Eocene *Ignacius*: Re-evaluation of the plesiadapiform-dermopteran link, *Am. J. phys. Anthrop.* **116**: 184–198.

Bloch, J. I., and Silcox, M. T., 2003, Comparative cranial anatomy and cladistic analysis of Paleocene primate *Carpolestes simpsoni* using ultra high resolution X-ray computed tomography, *Am. J. phys. Anthrop.* **120**(S1): 68.

Bloch, J. I., Silcox, M. T., 2006, Cranial anatomy of the Paleocene plesiadapiform *Carpolestes simpsoni* (Mammalia, Primates) using ultra high-resolution X-ray computed tomography, and the relationships of plesiadapiforms to Euprimates. *J. hum. Evol.* **50**: 1–35.

Bloch, J. I., Silcox, M. T., and Sargis, E. J., 2002, Origin and relationships of Archonta (Mammalia, Eutheria): Re-evaluation of Eudermoptera and Primatomorpha, *J. Vert. Paleontol.* **22** (supp. to No. 3): 37A.

Bloch, J. I., Boyer, D.M., and Houde, P., 2003. New skeletons of Paleocene-Eocene micromomyds (Mammalia, Primates): functional morphology and implications for euarchontan relationships, *J. Vert. Paleontol.* **23** (suppl. To 3): 35A.

Bloch, J. I., Silcox, M.T., Boyer, D. M., and Sargis, E. J. 2004. New hypothesis of primate supraordinal relationships and its bearing on competing models of primate origins: a test from the fossil record. *Am. J. phys. Anthrop.* **123**(S38): 64.

Bown, T. M., and Rose, K. D., 1976, New early Tertiary primates and a reappraisal of some Plesiadapiformes, *Folia primatol.* **26**: 109–138.

Boyer, D. M., Bloch J. I., and Gingerich, P. D., 2001, New skeletons of Paleocene paromomyids (Mammalia, ?Primates): Were they mitten gliders? *J. Vert. Paleontol* **21**(suppl.to 3): 35A.

Cantino, P. D., and De Queiroz, K., 2000, Athens, Ohio, Phylocode: A phylogenetic code of biological nomenclature (April 8, 2000); http://www.ohiou.edu/phylocode/printable.html

Cantino, P. D., Bryant, H. N., De Queiroz, K., Donoghue, M. J., Eriksson, T., Hillis, D. M., and Lee, M. S. Y., 1999, Species names in phylogenetic nomenclature, *Syst. Biol.* **48**: 790–807.

Cartmill, M., 1972, Arboreal adaptations and the origin of the order Primates, in: *The Functional and Evolutionary Biology of Primates*, R. Tuttle, ed., Aldine-Atherton, Chicago, pp. 97–122.

Cartmill, M., 1974, Rethinking primate origins, *Science* **184**: 436–443.

Cartmill, M., 1992, New views on primate origins, *Evol. Anthrop.* **1**(3): 105–111.

Cartmill, M., and MacPhee, R. D. E., 1980, Tupaiid affinities: The evidence of the carotid arteries and cranial skeleton, in: *Comparative Biology and Evolutionary Relationships of Tree Shrews*, W. P. Luckett, ed., Plenum Publishing, New York, pp. 95–132.

De Queiroz, K., 1994, Replacement of an essentialistic perspective on taxonomic definitions as exemplified by the definition of "Mammalia", *Syst. Biol.* **43**: 497–510.

De Queiroz, K., 1997, Misunderstandings about the phylogenetic approach to biological nomenclature: A reply to Lidén and Oxelman, *Zool. Scripta* **26**:67–70.

De Queiroz, K., and Gauthier, J., 1990, Phylogeny as a central principle in taxonomy: Phylogenetic definitions of taxon names, *Syst. Zool.* **39**: 307–322.

De Queiroz, K., and Gauthier, J., 1992, Phylogenetic taxonomy, *Annu. Rev. Ecol. Syst.* **23**: 449–480.

De Queiroz, K., and Gauthier, J., 1994, Toward a phylogenetic system of biological nomenclature, *Trends Ecol. Evol.* **9**: 27–31.

Ghiselin, M. T., 1984, "Definition", "Character" and other equivocal terms, *Syst. Zool.* **33**: 104–110.

Gidley, J. W., 1923, Paleocene primates of the Fort Union, with discussion of relationships of Eocene primates, *Proc. U. S. Natl. Mus.* **63**: 1–38.

Gingerich, P. D., 1973, First record of the Paleocene primate *Chiromyoides* from North America, *Nature* **244**: 517–8.

Gingerich, P. D., 1976, Cranial anatomy and evolution of early Tertiary Plesiadapidae (Mammalia, Primates), *Papers Paleontol. Univ. Michigan* **15**: 1–141.

Gingerich, P. D., 1986, *Plesiadapis* and the delineation of the order Primates, in: *Major Topics in Primate and Human Evolution*, B. Wood, L. Martin, and P. Andrews, eds., Cambridge University Press, Cambridge, pp. 32–46.

Gingerich, P. D., 1989, New earliest Wasatchian mammalian fauna from the Eocene of northwestern Wyoming: Composition and diversity in a rarely sampled high-floodplain assemblage, *Papers Paleontol. Univ. Michigan* **28**: 1–97.

Griffiths, G. C. D., 1976, The future of Linnean nomenclature, *Syst. Zool.* **25**: 168–173.

Gunnell, G. F., 1985, Systematics of early Eocene Microsyopinae (Mammalia, Primates) in the Clark's Fork Basin, Wyoming, *Contrib. Mus. Paleontol. Univ. Michigan* **27**: 51–71.

Gunnell, G. F., 1989, Evolutionary history of Microsyopoidea (Mammalia, ?Primates) and the relationship between Plesiadapiformes and Primates, *Papers Paleontol. Univ. Michigan* **27**: 1–157.

Hennig, W., 1966, *Phylogenetic Systematics*, University of Illinois Press, Urbana.

Hoffstetter, R., 1977, Phylogénie des primates, *Bull. Mém. Soc. Anthrop. Paris* t.4, série **XIII:** 327–346.

Hooker, J. J., Russell, D. E., and Phélizon, A., 1999, A new family of Plesiadapiformes (Mammalia) from the old world lower Paleogene, *Palaeontology* **42:** 377–407.

Jeffries, R. P. S., 1979, The origin of Chordates–a methodological essay, in: *The Origin of Major Invertebrate Groups*, M. R. House, ed., Systematics Association Special Volume 12, Academic Press, New York, pp. 443–477.

Ji, Q., and Ji, S. A., 1996, On the discovery of the earliest bird fossil in China and the origin of birds, *Chinese Geology* **233:** 30–33.

Kay, R. F., and Cartmill, M., 1977, Cranial morphology and adaptations of *Palaechthon nacimienti* and other Paromomyidae (Plesiadapoidea, ?Primates), with a description of a new genus and species, *J. hum. Evol.* **6:** 19–53.

Kay, R. F., Thorington R. W. Jr., and Houde, P., 1990, Eocene plesiadapiform shows affinities with flying lemurs not primates, *Nature* **345:** 342–344.

Kay, R. F., Thewissen, J. G. M., and Yoder, A. D., 1992, Cranial anatomy of *Ignacius graybullianus* and the affinities of the Plesiadapiformes, *Am. J. phys. Anthrop.* **89:** 477–498.

Kluge, A. G., 1989, A concern for evidence and a phylogenetic hypothesis of relationships among *Epicrates* (Boidae, Serpentes), *Syst. Zool.* **38:** 7–25.

Le Gros Clark, W. E., 1959, *The Antecedents of Man*, Edinburgh University Press, Edinburgh.

Leakey, L. S. B., Tobias, P. V., and Napier, J. R., 1964, A new species of the genus *Homo* from Olduvai Gorge, *Nature* **202:** 7–9.

Lee, M. S. Y., 1996, The phylogenetic approach to biological taxonomy: Practical aspects, *Zool. Scripta* **25:** 187–190.

Lidén, M., and Oxelman. B., 1996, Do we need phylogenetic taxonomy? *Zool. Scripta* **25:** 183–5.

Lidén, M., Oxelman, B., Backlund, A., Andersson, L., Bremer, B., Eriksson, R., Moberg, R., Nordal, I., Persson, K., Thulin, M., and Zimmer, B., 1997, Charlie is our darling, *Taxon* **46:** 735–738.

Liu, F.-G. R., and Miyamoto, M. M., 1999, Phylogenetic assessment of molecular and morphological data for eutherian mammals, *Syst. Biol.* **48:** 54–64.

Liu, F.-G. R., Miyamoto, M. M., Freire, N. P., Ong, P. Q., Tennant, M. R., Young, T. S., and Gugel, K. F., 2001, Molecular and morphological supertrees for eutherian (placental) mammals, *Science* **291:** 1786–1789.

MacPhee R. D. E., and Cartmill, M., 1986, Basicranial structures and primate systematics, in: *Comparative Primate Biology, Volume 1: Systematics, evolution, and anatomy*, D. R. Swisher and J. Erwin, eds., Alan R. Liss, New York, pp. 219–275.

MacPhee R. D. E., Cartmill, M., and Gingerich, P. D., 1983, New Paleogene primate basicrania and the definition of the order Primates, *Nature* **301**: 509–511.

MacPhee, R. D. E., Novacek, M. J., and Storch, G., 1988, Basicranial morphology of early Tertiary erinaceomorphs and the origin of Primates, *Am. Mus. Novit.* **2921**: 1–42.

MacPhee, R. D. E., Beard, K. C., Flemming, C., and Houde, P., 1995, Petrosal morphology of *Tinimomys graybulliensis* is plesiadapoid not microsyopid, *J. Vert. Paleontol.* **15** (Suppl. to 3): 42A.

Maddison, W. P., and Maddison, D. R., 1992, *MacClade 3, Program and Documentation*, Sinauer Associates, Inc., Sunderland MA.

Madsen, O., Scally, M., Douady, C. J., Kao, D. J., DeBry, R. W., Adkins, R. M., Amrine, H. M., Stanhope, M. J., de Jong, W. W., and Springer, M. S., 2001, Parallel adaptive radiations in two major clades of placental mammals, *Nature* **409**: 610–614.

Martin, R. D., 1968, Towards a new definition of Primates, *Man* **3**: 377–01.

Martin, R. D., 1986, Primates: A definition, in: *Major Topics in Primate and Human Evolution*, B. Wood, L. Martin, and P. Andrews, eds., Cambridge University Press, Cambridge, pp. 1–31.

Martin, R. D., 1993, Primate origins: Plugging the gaps, *Nature* **363**: 223–234.

Mayr, E., 1950, Taxonomic categories in fossil hominids, *Cold Spring Harb. Symp. Quant. Biol.* **15**: 109–118.

McKenna, M. C., 1966, Paleontology and the origin of the Primates, *Folia primatol.* **4**: 1–25.

McKenna, M. C., and Bell, S. K., 1997, *Classification of Mammals Above the Species Level*, Columbia University Press, New York.

Mivart, St. G., 1873, On *Lepilemur* and *Cheirogaleus*, and on the zoological rank of the Lemuroidea, *Proc. Zool. Soc. Lond.* **1873**: 484–510.

Miyamoto, M. M., Porter, C. A., and Goodman, M., 2000, *c-Myc* gene sequences and the phylogeny of bats and other eutherian mammals, *Syst. Biol.* **49**: 501–514.

Moore, G., 1998, A comparison of traditional and phylogenetic nomenclature, *Taxon* **47**: 561–579.

Murphy, W. J., Eizirik, E., Johnson, W. E., Zhang, Y. P., Ryder, O. A., and O'Brien, S. J., 2001a, Molecular phylogenetics and the origins of placental mammals, *Nature* **409**: 614–618.

Murphy, W. J., Eizirik, E., O'Brien, S. J., Madsen, O., Scally, M., Douady, C. J., Teeling, E. C., Ryder, O. A., Stanhope, M. J., de Jong, W. W., and Springer, M. S., 2001b, Resolution of the early placental mammal radiation using Bayesian phylogenetics, *Science* **294**: 2348–2351.

Napier, J. R., and Napier, P. H., 1967, *A Handbook of Living Primates*, Academic Press, London.

Nixon, K. C., and Carpenter, J. M., 2000, On the other "phylogenetic systematics," *Cladistics* **16**: 298–318.

Pennisi, E., 1996, Evolutionary and systematic biologists converge, *Science* **273**: 181–2.

Pumo, D. E., Finamore, P. S., Franek, W. R., Phillips, C. J., Tarzami, S., and Balzarano, D., 1998, Complete mitochondrial genome of a neotropical fruit bat, *Artibeus jamaicensis* and a new hypothesis of the relationships of bats to other eutherian mammals, *J. Mol. Evol.* **47**: 709–717.

Raup, D. M., 1992, *Extinction: Bad Genes or Bad Luck?* W. W. Norton, New York.

Rose, K. D., 1975, The Carpolestidae: Early Tertiary primates from North America, *Bull. Mus. Comp. Zool.* **147**: 1–74.

Rose, K. D., 1977, Evolution of carpolestid primates and chronology of the North American middle and late Paleocene, *J. Paleontol.* **51**: 536–542.

Rose, K. D., and Bown, T. M., 1996, A new plesiadapiform (Mammalia: Plesiadapiformes) from the early Eocene of the Bighorn Basin, Wyoming, *Ann. Carnegie Mus.* **65**: 305–321.

Rowe, T., 1987, Definition and diagnosis in the phylogenetic system, *Syst. Zool.* **36**: 208–211.

Rowe, T., 1988, Definition, diagnosis, and origin of Mammalia, *J. Vert. Paleontol.* **8**: 241–264.

Russell, D. E., 1959, Le crâne de, *Plesiadapis. Bull. Soc. Géol. Fr.* **4**: 312–4.

Russell, D. E., 1964, Les mammifères Paléocène d'Europe, *Mém. Mus. Hist. nat.* nouvelle série **13**: 1–324.

Russell, D. E. 1981, Un primate nouveau du Paléocène supérieur de France, *Géobios* **14**: 399–405.

Sánchez-Villagra, M. R., and Williams, B. A., 1998, Levels of homoplasy in the evolution of the mammalian skeleton, *J. Mammal. Evol.* **52**: 113–126.

Sargis, E. J., 2002a, Functional morphology of the forelimb of tupaiids (Mammalia, Scandentia) and its phylogenetic implications, *J. Morphol.* **253**: 10–42.

Sargis, E. J., 2002b, Functional morphology of the hindlimb of tupaiids (Mammalia, Scandentia) and its phylogenetic implications, *J. Morphol.* **254**: 149–185.

Sargis, E. J., 2002c, The postcranial morphology of *Ptilocercus lowii* (Scandentia, Tupaiidae): An analysis of Primatomorphan and Volitantian characters, *J. Mammal. Evol.* **9**: 137–160.

Sargis, E. J., (2006), The postcranial morphology of *Ptilocercus lowii* (Scandentia, Tupaiidae) and its implications for primate supraordinal relationships, in: Primate Origins and Adaptations: a Multidisciplinary Perspective, M.J. Ravosa and M. Dagosto, eds., Plenum Press, New York.

Schander, C., and Thollesson, M., 1995, Phylogenetic taxonomy–some comments, *Zool. Scripta* **24**: 263–268.

Schopf, T. J. M., 1982, Extinction of the dinosaurs: A 1982 understanding, *Geological Society of America Special Paper* **190**: 415–422.

Shipman, P., 1990, Primate origins up in the air again, *New Scientist* **126**: 57–60

Silcox, M. T., 2001, *A Phylogenetic Analysis of Plesiadapiformes and Their Relationship to Euprimates and other Archontans*, Ph.D. Dissertation, Johns Hopkins School of Medicine, Baltimore, Maryland (Unpublished)

Silcox, M. T., 2003, New discoveries on the middle ear anatomy of *Ignacius graybullianus* (Paromomyidae, Primates) from ultra high resolution X-ray computed tomography, *J. hum. Evol.* **44**: 73–86.

Silcox, M. T., Krause D. W., Maas M. C., and Fox R. C., 2001, New specimens of *Elphidotarsius russelli* (Mammalia, ?Primates, Carpolestidae) and a revision of plesiadapoid relationships, *J. Vert. Paleontol.* **21**: 132–152.

Simmons, N. B., 1995, Bat relationships and the origin of flight, *Symp. Zool. Soc. Lond.* **67**: 27–43.

Simmons, N. B., and Geisler, G. H., 1998, Phylogenetic relationships of *Icaronycteris, Archaeonycteris, Hassianycteris*, and *Palaeochiropteryx* to extant bat lineages, with comments on the evolution of echolocation and foraging strategies in Microchiroptera, *Bull. Am. Mus. Nat. Hist.* **235**: 1–182.

Simmons, N. B., and Quinn, T. H., 1994, Evolution of the digital tendon locking mechanism in bats and dermopterans: A phylogenetic perspective, *J. Mammal. Evol.* **2**: 231–254.

Simpson, G. G., 1928, A new mammalian fauna from the Fort Union of Southern Montana, *Am. Mus. Novit.* **297**: 1–15.

Simpson, G. G., 1935a, The Tiffany fauna, Upper Paleocene, II. Structure and relationships of *Plesiadapis, Am. Mus. Novit.* **816**: 1–30.

Simpson, G. G., 1935b, The Tiffany fauna, Upper Paleocene, III. Primates, Carnivora, Condylarthra, and Amblypoda, *Am. Mus. Novit.* **817**: 1–28.

Simpson, G. G., 1955, The Phenacolemuridae, new family of Early Primates, *Bull. Am. Mus. Nat. Hist.* **105**: 415–441.

Simpson, G. G., 1963, The meaning of taxonomic statements, in: *Classification and Human Evolution*, S. L. Washburn, ed., Aldine, Chicago, pp. 1–31.

Soligo, C. and Martin, R. D. in press. Adaptive origins of primates revisited. *J. hum. Evol.*

Springer, M.S., Murphy, W.J., Eizirik, E., Madsen, O., Scally, M., Douady, C.J., Teeling, E.C., Stanhope, M.J., de Jong, W.W., and O'Brien, S. J., (2006), A molecular classification for the living orders of placental mammals and the phylogenetic placement of Primates. in: *Primate Origins and Adaptations: A Multidisciplinary Perspective*, M. J. Ravosa and M. Dagosto, eds., Plenum Press, New York.

Stanhope, M. J., Waddell, V. G., Madsen, O., de Jon, W. W., Hedges, S. B., Cleven, G. C., Kao, D., and Springer, M. S., 1998, Molecular evidence for multiple origins of the Insectivora and for a new order of endemic African mammals, *Proc. Natl. Acad. Sci.* **95**: 9967–9972.

Swofford, D. L., 2001, *Phylogenetic Analysis Using Parsimony* (PAUP*) version 4.0β6*, program, Sinauer Associates, Inc., Sunderland, MA.

Szalay, F. S., 1969, Mixodectidae, Microsyopidae, and the insectivore-primate transition, *Bull. Am. Mus. Nat. Hist.* **140**: 195–330.

Szalay, F. S., 1972, Cranial morphology of the early Tertiary *Phenacolemur* and its bearing on primate phylogeny, *Am. J. Phys. Anthrop.* **361**: 59–76.

Szalay, F. S., 1975, Where to draw the nonprimate-primate taxonomic boundary, *Folia Primatol.* **23**: 158–163.

Szalay, F. S., Tattersall, I., and Decker, R. L., 1975, Phylogenetic relationships of *Plesiadapis*-postcranial evidence, in: *Approaches to Primate Paleobiology*, F. S. Szalay, F. S. ed., Karger, Basel, pp. 136–166.

Szalay, F. S., Rosenberger, A. L., and Dagosto, M., 1987, Diagnosis and differentiation of the order Primates, *Yearb. Phys. Anthrop.* **30**: 75–105.

Van Valen, L. M., 1994, The origin of the plesiadapid primates and the nature of *Purgatorius, Evolutionary Monographs* **15**: 1–79.

Waddell, P. J., Okada, N., and Hasegawa, M., 1999, Towards resolving the interordinal relationships of placental mammals, *Syst. Biol.* **48**: 1–5.

Wible, J. R., 1991, Origin of Mammalia: The craniodental evidence reexamined, *J. Vert. Paleontol.* **11**: 1–28.

Wible J. R., and Covert, H. H., 1987, Primates: Cladistic diagnosis and relationships, *J. hum. Evol.* **16**: 1–22.

Wiley, E. O., 1981, *Phylogenetics: The Theory and Practice of Phylogenetic Systematics*, Wiley Interscience, New York.

Williams, B. A., and Kay, R. F., 1995, The taxon Anthropoidea and the crown clade concept, *Evol. Anthropol.* **3**: 188–190.

Wood, B., and Collard, M., 1999, The human genus, *Science* **284**: 65–71.

Wyss, A. R., and Flynn, J. J., 1995, "Anthropoidea": A name, not an entity, *Evol. Anthropol.* **3**: 187–188.

Xu, X., Zhou, Z., Wang, X., Kuang, X., Zhang, F., and Du, X., 2003, Four-winged dinosaurs from China, *Nature* **421**: 335–340.

Zimmer, C., 1991, Family affairs, *Discover* **12**: 64–5

CHAPTER SIX

Jaw-Muscle Function and the Origin of Primates

Christopher J. Vinyard, Matthew J. Ravosa, Susan H. Williams, Christine E. Wall, Kirk R. Johnson, and William L. Hylander

INTRODUCTION

Anthropologists studying primate chewing have focused on the origins and evolution of the masticatory apparatus of anthropoids and humans. We know far less about the functional morphology and evolution of the masticatory apparatus in the earliest euprimates (e.g., Jablonski, 1986). A more complete understanding of masticatory apparatus function in the earliest primates would greatly benefit studies of chewing behavior in both strepsirrhines and haplorhines. We begin addressing this shortcoming in this chapter by asking, "To what extent do treeshrews share similar jaw-muscle activity patterns during chewing with living primates?" We use the small, nonprimate mammal, Belanger's treeshrew (*Tupaia belangeri*), as an extant model of jaw-muscle activity during chewing, or mastication, in early euprimates. By comparing

Christopher J. Vinyard • Department of Anatomy, Northeastern Ohio Universities College of Medicine. Rootstown, Ohio 44272-0095 **Matthew J. Ravosa** • Department of Cell and Molecular Biology, Northwestern University, Feinberg School of Medicine, Chicago, IL; Department of Zoology, Division of Mammals, Field Museum of Natural History, Chicago, IL **Susan H. Williams** • Department of Anatomy, Ohio University College of Osteopathic Medicine. Athens, Ohio 45701 **Christine E. Wall, Kirk R. Johnson, and William L. Hylander** • Department of Biological Anthropology and Anatomy, Duke University Medical Center, Durham, NC 27710

living primates to this treeshrew, we can infer whether the origin of primates involved significant changes in jaw-muscle activity patterns during chewing. Because we can make some basic functional links between jaw-muscle activity patterns and jaw form, our results will aid future interpretations of masticatory apparatus function from jaw form in living and fossil primates.

Functional Morphology of the Primate Masticatory Apparatus

What we know about how primates chew and the relationship between form and function of their masticatory apparatus comes from three seemingly disparate research agendas. The earliest and hence most enduring efforts have been made by scientists, dentists, and physicians focused on advancing applied dentistry and related medical fields (e.g., Ahlgren, 1966; Bennett, 1908; Carlsöö, 1952; DuBrul, 1988; Gibbs et al., 1971; Lindblom, 1960; Linden, 1998; Moller, 1966). A second group of researchers have studied the masticatory apparatus of humans, other primates, and more rarely nonprimates with the goal of understanding the evolution of human teeth and jaws (e.g., Ashton and Zuckermann, 1954; Biegert, 1956, 1963; Daegling and Grine, 1991; Demes and Creel, 1988; DuBrul, 1977; Gregory, 1920; Jolly, 1970; Kinzey, 1974; Rak, 1983; Robinson, 1956; Walker, 1981; Weidenreich, 1943; Wolpoff, 1973; Wood, 1981). Finally, a smaller group of researchers have been working to describe the patterns of variation in masticatory apparatus form and function among nonhuman primates (e.g., Beecher, 1977a,b, 1979; Bouvier, 1986; Daegling, 1989, 1992, 2001; Hylander, 1979a,b, 1988; Hylander et al., 1987, 1998, 2000; Kay, 1975, 1978; Luschei and Goodwin, 1974; McNamara, 1974; Ravosa, 1991, 1996; Ravosa and Hylander, 1994; Ross and Hylander, 2000; Smith, 1983; Teaford, 1994; Vinyard and Ravosa, 1998; Wall, 1999). The goal of many of the researchers in this third group is to understand the evolutionary history of the primate masticatory apparatus and how this history relates to the patterning of functional and morphological variation among primates.

We begin this review by briefly describing chewing in mammals. As part of this description, we provide several definitions frequently used by researchers studying chewing. We then consider the evidence linking jaw morphology and mechanical loads, that deform the jaw during chewing. Subsequently, we discuss the relationship between jaw loading and jaw-muscle activity patterns and by inference, the links between jaw-muscle activity and jaw form.

Understanding the associations among jaw loading, jaw-muscle activity and jaw form is fundamental to reconstructing the functional morphology of the masticatory apparatus of fossil primates. With this background in hand, we will then consider the masticatory apparatus in the earliest euprimates and the role that living treeshrews arguably play in studying its form and function.

Chewing

Chewing is the mechanical processing of foods in the oral cavity prior to swallowing. In mammals, chewing involves cyclic patterns of jaw movement during which foods are reduced between the opposing upper and lower postcanine dentition (Hiiemae and Crompton, 1985). A single movement circuit of the mandible is referred to as a chewing cycle, which can be divided into a closing stroke, a power stroke, and an opening stroke (Hiiemae, 1978) (Figure 1). A chewing sequence comprises multiple, sequential chewing cycles. A chewing sequence frequently will have multiple swallows interspersed among its chewing cycles. Finally, most mammals typically chew on one side of the jaw at a time.

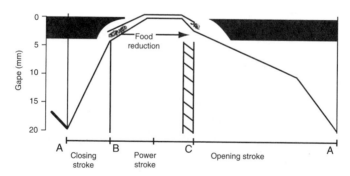

Figure 1. The mammalian chewing cycle is divisible into a closing stroke, a power stroke, and an opening stroke. In this idealized schematic gape (i.e., the distance between the upper and lower incisors) is depicted along the y-axis during these different segments of the chewing cycle. The closing stroke begins at maximum gape (A) Mandibular gape decreases during the closing stroke as the teeth move toward occlusal contact. The power stroke begins at tooth-tooth or tooth-food-tooth contact (B) Mechanical reduction of food occurs at this time. Minimum gape also takes place during the power stroke. The opening stroke begins as the teeth move inferiorly and out of contact (C) Gape increases during the opening stroke as the jaw opens prior to the next closing stroke (adapted from Hiiemae, 1976).

The opening stroke of the chewing cycle begins as the upper and lower teeth move out of occlusal (tooth–tooth) or tooth–food–tooth contact (Hiiemae, 1978) (Figure 1). The lower jaw moves away from the upper teeth, primarily inferiorly, during the opening stroke. After the jaw reaches maximum opening, the closing stroke begins with the rapid movement of the jaw superiorly and often laterally to align the teeth for the upcoming power stroke. The closing stroke ends at the commencement of occlusal or tooth–food–tooth contact. The power stroke begins at this contact as the jaw continues to move superiorly, medially, and often anteriorly, albeit at a slower rate (Hiiemae, 1978). Simultaneously, the masticatory apparatus experiences significant loads as foods are mechanically reduced during the power stroke (Hylander and Crompton, 1980, 1986). The next opening stroke begins when occlusal or tooth–food–tooth contact is lost. The extent of jaw movement and the magnitude of jaw loads in a chewing cycle are influenced by the structural and mechanical properties of the food, as well as the relative position of the chewing cycle in a chewing sequence (e.g., Agrawal et al., 1998; Chew et al., 1988; Fish and Mendel, 1982; Hiiemae, 1978; Hylander, 1979a; Luschei and Goodwin, 1974; Oron and Crompton, 1985; Thexton et al., 1980).

Masticatory apparatus form and jaw loads during chewing

We begin this section by stating our basic assumption that a significant component of the variation in mammalian jaw morphology is functionally related to differences in chewing behavior. This is not to say that all of the morphological variation in the masticatory apparatus is related to functional differences in chewing behaviors, but rather that some measurable component is. This explains our quantifying specific, mechanical aspects of chewing in order to correlate functional variation with morphological variation. Both forces and movements during chewing likely impact masticatory apparatus form, but in this chapter, we concentrate on jaw-muscle forces that occur primarily during the power stroke of mastication.

Numerous studies of primates, including *in vivo* and comparative morphological analyses point to an association between masticatory apparatus form on the one hand and jaw movements and forces during chewing on the other (e.g., Bouvier, 1986; Bouvier and Hylander, 1981; Daegling, 1992; Hylander, 1979a,b, 1984, 1988; Hylander et al., 1987, 1998, 2000; Kay, 1975, 1978; Ravosa, 1991, 1992; Ravosa et al., 2000; Strait, 1993; Taylor, 2002; Teaford and Walker, 1984; Vinyard and Ravosa, 1998; Wall, 1999).

Researchers, however, continue to discuss the strength of this relationship within and among primate species (Daegling, 2002; Daegling and McGraw, 2001; Hylander, 1979b; Smith, 1983, 1984; Wood, 1994). Undoubtedly, the morphology of an individual primate's masticatory apparatus reflects the influence of numerous behaviors that have nothing to do with chewing (e.g., anterior-tooth biting, display, or ingestion). Given what we know about the relationship between masticatory apparatus form and function, however, we anticipate that the functional role played by the masticatory apparatus during chewing does affect its form. Furthermore, we expect that these influences may be most easily discerned via interspecific comparisons of higher order groups of primates that differ in specific aspects of their chewing behaviors.

In vivo *analyses of facial bone strains during mastication.* Researchers possess several ways to study the functional morphology of the primate masticatory apparatus. It turns out that one of the most productive approaches for understanding masticatory forces is to attach strain gages directly to the facial bones of living animals and record their deformation or strain during chewing (Daegling and Hylander, 2000; Hylander, 1979a, 1985). This approach provides data showing how jaws are stressed or deformed during chewing.[1] For the mandible, these analyses have focused on the mandibular condyle, corpus, and symphysis—three regions of the jaw which experience significant internal forces, or internal loads, during chewing.[2] We define the phrase "significant loads" relative to an empirical observation that homotypic locations on load-bearing elements experience relatively similar strain magnitudes, as large as 1000–3000 $\mu\varepsilon$, during habitual loading across vertebrates of different body sizes and shapes (e.g., Biewener, 1982, 1989, 1990; Lanyon and Rubin, 1985; Rubin and Lanyon, 1984; Rubin et al., 1992, 1994).

The mandibular condyles on both the working side (chewing side) and balancing side (nonchewing side) of the jaw are typically loaded in compression

---

[1] When an external force is applied to a structure, the structure deforms as it resists this force. This deformation, or internal force, is measured by the stress created within the structure, while the displacement created within the structure is measured by its strain. Stress (a) is defined as force per unit area and is generally expressed in Newtons (N) per meter or Pascals (N/m). Strain (e) is a dimensionless unit measuring the amount of displacement or change in length (ΔL) divided by the original length (L) of a structure ($\Delta L/L$). Strain is often expressed in microstrain ($\mu\varepsilon$) or 1×10^{-6} mm/mm, (see e.g., Beer and Johnston, 1977; Biewener, 1992).

[2] Hylander and Johnson (1997) also demonstrated that the anterior portion of the zygomatic arch in macaques experiences significant strains during chewing. However, we are not discussing this region because there has not been a systematic analysis of zygomatic arch morphology among primates.

during the power stroke of mastication (Boyd et al., 1990; Brehnan et al., 1981; Hylander, 1979c; Hylander and Bays, 1979). The best way to resist a compressive force at the condyle is to increase the area of the articular surface resisting the load simply because stress is a function of a force per unit area (Bouvier, 1986a,b; Herring, 1985; Hylander, 1979b; Smith et al., 1983; Vinyard, 1999; Wall, 1999).

The working-side mandibular corpus in the molar region is sheared dorsoventrally: twisted about its long axis and bent in parasagittal and transverse planes during the power stroke (Hylander, 1979a; Hylander et al., 1987). Twisting of the corpus about its long axis appears to create the largest stresses on the working side of the jaw during mastication (Dechow and Hylander, 2000; Hylander, 1988). Generally, the most effective solution for resisting twisting loads is to increase the mediolateral width of the mandibular corpus (Daegling, 1992; Hylander, 1979a,b, 1988; Ravosa, 1991, 1996; Ravosa and Hylander, 1994).

On the balancing side, the mandibular corpus in the molar region is sheared dorsoventrally: twisted about its long axis and bent in a parasagittal plane during chewing (Hylander, 1979a). Parasagittal bending appears to create the largest loads along the balancing-side corpus during chewing (Hylander, 1988). The most effective way of providing greater resistance to parasagittal bending moments is to increase the vertical depth of the corpus (Daegling, 1992; Hylander, 1979a,b, 1988; Ravosa, 1991, 1996; Ravosa and Hylander, 1994).

The primate mandibular symphysis during chewing routinely experiences dorsoventral (DV) shearing, bending in a coronal plane and in some species lateral transverse bending, or wishboning (Hylander, 1984, 1985). In species that routinely wishbone their symphyses, this loading regime arguably generates the largest stresses, or internal forces, at the symphysis (Hylander, 1984, 1985, 1988; Hylander et al., 1998). Fusing the left and right halves of the mandible strengthens the symphysis by replacing relatively weaker ligaments with bone thereby providing increased ability to resist loads. Furthermore, increasing symphyseal area also provides greater resistance to DV shearing. Increasing the vertical (or dorsoventral) length of the symphysis offers an efficient means of providing more resistance to symphyseal bending in a coronal plane, while increasing the anteroposterior width of the symphysis provides an effective way of increasing the ability to resist wishboning (Daegling, 1992; Hylander, 1984, 1985, 1988; Ravosa and Hylander, 1994; Ravosa and Simons, 1994).

The morphological bottom line for resisting these loads is that larger and/or denser jaws, sometimes larger in a specific direction and other times larger in magnitude regardless of direction, offer increased resistance ability. One of the best examples linking jaw form and jaw load–resistance ability in primates is the comparison of strepsirrhines with mobile, unfused mandibular symphyses to living anthropoids all with fully ossified symphyses. Hylander (1979a) and Hylander et al. (1998) demonstrate that greater galagos, a strepsirrhine with an unfused symphysis, shows comparable levels of bone strain on the working side of the mandibular corpus during vigorous chewing as compared to macaques and owl monkeys—two anthropoids with fully fused symphyses (Table 1). Galagos, however, have much lower levels of corporal strain on the balancing side of the jaw during chewing as compared to these two anthropoids (Table 1). Thus, the ratio of working- to balancing-side (W/B) corporal strain is much higher in galagos as compared to macaques and owl monkeys (Table 1). This suggests that galagos, and perhaps all strepsirrhines with loose, mobile symphyses, recruit relatively less muscle force from the balancing side of the jaw during chewing when compared to anthropoids with fused symphyses (Hylander, 1977, 1979a,b; Hylander et al., 1998).

Primate suborder comparisons of jaw functional morphology. These differences in balancing-side corporal strains between galagos and macaques prompted a series of morphological analyses comparing jaw shapes between members of the two primate suborders (Hylander, 1979b; Ravosa, 1991;

Table 1. Comparison of average corporal bone strain and average masseter EMG W/B ratios in greater galagos, macaques, and owl monkeys during chewing of hard and/or tough foods

Species	Average Working-side Corporal Shear Strain ($\mu\varepsilon$)[a]	Average Balancing-side Corporal Shear Strain ($\mu\varepsilon$)[a]	W/B Corporal strain ratio[a,b]	Average W/B ratio for superficial masseter EMG[c]	Average W/B ratio for deep masseter EMG[c]
Galago	1197	216	7.00	2.2	3.9
Macaque	724	501	1.55	1.4	1.0
Owl monkey	1061	836	1.28	1.4	1.3

[a]Data from Hylander et al. (1998). Strain in microstrain ($\mu\varepsilon$).
[b]The reported W/B strain ratio is not the ratio of average working- and balancing-side shear strains reported here, but rather the average of experimental W/B strain ratios reported in Hylander et al. (1998).
[c]Data from Hylander et al. (2000).

Vinyard, 1999). Given that the higher balancing-side corporal strains in anthropoids are likely linked to increased parasagittal bending, anthropoids should have relatively deeper mandibular corpora. Figure 2A shows that for a given chewing moment arm length, anthropoids tend to have deeper mandibular corpora than strepsirrhines with unfused symphyses (Hylander, 1979b; Ravosa, 1991). We need to remind the reader that the converse of the earlier statement

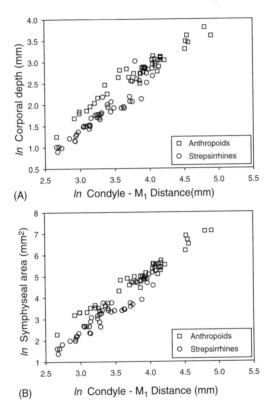

Figure 2. Comparison of load–resistance ability in strepsirrhine and anthropoid jaws. (A) Plot of *ln* corporal depth versus the *ln* chewing moment arm (condyle—M1 distance) in 44 anthropoids and 47 strepsirrhines. Only strepsirrhines with unfused symphyses are included. Anthropoids are visibly transposed above strepsirrhines for their corporal depth at a given condyle—M1 length. This transposition suggests that anthropoids can resist relatively greater amounts of corporal bending for a given moment arm as compared to strepsirrhines. (B) Plot of *ln* symphyseal area versus *ln* chewing moment arm length for anthropoids and strepsirrhines. Anthropoids are similarly transposed above strepsirrhines suggesting they have relatively greater load resistance ability at the symphysis.

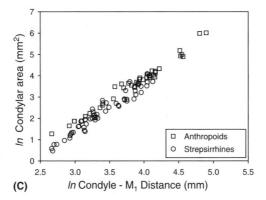

Figure 2. (*Continued*) (C) Plot of *ln* condylar area versus *ln* chewing moment arm length in anthropoids and strepsirrhines. The transposition of anthropoids above strepsirrhines again suggests that they have relatively greater ability to resist loads at the condyle as compared to strepsirrhines. In plots 2a-2c, indriids show considerable overlap with anthropoids. (See Vinyard, 1999 for further information).

is not necessarily equally valid. That is, corporal depth in strepsirrhines may not be linked to low balancing-side strains. This is true because, if strepsirrhines have significantly lower balancing-side as compared to working-side strains, then their corporal form should be related to working-side loads during chewing. Balancing-side strains need to be larger and/or have a different loading pattern than working-side strains so as to influence corporal form.

If higher balancing-side strains also indicate greater vertically-directed balancing-side jaw-muscle force recruitment (Hylander, 1977) and hence greater DV shear at the symphysis, then anthropoids are predicted to have relatively larger symphyseal areas and/or fused symphyses. While it is well known that living anthropoids have fused symphyses, Figure 2B indicates that they have also relatively large symphyseal areas for their moment arm length when compared with strepsirrhines (Ravosa, 1991).[3]

---

[3] A strong argument has been made that symphyseal fusion in anthropoids is related to resisting increased transversely, rather than vertically directed forces during chewing (Ravosa and Hylander, 1994; Hylander et al., 2000). Thus, fusion would be related to resisting wishboning rather than DV shear forces. Similarly, the relatively larger symphyses of anthropoids may reflect further structural buttressing to withstand wishboning forces during mastication (Hylander, 1984, 1985; Hylander et al., 2000; Ravosa and Hylander, 1994). The corporal strain data, however, suggest that strepsirrhines experience less DV shear than anthropoids, regardless of whether anthropoid symphyseal fusion is directly related to this loading regime. In other words, it is possible that if the anthropoid symphysis had never needed to fuse to resist wishboning, then it might have needed to fuse or become larger in area to resist increased DV shearing forces.

Finally, greater recruitment of the balancing-side jaw muscles increases reaction forces at the balancing-side condyle. Thus, anthropoids should have relatively larger condylar areas than strepsirrhines because of the increased forces at the balancing-side condyle. (This assumes that anthropoids and strepsirrhines experience relatively similar working-side condylar stresses.) Anthropoids do tend to have relatively larger condylar areas than strepsirrhines with unfused symphyses for a given chewing moment arm length (Vinyard, 1999) (Figure 2C). These comparisons collectively show that variation in jaw loading between the two suborders correlates with variation in jaw morphology. Arguably, these correlations reflect an underlying link between jaw stresses during chewing and jaw form among primates.

Jaw loading and jaw-muscle activity patterns during chewing

Even though *in vivo* strain gage studies provide unique insights into the stresses along both sides of the jaw during chewing, they do not offer direct information on how animals recruit specific jaw muscles. Because the jaw muscles create masticatory forces and jaw movements during chewing, we need to understand when the various jaw muscles are active and how strongly they are recruited. Biologists routinely use electromyography (EMG) to study jaw-muscle activity patterns during chewing (e.g., Ahlgren, 1966; Moller, 1966; Kallen and Gans, 1972; Luschei and Goodwin, 1974; de Vree and Gans, 1976; Clark et al., 1978; Hiiemae, 1978; Herring et al., 1979; Gorniak and Gans, 1980; Hannam and Wood, 1981; Weijs and Dantuma, 1981; Oron and Crompton, 1985; Hylander et al., 1987, 2000; de Gueldre and de Vree, 1988). Electromyographic approaches allow us to identify when these muscles are most active and in some cases allow us to speculate on the relative amount of muscle force recruited (e.g., Basmajian, 1978; Gans et al., 1978; Loeb and Gans, 1986).

The relationship between jaw-muscle EMG and forces during chewing. One of the first questions to ask when trying to link masticatory EMG studies to jaw form is, to what extent are jaw-muscle EMG magnitudes correlated with forces during chewing? Many researchers have demonstrated a positive correlation between the relative magnitudes of jaw-muscle EMG and submaximal isometric bite forces (see references in Hylander and Johnson, 1989). In these static biting situations, the jaw muscles fire under nearly isometric conditions (i.e., with minimal jaw movement). This means that EMG activity largely reflects the production of muscle and reaction forces (bite force and TMJ reaction forces) instead of some combination of masticatory force and jaw movements. Hylander and Johnson (1989) demonstrated that the relative magnitude of masseter EMG is highly correlated with the magnitude of *in vivo* strains in the zygomatic arch during chewing. This result supports a previous assertion by Weijs (1980) that relative EMG magnitude can be used as a relative estimate of muscle force during chewing. In this situation, we must model maximum force production during the power stroke as a "quasi-static" event. It appears reasonable to do this because jaw muscles shorten slowly at this time.

This link between force and jaw-muscle EMG magnitude is important in theory, but in practical terms we find it extremely difficult to identify an individual jaw muscle's contribution to chewing forces. This difficulty arises because multiple jaw muscles act both synchronously and asynchronously during chewing and the magnitude of any one muscle's EMG interference pattern is not directly comparable to that of other muscles (Loeb and Gans, 1986). One solution to this problem is to scale an electrode's EMG magnitude during a chewing cycle to the maximum value observed in that same electrode across all of the other chewing cycles in an experiment and then compare the ratio of scaled values between the working and balancing side of a muscle pair (Dessem and Druzinsky, 1992; Hylander et al., 1992, 2000). This working–balancing (W/B) ratio provides an estimate of the peak activity of the working-side muscle as compared to its balancing-side counterpart. We must remember, however, that the peak activities of these two muscles are not necessarily occurring at the same time.

There is some evidence that the W/B ratio correlates with overall force levels during chewing. Hylander et al. (1992) found that as masticatory forces increased, as estimated by zygomatic arch strains, masseter W/B EMG ratios tend to decrease during chewing in macaques. Thus, on average, larger forces correlate with lower W/B EMG ratios (Hylander et al., 1992). This suggests that macaques, and maybe other primates too, often increase force production during chewing by recruiting relatively greater amounts of balancing-side jaw-muscle force and thus lowering their W/B ratio. As a cautionary note, Hylander et al. (1992) point out that this pattern is variable as some animals show no relationship between zygomatic arch strain levels and W/B EMG ratios. Furthermore, there is greater variation in W/B ratio with foods that are easier to chew (i.e., structurally weaker and/or less tough) because an animal can recruit the necessary chewing force with many different muscle combinations (Hylander, 1979a). In conclusion, an association often exists between W/B EMG ratios and jaw-muscle forces, but this relationship is not an invariant one.

Suborder comparisons of W/B EMG ratios during chewing. We can return to our initial comparison of galagos to macaques and owl monkeys and ask whether the W/B EMG ratios for the superficial and deep masseters show the same interspecific pattern as seen in the W/B corporal strain ratios (Table 1). Like the W/B corporal strain ratios, galagos have higher average masseter W/B EMG ratios than macaques and owl monkeys. This suggests that compared to these anthropoids, galagos recruit their balancing-side masseters relatively less strongly than they do their working-side masseters (Hylander et al., 2000). The observed similarity between average corporal strain ratios and masseter EMG ratios provides some evidence linking relative EMG magnitude to force levels and by inference to differences in jaw form between the two suborders (see Table 1 and Figure 2).

Hylander et al. (2000, 2002) tested two hypotheses linking balancing-side masseter force and the presence or absence of symphyseal fusion in greater galagos, ring-tailed lemurs, macaques, baboons and owl monkeys. Greater galagos, with unfused symphyses, have higher average W/B ratios than the three anthropoid species for the superficial masseters, while ring-tailed lemurs are roughly comparable. However, the average superficial masseter W/B ratios for individual experiments in greater galagos overlap with anthropoid average W/B ratios across experiments. On the other hand, the deep masseter W/B ratios are significantly higher in greater galagos and ring-tailed lemurs as compared to anthropoids. Given the large transverse direction of pull in the deep masseter, these results support the hypothesis that fusion of the anthropoid symphysis relates primarily to transversely-directed forces from the deep masseter (Hylander et al., 2000, 2002).

We cannot completely rule out the hypothesis linking vertically-directed forces and symphyseal fusion based solely on masseter EMG if the balancing-side temporalis is the primary muscle generating DV shear at the symphysis (Hylander et al., n.d.). The linking of balancing-side temporalis activity to DV shear follows from the observations that (1) the balancing-side muscles may be creating most of the balancing-side corporal strain (Hylander, 1977), (2) peak balancing-side strains in the mandibular corpus occur after working-side corporal strains (Hylander et al., 1987), and (3) the balancing-side temporalis muscle peaks late in the chewing sequence (Hylander and Johnson, 1994; Weijs, 1994). These observations suggest that a comparison of temporalis W/B ratios also may be important for evaluating the link between vertically-directed forces and symphyseal form (Hylander et al., 2005).

The timing of peak EMG activity during the chewing cycle. Electromygraphic data also provide information about the timing of jaw-muscle force production during a chewing cycle. As might be expected, EMG studies have shown that all mammals do not fire their jaw muscles in the same sequence during chewing (Weijs, 1994). That having been said, several researchers suggest that there may be a common, primitive firing pattern found in many mammals including several primates (Gorniak, 1985; Hiiemae, 1978; Langenbach and van Eijden, 2001; Weijs, 1994). Most of the jaw-closing muscles are thought to fire in three, somewhat distinct, groups in this generalized model. These groups are: (1) a vertically oriented group of symmetric closers (VSC), (2) Triplet I, and (3) Triplet II. The VSC group includes the anterior and deep portions of the temporalis and the zygomaticomandibularis muscles on both the working and balancing sides. Triplet I muscles include the working-side posterior temporalis, balancing-side medial pterygoid and balancing-side superficial masseter. Triplet II muscles include the balancing-side posterior temporalis, working-side medial pterygoid and working-side superficial masseter. The VSC muscles are said to fire first in a chewing cycle and are thought to peak during the closing stroke. Triplet I muscles peak next near the start of the power stroke and Triplet II muscles peak after Triplet I muscles later during the power stroke (see Hylander et al., 2005 and Vinyard et al., 2005 for tests of this model in primates).

Hylander et al. (2000, 2002) also compared the timing of peak EMG activity of the masseters during chewing in greater galagos, ring-tailed lemurs and anthropoids. While the working-side superficial masseter acts as part of Triplet II, the deep masseters vary in their peak activity among species and are not consistently linked to either Triplet. In fact, previous work demonstrates that wishboning in the macaque mandibular symphysis largely relates to the activity of the deep masseter on the balancing side of the jaw (Hylander and Johnson, 1994; Hylander et al., 1987). Peak activity of the balancing-side deep masseter occurs late in the power stroke when most of the other jaw muscles, particularly the ones that could counter the laterally-directed pull of the deep masseter, have already peaked and are rapidly relaxing. To date, all other nonhominoid anthropoids show this firing pattern for the balancing-side deep masseter (Hylander et al., 2000; Vinyard et al., 2001). Greater galagos and ring-tailed lemurs, both with unfused symphyses, do not routinely exhibit this late-peak activity of the balancing-side deep masseter (Hylander et al., 2000, 2002). Because of its late peak activity, large transverse component of pull and significant relative recruitment in anthropoids when compared to the working-side muscle, the balancing-side deep masseter is considered to be a significant cause of wishboning of the anthropoid symphysis. These observations led Hylander et al. (2000, 2002) to hypothesize that symphyseal fusion in anthropoids relates to this wishboning-loading regime.

In summary, we can draw some clear links between primate jaw form, internal jaw forces and jaw-muscle activity patterns during chewing. Such links may prove particularly useful in comparisons of higher-level primate clades where it is possible to identify distinct morphological and functional differences among groups. Admittedly, we cannot presently derive irrefutable causal connections between masticatory apparatus form and chewing functions. We do, however, think that it is possible to speculate on the nature of these relationships based on what we know about the biomechanics of chewing in primates.

Interpretations of the Masticatory Apparatus in the First Primates

The topic of primate origins has a long history of discussion. We will not review this vast literature. Readers interested in such an appraisal should direct their attention to both earlier publications and recent reviews that consider the origin of primates in their appropriate historical context (e.g., Cartmill, 1974; Cartmill, 1992; Jones, 1917; Kay and Cartmill, 1977; Le Gros Clark, 1959a,b; Martin, 1993; McKenna, 1966; Smith, 1924; Sussman, 1991; Szalay, 1968). Instead, we will review references to the evolution of the primate masticatory apparatus during the origin of primates. We will categorize these publications into two hypotheses.

The "no adaptive change" in the masticatory apparatus hypothesis

Many discussions of primate origins offer little or no consideration of the evolution of the masticatory apparatus. For our purposes, we must argue from lack of evidence that these authors did not believe that the masticatory apparatus experienced noteworthy evolutionary changes during the origin of primates.

Some authors explicitly argued that the origin of primates did not involve any major adaptive changes in form (e.g., Cain, 1954; Davis, 1955; Simpson, 1955, Straus, 1949; Zuckerman, 1933; 1961). For example, Simpson (1955: 268) stated that, "the order Primates ... arose by adaptive improvement and not by any more or less clear-cut single basic adaptation." He went on to argue that "the most primitive primates are distinguished only arbitrarily from primitive Insectivora." It must be true that as we trace back a stem lineage the evolutionary changes between reproductively isolated sister groups disappear making taxonomic distinctions arbitrary. However, Simpson's view seems to envision evolution along the euprimate stem lineage as a process of accumulation of unremarkable changes in form that when added together allow us to designate something as a primate.

Washburn (1950) argued that the origin of primates involved a major adaptive shift in the form and function of the locomotor skeleton. He succinctly characterizes this viewpoint:

"The earliest primates were distinguished from other primitive mammals by the use of the hands and feet for grasping.... This basic adaptation has been the foundation of the whole history of primates.... The origin of primates was primarily a locomotor adaptation."

(Washburn, 1950: 68)

The unequivocal nature of this statement allows us to reasonably infer that Washburn did not envision significant changes in the masticatory apparatus associated with the origin of primates.

These two models of primate origins both predict relatively little evolutionary modification of the masticatory apparatus during the origin of primates. Thus, we would expect few changes in both the form and function of the masticatory apparatus between many living primates and other closely-related mammalian species. Given our intention to focus only on jaw-muscle activity patterns during chewing in this chapter, this hypothesis predicts significant overlap in jaw-muscle activity patterns during chewing in treeshrews and living primates, particularly strepsirrhines.

The 'herbivorous feeding adaptation' hypothesis. Szalay (1968, 1969, 1972, 1973) argued that the first primates differentiated from an ancestral stock through feeding adaptations in a burgeoning frugivorous and herbivorous arboreal niche.

"It is safe to presume, however, that the various features of the early prosimian dentition reflect a rather important shift in the nature of the *whole feeding mechanism.* Sporadic finds of primate skulls in the early Tertiary confirm this shift as a change from an insectivorous diet... to a herbivorous one. This change in diet that concomitantly affected the feeding mechanism was not an absolute one, as the insectivorous—carnivorous mode of life of many recent primates of any of the suborders testifies. Nevertheless, it is only an increasing occupation of feeding on fruits, leaves and other herbaceous matter that explains the first radiation of primates." (Italics added)

(Szalay, 1968:32)

Thus, according to Szalay, the origin of primates primarily involved adaptive modification of the masticatory apparatus due to an increased emphasis on plants as a food source. Despite this claim, no one has attempted a comprehensive comparison of masticatory apparatus form, beyond the teeth, or jaw-muscle activity patterns between primates and closely related nonprimates.

Szalay's discussion of the morphological changes in this feeding adaptation focused on the dentition. He noted that the teeth became better adapted for crushing and grinding. As part of this change, the cusps shortened and became more bulbous and the trigonid lowered (Szalay, 1968, 1973). He further argued that there were adaptive changes in the skull related to this change in diet. The facial skull shortened, although he also linked this change to evolutionary modifications of the neural, visual and olfactory systems. The zygomatic arches broadened and became more robust to allow larger attachment areas for jaw muscles. Finally, the primate masticatory apparatus evolved more transverse movement capabilities linked to an increased emphasis on "grinding" and "crushing" during chewing (see also Biegert, 1963; Hiiemae and Kay, 1972; Kay and Hiiemae, 1974; Szalay, 1968, 1972, 1973).

Szalay is not alone in suggesting that primates are marked by changes in the form and function of the masticatory apparatus. Harrison et al. (1977: 24) suggested that "the very origin of the Primates can be attributed, in the final analysis, to the presence of an arboreal food supply." Harrison et al. also cite changes in primate tooth form and function during this shift to a more herbivorous diet. Campbell (1974) suggested that primates, once freed from the need for grasping with the anterior teeth, improved chewing efficiency by posterior migration of the jaws. Biegert (1963) suggested that the masticatory apparatus became enlarged and specialized for grinding foods throughout primate evolution. Neither of these last two authors directly specified whether these changes in skull form occurred during the origin of primates.

An immediate issue with the herbivorous feeding adaptation hypothesis is that the diets of living primates and the reconstructed diets of extinct primates broadly overlap with those of closely related nonprimates (e.g., Covert, 1986; Emmons, 2000; Kay and Cartmill, 1977; Szalay, 1968; Van Valen, 1965). Szalay (1968, 1975) and Szalay and Delson (1979) acknowledged this overlap. We know, however, that the dietary categories commonly used by biologists, such as frugivory, folivory, insectivory, do not accurately describe the mechanical properties of the foods being consumed (e.g., Kay, 1975; Kay et al., 1978; Lucas, 1979; Lucas and Luke, 1984; Lucas and Peter, 2000; Lucas and Teaford, 1994; Rosenberger, 1992; Rosenberger and Kinzey, 1976; van Roosmalen, 1984; Yamashita, 1996, 1998). Obviously, it is the foods' mechanical and structural properties that relate to the internal and external forces during chewing rather than the broad classificatory nature of the food. Thus, while Kay and Cartmill (1977) correctly point out that similar dietary categories are found on both sides of the primate boundary, we cannot be certain that there was not a shift toward harder and/or tougher foods in these diets. Potential changes in the percentages of insects, fruits and leaves are particularly relevant to this question. At present, we simply lack this kind of mechanical information on the diets of living animals and likely can never collect these data for fossil taxa. Broad dietary overlap, therefore, cannot reject the spirit of Szalay's hypothesis suggesting that the first primates consumed harder and tougher foods. We might, however, expect to see evidence of these purported changes in mechanical properties of foods in the load-resistance ability of primate jaws and/or their jaw-muscle activity patterns in comparison to nonprimate species.

The herbivorous feeding adaptation hypothesis suggests that primates will have increased their relative force production during chewing, particularly transverse forces during the power stroke. The morphological implication of this hypothesis is that primates will have relatively robust jaws in comparison with closely related nonprimates so as to withstand these relatively larger loads during chewing. The herbivorous feeding adaptation hypothesis does not make a specific prediction regarding jaw-muscle recruitment patterns during chewing. Therefore, we cannot strictly test this hypothesis by comparing jaw-muscle EMG data from primates and treeshrews. The primary difficulty arises from the fact that differences in jaw-muscle activity patterns are not strictly linked to variation in diets. Based on what we know about *in vivo* jaw-muscle recruitment in primates, we suggest that this hypothesis would be supported if primates recruit relatively larger amounts of balancing-side jaw muscle force during chewing than treeshrews (i.e., primates will have lower W/B ratios).

Treeshrew Feeding Ecology and Jaw Morphology—A Reasonable Early Primate Model?

Treeshrews represent an enigmatic group of small mammals that have bewildered primatologists for over a century. This confusion persists in part because of our poor understanding of treeshrew ecology. Emmons (2000) has provided a much-needed field study of treeshrew behavioral ecology. Emmons (2000) emphasizes that the ecological diversity among treeshrew species makes it difficult to summarize the behavioral ecology of the order. Among treeshrews there are nocturnal and diurnal species, montane- and lowland-living taxa, terrestrial and arboreal groups as well as a variety of substrate specializations among the different species (Emmons, 2000). All treeshrew species eat fruits and insects, but not necessarily the same ones or in the same proportions (Emmons, 1991, 2000). As early Cenozoic primates are reconstructed to have broadly similar diets (e.g., Covert, 1986; Strait, 2001), treeshrews maybe useful as a living model for studying the chewing behaviors of the earliest primates.

We know more about treeshrew morphology as compared to their behavioral ecology. Several studies have focused on the morphology of the treeshrew masticatory apparatus, or at least specific parts of it (e.g., Butler, 1980; Fish, 1983; Gregory, 1910; Hiiemae and Kay, 1973; Kay and Hiiemae, 1974; Le Gros Clark, 1924, 1925; Mills, 1955, 1963, 1967). Treeshrews are small mammals, ranging in size from 45 to 350 g. Compared to primates, they have long jaws for their head size. These longer jaws are linked to their relatively longer snouts. Butler (1980) describes treeshrew molar teeth as dilambdodont and suggests that they "have not departed far from the primitive tribosphenic type" (Butler, 1980: 184). Hiiemae and Kay also state that treeshrew molars are well-suited for vertical shearing as opposed to crushing and grinding (Hiiemae and Kay, 1973; Kay and Hiiemae, 1974). They add that primate molar evolution is marked by a shift away from predominantly molar shearing toward an increased emphasis on crushing and grinding. This supports the argument that treeshrew teeth represent a reasonable functional model of early primate dentitions.

We are certainly not the first researchers to utilize treeshrews as a living functional model of early primates (e.g., Emmons, 2000; Le Gros Clark, 1959a,b; Sargis, 2001; Simpson, 1950, 1965; Tattersall, 1984). Numerous researchers have adopted treeshrews as a model for primitive primate chewing (see Biegert, 1956, 1963; Butler, 1980; Fish, 1983; Le Gros Clark, 1927, 1959a; Hiiemae and Kay, 1972, 1973; Jablonski, 1986; Kay and Hiiemae, 1974; Simpson, 1945; Mills, 1955). It certainly is possible, however, that treeshrews are a poor model for primitive primates (see e.g., Martin, 1990). We simply assume that treeshrews mimic the feeding behaviors of early primates, but we can only provide inferential evidence supporting this assumption. As with all such analogies, our use of *Tupaia belangeri* (or any other living species for that matter) as a functional model for early primate chewing should be viewed with caution.

Cineradiographic studies have documented complex jaw movements during chewing in *Tupaia glis* (Fish and Mendel, 1982; Hiiemae and Kay, 1973; Kay and Hiiemae, 1974). Three aspects of treeshrew jaw morphology may explain this capacity for complex jaw movements.

First, they have highly mobile, unfused mandibular symphyses that facilitate independent movement of the two hemimandibles during chewing (Fish, 1983, Fish and Mendel, 1982; Hiiemae and Kay, 1973; Kay and Hiiemae, 1974). Second, the morphology of their temporomandibular joint allows a range of complex jaw movements (Fish, 1983). Finally, the position and orientation of treeshrew jaw muscles facilitate mandibular movements in multiple directions (Fish, 1983).

We know very little about jaw-muscle activity patterns and force production during chewing in treeshrews. There also has been little comparative morphometric work examining the functional morphology of treeshrew jaws. Our goal here is to begin addressing these shortcomings by characterizing jaw-muscle activity patterns during chewing in treeshrews.

MATERIALS AND METHODS

Jaw-Muscle Electromyography

Subjects

We recorded jaw-muscle EMG patterns during chewing in five Belanger's treeshrews (*Tupaia belangeri*). All subjects were healthy adult males that weighed between 150 and 250 g. We habituated individuals to accepting food prior to recording EMG data. Our goal was to record EMG data from each treeshrew during three separate experiments. However, for reasons unrelated to our work, treeshrew 1 was available for only one experiment.

Electrodes and their placement

We placed fine-wire, bipolar indwelling electrodes in the left and right superficial and deep masseter muscles as well as the anterior and posterior portions of the left and right temporalis muscles (Figure 3). Prior to inserting electrodes, we sedated each treeshrew with ketamine (dosage: 40–50 mg/kg) and

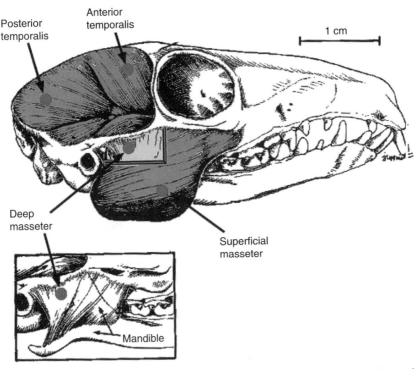

Figure 3. Lateral view of a treeshrew skull and jaw muscles. The dots depict the approximate placement of EMG electrodes in this study. In the inset, we have removed the superficial masseter in order to show the morphology of the treeshrew deep masseter. We inserted the deep masseter electrode by pushing the electrode through the superficial masseter and into the deep masseter. (Modified from Fish, 1983).

placed the subject in a restraining sling-suit that allowed free movement of the head and neck. To insert an electrode, we placed the tip of the electrode in a 27 gauge needle and inserted the needle in the appropriate muscle until the needle's tip contacted bone. We then removed the needle leaving the electrode in the muscle near the bone.

We placed electrodes in the superficial masseter by positioning the needle (with inserted electrode) midway between the muscle's anterior and posterior borders and inserting it until it contacted the lower border of the mandibular angle. The electrode tended to be located in the middle of the muscle, rather than more inferiorly, because the inferior portion of the superficial masseter wraps below the mandibular angle to attach on the jaw's medial surface just below the medial pterygoid (Fish, 1983). We placed deep masseter electrodes by inserting the needle just below the zygomatic arch midway between the mandibular condyle and the junction of the arch and postorbital bar. We inserted the needle at approximately a 30° downward angle relative to the ramus until it contacted bone. We inserted anterior temporalis electrodes approximately 5 mm above the zygomatic arch midway between the postorbital bar and the lateralmost extent of the temporal bone. We inserted needles downward and posteriorly, both at approximately 45° relative to horizontal, until contact with bone. We positioned posterior temporalis electrodes above and behind the ear and inserted them downward and anteriorly at approximately 45° until contact with bone. We did not verify electrode position via dissection because we did not kill the animals at the end of an experiment. Electrode construction and general placement procedures are described in more detail in Hylander and Johnson (1985, 1994) and Hylander et al. (2000).

Recording electromyography

We allowed the restrained animal to recover fully from sedation at which time we began feeding it dried apricots, dried raisins and crickets. We amplified and band-pass filtered (100–3000 Hz) EMG potentials from the jaw muscles during chewing of these foods and then recorded them onto a 14-channel FM tape recorder at a rate of 15 in./sec. Hylander et al. (2000) provide a detailed account of the recording procedure.

We continued feeding the treeshrew until we had collected sufficient data or the animal refused to eat any more. Following data collection, we removed the electrodes, freed the animal from its restraints and returned it to its cage. All recoveries from the experimental procedures were uneventful.

Quantifying electromyography

Processing EMG data followed the methods in Hylander et al. (2000). We selected chewing sequences for analysis by visually examining raw EMG signals on two multichannel, dual-beam oscilloscopes. We converted the raw analog EMG data from these selected chewing sequences to digital data sampling at 10 kHz and recorded the digital data to a microcomputer using LabView software. These recorded digitized EMG data were then filtered with a digital Butterworth band-pass filter (100–3000 Hz). We then calculated the root-mean-square (rms) of each digitized EMG signal using a 42-millisecond (ms) time constant (see Figure 2 in Hylander et al., 2000). All rms data for chewing sequences were calculated over 2 ms intervals (Hylander and Johnson, 1989; Hylander et al., 2000). We analyzed the rms EMG data.

Jaw-muscle recruitment: W/B ratios

The magnitude of EMG potentials cannot be directly compared across electrodes because of variation in signal quality and electrode position (e.g., Gans et al., 1978; Loeb and Gans, 1986). Therefore, we scaled each electrode's signal relative to its own peak signal in order to compare relative amounts of muscle activity among the jaw-closing muscles during chewing. To do this scaling, we identified the largest peak rms EMG potential for each muscle (i.e., an electrode) during a chewing cycle in an experiment regardless of whether we thought it acted as a working- or balancing-side muscle. We assigned the peak EMG a value of 1.0 and then re-scaled all other peak values for other chewing cycles in that electrode in a linear fashion relative to this largest peak.

We then divided the scaled peak value of the working-side muscle by the scaled peak value of the balancing-side muscle for each chewing cycle (Hylander et al., 1992, 2000). This working–balancing side (W/B) ratio measures the amount of scaled balancing-side muscle activity relative to that jaw muscle's working-side counterpart. For example, a W/B ratio of 1.0 indicates equal amounts of relative muscle activity, while a W/B ratio of 3.0

indicates three times more scaled muscle activity on the working as compared to the balancing side.[4]

We calculated the mean and standard deviation of W/B ratios across all chewing cycles in an experiment for the four pairs of electrodes in the masseter and temporalis muscles. W/B ratios tend to be right skewed with values above and below 1.0 (Hylander and Johnson, 1994). Log_{10} transformation of the individual W/B ratios helps to normalize these data (Sokal and Rohlf, 1997). Additionally, the largest peak EMG values for the left- and right-side electrodes in a muscle pair represent two, likely independent, scaling factors that affect a W/B ratio. In other words, the left muscle may have a peak rms value that represents a relatively greater amount of contraction for that muscle as compared to the peak rms value of the right-sided muscle (or vice versa). This possibility exists when an animal chews more forcefully on one side of the jaw than the other. Therefore, we first calculated the average of the logged W/B ratios for all left-sided chewing cycles separate from the average of right-sided chewing cycles. We then averaged the left and right W/B ratios to adjust for this side-related effect. Lastly, we antilogged this corrected average to provide the mean W/B ratio for that muscle pair in an experiment.

Jaw-muscle firing patterns: Timing

We examined when the various jaw muscles were active relative to each other during a chewing cycle by comparing the timing of their peak rms signals. To provide a standardized method for comparing jaw-muscle firing patterns, we compared the time of each muscle's EMG peak to the peak activity of the working-side superficial masseter (WSM) (Hylander and Johnson, 1994; Hylander et al., 2000). If a peak EMG for a muscle precedes the WSM peak, then we reported the number of milliseconds this muscle peaks prior to it as a positive value. Alternatively, if a muscle's peak EMG occurs after the peak of the WSM, then we reported the number of milliseconds this peak is trailing it as a negative value. We calculated the mean timing of each muscle's peak activity relative to the WSM across all chewing cycles in an experiment.

---

[4] We arbitrarily established a maximum W/B ratio of 10.0 for any single chew. Thus, if the relative working-side recruitment exceeded the relative balancing-side level by more than a 10:1 ratio, then we recorded the W/B ratio as 10.0 for that chew. We did this to reduce the likelihood that a few exceptionally large values would inflate the average W/B ratio (Hylander *et al.*, 2000).

Grand means and 95% confidence intervals

We report treeshrew grand means for W/B ratios and jaw-muscle peak firing times as the mean of 13 experimental averages (Hylander et al., 2000). For each W/B ratio, we averaged the Log_{10} values from the 13 experimental means and reported its antilog as the grand mean. We give the standard deviations of the 13 logged experimental means because of the known difficulties in antilogging a standard deviation (e.g., Finney, 1941; Laurent, 1963). We also calculated grand means and standard deviations for the jaw-muscle peak firing times based on the 13 experimental means.

We estimated the 95% confidence interval (CI) for these grand means using a bootstrapping approach (Efron, 1979; Manly, 1997) because the distributions of these variables are not fully understood. To calculate these confidence intervals, we pooled the 13 experimental means for each variable (for W/B ratios we used the logged means) and then resampled these values with replacement to create 1000 new samples each with 13 values. We then ranked the averages of these 1000 bootstrapped samples from smallest to largest. We took the values at the 2.5th and 97.5 percentiles of this ranked distribution as the 95% CI for that variable.[5]

Comparative Primate Electromyographic Data

We compared treeshrew EMG data to primate EMG data collected from greater galagos (*Otolemur crassicaudatus* and *O. garnetti*), ring-tailed lemurs (*Lemur catta*), owl monkeys (*Aotus trivirgatus*), callitrichids (*Callithrix jacchus* and *Sagunius fuscicollis*), macaques (*Macaca fascicularis* and *M. fuscata*) and baboons (*Papio anubis*) (Hylander et al., 2000, 2002, 2005; Vinyard et al., 2001). All primate data were collected and analyzed in a similar fashion to the treeshrew EMG data following the methods outlined in Hylander et al. (2000). We calculated grand means for primate W/B ratios and peak firing times from these sources following the same methods described above for treeshrews. All comparisons between primates and treeshrews are based on these grand means.

---

[5] The 95% confidence intervals based on this bootstrapping approach gave very similar results to the 95% confidence intervals for the grand means calculated using a typical parametric approach. The coincidence of these values supports the use of the bootstrapped confidence intervals (Manly, 1997).

RESULTS

Treeshrew Electromyography

Masseter W/B ratios

The grand means for the superficial and deep masseter W/B ratios are similar in treeshrews (2.7 and 2.8, respectively) (Table 2). On average, treeshrews recruit almost three times as much relative working-side masseter activity when compared to their balancing-side masseter. The confidence intervals (CIs) for the grand means for both W/B ratios range from a little over 2 to about 3.5 (Table 2). Individual treeshrews clearly vary in their average superficial and deep masseter W/B ratios. Specifically, treeshrew 1 tends to have much higher levels of relative balancing-side superficial and deep masseter recruitment (i.e., W/B ratios of 1.5 and 1.7, respectively) than treeshrew 2 who recruits the working-side muscles up to five times more than the balancing-side muscles (i.e., W/B ratios of 5.0 and 4.6, respectively) (Table 2).

Temporalis W/B ratios

The grand means of the W/B ratios for the anterior temporalis and posterior temporalis are also quite similar to each other (2.1 and 2.0, respectively).

Table 2. Treeshrew jaw muscle average W/B ratios

Subject	N	Superficial masseter		Deep masseter		Anterior temporalis		Posterior temporalis	
		Mean	Log_{10} SD	Mean	Log_{10} SD	Mean	Log_{10} SD	Mean	Log_{10} SD
Treeshrew 1									
Subject mean	202	1.5	–	1.7	–	1.1	–	1.2	–
Treeshrew 2									
Subject mean	257	5.0	0.04	4.6	0.18	3.0	0.20	2.5	0.02
Treeshrew 3									
Subject mean	295	2.0	0.08	1.9	0.08	1.4	0.11	1.6	0.03
Treeshrew 4									
Subject mean	214	2.6	0.10	2.4	0.10	2.3	0.16	1.9	0.17
Treeshrew 5									
Subject mean	324	2.4	0.21	3.2	0.10	2.3	0.27	2.4	0.07
Grand mean		2.7	0.20	2.8	0.19	2.1	0.22	2.0	0.13
95%CI for— grand mean		2.13–3.37		2.25–3.50		<1.62–2.69		1.71–2.32	

N = number of chewing cycles; Log_{10} SD = standard deviation of Log_{10} transformed ratios.
95% CI = Confidence Interval for the Grand mean based on bootstrapping the 13 experimental means.

Both are lower on average than the grand means for the two masseter W/B ratios (Table 2). The confidence interval (CI) for the grand mean of the anterior temporalis W/B ratio ranges from 1.6 to 2.7. The CI for the posterior temporalis is slightly narrower ranging from 1.7 to 2.3. Treeshrews 1 and 2 also set the range of variation in averages among treeshrews (Table 2). Similar to the pattern seen with the masseters, treeshrew 1 temporalis W/B ratios indicate a near equivalence in amounts of relative working-versus balancing-side muscle recruitment (W/B ratios of 1.1 and 1.2, respectively).

Alternatively, treeshrew 2 tends to have the highest levels of relative working-side recruitment for the temporalis (W/B ratios of 3.0 and 2.5, respectively) (Table 2).

Jaw-muscle firing patterns

The treeshrew working-side deep masseter (WDM), balancing-side superficial masseter (BSM) and balancing-side deep masseter (BDM) each peak on average before the peak activity of the working-side superficial masseter (WSM) (Table 3). The BSM and BDM both peak about 10 ms prior to the WSM. The WDM peaks on average only 3 ms prior to the WSM peak. The confidence interval (CI) for each of these muscles support this observation in that the CIs for the BSM and BDM only shows slight overlap with the WDM confidence interval and the CI of the WDM does not include zero. Individual treeshrew averages generally uphold this timing pattern of peak EMG activity where the BSM peaking first followed by the BDM, then the WDM and finally the WSM.

Peak firing in the working-side anterior temporalis (WAT) and posterior temporalis (WPT) muscles occurs on average at about the same time (Table 3). Both peak before the WSM. The grand mean and confidence intervals for the WAT and WPT are essentially the same. The grand mean for the time of peak activity in the balancing-side anterior temporalis (BAT) precedes the average peak of the balancing-side posterior temporalis (BPT) by about two milliseconds. Even though this two millisecond difference is small, the CIs for the two grand means show little overlap suggesting that there may be a general tendency for the BAT to peak just prior to the BPT (Table 3). The balancing-side anterior temporalis and BPT both reach peak activity after the WSM on average. Furthermore, comparison of individual treeshrew means tends to support this pattern of the BAT peaking after the WSM, but prior to the BPT.

The two working-side temporalis muscles are the first muscles on average to show peak activity during a chewing cycle (Table 3). Alternatively, the BAT

Table 3. Treeshrew timing differences in ms between peak jaw-muscle activity relative to the working-side superficial masseter

Subject	N	WDM		BSM		BDM		WAT		BAT		WPT		BPT	
		Mean	SD	Mean	SD	Mean	SD	Mean	SD	Mean	SD	Mean	SD	Mean	SD
Treeshrew 1															
Subject mean	202	−1	–	6	–	1	–	11	–	−2	–	17	–	−4	–
Treeshrew 2															
Subject mean	257	1	2	7	6	6	7	9	3	−5	3	11	5	−6	1
Treeshrew 3															
Subject mean	295	6	5	6	4	8	5	13	4	−1	2	11	1	−3	3
Treeshrew 4															
Subject mean	214	2	7	13	4	9	7	13	3	−3	1	11	2	−3	2
Treeshrew 5															
Subject mean	324	5	4	14	3	12	3	13	2	0	3	15	3	−5	1
Grand mean		3	5	10	5	8	6	12	3	−2	3	12	3	−4	2
95% CI for the Grand mean		0.9–5.5		7.3–12.5		5.4–11.1		10.5–13.7		−1.0–(−3.5)		10.5–13.7		−3.3–(−5.3)	

Positive values indicate that peak EMG activity of the muscle precedes peak EMG activity of the working-side superficial masseter.

Negative values indicate the reverse condition.

N = number of chewing cycles; SD = standard deviation.

95% CI = Confidence Interval for the Grand mean based on bootstrapping the 13 experimental means.

and BPT are the last muscles to exhibit peak activity during a chewing cycle. Thus of all of the muscles analyzed, the temporalis muscles show the greatest amount of offset in timing of peak activity between the working- and balancing-side muscles during chewing in treeshrews.

Figure 4 shows the progression of average peak firing times among the treeshrew jaw-closing muscles. The WAT, WPT, BSM, and to some extent the BDM, form a group of jaw-closing muscles that tend to fire first during a chewing cycle (Triplet I). The BAT, BPT and WSM, and arguably the WDM, form a second group of jaw-closing muscles that show peak activity later in the power stroke (Triplet II).

Comparison of Treeshrew and Primate Electromyography

W/B Ratios

Compared to anthropoids, strepsirrhines with unfused symphyses tend to have higher average W/B ratios for their jaw muscles during chewing (Hylander et al., 2000, 2002). The deep masseter clearly shows the most

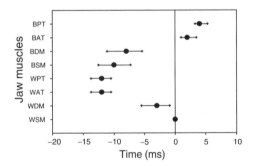

Figure 4. Plot of average peak jaw-muscle activities during chewing in treeshrews. On average, most treeshrew jaw muscles peak in two separate groups. The working-side anterior temporalis (WAT), working-side posterior temporalis (WPT), balancing-side superficial masseter (BSM), and balancing-side deep masseter (BDM) tend to peak before the remaining muscles. The balancing-side anterior temporalis (BAT), balancing-side posterior temporalis (BPT), and working-side superficial masseter (WSM) (represented by the dot on the solid line at 0 on the x-axis) peak later in the chewing cycle. The working-side deep masseter is intermediate between these two groups. The dot represents the average peak activity for a muscle, while the bar passing through the dot demonstrates the 95% CI of the mean.

Table 4. Summary comparisons of average jaw muscle W/B ratios among treeshrews and primates

Species	Superficial masseter	Deep masseter	Anterior temporalis	Posterior temporalis
Treeshrew	2.7	2.8	2.1	2.0
Galago	2.2	3.9	4.4	2.4
Lemur	1.7	2.4	1.5	2.0
Baboon	1.9	1.0	1.2	1.0
Macaque	1.4	1.0	1.2	1.2
Owl monkey	1.4	1.3	1.4	1.4
Callitrichid	1.9	1.3	1.3	1.2

differences in W/B ratios between these strepsirrhines and anthropoids. The W/B ratios for the remaining muscles, in particular the superficial masseter, are more similar between strepsirrhines and anthropoids. Treeshrew average W/B ratios are more similar to those for ring-tailed lemurs and greater galagos as compared to the lower W/B ratios found in the four anthropoid groups (Table 4). Thus, treeshrews appear more like strepsirrhines with unfused symphyses in showing lower levels of balancing-side jaw-muscle activity relative to the working-side muscles (cf Hylander et al., 2000).

Jaw-muscle firing patterns

The posterior temporalis and superficial masseters in strepsirrhines and anthropoids both generally follow the Triplet hypothesis with Triplet I jaw muscles peaking before Triplet II muscles during a chewing cycle (Weijs, 1994). The vertically oriented primate anterior temporalis fires along with its same-side posterior temporalis (Hylander et al., 2005) rather than peaking prior to Triplets I and II as suggested by Weijs (1994). Here we follow Hylander et al. (2005) by including the primate working-side anterior temporalis with Triplet I and the balancing-side anterior temporalis as part of Triplet II. Platyrrhines may be an exception to this generalized Triplet pattern given that their balancing-side superficial masseter tends to peak at about the same time as their working-side superficial masseter on average and hence is not firing with the working-side temporalis as part of Triplet I (Table 5).

The firing patterns of the deep masseter show the most variation among primate jaw-closing muscles. Anthropoids differ from those strepsirrhines with mobile, unfused symphyses by peaking their balancing-side deep masseter (BDM) after their working-side superficial masseter (WSM) (Hylander et al.,

Christopher J. Vinyard et al.

Table 5. Summary comparisons of average timing differences between peak EMG activity relative to the working-side superficial masseter for treeshrews and primates

Species	WDM	BSM	BDM	WAT	BAT	WPT	BPT
Treeshrew	3	10	8	12	−2	12	−4
Galago	11	21	23	13	−3	14	−3
Lemur	37	14	24	23	6	17	0
Baboon	47	17	−6	16	−6	20	−19
Macaque	65	17	−20	8	−6	7	−16
Owl monkey	13	−1	−11	13	−14	13	−15
Callitrichid	10	−5	−12	3	−13	11	−15

Positive values indicate that peak EMG activity of the muscle precedes peak EMG activity of the working-side superficial masseter. Negative values indicate the reverse condition.

2000, 2002) (Table 5). In contrast, the BDM peaks well before the WSM in greater galagos and ring-tailed lemurs. The working-side deep masseter (WDM) is also quite variable in its timing of peak activity among primates. The WDM is the first muscle to peak, at a time well before the muscles in Triplet I, in lemurs, macaques and baboons (Table 5). Alternatively, the WDM peaks with the other muscles of Triplet I in galagos and platyrrhines (Table 5).

Treeshrews share similar firing patterns for their jaw-closing muscles during chewing with strepsirrhines, and are arguably most similar to galagos. Treeshrews also appear to fire their jaw-closing muscles in two triplets; with the working-side anterior temporalis (WAT), working-side posterior temporalis (WPT) and balancing-side superficial masseter (BSM) firing as an initial group followed by the working-side superficial masseter (WSM), balancing-side anterior temporalis (BAT) and balancing-side posterior temporalis (BPT) (Tables 3, 5). Given that we do not have EMG data from the medial pterygoids during chewing (i.e., the third muscle in both Triplet I and Triplet II), we cannot unequivocally state that treeshrew jaw-muscle activity patterns fire in two Triplets. The main difference between galagos and treeshrews for Triplet I is that the BSM and the balancing-side deep masseter (BDM) tend to peak before the working-side temporalis in galagos as compared to the reverse condition in treeshrews. Compared to anthropoids, treeshrews are similar to strepsirrhines in that their BDM peaks before their WSM. This difference in the timing of the peak activity of the BDM clearly distinguishes treeshrews and strepsirrhines with mobile, unfused symphyses from anthropoids (Table 5).

DISCUSSION

Jaw-Muscle EMG and Jaw Morphology in Treeshrews and Primates

Belanger's treeshrews, greater galagos and ring-tailed lemurs are more similar to each other in jaw-muscle EMG activity patterns as compared to this anthropoid sample. Given this broad resemblance, we find it worthwhile to ask whether treeshrew jaw morphology is also more similar to strepsirrhines than anthropoids. If so, then we can extend the associations among jaw-muscle activity patterns, inferentially internal jaw forces, and jaw form observed in comparisons of the primate suborders to include treeshrews (Hylander, 1979a,b; Hylander et al., 1998, 2000; Ravosa, 1991; Ravosa et al., 2000; Vinyard, 1999).

In Figure 5, we have added 8 treeshrew species to our previous comparisons of anthropoids to strepsirrhines with unfused symphyses. Treeshrews appear more similar to strepsirrhines than anthropoids in their relative corporal depth (Figure 5A), symphyseal area (Figure 5B) and condylar area (Figure 5C). Thus, treeshrews are more like strepsirrhines in both their jaw morphology and jaw-muscle activity patterns during chewing. This result further substantiates the link between jaw form and its load-bearing function during chewing. Based on the EMG and morphological comparisons, we predict that treeshrews will have relatively higher W/B corporal strain ratios than anthropoids during chewing.

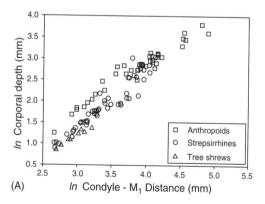

(A)

Figure 5. Comparison of load–resistance ability in the jaws of strepsirrhines with unfused symphyses, anthropoids, and treeshrews. (A) Plot of *ln* corporal depth versus the *ln* chewing moment arm (condyle—M1 distance) in 44 anthropoids, 47 strepsirrhines, and 8 treeshrew species.

(*Continued*)

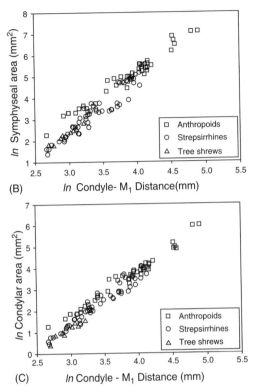

Figure 5. (*Continued*) (B) Plot of *ln* symphyseal area versus *ln* chewing moment arm length. (C) Plot of *ln* condylar area versus *ln* chewing moment arm length. treeshrews appear more similar to strepsirrhines with unfused symphyses than either does to anthropoids in all three plots (5a–5c). This result suggests that anthropoids can resist relatively greater chewing forces for a given chewing moment arm length when compared to both strepsirrhines and treeshrews.

Despite a highly plausible functional argument, we are careful to point out that this relationship is still only a correlation between jaw form and jaw loading patterns during chewing. Furthermore, this link may only be apparent at higher taxonomic levels and disappear when comparing more closely related species. It has been convincingly argued and empirically demonstrated that form and function need not share a law-like relationship in animals (Bock, 1977, 1989; Lauder, 1995, 1996). Without this invariant link, we must bear in mind that any inference of function from form in species lacking the appropriate *in vivo* data is a significant assumption that may not hold up on further analysis.

Establishing this correlation between jaw form and jaw-muscle activity during chewing across treeshrews and primates raises the question of whether we can extend these form–function relationships to include additional mammalian groups. While researchers have collected EMG data from the jaw muscles of other small mammals (e.g., Crompton et al., 1977; de Gueldre and de Vree, 1988; Dotsch and Dantuma, 1989; Kallen and Gans, 1972; Oron and Crompton, 1985), these data are not directly comparable to our results because of methodological differences in EMG data collection and analysis. More importantly, no one has systematically compared jaw morphologies of primates to those of other small mammals. Such comparisons would help us to better understand whether, and if so how, the functional morphology of strepsirrhine and anthropoid jaws differ from jaw forms in these other mammalian groups.

The overall similarity of treeshrew and strepsirrhine jaw-muscle activity patterns during chewing as compared to anthropoids mirrors observed differences in the morphology of their mandibular symphyses. Treeshrews, greater galagos and ring-tailed lemurs have relatively weak, unfused mandibular symphyses as compared to the fused symphyses of the anthropoid sample (Beecher, 1977b). Although we have described the symphysis as either fused or unfused throughout this chapter and we can reasonably interpret the relative strength and stiffness of the joint using such categorical terms, this characterization masks a continuous range of variation in the mechanical properties of the symphysis for resisting loads in various directions (Beecher, 1977a,b, 1979). We presently lack such mechanical information for primate mandibular symphyses.

Beecher (1977a) estimated the relative stress-resisting ability of the symphyses of several strepsirrhine species with either unfused or partially fused symphyses. We have scaled his estimates by the moment-arm length for chewing at the M_1 to compare these relative symphyseal strength estimates to the average W/B ratios for the masseters and temporalis in treeshrews, greater galagos and ring-tailed lemurs (Figure 6). We also have included the representative anthropoids in this plot by making the assumption that their fused symphyses are relatively stronger than the unfused symphyses in these strepsirrhines and treeshrews. Variation in the anthropoid values reflects the area of their symphyses scaled by this same chewing moment-arm length.

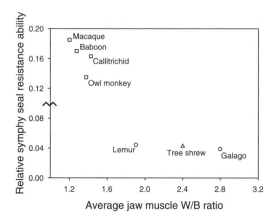

Figure 6. Plot of relative symphyseal resistance ability versus the average jaw-muscle W/B ratios for treeshrews and primates. There is a near stepwise relationship between increasing resistance ability and decreasing W/B ratios, both within anthropoid species and among treeshrews and strepsirrhines. This association suggests that increasing balancing-side muscle recruitment may be related to symphyseal strength among primates and treeshrews. Relative symphyseal resistance ability for strepsirrhines and treeshrews is based on data from Beecher (1977a) that has been scaled by the chewing moment arm length. In anthropoids, this estimate is calculated by dividing (symphyseal area) by the chewing moment arm length. The broken line along the y-axis signifies that the values above and below the break cannot be compared except at a categorical level. Thus, we assume that anthropoids have greater resistance ability than treeshrews and strepsirrhines with unfused symphyses. The x-axis is the average of the species means of W/B ratios for the four jaw muscles from Table 4.

There appears to be a consistent stepwise relationship between these two variables both across all taxa as well as within those species with fused and unfused symphyses. In fact, the rank correlation between these two variables is −0.92 across the seven groups (It is important to remember that there is a break in the y-axis between anthropoids and the remaining species in their relative strength estimates. A rank correlation is appropriate here if we assume that for a given size fused symphyses are stronger than unfused ones). This association provides further evidence linking relative balancing-side muscle forces to the strength of the mandibular symphysis. Furthermore, this plot fuels speculation that jaw-muscle force production and symphysis form may share a quantitative relationship across primates and treeshrews. We are currently measuring the strength and stiffness of primate and treeshrew symphyses under various loading regimes. We intend to use these quantitative strength estimates to determine how tightly these symphyseal properties are correlated with jaw-muscle EMG patterns. If the results of these analyses corroborate the pattern in Figure 6, then we suggest this correlation offers strong evidence linking symphyseal form, and hence strength, to relative balancing-side jaw-muscle forces during chewing.

Regardless of the outcome of these future tests, there is a categorical relationship between symphyseal fusion and jaw loading during chewing. This association between jaw-muscle EMG data and symphyseal morphology allows us to speculate on chewing behaviors in the earliest euprimates. If the earliest primates were small, had mobile unfused symphyses, and ate insects and fruits, then they probably chewed more like treeshrews and strepsirrhines than living anthropoids. More specifically, they likely recruited relatively less muscle force from their balancing-side jaw muscles during chewing and they did not elicit large amounts of transverse muscle force from their balancing-side deep masseters late in the power stroke. As our interpretation is based on the correlation between jaw form and jaw-muscle functions it arguably remains valid even if treeshrews turn out to not be the sister taxa of primates (see Kupfermann et al., 1999; Liu et al., 2001; Madsen et al., 2001; Murphy et al., 2001a,b; Scally et al., 2001). What is important to our argument is that treeshrews share broadly similar diets and jaw form with the earliest euprimates.

Jaw-Muscle EMG and the Conservation of Primate Masticatory Behaviors

In the early 1970s, Hiiemae and Kay analyzed jaw movements during chewing, tooth occlusal morphology and tooth wear in treeshrews, galagos, squirrel monkeys and spider monkeys (Hiiemae and Kay, 1972, 1973; Kay and Hiiemae, 1974). They concluded that primates share a similar pattern of jaw movement during chewing with a trend toward increased crushing and grinding in anthropoids.[6] Hiiemae (1978) went on to argue that mammals, including primates, share a broadly similar pattern of jaw-muscle activation during chewing (cf. Gorniak, 1985; Weijs, 1994). Langenbach and van Eijden (2001)

---

[6] These authors did stress, however, that jaw-movement patterns varied with the structural and mechanical properties of the foods as well a sthe position of a chewing cycle in a chewing sequence.

also have argued for a general uniformity of jaw-muscle activity patterns during chewing. Their argument implies that the structural changes in the jaws and teeth of recent primates have taken place within a behavioral framework established early in primate evolution (Hiiemae, 1984; Hiiemae and Kay, 1972, 1973; Kay and Hiiemae, 1973).

We are now beginning to accumulate jaw-muscle EMG data from enough primate species to realistically evaluate whether primates share broadly similar jaw-muscle activation patterns during chewing. Prior to discussing this hypothesis, we need to make clear that similar jaw movements during chewing may be created by different jaw-muscle activity patterns. Thus, even if primates move their jaws in the same way (a proposal requiring further testing), we still could see different jaw-muscle activity patterns creating these movements. We also want to clarify that primates move their jaws in orbits during chewing. At this level all primates are alike. This shared similarity, however, is likely a physical requirement for repeated chewing cycles and as such is not particularly interesting for studying the evolution of primate chewing.

Examination of the W/B ratios indicates that treeshrews and primates show marked variation among species in the relative magnitudes of jaw-muscle activity on the working-versus balancing-side during chewing (Table 4). The systematic differences in W/B ratios in anthropoids versus strepsirrhines and treeshrews suggest that this variation may characterize larger primate clades. Given that we have discussed W/B ratios at length above, we will not consider them further except to emphasize that all primates do not appear to recruit similar magnitudes of relative working-and balancing-side jaw-muscle forces during chewing.

The temporalis muscles show the most consistency across primates and treeshrews in peak firing times. The working-side temporalis peaks on average before the balancing-side temporalis in all of the taxa in Table 5. The working-side temporalis typically peaks at about the same time as the balancing-side superficial masseter with the possible exception of platyrrhines. The balancing-side temporalis usually peaks late in the power stroke typically after the peak of the working-side superficial masseter. Again with the potential exception of the South American primates, the balancing-side superficial masseter typically peaks before the working-side superficial masseter in primates and treeshrews. The consistency of temporalis peak firing times across primates is interesting given that it is also the largest jaw-closing muscle in primates (Cachel, 1979; Turnbull, 1970).

Callitrichids and owl monkeys appear to have a jaw-muscle firing pattern that does not always follow this triplet pattern described above. In some individuals, the balancing-side superficial masseter (BSM) peaks on average slightly after the working-side superficial masseter (WSM). These peaks, however, are quite close in time suggesting that a more realistic interpretation is that on average the two muscles are peaking at about the same time. We cannot say presently whether the triplet pattern has been altered so that the BSM is delayed or the WSM is advanced in these animals. If this pattern is present in other platyrrhines, then it suggests that platyrrhines have evolved a distinct jaw-muscle firing pattern.

The deep masseter shows the greatest variation in peak firing times across primates and treeshrews (Table 5). The working-side deep masseter (WDM) peaks quite early in ring-tailed lemurs and cercopithecoids. The WDM peaks much closer to the working-side superficial masseter (WSM) in galagos and treeshrews and close to the working-side temporalis in platyrrhines. Furthermore, the balancing-side deep masseter (BDM) is the first muscle to peak on average in galagos. Although the BDM does not peak first in treeshrews and lemurs, it does on average peak well before the WSM in these species. In contrast, the BDM peaks after both the working-side and balancing-side superficial masseters in baboons, macaques, owl monkeys and callitrichids as part of the "wishboning" firing pattern (Hylander and Johnson, 1994; Hylander et al., 2000). Finally, humans do not appear to show this relatively late peak of the balancing-side deep masseter (van Eigden et al., 1993).

Mammals appear to be extremely labile in their deep masseter firing patterns relative to the other jaw-closing muscles. If we assume that the BDM peaked before the WSM in primitive mammals, then the late peak activity of the BDM appears to have evolved at least six times in mammals. EMG data indicate that rabbits (Weijs and Dantuma, 1981), pigs (Herring and Scapino, 1973; Huang et al., 1994), alpacas (Williams et al., 2003), horses (Williams et al., 2003), anthropoids (Hylander and Johnson, 1994; Hylander et al., 2000), and sifakas (Hylander et al., 2003) exhibit this late-peak activity in the BDM. Based on the EMG data from van Eijden et al. (1993), humans may have evolved an early average peak of the BDM (i.e., before the WSM) from an anthropoid ancestor exhibiting a late peaking BDM. The alternative interpretation of the human deep masseter EMG data is that New World and Old World monkeys convergently evolved a late-peaking BDM (cf. Ravosa, 1999).

The interspecific variation in firing times for the deep masseters as well as the convergent evolution of the late peak activity of the BDM is likely related to transverse jaw forces during the power stroke (Herring and Scapino, 1979; Hylander and Johnson, 1994; Hylander et al., 2000; Ravosa et al., 2000; Weijs, 1994). The orientation of the deep masseters tends to have a relatively large transverse component of pull during contraction as compared to most of the other jaw adductors. Thus, the deep masseters may be more effective in creating transverse jaw movements and/or transverse occlusal forces during chewing. Alternatively, the deep masseters may be freer to vary their recruitment patterns than the other jaw adductors. If so, then the remaining jaw adductors may be constrained to produce vertical and/or transverse forces and jaw movements at specific times during the chewing cycle. The deep masseters may be acting to modify existing jaw movements and/or occlusal forces during chewing, particularly those in a transverse plane (e.g., Hylander and Johnson, 1994; Ravosa et al., 2000).

If we accept treeshrew jaw-muscle activity patterns as primitive for primates, then we can reconstruct early primates as firing their superficial and deep masseters as a more or less single unit. Thus, primates may have evolved several modifications of their deep masseter firing patterns. The early peak activity of the WDM may be functionally related to positioning the mandibular molars for the upcoming power stroke. The early peak activity of the WDM in ring-tailed lemurs indicates that this WDM pattern can evolve in species with unfused symphyses.

Functionally, this observation suggests that transverse jaw movements during the closing stroke may not involve significant stresses at the mandibular symphysis. In anthropoids, the changes in the deep masseter-firing pattern have been argued to reflect an increased emphasis on transverse jaw movements and/or force production during the power stroke (Hylander and Johnson, 1994; Hylander et al., 2000; Ravosa et al., 2000). Regardless of the functional interpretation, the timing of peak activity in the primate deep masseter appears to be easily and routinely uncoupled from the superficial masseter and the other jaw adductors throughout primate evolution. This observation may be characteristic of mammals in general.

In summary, there are distinct differences in jaw-muscle activity patterns among these primate species. We anticipate that studying additional primate species will likely add to this variation. Based on the existing EMG data, we find it difficult to conclude that primates have similar jaw-muscle activity patterns during chewing and that all primates chew in a similar way. Having said this, it is clear that certain muscles, such as the temporalis, show broadly similar activity patterns during chewing across primates and treeshrews. Hiiemae and Kay's (1972, 1973) assertion may be correct for specific aspects of primate chewing behaviors—such as this similarity in temporalis timing patterns. It is probably misleading, however, to characterize primate-chewing behavior as similar because it masks the observed interspecific variation in EMG activity patterns in other muscles. We must appreciate this variation in order to understand the evolution of chewing behaviors in specific primate clades. Furthermore, it remains to be seen whether these behavioral modifications are correlated with morphological changes in primate jaws.

Mastication in the First Primates: *In vivo* Evidence from Treeshrews and Primates

We speculated above based on an observed association between jaw form and function that if primitive primates were small, had unfused symphyses, and ate insects and fruits, then they probably chewed more like these strepsirrhines and treeshrews than living anthropoids. The earliest adapids, such as *Donrussellia,* and the earliest omomyids, such as *Teilhardina,* appear to have been small, fruit and insect eating animals with unfused symphyses (e.g., Covert, 1986; Covert and Williams, 1994; Rose et al., 1994; Strait, 2001). Thus, we hypothesize that the earliest euprimates fit the morphological and behavioral pattern found in the masticatory apparatus of treeshrews and galagos.

Our *in vivo* data on jaw-muscle activity during chewing support the hypothesis that the origin of primates did not involve major adaptive changes in the masticatory apparatus. Indirectly, this result supports the arguments that the origin of primates involved evolutionary changes in the locomotor skeleton (Washburn, 1950) or that it did not involve major adaptive changes in primate form (Cain, 1954; Davis, 1955; Simpson, 1955, 1961; Straus, 1949; Zuckerman, 1933). In other words, we presently see little evidence for a major shift in chewing behavior between treeshrews and the two strepsirrhine taxa. We base this conclusion on the observation that Belanger's treeshrews are more similar to greater galagos and ring-tailed lemurs than anthropoids in their jaw-muscle activity patterns during chewing. Additionally, treeshrew jaw form appears more like that of strepsirrhines with unfused symphyses than anthropoids (Figure 5).

Conversely, our results do not clearly support Szalay's (1968, 1969, 1972, 1973) hypothesis that the origin of primates marked an adaptive shift toward increased feeding on fruits and plants. If we had observed a systematic difference in jaw-muscle activity patterns between treeshrews and primates, then this hypothesis would have been strongly supported. The similarity of both jaw-muscle W/B ratios and peak firing times in treeshrews and the two strepsirrhines suggests that the origin of primates (or at least strepsirrhines) did not involve significant changes in jaw-muscle activity patterns. As stated earlier, the comparison of jaw-muscle activity patterns does not offer a direct test of Szalay's hypothesis because variation in diets is not directly linked to variation in jaw-muscle EMGs. Thus, we are not rejecting Szalay's hypothesis, but rather suggesting that if an adaptive shift toward increased consumption of plant foods occurred at the origin of primates, then this dietary change likely took place without major changes in jaw-muscle activity patterns during chewing. Because our results cannot directly address potential changes in jaw form at the origin of primates, Szalay's hypothesis deserves further testing by comparing jaw morphology in primates and closely related nonprimates.

The two most recent adaptive explanations of primate origins, the visual predation hypothesis (Cartmill, 1972, 1974) and the angiosperm coevolution hypothesis (Sussman, 1991), both involve feeding adaptations, but deal more with acquiring rather than chewing foods. The visual predation hypothesis focuses on spying and catching quick, agile insects on slender branches, while the angiosperm coevolution hypotheses concentrate on early primates' ability to feed on the rich repository of plant materials in the terminal branches of trees. Neither hypothesis stresses dietary shifts in the origin of primates, but rather each emphasizes either insects or plants over the other. Our results suggest that if these predicted changes in food acquisition occurred in the earliest primates (e.g., Cartmill, 1972, 1974; Sussman, 1991), then they did so without significant changes in chewing behavior. Furthermore, our results are in agreement with the scenario provided by Rasmussen (1990) where early primates fed on both plants and insects in the terminal branches.

The differences in jaw-muscle activity patterns and jaw form between living anthropoids as compared to treeshrews and strepsirrhines suggest that the origin of crown anthropoids is associated with significant changes in chewing behavior and jaw morphology. Thus, a major evolutionary change in chewing appears to have occurred along the lineage leading to living anthropoids rather than during the origin of primates. We are not claiming that specific strepsirrhine clades, such as the subfossil lemurs or indriids, did not undergo significant changes in chewing behaviors during their evolution. Rather it appears that among the two primate suborders, the origin of living anthropoids is more likely to have involved major restructuring of the masticatory apparatus (e.g., Hylander, 1979b; Ravosa, 1991; Ravosa et al., 2000; Rosenberger, 1986). Future work will determine whether these changes occurred in parallel among platyrrhines and catarrhines and to what extent this restructuring of the masticatory apparatus is related to functional and morphological changes in other organ systems housed in the skull.

CONCLUSIONS

Feeding behaviors have played a major role in the leading hypotheses of primate origins. Given this, we find it surprising that no one has systematically compared primate chewing behaviors and masticatory apparatus form, outside of the teeth, to closely related nonprimates. We find that treeshrew jaw-muscle activity patterns are more similar to representative strepsirrhines than either is to anthropoids. We argue, based on this similarity, that there is little evidence for a major shift in chewing behavior at the origin of primates. This finding suggests that if the origin of primates involved the evolution of derived traits for exploiting new foods in terminal branches (e.g., Sussman, 1991) and/or new food gathering techniques (Cartmill, 1972, 1974), then these behavioral changes occurred without altering the chewing behaviors of the earliest primates.

ACKNOWLEDGMENTS

We thank D. Fitzpatrick and M. Pucak for allowing us to work with the treeshrews under their care and Shannon Mangum for her daily efforts in helping us handle the animals. We thank the following institutions and individuals for providing access to skeletal materials: Field Museum of Natural History – L. Heaney, B. Patterson, W. Stanley; National Museum of Natural History – L. Gordon, R. Thorington; American Museum of Natural History – R. MacPhee; Natural History Museum, London – P. Jenkins; Museum National d'Histoire Naturelle – J. Cuisin; Naturhistorisches Museum Basel – M. Sutermeister, F. Weidenmayer. We thank P. Lemelin, D. Schmitt and P. Vinyard for numerous helpful discussions and comments that greatly improved this manuscript. Funding for the EMG data collection was provided by NSF (SBR-9420764; BCS-0094666; BCS-01-38565). Funding for the morphometric data collection was provided by Sigma-Xi, the Boise Fund, the AMNH, NSF (SBR-9701425) and the L.S.B. Leakey Foundation. This is Duke University Primate Center Publication #785.

REFERENCES

Agrawal, K. R., Lucas, P. W., Bruce, I. C., and Prinz, J. F., 1998, Food properties that influence neuromuscular activity during human mastication, *J. Dent. Res.* **77**: 1931–1938.

Ahlgren, J., 1966, Mechanism of mastication: A quantitative cinematographic and electromyographic study of masticatory movements in children, with special reference to occlusion of the teeth, *Acta Odontol. Scand.* **24**: 1–109.

Ashton, E. H., and Zuckerman, S., 1954, The anatomy of the articular fossa (fossa mandibularis) in man and apes, *Am. J. Phys. Anthropol.* **12**: 29–62.

Basmajian, J. V., 1978, *Muscles Alive,* Williams & Wilkins, Baltimore.

Beecher, R. M., 1977a, *Functional Significance of the Mandibular Symphysis,* Ph. D. Thesis, Duke University.

Beecher, R. M., 1977b, Function and fusion of the mandibular symphysis, *Am. J. Phys. Anthropol.* **47**: 325–336.

Beecher, R. M., 1979, Functional significance of the mandibular symphysis, *J. Morphol.* **159**: 117–130.

Beer, F. P., and Johnston, E. R., 1977, *Vector Mechanics for Engineers,* McGraw-Hill, New York.

Bennett, N. G., 1908, A contribution to the study of the movements of the mandible, *Proc. Roy. Soc. Med. (Odontol. Sect.),* **1**: 79–98.

Biegert, J., 1956, Das Kiefergelenk der Primaten, *Geg. Morphol. Jahrb.* **97**: 249–404.

Biegert, J., 1963, The evaluation of characteristics of the skull, hands, and feet for primate taxonomy, in: *Classification and Human Evolution,* S. L. Washburn, ed., Wenner-Gren Foundation, New York, pp. 116–145.

Biewener, A. A., 1982, Bone strength in small mammals and bipedal birds: Do safety factors change with body size? *J. Exp. Biol.* **98**: 289–301.

Biewener, A. A., 1989, Scaling body support in mammals: Limb posture and muscle mechanics, *Science* **245**: 45–48.

Biewener, A. A., 1990, Biomechanics of mammalian terrestrial locomotion, *Science* **250**: 1097–1103.

Biewener, A. A., 1992, Overview of structural mechanics, in: *Biomechanics: A Practical Approach,* A. A. Biewener, ed., OIRL Press, Oxford, pp. 1–20.

Bock, W. J., 1977, Adaptation and the comparative method, in: *Major Patterns in Vertebrate Evolution,* M. K. Hecht, P. C. Goody, and B. M. Hecht, eds., Plenum Press, New York, pp. 57–82.

Bock, W. J., 1989, Principles of biological comparison, *Acta Morphol. Neerl. Scand.* **27:** 17–32.

Bouvier, M., 1986, A biomechanical analysis of mandibular scaling in Old World monkeys, *Am. J. Phys. Anthropol.* **69:** 473–482.

Bouvier, M., and Hylander, W. L., 1981, Effect of bone strain on cortical bone structure in macaques (*Macaca mulatta*), *J. Morphol.* **167:** 1–12.

Boyd, R. L., Gibbs, C. H., Mahan, P. E., Richmond, A. F., and Laskin, J. L., 1990, Temporomandibular joint forces measured at the condyle *of Macaca arctoides, Am. J. Orthod. Dentofac. Orthop.* **97:** 472–479.

Brehnan, K., Boyd, R. L., Gibbs, C. H., and Mahan, P., 1981, Direct measurement of loads at the temporomandibular joint in *Macaca arctoides, J. Dent. Res.* **60:** 1820–1824.

Butler, P. M., 1980, The tupaiid dentition, in: *Comparative Biology and Evolutionary Relationships of treeshrews,* W. P. Luckett, ed., Plenum Press, New York, pp. 171–204.

Cachel, S. M., 1979, A functional analysis of the primate masticatory system and the origin of the anthropoid post-orbital septum, *Am. J. Phys. Anthropol.* **50:** 1–18.

Cain, A. J., 1954, *Animal Species and Their Evolution,* Hutchinson, London.

Campbell, B., 1974, *Human Evolution,* Aldine, New York.

Carlsöö, S., 1952, An electromyographic study of the activity, and an anatomic analysis of the mechanics of the lateral pterygoid muscle, *Acta Anat.* **26:** 339–348.

Cartmill, M., 1972, Arboreal adaptations and the origin of the Order Primates, in: *The Functional and Evolutionary Biology of Primates,* R. Tuttle, ed., Aldine-Atherton, Chicago, pp. 97–122.

Cartmill, M., 1974, Rethinking primate origins, *Science* **184:** 436–443.

Cartmill, M., 1992, New views on primate origins, *Evol. Anthropol.* **1:** 105–111.

Chew, C. L., Lucas, P. W., Tay, D. K. L., Keng, S. B., and Ow, R. K. K., 1988, The effect of food texture on the replication of jaw movements in mastication, *J. Dent.* **16:** 210–214.

Clark, R. W., Luschei, E. S., and Hoffman, D. S., 1978, Recruitment order, contractile characteristics, and firing patterns of motor units in the temporalis muscle of monkeys, *Expl. Neurol.* **61:** 31–52

Covert, H. H., 1986, Biology of early Cenozoic primates, in: *Comparative Primate Biology: Systematics, Evolution, and Anatomy,* D. R. Swindler and J. Erwin, eds., vol. 1, pp. 335–359, Alan R. Liss, New York.

Covert, H. H., and Williams, B. A., 1994, Recently recovered specimens of North American Eocene omomyids and adapids and their bearing on debates about anthropoid origins, in: *Anthropoid Origins,* J. G. Fleagle and R. F. Kay, eds., Plenum Press, New York, pp. 29–54.

Crompton, A. W., Thexton, A. J., Parker, P., and Hiiemae, K., 1977, The activity of the jaw and hyoid musculature in the Virginia opossum, *Didelphis virginiana,* in: *The Biology of Marsupials,* B. Stonehouse and D. Gilmore, eds., University Park Press, Baltimore, pp. 287–305.

Daegling, D. J., 1989, Biomechanics of cross-sectional size and shape in the hominoid mandibular corpus, *Am. J. Phys. Anthropol.* **80**: 91–106.

Daegling, D. J., 1992, Mandibular morphology and diet in the genus *Cebus, Int. J. Primatol.* **13**: 545–570.

Daegling, D. J., 2001, Biomechanical scaling of the hominoid mandibular symphysis, *J. Morphol.* **250**: 12–23.

Daegling, D. J., 2002, Bone geometry in cercopithecoid mandibles, *Arch. Oral Biol.* **47**: 315–325.

Daegling, D. J., and Grine, F. E., 1991, Compact bone distribution and biomechanics of early hominid mandibles, *Am. J. Phys. Anthropol.* **86**: 321–339.

Daegling, D. J., and Hylander, W. L., 2000, Experimental observation, theoretical models, and biomechanical inference in the study of mandibular form, *Am. J. Phys. Anthropol.* **112**: 541–551.

Daegling, D. J., and McGraw, W. S., 2001, Feeding, diet, and jaw form in West African *Colobus* and *Procolobus, Int. J. Primatol.* **22**: 1033–1055.

Davis, D. D., 1955, Primate evolution from the viewpoint of comparative anatomy, in: *The Nonhuman Primates and Human Evolution*, J. A. Gavan, ed., Wayne State University Press, Detroit, pp. 33–41.

Dechow, P. C., and Hylander, W. L., 2000, Elastic properties and masticatory bone stress in the macaque mandible, *Am. J. Phys. Anthropol.* **112**: 553–574.

de Gueldre, G., and de Vree, F., 1988, Quantitative electromyography of the masticatory muscles of *Pteropus giganteus* (Megachiroptera), *J. Morphol.* **196**: 73–106.

de Vree, F., and Gans, C., 1974, Mastication in Pygmy goats *"Capra hircus,"* Ann. *Soc. Roy. Zool. Belg.* **105**: 255–306.

Demes, B., and Creel, N., 1988, Bite force, diet, and cranial morphology of fossil hominids, *J. Hum. Evol.* **17**: 657–670.

Dessem, D., and Druzinsky, R. E., 1992, Jaw-muscle activity in ferrets, *Mustela putorius furo, J. Morphol.* **213**: 275–286.

Dotsch, C., and Dantuma, R., 1989, Electromyography and masticatory behavior in shrews (Insectivora), in: *Progress in Zoology: Trends in Vertebrate Morphology,* H. Splechtna and H. Hilgers, eds., Gustav Fischer, New York, pp. 146–147.

DuBrul, E. L., 1977, Early hominid feeding mechanisms, *Am. J. Phys. Anthropol.* **47**: 305–320.

DuBrul, E. L., 1988, *Sicker and DuBrul's Oral Anatomy*, Ishiyaku Euro America, St. Louis.

Efron, B., 1979, Bootstrap methods: Another look at the jacknife. *Ann Stats.* **7:** 1–26.

Emmons, L. H., 1991, Frugivory in treeshrews (*Tupaia*), *Am. Nat.* **138:** 642–649.

Emmons, L. H., 2000, *Tupai: A Field Study of Bornean Treeshrews*, University of California Press, Berkeley.

Finney, D. J., 1941, On the distribution of a variate whose logarithm is normally distributed, *J. Roy. Stat. Soc. Lond.* **222A:** 309–368.

Fish, D. R., 1983, Aspects of masticatory form and function in common treeshrews, *Tupaia glis, J. Morphol.* **176:** 15–29.

Fish, D. R., and Mendel, F. C., 1982, Mandibular movement patterns relative to food types in common treeshrews (*Tupaia glis*), *Am. J. Phys. Anthropol.* **58:** 255–269.

Gans, C., de Vree, F., and Gorniak, G. C., 1978, Analysis of mammalian masticatory mechanisms: Progress and problems, *Zbl. Vet. Med. C. Anat. Histol. Embryol.* **7:** 226–244.

Gibbs, C. H., Meserman, T., Reswick, J. B., and Derda, H. J., 1971, Functional movements of the mandible, *J. Prosth. Dent.* **26:** 604–620.

Gorniak, G. C., 1985, Trends in the actions of mammalian masticatory muscles, *Am. Zool.* **25:** 331–337.

Gorniak, G. C., and Gans, C., 1980, Quantitative assay of electromyograms during mastication in domestic cats (*Felis catus*), *J. Morphol.* **163:** 253–281.

Gregory, W. K., 1910, The orders of mammals, *Bull. Am. Mus. Nat. Hist.* **27:** 1–524.

Gregory, W. K., 1920, The origin and evolution of the human dentition, Parts I–IV, *J. Dent. Res.* **2:** 89–186, 215–284, 357–466, 607–718.

Hannam, A. G., and Wood, W. W., 1981, Medial pterygoid muscle activity during the closing and compressive phases of human mastication, *Am. J. Phys. Anthropol.* **55:** 359–367.

Harrison, G. A., Weiner, J. S., and Reynolds, V., 1977, Human evolution, in: *Human Biology*, G. A. Harrison, J. S. Weiner, J. M. Tanner, and N. A. Barnicot, eds., Oxford University Press, London, pp. 3–96.

Herring, S. W., 1985, Morphological correlates of masticatory patterns in peccaries and pigs, *J. Mammal.* **66:** 603–617.

Herring, S. W., and Scapino, R. P., 1973, Physiology of feeding in miniature pigs, *J. Morphol.* **141:** 427–460.

Herring, S. W., Grimm, A. F., and Grimm, B. R., 1979, Functional heterogeneity in a multipinnate muscle, *Am. J. Anat.* **154:** 563–576.

Hiiemae, K. M., 1976, Masticatory movements in primitive mammals, in: *Mastication*, D. J. Anderson and B. Matthews, eds., Wright and Sons, Bristol, pp. 105–118.

Hiiemae, K. M., 1978, Mammalian mastication: A review of the activity of the jaw muscles and the movements they produce in chewing, in: *Development, Function and Evolution of Teeth*, P. M. Butler and K. A. Joysey, eds., Academic Press, New York, pp. 359–398.

Hiiemae, K., 1984, Functional aspects of primate jaw morphology, in: *Food Acquisition and Processing in Primates*, D. J. Chivers, B. A. Wood and A. Bilsborough, eds., Plenum Press, New York, pp. 257–281.

Hiiemae, K. M., and Crompton, A. W., 1985, Mastication, food transport, and swallowing, in: *Functional Vertebrate Morphology*, D. B. M. Hildebrand, K. Liem, and D. Wake, eds., Harvard University Press, Cambridge, pp. 262–290.

Hiiemae, K. M., and Kay, R. F., 1972, Trends in the evolution of primate mastication, *Nature* **240**: 486–487.

Hiiemae, K. M., and Kay, R. F., 1973, Evolutionary trends in the dynamics of primate mastication, *Craniofacial Biology of Primates*, Symposia of the 4th International Congress of Primatology, vol. 3, S. Karger, Basel, pp. 28–64.

Huang, X., Zhang, G., and Herring, S. W., 1994, Age changes in mastication in the pig. *Comp. Biochem. Physiol.* **107A**: 647–654.

Hylander, W. L., 1977, In vivo bone strain in the mandible *of Galago crassicaudatus*, *Am. J. Phys. Anthropol.* **46**: 309–326.

Hylander, W. L., 1979a, Mandibular function in *Galago crassicaudatus and Macaca fascicularis*: An in vivo approach to stress analysis, *J. Morphol.* **159**: 253–296.

Hylander, W. L., 1979b, The functional significance in primate mandibular form, *J. Morphol.* **160**: 223–240.

Hylander, W. L., 1979c, An experimental analysis of temporomandibular joint reaction forces in macaques, *Am. J. Phys. Anthropol.* **51**: 433–456.

Hylander, W. L., 1984, Stress and strain in the mandibular symphysis of primates: A test of competing hypotheses, *Am. J. Phys. Anthropol.* **61**: 1–46.

Hylander, W. L., 1985, Mandibular function and biomechanical stress and scaling, *Am. Zool.* **25**: 315–330.

Hylander, W. L., 1988, Implications of *in vivo* experiments for interpreting the functional significance of 'robust' australopithecines jaws, in: *Evolutionary History of the Robust Australopithecines*, F. E. Grine, ed., Gruyter Press, New York, pp. 55–83.

Hylander, W. L., and Bays, R., 1979, An *in vivo* strain-gauge analysis of the squamosal-dentary joint reaction force during mastication and incisal biting in *Macaca mulatta and Macaca fascicularis*, *Arch. Oral Biol.* **24**: 689–697.

Hylander, W. L. and Crompton, A. W., 1980, Loading patterns and jaw movement during the masticatory power stroke in macaques, *Am. J. Phys. Anthropol.* **52**: 239.

Hylander, W. L., and Crompton, A. W., 1986, Jaw movements and patterns of mandibular bone strain during mastication in the monkey *Macaca fascicularis*, *Arch. Oral Biol.* **31**: 841–848.

Hylander, W. L., and Johnson, K. R., 1985, Temporalis and masseter muscle function during incision in macaques and humans, *Int. J. Primatol.* **6**: 289–322.

Hylander, W. L., and Johnson, K. R., 1989, The relationship between masseter force and masseter electromyogram during mastication in the monkey *Macaca fascicularis, Arch. Oral Biol.* **34**: 713–722.

Hylander, W. L., and Johnson, K. R., 1994, Jaw muscle function and wishboning of the mandible during mastication in macaques and baboons, *Am. J. Phys. Anthropol.* **94**: 523–547.

Hylander, W. L., and Johnson, K. R., 1997, In vivo bone strain patterns in the zygomatic arch of macaques and the significance of these patterns for functional patterns of craniofacial form, *Am. J. Phys. Anthropol.* **102**: 203–232.

Hylander, W. L., Johnson, K. R., and Crompton, A. W., 1987, Loading patterns and jaw movements during mastication in *Macaca fascicularis:* A bone-strain, electromyographic, and cineradiographic analysis, *Am. J. Phys. Anthropol.* **72**: 287–314.

Hylander, W. L., Johnson, K. R., and Crompton, A. W., 1992, Muscle force recruitment and biomechanical modeling: An analysis of masseter muscle function during mastication in *Macaca fascicularis, Am. J. Phys. Anthropol.* **88**: 365–387.

Hylander, W. L., Ravosa, M. J., Ross, C. F., and Johnson, K. R., 1998, Mandibular corpus strain in primates: Further evidence for a functional link between symphyseal fusion and jaw-adductor muscle force, *Am. J. Phys. Anthropol.* **107**: 257–271.

Hylander, W. L., Ravosa, M. J., Ross, C. F., Wall, C. E., and Johnson, K. R., 2000, Symphyseal fusion and jaw-adductor muscle force: An EMG study, *Am. J. Phys. Anthropol.* **112**: 469–492.

Hylander, W. L., Vinyard, C. J., Wall, C. E., Williams, S. H., and Johnson, K. R., 2002, Recruitment and firing patterns of jaw muscles during mastication in ring-tailed lemurs, *Am J. Phys. Anthropol.* **34** (Suppl.): 88

Hylander W. L., Vinyard, C. J., Wall C. E., Williams S. H., and Johnson, K. R., 2003, Convergence of the "wishboning" jaw-muscle activity pattern in anthropoids and strepsirrhines: The recruitment and firing of jaw muscles in *Propithecus verreauxi. Am. J. Phys. Anthropol. Suppl.* **36**: 120.

Hylander, W. L., Wall, C. E., Vinyard, C. J., Ross, C. F., Ravosa, M. J., Williams, S. H., and Johnson, K. R., 2005, Temporalis function in anthropoids and strepsirrhines: An EMG study. *Am J. Phys. Anthropol.* **128**: 35–56.

Jablonski, N. G., 1986, A history of form and function in the primate masticatory apparatus from the ancestral primate through the Strepsirhines, in: *Comparative Primate Biology: Systematics, Evolution, and Anatomy,* D. R. Swindler and J. Erwin, eds., vol. 1, Alan R. Liss, New York, pp. 537–558.

Jolly, C. J., 1970, The seed-eaters: A new model of hominid differentiation based on a baboon model, *Man* **5**: 5–26.

Jones, F. W., 1917, *Arboreal Man,* Edward Arnold, London.

Kallen, F. C., and Gans, C., 1972, Mastication in the little brown bat, *Myotis lucifugus, J. Morphol.* **136:** 385–420.

Kay, R. F., 1975, The functional adaptations of primate molar teeth, *Am. J. Phys. Anthropol.* **43:** 195–216.

Kay, R. F., 1978, Molar structure and diet in extant Cercopithecidae, in: *Development, Function and Evolution of Teeth,* P. M. Butler and K. A. Joysey, eds., Academic Press, New York, pp. 309–339.

Kay, R. F., and Cartmill, M., 1977, Cranial morphology and adaptations *of Palaecthon nacimienti* and other Paromomyidae (Plesiadapoidea, ? Primates), with a description of new genus and species, *J. Hum. Evol.* **6:** 19–53.

Kay, R. F., and Hiiemae, K. M., 1974, Jaw movement and tooth use in recent and fossil primates, *Am. J. Phys. Anthropol.* **40:** 227–256.

Kay, R. F., Sussman, R. W., Tattersall, I., 1978, Dietary and dental variation in the genus *Lemur,* with comments concerning dietary-dental correlations among Malagasy primates, *Am. J. Phys. Anthropol.* **49:** 119–128.

Kinzey, W. G., 1974, Ceboid models for the evolution of hominoid dentition, *J. Hum. Evol.* **3:** 193–203.

Kupfermann, H., Satta, Y., Takahata, N., Tichy, H., and Klein, J., 1999, Evolution of *Mhc-DRB* introns: Implications for the origin of primates. *J. Mol. Evol.* **48:** 663–674.

Langenbach, G. E. J., and van Eijden, T. M. G. J., 2001, Mammalian feeding motor patterns, *Am. Zool.* **41:** 1338–1351.

Lanyon, L. E., and Rubin, C. T., 1985, Functional adaptation in skeletal structures, in: *Functional Vertebrate Morphology,* D. B. M. Hildebrand, K. Liem, and D. Wake, eds., Harvard University Press, Cambridge, pp. 1–25.

Lauder, G. V., 1995, On the inference of function from structure, in: *Functional Morphology in Vertebrate Paleontology,* J. J. Thomason, ed., Cambridge University Press, Cambridge, pp. 1–18.

Lauder, G. V., 1996, The argument from design, in: *Adaptation,* M. R. Rose and G. V. Lauder, eds., Academic Press, New York, pp. 55–91.

Laurent, A. G., 1963, The lognormal distribution and the translation method: Description and estimation problems, *J. Am. Stat. Assoc.* **58:** 231–235.

Le Gros Clark, W. E., 1924, The myology of the tree-shrew (*Tupaia minor*), *Proc. Zool. Soc. Lond.* **1924:** 1053–1074.

Le Gros Clark, W. E., 1925, On the skull *of Tupaia, Proc. Zool. Soc. Lond.* **1925:** 559–567.

Le Gros Clark, W. E., 1927, On the treeshrew, *Tupaia minor, Proc. Zool. Soc. Lond.* **1927:** 254–256.

Le Gros Clark, W. E., 1959a, *The Antecedents of Man,* Harper & Row, New York.

Le Gros Clark, W. E., 1959b, *History of the Primates,* University of Chicago Press, Chicago.

Lindblom, G., 1960, On the anatomy and function of the temporomandibular joint, *Acta Odontol. Scand. Suppl.* **28**: 1–287.

Linden, R. W., 1998, The Scientific Basis of Eating: Taste and Smell, Salivation, Mastication and Swallowing and their Dysfunctions, Karger, New York.

Liu, E. G., Miyamoto, M. M., Freire, N. P., Ong, P. Q., Tennant, M. R., Young, T. S., and Gugel, K. F., 2001, Molecular and morphological supertrees for eutherian (placental) mammals, *Science* **291**: 1786–1789.

Loeb, G. E., and Gans, C., 1986, *Electromyography for Experimentalists,* University of Chicago Press, Chicago.

Lucas, P. W., 1979, The dental-dietary adaptations of mammals, *N. Jahr. Geol. Palaont. Mh.* **8**: 486–512.

Lucas, P. W., and Luke, D. A., 1984, Chewing it over: Basic principles of food breakdown, in: *Food Acquisition and Processing in Primates,* D. J. Chivers, B. A. Wood, and A. Bilsborough, eds., Plenum Press, New York, pp. 283–301.

Lucas, P. W, Peter, C. R., 2000, Function of postcanine tooth crown shape in mammals, in: *Development, Function and Evolution of Teeth,* M. F. Teaford, M. M. Smith, and M. W. J. Ferguson, eds., Cambridge University Press, Cambridge, pp. 282–289.

Lucas, P. W., and Teaford, M. F., 1994, Functional morphology of colobine teeth, in: *Colobine Monkeys: Their Ecology, Behavior and Evolution,* A. G. Davies and J. F. Gates, eds., Cambridge University Press, Cambridge, pp. 173–203.

Luschei, E. S., and Goodwin, G. S., 1974, Patterns of mandibular movement and jaw muscle activity during mastication in the monkey, *J. Neurophysiol.* **37**: 954–966.

Madsen, O., Scally, M., Douady, C. J., Kao, D. J., DeBry, R. W., Adkins, R., Amrine, H. M., Stanhope, M. J., de Jong, W. W., and Springer, M. S., 2001, Parallel adaptive radiations in two major clades of placental mammals, *Nature* **409**: 610–614.

Manly, B. F. J., 1997, *Randomization, Bootstrap and Monte Carlo Methods in Biology,* Chapman & Hall, London.

Martin, R. D., 1990, *Primate Origins and Evolution: A Phylogenetic Reconstruction,* Princeton University Press, Princeton.

Martin, R. D., 1993, Primate origins: Plugging in the gaps, *Nature* **363**: 223–234.

McKenna, M. C., 1966, Paleontology and the origin of primates, *Folia Primatol.* **4**: 1–25.

McNamara, J. A., 1974, An electromyographic study of mastication in the rhesus monkey (*Macaca mulatta*), *Arch. Oral Biol.* **19**: 821–823.

Mills, J. R. E., 1955, Ideal dental occlusion in the primates, *Dent. Pract.* **6**: 47–63.

Mills, J. R. E., 1963, Occlusion and malocclusion of the teeth of primates, in: *Dental Anthropology,* D. R. Brothwell, ed., Pergamon Press, New York, pp. 29–52.

Mills, J. R. E., 1967, A comparison of lateral jaw movements in some mammals from wear facets on the teeth, *Arch. Oral Biol.* **12**: 645–661.

Moller, E., 1966, The chewing apparatus, *Acta Physiol. Scand.* **69** (Suppl. 280): 1–229.

Murphy, W. J., Eizirk, E., Johnson, W. E., Zhang, Y. P., Ryder, O. A., and O'Brien, S. J., 2001a, Molecular phylogenetics and the origins of placental mammals, *Nature* **409**: 614–618.

Murphy, W. J., Eizirik, E., O'Brien, S. J., Madsen, O., Scally, M., Douady, C. J., Teeling, E., Ryder, O. A., Stanhope, M. J., de Jong, W. W., and Springer, M. S., 2001b, Resolution of the early placental mammal radiation using Bayesian phylogenetics, *Science* **294**: 2348–2351.

Oron, U., and Crompton, A. W., 1985, A cineradiographic and electromyographic study of mastication in *Tenrec ecaudatus, J. Morphol.* **185**: 155–182.

Rak, Y., 1983, *The Australopithecine Face,* Academic Press, New York.

Rasmussen, D. T., 1990, Primate origins: Lessons from a Neotropical marsupial. *Am. J. Primatol.* **22**: 263–277.

Ravosa, M. J., 1991, Structural allometry of the prosimian mandibular corpus and symphysis, *J. Hum. Evol.* **20**: 3–20.

Ravosa, M. J., 1992, Allometry and heterochrony in extant and extinct Malagasy primates, *J. Hum. Evol.* **23**: 197–217.

Ravosa, M. J., 1996, Mandibular form and function in North American and European Adapidae and Omomyidae, *J. Morphol.* **229**: 171–190.

Ravosa, M. J., 1999, Anthropoid origins and the modern symphysis, *Folia Primatol.* **70**: 65–78.

Ravosa, M. J., and Hylander, W. L., 1994, Function and fusion of the mandibular symphysis in primates: Stiffness or strength? in: *Anthropoid Origins,* J. G. Fleagle, and R. F. Kay, eds., Plenum Press, New York, pp. 447–468.

Ravosa, M. J., and Simons, E. L., 1994, Mandibular growth and function in *Archaeolemur, Am. J. Phys. Anthropol.* **95**: 63–76.

Ravosa, M. J., Vinyard, C. J., Gagnon, M., and Islam, S. A., 2000, Evolution of anthropoid jaw loading and kinematic patterns, *Am. J. Phys. Anthropol.* **112**: 493–516.

Robinson, J. T., 1956, The dentition of the Australopithecinae, *Mem. Transvl. Mus.* **9**: 1–179.

Rose, K. D., Godinot, M., and Bown, T. M., 1994, The early radiation of Euprimates and the initial diversification of Omomyidae, in: *Anthropoid Origins,* J. G. Fleagle and R. F. Kay, eds., Plenum Press, New York, pp. 1–27.

Rosenberger, A. L., 1986, Platyrrhines, catarrhines and the anthropoid transition, in *Major Topics in Primate and Human Evolution,* B. A. Wood, L. Martin, and P. Andrews, (eds,), Cambridge University Press, Cambridge, pp. 66–88.

Rosenberger, A. L., 1992, Evolution of feeding niches in New World monkeys, *Am. J. Phys. Anthropol.* **88**: 525–62.

Rosenberger, A. L., and Kinzey, W. G., 1976, Functional patterns of molar occlusion in platyrrhine primates, *Am. J. Phys. Anthropol.* **45**: 281–98.

Ross, C. F., and Hylander, W. L., 2000, Electromyography of the anterior temporalis and masseter muscles of owl monkeys (*Aotus trivirgatus*) and the function of the postorbital septum, *Am. J. Phys. Anthropol.* **112**: 455–468.

Rubin, C. T., and Lanyon, L. E., 1984, Dynamic strain similarity in vertebrates; An alternative to allometric limb bone scaling, *J. Theor. Biol.* **107**: 321–327.

Rubin, C. T., Gross, T., Donahue, H., Guilak, F., and McLeod, K., 1994, Physical and environmental influences on bone formation, in: *Bone Formation and Regeneration,* C. Brighton, G. Friedlander, and J. Lane, eds., American Academy of Orthopedic Surgeons, New York, pp. 61–78.

Rubin, C. T., McLeod, K. J., Gross, T. S., and Donahue, H. J., 1992, Physical stimuli as potent determinants of bone morphology, in: *Bone Biodynamics in Orthodontic and Orthopedic Treatment,* Craniofacial Growth Series, Ann Arbor, pp. 75–91.

Sargis, E. J., 2001, The grasping behaviour, locomotion and substrate use of the treeshrews *Tupaia minor* and *T. tana* (Mammalia, Scandentia), *J. Zool,* London, **253**: 485–490.

Scally, M., Madsen, O., Douady, C. J., de Jong, W. W., Stanhope, M. J., and Springer, M. S., 2001, Molecular evidence for the major clades of placental mammals, *J. Mamm. Evol.* **8**: 239–277.

Simpson, G. G., 1945, The principles of classification and a classification of mammals, *Bull. Am. Mus. Nat. Hist.* **85**: 1–350.

Simpson, G. G., 1950, *The Meaning of Evolution,* Yale University Press, New Haven.

Simpson, G. G., 1955, The nature and origin of supraspecific taxa, *Cold Spring Harbor Symp. Quant. Biol.* **24**: 255–271.

Simpson, G. G., 1961, *Principles of Animal Taxonomy,* Columbia University Press, New York.

Simpson, G. G., 1965, Long-abandoned views, *Science* **147**: 1397.

Smith, G. E., 1924, *The Evolution of Man,* Oxford University Press, New York.

Smith, R. J., 1983, The mandibular corpus of female primates: Taxonomic, dietary, and allometric correlates of interspecific variations in size and shape, *Am. J. Phys. Anthropol.* **61**: 315–330.

Smith, R. J., 1984, Comparative functional morphology of maximum mandibular opening (gape) in primates, in: *Food Acquisition and Processing in Primates,* D. J. Chivers, B. A. Wood, and A. Bilsborough, eds., Plenum Press, New York, pp. 231–255.

Smith, R. J., Petersen, C. E., and Gipe, D. P., 1983, Size and shape of the mandibular condyle in primates, *J. Morphol.* **177**: 59–68.

Sokal, R. R., and Rohlf, F. J., 1997, *Biometry,* Freeman and Company, New York.

Strait, S. G., 1993, Differences in occlusal morphology and molar size in frugivores and faunivores, *J. Hum. Evol.* **25**: 471–484.

Strait, S. G., 2001, Dietary reconstruction of small-bodied omomyoid primates, *J. Vert. Paleo.* **21**: 322–334.

Sussman, R. W., 1991, Primate origins and the evolution of angiosperms. *Am. J. Primatol.* **23**: 209–223.

Szalay, F. S., 1968, The beginnings of primates, *Evolution* **22**: 19–36.

Szalay, F. S., 1969, Mixodectidae, Microsyopidae, and the insectivore-primate transition, *Bull. Am. Mus. Nat. Hist.* **140**: 193–330.

Szalay, F. S., 1972, Paleobiology of the earliest primates, in: *The Functional and Evolutionary Biology of Primates,* R. Tuttle, ed., Aldine, Chicago, pp. 3–35.

Szalay, F. S., 1973, New Paleocene primates and a diagnosis of the new suborder Paromomyiformes, *Folia Primatol.* **19**: 73–87.

Szalay, F. S., 1975, Where to draw the nonprimate-primate taxonomic boundary, *Folia Primatol.* **23**: 158–163.

Szalay, F. S., and Delson, E., 1979, *Evolutionary History of the Primates,* Academic Press, New York.

Tattersall, I., 1984, The tree-shrew, *Tupaia:* A 'living model' of the ancestral primate? in: *Living Fossils,* N. Eldredge and S. M. Stanley, eds., Springer-Verlag, New York, pp. 32–37.

Taylor, A. B., 2002, Masticatory form and function in African apes, *Am. J. Phys. Anthropol.* **117**: 133–156.

Teaford, M. F., 1994, Dental microwear and dental function, *Evol. Anthropol.* **3**: 17–30.

Teaford, M. F., and Walker, A. C., 1984, Quantitative differences in dental microwear between primate species with different diets and a comment on the presumed diet of *Sivapithecus, Am. J. Phys. Anthropol.* **64**: 191–200.

Thexton, A. J., Hiiemae, K. M., and Crompton, A. W., 1980, Food consistency and bite size as regulators of jaw movement during feeding in the cat, *J. Neurophysiol.* **44**: 456–474.

Turnbull, W. D., 1970, The mammalian masticatory apparatus, *Fieldiana: Geol.* **18**: 148–356.

van Eijden, T. M. G. J., Blanksma, N. G., and Brugman, P., 1993, Amplitude and timing of EMG activity in the human masseter muscle during selected motor tasks, *J. Dent. Res.* **72**: 599–606.

van Roosmalen, M. G. M., 1984, Subcategorizing foods in primates, in: *Food Acquisition and Processing in Primates,* D. J. Chivers, B. A. Wood, and A. Bilsborough, eds., Plenum Press, New York, pp. 167–175.

van Valen, L., 1965, treeshrews, primates and fossils, *Evolution* **19**: 137–151.

Vinyard, C. J., 1999, *Temporomandibular Joint Morphology and Function in Strepsirhine and Eocene Primates,* Ph. D. Thesis, Northwestern University.

Vinyard, C. J., and Ravosa, M. J., 1998, Ontogeny, function, and scaling of the mandibular symphysis in papionin primates, *J. Morphol.* **235:** 157–175.

Vinyard, C. J., Williams, S. H., Wall, C. E., Johnson, K. R., and Hylander, W. L., 2001, Deep masseter recruitment patterns during chewing in callitrichids, *Am. J. Phys. Anthropol.* **32** (Suppl.): 156.

Vinyard, C. J., Williams, S. H., Wall, C. E., Johnson, K. R., and Hylander, W. L., 2005, Jaw-muscle electromyography during chewing in Belanger's treeshrew (*Tupaia belangeri*), *Am. J. Phys. Anthropol.* **127:** 26–45.

Walker, A., 1981, Diet and teeth: Dietary hypotheses and human evolution, *Phil. Trans. Roy. Soc. Lond.* **292:** 57–64.

Wall, C. E., 1999, A model of TMJ function in anthropoid primates based on condylar movements during mastication, *Am. J. Phys. Anthropol.* **109:** 67–88.

Washburn, S. L., 1950, The analysis of primate evolution with particular reference to the origin of man, *Cold Spring Harbor Symp. Quant. Biol.* **50:** 67–78.

Weidenreich, F., 1943, The skull of *Sinanthropus pekinensis*: A comparative study on a primitive hominid skull, *Palaeont. Sin. D.* **10:** 1–484.

Weijs, W. A., 1980, Biomechanical models and the analysis of form: A study of the mammalian masticatory apparatus, *Am. Zool.* **20:** 707–719.

Weijs, W. A., 1994, Evolutionary approach of masticatory motor patterns in mammals, *Adv. Comp. Env. Phys.* Springer-Verlag, Berlin, pp. 282–320.

Weijs, W. A., and Dantuma, R., 1981, Functional anatomy of the masticatory apparatus in the rabbit (*Oryctolagus cuniculus* L.), *Neth. J. Zool.* **31:** 99–147.

Williams, S. H., Vinyard, C. J., Wall, C. E. and Hylander, W. L., 2003, Symphyseal fusion in anthropoids and ungulates: A case of functional convergence? *Am. J. Phys. Anthropol.* **36** (Suppl.): 226.

Wolpoff, M. H., 1973, Posterior tooth size, body size and diet in South African gracile australopithecines, *J. Phys. Anthropol.* **39:** 375–394.

Wood, B. A., 1981, Tooth size and shape and their relevance to studies of human evolution, *Phil. Trans. R. Soc. Lond. B.* **292:** 65–76.

Wood, C., 1994, The correspondence between diet and masticatory morphology in a range of extant primates. *Z. Morphol. Anthropol.* **80:** 19–50.

Yamashita, N., 1996, Seasonality and site specificity of mechanical dietary patterns in two Malagasy lemur families (Lemuridae and Indriidae), *Int. J. Primatol.* **17:** 355–387.

Yamashita, N., 1998, Functional dental correlates of food properties in five Malagasy lemur species, *Am. J. Phys. Anthropol.* **106:** 169–88.

Zuckerman, S., 1933, *Functional Affinities of Man, Monkeys, and Apes,* Harcourt, Brace & Co., New York.

CHAPTER SEVEN

Were Basal Primates Nocturnal? Evidence from Eye and Orbit Shape

Callum F. Ross, Margaret I. Hall, and Christopher P. Heesy

INTRODUCTION

The adaptations of basal primates are of interest to paleoprimatologists because they give insight into the context in which primates diverged from other mammals. In addition, these basal adaptations may have biased the evolutionary trajectories taken by the lineages leading to extant primates. Diel activity pattern is an important component of an animal's ecology because it has pervasive influence on many aspects of primate morphology and behavior, including body size, diet, substrate preference, communication, and adaptations of the sensory systems (Charles-Dominique, 1975; Heesy and Ross, 2001, 2004; Martin, 1979).

Early explanations for primate origins did not specify the activity pattern of basal primates (Cartmill, 1970, 1972, 1974; Elliot Smith, 1924; Le Gros Clark, 1959; Wood Jones, 1916), with one possible exception. Writing in the spirit of the "Primatological Synthesis" (*sensu* Cartmill, 1982), in which

Callum F. Ross • Organismal Biology & Anatomy, University of Chicago, Chicago, IL 60637
Margaret I. Hall • Department of Anatomical Sciences, Health Sciences Center, Stony Brook University, Stony Brook, NY 11794-8081 Christopher P. Heesy • · Department of Anatomy, New York College of Osteopathic Medicine, Old Westbury, NY 11568

primates were defined by a set of pervasive trends, Polyak (1957) argued that trends toward diurnality, high-visual acuity, and color vision, culminating in the higher primates, suggested continuity of diurnal "potential" through the mammalian stem lineage, up through primates. According to Polyak (1957: 968–969), nocturnal strepsirrhines are divergent from the diurnal mainstream of primate evolution.

By the beginning of the 1970s, however, P. Charles-Dominique and R. D. Martin's field studies had revealed many ecological and behavioral similarities between *Microcebus murinus* and "*Galago demidovii*," including nocturnality, leading them to suggest that many of these aspects are likely to be both ancestral strepsirrhine and ancestral primate characteristics (Charles-Dominique and Martin, 1970). Martin (1973) bolstered the argument that the ancestral strepsirrhines were nocturnal by noting that many diurnal lemurid species possess a tapetum. Tapeta are usually found in nocturnal animals (Walls, 1942; see Ross, 2004, for a review), suggesting that their presence in extant diurnal strepsirrhines is due to "primitive retention." Extending this "primitive retention" argument to explain the *absence* of a tapetum in *Aotus* and *Tarsius*, Martin suggested that these animals were descended from diurnal ancestors, retaining a diurnal adaptation into a nocturnal environment (Martin, 1973). This argument was reinforced by the observation that *Tarsius* also possesses a retinal fovea (Le Gros Clark, 1959)—a traditionally diurnal adaptation (Ogden, 1974; Polyak, 1957; Ross, 2004; Walls, 1942). In contrast with Martin's "primitive retention" explanation for the imperfect correlations between primate morphology and activity pattern, Charles-Dominique (1975: 86) suggested that the last common ancestor of primates "had an eye slightly differentiated for both nocturnal and diurnal vision," capable of evolving into an anthropoid or a strepsirrhine eye.

By the late 1970s, the issue of the activity pattern of basal primates was independently addressed by a number of workers. Martin (1979) marshaled evidence, including body size, relative size of the olfactory bulb, and the presence or absence of tapeta, to suggest "that nocturnal life involving at least some predation on small animals is a primitive feature for the lemurs and lorises, and possibly for the primates as a whole" (Martin, 1979: 72). Supporting Martin's argument were functional interpretations of two features assumed to have characterized basal primates: high degrees of orbital convergence and relatively large orbital apertures.

ORBITAL CONVERGENCE

Primates have long been noted to have more convergent orbital apertures than most other mammals. Early explanations related convergence to arboreality (Elliot Smith, 1924; Le Gros Clark, 1959; Wood Jones, 1916). However, comparisons with other animals suggested to Cartmill (1970, 1972) that convergent orbits facilitated visual predation on insects in the fine branches of the shrub layer of tropical rainforests. Cartmill argued that "Stereoptic integration of the two visual fields improves the accuracy of the final strike; increase in visual-field overlap facilitates compensation for evasive movements of the prey" (1972: 113). Cartmill's hypothesis did not specify whether these first primates were diurnal, nocturnal, crepuscular, or cathemeral, and it was left to Jack Pettigrew and John Allman to round out the visual predation hypothesis, specifying nocturnality as an essential part of the argument (Cartmill, 1992).

Pettigrew (cited by Allman, 1977: p. 29; Pettigrew, 1978) and Allman (1977) pointed out that the dioptric benefits of orbital convergence accrue to nocturnal rather than diurnal animals. The optical axis is the axis of the dioptric apparatus of the eye (i.e., lens and cornea), around which image quality is highest, whereas the visual axis is the "physiological line of fixation" (Walls, 1942: 292), approximated by a line passing through the center of the pupil and the retinal fovea or area centralis. Thus, alignment of the optic axis with the visual axis maximizes image quality in the fovea or area centralis (Figure 1). The Allman-Pettigrew model posits that orbital convergence is correlated with convergence of the optic axes on the visual axes (Figure 2), providing improved image quality in nocturnal primates. Another way to ensure high-image quality across the retina is to restrict incoming images to the paraxial region of the dioptric apparatus. This can be achieved by decreasing diameter of the pupil, but this option is not available to nocturnal animals that must maintain large pupil sizes in order to maintain image brightness. Consequently, nocturnal animals can only improve image quality in the area of visual field overlap by optic convergence (Figure 2). This suggested to Allman (1977) that if the first primates had high degrees of orbital convergence, then they were probably nocturnal.

Convergence of the optic axes on each other increases the size of the region of visual field overlap (Heesy, 2004; Ross, 2000), something that Cartmill hypothesized was advantageous in the pursuit of evasive prey.

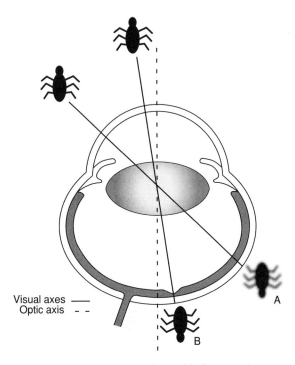

Figure 1. Diagram of eye illustrating the relationship between image quality and the orientation of the optic and visual axes. The optic axis is the axis of the dioptric apparatus (i.e., the primary refracting surfaces of the eye, the cornea, and the lens). The visual axis is the "physiological line of fixation" (Walls, 1942: 292), approximated by a line intersecting the center of the pupil and the center of the region of the retina which the animal directs at objects of interest, and in which the animal has the highest visual acuity. In all mammals this region is the area centralis, and in haplorhine primates this region also contains a fovea. Image quality is highest for paraxial images or images close to the optic axis and deteriorates with distance from the optic axis. Image quality and visual acuity are maximized when the visual axis is close to the optic axis (A) Image quality and visual acuity are decreased when the visual axis is divergent from the optic axis. (B) Image quality and visual acuity are increased when the visual axis is close to the optic axis.

However, nocturnal animals also benefit from visual field overlap because it increases sensitivity to low-light levels, the eyes effectively having double the chance of registering a photon from any part of the binocular visual field. This improves the signal-to-noise ratio of the image, and improves contrast detection (Lythgoe, 1979; Pettigrew, 1986). Thus, optic (and presumably orbital) convergence provides two advantages to nocturnal animals: improved image quality, and increased image brightness in the binocular visual field.

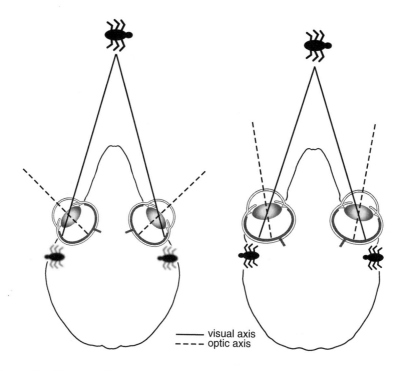

Figure 2. Diagram illustrating relationship between orientation of optic axis, visual axis, and orbits in a plesiadapiform-like animal with divergent orbits (left), and a primate with convergent orbits (right). The retinal images of objects in the binocular visual field (in front of the snout) are of lower quality in the animal with divergent orbits than in the primate. This is because the visual axis is less closely aligned with the optic axis in the animal with divergent orbits.

ORBIT SIZE AND SHAPE

Small-bodied nocturnal primates have long been noted to have larger orbital apertures than small-bodied primates of the same skull length (Kay and Cartmill, 1977; Heesy and Ross, 2001; Kay and Kirk, 2000; Walker, 1967). These differences in orbital aperture dimensions may reflect several aspects of eye size and shape (Kirk, 2004; Ross, 2000). The dioptric principles underlying the relationship between eye shape and activity pattern are illustrated in Figure 3. Nocturnal animals live in scotopic, or low light level conditions. Low light levels compromise image quality by providing poor sampling of the visual environment. Consequently, visually dependent nocturnal animals

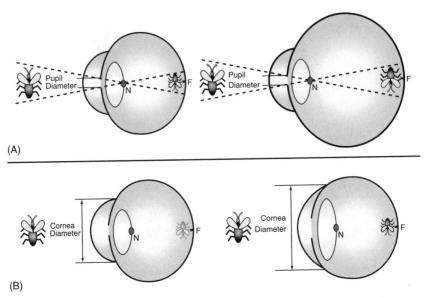

Figure 3. Diagrams illustrating hypothesized relationship between eye shape and activity pattern. N = nodal point. Distance from nodal point to retina is focal length. Visually dependent diurnal animals are expected to have large focal lengths relative to pupil diameter in order to maximize image size and, hence, the resolution of the image (A). Visually dependent nocturnal animals are expected to have large pupil diameters relative to focal length in order to maximize image brightness (B). Nocturnal animals can also increase image brightness by increasing eye size while also increasing the effective size of the photoreceptor units (Land and Nilsson, 2002).

exhibit adaptations for increasing the number of photons captured from the visual field, or image brightness. Image brightness is directly related to the area of the pupil and to the solid angle in space from which each receptor samples light (Land and Nilsson, 2002). The first determines the number of photons that can enter the eye simultaneously, the second determines the number of photons that can enter each photoreceptor (Land, 1981). The angle of acceptance of a photoreceptor is inversely related to the focal length of the eye (or posterior nodal distance), roughly the distance from the lens to the retina (see Figure 4), so the longer the focal length of the eye, the dimmer the image. An intuitive sense of this can be achieved by shining a flashlight on a wall: the closer the flashlight is to the wall, the smaller and brighter the image. The same is true of the eye: the shorter the focal length of the eye, the smaller and brighter the image on the retina. Hence, visually dependent

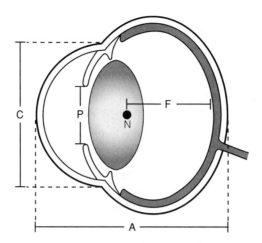

Figure 4. Diagram illustrating measures of eye shape used in this study and the variables they estimate. N=Nodal point; F=Focal length or posterior nodal distance; P=Pupil diameter; C=Cornea diameter; A=Axial diameter or length of the eye. Cornea diameter is a reasonable estimator of pupil diameter (see Ross, 2000). Axial diameter of the eye is assumed to be a sufficiently good estimator of focal length of the eye, following Hughes (1977).

nocturnal animals are predicted to have large corneas relative to their focal length regardless of body or eye size (Figure 3B).

In addition to these effects on eye shape, nocturnality is also predicted to be associated with increased eye size. The pupil obviously cannot be larger than the eye, so eye size limits the amount that the pupil can be expanded. If pupil diameter is increased by making the eye bigger, focal length will also increase, reducing image brightness by decreasing the angle of acceptance of the photoreceptors. However, this latter effect can be compensated for by increasing the effective size of the photoreceptors. In vertebrates this is done by pooling many receptors into one functional unit by connecting many of them up to a single ganglion cell—a phenomenon widespread among vertebrates, including primates (Kay and Kirk, 2000; Rohen and Castenholtz, 1967). Thus, photoreceptor pooling allows image brightness to be increased purely by increasing eye size, so it is expected that nocturnal animals will have larger eyes than similarly sized diurnal animals.

It has also been argued that photoreceptor pooling has the added benefit of widening the range of "image brightness," or luminance, to which an animal is sensitive (Martin, 1999). This hypothesis assumes that the size of the photoreceptor (rod) pools of nocturnal animals is flexible at the retinal level, allowing changes in the size of the pool in response to different luminance levels. This would be valuable in nocturnal environments that are characterized by a much wider range of luminance levels than diurnal environments (Lythgoe, 1979; Martin, 1990, 1999). The hypothesis argues that increases in eye size would augment the number of photoreceptor pools, creating the possibility of greater flexibility to the greater range of light levels in nocturnal environments (Martin, 1990, 1999). The validity of this hypothesis remains in doubt until mechanisms for adjusting receptive field size in scotopic conditions are demonstrated. Primates lack rod–rod coupling that might be one mechanism to accomplish this (Djamgoz et al., 1999), but there do appear to be up to three pathways for information to pass from rods to the inner retina, and these pathways may operate under different ambient light conditions (Bloomfield and Dacheux, 2001).

Diurnal animals, in contrast, are not constrained by the need to shorten focal length or enlarge the pupil, because image brightness is not a problem in the photopic or light rich environment. Consequently, visually dependent diurnal animals are able to have long focal lengths, thereby decreasing the acceptance angle of each photoreceptor in the retina and increasing visual resolution (Figure 3A). Another way of saying this is that increased focal length spreads the image over a larger number of photoreceptors, increasing visual resolution. Moreover, because diurnal animals do not need to enhance image brightness, they are predicted to have small pupils relative to focal length.

RECONSTRUCTIONS OF ORBIT SIZE AND SHAPE IN BASAL PRIMATES

Functional interpretations of orbital convergence and enlargement only suggest that basal primates were nocturnal if these features were present in basal primates. The last common ancestor of primates is not known from fossil evidence, and nor are their immediate outgroups. Consequently, the assumption that basal primates had orbital apertures that were convergent and enlarged rests on interpretation of the available evidence from fossils and extant taxa.

The objectives of this study are: (a) to document the relationship between eye size and shape, and activity patterns in extant primates; (b) to document the relationships between eye size and shape, and activity patterns in extant amniotes and use these data to interpret primate eye shape; and (c) to use orbit shape data to reconstruct activity pattern in fossil primates.

MATERIALS AND METHODS

Eye Shape Measures

On the basis of the dioptric principles outlined in the Introduction, visually dependent nocturnal animals are predicted to have large pupils relative to focal length, and visually dependent diurnal animals are predicted to have small pupil diameters for their focal lengths. These dimensions cannot be measured accurately in preserved eyes and are known for only a small number of vertebrates (e.g., Arrese, 2002; Hughes, 1977; Martin, 1999). Here, we use the axial diameter of the eye as a surrogate for focal length, and cornea diameter as a surrogate for pupil diameter (Figure 4). Hughes (1977, Figure 9B) has shown that, across a range of vertebrates of differing activity patterns, focal length is approximately 0.6 axial diameter of the eye. Assuming that this relationship is constant across vertebrates, we use axial diameter of the eye as a surrogate for focal length. Cornea diameter is a reasonable surrogate for pupil diameter as there is no obvious reason to have a cornea that is significantly larger than the pupil.

To investigate scaling relationships of cornea diameter and axial length, head and body length is used as a measure of body size for comparisons across different groups of amniotes.

Orbit Shape Measures

The relationship between relative size of the orbital aperture and activity pattern has long been of interest to paleoprimatologists because it provides a method for reconstructing activity pattern in moderately well-preserved fossils (Beard et al., 1991; Heesy and Ross, 2001; Kay and Cartmill, 1977; Kay and Kirk, 2000; Martin, 1990; Rasmussen and Simons, 1992; Walker, 1967). Given the relationship between eye shape and activity pattern predicted above, we also predict that there will be a relationship between orbit shape and activity pattern. Specifically, we predict that the size of the orbital aperture relative to the axial depth of the orbit should be correlated with activity pattern. The underlying assumptions, that pupil or cornea area is correlated with orbital aperture area, and that the axial length of the eye is correlated with the axial depth of the orbit, have not yet been evaluated using measures of orbit and eye shape taken from the same individuals.

Orbital aperture size is estimated by the diameter of the orbital aperture, measured from orbitale inferius to the orbitale superius (Cartmill, 1970). Orbit depth is calculated as the distance from the midpoint of orbitale superius–orbitale inferius chord to the superiormost point along the rim of the optic canal. These measures were extracted using customized macros in Microsoft Excel, from 3D coordinates of points collected using a Microscribe 3DX digitizer (Immersion Corp., San Jose, CA).

Orbit diameter data for *Cantius abditus* are from Heesy and Ross (2001). The optic canal of *Cantius abditus* (USNM 494881: Rose et al., 1999) is not preserved. A minimum estimate of the axial depth of the orbit in *C. abditus* was obtained by combining the length of the orbit floor in *C. abditus* (USNM 494881) with that of another specimen of this taxon (USNM 93938). Comparisons of *C. abditus* (USNM 494881) orbit and preserved braincase with similar-sized extant strepsirrhines (e.g., *Otolemur crassicaudatus*) suggest that the orbit depth was not substantially longer than this estimate.

The majority of the eye shape data were derived from Ritland (1982). Data from nonadult individuals were excluded. The primate eye shape data were derived in part from Ritland (1982), as well as from unpublished observations made by C. F. Ross. Additional data on bats and birds were collected by M. I. Hall (Hall, 2005). The orbit shape data were collected from different specimens than the eye shape data.

Reduced major axis (RMA) regression equations for all Aves, Primates, and Mammalia were calculated from least squares equations generated by SPSS 12.0 for Windows. The RMA slopes, intercepts, and correlation coefficients are given in Table 1.

RESULTS

Eye Size and Shape

Figure 5 plots axial eye diameter against head and body length in extant primates and other mammals. Nocturnal primate eyes have larger axial diameters than most similarly sized nonprimate mammals. Tarsiers have longer eyes relative to body size than any other mammals. Axial diameter of the eye scales with negative allometry across all diurnal and nocturnal mammals, except

Table 1. Reduced major axis (RMA) regression statistics

Variables	Activity	n	RMA slope	RMA intercept	r
CD vs. AD					
Aves	Nocturnal	129	0.967	−0.100	0.951
	Diurnal	166	0.930	−0.174	0.971
Mammals	Nocturnal	174	0.912	0.0194	0.979
	Diurnal	103	1.138	−0.332	0.921
Primates	Nocturnal	30	0.765	0.179	0.971
	Diurnal	49	0.963	−0.192	0.892
CD vs. HBL					
Aves	Nocturnal	146	0.875	−0.799	0.864
	Diurnal	193	0.678	−0.590	0.826
Mammals	Nocturnal	174	0.802	−0.941	0.699
	Diurnal	103	0.576	−0.480	0.852
Primates	Nocturnal	30	0.417	0.103	0.539
	Diurnal	49	0.450	−0.182	0.895
AD vs. HBL					
Aves	Nocturnal	129	0.906	−0.726	0.85
	Diurnal	166	0.729	−0.446	0.837
Mammals	Nocturnal	197	0.882	−1.059	0.724
	Diurnal	107	0.507	−0.132	0.881
Primates	Nocturnal	30	0.546	−0.100	0.603
	Diurnal	56	0.467	0.011	0.903

AD = axial diameter; CD = cornea diameter; HBL = head and body length.

those at very small body sizes (Kiltie, 2000) and all diurnal and nocturnal primates. However, as noted elsewhere (Ross, 2000), the relationship between these variables across all mammals is not linear, being positively allometric at small body sizes and negatively allometric at large body sizes. Because small primates have relatively larger eyes than other mammals, the primate slopes are significantly less steep than those for all mammals combined.

Figure 6 plots axial diameter of the eye against head and body length in extant primates and birds. Tarsiers fall within the range of nocturnal birds, i.e., Strigiformes (owls), Caprimulgidae (nightjars), and Podargidae (frogmouths). Most of the diurnal birds lying between the nocturnal bird and diurnal primate distributions are falconiforms (Hall, 2005). The rest of the diurnal birds are parrots, pigeons, and procellariform sea-birds, which fall among and below the primate distributions. Across nocturnal and diurnal birds, eye axial diameter is negatively allometric. However, the 95% confidence limits for the slope across nocturnal birds almost include 1.00 (= 0.99), a slope significantly steeper than that of primates.

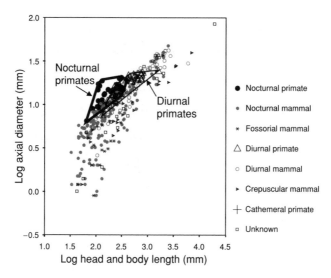

Figure 5. Bivariate plot of axial diameter ($\log_{10}$) against head and body length ($\log_{10}$) across all mammals. Minimum spanning polygons for nocturnal primates and diurnal primates are added. Nocturnal primates have longer axial lengths for their body size than similarly sized nonprimate mammals.

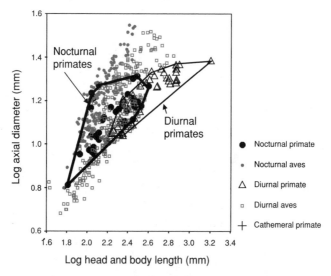

Figure 6. Bivariate plot of axial diameter ($\log_{10}$) against head and body length ($\log_{10}$) across birds and primates. Minimum spanning polygons for nocturnal primates and diurnal primates are added. The nocturnal primates with the longest axial lengths for their body size are tarsiers, which plot with the nocturnal birds represented here; i.e., Strigiformes (owls) Caprimulgidae (nightjars), and Podargidae (frogmouths).

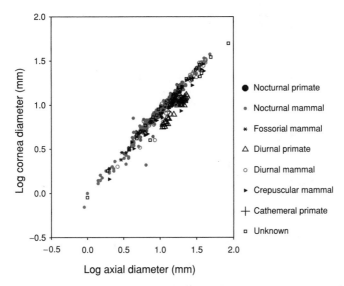

Figure 7. Bivariate plot of corneal diameter ($\log_{10}$) against axial diameter ($\log_{10}$) across mammals.

Figure 7 plots cornea diameter against axial diameter in extant primates and other mammals. Nocturnal primates, as well as diurnal and cathemeral strepsirrhines, fall with other mammals on a plot of cornea diameter versus axial diameter of the eye (see also Ross, 2000). Diurnal anthropoids differ from all other mammals in having small corneas relative to axial length, or longer eyes relative to cornea diameter. The separation between nocturnal and diurnal primates extends to all body sizes sampled. The RMA slopes for nocturnal and diurnal primates are not significantly different, and the intercept for diurnal primates is significantly lower than that for nocturnal primates. Eye shape does not sort nocturnal and diurnal nonprimate mammals. Cornea diameter scales isometrically with axial diameter in diurnal primates, with positive allometry across all diurnal mammals, with negative allometry across nocturnal primates, and with slight negative allometry across all nocturnal mammals.

Figure 8 plots cornea diameter against axial diameter in extant primates and birds. Nocturnal primates and diurnal strepsirrhines plot with nocturnal birds in relative cornea and axial diameter, whereas diurnal anthropoids plot amongst diurnal birds. Eye shape sorts nocturnal and diurnal birds from each other, and diurnal anthropoids from other primates. Cornea diameter scales

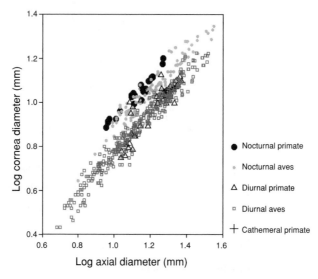

Figure 8. Bivariate plot of corneal diameter ($\log_{10}$) against axial diameter ($\log_{10}$) across birds and primates.

isometrically with axial diameter in nocturnal birds, and with slight negative allometry in diurnal birds.

The RMA slopes of corneal diameter against axial diameter are not significantly different from each other in diurnal primates, diurnal birds, nocturnal birds, and nocturnal mammals. Nocturnal primates have a significantly shallower slope than diurnal primates, and diurnal mammals have a significantly steeper slope than all other groups.

Figure 9 diagrams the distributions of cornea and axial diameters in mammals and birds with minimum spanning polygons. There is surprisingly little variability in eye shape across these clades. Diurnal anthropoids and diurnal birds have longer axial lengths for their cornea diameters than other birds and mammals, whereas nocturnal birds, nocturnal primates, and diurnal strepsirrhines (asterisks) plot among nonprimate mammals, without regard for their activity pattern.

Orbit Size and Shape

Figure 10 plots orbit aperture diameter against orbital depth in extant primates and some fossils. Orbital aperture diameter is positively correlated with orbit depth in both nocturnal and diurnal taxa, and within each activity group there is little variation in orbit diameter at any given orbit depth. In this

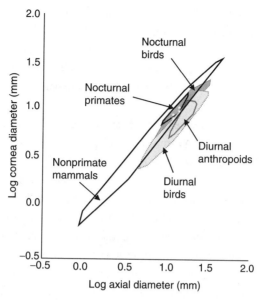

Figure 9. Bivariate plot of corneal diameter ($\log_{10}$) against axial diameter ($\log_{10}$) across birds, nonprimate mammals, and primates. Minimum spanning polygons are illustrated, excluding two outliers for nonprimate mammals. White asterisks on nocturnal primate polygon are diurnal strepsirrhine primates.

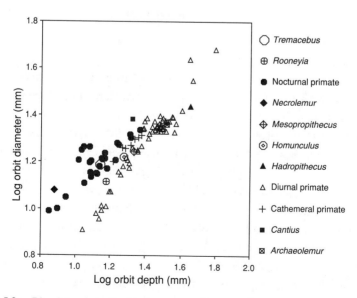

Figure 10. Bivariate plot of orbital aperture diameter ($\log_{10}$) against axial depth of the orbit ($\log_{10}$) across primates.

respect this plot resembles the plot of cornea diameter against axial length of the eye in birds and primates (Figure 7). However, the slopes of the two distributions differ, so that at small body sizes nocturnal primates have larger orbital apertures than diurnal primates with the same axial depths, but above an axial depth = 1.3 ($\log_{10}$) the nocturnal and diurnal distributions begin to overlap.

The activity patterns of several fossil taxa can be reconstructed by plotting them on this distribution. As noted by others, *Necrolemur* was almost certainly nocturnal. *Tremacebus* and *Homunculus* also plot as nocturnal, and *Mesopropithecus* plots as diurnal. *Rooneyia* plots closest to the diurnal primates, and was probably diurnal, although it has a slightly larger orbit diameter than extant diurnal primates. At orbit depths greater than that of *Mesopropithecus* there are no nocturnal extant primates, so it is not possible to estimate the activity patterns of larger fossil forms, such as *Hadropithecus*. *Cantius* falls above the distribution of extant nocturnal primates, suggesting that it was nocturnal, confirming the results of Heesy and Ross (2001). However, because of the uncertainty of orbit depth in this taxon, this conclusion must be regarded as preliminary.

DISCUSSION

The Eyes of Basal Primates

There is debate as to the precise ecological significance of the relatively high levels of orbital convergence seen in primates. Cartmill attributes it to selection for nocturnal visual predation (Cartmill, 1992), whereas others associate it with manual manipulation or visual detection of small objects, including fruits, insects, and small branches in a nocturnal rainforest environment (Crompton, 1995; Rasmussen, 1990; Sussman, 1991, 1995). Common to all these models is the relationship between orbital convergence and nocturnality. Although relatively high degrees of orbital convergence have not been extrapolated down to basal primates using rigorous character optimization methods, all primates, living and fossil, with only one exception (*Megaladapis*) have more highly convergent orbits (Heesy, 2003; Ni et al., 2004; Ravosa and Savokova, 2004) than seen in nonprimate mammals, including plesiadapiforms. It, therefore, seems probable that basal primates also had relatively high degrees of orbital convergence, and hence were nocturnal.

This conclusion is congruent with character optimization studies of the evolution of activity pattern and chromacy in primates, and their relatives (Heesy and Ross, 2001, 2004), but runs counter to recent claims that basal primates were diurnal (Li, 2000; Ni et al., 2004; Tan and Li, 1999). We have discussed our objections to Tan and Li's arguments elsewhere (Heesy and Ross, 2001, 2004).

The eye shape data presented here suggest that the eyes of these basal primates were probably not distinguished from those of their ancestors on the basis of shape, as anthropoids are the only mammals with a distinctive eye shape (Figure 7). The reason for the lack of correlation between eye shape and diel activity pattern in nonprimate mammals is not obvious. One possibility is that the nocturnality generally assumed for the mammalian stem lineage resulted in a nocturnal-shaped eye (i.e., with a large cornea relative to axial length), and that nocturnality and its characteristic eye shape persisted in the lineage leading from basal mammals to basal primates. Of course, this does not explain why all nonanthropoid diurnal mammals possess a "nocturnal eye shape," including many visually dependent diurnal mammals. Another possibility is that the measures of eye shape used here are poor indicators of image brightness; in particular, axial diameter of the eye may not accurately reflect focal length in mammals. Future work should evaluate this possibility.

In contrast with these conclusions regarding eye shape, it can be hypothesized that basal primates, if they were nocturnal, were distinguished from their ancestors by larger eye size (Figure 5): extant nocturnal primate eyes have larger axial diameters than similarly sized nonprimate mammals. As noted earlier, when accompanied by photoreceptor pooling, increase in eye size increases image brightness (Land and Nilsson, 2002). Increase in axial length of the eye in basal primates will also increase visual acuity in a nocturnal environment, the same way as it increases acuity for diurnal animals (i.e., by enlarging the image and spreading it over a greater number of photoreceptors). Of course, this would make the image dimmer if the cornea and pupil did not also increase in size to maintain image brightness, but image brightness is maintained in primates (Figure 7) regardless of differences in eye size (Figure 5).

Increased visual acuity is also expected in the context of the increased orbital convergence that also characterized basal primates. One effect of increased orbital convergence is to improve image quality along the visual axis (by aligning optic and visual axes), so it seems reasonable to expect that the eye would be altered to take advantage of the improved image quality. Increasing axial length and spreading the image over a greater number of photoreceptors is one way to do this.

If the basal primate eye was characterized by features functioning to increase visual acuity in a nocturnal environment, this acuity could have been put to a number of uses, including visual predation on insects (Cartmill, 1992), detection of small fruits, and locomotion in the terminal branches (Crompton, 1995; Sussman, 1995). Several workers have criticized the "nocturnal visual predation" model of primate origins by pointing out that many nocturnal primates use their auditory sense to detect prey, suggesting that this weakens the link between orbital convergence and visual predation (Crompton, 1995; Rasmussen, 1991; Sussman, 1991, 1995;). Clearly, however, basal primates could well have been using both senses to find their prey. R.S. Heffner and H.E. Heffner provide evidence that in extant mammals, increased acuity in sound localization is positively correlated with both increased width of the binocular visual field (1985) and a narrowing of the field of highest visual acuity, estimated by the width of the area centralis (in degrees) (1992). The sound localization threshold is a measure of acuity, such that the lower the threshold, the smaller the difference in the angular position of a sound source that can be detected. Hence, animals with large binocular visual fields and narrow fields of high-acuity vision tend to have the highest auditory acuity. Heffner and Heffner argue that auditory and visual acuity are correlated because hearing is used to guide the eyes toward the target more precisely. Indeed, they go so far as to suggest that "it is the function of sound localizing, i.e., directing the attention of other senses toward the sound-producing object . . . which underlies the variation in mammalian sound-localizing acuity" (R. S. Heffner and H. E. Heffner, 1992: 711). This suggests that if basal primates exhibited adaptations for prey localization, these adaptations probably were found in both the hearing and visual systems.

Heffner and Heffner's data do not include many primates, but support for a link between visual and auditory acuity, and degree of insectivory among primates is found in Tetreault et al.'s (2004) study of retinal ganglion cell densities in *Cheirogaleus* and *Microcebus*. *Microcebus* has a higher retinal ganglion cell density than *Cheirogaleus*, is more insectivorous, and has larger more mobile pinnae, an important determinant of sound localizing ability (Brown, 1994; Coleman and Ross, 2004; Heffner and Heffner, 1992). Their data also suggest that *Microcebus* has a narrower field of high-acuity vision than *Cheirogaleus*. Clearly more research is needed into the sound localizing and visual acuity of strepsirrhines, as well as the interactions between the two systems.

The Eyes of Haplorhines

The increase in orbital convergence in anthropoid primates over and above that of most prosimians (Ross, 1995), combined with the decreased pupil diameter associated with diurnality, probably further improved image quality along the visual axis of anthropoids. In this context it would have been worthwhile to both further increase image size by increasing axial length of the eye (producing the unusually long eyes of extant anthropoids), and add a retinal fovea to the visual axis. It is noteworthy in this regard that diurnal anthropoids fall with diurnal birds on the plot of cornea diameter and axial diameter (Figure 9), and most diurnal birds have retinal foveae as well (Ross, 2004).

The tarsier eye exhibits adaptations for increased acuity in a nocturnal environment over and above those predicted for basal primates. The orbits of tarsiers are highly convergent for their size, suggesting that tarsiers are maximizing convergence as much as possible to improve image quality on the retina. The eyes of tarsiers are longer in axial length than any other mammals, plotting with strigiform and caprimulgiform birds. This may reflect increases in overall eye size, to increase either image brightness or the range of light levels over which their eyes are sensitive. It may also be an attempt to increase visual acuity. Tarsiers possess a retinal fovea characterized by a high density of photoreceptors and ganglion cells (Hendrickson et al., 2000), and the exclusion of blood vessels from the center of the fovea, or foveola (Hendrickson et al., 2000; Polyak, 1957; Ross, 2004). Tarsiers lack a tapetum (Hendrickson et al., 2000; Martin, 1973), also probably an adaptation for increased acuity (Cartmill, 1980; Ross, 2004) and possess a postorbital septum to insulate their fine-grained retina against movements in the temporal fossa during mastication (Cartmill, 1980, Heesy et al., this volume; Ross, 1996). In most of these features, tarsiers resemble owls, animals with similar relative axial diameters of the eye (Figure 6), supporting Niemitz' (1985) suggestion of ecological convergence between the two.

Tarsiers and anthropoids share several features of the visual system that are divergent from the basal primate condition. They both possess retinal foveae and lack tapeta, even when nocturnal, and their eyes exhibit large axial diameters. Their orbits are highly convergent for their size and are characterized by a postorbital septum. One explanation is that these shared features are adaptations to diurnality that have been retained by the tarsier lineage when it adopted nocturnal habits (Cartmill, 1980; Ross, 2000). These features of the tarsier eye may also be adaptations for high acuity in a nocturnal environment. Parsimony suggests that the last common ancestor of extant haplorhines was nocturnal (Heesy and Ross, 2001, 2004; Ross, 2004), but definitive resolution of this question must await discovery of fossils closer to the ancestral haplorhine node.

CONCLUSIONS

The origin of primates was accompanied by increases in eye size and orbital convergence. These changes almost certainly occurred in a nocturnal lineage and likely functioned to improve image brightness and visual acuity in a nocturnal environment. The exact use to which this increased acuity was put cannot be determined from eye shape and size alone. Comparative studies of nonprimate mammals suggest that increased visual acuity was associated with increased auditory acuity as well (Heffner and Heffner, 1992).

The changes in the visual system at the origin of primates were similar in kind to but less in degree than those that took place along the anthropoid stem lineage; i.e., anthropoids exhibit a further increase in orbital convergence, axial diameter of the eye, and visual acuity. Selections for these changes in the anthropoid visual system are most likely to have occurred in the context of the changes in visual system anatomy put in place at the origin in primates.

ACKNOWLEDGMENTS

We thank Matt Ravosa and Marian Dagosto for inviting us to participate in the Primate Origins Conference, and in this volume. This research was supported by a grant from NSF Physical Anthropology to C.F. Ross (SBR 9706676) and a grant from the L.S.B. Leakey Foundation to C.P. Heesy. Mark Coleman alerted us to important papers documenting the relationship between auditory and visual acuity, and read and provided comments on the manuscript.

REFERENCES

Allman, J., 1977, Evolution of the visual system in the early primates, *Prog. Psychobiol. Physiol. Psychol.* **7**: 1–53.

Arrese, C., 2002, Pupillary mobility in four species of marsupials with differing lifestyles, *J. Zool.* (Lond). **256**: 191–197.

Beard, K. C., Krishtalka, L., and Stucky, R. K., 1991, First skulls of the Early Eocene primate *Shoshonius cooperi* and the anthropoid-tarsier dichotomy, *Nature* **349**: 64–67.

Bloomfield, S. A., and Dacheux, R. F., 2001, Rod vision: Pathways and processing in the mammalian retina, *Prog. Ret. Eye Res.* **20**: 351–384.

Brown, C., 1994, Sound localization, in: *Comparative Hearing in Mammals*, R. R. Fay and A. N. Popper, eds., Springer-Verlag, New York, pp. 57–96.

Cartmill, M., 1970, *The Orbits of Arboreal Mammals: A Reassessment of the Arboreal Theory of Primate Evolution*, Unpublished Ph.D. dissertation, University of Chicago, Chicago, IL.

Cartmill, M., 1972, Arboreal adaptations and the origin of the Order Primates, in: *The Functional and Evolutionary Biology of Primates*, R. Tuttle, ed.,. Aldine, Chicago, pp. 97–122.

Cartmill, M., 1974, Rethinking primate origins, *Science* **184**: 436–443.

Cartmill, M., 1980, Morphology, function and evolution of the anthropoid postorbital septum, in: *Evolutionary Biology of the New World Monkeys and Continental Drift*, R. L. Ciochon and A. B. Chiarelli, eds., Plenum Press, New York, pp. 243–274.

Cartmill, M., 1982, Basic primatology and prosimian evolution, in: *Fifty Years of Physical Anthropology in North America*, F. Spencer, ed.,. Academic Press, New York, pp. 147–186.

Cartmill, M., 1992, New views on primate origins, *Evol. Anthropol.* **3**: 105–111.

Charles-Dominique, P., 1975, Nocturnality and diurnality: An ecological interpretation of these two modes of life by an analysis of the higher vertebrate fauna in tropical forest ecosystems, in: *Phylogeny of the Primates: A Multidisciplinary Approach*, W. P. Luckett and F. S. Szalay,eds., Plenum Press, New York, pp. 69–88.

Charles-Dominique, P., and Martin, R. D., 1970, Evolution of lorises and lemurs, *Nature* **227**: 257–260.

Coleman, M.N. & Ross, C. F. 2004 Primate auditory diversity and its influence on hearing perfomance. *Anat. Rec* **281A**, 1123–1137.

Crompton, R. H., 1995, "Visual predation," habitat structure, and the ancestral primate niche, in: *Creatures of the Dark: The Nocturnal Prosimians*, L. Alterman, G. A. Doyle, and M. K. Izard,eds., Plenum Press, New York, pp. 11–30.

Djamgoz, M. B. A., Vallerga, S., and Wagner, H.-J., 1999, Functional organization of the outer retina in aquatic and terrestrial vertebrates: Comparative aspects and possible significance to the ecology of vision, in: *Adaptive Mechanisms in the Ecology of*

Vision, S. N. Archer, M. B. A. Djamgoz, E. R. Loew, J. C. Partridge, and S. Vallerga, eds., Kluwer Academic Publishers, Dordrecht, pp. 329–382.

Elliot Smith, G. E., 1924, *The Evolution of Man*, Oxford University Press, London and New York.

Frishman, L. J., and Robson, J. G., 1999, Inner retinal signal processing: Adaptation to environmental light, in: *Adaptive Mechanisms in the Ecology of Vision*, S. N. Archer, M. B. A. Djamgoz, E. R. Loew, J. C. Partridge, and S. Vallerga, eds., Kluwer Academic Publishers, Dordrecht, pp. 383–412.

Hall, M. I. 2005 The Roles of Function and Phylogeny in the Morphology of the Diapsid Visual system. In *Anatomical sciences*, vol. Ph.D. Stony Brook: Stony Brook University.

Heesy, C. P. 2003 The Evolution of Orbit Orientation in Mammals and the Function of the Primate Postorbital Bar. *In Interdepartmental Doctoral Program in Anthropological Sciences*, vol. Ph.D. Stony Brook: Stony Brook University.

Heesy, C. P., 2004, On the relationship between orbit orientation and binocular visual field overlap in mammals, *Anat. Rec.* **281A**: 1104–1110.

Heesy, C. P., and Ross, C. F., 2001, Evolution of activity patterns and chromatic vision in primates: Morphometrics, genetics and cladistics, *J. Hum. Evol.* **40**: 111–149.

Heesy, C. P., and Ross, C. F., 2004, Mosaic evolution of activity pattern, diet, and color vision in haplorhine primates, in: *Anthropoid Origins: New Visions*, C. F. Ross and R. F. Kay, eds., Kluwer Academic/Plenum Publishers, New York, pp. 665–698.

Heffner, R. S., and Heffner, H. E., 1985, Auditory localization and visual fields in mammals, *Neurosci. Abstracts* **11**: 547.

Heffner, R. S., and Heffner, H. E., 1992, Evolution of sound localization in mammals. in: *The Evolutionary Biology of Hearing*, D. B. Webster, R. R. Fay, and A. N. Popper, eds., Springer-Verlag, New York, pp. 691–715.

Hendrickson, A. E., Djajadi, H. R., Nakamura, L., Possin, D. E., and Sajuthi, D., 2000, Nocturnal tarsier retina has both short and long/medium-wavelength cones in an unusual topography, *J. Comp. Neurol.* **424**: 718–730.

Hughes, A., 1977, the topography of vision in mammals of contrasting life style: Comparative optics and retinal organization, in: *The Visual System in Vertebrates*, F. Crescitelli, ed., Springer-Verlag, Berlin, pp. 613–756.

Kay, R. F., and Cartmill, M., 1977, Cranial morphology and adaptations of *Palaechthon nacimienti* and other Paromomyidae (Plesiadapoidea,? Primates), with a description of a new genus and species, *J. Hum.Evol.* **6**: 19–35.

Kay, R. F., and Kirk, E. C., 2000, Osteological evidence for the evolution of activity pattern and visual acuity, *Am. J. Phys. Anthropol.* **113**: 235–262.

Kirk, E. C. 2004 Comparative morphology of the eye in primates. *Anatomical Record A* **281A**, 1095–1103.

Land, M. F., 1981, Optics and vision in invertebrates. In: H. Autrum, ed. *Comparative physiology and evolution of vision in invertebrates. B: Invertebrate visual centers and behavior I. Handbook of sensory physiology VII/6B*, Springer Verlag, New York, pp. 471–592.

Land, M. F., and Nilsson, D. E., 2002, *Animal Eyes*, Oxford University Press, Oxford.

Le Gros Clark, W. E., 1959, *The Antecedents of Man*, Harper, New York.

Li, W. H., 2000, Genetic systems of color vision in primates, *Am. J. Phys. Anthropol.* **30** (Suppl.): 318.

Lythgoe, J. N., 1979, *The Ecology of Vision*, Clarendon Press, Oxford.

Martin, R. D., 1990, *Primate Origins and Evolution: A Phylogenetic Reconstruction*, Princeton University Press, Princeton, New Jersey.

Martin, G. R., 1990, *Birds by Night*, T & A.D. Poyser, London.

Martin, G. R., 1999, Optical structure and visual fields in birds: Their relationship with foraging behaviour and ecology, in: *Adaptive Mechanisms in the Ecology of Vision*, S. N. Archer, M. B. A. Djamgoz, E. R. Loew, J. C. Partridge, and S. Vallerga, eds., Kluwer Academic Publishers, Dordrecht, pp. 485–508.

Martin, R. D., 1973, Comparative anatomy and primate systematics, *Symp. Zool. Soc. Lond.* **33**: 301–337.

Martin, R. D., 1979, Phylogenetic aspects of prosimian behavior, in: *The Study of Prosimian Behavior*, (ed. G. A. Doyle & R. D. Martin) pp. 45–78. New York: Academic Press.

Ni, X., Wang, Y., Hu, Y., and Li, C., 2004, A euprimate skull from the early Eocene of China, *Nature* **427**: 65–68.

Niemitz, C., 1985, Can a primate be an owl? Convergences in the same ecological niche, in: *Functional Morphology in Vertebrates. Proceedings of the 1st International Symposium on Vertebrate Morphology, Giessen, 1983*, H.-R. Duncker and G. Fleischer, eds., Gustav Fischer Verlag, Stuttgart, pp. 667–670.

Ogden, T. E., 1974, The morphology of retinal neurones of the owl monkey *Aotes*, *J. Comp. Neurol.* **153**: 399–428.

Pettigrew, J. D., 1978, Comparison of the retinotopic organization of the visual wulst in nocturnal and diurnal raptors, with a note on the evolution of frontal vision, in: *Frontiers of Visual Science*, S. J. Cool and E. L. Smith, eds., Springer-Verlag, New York, pp. 328–335.

Pettigrew, J. D., 1986, Evolution of binocular vision, in: *Evolution of the Eye and Visual System*, J. R. Cronly-Dillon and R. L. Gregory, eds., Macmillan Press, New York, pp. 271–283.

Polyak, S., 1957, *The Vertebrate Visual System*, University of Chicago Press, Chicago.

Rasmussen, D. T., 1990, Primate origins: Lessons from a neotropical marsupial, *Am. J. Primatol.* **22**: 263–277.

Rasmussen, D. T., and Simons, E. L., 1992, Paleobiology of the oligopithecines, the earliest known anthropoid primates, *Int. J. Primatol.* **13**(5): 477–508.

Ravosa, M. J., and Savakova, D. G., 2004, Euprimate origins: The eyes have it, *J. Hum. Evol.* **46**: 355–362.

Ritland, S., 1982, *The Allometry of the Vertebrate Eye*, Unpublished Ph.D. dissertation, Department of Biology, University of Chicago, Chicago.

Rohen, J. W., and Castenholz, A., 1967, Über die Zentralisation der Retina bei Primaten, *Folia Primatol.* **5**: 92–147.

Rose, K. D., MacPhee, R. D. E., and Alexander, J. P., 1999, Skull of early Eocene *Cantius abditus* (Primates: Adapiformes) and its phylogenetic implications, with a reevaluation of "*Hesperolemur*" *actius, Am. J. Phys. Anthropol.* **109**: 523–539.

Ross, C. F., 1995, Allometric and functional influences on primate orbit orientation and the origins of the Anthropoidea, *J. Hum. Evol.* **29**: 201–227.

Ross, C. F., 2000, Into the light: The origin of Anthropoidea, *Annu. Rev. Anthropol.* **29**: 147–194.

Ross, C. F., 2004, The tarsier fovea: Functionless vestige or nocturnal adaptation? in: *Anthropoid Origins: New Visions*, C. F. Ross and R. F. Kay, eds., Kluwer Academic/Plenum Publishers, New York, pp. 437–577.

Sussman, R. W., 1991, Primate origins and the evolution of angiosperms, *Am. J. Primatol.* **23**: 209–223.

Sussman, R. W., 1995, How primates invented the rainforest and vice versa, in: *Creatures of the Dark: The Nocturnal Prosimians*, L. Alterman, G. A. Doyle, and M. K. Izard, eds., Plenum Press, New York, pp. 1–10.

Tan, Y., and Li, W.-H., 1999, Trichromatic vision in prosimians, *Nature* **402**: 36.

Tetreault, N., Hakeem, H., and Allman, J., 2004, The distribution and size of retinal ganglion cells in *Cheirogaleus medius* and *Tarsius syrichta*: Implications for the evolution of sensory systems in Primates, in: *Anthropoid Origins: New Visions*, C. F. Ross and R. F. Kay, eds., Kluwer Academic/Plenum Publishers, New York.

Walker, A. C., 1967, Patterns of extinction among the subfossil Madagascan lemuroids, in: *Pleistocene Extinctions. The Search for a Cause*, P. S. Martin and H. E. Wright, eds., Yale University Press, New Haven, pp. 425–432.

Walls, G. L., 1942, *The Vertebrate Eye and Its Adaptive Radiation*, Hafner, New York.

Wood Jones. F., 1916, *Arboreal Man*, Arnold, London.

CHAPTER EIGHT

Oculomotor Stability and the Functions of the Postorbital Bar and Septum

Christopher P. Heesy, Callum F. Ross, and Brigitte Demes

INTRODUCTION

The postorbital bar and septum are circumorbital structures that are important to adaptive hypotheses for the origins of primates and haplorhines, respectively. All primates possess complete postorbital bars, bony arches formed by processes of the frontal and zygomatic bones that encompass the lateral aspect of the eye. Postorbital septa, bony walls formed by the frontal, zygomatic and alisphenoid bones, walling off the orbit from the anterior temporal fossa, are limited to tarsiers and anthropoids.

Numerous functional hypotheses have been advanced for postorbital bars and septa. Many of these hypotheses can easily be rejected (Cartmill, 1970, 1972, 1980; Ravosa, 1991a,b; Ravosa et al., 2000a,b; Ross, 1994, 1995a,b, 1996, 2000, 2001; Ross and Hylander, 1996; see Heesy, 2003). Cartmill (1970, 1972, 1980; see also Collins, 1921) suggested that in therian mammals with large eyes, relatively small temporal fossae, and derived orbit

Christopher P. Heesy • Department of Anatomy, New York College of Osteopathic Medicine, Old Westbury, NY 11568 Callum F. Ross • Organismal Biology & Anatomy, University of Chicago, 1027 East 57th Street, Chicago, IL 60637 Brigitte Demes • Department of Anatomical Sciences, Health Sciences Center, Stony Brook University, Stony Brook, NY 11794-8081

convergence (orbits facing in the same direction), the plane of the bony orbit would deviate from the "plane" of the temporal fossa. Cartmill proposed that increasing orbital convergence "drags" the anterior temporalis muscle and temporalis fascia from a posterior position to a lateral position (Figure 1). In such taxa, including Primates, Cartmill suggested that contractions of the masticatory musculature, particularly the anterior temporalis muscle, would be likely to distort the lateral orbital margin, potentially disrupting oculomotor precision. Replacement of the postorbital ligament with an osseous postorbital bar should stiffen the lateral bony orbit and prevent oculomotor disruption.

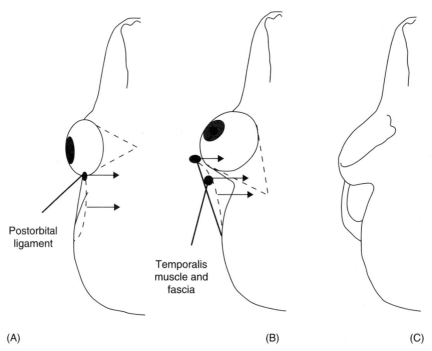

(A)   (B)   (C)

Figure 1. Cartmill's hypothesized effects of orbit orientation on postorbital ossification. In many mammals, the postorbital ligament and temporalis fascia sit posterior to the eye (A). Cartmill hypothesized that increasing orbital convergence drags the ligament anterolaterally (B), where it can be deformed during anterior temporalis contraction and temporalis fascia tension (indicated by arrows). Preventing disruptive eye movements during mastication requires the evolution of postorbital processes or a bar to prevent the fascia from encroaching on the eye (C).

Cartmill's hypothesis predicts that increasing the angular or planar deviation between the orbit and the temporalis fascia should lead to increased replacement of the postorbital ligament with postorbital processes and other stiffening structures in order to prevent disruption of oculomotor coordination. Taxa such as megachiropterans and small-bodied herpestid carnivorans with slightly lesser degrees of orbital convergence than most strepsirrhines (Cartmill, 1970) have well-developed postorbital processes, corroborating Cartmill's hypothesis (Noble et al., 2000; Ravosa et al., 2000a). Ravosa and colleagues (Noble et al., 2000; Ravosa et al., 2000a), in an extensive analysis of megachiropterans, herpestid, and felid carnivorans, found that frontation (relative vertical orbit orientation) as well as relative encephalization and relative orbit size (a factor identified by Cartmill) are contributing factors to the evolution of bars among these taxa. The association between bone strain patterns and anterior temporalis contraction has not been directly evaluated in strepsirrhines. However, *in vivo* bone strain data collected on the lateral aspect of the postorbital bar of *Otolemur* indicate that it experiences nontrivial levels of strain during mastication (Ravosa et al., 2000a,b). If the strepsirrhine skull is twisting (e.g., Greaves, 1985), or the palate is "rocking" on the interorbital region (Ross, 2001; Ross and Hylander, 1996) during mastication, the bar would be deformed. These results suggest that if a postorbital ligament were in place of the bar, the lateral orbit would experience substantial deformation.

The evolution of postorbital septa in haplorhines is related to Cartmill's hypothesis for postorbital bars in the following ways. Anthropoids have higher orbit convergence and frontation than strepsirrhines (Cartmill, 1974; Ross, 1995a, 2000). These high degrees of orbit convergence and frontation in anthropoids are also unique among mammals (Cartmill, 1974). The postorbital septum separates the anterior temporal fossa contents, which are posterolateral to the orbit, from the eye and orbital cone (Cartmill, 1994; Ross, 1995b). Just as increasing orbital convergence "drags" the anterior temporalis muscle and temporalis fascia from a posterior position to a lateral position in other mammals, the continuation of this to the uniquely high convergence and frontation in anthropoids probably accounts for a further lateral placement of these tissues (Cartmill, 1980, 1994; Ross, 1995b). In this new anatomical configuration, the anterior temporalis muscle would be in a position to directly impinge on the eye and orbital contents if the postorbital septum were not in place to prevent this (Ross, 1995a,b, 1996, 2000).

Cartmill hypothesized that the function of the postorbital septum was to prevent impingement on the eye and orbital contents in order to improve visual acuity (Cartmill, 1980). Cartmill related the importance of visual acuity to the evolution of the haplorhine retinal fovea (a depression in the retina with a concentration of cones that is associated with high visual acuity), suggesting that the septum was required to insulate the foveate eye (Cartmill, 1980; Ross, 2004; Walls, 1942). Ross' hypothesis for the function of the septum differs from Cartmill's in that he emphasized the shift in orbit orientation and the possible disruption of normal oculomotor function as the predominant factors related to the evolution of the septum, regardless of whether the lineage that first evolved it had high acuity (Ross, 1995a,b, 1996; Ross and Kay, 2004).

From the preceding, it is clear that the postorbital bar and septum have been suggested to perform two interrelated functions: (a) maintain the shape of the orbital margin, and (b) prevent disruptive movement of the eye. Two questions remain: (a) Would deformation of the orbital margin disrupt binocular vision and, if so, how? (b) Does a bony orbital margin prevent the masticatory muscles from displacing the eye and disrupting vision during mastication? In the remainder of the introduction, we will relate these hypotheses to what is currently known about the neurological and morphological systems responsible for oculomotor coordination.

Neurological and Morphological Maintenance of Eye Position

The hypotheses discussed above for the functions of the postorbital bar and septum relate to the function and maintenance of oculomotor control. In this section we provide a brief overview of the mammalian oculomotor system, describe how minor disruptive eye movements are corrected, and relate these to the anatomy of the orbit in taxa with and without postorbital bars.

The crucial function of oculomotor control is to prevent disruptive image blur across the retina during visual targeting and locomotion (Land and Nilsson, 2002; Walls, 1962). To achieve this, the oculomotor system generates precise, coordinated movements of the eyes to maintain image position on the retinae during movement of either the object or the observer (Goldberg et al., 1991). In addition to primary oculomotor commands (i.e., brainstem and cortically-based commands, reviewed in Robinson, 1975), oculomotor control is maintained by a collateral discharge system under which collateral axons from extraocular interneurons provide information on the timing and magnitude of contractions to other extraocular motoneurons (e.g., Delgado-Garcia et al., 1977; Evinger et al., 1981; Guthrie et al., 1983; Highstein and Baker, 1978; Highstein et al., 1982; Matin et al., 1982). Since significant forces other than those imparted by the extraocular muscles typically do not act on the eye, set contractions of extraocular muscles should predictably position the eye (Goldberg et al., 1991; Guthrie et al., 1983; Ruskell, 1999). These contraction and eye position data are "stored" such that for subsequent eye movements, extraocular motoneurons are recruited based upon prior eye position (Kandel et al., 2000).

Several mechanisms have been shown to effectively correct for small movements at low frequencies in humans and may be present in other taxa as well. For example, Ilg et al. (1989) found that small magnitude movements, less than 1 degree, and frequencies less than 1 Hz are corrected for, probably by compensatory extraocular muscle firing. Additional studies have found that movements due to bone transduction of the forces generated during chewing, or simply those associated with eye movements or fixation (i.e., retinal slip, the image blur traveling across the retina proportional to the rotation speed of the eyes) are compensated for by cortical calculation of perceived relative movement of images across the retina (Murakami and Cavanagh, 1998, 2001; Sasaki et al., 2002). However, these corrective adjustments of eye position seem to operate optimally at low magnitude eye displacements, and fail for large magnitude and high frequency eye movements (Ilg et al., 1989; Rashbass and Westheimer, 1961; Velay et al., 1997). What these studies do demonstrate is that neurological corrective mechanisms exist to correct for displacement of the eye in humans. If these same mechanisms exist in other primate and mammalian taxa, then small magnitude and low frequency movements of the eye due to bone transduction of masticatory forces, or disruptive movements generated by the anterior temporalis tissues may be corrected, possibly with compensatory firing of the extraocular muscles.

Several aspects of orbital anatomy are crucial for maintaining oculomotor stability. Principal among these are the ligamentous attachments that suspend the eye within the orbit and maintain the normal or rest position. In humans, connective tissue extends from Tenon's capsule (the fibrous membrane that envelops the eye from the cornea to the optic nerve) to the periorbita (Koornneef, 1992; Wolff, 1948). The check ligaments of the medial and lateral recti, which attach to the medial and lateral palpebral ligaments as well as the lacrimal and lateral orbital rim respectively in humans, have long been thought to be collagenous extensions of the recti (Bannister et al., 1995; Lockwood, 1886). It has been demonstrated that not only do they contain innervated muscular components but that these so-called orbital heads function as specialized pulleys of the extraocular muscles, and have been found in humans, macaques, rabbits, and rats (e.g., Briggs and Schachat, 2002; Clark et al., 1997, 1998; Demer et al., 1995, 1996; Khanna and Porter, 2001; Oh et al., 2001). The medial and lateral extraocular muscle pulleys are believed to function to maintain linear position of the eye and influence the position of its rotational axis (Demer, 2002; Demer et al., 2000; Haslwanter, 2002). The orbital heads may also contribute to eye movement during saccadic (foveating) movements (Briggs and Schachat, 2002; Demer, 2002).

Displacement of these extraocular muscle pulleys in humans due to abnormal development, surgery or orbital "blow out" fractures causes misalignment and improper rotation, and is associated with strabismus, which is a deviation of the visual axis of the affected eye during normal vision (Abramoff et al., 2002; Clark et al., 1998; Koornneef, 1992; Miller and Demer, 1992). The lateral pulley in particular is important when considering the functional implications of Cartmill's model. In rats and rabbits, the pulley of the lateral rectus attaches to the postorbital ligament, whereas in macaques and humans it attaches to the lateral orbital rim (Demer et al., 2000; Eglitis, 1964; Khanna and Porter, 2001). In the type of orbit deformation described by Cartmill, contraction of the anterior temporalis muscle and tension in the temporalis fascia (of which the postorbital ligament is the anteriormost portion) would deform the postorbital ligament, and also displace the pulleys of the lateral and superior recti. Displacement and the associated change in tension of the pulleys during mastication would probably cause symptoms similar to those experienced by humans with congenital dysfunction or injuries. Therefore, whereas there are neural circuits that probably maintain eye position during unavoidable displacement, these depend on maintaining the integrity of the orbit because corrective eye movements are made by the extraocular muscles, which require predictable orbit position. If the orbit were deformed during such corrective contractions of the extraocular muscles, then the subsequent eye movements would cause misalignment.

What Are the Consequences of Disruption of Oculomotor Coordination?

The disruption of oculomotor precision would have several effects on visual perception, which in turn would affect an animal's ability to interact with its environment. In taxa like primates with convergent orbits, and significant binocular field overlap and stereopsis (the cortically driven perception of depth and solidity based on binocular cues), there is a substantial field of points in space that projects images to corresponding points on each retina based on the intersection of the visual axes (point F in Figure 2). This field in visual space is called the *horopter* (Figure 2; see Howard and Rogers, 1995). Objects situated within the horopter can be fused into single images with associated depth and solidity visual information. Displacement of one eye leads to a shift in retinal horizontal disparity and the relative position of the horopter. If the object of interest did not shift position, the shift in retinal horizontal disparity would result in misalignment of the visual axes, thereby causing images to fall on noncorresponding areas of the two retinae (Leigh and Zee, 1991). This is schematically illustrated in Figure 2B in which the two visual axes no longer intersect at the object of interest, point F, but beyond it. This would lead to the perception of two versions of the same object, a phenomenon known as *diplopia*. Mastication is a cyclical behavior occurring at 1–3 Hz in primates (Ross, unpublished data), and the impingement on the eye would also be cyclically correlated with each power stroke. This oscillating misalignment of the optic axes during mastication would result in *oscillopsia* (the apparent movement of objects and the environment) (Duke-Elder and Wybar, 1973). Oscillopsia and diplopia lead to visual confusion, vertigo, and nausea in humans (Duke-Elder and Wybar, 1973). These symptoms are experienced in humans with tumors deep to the anterior temporalis muscle that pierce the postorbital septum and impart a medially directed force on the lateral rectus muscle during temporalis contraction (Crone, 1973; Emerick et al., 1997; Knight et al., 1984). For nonconjugate eye movements to result in diplopia, the visual axis of one eye must move beyond the bounds of the disparity threshold, the largest retinal disparity between the two images presented to each eye that can be fused into a single image (Howard and Rogers, 1995). The required shift in horizontal disparity for diplopia to occur differs among taxa. For example, in *Felis*, lateral eye movements leading to horizontal disparities beyond approximately 1° of arc result

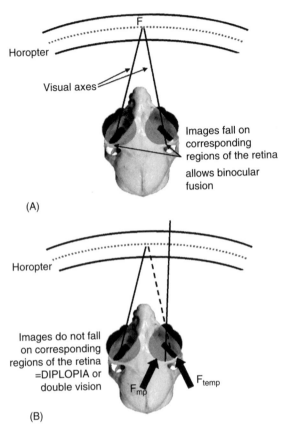

Figure 2. Schematic of disruptive movement of the right eye during either right temporalis or medial pterygoid contraction. The horopter is a substantial field of points in space that project images to corresponding points on each retina based on the intersection of the visual axes. The visual axis is defined as a line joining the object or fixation point, the center of the pupil, and the fovea centralis in anthropoids. An equivalent definition is used here for taxa without a fovea. (A) The object of interest, point F, lies within the horopter and is normal fused into a single image. (B) When displacement of the right eye occurs due to muscle contraction (the arrows indicate the presumed direction of anterior temporalis (F_{temp}) and medial pterygoid (F_{mp}) forces acting on the orbital contents), the visual axis of the right eye falls beyond the horopter F. Point F cannot be fused and is presented in different positions in space to each eye and retina. This would cause diplopia.

Table 1. Binocular disparity thresholds

Taxon	Binocular disparity threshold	Eye radius[a]	Diplopia arc length[b]	References
Equus	15 min arc[c]	20 mm	0.087 mm	Timney and Keil (1994, 1999)
Homo	10 min arc[c]	12 mm	0.035 mm	Mitchell (1966)[d]
Macaca	6–18 min arc[c]	11 mm	0.019 mm	Poggio and Poggio (1984)
Felis	30–50 min arc[c]	10.8 mm	0.094 mm	Packwood and Gordon (1975)
Suricata	15–20 min arc[c]	5.25 mm	0.031 mm	Moran et al. (1983)
Otolemur	?	7.9 mm	0.069 mm[f]	Ordy and Samorajski (1968)
Ovis	3 degrees[c]	15 mm	0.785 mm	Clarke et al. (1976) Pettigrew et al. (1984) Ramachandran et al. (1977)

Relationship between binocular disparity, eye size and diplopia in mammals. The binocular disparity threshold is the largest retinal disparity between the images presented to each eye that can be fused into a single image. Differential eye movements beyond this result in double vision. The minimum values for binocular disparity thresholds can be combined with mean eye radii to estimate the minimum distance one visual axis would need to move to generate diplopia.

[a] Eye diameter data for *Equus, Felis, Macaca, Suricata,* and *Ovis* are taken from Ritland (1982). Eye diameter data for *Homo* are from Williams et al. (1980). Data on *Otolemur crassicaudatus* were provided by C. F. Ross (unpublished data).

[b] The arc lengths were calculated by the formula $S = R\theta$, where R is the radius of the eye, θ (in radians) is the angular excursion. The minimum disparity threshold values were used for these estimates.

[c] Data derived from psychophysical or behavioral studies.

[d] Two studies have made the claim that images with disparities as high as 2 can still be fused. Howard and Rogers (1995) point out several flaws with these two studies, such as the fact that the criteria for perceiving fusion without depth are not described. Nor have these results been replicated by numerous other studies. For this reason, we use the Mitchell (1966) data.

[e] Data derived from electrophysiological studies.

[f] The diplopia arc length for *Otolemur* was computed using the *Felis* binocular disparity threshold value (see text).

in diplopia (Packwood and Gordon, 1975), whereas in humans disparities beyond 5–10° of arc result in diplopia (Mitchell, 1966; also see Table 1). Protruding movements of the eye would also lead to disparity and blur in a way similar to mediolateral movements. This point is listed in Figure 2B in which displacement of the right visual axis shifts the point of fixation, F, beyond the horopter and object of fixation. Protrusion in this study is defined as relative to the orbital plane and as such is anterolateral relative to the skull. This would shift the position of the image on the right retina to one that does not correspond to the position on the left retina. Depending on the magnitude of the movement and the size of the eye, protrusion could also lead to diplopia.

There are several potential behavioral and ecological consequences of oscillopsia and diplopia. While the animal is chewing, accurate location of potential predators or prey would be impossible, and likely the time both during and slightly after chewing would be one of visual confusion, as it is in humans that experience oscillopsia and diplopia. An arboreal animal would be unable to effectively navigate in its environment. Data on *Loris tardigradus* and the didelphid marsupial *Caluromys derbianus* (Nekaris and Rasmussen, 2001, 2003; Rasmussen, 1990), both nocturnal visually directed predators in the terminal branches, indicate that these taxa opportunistically capture flying and nonflying insect prey while feeding on flowers and fruits in the terminal branches. Diplopia and oscillopsia would prevent this feeding strategy.

For taxa like perissodactyls and artiodactyls with panoramic visual fields that spend a large component of the day masticating, oscillopsia would make the location of predators difficult. Presumably the loss of stereopsis for these animals would not be critical because the stereoscopic field is not large in these taxa (e.g., Hughes, 1977; Hughes and Whitteridge, 1973; Pettigrew et al., 1984), although the stereoscopic portion of the visual field may be critical to taxa that locomote on cliffs and other precarious substrates (Ramachandran et al., 1977).

Focus of This Study

Whereas the biomechanics of the bony orbit has received extensive attention (Hylander et al, 1991; Ravosa et al., 2000a,b; Ross, 2001; Ross and Hylander, 1996), one question common to hypotheses for the functions of postorbital bars and septa has not been directly evaluated, and that is whether the masticatory muscles displace and disrupt the eye during mastication. In this study we use ocular kinematic methods in anesthetized subjects to test specifically whether the masticatory muscles can disrupt eye position. We further distinguish between eye movement caused by anterior temporalis and medial pterygoid muscles as well as compare the magnitudes of the eye movements they cause. The goal is to characterize the eye movements, if any, caused by masticatory muscle contraction, compare these with known or estimated binocular disparity thresholds, and then reevaluate hypotheses for primate lateral orbital wall function.

METHODS

As described above, several possible neural mechanisms exist to correct for minor eye displacements. These mechanisms, when combined with voluntary and reflexive eye and head movements in a conscious subject, pose problems for measuring eye movements due to masticatory muscle contractions. We attempted to circumvent these difficulties by measuring eye movements in anesthetized subjects.

The Institutional Animal Care and Use Committee at Stony Brook University approved animal procedures. The magnitudes of eye movements that anterior temporalis and medial pterygoid are able to cause were measured in one *Otolemur*, and three *Felis* subjects sedated with intramuscular injections of ketamine and acepromazine, then anesthetized with inhalant isoflurene. Two of the *Felis* subjects had postorbital processes and short postorbital ligaments, and one had bilaterally complete postorbital bars, as verified by radiographs. Two small incisions (approximately 1–1.5 cm) were made, one in the infraorbital region and one above the center of the orbit. Small markers (consisting of flat-based metal posts) were glued onto the bone with cyanoacrylate adhesive. A high contrast cotton marker was applied to the cornea and sclera. The purpose of the scleral/corneal and bone markers was to allow calculation of eye displacement due to masticatory muscle contraction while simultaneously subtracting ancillary head movements. Indwelling bipolar fine-wire EMG electrodes were inserted into the anterior temporalis and medial pterygoid muscles, and connected to a stimulator. These electrodes consisted of two nickel–chromium alloy fine wires (0.05 mm diameter) with the insulation stripped off at the tips, and fed through a 25-gauge hypodermic needle. Two Sony DCR-PC110 high-resolution megapixel digital camcorders equipped with 120x zoom lenses were used to videotape eye movement. The lenses of the two cameras were positioned parallel and perpendicular to the orbital plane, respectively, in order to quantify mediolateral displacement or rotation and protrusion of the eye during muscle stimulation. The parallel camera was positioned anterolaterally to "face" the eye, and the perpendicular camera was positioned superiorly. We present data on movements of the eyeball marker, parallel and perpendicular to the orbital plane. A grid with 1 mm increments was placed within the plane of the tips of the bone and eye markers and videotaped for calibration. At the full macro option, much of the eye and orbit filled the screen. The anterior temporalis and medial pterygoid muscles were stimulated by applying currents (250–450 μ amp at 50 Hz) through the electrodes (Stern and Susman, 1981; Susman et al., 1982). Muscles were stimulated individually to tetanic contraction and the eye movements were recorded on digital videotape.

Eye movement data were analyzed by playing the videotape and capturing images of the calibration grid and of the eye both before and during the muscle stimulation. The video images were imported into SigmaScan Image Measurement Software (Jandel Scientific Software, San Rafael, CA). The calibration grid was used to scale measurements. Resolution was evaluated by measuring known distances and was determined to be 2.0×10^{-2} mm/pixel. Eye movement was measured by drawing calibration lines on the images between the two bone markers and measuring the linear distances between these lines and the marker on the eyeball. These linear distances, after calibration for each camera, were used to compute the translation of the eye markers parallel and perpendicular to the orbital plane relative to the orbital markers. Note that linear motion will be less than an arc motion, and in the cases of rotation of the eye, may underestimate the amount of movement generated.

RESULTS

The magnitudes of movement generated during anterior temporalis and medial pterygoid stimulation in *Otolemur* are presented in Table 2 and shown schematically in Figure 3. For *Otolemur*, movements detected by the orbital view camera were recorded for simultaneous anterior temporalis and medial pterygoid stimulation, and for anterior temporalis stimulation alone. For *Otolemur*, mediolateral eye movements caused by medial pterygoid alone are unavailable for the orbital view camera. The magnitudes of mediolateral eye movements generated during anterior temporalis stimulation in general exceed the diplopia arc length computed for *Otolemur* (see Table 1; Table 2, Orbital View Camera; Figure 3A, left). The average eye movements resulting from the simultaneous stimulation of the anterior temporalis and medial pterygoid do not exceed the estimated linear value of the disparity threshold value of *Otolemur* with the exception of one movement, which was the first muscle stimulation of the entire experiment. There was also variation in the directions of movements during these simultaneous stimulations: four made the eye translate medially, two laterally. Movements generated by isolated medial pterygoid stimulation were captured with the superior view camera

Table 2. Descriptive statistics for ocular kinematic data in *Otolemur garnettii* and *Felis catus*

	Superior view camera						Orbital view camera					
	N	Mean	SD	Max	Min	Direction	N	Mean	SD	Max	Min	Direction
Otolemur												
Anterior temporalis		No data from this camera view					5	6.8×10^{-2}	4.8×10	0.12	6.0×10^{-2}	Medial
Medial pterygoid	5	9.6×10^{-2}	5.1×10^{-2}	0.18	4.0×10^{-2}	Protrude		No data from this camera view				
AT+MP[a]		No data from this camera view					6	5.6×10^{-2}	7.1×10	$5.0 \times 10^{-}$	5.0×10^{-2}	4 med. /2lat.
Cat 1												
Anterior temporalis	8	0.38	0.10	0.45	0.22	Protrude	10	0.31	0.14	0.60	0.14	Lateral
Medial pterygoid	6	0.97	0.34	1.6	0.70	Protrude	10	0.95	0.16	1.16	0.72	Lateral
Cat 2												
Medial pterygoid		No data from this camera view					10	5.0×10^{-2}	3.1×10^{-2}	0.11	1.3×10^{-2}	Lateral
Cat 3												
Anterior temporalis	1	0.35	0.15	0.56	0.09	Protrude	10	0.51	0.13	0.68	0.29	Lateral
Medial pteryoid	1	0.66	0.13	0.75	0.38	Protrude	10	0.86	0.11	0.99	0.68	Lateral

Data are in millimeters.

[a]AT + MP is a series of simultaneous stimulations of the anterior temporalis and medial pterygoid muscles.

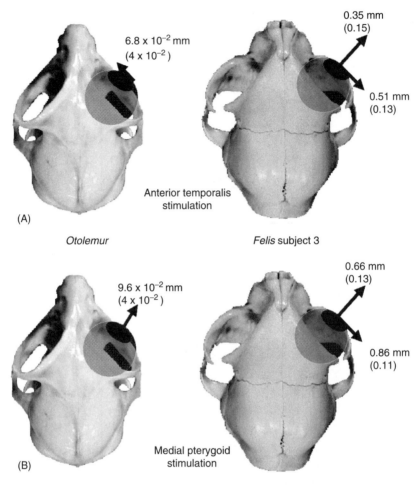

Figure 3. Eye movements during anterior temporalis and medial pterygoid stimulation. Schematics for the movements generated during anterior temporalis stimulation (A) in the *Otolemur* (left) and *Felis* subject 3 (right), and the medial pterygoid stimulation (B). The arrows indicate the directions of eye movements, the numbers are magnitudes (mean and, in parentheses, standard deviations).

(Table 2). When the medial pterygoid was stimulated in isolation, the mean amount of protrusion was nearly 0.1 mm (Figure 3B, left).

In *Felis*, anterior temporalis stimulation caused both protrusion and lateral displacement of the eye in subjects 1 and 3 (we do not have anterior temporalis data for subject 2) (Table 2, Figure 3A, right). The mean lateral displacement substantially exceeds the magnitude required to cause diplopia in *Felis* (see Table 1). Medial pterygoid stimulation caused both protrusion in subjects 1 and 3 (no data from the superior view camera for subject 2) and lateral displacement in all three subjects (Table 2). The mean lateral displacement for subject 2 was below the diplopia threshold, but the maximum displacement in this subject did exceed the estimated linear disparity threshold (Table 2). The mean lateral displacement exceeds the magnitude required to cause diplopia in two *Felis* subjects by nearly an order of magnitude. Displacement due to anterior temporalis stimulation as well as displacement and protrusion due to medial pterygoid are both responsible for disruptive-level eye movement magnitudes.

Data for simultaneous anterior temporalis and medial pterygoid stimulations were not recorded in the cats. As these two muscles in isolation cause similar eye movements, it is unlikely that synchronous effects would cancel each other out.

DISCUSSION AND CONCLUSIONS

Discussion of the Ocular Kinematic Results

The ocular kinematic data collected in *Felis* and *Otolemur* demonstrate that both the anterior temporalis and medial pterygoid muscles can cause displacements of the eye during muscle contraction. Stimulation of the anterior temporalis caused mediolateral movements in both taxa, although the directions were opposite in *Otolemur* and *Felis*. Identifying a basis for the difference in direction between taxa is not possible from the data at hand. Probable causes include differences in the position or size of the anterior temporalis and eye size and position between these taxa. Orbit orientation is not a factor because orbit convergence is not significantly different between *Otolemur* and small-bodied felids (Heesy, 2003). Also of interest is the protrusion generated by both anterior temporalis and medial pterygoid stimulation in *Felis*, and medial pterygoid stimulation in *Otolemur* (we do not have video data to evaluate protrusion during anterior temporalis stimulation in *Otolemur*). Protruding movements of the eye could also lead to disparity and blur.

In several of the simultaneous stimulations of the anterior temporalis and medial pterygoid in *Otolemur*, the resulting movements approached but did not exceed the diplopia threshold. However it is important to note that both eyes would be affected during mastication, and the movements of the eyes and visual axes would be additive (really doubled). This would pull the visual axes out of fixation, leading to diplopia.

The mean magnitudes of the movements generated in cat subjects 1 and 3 exceed those found in the *Otolemur* and cat subject 2 by an order of magnitude. We view the data collected for the second cat subject with some caution because we had difficulty stimulating either muscle during the experiment. In all cases for subject 2, the contractions were weak, and this is probably related to the difference in magnitudes found. Reasons for the order of magnitude difference between the *Otolemur* and the other two cat subjects could be related to, as above, differences between these taxa in the position or size of the anterior temporalis and eye size, and position or differences in degree of contraction due to electrode position. One reason, the functional difference between large postorbital processes and bars, can be addressed. Cat subjects 1 and 3 varied in lateral orbital wall morphology: subject 1 had a postorbital ligament whereas 3 had a complete bar. Yet the movements generated are not demonstrably different in orientation or magnitude between subjects 1 and 3 (Table 2). For this reason, it is unlikely that large postorbital processes are different from bars in insulating the eye and orbit from masticatory muscles. In addition, the fact that movements were generated in two animals with bars, *Otolemur* and one *Felis*, weakens the argument that the presence of bars alone prevents potentially disruptive eye movements.

Oculomotor Stability and the Function of the Postorbital Bar

The ocular kinematic data are relevant to explaining the function of postorbital bars in primates as well as bars and postorbital processes in other mammals. These data show that the contraction of masticatory muscles has the potential to disruptively move the eye in subjects with complete postorbital bars. Therefore, the presence of postorbital bars is not sufficient to maintain normal oculomotor function during mastication unless the movements caused by masticatory muscle contractions are corrected for by extraocular muscle activity in the awake animal. As reviewed in the earlier section entitled Neurological and Morphological Maintenance of Eye Position, the lateral orbit is the site of attachment of orbital septa and the lateral rectus pulleys. Stability of the lateral orbit is crucial because the orbital head and pulley of the lateral rectus, and possibly the superior rectus muscle as well as these orbital septa maintain eye position. Without this stability the corrective linear movements provided by the orbital head and pulley of the lateral rectus would fail. Additionally, normal oculomotor function would fail because the corollary discharge system would no longer have reliable eye and extraocular muscle position upon which to coordinate subsequent movements. As a stiff bony structure, the postorbital bar prevents substantial deformation of the lateral orbit during mastication in galagonids and presumably other mammals as well (Ravosa et al., 2000a,b), and serves the function that Cartmill first suggested, preventing orbit deformation.

Nevertheless, it is reasonable to question the relevance of a stiff lateral orbit if the anterior temporalis and medial pterygoid muscles can directly displace the eye during muscle contraction. Based on our review of the literature as presented in the introduction, we believe that several possible neural reflexes or mechanisms exist to correct displacements due to masticatory muscle contraction, all of which rely on stable orbit position. Retinocortical mechanisms have been studied in humans that correct for eye displacement or induced visual jitter (Ilg et al., 1989; Murakami and Cavanagh, 1998, 2001; see also Bridgeman and Delgado, 1984; Stark and Bridgeman, 1983). These corrective movements function to maintain the point of fixation and binocular image fusion by compensatory eye rotation (Ilg et al., 1989; see also Rine and Skavenski, 1997). These displacements can be as great as one degree (Ilg et al., 1989), which is substantially larger than disparity sensitivity based on passive retinal correspondence (Table 1). However, in humans these mechanisms seem to operate optimally for low frequency eye displacements (<1 Hz) and fail for higher frequency (>1–1.5 Hz) and magnitude (≈1°) eye movements (Ilg et al., 1989; Rashbass and Westheimer, 1961; Velay et al., 1997). Chewing frequency and associated masticatory muscle firing in many mammals exceeds this frequency (Druzinsky, 1993), and for this reason the corrective mechanism found in humans might be expected to fail in other mammals. Early primates were probably very small-bodied (Gebo, 2000; Ross, 2000). As small mammals chew at high frequencies (Druzinsky, 1993), oculomotor failure may have occurred. On the other hand, humans, like other anthropoids, have higher visual acuity than any other mammals tested (Ross, 2000), and may therefore have much less tolerance for moderate to high frequency and magnitude eye displacements. Other taxa, including strepsirrhines, may have greater tolerance than do anthropoids (Table 1). The data at hand do not serve to resolve this issue. We consider compensation to be the likely explanation for maintenance of oculomotor stability. The function of the postorbital bar is to maintain a stiff lateral orbit to prevent gross deformation of the orbital margin. This achieves two things. First, as suggested by Cartmill (1970, 1972) it ameliorates gross eye movements caused by deformation of the orbit. Second, it provides a stable substrate from which the extraocular muscle system can compensate for the remaining small scale eye movements identified by the present study as being caused by contraction of the anterior temporalis and medial pterygoid. Another alternative, which we consider to be much less probable, is that eye displacements and visual disruption occur during mastication in nonanthropoid mammals.

It is important to consider the possibility that our method for stimulating the anterior temporalis and medial pterygoid does not adequately simulate normal muscle contraction during mastication. For example, our method may stimulate more motor units than are normally contracting at any one time during chewing. This in turn may cause eye displacements that exceed those found in awake masticating animals. This possibility suggests that disruptive eye movements are dependent on degree of contraction, with low to moderate masticatory contractions not disrupting vision whereas powerful mastication may. One way, therefore, for an animal to prevent oscillopsia is simply not to masticate powerfully. Whereas our data do establish that these two muscles can potentially displace the eye, further data are required to establish the degree to which they do during mastication. Ideally, in order to address this question one would require simultaneous masticatory muscle EMG and eye movement data collected in conjunction with data that distinguish voluntary and involuntary eye movement commands from ancillary eye movements.

Finally, it is interesting to mention that certain neural compensatory mechanisms may be limited to primates. Using fMRI, Sasaki and colleagues (2002) found that area MT+ (in the visual cortex) was active in compensating for visual jitter. If MT+ is involved in compensating for jitter, such as that generated by bone transduction of masticatory forces (Murakami and Cavanagh, 1998, 2001), then the possibility exists that primates might be better at correcting for jitter than other taxa because they have a specialized or unique area MT (Allman and Kaas, 1971; Allman et al., 1973; Kaas, 1978; Kaas and Preuss, 1993). This cortical area, which is known to be responsible for the analysis of movement in visual images (e.g., Allman, 1999), has not been unequivocally identified in any group other than primates (Allman and Kaas, 1971; Allman et al., 1973; Kaas, 1978; Kaas and Preuss, 1993). An equivalent zone has been suggested to be present in tree shrews and megachiropterans, but without the morphology, input, and mapping of the contralateral visual field characteristic of the primate area MT (Kaas and Preuss, 1993; Sesma et al., 1984). If area MT in primates compensates for jitter as the human area MT+ (Sasaki et al., 2002), then such improved oculomotor stability may be unique to primates.

Oculomotor Stability and the Function of the Haplorhine Septum

The haplorhine postorbital septum is a bony wall that has been hypothesized to insulate the eye and orbital contents from impingement by the anterior temporalis muscle and fascia (Cartmill, 1980; Ross, 1995a,b, 1996). In haplorhines, the anterior temporalis partly curves around the septum unlike in strepsirrhines and other taxa without septa, where this muscle is merely laterally adjacent to the orbital contents. (Cartmill, 1980, 1994; Ross, 1995b). In general, our ocular kinematic results are consistent with the suggestion that the anterior temporalis muscle could impinge on the eye if the septum were not in place to prevent this. In taxa with postorbital processes or bars, masticatory muscle contraction is capable of generating disruptive movements of the eye, which a septum could potentially alleviate. This may not be true of *Tarsius*, however, because the orbital head of the medial pterygoid originates from the orbital wall and travels through the inferior orbital fissure out of the orbit (Cartmill, 1978, 1980; Fiedler, 1953; Ross, 1995b). The site of origin of this muscle is roughly equivalent in galagonids, although not as rostral or dorsal as in *Tarsius* (C.F. Ross, personal observation), and it is possible that protruding movement would be generated in *Tarsius* during medial pterygoid stimulation as was found for *Otolemur garnettii* in this study. The medial pterygoid stimulation results in *Otolemur* are certainly consistent with Cartmill's suggestion that medial pterygoid contractions are transmitted to the orbital contents in *Tarsius* (Cartmill, 1980). It has been alternatively suggested that the dense taut periorbita in *Tarsius* may be sufficient to prevent vibrations due to medial pterygoid contraction (Ross, 1995b).

SUMMARY

Based on the analyses and discussion above, we conclude the following:

1. A stable position of the eye is required for normal oculomotor function. Minor low magnitude, low frequency displacements of eye position can be compensated for by several neural circuits. However, based on experimental and clinical data, moderate to high magnitude and frequency displacements cannot be compensated for in humans, and presumably other mammals.

2. Anterior temporalis and medial pterygoid muscle stimulation displaces the eye in anesthetized *Otolemur* and *Felis*. The magnitudes of these displacements exceed the magnitudes required to cause diplopia in awake animals. It is possible that diplopia may occur in awake masticating animals.

3. The presence of large postorbital processes in two *Felis*, and postorbital bars in one *Otolemur* and one *Felis* did not prevent eye displacement during isolated contractions of the anterior temporalis and medial pterygoid muscles. These data suggest that postorbital bars do not insulate the eye and orbital contents from the actions of the masticatory muscles in small-bodied primates and carnivorans.

4. Based on the anatomical and clinical literature, it is probable that the postorbital bar functions to maintain rigidity of the orbit required for the extraocular muscles to position and reposition the eye, and potentially compensate for movements due to masticatory muscle activity. The function of the postorbital bar is to maintain a stiff lateral orbit to prevent gross deformation of the orbital margin. This achieves two things. First, as suggested by Cartmill (1970, 1972) it ameliorates gross eye movements caused by deformation of the orbit. Second, it provides a stable substrate from which the extraocular muscle system can compensate for the remaining small scale eye movements identified by the present study as being caused by contraction of the anterior temporalis and medial pterygoid.

5. These ocular kinematic data collected during anterior temporalis and medial pterygoid muscle stimulation suggest that the haplorhine postorbital septum, the bony wall that largely insulates the orbit from the anterior temporal fossa, would prevent the masticatory muscles from disrupting eye position. This supports the hypotheses advanced for the function of the postorbital septum by Cartmill (1980) and Ross (1995, 1996), although these data do not allow us to distinguish between these hypotheses.

ACKNOWLEDGMENTS

We thank Matt Ravosa and Marian Dagosto for inviting us to participate in the *Primate Origins and Adaptations* conference, and appreciate their patience. This work would not have been possible without Sue Van Horn and the Sherman Lab group of the Department of Neurobiology and Behavior, Stony Brook University, which lent us the three cat subjects. We are grateful for the assistance of Paul Cameau of the Stony Brook Division of Laboratory Animal Research. This research was supported by grants from L.S.B. Leakey Foundation, Sigma Xi, and IDPAS – Stony Brook University to C. P. Heesy, NSF Physical Anthropology to C. F. Ross (SBR 0109130 & 9706676), and NSF Physical Anthropology to B. Demes (SBR 9209061).

REFERENCES

Abramoff, M. D., Kalmann, R., de Graaf, M. E. L., Stilma, J. S., and Mourits, M. P., 2002, Rectus extraocular muscle paths and decompression surgery for Graves Orbitopathy: Mechanism of motility disturbances, *Invest. Ophthalmol. Vis. Sci.* **43:** 300–307.

Allman, J. M., 1999, *Evolving Brains*, W.H Freeman and Company, New York.

Allman, J. M., and Kaas, J. H., 1971, A representation of the visual field in the posterior third of the middle temporal gyrus of the owl monkey (*Aotus trivirgatus*), *Brain Res.* **31:** 85–105.

Allman, J. M., Kaas, J. H., and Lane, R. H., 1973, The middle temporal visual area (MT) in the bush baby (*Galago senegalensis*), *Brain Res.* **57:** 197–202.

Bannister, L. W., Berry, M. M., Collins, P., Dyson, M., Dussek, J. E., and Ferguson, M. W. J., eds., 1995, *Gray's Anatomy: The Anatomical Basis of Medicine and Surgery*, 38th edn., Churchill Livingstone, Edinburgh.

Bridgeman, B., and Delgado, D., 1984, Sensory effects of eyepress are due to efference. *Percep. Psychophys.* **36:** 482–484.

Briggs, M. M., and Schachat, F., 2002, The superfast extraocular myosin (MYH13) is localized to the innervation zone in both the global and orbital layers of rabbit extraocular muscle, *J. Exp. Biol.* **205:** 3133–3142.

Cartmill, M., 1970, *The Orbits of Arboreal Mammals: A Reassessment of the Arboreal Theory of Primate Evolution*, Unpublished Ph.D. dissertation, University of Chicago.

Cartmill, M., 1972, Arboreal adaptations and the origin of the Order Primates, in: *The Functional and Evolutionary Biology of Primates*, R. Tuttle,. ed.,. Aldine, Chicago, pp. 97–122.

Cartmill, M., 1974, Rethinking primate origins, *Science* **184**: 436–443.

Cartmill, M., 1978, The orbital mosaic in prosimians and the use of variable traits in systematics, *Folia Primatol.* **30**: 89–114.

Cartmill, M., 1980, Morphology, function, and evolution of the anthropoid postorbital septum, in: *Evolutionary Biology of the New World Monkeys and Continental Drift*, R. L. Ciochon and A. B. Chiarelli, eds., Plenum, New York, pp. 243–274.

Cartmill, M., 1994, Anatomy, antinomies, and the problem of anthropoid origins, in: *Anthropoid Origins*, J. G. Fleagle and R. F. Kay, eds., Plenum Press, New York, pp. 549–566.

Clark, R. A., Miller, J. M., and Demer, J. L., 1997, Location and stability of rectus muscle pulleys: Muscle paths as a function of gaze, *Invest. Ophthalmol. Vis. Sci.* **38**: 227–240.

Clark, R. A., Miller, J. M., Rosenbaum, A. L., and Demer, J. L., 1998, Heterotopic muscle pulleys or oblique muscle dysfunction? *J. A.A.P.O.S.* **2**: 17–25.

Clarke, P. G. H., Donaldson, I. M. L., and Whitteridge, D., 1976, Binocular visual mechanism in cortical areas I and II of the sheep, *J. Physiol.* **256**: 509–526.

Collins, E. T., 1921, Changes in the visual organs correlated with the adoption of arboreal life and with the assumption of the erect posture, *Trans. Ophthal. Soc.*, UK **41**: 10–90.

Crone, R. A., 1973, *Diplopia*, American Elsevier Publishing Co, New York.

Delgado-Garcia, J., Baker, R., and Highstein, S. M., 1977, The activity of internuclear neurons identified within the abducens nucleus of the alert cat, in: *Control of Gaze by Brain Stem Neurons*, R. Baker and A. Berthoz, eds., Elsevier/North-Holland Biomedical Press, New York, pp. 291–300.

Demer, J. L., 2002, The orbital pulley system: A revolution in concepts of orbital anatomy, *Ann. NY Acad. Sci.* **956**: 17–32.

Demer, J. L., Miller, J. M., Poukens, V., Vinters, H. V., and Glasgow, B. J., 1995, Evidence for fibromuscular pulleys of the recti extraocular muscles, *Invest. Ophthalmol. Vis. Sci.* **36**: 1125–1136.

Demer, J. L., Miller, J. M., and Poukens, V., 1996, Surgical implications of the rectus extraocular muscle pulleys, *J. Pediatr. Ophthalmol. Strabismus* **33**: 208–218.

Demer, J. L., Oh, S. Y., and Poukens, V., 2000, Evidence for active control of rectus extraocular muscle pulleys, *Invest. Ophthalmol. Vis. Sci.* **41**: 1280–1290.

Demer, J. L., Poukens, V., and Micevych, P., 1997, Innervation of extraocular pulley smooth muscle in monkey and human, *Invest. Ophthalmol. Vis. Sci.* **38**: 1774–1785.

Druzinsky, R. E., 1993, The time allometry of mammalian chewing movements: Chewing frequency scales with body mass in mammals, *J. theor. Biol.* **160**: 427–440.

Duke-Elder, S., and Wybar, K., 1973, *System of Ophthalmology. Volume VI. Ocular Motility and Strabismus*, The C.V. Mosby Co, St. Louis.

Eglitis, I., 1964, The orbital fascia, in: *The Rabbit in Eye Research*, J. H. Prince, ed., Springfield, Illinois: Charles C. Thomas, pp. 28–37.

Emerick, G. T., Shields, C. L., Shields, J. A., Eagle Jr., R. C., De Potter, P., and Markowitz, G. L., 1997, Chewing-induced visual impairment from a dumbbell dermoid cyst, *Ophthal. Plastic Recon. Surg.* **13**: 57–61.

Evinger, C., Baker, R., McCrea, R. A., and Spencer, R., 1981, Axon collaterals of oculomotor nucleus motoneurons, in: *Progress in Oculomotor Research*, A. F. Fuchs and W. Becker, eds., Elsevier/North-Holland, New York, pp. 263–270.

Fiedler, W., 1953, Die Kaumuskulatur der Insectivora, *Acta Anat.* **18**: 101–175.

Gebo, D. L., Dagosto, M., Beard, K. C., and Qi, T., 2000, The smallest primates, *J. hum. Evol.* **38**: 585–594.

Goldberg, M. E., Eggers, H. M., and Gouras, P., 1991, The ocular motor system, in: *Principles of Neural Science*, E. R. Kandel, J. H. Schwartz, and T. M. Jessell, eds., 3rd edn., Appleton and Lange, Norwalk, Connecticut, pp. 660–678.

Greaves, W. S., 1985, The mammalian postorbital bar as a torsion-resisting helical strut, *J. Zool., London* **207**: 125–136.

Guthrie, B. L., Porter, J. D., and Sparks, D. L., 1983, Corollary discharge provides accurate eye position information to the oculomotor system, *Science* **221**: 1193–1195.

Haslwanter, T., 2002, Mechanics of eye movements: Implications of the "orbital revolution," *Ann. NY Acad. Sci.* **956**: 33–41.

Heesy, C. P. (2003). *The Evolution of Orbit Orientation in Mammals and the Function of the Primate Postorbital Bar*, Ph.D. dissertation, Stony Brook University, Stony Brook, New York.

Highstein, S. M., and Baker, R., 1978, Excitatory termination of abducens internuclear neurons on medial rectus motoneurons: Relationship to syndrome of internuclear opthalmoplegia, *J. Neurophysiol.* **41**: 1647–1661.

Highstein, S. M., Karabelas, A., Baker, R., and McCrea, R. A., 1982, Comparison of the morphology of physiologically identified abducens motor and internuclear neurons in the cat: A light microscopic study employing the intracellular injection of horseradish peroxidase, *J. Comp. Neuro.* **208**: 369–381.

Howard, I. P., and Rogers, B. J., 1995, *Binocular Vision and Stereopsis*, Oxford University Press, New York.

Hughes, A., 1977, The topography of vision in mammals of contrasting lifestyle: Comparative optics and retinal organization, in: *The Visual System in Vertebrates*, F. Crescitelli, ed., Springer-Verlag, New York, pp. 613–756.

Hughes, A., and Whitteridge, D., 1973, The receptive fields and topographical organization of goat retinal ganglion cells, *Vision Res.* **13**: 1101–1114.

Hylander, W. L., Picq, P. G., and Johnson, K. R., 1991, Masticatory-stress hypotheses and the supraorbital region of primates, *Am. J. Phys. Anthropol.* **86**: 1–36.

Ilg, U. J., Bridgeman, B., and Hoffmann, K. P., 1989, Influence of mechanical disturbance on oculomotor behavior, *Vision Res.* **29**: 545–551.

Kaas, J. H., 1978, The organization of visual cortex in primates, in: *Sensory Systems of Primates*, C. R. Noback, ed., Plenum Press, New York, pp. 151–179.

Kaas, J. H., and Preuss, T. M., 1993, Archontan affinities as reflected in the visual system, in: *Mammal Phylogeny: Placentals*, F. S. Szalay, M. J. Novacek, and M. C. McKenna, eds., Springer-Verlag, New York, pp. 115–128.

Kandel, E. R., Schwartz, J. H., and Jessell, T. M., 2000, The control of gaze, in: *Principles of Neural Science*, E. R. Kandel, J. H. Schwartz, and T. M. Jessell, eds., 4th Ed., McGraw-Hill, New York, pp. 782–800.

Khanna, S., and Porter, J. D., 2001, Evidence for rectus extraocular muscle pulleys in rodents, *Invest. Ophthalmol. Vis. Sci.* **42**: 1986–1992.

Knight, R. T., St. John, J. N., and Nakada, T., 1984, Chewing oscillopsia: A case of voluntary illusions of movement, *Arch. Neurol.* **41**: 95–96.

Koornneef, L., 1992, Orbital connective tissue, in: *Duane's Foundations of Clinical Ophthalmology*, W. Tasma, ed., J. S., Lippincott, Co., Philadelphia, pp. 1–23.

Land, M. F., and Nilsson, D.-E., 2002, *Animal Eyes*, Oxford University Press, New York.

Leigh, R. J., and Zee, D. S., 1991, *The Neurology of Eye Movements*, 2nd edn., F. A., Davis Company, Philadelphia.

Lockwood, C. B., 1886, The anatomy of the muscles, ligaments, and fasciae of the orbit, including an account of the capsule of Tenon, the check ligaments of the recti, and of the suspensory ligament of the eye, *Journal of Anatomy and Physiology* **20**: 1–25.

Matin, L., Picoult, E., Stevens, J. K., Edwards Jr., M. W., Young, D., and MacArthur, R., 1982, Oculoparalytic illusion: Visual-field dependent spatial mislocalizations by humans partially paralyzed with curare, *Science* **216**: 198–201.

Miller, J. M., and Demer, J. L., 1992, Biomechanical analysis of strabismus, *Binoc. Vis. Eye Muscle Surg. Q.* **7**: 233–248.

Mitchell, D. E., 1966, Retinal disparity and diplopia, *Vision Res.* **6**: 441–451.

Moran, G., Timney, B., Sorenson, L., and Desrochers, B., 1983, Binocular depth perception in the Meerkat (*Suricata suricatta*), *Vision Res.* **23**: 965–969.

Murakami, I., and Cavanagh, P., 1998, A jitter after-effect reveals motion-based stabilization of vision, *Nature* **395**: 798–801.

Murakami, I., and Cavanagh, P., 2001, Visual jitter: Evidence for a visual-motion-based compensation of retinal slip due to small eye movements, *Vision Res.* **41**: 173–186.

Nekaris, K. A. I., and Rasmussen, D. T., 2001, The bug-eyed slender loris: Insect predation and its implications for primate origins, *Am. J. phys. Anthropol. Suppl.* **32**: 112.

Nekaris, K. A. I., and Rasmussen, D. T., 2003, Diet and feeding behavior of Mysore slender lorises, *Int. J. Primatol.* **24**: 33–46.

Noble, V. E., Kowalski, E. M., and Ravosa, M. J., 2000, Orbit orientation and the function of the mammalian postorbital bar, *J. Zool.,London* **250**: 405–418.

Oh, S. Y., Poukens, V., and Demer, J. L., 2001, Quantitative analysis of rectus extraocular muscle layers in monkey and humans, *Invest. Ophthalmol. Vis. Sci.* **42**: 10–16.

Ordy, J. M., and Samorajski, T., 1968, Visual acuity and ERG-CFF in relation to the morphologic organization of the retina among diurnal and nocturnal primates, *Vision Res.* **8**: 1205–1225.

Packwood, J., and Gordon, B., 1975, Stereopsis in normal domestic cat, siamese cat, and cat raised with alternating monocular occlusion, *J. Neurophysiol.* **38**: 1485–1499.

Pettigrew, J. D., Ramachandran, V. S., and Bravo, H., 1984, Some neural connections subserving binocular vision in ungulates, *Brain, Behav. Evol.* **24**: 65–93.

Poggio, G. F., and Poggio, T., 1984, The analysis of stereopsis, *Ann. Rev. Neurosci.* **7**: 379–412.

Ramachandran, V. S., Clarke, P. G. H., and Whitteridge, D., 1977, Cells selective to binocular disparity in the cortex of newborn lambs, *Nature* **268**: 333–335.

Rashbass, C., and Westheimer, G., 1961, Disjunctive eye movements, *J. Physiol.* **159**: 339–360.

Rasmussen, D. T., 1990, Primate origins: Lessons from a neotropical marsupial, *Am. J. Primatol.* **22**: 263–277.

Ravosa, M. J., 1991a, Interspecific perspective on mechanical and nonmechanical models of primate circumorbital morphology, *Am. J. Phys. Anthropol.* **86**: 369–396.

Ravosa, M. J., 1991b, Ontogenetic perspective on mechanical and non-mechanical models of primate circumorbital morphology, *Am. J. Phys. Anthropol.* **85**: 95–112.

Ravosa, M. J., Noble, V. E., Hylander, W. L., Johnson, K. R., and Kowalski, E. M., 2000a, Masticatory stress, orbital orientation and the evolution of the primate postorbital bar, *J. hum. Evol.* **38**: 667–693.

Ravosa, M. J., Johnson, K. R., and Hylander, W. L., 2000b, Strain in the galago facial skull, *J. Morph.* **245**: 51–66.

Rine, R. M., and Skavenski, A. A., 1997, Extraretinal eye position signals determine perceived target location when they conflict with visual cues, *Vision Res.* **37**: 775–787.

Ritland, S., 1982, *The Allometry of the Vertebrate Eye*, Unpublished Ph.D. dissertation, Department of Biology, University of Chicago, Chicago.

Robinson, D. A., 1975, Oculomotor control signals, in: *Basic Mechanisms of Ocular Motility and Their Clinical Implications*, G. Lennerstrand and P. Bach-y-Rita, eds., Pergamon Press, Oxford, pp. 337–378.

Ross, C., 1994, The craniofacial evidence for anthropoid and tarsier relationships, in: *Anthropoid Origins*, J. G. Fleagle and R. F. Kay, eds., Plenum Press, New York, pp. 469–547.

Ross, C. F., 1995a, Allometric and functional influences on primate orbit orientation and the origins of the Anthropoidea, *J. hum. Evol.* **29**: 201–227.

Ross, C. F., 1995b, Muscular and osseous anatomy of the primate anterior temporal fossa and the functions of the postorbital septum, *Am. J. phys. Anthropol.* **98**: 275–306.

Ross, C. F., 1996, Adaptive explanation for the origins of the Anthropoidea (Primates), *Am. J. Primatol.* **40**: 205–230.

Ross, C. F., 2000, Into the light: The origin of Anthropoidea, *Ann. Rev. Anthrop.* **29**: 147–194.

Ross, C. F., 2001, In vivo function of the craniofacial haft: The interorbital "pillar," *Am. J. phys. Anthropol.* **116**: 108–139.

Ross, C. F., 2004, The tarsier fovea: Functionless vestige or nocturnal adaptation? in: *Anthropoid Origins: New Visions*, C. F. Ross and R. F. Kay, eds., Kluwer Academic/Plenum Publishers, New York, pp. 477–537.

Ross, C. F., and Hylander, W. L., 1996, In vivo and in vitro bone strain in the owl monkey circumorbital region and the function of the postorbital septum, *Am. J. phys. Anthropol.* **101**: 183–215.

Ross, C. F., and Kay, R. F., 2004, Evolving perspectives of Anthropoidea, in: *Anthropoid Origins: New Visions*, C. F. Ross and R. F. Kay, eds., Kluwer Academic/Plenum Publishers, New York, pp. 3–41.

Ruskell, G. L., 1999, Extraocular muscle proprioceptors and proprioception, *Prog. Retin. Eye Res.* **18**: 269–291.

Sasaki, Y., Murakami, I., Cavanaugh, P., and Tootell, R. H. B., 2002, Human brain activity during illusory visual jitter as revealed by functional magnetic resonance imaging, *Neuron* **35**: 1147–1156.

Sesma, M. A., Casagrande, V.A., and Kaas, J. H., 1984, Cortical connections of area 17 in tree shrews, *J. Comp. Neurol.* **228**: 337–351.

Stark, L., and Bridgeman, B., 1983, Role of the corollary discharge in space constancy, *Percep. Psychophys.* **34**: 371–380.

Stern Jr., J. T., and Susman, R. L., 1981, Electromyography of the gluteal muscles in *Hylobates, Pongo,* and *Pan*: Implications for the evolution of hominid bipedality, *Am. J. phys. Anthropol.* **55**: 153–166.

Susman, R. L., Jungers, W. L., and Stern Jr., J. T., 1982, The functional morphology of the accessory interosseous muscle in the gibbon hand: Determination of locomotor and manipulatory compromises, *J. Anat.* **134**: 111–120.

Timney, B., and Keil, K., 1994, Local and global stereopsis in the horse, *Invest. Ophthalmol. Vis. Sci.* **35**: 2110.

Timney, B., and Keil, K., 1999, Local and global stereopsis in the horse, *Vision Res.* **39**: 1861–1867.

Velay, J. L., Allin, F., and Bouquerel, A., 1997, Motor and perceptual responses to horizontal and vertical eye vibration in humans, *Vision Res.* **37**: 2631–2638.

Walls, G. L., 1942, *The Vertebrate Eye and Its Adaptive Radiation*, Hafner, New York.

Walls, G. L., 1962, The evolutionary history of eye movements, *Vision Res.* **2**: 69–80.

Williams, P. L., Warwick, R., Dyson, M., and Bannister, L. H., eds., 1980, *Gray's Anatomy*, 36th edn., Churchill Livingstone, New York.

Wolff, E., 1948, *The Anatomy of the Eye and Orbit.*, 3rd edn., Blakiston Co, Philadelphia.

CHAPTER NINE

Primate Origins and the Function of the Circumorbital Region: What Is Load Got to Do with It?

Matthew J. Ravosa, Denitsa G. Savakova, Kirk R. Johnson, and William L. Hylander

INTRODUCTION

Due to the wide range of morphological variability within and among major primate clades, the circumorbital region has long been the focus of functional and phylogenetic investigations. As is well known, all euprimates differ from their putative ancestors in having the more derived character state of a bony postorbital bar along the lateral orbital margins extending between the frontal and jugal bones (Cartmill, 1970, 1972, 1974, 1992; Fleagle, 1999;

Matthew J. Ravosa • Department of Cell and Molecular Biology, Northwestern University, Feinberg School of Medicine, Chicago, IL; Department of Zoology, Division of Mammals, Field Museum of Natural History, Chicago, IL, **Denitsa G. Savakova** • Department of Cell and Molecular Biology, Northwestern University, Feinberg School of Medicine, Chicago, IL, **Kirk R. Johnson** • Department of Biological Anthropology and Anatomy, Duke University Medical Center, Durham, NC, **William L. Hylander** • Department of Biological Anthropology and Anatomy, Duke University Medical Center, Durham, NC

Martin, 1986, 1990, 1993; Szalay and Delson, 1979; Szalay et al., 1987; Wible and Covert, 1987). Compared to basal euprimates and strepsirrhines, anthropoids are further derived in possessing an orbital cavity largely walled off from the temporal fossa by a bony postorbital septum. Therefore, information on circumorbital function in strepsirhines, and by inference basal euprimates, is of added importance for understanding the origin of anthropoid cranial adaptations.

The purpose of this chapter is to review personal work regarding the functional significance of the postorbital bar and circumorbital region. In particular, we address two long-standing and influential hypotheses regarding the primary function of the euprimate postorbital bar: (1) that it resists facial torsion associated with masticatory stresses transmitted across the temporal fossa from the maxilla to the braincase during unilateral molar chewing and biting (Greaves, 1985, 1991, 1995); and (2) that it provides rigidity to the lateral orbital margins in order to prevent excessive ocular movements and thus maintain a high degree of visual acuity during nocturnal predation on small vertebrates and invertebrates (Cartmill, 1970, 1972). To test the facial torsion model, we analyzed *in vivo* bone-strain data so as to determine mandibular and circumorbital loading patterns in representative primates with a postorbital bar and masticatory apparatus similar to the first modern primates (Ravosa et al., 2000a–c). To investigate the nocturnal visual predation hypothesis (NVPH), we collected metric data on orbit orientation in living and fossil euprimate sister taxa, 12 strepsirrhine, and anthropoid postnatal growth series, as well as several clades of mammalian visual predators and foragers that vary interspecifically in postorbital bar formation (Noble et al., 2000; Ravosa et al., 2000a,b; Ravosa, unpublished). By integrating and evaluating experimental, comparative and ontogenetic evidence in a phylogenetic framework, we then attempt a more comprehensive characterization of adaptive transformations in skull form during the origin of Euprimates (cf., Cartmill, 1972, 1974, 1992; Fleagle, 1999; Martin, 1990, 1993; Rasmussen, 1990; Sussman, 1991).

MASTICATORY STRESS AND CIRCUMORBITAL FORM

Mammalian circumorbital features such as the supraorbital rim (browridge), postorbital bar, interorbital pillar, and postorbital septum are purportedly adaptations to resist torsion of the facial skull relative to the neurocranium during unilateral mastication (Greaves, 1985, 1991, 1995; also Rosenberger, 1986; for alternative masticatory explanations for circumorbital structures see Bookstein et al., 1999; Endo, 1966; Hilloowala and Trent, 1988a,b; Lahr and Wright, 1996; Oyen et al., 1979; Rak, 1983; Rangel et al., 1985; Russell, 1985; Tattersall, 1995; Wolpoff, 1996). This torsion of the craniofacial "cylinder" results from molar bite forces, which twist the working-side (WS) toothrow and face about a central anteroposterior axis in one direction, and relatively high balancing-side (BS) condylar reaction forces, which twist the cranial vault in the opposite direction. Jaw-adductor and nuchal forces on the working and balancing sides also produce significant axial torques during mastication and it is unlikely that their moments cancel each other out (Figure 1) (Hylander et al., 1991a,b). In turn, the facial torsion model predicts that twisting stresses are oriented 45° relative to the long axis of the skull, with the BS postorbital bar loaded axially in tension and the WS postorbital bar loaded axially in compression (Greaves, 1985, 1991, 1995).

In testing any hypothesis of craniofacial function, it is logical to inquire what constitutes a sufficiently high level of strain, stress, or load for a cranial element to be considered a "functional adaptation" to a masticatory loading regime. If a structure is optimized for bearing masticatory stresses, it is necessary to demonstrate that the observed safety factor (strain level at yield/observed strain) is no larger than 4 or 5 as this indicates that the structure is arguably optimized for resisting routine masticatory loads (i.e., it exhibits maximum strength with minimum tissue) (cf., Hylander et al., 1991a,b).[1]

Prior work demonstrates that peak-strain magnitudes from the anthropoid mandible, maxilla, and zygoma during unilateral mastication (Hylander, 1979a,b; Hylander and Ravosa, 1992; Hylander and Johnson, 1992, 1997a,b; Hylander et al., 1991a,b, 1998; Ross and Hylander, 1996) fall within the range of values at midshaft for vertebrate postcranial elements, which experience high stresses during locomotion (cf., Biewener, 1993;

---

[1] Strain is a dimensionless unit that equals the change in length (L) of an object relative to its original length (L). It is measured in microstrain (με), e.g., 1×10^{-6} cm/cm. The largest tensile strain is the maximum principal strain (ε_1), whereas the largest compressive strain is the minimum principal strain (ε_2). Tensile strains are positive and compressive strains negative. Shear strains (γmax = $\varepsilon_1 - \varepsilon_2$) are used as overall descriptors of peak masticatory forces along the facial skull. The angular value of ε_1 is measured versus the long (A) axis of one of the delta-rosette gage elements. Positive values are measured counterclockwise to the A axis and negative values are clockwise to the A axis. The angular value of ε_2 is determined by adding or subtracting 90° to the value of ε_1.

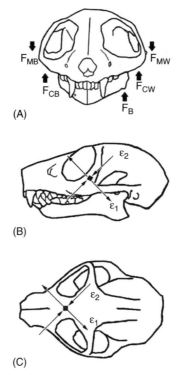

(A)

(B)

(C)

Figure 1. Frontal (A), lateral (B), and dorsal (C) views of an adult greater galago skull during left-sided unilateral mastication (adapted from Ravosa et al., 2000a). Extrinsic forces causing facial torsion about a central cranial axis are the bite force (F_B), working-(F_{CW}) and balancing-side (F_{CB}) condylar reaction forces, and working-(F_{MW}) and balancing-side (F_{MB}) jaw-adductor forces (A). Although not depicted, facial torsion is also likely to be affected by working- and balancing-side nuchal forces. In this case, chewing results in a counterclockwise rotation of the facial skull versus the cranial vault (A). Expected strains during left-sided mastication are depicted for the left postorbital bar (B) and interorbit (C). Both ε_1 (tension) and ε_2 (compression) are oriented 45° versus the cranial axis. When chewing changes from the left side of the dental arcade to the right, principal-strain directions (ε_1 and ε_2) reverse due to a shift in the direction of torsion. Predicted strain directions for the WS postorbital bar are similar in the facial torsion model and NVPH.

Lanyon and Rubin, 1985). In contrast, peak-strain levels during mastication for the anthropoid circumorbital region are much lower than those values recorded at the mandible, maxilla, and zygomatic arches (Bouvier and Hylander, 1996a,b; Hylander and Johnson, 1992, 1997a,b; Hylander and

Ravosa, 1992; Hylander et al., 1991a,b, 1998; Ross and Hylander, 1996). This presence of a significant strain gradient indicates that circumorbital structures are routinely overbuilt for stresses encountered during routine biting and chewing. Thus, the amount of circumorbital bone mass could be decreased significantly without causing it to experience structural failure during normal masticatory behaviors (Hylander et al., 1991a,b).

On the other hand, if circumorbital peak strains were large, then a considerable reduction in the amount of cortical bone at the postorbital bar, interorbital pillar, and browridge would likely result in dangerously high strains (as would occur with the mandibular corpus and symphysis). This in turn would suggest that circumorbital structures are functional adaptations to resist masticatory stresses since they appear designed to maximize strength with a minimum of material.

Until recently all studies of primate circumorbital strains were of anthropoids, taxa which possess the derived condition of a bony postorbital septum along the lateral orbital wall. Compared to anthropoids, greater galagos (Strepsirrhini, Primates) recruit relatively less BS jaw-adductor force during unilateral mastication, which explains why they possess unfused mandibular symphyses (Hylander, 1979a,b; Hylander et al., 1998, 2000, 2004; Ravosa and Hogue, 2004; Ravosa et al., 2000a–c; Vinyard et al., this volume). Moreover, as the first modern primates had unfused symphyses and postorbital bars, galagos likely provide a good extant analog for the masticatory complex of basal euprimates (Ravosa, 1991a, 1996, 1999; Ravosa and Hylander, 1994). As there are no circumorbital strain data for an alert mammal with only a postorbital bar (but no postorbital septum), the galago *in vivo* analyses are of further importance for interpreting circumorbital form in other mammals with this more common character state (Cartmill, 1970, 1972; Noble et al., 2000; Pettigrew et al., 1989; Ravosa et al., 2000a,b). In evaluating the galago experimental evidence vis-à-vis the facial torsion model, peak-strain magnitude data are considered first and the principal-strain direction data are discussed next (Ravosa et al., 2000a,b).

Galago Circumorbital Peak-Strain Magnitudes

The mean peak shear strain (γ_{max}) recorded from the galago left postorbital bar during powerful unilateral mastication of tough foods is 534 $\mu\epsilon$ when chewing is on the left (WS) side and 174 $\mu\epsilon$, when chewing is on the right

Table 1. Greater galago peak shear-strain magnitudes during powerful mastication of tough foods

Gage site	Power strokes, n	γ_{max} mean, $\mu\epsilon$	γ_{max} range, $\mu\epsilon$
WS postorbital bar	110	534	87–1320
BS postorbital bar	113	174	36–587
Dorsal interorbit	469	420	44–1221
WS mandibular corpus	142	1197	168–2653
BS mandibular corpus	119	216	68–746

(BS) side (Table 1; Figure 2) (Ravosa et al., 2000a,b).[2] Observed differences in WS/BS postorbital bar shear strain ratios mirror WS/BS variation in mandibular corpus strain ratios (Table 1). This appears due to the fact that galagos recruit relatively low levels of BS jaw-adductor force (Hylander, 1979a,b; Hylander et al., 1998, 2000, 2004; Ravosa et al., 2000a,b; Vinyard et al., this volume). As left- and right-sided powerful chews are equivalent for the dorsal interorbit, galagos exhibit a mean peak shear strain of 420 $\mu\epsilon$ during forceful molar biting and chewing (Table 1; Figures 2 and 3)(Ravosa et al., 2000a,b). Lastly, interorbital and WS postorbital bar strains are of similar magnitude and peak values for both sites are significantly lower than the mean peak shear strain of 1197 $\mu\epsilon$ for the WS corpus during mastication (Table 1; Figures 2 and 3).

As the galago postorbital bar and dorsal interorbit are overbuilt for loads encountered during the vigorous molar processing of tough/hard foods, the principal function of circumorbital structures cannot be to resist facial torsion during mastication (contra Greaves, 1985, 1991, 1995; Rosenberger, 1986). As the anthropoid skull is also characterized by a significant strain gradient (Bouvier and Hylander, 1996a,b; Hylander and Johnson, 1992, 1997a,b; Hylander and Ravosa, 1992; Hylander et al., 1991a,b; Ross and Hylander, 1996), there are no experimental data supporting the hypothesis that primate circumorbital features are functionally adapted to counter routine masticatory stress (contra Bookstein et al., 1999; Endo, 1966; Hilloowala and Trent,

---

[2] Procedures for bonding three-element 120-ohm stacked delta-rosette gages, for recording bone strain, and for analyzing cranial strain data are detailed elsewhere (Hylander, 1979a; Hylander et al., 1991a,b, 1998). Seven ACUC-approved experiments were performed on three adult male greater galagos, such that each was used more than once (Ravosa et al., 2000b). In all cases, an attempt was made to align the A element of each rosette the following way: dorsal interorbit – along the midsagittal plane, postorbital bar – along the bar's long axis, and mandibular corpus – parallel to its long axis. Chewing side during mastication of tough/hard foods (prune nut, dried prune, dried apricot, dried gummi bear) was identified via monitoring of surface electromyograms (EMGs) for both superficial masseters, i.e., WS EMG is relatively higher than BS EMG (Hylander et al., 2000).

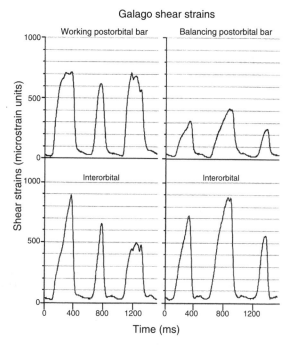

Figure 2.   Plot of peak shear strains recorded simultaneously from the dorsal interorbit and left postorbital bar of a greater galago during powerful mastication on both sides of the face (Ravosa et al., 2000b) (Table 1). Postorbital bar strains for left-sided chews (working side) are much higher than right-sided chews (balancing side). Due to the midsagittal location of the dorsal interorbital gage, such strains are similar in level regardless of chewing side. Interorbital and WS postorbital bar strains are also similar in peak magnitude.

1988a,b; Lahr and Wright, 1996; Oyen et al., 1979; Rak, 1983; Rangel et al., 1985; Russell, 1985; Tattersall, 1995; Wolpoff, 1996).

It is thus evident that only certain primate cranial elements experience relatively high-peak strains during powerful biting and chewing—mandible (corpus and symphysis), maxilla, anterior portion of the zygomatic arch, the anterior root of the zygoma, or the infraorbital region. Such structures appear designed to minimize cortical bone and maximize strength for countering routine, cyclical masticatory loads (Hylander and Johnson, 1997a,b). This regional disparity in facial strains highlights the underlying nature of safety factors throughout the skull. When analyzing masticatory related features, it

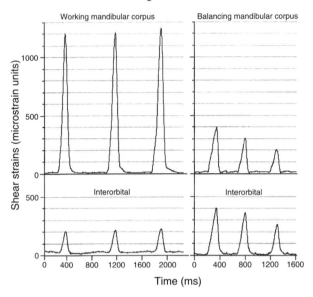

Figure 3. Plot of peak shear strains recorded simultaneously from the dorsal interorbit and right mandibular corpus during forceful unilateral mastication (Ravosa et al., 2000b) (Table 1). Corpus strains for right-sided (WS) chews are much higher than left-sided (BS) chews, while interorbital strains are similar in magnitude regardless of chewing side. Interorbital strains are considerably lower than WS corpus strains, which indicates a significant strain gradient along the facial skull.

is correct to base estimates of safety factors on the fact that cortical bone undergoes monotonic yield failure at 6800 με or fatigue failure at 3000 με (following 10^6 loading cycles – Currey, 1984). However, in analyses of the circumorbital region, it seems more appropriate to estimate safety factors based on the fact that cortical bone experiences ultimate failure at 16,000 με—a load indicative of the magnitude required for an accidental or traumatic force to significantly reduce the fitness of an individual (Hylander and Johnson, 1997a,b; Ravosa et al., 2000b,d). This variation also suggests that epigenetic and genetic control of craniofacial form results in circumorbital and neurocranial safety factors for traumatic loads, and thus safety factors for masticatory stresses significantly higher (10–20 times) than for elements such as the mandible, anterior root of the zygoma and zygomatic arch. A similar pattern is observed in comparisons of rat ulnar strains (large in magnitude) during locomotion with rat calvarial strains (low in magnitude) during biting and chewing (Rawlinson et al., 1995).

Galago Circumorbital Principal-Strain Directions

The average angular direction of peak ε_1 along the left postorbital bar is 53° relative to the A element during left-sided mastication and −44° during chewing on the right (Table 2; Figure 4) (Ravosa et al., 2000a,b). Depending on chewing side, there is a characteristic reversal pattern in the angle of ε_1 relative to the skull's long axis (±49°), much as predicted by the facial torsion model (Greaves, 1985, 1991, 1995). During left-sided chewing peak ε_1 is directed anterosuperiorly, whereas during right-sided mastication peak ε_1 is oriented posterosuperiorly. Contrary to the facial torsion model, however, peak principal strains are not oriented orthogonal and parallel to the axis of the postorbital bar. Thus, the primate postorbital bar is not compressed axially on the working side, nor does it experience axial tension on the balancing side (Ravosa et al., 2000a,b).

The average angular direction of peak ε_1 along the dorsal interorbit is −49° relative to the long axis of the skull during powerful left-sided mastication and 63° during chewing and biting on the right (Table 2; Figure 5) (Ravosa et al., 2000a,b). Thus, similar to the postorbital bar, there is a stereotypical reversal or flip-flop in the direction of ε_1, such that when mastication occurs on the left peak ε_1 is oriented posterolaterally to the left, while during chewing on the right peak ε_1 is directed posterolaterally to the right. Assuming symmetry of left- and right-sided loading patterns for a midsagittal structure, such as the interorbital pillar, the mean angle of ε_1 (±56°) is close to, but somewhat larger than, the 45-degree predictions of the facial torsion model (Greaves, 1985, 1991, 1995). The most probable cause why the direction of ε_1 at the dorsal interorbit is typically larger than this prediction is that, in addition to facial twisting, the galago interorbit also likely experiences bending in the frontal plane during unilateral mastication. This circumorbital bending regime occurs in anthropoids and is due likely to bilateral masseter contraction and the consequent inferior deflection of the zygomatic arches and, in turn, the postorbital bars (Hylander and Johnson, 1997a; Hylander et al., 1991a,b).

Table 2. Angular direction (°) of maximum principal strain (peak ε_1) during powerful mastication

Chew side[a]	Power strokes, n	Dorsal interorbit, ε_1 mean (min/max)	Left postorbital bar, ε_1 mean (min/max)
Galago 1: Experiment A			
Left	22	−14 (−4/−18)	−
Right	31	55 (47/60)	−
Mean	53	±35	−
Galago 1: Experiment B			
Left (WS)	43	−	57 (54/60)
Right (BS)	49	−	−40 (−34/−49)
Mean	92	−	±49
Galago 1: Experiment C			
Left	46	−67 (−60/−70)	−
Right	61	65 (58/77)	−
Mean	107	±66	−
Galago 2: Experiment A			
Left (WS)	32	−58 (−43/−80)	46 (23/86)
Right (BS)	22	74 (59/116)	−80 (−76/−87)
Mean	54	±66	±63
Galago 2: Experiment B			
Left	38	−66 (−63/−77)	−
Right	38	83 (34/95)	−
Mean	76	±75	−
Galago 3: Experiment A			
Left (WS)	35	−45 (−38/−53)	57 (39/80)
Right (BS)	42	49 (43/63)	−11 (−18/6)
Mean	77	±47	±34
Galago 3: Experiment B			
Left	44	−46 (−35/19)	−
Right	58	51 (38/57)	−
Mean	102	±49	−
Grand Mean: Total	469	±56	−
Left	217	−49	−
Right	252	63	−
Grand Mean: Total	223	−	±49
Left	110	−	53
Right	113	−	−44

[a]WS = working-side direction of tension; and BS = balancing-side value. As before, it is unnecessary to note the chewing side for the dorsal interorbital gage.

Facial Torsion and the Evolution of the Primate Postorbital Bar

Based on the galago strain-magnitude data, it appears that the primate postorbital bar is overbuilt to resist masticatory stresses. Thus, there is more than enough cortical bone to provide rigid lateral orbital margins. Perhaps once ossified (for whatever nonmasticatory function), the postorbital bar must

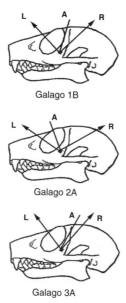

Galago 1B

Galago 2A

Galago 3A

Figure 4. Average directions of ε_1 for the left postorbital bar during mastication on the working (L) and balancing (R) side (adapted from Ravosa et al., 2000b) (Table 2). The line "A" marks the orientation of the A element of the delta-rosette gage in each of six experiments. The grand mean ($\pm 49°$) is close to the predictions of the facial torsion model; however, this is opposite what should be found in galagos. Only WS postorbital bar stain patterns are as predicted by the NVPH.

then be constructed of an amount of tissue sufficient to counter accidental, traumatic loads to its outwardly flared margins (Hylander and Johnson, 1992, 1997a,b; Hylander and Ravosa, 1992; Hylander et al., 1991a,b; Ravosa et al., 2000a,b). As argued earlier, circumorbital structures are probably not specially designed to counter masticatory stresses because bone in this area could be reduced considerably without risking structural failure associated with forceful chewing and biting (Bouvier and Hylander, 1996a,b; Hylander and Johnson, 1992, 1997; Hylander and Ravosa, 1992; Hylander et al., 1991a,b; Ravosa et al., 2000a,b,d; Ross and Hylander, 1996). To do so, however, increases the risk of fracturing the circumorbital region due to accidental, traumatic external forces of a nonmasticatory nature (e.g., excessive loads during falls), and this would significantly reduce the fitness of such an organism (Hylander and Johnson, 1992, 1997a,b; Hylander and Ravosa, 1992;

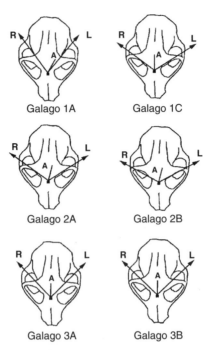

Figure 5. Average directions of ε_1 during left-sided (L) and right-sided (R) mastication along the dorsal interorbit of greater galagos (adapted from Ravosa et al., 2000b) (Table 2). The line marked "A" represents the orientation of the A element of the delta-rosette strain gage during each of the three experiments. The direction of peak maximum principal strain (ε_1) is always given relative to the A element. The grand mean of $\pm 56°$ is close to predictions of the facial torsion model, although this may be opposite what should be observed in galagos (see text).

Hylander et al., 1991a,b; Ravosa et al., 2000a,b,d). Therefore, selection for circumorbital and neurocranial safety factors of higher magnitude than those for masticatory elements is especially critical for any skeletal structure that houses and/or protects special sense organs.

　　If the evolution of a novel circumorbital structure such as the postorbital bar has been a two-step process, this has direct implications for the facial torsion model. For instance, it could be argued that the postorbital bar of the first primates was initially an adaptation to resist facial twisting during mastication and that in subsequent basal taxa, the amount of cortical bone along the lateral orbital margin was increased so as to ensure an adequate safety factor for accidental nonmasticatory forces. That is, once of adequate size (to counter traumatic loads), the postorbital bar of modern primates now experiences negligible strains during postcanine biting and chewing.[3]

Nevertheless, several lines of evidence directly refute the importance of the facial torsion model. First, and perhaps most importantly, neither the postorbital bars nor browridges of primates are oriented 45° relative to the cranial long axis (Hylander and Ravosa, 1992; Ravosa, 1991a,b; Ravosa et al., 2000a,b). Second, our analysis of jaw-adductor activity patterns and condylar reaction forces indicates that, while the galago circumorbital region is likely twisted, the underlying causative forces differ from the predictions of the facial torsion model. Finally, circumorbital strain directions for three anthropoids also contradict the facial torsion model (Hylander et al., 1991a,b; Ross and Hylander, 1996)—a finding which undermines the applicability of this model to primates.

Interestingly, the galago strain-direction pattern contrasts with that for the anthropoid dorsal interorbital region. Whereas papionins and owl monkeys also show a reversal pattern, interorbital strain directions are opposite those for galagos and thus, contrary to predictions of the facial torsion model (Hylander and Johnson, 1992; Hylander and Ravosa, 1992; Hylander et al., 1991a,b; Ross and Hylander, 1996). There are two interrelated causes of this apparent suborder difference in circumorbital principal-strain directions. In contrast to anthropoids (and the facial torsion model), condylar reaction forces and jaw-adductor forces during mastication are largely localized to the working side of the galago face (and presumably that of all strepsirhines and "prosimians" with unfused symphyses – Hylander, 1979a,b; Hylander et al., 1998, 2000, 2004; Ravosa and Hylander, 1994; Ravosa and Hogue, 2004; Ravosa et al., 2000a; Vinyard et al., this volume). This is reflected in the large disparity in WS/BS peak-strain magnitude ratios for bilateral structures such as the postorbital bar and mandibular corpus (Table 1). On the other hand, anthropoids recruit relatively higher BS jaw-adductor forces and, in turn, experience less variation in cranial peak strains between working and balancing sides (Hylander, 1979a; Hylander et al., 1991a,b, 1998; Ross and Hylander, 1996). Therefore, despite the fact that primate circumorbital structures are not designed to resist masticatory stress, it is now evident that suborder differences in skull form

---

[3] Our explanation differs from the argument that a bony postorbital bar serves to protect the lateral aspect of the eye from injury when a greater proportion of the lateral orbital margin lies exposed to branches during locomotion (Prince, 1953; Simons, 1962). This latter suggestion is flawed because an ossified postorbital bar would characterize a significantly greater number of mammalian clades if its function were solely protective (Cartmill, 1970, 1972).

significantly influence suborder patterns of stress along the circumorbital region. Indeed, as galago forces during mastication differ from expectations or assumptions of the facial torsion model, but nonetheless result in circumorbital strain directions in support of this model; it is unlikely that the strepsirhine (Ravosa et al., 2000a,b) or anthropoid (Hylander et al., 1991a,b) skull functions as a simple hollow cylinder. This highlights the considerable benefit of modeling complex biological systems with a broad, phylogenetic characterization of *in vivo* patterns of functional variation (Hylander et al., 1998, 2000, 2004; Lauder, 1995; Ravosa et al., 2000a,b).

NOCTURNAL VISUAL PREDATION AND CIRCUMORBITAL FORM

The Nocturnal Visual Predation Hypothesis (NVPH) argues that in the first modern primates, a shift to nocturnal visual predation on small invertebrates and vertebrates necessitated more anteriorly directed and medially approximated orbital apertures and eyeballs (Figure 6A) (Cartmill, 1970, 1972, 1974, 1992; also Collins, 1921). This increased orbital convergence in turn results in a larger binocular field for greater stereoscopic vision, as well as an exceptionally clear retinal image during nocturnal prey location and capture at short distances (Allman, 1977, 1982; Cartmill, 1970, 1972, 1974, 1992). The NVPH further explains the relatively larger orbits and grasping hands/feet possessing digits with nails as basal euprimate adaptations to nocturnality in an arboreal, terminal-branch milieu (Cartmill, 1970, 1972, 1974, 1992; Covert and Hamrick, 1993; Dagosto, 1988; Hamrick, 1998, 1999; Heesy and Ross, 2001; Kay and Cartmill, 1974, 1977; Kay and Kirk, 2000; Lemelin, 1999). In contrast, putative ancestors had the primitive mammalian condition of relatively small and less convergent orbits, larger olfactory complexes, digits with claws, and more terrestrial locomotor specializations (Cartmill, 1970, 1972, 1974, 1992; Kay and Cartmill, 1974, 1977; but see Bloch and Boyer, 2002; Bloch et al., this volume). Aside from alternative and less compelling claims that convergent orbits and grasping appendages are adaptations for nocturnal visual foraging on fruits and flowers in terminal branches (Crompton, 1995; Rasmussen, 1990; Sussman, 1991), the NVPH has become a well-accepted model of euprimate origins (Fleagle, 1999; Martin, 1990, 1993; Ravosa et al., 2000a).

The NVPH also addresses the correlated effects of changes in orbital form on the function of the circumorbital region. During molar chewing and bit-

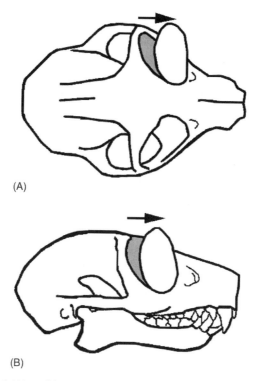

(A)

(B)

Figure 6. Dorsal (A) and lateral (B) views of an adult greater galago skull (adapted from Ravosa et al., 2000a). Orbital convergence (A) refers to the extent the orbital margins face forward, such that orbits directed more anteriorly are convergent (arrow). This morphology is posited to facilitate binocular stereoscopic acuity and increased depth perception. Orbital frontation (B) refers to the degree of verticality of the margins, such that orbits more orthogonal to the cranial long axis are frontated (arrow). This condition appears linked primarily to greater basicranial flexion and ultimately increased levels of encephalization. Basal primates are derived relative to plesiadapiforms and other sister taxa in exhibiting greater convergence and increased frontation.

ing, the lateral orbital margins on both sides of the face, thought to be pulled posteroinferiorly by the masseter and temporalis (Cartmill, 1970, 1972)—a loading pattern which is supported by the galago WS (but not BS) postorbital bar strain data (Figure 1B) (Ravosa et al., 2000a,b). In species with increased convergence, the orbital apertures are directed more out of the plane of the temporal fossa, and the above-circumorbital loading regime is posited to entail greater disruption of the orbital contents. Thus, the lateral orbital margins of taxa, like euprimates, are compressed more along the optical axis instead of along the orbital aperture as in more primitive mammals with lower convergence levels (Cartmill, 1970, 1972). It follows that to ensure a high degree of stereoscopic acuity in a nocturnal organism that hunts and forages while processing its prior meal,[4] the postorbital bar of visual predators arguably functions to stiffen the lateral orbital margins and thus resist ocular deformation during mastication (Cartmill, 1970, 1972).

Though more vertical, frontated orbital apertures also deviate from the plane of the temporal fossa (Figure 6B), the role of orbital frontation has not figured into discussions on the origin of the primate postorbital bar (Cartmill, 1970, 1972). This dichotomy between the effects of convergence versus frontation on postorbital bar development and ocular movements is arguably unnecessary. Variation in one or both orbital parameters could influence postorbital bar formation. In fact, the visually oriented frugivore *Caluromys* differs from other didelphimorphs in exhibiting increased frontation (but not greater convergence – see in a later section), a relatively larger brain and larger postorbital processes of the frontal and jugal bones (Cartmill, 1970, 1972, 1974, 1992; Rasmussen, 1990).[5] Due to both elevated convergence and frontation, the anthropoid postorbital septum is also posited to function in dampening ocular oscillations (Cartmill, 1980; Ross, 1995). In addition to higher convergence levels, basal primates were more encephalized and frontated than plesiadapiformes (Cartmill, 1992; Fleagle, 1999; Martin, 1990; Simons, 1962; Szalay and Delson, 1979).

To investigate the relationship between orbital orientation and the presence of a postorbital bar, three clades varying interspecifically in postorbital bar formation were examined (pteropodids, herpestids, felids – Noble et al., 2000; Ravosa et al., 2000a). To assess the link between increased orbital convergence and visual strategy, as well as phylogenetic and size-related patterns of orbital form in primate and nonprimate mammals, a series of adult interspecific analyses were performed (Ravosa and Savakova, 2004). As the presence of

---

[4] One possible shortcoming of the NVPH is that it is unclear whether an organism would actively predate and/or forage while simultaneously chewing. If most animals are sedentary while eating, then why would a rigid postorbital bar be necessary for a high level of visual acuity (when predation is temporarily interrupted)? One related observation is that, unlike squirrels that return to the center of a tree when feeding, basal euprimate and nonprimate analogs tend to remain on terminal branches and thus apparently maintain elevated activity levels (cf. Cartmill, 1992).

[5] In a didelphimorph marsupial sample with species means of 30–38 mm for the nasion-inion chord (n = 7 taxa, 68 adults), three sister taxa of the genus *Caluromys* (mean = 44.7°, range = 42.4–48.9°) are significantly more frontated than four species of the genera *Philander*, *Chironectes*, and *Metachirus* (mean = 32.2°, range = 25.3–37.7°) (Mann-Whitney U test, p = 0.034).

intrafamilial variation in activity cycle is vitally important for evaluating competing explanations for forward-facing orbits, phylogenetically restricted comparisons were performed in seven didelphimorphs (n=68),[6] 11 procyonids (n=88), 28 herpestids (n=183), and five tupaiids (n=21). To examine relative and absolute levels of orbital convergence in the earliest euprimates, four omomyids (n=8, nocturnal predators) and four adapids (n=12, diurnal foragers) were compared to functional analogs and sister taxa: 11 extant primate nocturnal predators (n=48),[7] 31 felids (n=208, nocturnal predators), 28 herpestids (n=183, mostly diurnal predators), a dermopteran (n=6, nocturnal foragers), five tupaiids (n=21, mostly diurnal predators), 64 pteropodids (n=277, nocturnal foragers), and two plesiadapiforms (n=2, diurnal foragers) (Ewer, 1973; Fleagle, 1999; Martin, 1990, 1993; Nowak, 1999; Richard, 1985; Simons, 1962; Zeveloff, 2002). A benefit of this last, higher-level comparison over prior work (Ravosa et al., 2000a) is the inclusion of data for the most appropriate analogs based on body size, activity cycle, and feeding behavior for the first modern primates: living primate nocturnal predators.

The allometry of orbital orientation was further investigated for the postnatal ontogeny of six strepsirhines: *Propithecus verreauxi* (11 adults and 24 nonadults), *Eulemur fulvus* (10 adults and 27 nonadults), *Hapalemur griseus* (11 adults and 20 nonadults), *Otolemur crassicaudatus* (11 adults and 21 nonadults), *Nycticebus coucang* (12 adults and 31 nonadults), and *Perodicticus potto* (12 adults and 34 nonadults) (Ravosa and Savakova, 2004). These primates were selected because they represent both extant strepsirrhine infraorders and most extant superfamilies, equal the interspecific range of variation in primate convergence values, and closely approximate the skull morphology of basal euprimates. In all taxa, at least five adults of each sex were examined.

Orbital Form and Patterns of Covariation

Herpestids, felids, and pteropodids—all exhibit positive correlations between orbital convergence and skull size (also procyonids and tupaiids – Table 3).

---

[6] To eliminate the chance that an enlarged masticatory complex in folivorous didelphimorphs would differentially influence the position of the lower orbital margin, and thus an evaluation of variation in orbital convergence across this clade (Cartmill, 1972), no such species are included.

[7] Orbital convergence species means for four cheirogaleids and two tarsiids are from Ross (1995). Means for three galagids and two lorisids are based on adult data collected by the authors or Ross (1995). Linear dimensions for all nocturnal primate faunivores were taken by the authors.

Table 3. Bivariate correlations (r) for mammalian interspecific series[a]

Family, n	Orbital convergence versus nasion-inion chord[b]	Orbital frontation versus nasion-inion chord[b]	Orbital convergence versus orbital frontation
Felidae (31 taxa, 208 adults)	0.257*	−0.479****	0.261*
Herpestidae (28 taxa, 183 adults)	0.415***[c]	−0.041ns	0.309**
Pteropodidae (64 taxa, 277 adults)	0.644****	−0.207**[d]	0.112ns
Procyonidae (11 taxa, 88 adults)	0.560****[e]	−0.662***[f]	0.419*
Tupaiidae (5 taxa, 21 adults)	0.770***[g]	−0.409*	0.151ns

[a]Significance levels: **** = $p < 0.01$; *** = $p < 0.05$; ** = $p < 0.10$; * = $p < 0.15$; ns = $p < 0.15$.
[b]Comparisons versus other cranial measures differ little.
[c]When restricted to only diurnal species ($n = 27$) and thus a single activity cycle, $r = 0.556$ at $p < 0.01$.
[d]Due to two larger-sized outlier species, pteropodids exhibit a negative correlation between orbital frontation and size. With these data excluded, the correlation is no longer significant ($p > 0.15$).
[e]When restricted to only nocturnal species ($n = 8$) and thus a single activity cycle, $r = 0.838$ at $p < 0.01$.
[f]This represents the value with one large-sized outlier species eliminated from the sample ($n = 10$). Analysis of all 11 sister taxa results in a nonsignificant correlation ($r = -0.273$; $p > 0.15$).
[g]When restricted to only diurnal species ($n = 4$) and thus a single activity cycle, $r = 0.989$ at $p < 0.01$.

Of these three clades, only larger-sized pteropodids and herpestids with greater convergence tend to have postorbital bars (Figure 7). Contrary to predictions of the NVPH (Cartmill, 1970, 1972), felids do not show a link between postorbital bar formation and orbital convergence (Noble et al., 2000; Ravosa et al., 2000a).

Felids also exhibit a negative correlation between orbital frontation and skull size (also procyonids and tupaiids – Table 3), with postorbital bar formation bars tending to characterize only smaller, more frontated cats (Figure 8A). Herpestids with greater orbital frontation also tend to possess bony postorbital bars (Figure 8B). Pteropodids, however, do not exhibit this pattern (Noble et al., 2000; Ravosa et al., 2000a). In addition, size-related decreases in orbital frontation are not observed in the less-encephalized mongooses and fruit bats (cf., Figures 8B and 9).

Felids, basal euprimates, and extant primate nocturnal predators exhibit significantly greater orbital convergence than plesiadapiforms, tupaiids, dermopterans, pteropodids, and herpestids (Figure 10; significant Y-intercept difference or transposition of the euprimate/felid LS line above that for the remaining mammalian clades – ANCOVA, $p < 0.001$) (Ravosa and Savakova, 2004; Ravosa et al., 2000a). In tupaiids, procyonids, and herpestids, nocturnal predators (Ewer, 1973; Nowak, 1999; Zeveloff, 2002) display higher

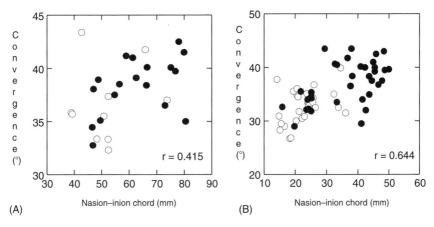

(A)

(B)

Figure 7. Orbital convergence versus skull size in herpestids (A) and pteropodids (B) (adult species means -Noble et al., 2000; Ravosa et al., 2000a). Both mammal families show allometric increases in convergence (Table 3). Convergence and postorbital bar formation are linked, such that bony bars occur primarily in the larger taxa of each clade. A skull is described as possessing a postorbital bar if at least a near-continuous strut is present; complete, and near-complete states are assumed to afford similar levels of rigidity. (Open circle: ligament; Solid circle: bony bar.)

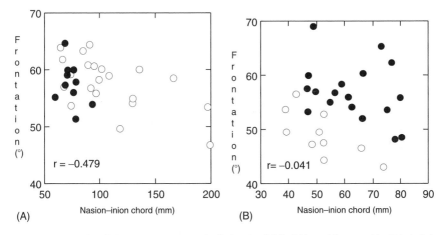

(A)

(B)

Figure 8. Orbital frontation versus skull size in felids (A) and herpestids (B) (adult species means—Noble et al., 2000; Ravosa et al., 2000a). In the former, clade frontation decreases significantly with size, while in the latter, no such pattern exists (Table 3). More frontated sister taxa in both families tend to exhibit bony postorbital bars. In felids and basal primates, this is restricted to small sizes due to allometric decreases in relative brain size and thus increased orbital frontation. (Key: same as Fig. 7.)

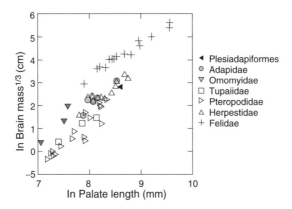

Figure 9. Brain mass$^{(0.33)}$ versus face size in plesiadapiforms, omomyids, adapids, tupaiids, felids, herpestids, and pteropodids (based on adult species means in Stephan and Pirlot, 1970; Stephan et al., 1981; Gittleman, 1986, 1991; Martin, 1990; Ravosa et al., 2000a). Versus the other clades, felids, and omomyids are significantly more encephalized for a given face size, thus suggesting the presence of increased basicranial flexion and greater frontation of the orbital apertures (Ravosa, 1991b,c, unpublished; Ross and Ravosa, 1993; Lieberman et al., 2000; Ravosa et al., 2000a,c).

convergence levels than diurnal sister taxa (Figures 10 and 11; also in a later section). Interestingly, the nocturnal, more frugivorous kinkajou (*Potos*) has a degree of orbital convergence similar to other nocturnal procyonids (Figure 11). In similar-sized didelphimorphs, four nocturnal faunivores of the genera *Philander, Chironectes,* and *Metachirus*—all possess more convergent orbits (mean = 53.5°; range = 50.4–57.8°) than three nocturnal arboreal frugivores of the genus *Caluromys* (mean = 42.4°; range = 42.0–43.0°) (Mann-Whitney U test among species means of 30–38 mm for the nasion-inion chord, p = 0.034). Therefore, as compared to nocturnal frugivores and diurnal faunivores, only nocturnal predators appear to display marked stereoscopic visual acuity and depth perception (cf., Allman, 1977, 1982; Cartmill, 1972, 1974, 1992).

Analyses of ontogenetic series for 12 primate species (six strepsirrhines, six haplorhines) provide additional evidence regarding covariation in orbital orientation (Ravosa, unpublished). On the one hand, all 12 taxa are characterized by size-related decreases in orbital frontation during growth, such that infants are more frontated than adults (Table 4). Furthermore, strepsirrhines uniformly exhibit ontogenetic increases in orbital convergence, but their

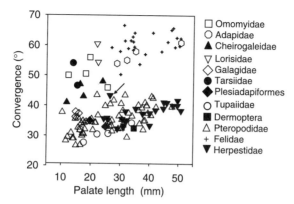

Figure 10. Orbital convergence versus face size in Eocene primates, primate sister taxa, extant primate nocturnal predators and other putative analogs (adult species means – adapted from Ravosa and Savakova, 2004). The first primates (omomyids, adapids) are similar to primate nocturnal predators (cheirogaleids, lorisids, galagids, tarsiids) and felids in exhibiting relatively higher convergence levels. The remaining mammals display more divergent orbits—tupaiids, plesiadapiforms, dermopterans, pteropodids, and herpestids. In contrast to its largely diurnal sister taxa, the herpestid nocturnal faunivore *Dologale* (arrow) displays a relatively greater degree of orbital convergence. These and other comparisons suggest that the derived presence of forward-facing orbits in basal primates was an adaptation for nocturnal visual predation (cf., Figure 11). Elevated convergence in diurnal adapids (cf., Cartmill, 1974; Martin, 1990) is due to the retention of basal primate (=omomyids-like) levels of convergence coupled with subsequent evolutionary increases in body size.

infants exhibit less convergence than adults (Ravosa and Savakova, 2004). On the other hand, anthropoids are characterized by postnatal increases, decreases, and isometry of orbital convergence (Table 4). Not surprising, strepsirrhines exhibit a common growth pattern, whereby orbital convergence and frontation are negatively correlated, while anthropoids show isometric or positive ontogenetic relations between convergence and frontation. Therefore, it is unlikely that a single explanation can explain both positive and negative correlations between convergence and frontation during ontogeny (Table 4) or across a clade (Table 3).

Another way the two suborders differ is in the relative level of orbital convergence and orbital frontation during postnatal ontogeny (Table 4). Anthropoid infants typically exhibit higher convergence levels than strepsirrhines

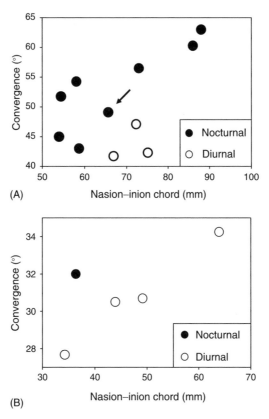

Figure 11. Orbital convergence versus skull size in procyonids (A) and tupaiids (B) (adult species means – adapted from Ravosa and Savakova, 2004). In the former clade, the nocturnal, highly frugivorous kinkajou (*Potos*-arrow) has an average degree of orbital convergence versus other nocturnal sister taxa (A); at about the procyonid median skull size, kinkajous have the median convergence value. This is opposite claims that elevated convergence and stereoscopic visual acuity are related to nocturnal arboreal frugivory (Rasmussen, 1990; Sussman, 1991; Crompton, 1995). In support of the NVPH, procyonid (A) and tupaiid (B) nocturnal predators (solid circles) exhibit relatively higher convergence levels than diurnal predators and foragers (open circles). Similar to other mammals, convergence increases allometrically in these clades (Tables 3 and 4).

and this difference is maintained into adulthood. In addition, anthropoid infants (except in the smaller-brained *Alouatta*) exhibit relatively greater frontation, and this pattern likewise characterizes suborder variation in adult levels of orbital frontation (Table 4).

Table 4. Bivariate correlations (r) for primate ontogenetic series[a]

Species, n	Orbital convergence versus palate length[b] (angle range)	Orbital frontation versus palate length[b] (angle range)	Orbital convergence versus orbital frontation
Propithecus verreauxi	0.692**	−0.641**	−0.522**
(11 adults and 24 nonadults)	(32.5–63.5)	(65.5–88.5)	
Eulemur fulvus	0.711**	−0.654**	−0.354*
(10 adults and 27 nonadults)	(30.0–58.0)	(50.0–84.0)	
Hapalemur griseus	0.416*	−0.830**	−0.257ns
(11 adults and 20 nonadults)	(39.0–57.5)	(49.0–79.5)	
Nycticebus coucang	0.832**	−0.367*	−0.319*
(12 adults and 31 nonadults)	(33.0–68.0)	(45.0–70.0)	
Perodicticus potto	0.677**	−0.527**	−0.498**
(12 adults and 34 nonadults)	(30.5–61.5)	(38.0–86.5)	
Otolemur crassicaudatus	0.653**	−0.700**	−0.243ns
(11 adults and 21 nonadults)	(26.0–51.0)	(40.0–75.0)	
Alouatta palliata	0.305*	−0.785**	−0.165ns
(10 adults and 53 nonadults)	(57.5–77.5)	(47.5–82.0)	
Macaca fascicularis	0.081ns	−0.695**	0.011ns
(12 adults and 51 nonadults)	(67.5–90.0)	(81.0–97.5)	
Macaca nemestrina	−0.006ns	−0.638**	−0.084ns
(2 adults and 46 nonadults)	(67.5–88.5)	(73.5–101.5)	
Nasalis larvatus	−0.494**	−0.728**	0.464**
(4 adults and 35 nonadults)	(66.0–82.5)	(77.5–101.5)	
Pongo pygmaeus	−0.323*	−0.895**	0.347*
(6 adults and 40 nonadults)	(73.5–90.0)	(55.0–98.5)	
Pan troglodytes	−0.081ns	−0.665**	0.009ns
(2 adults and 62 nonadults)	(80.0–90.0)	(78.0–108.0)	

[a]Significance levels: ** = p < 0.001; * = p ≤ 0.05; ns = p > 0.05.
[b]Comparisons versus other cranial measures differ little.

Nocturnal Visual Predation and the Evolution of Orbit Orientation and the Postorbital Bar

The interspecific analyses indicate that the presence of a postorbital bar is correlated with greater orbital convergence in herpestids and pteropodids, and with increased orbital frontation in felids and herpestids. Therefore, moderate support is provided for the NVPH's prediction about orbital convergence, as well as our suggestion regarding the influence of orbital frontation. As both orbital parameters increase during euprimate origins (Cartmill, 1970, 1972, 1974, 1992; Fleagle, 1999; Martin, 1990; Simons, 1962), it is reasonable to infer that both changes in orbital morphology are implicated in the development of the primate postorbital bar (Ravosa et al., 2000a; see also later section). In fact, the coevolution of these orbital parameters characterized the origin of anthropoids and the evolution of a postorbital septum is also thought to dampen ocular movements (cf., Cartmill, 1980; Ravosa, 1991b,c, unpublished; Ross, 1995; Ross and Ravosa, 1993) (Table 4).

The allometry of orbital parameters also appears important to a consideration of postorbital bar formation. For example, orbital convergence increases with size in herpestids and pteropodids (Table 3) and larger taxa in each clade evince a higher occurrence of a postorbital bar (Figure 7). Due to the negative scaling of brain size across felids—a pattern common to mammalian clades (Gould, 1975; Martin, 1990; Shea, 1987)—smaller cats are more frontated and only such diminutive species tend to exhibit postorbital bars (Figure 8A) (Ravosa et al., 2000a). Therefore, orbital frontation in felids appears to be proportional to relative brain size, with larger, more anteriorly located frontal lobes displacing the anterior cranial base and superior orbital margins rostrally so that the orbital apertures are more vertical (Cartmill, 1970, 1972, 1980; Radinsky, 1968). While herpestids with postorbital bars are more frontated, the fact that they are not as encephalized as felids may explain the lack of allometric decreases in orbital frontation in this clade (Figure 8B) (Ravosa et al., 2000a).

Levels of orbital frontation are likewise elevated in more encephalized primates (Cartmill, 1980, 1992; Ravosa, 1991b,c, unpublished; Ross, 1995; Ross and Ravosa, 1993). Contrary to a recent suggestion (Heesy, 2005), added support for this structural pattern is provided by suborder comparisons of the primate growth data. Due to greater encephalization and increased basicranial flexion (Ross and Ravosa, 1993), anthropoids typically exhibit relatively elevated levels of orbital frontation throughout ontogeny (Table 4). Furthermore, all 12 primates exhibit age-related decreases in orbital frontation (Table 4)—a pattern which strongly belies the negative scaling of brain size and basicranial flexion common to the postnatal ontogeny of a wide variety of mammals (Gould, 1975; Lieberman et al., 2000; Martin, 1990; Shea, 1987).

One long-standing controversy regarding the craniodental adaptations and behavior of basal euprimates centers on the mammal clade(s) used to elucidate the functional underpinnings of an important euprimate synapomorphy (forward-facing orbits). On one hand, the NVPH emphasizes that felid-like nocturnal visual predation is critical for understanding the adaptive significance of increased convergence levels and binocular visual acuity during euprimate origins (Allman, 1977, 1982; Cartmill, 1972, 1974, 1992). Alternative scenarios regarding the evolution of the euprimate circumorbital region posit that better functional analogs are to be found among nocturnal frugivores such as pteropodids and didelphimorh marsupials (Crompton, 1995; Pettigrew et al., 1989; Rasmussen, 1990; Sussman, 1991; Sussman and Raven, 1978). While the principal source of disagreement centers on the extant clade(s) selected to elucidate the functional underpinnings of orbital character states, this unresolved debate is further complicated by the lack of a broad-based empirical analysis of factors posited to influence variation in orbital orientation. This is especially surprising given that elevated levels of orbital convergence are also purportedly linked to the presence of relatively smaller orbital diameters (Cartmill, 1980), so that convergence increases with skull size due to the negative scaling of eye/orbit size (Kay and Cartmill, 1977; Martin, 1990).

The recent discovery of an exceptionally well-preserved Paleocene plesiadapiform (*Carpolestes*) with a unique constellation of skeletal features has rekindled debate regarding the patterning of morphological and adaptive transformations during the origin of archaic (Plesiadapiformes) and modern (Euprimates) primates (Bloch and Boyer, 2002, 2003; Kirk et al., 2003; Sargis, 2002). In positing that a series of manual and pedal features shared between *Carpolestes* and basal euprimates are homologous and derived (rather than simply a case of functional convergence), Bloch and Boyer (2002, 2003) argue that grasping appendages and terminal-branch feeding *preceded* the evolution of increased levels of orbital convergence and stereoscopic visual acuity characteristic of the earliest modern primates. With *Carpolestes* reconstructed as frugivorous (and having low levels of orbital convergence), grasping extremities in carpolestids and the ancestors of euprimates are inferred to be adaptations for terminal-branch foraging on fruits, flowers, and buds (Bloch and Boyer, 2002, 2003; Sargis, 2002). Accordingly, the phylogenetic independence of grasping and forward-facing orbits is incompatible with the NVPH's version of euprimate origins (Cartmill, 1972, 1974, 1992; Kay and Cartmill, 1974, 1977; Kirk et al., 2003). Instead, grasping capabilities in basal euprimates are interpreted as exaptations for nocturnal visual predation, supporting an alternative scenario regarding the sequence of acquisition and function of important primate postcranial synapomorphies (cf., Rasmussen, 1990; Sussman, 1991). Bloch and Boyer's (2002, 2003) study also suggests that grasping adaptations occurred prior to the emphasis on leaping behaviors as a component of the locomotor repertoire in basal euprimates (Dagosto, 1988; Szalay, 1973, 1981; Szalay et al., 1987).

In differentially focusing on the phylogenetic implications of the postcranial features, Bloch and Boyer (2002, 2003) are equivocal regarding an unresolved claim that nocturnal arboreal frugivores are the most appropriate functional analogs for understanding the evolution of euprimate visual acuity and forward-facing orbits (Rasmussen, 1990; Sussman, 1991; also Crompton, 1995; Pettigrew et al., 1989). Three sets of comparisons bear on the unresolved questions regarding the functional significance of forward-facing orbits. First, nocturnal arboreal frugivores, such as pteropodids, exhibit relatively lower amounts of orbital convergence similar to that in tupaiids, plesiadapiforms, dermopterans, and herpestids (Figure 10). Apart from representing the plesiomorphic state for euprimate sister taxa and presumably all eutherians, more divergent orbits characterize mammals with widely disparate activity patterns, including diurnal predators, as well as nocturnal and diurnal foragers (Ravosa and Savakova, 2004; Ravosa et al., 2000a). In contrast, the earliest euprimates are similar to nocturnal primate and felid faunivores in uniquely possessing enlarged (Kay and Cartmill, 1977; Martin, 1990) and relatively convergent orbits (Figure 10). Further comparisons between similarly sized nocturnal omomyids and extant primate nocturnal faunivores indicate no significant group differences in levels of orbital convergence (Mann-Whitney U test among species means of 12–26 mm for palate length, $p = 0.240$). Therefore, the derived presence of relatively larger eyes and forward-facing orbits in the first modern primates suggests that consequent increases in stereoscopic acuity and image clarity at close range were functionally linked to nocturnal stalking and capturing of mobile prey (Allman, 1977, 1982; Cartmill, 1972, 1974, 1992; Ravosa and Savakova, 2004; Ravosa et al., 2000a).

More phylogenetically restricted comparisons further demonstrate the lack of an association between nocturnal frugivory and marked orbital convergence (Ravosa and Savakova, 2004). In procyonids, the more frugivorous, nocturnal kinkajou exhibits an amount of orbital convergence equivalent to its nocturnal sister taxa (Figure 11A). Interestingly, kinkajous and other nocturnal predators display relatively greater levels of convergence than diurnal predators (Mann-Whitney U test of residuals from the procyonid LS line, $p = 0.001$). This suggests that the small vertebrate and insect component of the kinkajous diet—a proclivity shared with all nocturnal procyonids (Ewer, 1973; Nowak, 1999; Zeveloff, 2002), underlies variation in orbital form within and between nocturnal and diurnal members of this clade. Likewise, the nocturnal faunivore

Ptilocercus exhibits relatively greater convergence than diurnal tupaiid sister taxa (Figure 11B; studentized residual of 5.974 from the tupaiid LS line). In didelphimorphs of similar size, four nocturnal faunivores of the genera *Philander*, *Chironectes*, and *Metachirus* (mean = 53.5°; range = 50.4–57.8°)—all possess significantly more convergent orbits than three nocturnal arboreal frugivores of the genus *Caluromys* (mean = 42.4°; range = 42.0–43.0°) (Mann-Whitney U test among species means of 30–38 mm for the nasion-inion chord, p = 0.034). Compared to other largely diurnal herpestids, the diminutive nocturnal predator *Dologale* also displays an elevated level of convergence (Figure 10; studentized residual of 3.057 from the herpestid LS line). In support of the NVPH, these independent analyses clearly indicate that the presence of forward-facing orbits in extant and extinct mammals are related to an adaptive strategy of nocturnal visual predation.

One can nonetheless identify apparent support for the relationship between elevated orbital convergence and frugivory. In nocturnal pteropodids, larger-bodied species are more frugivorous and exhibit greater convergence than smaller, more insectivorous sister taxa (Nowak, 1999; Table 3). This pattern also characterizes orbital and dietary variation between adapids and the smaller-bodied omomyids (Figure 10). However, there are several reasons why these correlations are spurious and due rather to the independent scaling of dietary preference and orbit orientation. Controlling for activity cycle, and given the preponderance of faunivory (Ewer, 1973; Nowak, 1999), orbital convergence nevertheless increases with size in diurnal herpestids, nocturnal procyonids, diurnal tupaiids, and felids (Table 3; Figures 10 and 11). Postnatal development in three lorisiformes and three lemuriformes, taxa arguably most similar to basal euprimates in skull form, is uniformly characterized by size-related increases in convergence (Table 4). Moreover, increases in orbital convergence among these six strepsirrhines continue long after the early postnatal shift to weaning (inferred from the eruption of the first permanent molar – Smith et al., 1994) and relatively invariant adult feeding and chewing behaviors (see Watts, 1985 and review in Ravosa and Hogue, 2004). Such findings, particularly the ontogenetic evidence, offer strong empirical support for the argument that the allometric patterning of orbital convergence is simply a structural consequence of the negative scaling of orbital aperture diameter both during ontogeny and across a size series of close relatives (Cartmill, 1980; Ravosa and Savakova, 2004).

Contrary to previous assertions (Crompton, 1995; Pettigrew et al., 1989; Rasmussen, 1990; Sussman, 1991; Sussman and Raven, 1978), phylogenetically and allometrically controlled analyses highlight the unmistakable imprints of nocturnality and visual predation on the evolution of the skull and sensory system in the first primates of modern aspect (Allman, 1977, 1982; Cartmill, 1972, 1974, 1992; Kay and Cartmill, 1974, 1977; Kirk et al., 2003; Ravosa and Savakova, 2004; Ravosa et al., 2000a). Since basal euprimates perhaps were no larger than the 100-g *Carpolestes* (Bloch and Boyer, 2002, 2003; Fleagle, 1999; Martin, 1990, 1993; Sargis, 2002), allometry cannot be invoked to explain the derived presence of relatively higher levels of orbital convergence in this clade (as would be the case if a descendant were larger in body size than its ancestor). Thus, while the evidence of *Carpolestes* supports earlier studies regarding the more mosaic sequence of acquisition of euprimate synapomorphies (Rasmussen, 1990; Sussman, 1991), accompanying ecomorphological models positing the importance of nocturnal arboreal frugivory as a basis for elevated orbital convergence are unfounded (Ravosa and Savakova, 2004). Conversely, although the NVPH's explanation for the functional significance of increased visual acuity is well supported, because certain grasping features arguably predate the origin of Euprimates (Bloch and Boyer, 2002, 2003; Sargis, 2002), the shift to predatory and leaping behaviors in this clade occurred in an ancestor already frequenting a terminal-branch milieu (Dagosto, 1988; Szalay, 1973, 1981; Szalay et al., 1987). This is at odds with the NVPH's adaptive scenario for the coevolution of a wider range of euprimate cranial and postcranial synapomorphies from an ancestor with minimal grasping capabilities (Cartmill, 1972, 1974, 1992; Kay and Cartmill, 1974, 1977; Kirk et al., 2003; Soligo and Martin, 2006).

Phylogenetic and functional similarities between the pteropodid and primate skull have been further disputed. For instance, putative neuroanatomical synapomorphies of the visual system (Pettigrew et al., 1989) have been refuted by more comprehensive analyses (Johnson and Kirsch, 1993; Thiele et al., 1991). Our data also point to different influences on postorbital bar formation in primates and megabats: frontation and convergence in the former clade and only convergence in the latter. Moreover, whereas primates first evolved a postorbital bar at very small body sizes, only larger megabats exhibit this derived condition. Contrary to Pettigrew et al. (1989), it is unlikely that the postorbital bar of megachiropterans and primates is a synapomorphy (Ravosa et al., 2000a).

In addition to highly convergent orbits linked to nocturnal visual predation, basal primates and felids share the following features/trends: neural specializations of the visual system (Allman, 1977, 1982; Cartmill, 1970, 1972, 1974, 1992); relatively larger brains (Figure 9); enlarged orbits related to nocturnality (Ravosa et al., 2000a); postorbital bar development at small skull sizes (Figure 8A); and somewhat reduced olfactory bulbs presumably associated with a diminished emphasis on olfaction (Ravosa et al., 2000a using data from Gittleman, 1991). Such evidence supports claims of the NVPH that many early primate cranial adaptations approximate those of felids (Allman, 1977, 1982; Cartmill, 1970, 1972, 1974, 1992; Kay and Cartmill, 1974, 1977; Ravosa et al., 2000a).

As alluded to the earlier section, several of these features appear to be uniquely critical for explaining shared patterns of circumorbital covariation in the felid and basal euprimate skull (i.e., allometric decreases in orbital frontation and the evolution of a postorbital bar at small sizes). Felids are similar to the first (omomyid-like) euprimates in exhibiting greater encephalization likely related to a nocturnal visual predation strategy (i.e., a relatively larger visual cortex) (Barton, 1998; Cartmill, 1992). Felids and omomyids are also alike in exhibiting relatively larger, more convergent orbits associated with a nocturnal predatory lifestyle (Figure 10) (Cartmill, 1970, 1972, 1974, 1992; Covert and Hamrick, 1993; Heesy and Ross, 2001; Kay and Cartmill, 1977; Kirk and Kay, 2000; Martin, 1990; Noble et al., 2000; Ravosa et al., 2000a). Increases in the relative size of these adjacent structures appear to have created a spatial packing problem in which the orientation of the orbits (and thus supraorbital rims) are more highly affected by the position of the anterior cranial fossae and, in particular, relative brain size and shape (cf., Cartmill, 1970, 1972, 1980; Lieberman et al., 2000; Radinsky, 1968; Ravosa, 1991b,c, unpublished; Ross, 1995; Ross and Ravosa, 1993). Thus, due to the negative allometry of neural and orbital size, this structural constraint is especially pronounced in smaller taxa—exactly the range of skull sizes in which felids and basal euprimates exhibit greater orbital frontation and develop postorbital bars (Ravosa et al., 2000a).

In sum, our reformulation of the NVPH uniquely emphasizes the role of encephalization on patterns of covariation in circumorbital form and function during the origin of euprimates—due to increased relative brain size, greater orbital frontation (and in turn postorbital bar formation)—is a structural consequence of a nocturnal visual predation adaptive strategy. This explanation does not preclude the possibility that increased relative brain size among basal euprimates is also related to their unique combination of arboreality, precociality, and small body size (Shea, 1987, this volume).

Although the comparative data indirectly support the argument that more convergent and/or more frontated visual predators or foragers may require rigid lateral orbital margins, obviously, the most direct test of this prediction is to investigate experimentally the functional relations among orbital orientation, ocular oscillations, and postorbital bar formation (see also Heesy et al., this volume). In this regard, we briefly discuss the implications of the galago *in vivo* data vis-à-vis the NVPH and the evolution of the primate postorbital bar.

Perhaps the best insight offered by the galago strain data centers on the significant disparity between loading patterns along the working and balancing sides of the facial skull. As predicted by the NVPH, the galago WS postorbital bar encounters posteroinferiorly directed tension during mastication. However, opposite the NVPH, tensile strains at the BS postorbital bar are oriented posterosuperiorly (Table 2). This indicates that only the WS lateral orbital margin in basal primates was likely to have been compressed more along the optical axis instead of along the plane of the orbital aperture as inferred for putative sister taxa such as plesiadapiforms.

Other differences exist in galago WS/BS loading patterns. Plesiadapiforms and basal primates had unfused mandibular symphyses (Beecher, 1977, 1979, 1983; Ravosa, 1991a, 1996, 1999; Ravosa and Hylander, 1994; Ravosa et al., 2000c), and it is now well documented that primates with this character state recruit less BS jaw-adductor force (especially the transverse component) and exhibit correspondingly low BS corpus strains during mastication (Hylander, 1979a,b; Hylander et al., 1998, 2000, 2004; Ravosa and Hogue, 2004; Vinyard et al., this volume). This explains why galago peak strains are much higher along the WS postorbital bar and corpus (Table 1; Figures 2 and 3) (Ravosa et al., 2000a,b). In taxa with unfused symphyses and only a postorbital ligament, a galago-like recruitment pattern would result in an asymmetrical loading pattern in which elevated levels of orbital/ocular deformation occur along the chewing side of the face. As BS ocular acuity would be minimally affected, the corresponding asymmetry in visual disruption presumably poses deleterious consequences for effective binocular stereoscopic acuity, and this constitutes another reason why a rigid postorbital bar is important in the evolution of a nocturnal visual predation strategy in basal primates (Ravosa et al., 2000a).

There is also limited evidence from both extant primate suborders indicating allometric increases in the cranial strain gradient. During powerful mastication, a 2-kg *Otolemur crassicaudatus* exhibits lower peak-strain magnitudes at the dorsal interorbit and WS postorbital bar than a 1-kg *O. garnettii* (Ravosa et al., 2000a,b). In papionins, 14-kg baboons possess lower interorbital strain levels during chewing than 4-kg macaques (Hylander et al., 1991a). This positive allometry of the cranial strain gradient highlights a possible design criterion requiring the circumorbital region of larger forms to be increasingly overbuilt for masticatory loads—an interpretation consistent with the positive scaling of primate postorbital bar and supraorbital torus proportions (Hylander and Ravosa, 1992; Ravosa, 1988, 1991a–c; Vinyard and Smith, 1997, 2001). Such size-related decreases in circumorbital strains are also in contrast to an apparent pattern of strain similarity along the WS mandibular corpus across an interspecific primate size series (Hylander, 1979a; Hylander et al., 1998; see also Hylander, 1985; Vinyard and Ravosa, 1998, regarding stress similarity).

Given that the cranial strain gradient appears to scale positively, smaller taxa and individuals experience relatively higher strains along the WS lateral orbital margins that are variably lower than WS mandibular levels. This scaling pattern suggests that the first basal primates, which were quite small in body size, would have experienced relatively higher levels of WS ocular deformation coupled with a pronounced asymmetry in the amount of deformation between WS and BS lateral orbital margins.

Finally, as a close correspondence between orbital and ocular size characterizes only small-bodied species (Kay and Cartmill, 1977; Schultz, 1940), deformation of the lateral orbital margins is more likely to compromise ocular acuity in the earliest primates as they were quite small (100–300 g: Fleagle, 1999). Therefore, the development of a rigid postorbital bar may have been especially critical for maintaining high-effective levels of nocturnal visual acuity in small animals.

As anthropoids are further derived in having an orbital cavity mostly walled-off from the temporal fossa by a postorbital septum, data on strepsirhine masticatory and circumorbital function is important for understanding anthropoid evolution. For instance, the first anthropoid with a bony postorbital septum was small (700–900 g) and had only partial symphyseal fusion (Simons, 1989, 1992) and presumably a jaw-adductor activity pattern like galagos (Ravosa, 1999; Ravosa et al., 2000a–c). Thus, our argument regarding circumorbital-loading asymmetry and negative scaling of ocular size also may apply to the origin of the postorbital septum in stem anthropoids, especially since WS/BS asymmetry characterizes the temporalis and masseter of primates with unfused joints (Vinyard et al., this volume). This in turn suggests that functional and adaptive investigations of the anthropoid postorbital septum should account for the fact that the circumorbital region of extant anthropoids is loaded differently than in basal forms. Moreover, such studies should integrate data on orbital orientation in stem taxa so as to better estimate the extent to which the eye was disrupted along the orbital axis, as well as information on the position of the anterior temporalis versus the orbital contents so as to better estimate the proximity and direction of disruptive adductor forces.

In terms of the NVPH and postorbital bar function, there is recent evidence that in cats with high levels of orbital convergence and frontation, a complete bony bar does not prevent ocular movements during bilateral tetanic stimulation of the jaw adductors (Heesy et al., this volume). While this experimental information suggests that visual acuity may be compromised in mammals with postorbital bars, such ocular movements are documented under conditions hardly mirroring the biological role of alert cats during biting and chewing (cf., Gorniak and Gans, 1980). In fact, to evaluate if a postorbital bar does *or* does not serve to stiffen the lateral orbital margins, and thereby influence ocular movements, it is critical to determine if the magnitude of ocular movement in *alert* animals with and without bony bars is sufficient enough to inhibit effective stereoscopic vision. Furthermore, perhaps during the normal recruitment of the jaw-closing muscles, a neuromuscular mechanism may exist to counter ocular oscillations during mastication (thus obviating the need for a rigid postorbital bar). Finally, behavioral data on the extent to which arboreal taxa actively forage and/or predate while simultaneously chewing would directly test a fundamental premise of the NVPH regarding food procurement, activity level, visual acuity, and postorbital bar function.

Phylogenetic Evidence Regarding the NVPH and the Evolution of Circumorbital Form

Unlike the facial torsion model, which focuses exclusively on the "current utility" of the postorbital bar, a significant component of the NVPH is its dependence on the morphology of the sister taxon with which basal primates are compared (and thus the polarity of the ancestral condition from whence they arose – see also Dagosto, this volume). Due to certain basicranial, dental, and postcranial similarities with primates (Gingerich, 1976; Szalay, 1972), plesiadapiforms have figured heavily in this scenario (regardless of putative dermopteran affinities with this latter group – Beard, 1993; Kay et al., 1992). As they retain a primitive mammalian skull morphology, the use of plesiadapiforms as the sister taxon of primates allows one to relate derived changes in primate orbital form, postorbital bar formation, and encephalization to the evolution of nocturnal visual predation (Cartmill, 1970, 1972, 1974). Recent systematic analyses appear to provide further support for this phylogenetic hypothesis (Silcox, this volume; Springer et al., this volume). As dermopterans also exhibit low levels of orbital convergence and do not possess postorbital bars, suggestions that a clade with dermopterans and some plesiadapiforms as the sister taxon to primates (Beard, 1993; Kay et al., 1992) would also pose no problem regarding the NVPH of primate circumorbital and orbital evolution. Several lines of evidence from the postcranium also appear to support the NVPH (Dagosto, 1988, this volume; Hamrick, 1998, 1999; Lemelin, 1999; Lemelin and Schmitt, this volume); however, the fossil data arguably suggest more arboreal grasping behaviors for the ancestor of modern primates (Bloch and Boyer, 2002, 2003; Bloch et al., this volume).

Finally, if scandentians are the sister taxon to primates (Wible and Covert, 1987; Wible and Martin, 1993), certain aspects of the NVPH need to be reassessed—most important of which would be the link between postorbital bar development and increased orbital convergence and/or greater orbital frontation. If living tupaiids do approximate the condition of the ancestor to primates, then a postorbital bar clearly evolved prior to the split of scandentians and primates in a (presumably) diurnal, small-brained animal with low levels of orbital convergence and frontation.

CONCLUSIONS

The primary purpose of our chapter was to examine two long-standing hypotheses regarding circumorbital function, and then discuss the implications of these data for understanding adaptive transformation in skull form during primate origins. As greater galagos retain the primitive primate condition of a postorbital bar and an unfused symphysis, an understanding of masticatory function in such a representative strepsirhine is of considerable importance for interpreting circumorbital form in basal primates and other mammals (cf., Cartmill, 1970, 1972; Noble et al., 2000; Pettigrew et al., 1989; Ravosa et al., 2000a,b).

The presence of a significant strain gradient along the strepsirhine *and* anthropoid facial skull provides no support for the claim that mammalian circumorbital structures are functional adaptations to counter routine masticatory stresses (Bouvier and Hylander, 1996a,b; Hylander and Johnson, 1992, 1997a,b; Hylander and Ravosa, 1992; Hylander et al., 1991a,b; Ravosa et al., 2000a,b,d; Ross and Hylander, 1996). Galago circumorbital principal-strain directions during unilateral mastication are close to 45° relative to the skull's anteroposterior axis, much as predicted by the facial torsion model. Contrary to Greaves' model, neither the postorbital bars nor the supraorbital tori are oriented 45° relative to the cranial long axis in primates (Hylander and Ravosa, 1992; Ravosa, 1991b,c). Furthermore, as galago masticatory forces during biting and chewing differ from those of the facial torsion model but nonetheless result in circumorbital strain directions much as predicted, it is likely inappropriate to model the skull of primates and other mammals as a simple hollow cylinder loaded in axial torsion (Hylander et al., 1991a,b; Ravosa et al., 2000a,b).

Analyses of several mammalian clades suggest that the presence of a bony postorbital bar is correlated with higher levels of orbital convergence and/or frontation (Noble et al., 2000; Ravosa et al., 2000a). Consideration of the interspecific and ontogenetic evidence suggests that *relative* increases in these two orbital parameters during primate origins appear linked, respectively, to a shift to nocturnal visual predation and increased encephalization. Therefore, support is provided for the NVPH regarding postorbital bar and orbital convergence, as well as for our emphasis on the role of encephalization and orbital frontation in postorbital bar formation. These and several other aspects of the cranial bauplan of basal primates underscore the importance of small size on postorbital bar function (i.e., relatively larger brain and relatively larger, more convergent orbits). In this regard, our study complements a prior suggestion that a suite of life-history features unique to basal primates is associated with small body size (Shea, 1987).

These analyses contribute to an understanding of the broader influence of orbital frontation on other aspects of primate circumorbital form (e.g., browridge formation – Hylander and Ravosa, 1992; Moss and Young, 1960; Ravosa, 1988, 1991b,c; Ravosa et al., 2000d; Shea, 1986; Vinyard and Smith,

2001). Further support for this structural relationship is provided by within-species and between-suborder comparisons of the primate ontogenetic data (see an earlier section). Such postnatal growth data clearly demonstrate the importance of structural and allometric affects on orbital orientation in the evolution of basal primate and basal anthropoids. The interspecific and, especially, ontogenetic comparisons underscore the importance of controlling for size in assessing the functional and phyletic significance of forward-facing orbits. Indeed, selection for higher levels of convergence at small body sizes has to overcome the tendency for smaller sister taxa to develop more divergent orbits due to the presence of relatively large eyes. Such countervailing factors are presumably further pronounced if a given morphological transformation is coupled with a shift in activity cycle (i.e., the evolution of a relatively large-eyed nocturnal descendant from a smaller-eyed diurnal ancestor). In anthropoids, it has been argued that the origin of pronounced levels of convergence was linked to a shift to diurnality at small body sizes (Cartmill, 1980)—a pattern variably supported in interspecific analyses (Ross, 1995). Much stronger support for this prediction is demonstrated by the negative correlation between orbital convergence and relative orbit size during the postnatal development of six diverse strepsirrhines, and by the relatively elevated levels of convergence throughout anthropoid ontogeny (Table 4).

In evaluating the galago experimental data vis-à-vis the NVPH, we identify two factors underlying why basal primates may have evolved a rigid postorbital bar: loading asymmetry along the facial skull and negative scaling of ocular on orbital size. On the other hand, a recent experimental study suggests that visual acuity in domestic cats with postorbital bars may be compromised during bilateral tetanic bilateral stimulation of the jaw adductors (Heesy et al., this volume). Therefore, while the presence of a bony bar is linked to increased orbital convergence and/or frontation, it is conceivable we have yet to identify an adequate functional explanation for such a relationship, much as is the case for the postorbital bar of large-bodied taxa such as bovids. Furthermore, the presence of intraspecific variation in postorbital bar formation among certain felids, herpestids, and pteropodids (Noble et al., 2000) underscores the need for laboratory and field studies of alert organisms so as to properly address arguments regarding visual acuity, orbital orientation, and postorbital bar development. Indeed, additional testing of this and other functional models in a broader variety of clades would greatly improve our understanding of the evolutionary morphology of the masticatory apparatus and circumorbital region in primates and other mammals.

ACKNOWLEDGMENTS

The following kindly provided access to cranial specimens: L. Heaney, B. Patterson, and W. Stanley (Field Museum of Natural History); C. Beard (Carnegie Museum of Natural History); R. MacPhee (American Museum of Natural History); C. Smeenk and D. Reider (Rijksmuseum van Natuurlijke Historie); R. Angermann (Humboldt Universität Museum fuer Naturkunde); R. Thorington, L. Gordon, R. Emry, and R. Purdy (National Museum of Natural History); D. Gebo (Northern Illinois University); M. Tranier, D. Robineau, J. Cuisin, F. Renoult, M. Godinot, B. Senut, and C. Berge (Muséum National d'Histoire Naturelle); P. Jenkins (British Museum of Natural History); B. Latimer and L. Jellema (Cleveland Museum of Natural History); M. Rutzmoser (Harvard Museum of Comparative Zoology); and R. Kay, C. Vinyard (Duke University Medical Center). The authors thanked M. Cartmill for the loan of a dihedral goniometer. M. Dagosto, P. Freeman, S. Ghosh, M. Hamrick, A. Hogue, W. Kimbel, E. Kowalski, M. Morales, V. Noble, B. Shea, M. Silcox, S. Stack, and Y. Wu offered comments and assistance. For aid with and access to galagos, we thank I. Diamond, D. Schmechel, R. Ange, and K. Formo. This study was funded by the NSF (BCS-9709587 to MJR, SBR-9420764 to WLH), NIH (DE-05595 to MJR, DE-04531 to WLH), Leakey Foundation (MJR), Northwestern University (MJR), and Duke University (MJR and WLH). The NSF (BCS-0129349), Wenner-Gren Foundation for Anthropological Research, Field Museum and Northwestern University supported the international conference in which the findings of this chapter were presented.

REFERENCES

Allman, J., 1977, Evolution of the visual system in early primates, in: *Progress in Psychobiology, Physiology, and Psychology,* J. M. Sprague and J. M. Epstein, eds., vol. 7, pp. 1–53, Academic Press, New York.

Allman, J., 1982, Reconstructing the evolution of the brain in primates through the use of comparative neurophysiological and neuroanatomical data, in: *Primate Brain Evolution,* E. Armstrong and D. Falk, eds., Plenum Press, New York, pp. 13–28.

Barton, R. A., 1998, Visual specialization and brain evolution in primates, in: *Proc. R. Soc. Lond. B* **265**: 1933–1937.

Beard, K. C., 1993, Phylogenetic systematics of the Primatomorpha, with special reference to Dermoptera, in: *Mammal Phylogeny: Placentals,* F. S. Szalay, M. J. Novacek, and M. C. McKenna, eds., Springer-Verlag, New York, pp. 129–150.

Beecher, R. M., 1977, Function and fusion at the mandibular symphysis, *Am. J. Phys. Anthropol.* **47**: 325–336.

Beecher, R. M., 1979, Functional significance of the mandibular symphysis, *J. Morphol.* **159**: 117–130.

Beecher, R. M., 1983, Evolution of the mandibular symphysis in Notharctinae (Adapidae, Primates), *Int. J. Primatol.* **4**: 99–112.

Biewener, A. A., 1993, Safety factors in bone strength, *Calcif. Tissue Int.* **53**: 568–574.

Bloch, J. I., and Boyer, D. M., 2002, Grasping primate origins, *Science.* **298**: 1606–1610.

Bloch, J. I., and Boyer, D. M, 2003, Response to Kirk et al., *Science.* **300**:741; www.sciencemag.org/cgi/content/full/300/5620/741c

Bloch, J. I., & Boyer, D. M., 2006, New skeletons of Paleocene-Eocene Plesiadapiformes: A diversity of arboreal positional behaviors in early primates. In M. J. Ravosa & M. Dagosto (Eds.): Primate Origins: Adaptations and Evolution. New York: Springer Publishers, pp. 535–581.

Bookstein, F., Schafer, K., Prossinger, H., Seidler, H., Fieder, M., Stringer, C., et al., 1999, Comparing frontal cranial profiles in archaic and modern *Homo* by morphometric analysis, *Anat. Rec.* **257**: 217–224.

Bouvier, M., and Hylander, W. L., 1996a, The mechanical or metabolic function of secondary osteonal bone in the monkey, *Macaca fascicularis, Arch. Oral Biol.* **41**: 941–950.

Bouvier, M., and Hylander, W. L., 1996b, Strain gradients, age, and levels of modeling and remodeling in the facial bones of *Macaca fascicularis,* in: *The Biological Mechanisms of Tooth Movement and Craniofacial Adaptation,* Z. Davidovitch and L. A. Norton, eds., Harvard Society for the Advancement of Orthodontics, Boston, pp. 407–412.

Cartmill, M., 1970, *The Orbits of Arboreal Mammals: A Reassessment of the Arboreal Theory of Primate Evolution,* Ph.D. Dissertation, University of Chicago.

Cartmill, M., 1972, Arboreal adaptations and the origin of the order primates, in: *The Functional and Evolutionary Biology of Primates,* R. H. Tuttle, ed., Aldine de Gruyter, New York, pp. 97–122.

Cartmill, M., 1974, Rethinking primate origins, *Science* **184**: 436–443.

Cartmill, M., 1980, Morphology, function, and evolution of the anthropoid postorbital septum, in: *Evolutionary Biology of the New World Monkeys and Continental Drift,* R. L. Ciochon and A. B. Chiarelli, eds., Plenum Press, New York, pp. 243–274.

Cartmill, M., 1992, New views on primate origins, *Evol. Anthropol.* **1**: 105–111.

Collins, E. T., 1921, Changes in the visual organs correlated with the adoption of arboreal life and with the assumption of erect posture, *Trans. Ophthalmol. Soc. U.K.* **41**: 10–90.

Covert, H. H., and Hamrick, M. W., 1993, Description of new skeletal remains of the early Eocene anaptomorphine primate *Absarokius* (Omomyidae) and a discussion about its adaptive profile, *J. Hum. Evol.* **25**: 351–362.

Crompton, R. H., 1995, (Visual predation), habitat structure, and the ancestral primate niche, in: *Creatures of the Dark: The Nocturnal Prosimians*, L. Alterman, G. A. Doyle, and M. K. Izard, eds., Plenum Press, New York, pp. 11–30.

Currey, J. D., 1984, *The Mechanical Adaptations of Bones*, Princeton University Press, Princeton.

Dagosto, M., 1988, Implications of postcranial evidence for the origin of euprimates. *J. Hum. Evol.* **17**: 35–56.

Dagosto, M., 2006, The postcranial morphotype of primates. In M. J. Ravosa & M. Dagosto (Eds.): Primate Origins: Adaptations and Evolution. New York: Springer Publishers, pp. 489–534.

Endo, B., 1966, Experimental studies on the mechanical significance of the form of the human facial skeleton, *J. Faculty Science University Tokyo, Section V,* **3**(1):1–106.

Ewer, R. F., 1973, *The Carnivores*, Cornell University Press, Ithaca.

Fleagle, J. G., 1999, *Primate Adaptation and Evolution*, 2nd Ed., Academic Press, New York.

Gingerich, P. D., 1976, Cranial anatomy and evolution of early Tertiary Plesiadapidae (Mammalia, Primates), *Museum Paleontology University Michigan Papers Paleontology* **15**: 1–140.

Gittleman, J. L., 1986, Carnivore brain size, behavioral ecology, and phylogeny, *J. Mammalogy* **67**: 23–36.

Gittleman, J. L., 1991, Carnivore olfactory bulb size: Allometry, phylogeny, and ecology, *J. Zoolog. Lond.* **225**: 253–272.

Gorniak, G. C., and Gans, C., 1980, Quantitative assay of electromyograms during mastication in domestic cats (*Felis catus*), *J. Morphol.* **163**: 253–281.

Gould, S. J., 1975, Allometry in primates, with emphasis on scaling and the evolution of the brain, in: *Approaches to Primate Paleobiology. Contributions to Primatology*, F. S. Szalay, ed., vol. 5, pp. 244–292, S. Karger, Basel.

Greaves, W. S., 1985, The mammalian postorbital bar as a torsion-resisting helical strut, *J. Zoolog., Lond.* **207**: 125–136.

Greaves, W. S., 1991, A relationship between premolar loss and jaw elongation in selenodont artiodactyls, *Zoolog. J. Linnean Soc.* **101**: 121–129.

Greaves, W. S., 1995, Functional predictions from theoretical models of the skull and jaws in reptiles and mammals, in: *Functional Morphology in Vertebrate Paleontology*, J. J. Thomason, ed., Cambridge University Press, Cambridge, pp. 99–115.

Hamrick, M. W., 1998, Functional and adaptive significance of primate pads and claws: Evidence from New World anthropoids, *Am. J. Phys. Anthropol.* **106:** 113–127.

Hamrick, M. W., 1999, Pattern and process in the evolution of primate nails and claws, *J. Hum. Evol.* **37:** 293–297.

Heesy, C. P., and Ross, C. F., 2001, Evolution of activity patterns and chromatic vision in primates: Morphometrics, genetics and cladistics, *J. Hum. Evol.* **40:** 111–149.

Heesy, C. P., 2005, Function of the mammalian postorbital bar. *J. Morphol.* **264:**363–380.

Heesy, C. P., Ross, C. F., & Demes, B. 2006, Oculomotor stability and the functions of the postorbital bar and septum. In M. J. Ravosa & M. Dagosto (Eds.): Primate Origins: Adaptations and Evolution. New York: Springer Publishers, pp. 257–284.

Hilloowala, R. A., and Trent, R. B., 1988a, Supraorbital ridge and masticatory apparatus I: Primates, *Hum. Evol.* **3:** 343–350.

Hilloowala, R. A., and Trent, R. B., 1988b, Supraorbital ridge and masticatory apparatus II: Humans (Eskimos), *Hum. Evol.* **3:** 351–356.

Hylander, W. L., 1979a, Mandibular function in *Galago crassicaudatus* and *Macaca fascicularis*: An in vivo approach to stress analysis of the mandible, *J. Morphol.* **159:** 253–296.

Hylander, W. L., 1979b, The functional significance of primate mandibular form, *J. Morphol.* **160:** 223–240.

Hylander, W. L., 1985, Mandibular function and biomechanical stress and scaling, *Am. Zoolog.* **25:** 315–30.

Hylander, W. L., and Johnson, K. R., 1992, Strain gradients in the craniofacial region of primates, in: *The Biological Mechanisms of Tooth Movement and Craniofacial Adaptation*, Z. Davidovitch, ed., Ohio State University College of Dentistry, Columbus, pp. 559–569.

Hylander, W. L., and Johnson, K. R., 1997a, In vivo bone strain patterns in the zygomatic arch of macaques and the significance of these patterns for functional interpretations of craniofacial form, *Am. J. Phys. Anthropol.* **102:** 203–232.

Hylander, W. L., and Johnson, K. R., 1997b, In vivo bone strain patterns in the craniofacial region of primates, in: *Science and Practice of Occlusion*, C. McNeill, ed., Quintessence Publishing, Chicago, pp. 165–178.

Hylander, W. L., and Ravosa, M. J., 1992, An analysis of the supraorbital region of primates: A morphometric and experimental approach, in: *Structure, Function and Evolution of Teeth*, P. Smith and E. Tchernov, eds., Freund Publishing, Tel Aviv, pp. 223–255.

Hylander, W. L., Picq, P. G., and Johnson, K. R., 1991a, Masticatory-stress hypotheses and the supraorbital region of primates, *Am. J. Phys. Anthropol.* **86:** 1–36.

Hylander, W. L., Picq, P. G., and Johnson, K. R., 1991b, Function of the supraorbital region of primates, *Arch. Oral Biol.* **36:** 273–281.

Hylander, W. L., Ravosa, M. J., Ross, C. F., and Johnson, K. R., 1998, Mandibular corpus strain in primates: Further evidence for a functional link between symphyseal fusion and jaw-adductor muscle force, *Am. J. Phys. Anthropol.* **107:** 257–271.

Hylander, W. L., Ravosa, M. J., Ross, C. F., Wall, C. E., and Johnson, K. R., 2000, Symphyseal fusion and jaw-adductor muscle force: An EMG study, *Am. J. Phys. Anthropol.* **112:** 469–492.

Hylander, W. L., Vinyard, C. J., Ravosa, M. J., Ross, C. F., Wall, C. E., & Johnson, K. R., 2004, Jaw adductor force and symphyseal fusion. In F. Anapol, R. German & N. Jablonski (Eds.): *Shaping Primate Evolution. Form, Function and Behavior.* Cambridge: Cambridge University Press, pp. 229–257.

Johnson, J. I., and Kirsch, J. A. W., 1993, Phylogeny through brain traits: Interordinal relationships among mammals including Primates and Chiroptera, in: *Primates and Their Relatives in Phylogenetic Perspective*, R. D. E. MacPhee, ed., Plenum Press, New York, pp. 251–292.

Kay, R. F., and Cartmill, M., 1974, Skull of *Palaechthon nacimienti*, *Nature* **252:** 37–38.

Kay, R. F., and Cartmill, M., 1977, Cranial morphology and adaptations of *Palaechthon nacimienti* and other Paromomyidae (Plesiadapoidea,? Primates), with a description of a new genus and species, *J. Hum. Evol.* **6:** 19–35.

Kay, R. F., and Kirk, E. C., 2000, Osteological evidence for the evolution of activity pattern and visual acuity in primates, *Am. J. Phys. Anthropol.* **113:** 235–262.

Kay, R. F., Thewissen, J. G. M., and Yoder, A. D., 1992, Cranial anatomy of *Ignacius graybullianus* and the affinities of the Plesiadapiformes, *Am. J. Phys. Anthropol.* **89:** 477–498.

Kirk, E. C., Cartmill, M., Kay, R. F., and Lemelin, P., 2003, Comment on Bloch and Boyer, *Science* **300:**741; www.sciencemag.org/cgi/content/full/300/5620/741b

Lahr, M. M., and Wright, R. V. S., 1996, The question of robusticity and the relationship between cranial size and shape in *Homo sapiens*, *J. Hum. Evol.* **31:** 157–191.

Lanyon, L. E., and Rubin, C. T., 1985, Functional adaptation in skeletal structures, in: *Functional Vertebrate Morphology*, M. Hildebrand, D. M. Bramble, K. F. Liem, and D. B. Wake, eds., Harvard University Press, Cambridge, pp. 1–25.

Lauder, G. V., 1995, On the inference of function from structure, in: *Functional Morphology in Vertebrate Paleontology*, J. J. Thomason, ed., Cambridge University Press, Cambridge, pp. 1–18.

Lemelin, P., 1999, Morphological correlates of substrate use in didelphid marsupials: Implications for primate origins, *J. Zoolog. Lond.* **247:** 165–175.

Lemelin, P. & Schmitt, D., 2006, Origins of grasping and locomotor adaptations in primates: Comparative and experimental approaches using an opossum model. In

M. J. Ravosa & M. Dagosto (Eds.): Primate Origins: Adaptations and Evolution. New York: Springer Publishers, pp. 329–380.

Lieberman, D. E., Ross, C. F., and Ravosa, M. J., 2000, The primate cranial base: Ontogeny, function and integration, *Yearbook Phys. Anthropol.* **43**: 117–169.

Martin, R. D., 1986, Primates: A definition, in: *Major Topics in Primate and Human Evolution*, B. A. Wood, L. Martin, and P. Andrews, eds., Cambridge University Press Cambridge, pp. 1–31.

Martin, R. D., 1990, *Primate Origins and Evolution*, Princeton University Press, Princeton.

Martin, R. D., 1993, Primate origins: Plugging the gaps, *Nature* **363**: 223–234.

Moss, M. L., and Young, R. W., 1960, A functional approach to craniology, *Am. J. Phys. Anthropol.* **18**: 281–292.

Noble, V. E., Kowalski, E. M., and Ravosa, M. J., 2000, Orbital orientation and the function of the mammalian postorbital bar, *J. Zoolog. Lond.* **250**: 405–418.

Nowak, R. M., 1999, *Walker's Mammals of the World*, 6th Ed., The Johns Hopkins University Press, Baltimore.

Oyen, O. J., Walker, A. C., and Rice, R. W., 1979, Craniofacial growth in olive baboons (*Papio cynocephalus anubis*): Browridge formation, *Growth* **43**: 174–187.

Pettigrew, J. D., Jamieson, B. G. M., Robson, S. K., Hall, L. S., McNally, K. I., and Cooper, H. M., 1989, Phylogenetic relations between microbats, megabats and primates (Mammalia: Chiroptera and Primates), *Philoso. Trans. R. Soc. Lond.* **325**: 489–559.

Prince, J. H., 1953, Comparative anatomy of the orbit, *Brit. J. Physiol. Opt., N.S.* **10**: 144–154.

Radinsky, L. R., 1968, A new approach to mammalian cranial analysis, illustrated by examples of prosimian primates, *J. Morphol.* **124**: 167–180.

Rak, Y., 1983, *The Australopithecine Face*, Academic Press, New York.

Rangel, R. D., Oyen, O. J., and Russell, M. D., 1985, Changes in masticatory biomechanics and stress magnitude that affect growth and development of the facial skeleton, in: *Normal and Abnormal Bone Growth: Basic and Clinical Research*, A. D. Dixon and B. G. Sarnat, eds., A.R. Liss, New York, pp. 281–293.

Rasmussen, D. T., 1990, Primate origins: Lessons from a neotropical marsupial, *Am. J. Primatol.* **22**: 263–277.

Ravosa, M. J., 1988, Browridge development in Cercopithecidae: A test of two models, *Am. J. Phys. Anthropol.* **76**: 535–555.

Ravosa, M. J., 1991a, Structural allometry of the mandibular corpus and symphysis in prosimian primates, *J. Hum. Evol.* **20**: 3–20.

Ravosa, M. J., 1991b, Ontogenetic perspective on mechanical and nonmechanical models of primate circumorbital morphology, *Am. J. Phys. Anthropol.* **85**: 95–112.

Ravosa, M. J., 1991c, Interspecific perspective on mechanical and nonmechanical models of primate circumorbital morphology, *Am. J. Phys. Anthropol.* **86:** 369–396.

Ravosa, M. J., 1996, Mandibular form and function in North American and European Adapidae and Omomyidae, *J. Morphol.* **229:** 171–190.

Ravosa, M. J., 1999, Anthropoid origins and the modern symphysis, *Folia Primatol.* **70:** 65–78.

Ravosa, M. J., and Hylander, W. L., 1994, Function and fusion of the mandibular symphysis in primates: Stiffness or strength? in: *Anthropoid Origins*, J. G. Fleagle and R.F. Kay, eds., Plenum Press, New York, pp. 447–468.

Ravosa, M. J., & Hogue, A. S., 2004, Function and fusion of the mandibular symphysis in mammals: A comparative and experimental perspective. In C. F. Ross & R. F. Kay (Eds.): Anthropoid Evolution. New Visions. New York: Kluwer Academic/Plenum Publishers, pp. 413–462.

Ravosa, M. J., and Savakova, D. G., 2004, Euprimate origins: The eyes have it, *J. Hum. Evol.* **46:** 355–362.

Ravosa, M. J., Noble, V. E., Hylander, W. L., Johnson, K. R., and Kowalski, E. M., 2000a, Masticatory stress, orbital orientation and the evolution of the primate postorbital bar, *J. Hum. Evol.* **38:** 667–693.

Ravosa, M. J., Johnson, K. R., and Hylander, W. L., 2000b, Strain in the galago facial skull, *J. Morphol.* **244:** 51–66.

Ravosa, M. J., Vinyard, C. J., Gagnon, M., and Islam, S. A., 2000c, Evolution of anthropoid jaw loading and kinematic patterns, *Am. J. Phys. Anthropol.* **112:** 493–516.

Ravosa, M. J., Vinyard, C. J., and Hylander, W. L., 2000d, Stressed out: Masticatory forces and primate circumorbital form, *Anat. Rec. (New Anatomist)*, **261:** 173–175.

Rawlinson, S. C. F., Mosley, J. R., Suswillo, R. F. L., Pitsillides, A. A., and Lanyon, L. E., 1995, Calvarial and limb bone cells in organ and monolayer culture do not show the same early responses to dynamic mechanical strain, *J. Bone Miner. Res.* **10:** 1225–1232.

Richard, A. F., 1985, *Primates in Nature*, W. H. Freeman and Company, New York.

Rosenberger, A. L., 1986, Platyrrhines, catarrhines and the anthropoid transition, in: *Major Topics in Primate and Human Evolution*, B. A. Wood, L. Martin, and P. Andrews, eds., Cambridge University Press, Cambridge, pp. 66–88.

Ross, C. F., 1995, Allometric and functional influences on primate orbit orientation and the origins of the Anthropoidea, *J. Hum. Evol.* **29:** 201–227.

Ross, C. F., and Hylander, W. L., 1996, In vivo and in vitro bone strain in the owl monkey circumorbital region and the function of the postorbital septum, *Am. J. Phys. Anthropol.* **101:** 183–215.

Ross, C. F., and Ravosa, M. J., 1993, Basicranial flexion, relative brain size, and facial kyphosis in nonhuman primates, *Am. J. Phys. Anthropol.* **91:** 305–324.

Russell, M. D., 1985, The supraorbital torus: A most remarkable peculiarity, *Curr. Anthropol.* **26**: 337–360.

Sargis, E. J., 2002, Primate origins nailed, *Science* **298**: 1564–1565.

Schultz, A. H., 1940, The size of the orbit and of the eye in primates, *Am. J. Phys. Anthropol.* **25**: 398–408.

Shea, B. T., 1986, On skull form and the supraorbital torus in primates, *Curr. Anthropol.* **27**: 257–259.

Shea, B. T., 1987, Reproductive strategies, body size, and encephalization in primate evolution, *Int. J. Primatol.* **8**: 139–156.

Shea, B. T., 2006, Start small and live slow: Encephalization, body size and life history strategies in primate origins and evolution. In M. J. Ravosa & M. Dagosto (Eds.): Primate Origins: Adaptations and Evolution. New York: Springer Publishers, pp. 583–623.

Silcox, M. T., 2006, Primate taxonomy, plesiadapiforms, and approaches to primate origins. In M. J. Ravosa & M. Dagosto (Eds.): Primate Origins: Adaptations and Evolution. New York: Springer Publishers, pp. 143–178.

Simons, E. L., 1962, Fossil evidence relating to the early evolution of primate behavior, *Ann. N Y Acad. Sci.* **102**: 282–294.

Simons, E. L., 1989, Description of two genera and species of Late Eocene Anthropoidea from Egypt, *Proc. Natl. Acad. Sci., U.S.A.* **86**: 9956–9960.

Simons, E. L., 1992, Diversity in the early Tertiary anthropoidean radiation in Africa, *Proc. Natl. Acad. Sci., U.S.A.* **89**: 10743–10747.

Smith, B. H., Crummett, T. L., and Brandt, K. L., 1994, Ages of eruption of primate teeth: A compendium for aging individuals or comparing life histories, *Ybk. Phys. Anthropol.* **37**: 177–231.

Soligo, C. and R. D. Martin, 2006, Adaptive origins of primates revisited. *J. Hum. Evol.* **50**:414–430.

Springer, M. S., Murphy, W. J., Eizirik, E., Madsen, O., Scally, M., Douady, C. J., Teeling, E. C., Stanhope, M. J., de Jong, W. W., O'Brien, S. J., 2006, A molecular classification for the living orders of placental mammals and the phylogenetic placement of primates. In M. J. Ravosa & M. Dagosto (Eds.): Primate Origins: Adaptations and Evolution. New York: Springer Publishers, pp. 1–28.

Stephan, H., and Pirlot, P., 1970, Volumetric comparisons of brain structures in bats, *Z. Zool. Syst. Evol.* **8**: 200–236.

Stephan, H., Nelson, J. E., and Frahm, H. D., 1981, Brain size comparison in Chiroptera, *Z. Zool. Syst. Evol.* **19**: 195–222.

Sussman, R. W., 1991, Primate origins and the evolution of angiosperms, *Am. J. Primatol.* **23**: 209–223.

Sussman, R. W., and Raven, P. H., 1978, Pollination by lemurs and marsupials: An archaic coevolutionary system, *Science* **200**: 731–736.

Szalay, F. S., 1972, Paleobiology of the earliest primates, in: *The Functional and Evolutionary Biology of Primates*, R. H. Tuttle, ed., Aldine de Gruyter, Chicago, pp. 3–35.

Szalay, F. S., 1973, New Paleocene primates and a diagnosis of the new suborder Paromomyiformes, *Folia Primatol.* **19:** 73–87.

Szalay, F. S., 1981, Phylogeny and the problem of adaptive significance: The case of the earliest primates, *Folia Primatol.* **36:** 157–182.

Szalay, F. S., and Delson, E., 1979, *Evolutionary History of the Primates*, Academic Press, New York.

Szalay, F. S., Rosenberger, A. L., and Dagosto, M., 1987, Diagnosis and differentiation of the Order Primates, *Ybk. Phys. Anthropol.* **30:** 75–105.

Tattersall, I., 1995, *The Last Neanderthal*, Macmillan, New York.

Thiele, A., Vogelsang, M., and Hoffman, K. P., 1991, Pattern of retinotectal projection in the megachiropteran bat *Rousettus aegyptiacus*, *J. Comp. Neurol.* **314:** 671–683.

Vinyard, C. J., and Ravosa, M. J., 1998, Ontogeny, function, and scaling of the mandibular symphysis in papionin primates, *J. Morphol.* **235:** 157–175.

Vinyard, C. J., and Smith, F. H., 1997, Morphometric relationships between the supraorbital region and frontal sinus in Melanesian crania, *Homo* **48:** 1–21.

Vinyard, C. J., and Smith, F. H., 2001, Morphometric testing of structural hypotheses of the supraorbital region in modern humans, *Zeitschrift Morphologie Anthropologie* **83:** 23–41.

Vinyard, C. J., Ravosa, M. J., Williams, S. H., Wall, C. E., Johnson, K. R., & Hylander, W. L., 2006, Jaw muscle function and the origin of primates. In: M. J. Ravosa & M. Dagosto (Eds.): Primate Origins: Adaptations and Evolution. New York: Springer Publishers, pp. 179–231.

Watts, D. P., 1985, Observations on the ontogeny of feeding behavior in mountain gorillas (*Gorilla gorilla beringei*), *Am. J. Primatol.* **8:** 1–10.

Wible, J. R., and Covert, H. H., 1987, Primates: Cladistic diagnosis and relationships, *J. Hum. Evol.* **16:** 1–22.

Wible, J. R., and Martin, J. R., 1993, Ontogeny of the tympanic floor and roof in archontans, in: *Primates and Their Relatives in Phylogenetic Perspective*, R. D. E. MacPhee, ed., Plenum Press, New York, pp. 111–148.

Wolpoff, M. H., 1996, *Human Evolution*, McGraw-Hill, New York.

Zeveloff, S. I., 2002, *Raccoons: A Natural History*, Smithsonian Institution Press, Washington, DC.

CHAPTER TEN

Origins of Grasping and Locomotor Adaptations in Primates: Comparative and Experimental Approaches Using an Opossum Model

Pierre Lemelin and Daniel Schmitt

The earliest primates were distinguished from other primitive mammals by the use of the hands and feet for grasping....The origin of primates was primarily a locomotor adaptation.

—Washburn (1951: 68)

INTRODUCTION

Since the turn of the 20th century, most anthropologists agreed on one fundamental notion: the origin and evolution of the order Primates was closely tied with life in the trees. This view is founded on the obvious observation that the vast majority of extant primates live in the trees and have colonized many different arboreal habitats. Smith (1912) and Jones (1916) were among

Pierre Lemelin • Division of Anatomy, Faculty of Medicine and Dentistry, University of Alberta, Edmonton, Alberta, Canada, T6G 2H7 **Daniel Schmitt** • Department of Biological Anthropology and Anatomy, Duke University Medical Center, Durham, NC 27710

the first to relate some of the unique anatomical and behavioral characteristics of primates with arboreal life. Their views were promoted by LeGros Clark (1959), but later challenged and refined by Cartmill (1972, 1974a,b) who suggested that the forward-facing eyes and grasping extremities of primates can be interpreted as adaptations to cautious foraging for insect prey on thin, flexible branches. At the same time, Jenkins (1974: 112) suggested that "The adaptive innovation of ancestral primates was therefore not the invasion of the arboreal habitat, but their successful restriction to it." However, there are several extant mammal species other than primates that are restricted to an arboreal environment, particularly in which thin and flexible branches abound. As Cartmill (1972, 1974a,b) and Ramussen (1990) stressed, those nonprimate mammals offer great potential in addressing the problem of primate origins.

The views of Jenkins and Cartmill had a profound influence on the adaptive explanations of the postcranial and locomotor features that define primates as a group. Several primate postcranial and locomotor characteristics, rare in other mammals, are now being interpreted as evidence of an invasion and restriction to a fine-branch, arboreal niche by the earliest primates. For example, primates have prehensile hands and feet that bear nails instead of sharp claws (Cartmill, 1970, 1972, 1974a,b, 1985; Jones, 1916, 1929; LeGros Clark, 1959; Lemelin, 1996; Martin, 1968, 1986, 1990; Mivart, 1873; Napier, 1961, 1993; Napier and Napier, 1967; Szalay and Dagosto, 1988; Szalay et al., 1987) and relatively long limbs (Alexander et al., 1979; Polk et al., 2000) with more mobile joints, particularly in the forelimbs (Reynolds, 1985b). In addition to these postcranial features, most primates share three locomotor characteristics that are unusual or unique compared to other mammals (Larson, 1998). During quadrupedal walking, primates are characterized by: (a) an almost exclusive use of diagonal-sequence (DS) walking gaits (i.e., each hind footfall is followed by the contralateral fore footfall) (Cartmill et al., 2002; Hildebrand, 1967, 1985; Rollinson and Martin, 1981; Vilensky and Larson, 1989); (b) a protracted arm position at forelimb touchdown (i.e., arm greater than 90° relative to horizontal body axis) (Larson, 1998; Larson et al., 2000, 2001); (c) relatively lower peak vertical substrate reaction forces (V_{pk}) on the forelimbs compared to the hindlimbs (Demes et al., 1994; Kimura et al., 1979; Reynolds, 1985b); and (d) forelimb compliance (Larney and Larson 2004; Schmitt, 1998, 1999, 2003a,b; Schmitt and Hanna, 2004).

What has been lacking is a clear demonstration that mammals restricted to a fine-branch environment possess similar postcranial and locomotor characteristics that are functionally linked to moving and foraging on thin arboreal supports. In this chapter, we present the results of comparative and experimental studies that test the relationship between the presence of primate-like features and fine-branch arborealism using ecological convergence between didelphid marsupials and prosimian primates. Following a review of various models of primates, we present morphometric and behavioral data for opossums and primates that test specifically the functional link between the presence of more grasping, primate-like cheiridia and movement on thin branches. In the second part, we report experimental results that specifically test for the presence of three gait characteristics typical of most primates in a fine-branch arborealist, the woolly opossum (*Caluromys philander*). In the last part of this chapter, we discuss how these data accord with current theories of primate origins and assess the relevance of an opossum model in inferring the locomotor profile and ecological niche of the earliest primates.

MODELS OF PRIMATE ORIGINS: A REVIEW

For most of the last century, some of the postcranial characteristics that distinguish primates from other mammals have been linked to arboreality. For example, Smith (1912) proposed that with the adoption of a more arboreal lifestyle, the hand of primates became more prehensile and sensitive, thus replacing the role of the mouth for food collection. In the same vein, Jones (1916) suggested that the retention of an arboreal mode of life from primitive mammals led to a functional differentiation of the limbs of primates, with the hindlimbs assuming a more propulsive role while the forelimbs are used for reaching and manipulation. These ideas were essentially preserved in what became known as the "arboreal theory of primate evolution" (LeGros Clark, 1959). As LeGros Clark (1959: 43) pointed out: "The evolutionary trends observed in primates are a natural consequence of an arboreal habitat, a mode of life which among other things demands or encourages prehensile functions of the limbs ...." This arboreal theme is still predominant in more recent models of primate origins, such as the grasp-leaping model (Szalay and Dagosto, 1988; Szalay et al., 1987), which provides a detailed argument for the origins and evolution of the grasping foot in euprimates.

Cartmill (1970, 1972, 1974a,b) directly challenged the arboreal theory in arguing that the characteristic primate traits, including prehensile extremities, cannot be explained simply as adaptations to arboreal life in general. His thesis was based on two fundamental observations: first, many mammals, such as gray squirrels, are successful arborealists and lack all the morphological characteristics typical of primates. Second, and more importantly, Cartmill noted that many small prosimian primates and marsupials resemble each other in some aspects of their morphology and ecology. Specifically, Cartmill (1974b: 442) pointed out that "Visually directed predation on insects...is characteristic of many living prosimians, and also of small marsupials... This suggests that grasping extremities were evolved because they facilitate cautious well-controlled movements in pursuit of prey on slender supports ...."

In a more recent model of primate origins, Sussman (1991, 1995) stressed that searching and collecting angiosperm products on terminal branches were more determinant factors in shaping primate visual and postcranial specializations. Cartmill (1992) criticized this model arguing that the visual apparatus of primates is better suited for nocturnal animals trying to locate mobile prey rather than motionless fruits. In turn, Crompton (1995) provided a series of arguments against the visual predation model of Cartmill. He concluded that orbital convergence for animals like primates is as important for judging distances and avoiding obstacles when engaged in rapid, saltatory locomotion in a fine-branch niche as it is for visually discriminating fruits and insects from the arboreal background.

Aside from these differences of opinion on the reasons why primates evolved orbital convergence, which are beyond the scope of this chapter, both Cartmill and Sussman agree on one important factor: a similar kind of arboreality (mainly thin and flexible substrates) to explain the presence of prehensile hands and feet in primates. In other words, moving and foraging in a fine-branch niche were important factors in the evolution of prehensile extremities in primates.

Cartmill (1970) and Charles-Dominique and Martin (1970) proposed independently that ancestral primates originated in an arboreal environment in which supports of small diameter predominated. This concept of "fine branch niche"—later coined by Martin (1972a) to describe the physical environment in which the mouse lemur lives—has been invoked by several physical anthropologists to explain the origins of some of the locomotor peculiarities of primates described above. Since these morphological and locomotor characteristics appear to represent the primitive condition for the entire primate order, other mammals have to be considered to understand the potential link between these characteristics and movement on thin arboreal supports.

In his model of primate origins, Cartmill emphasized the importance of considering other mammal groups in addressing the problem of primate origins. On the basis of strong convergence in morphology, behavior, and ecology between small marsupials and small prosimians, Cartmill inferred that the last common ancestor of primates was a visual predator with prehensile extremities, which were useful for grabbing insects and moving cautiously on slender branches. In a landmark study, Rasmussen (1990) used a similar case of convergence to address the problem of primate origins. The rationale for his study on the behavioral ecology of two didelphid marsupials was that "... the behavior of nonprimate mammals that have convergently evolved primate like structure may be used as an adaptive analogy, or in effect, a sort of independent 'evolutionary experiment'" (Rasmussen, 1990: 265) to test or refine hypotheses of primate origins. First, Rasmussen noted that the woolly opossum (*Caluromys derbianus*) is more primate-like in skull morphology, eye size, encephalization quotient, and life-history patterns compared to other opossums. Then, he reported that, unlike other more terrestrial opossums like *Didelphis marsupialis*, woolly opossums rely heavily on terminal branches while moving and foraging for fruits and insects, just like many small-bodied prosimians do. Rasmussen (1990) argued that this case of convergence between arboreal opossums and primates establishes a link between the presence of primate-like traits and making a living in an arboreal environment of the sort advocated by Cartmill and Sussman in their models of primate origins. Rasmussen (1990: 275) concluded that "*Caluromys derbianus* makes a good model of what primitive prosimians may have been like" and added that "Studies of *C. derbianus* may be useful for testing hypotheses of primate origins, and may serve to revise and clarify theoretical perspectives on the issue." This conclusion is in agreement with the earlier suggestion of Hunsaker and Shupe (1977: 281), who stated that "These contrasts in behavior leave little doubt that the didelphids represent a virtually untouched group which has great potential for studies in convergent evolution and comparative behavior."

In the following section, we present the results of a comparative study that relies on convergence between opossums and primates to test the functional link between more prehensile extremities and facilitating movement and foraging on thin branches.

CONVERGENCE BETWEEN OPOSSUMS AND PRIMATES: COMPARATIVE STUDIES

The choice of an appropriate comparative strategy is critical for appropriate testing of adaptive hypotheses. Two major comparative methods can be employed to show evidence of adaptation (Brooks & McLennan, 1991): (a) comparing traits associated with an ecological role in closely related species that have diverged ecologically and (b) comparing traits associated with an ecological role in distantly related species that have converged on similar ecologies. While the first comparative method is frequently employed to understand associations between anatomy and behavior among primates (Daegling, 1992; Fleagle, 1977a,b, 1979; Fleagle and Meldrum, 1988; Rodman, 1979; Ward and Sussman, 1979), the second one is less commonly used by physical anthropologists. There are exceptions, such as Erikson's (1963) study showing similarities in forelimb morphology between *Ateles* and *Hylobates* associated with suspensory locomotion and posture, or Cartmill's (1974c) study pointing out similarities in skull and digit morphology between *Daubentonia* and *Dactylopsila* associated with grub extraction from tree bark. The convergent evolution of similar traits in similar environments in distantly related lineages has been argued as one of the strongest sources of evidence for adaptation (Brooks and McLennan, 1991; Biewener, 2002; Larson and Losos, 1996; Vogel, 1998). Such comparative strategy is especially well suited to understand the adaptive significance of traits linked with the origin of primates.

Since prehensile hands and feet are most likely to have characterized the last common ancestor of all primates, a comparison aiming to test the functional link between the presence of more grasping extremities and fine-branch arborealism *within* primates only is inappropriate. Instead, distantly related mammals that have converged on similar fine-branch niches need to be considered. In doing so, a stronger adaptive model that is general enough to explain the presence of prehensile hands and feet in all mammals inhabiting fine-branch habitats can be built. In the following section, we present the results of morphometric analyses and behavioral observations in laboratory conditions using both comparative methods described earlier: (a) comparisons within a closely related group of mammals that is ecologically diverse; in this case, five species of didelphid marsupials and (b) comparisons between two distantly related groups of mammals that have converged ecologically; in this case, some didelphid marsupials and cheirogaleid primates.

A Review of Didelphid and Cheirogaleid Ecology and Behavior

The family Didelphidae—commonly called the American opossums—is a group of marsupial mammals that is distributed throughout the New World, mainly in Mexico, Central America, and most of South America (Eisenberg, 1989; Nowak, 1999). Only the Virginia opossum (*Didelphis virginiana*) is found in the United States and parts of Canada (Eisenberg, 1989; McManus, 1974; Nowak, 1999). Traditionally, the family Didelphidae is considered to be a natural group that includes the genus *Caluromys*, the woolly opossum (Archer, 1984; Kirsch, 1977; Marshall et al., 1990; Reig et al., 1987). However, recent molecular evidence has shown that *Caluromys* may be more distinct from other didelphids and should be included in its own family Caluromyidae (Kirsch et al., 1997) or subfamily Caluromyinae (Jansa and Voss, 2000). In any event, since Didelphidae and Caluromyidae are sister families, we have retained the term "didelphids" throughout this chapter for practical reasons.

Like primates, didelphids are pentadactyl and show strong differentiation in body size, ecology, and behavior (Charles-Dominique, 1983; Charles-Dominique et al., 1981; Collins, 1973; Eisenberg and Wilson, 1981; Hall and Dalquest, 1963; Hunsaker, 1977; Hunsaker and Shupe, 1977; Figure 1). Five genera were considered for this comparative study: *Monodelphis, Didelphis, Philander, Marmosa,* and *Caluromys.*

Monodelphis, the short-tailed opossum, is among the smallest didelphids (about 84 g for *Monodelphis brevicaudata,* Eisenberg, 1989). It is restricted to the ground where it feeds largely on small prey and some fruits (Charles-Dominique et al., 1981; Eisenberg, 1989; Hunsaker, 1977; O'Connell, 1979). Cartmill (1972) described *Monodelphis* as a forest-floor predator that detects prey by smell, hearing, and vibrissal contact and captures them using its mouth; Hershkovitz (1969) characterized it as shrew-like in habits.

At the opposite end of the size spectrum, *Didelphis,* the large American opossum, weighs between 1 and 2 kg on an average (Eisenberg and Wilson, 1981). Like *Monodelphis, Didelphis* spends a large amount of its travel and feeding time budget on the ground (Cunha and Vieira, 2002; Eisenberg,

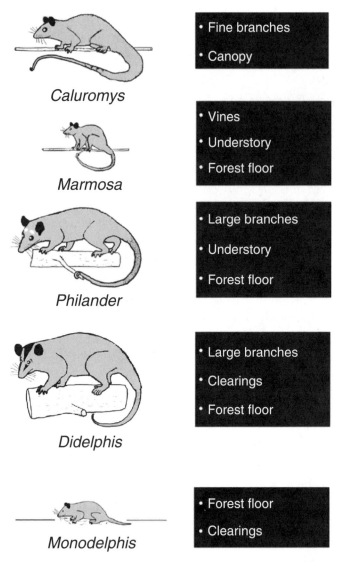

Figure 1. Review of didelphid ecology. For each genus, the forest level where it spends most of its time and types of substrates most commonly used are given. The information relative to *Marmosa* applies for *M. robinsoni*. Scale differences between animals are accurate. Figures of the opossums are modified after Eisenberg (1989).

1989; Hunsaker, 1977; McManus, 1974). In fact, *D. marsupialis* spends about 70% of its travel and feeding time on the ground (Charles-Dominique et al., 1981; Rasmussen, 1990). Nonetheless, *Didelphis* can also climb well, usually on large vertical trunks (Eisenberg, 1989; Hunsaker, 1977; Hunsaker and Shupe, 1977; McManus, 1970, 1974), and branches account for about 25% of the arboreal supports used by *D. marsupialis* (Charles-Dominique et al., 1981; Rasmussen, 1990). *Didelphis* feeds on a wide range of prey and fruits (Eisenberg, 1989; Hunsaker, 1977; Liete et al., 1996; McManus, 1974), but consumes more fruits when found in tropical forests compared to other environments (Atramentowicz, 1988; Hall and Dalquest, 1963).

Likewise, *Philander,* the "four-eyed" opossum, moves primarily on the forest floor (Charles-Dominique, 1983; Charles-Dominique et al., 1981; Cunha and Vieira, 2002; Eisenberg, 1989; Hunsaker, 1977), more often than *Didelphis* (Atramentowicz, 1988; Enders, 1935; Passamani, 1995). *Philander opossum* weighs on average between 330 g (Eisenberg, 1989) and 450 g (Atramentowicz, 1986, 1988), and forages almost exclusively on the ground for prey, ripe fruits, fallen off trees, nectars, and flowers (Atramentowicz, 1988; Charles-Dominique, 1983; Charles-Dominique et al., 1981).

Marmosa, the mouse opossum, is a small-bodied didelphid (*Marmosa robinsoni* averages 116 g, Eisenberg and Wilson, 1981) found in a wide variety of environments (Eisenberg, 1989; Hunsaker, 1977; Nowak, 1999). *M. robinsoni* prefers secondary forests and clearings where it spends most of its time on vines and bushes, but also descends to the ground (Eisenberg, 1989; Enders, 1935; Hall and Dalquest, 1963; Hunsaker, 1977; O'Connell, 1983). It feeds primarily on insects and uses its hands extensively to manipulate them (Eisenberg and Leyhausen, 1972; Enders, 1935; Hunsaker, 1977; Hunsaker and Shupe, 1977). Fruits also represent an important dietary component of *M. robinsoni* (O'Connell, 1983). Cartmill (1972) described *Marmosa* as a shrub-layer insectivore relying heavily on visual detection for stalking prey among the complex network of fine branches and capturing insects with the hands. Very often, *M. robinsoni* anchors itself on a thin support using its feet, while the hands are used to restrain and manipulate a food object (Hunsaker and Shupe, 1977).

Finally, *Caluromys,* the woolly opossum, is among the most arboreal didelphids. It travels and forages on small, terminal branches high in the canopy and very rarely descends to the ground (less than 1% of the time) (Atramentowicz, 1982; Bucher and Hoffman, 1980; Charles-Dominique et al., 1981; Eisenberg, 1989; Hall and Dalquest, 1963; Hunsaker, 1977; Liete et al., 1996; Malcolm, 1991; O'Connell, 1979; Rasmussen, 1990). *Caluromys philander* weighs on average 300 g and feeds primarily on ripe fruits, gums, and nectar (75% of the total diet), as well as invertebrates (25% of the total diet) (Atramentowicz, 1982, 1986, 1988; Charles-Dominique et al., 1981; Charles-Dominique, 1983; Julien-Laferrière, 1995, 1999; Liete et al., 1996). The locomotion of

Caluromys involves quick, agile, and acrobatic quadrupedalism on terminal branches by efficient grasping with the extremities; suspension, cantilevering, and short leaps have been also observed (Rasmussen, 1990). When feeding on terminal branches, upside-down postures by the hindlimbs and tail are frequent (Lemelin, 1996, 1999; Rasmussen, 1990), and the hands are used extensively for manipulation when consuming fruits and insects (Hunsaker and Shupe, 1977; Lemelin, 1996, 1999; Rasmussen, 1990).

The didelphid marsupials included in this study are characterized by two broad adaptive strategies: *Monodelphis, Didelphis,* and *Philander* spend a large portion of their activity budget on the ground. When venturing arboreally, they tend to be on large diameter supports. In contrast, *Marmosa* has a strong preference for understory vines despite spending some time on the ground, and *Caluromys* is fully committed to the canopy where it relies mainly on terminal branches. Both *Marmosa* and *Caluromys* use their hands more often to manipulate fruits and prey, especially insects. It is also very important to point out that the more frequent use of thin arboreal supports is not size related among didelphids. Indeed, there are smaller (*Marmosa*) and larger (*Caluromys*) fine-branch arborealist opossums, as well as smaller (*Monodelphis*) and larger (*Philander* and *Didelphis*) more ground-dwelling opossums.

The primates included in this comparative study are all members of the family Cheirogaleidae. Like *Marmosa* and *Caluromys*, cheirogaleids are agile quadrupeds moving on arboreal supports in search of fruits, gums, nectar, and insects (Hladik, 1979; Hladik et al., 1980; Martin, 1972a,b, 1973; Tattersall, 1982; Walker, 1979). They range in body mass from about 30 g for *Microcebus myoxinus* and 60 g for *Microcebus murinus* to about 400 g for *Cheirogaleus major,* with *Cheirogaleus medius* falling somewhere in between (Smith and Jungers, 1997). Of all cheirogaleids, mouse lemurs are the most committed fine-branch arborealists, moving and foraging frequently on small branches, twigs, or vines in the undergrowth and lower forest levels (Martin, 1972a,b, 1973). They are also the most predatory cheirogaleids, with prey accounting for about 40% of their diet (Hladik et al., 1980). Mouse lemurs use swift movements of the hands to capture fast-moving insect prey (Bishop, 1964; Lemelin, 1996). During feeding, cantilevering and upside-down postures are used by all cheirogaleids and the hands are used to hold food objects (Lemelin, 1996, 1999; Martin, 1972a; Gebo, 1987). Cheirogaleids share with *Marmosa* and *Caluromys* behavioral and ecological similarities that are believed to represent the primitive condition for the order Primates (Cartmill, 1972, 1974a,b;

Charles-Dominique, 1983; Charles-Dominique and Martin, 1970; Eisenberg and Wilson, 1981; Hershkovitz, 1969; Hunsaker and Shupe, 1977; Lemelin, 1996, 1999; Martin, 1972b; Rasmussen, 1990).

On the basis of this review, it is quite clear that these contrasts among didelphids and similarities between *Marmosa*, *Caluromys*, and cheirogaleids in ecology and behavior have potential to elucidate the functional role and adaptiveness of grasping hands and feet with regard to the problem of primate origins. In the next section, we present a functional model that explains why more prehensile hands and feet should be expected in mammals that travel and forage on thin arboreal supports compared to those that spend more time on the ground or larger arboreal substrates.

Substrate Use and Cheiridial Morphology: A Functional Model

Walking quadrupedally on thin and flexible arboreal supports presents considerable mechanical challenges. While standing on top of a relatively thin cylindrical support, the tendency of the body of the animal is to pitch or roll (Cartmill, 1985). Pitching and rolling will have the tendency to move the body's center of mass away from the margins of the narrow substrate, toppling the animal from the support (Figure 5.1 in Cartmill, 1985). One way to counteract pitching and rolling effects on a thin substrate is to generate torques equal and opposite using grasping extremities (Cartmill, 1974a, 1985; Napier, 1967; Preuschoft et al., 1995). The ability to grasp a thin support with the hands and feet is proportional to the length of the digits relative to the palm or sole (Lemelin, 1996, 1999; Napier, 1993; Washburn, 1951). Longer proximal and middle phalanges relative to the metapodials enable the digital portion of the ray to encircle completely a thin arboreal support by the action of the long flexor tendons and intrinsic muscles, thus providing a firm, powerful grip necessary to keep in check the unwanted torques described above (Lemelin, 1996, 1999; Fig. 2a). Similarly, longer phalanges relative to the metacarpals increase the potential of the hand to achieve prehensile grips (i.e., the ability to hold and retain an object from the pull of gravity with a single hand (Napier, 1961)) (Lemelin, 1996, 1999; Lemelin and Grafton, 1998).

Unlike fine-branch arborealists, maintaining balance is less problematic in mammals walking quadrupedally on the ground or relatively large arboreal supports. The digits of these mammals are probably under a much different mechanical environment. During the support phase of quadrupedal walking on

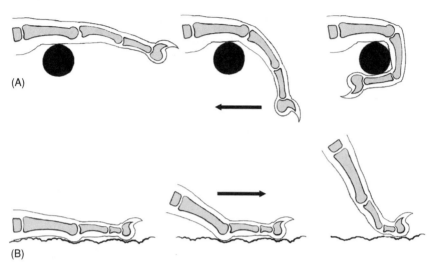

(A)

(B)

Figure 2. Digit kinematics on a thin branch (A) and on the ground (B) during a quadrupedal walking step in opossums. The top figures show a posterior view of pedal ray V, from the moment the sole strikes the thin pole to about mid-support (based on video data of *Caluromys philander* collected by the authors). The arrow in A indicates the movement of flexion at the metatarsophalangeal and proximal interphalangeal joints necessary to achieve a firm, powerful grip by completely encircling of the thin support. The bottom figures show a lateral view of the manual ray III, from the moment the palm strikes the flat surface to about mid-support (based on video data of *Monodelphis domestica* collected by the authors and cineradiographic data of *Didelphis marsupialis* published by Jenkins (1971)). The arrow in B shows the direction of travel during which movement of extension (and hyperextension) occurs at the metacarpophalangeal and interphalangeal joints as the digit is lifted from the substrate.

a flat surface (i.e., relatively large arboreal support or the ground), the rays of a plantigrade mammal undergo notable extension (and hyperextension) at the metapodiophalangeal joints and interphalangeal joints (Jenkins, 1971; McClearn, 1992; Fig. 2b). In theory, the resulting torques at these joints and bending moments on the phalanges are proportional in magnitude with digit length (Nieschalk and Demes, 1993; Preuschoft, 1969, 1970; Strasser, 1992, 1994), especially for the digit falling closest to the path of travel. This model is supported by observations within closely related mammal groups, such as cercopithecids (Etter, 1973; Gabis, 1960; Napier and Napier, 1967; Schultz, 1963; Strasser, 1992, 1994) and procyonids (Lemelin and Grafton, 1998;

McClearn, 1992), which show that more terrestrial taxa sport relatively shorter digits compared to more arboreal taxa, as well as the more general observation of relative digit length reduction across mammals with the adoption of more cursorial locomotor habits (Howell, 1944; Smith and Savage, 1956). On the basis of the functional model presented above, as well as the behavioral and ecological data presented in the previous section, we can make the following predictions: *Marmosa* and *Caluromys*, which spend more time moving and foraging on thin branches, should have more prehensile extremities (i.e., longer digits relative to the palm or sole) compared to *Monodelphis*, *Didelphis*, and *Philander*, which spend more time moving and foraging on the ground. When comparing hand and foot proportions, *Marmosa* and *Caluromys* should fall closer to cheirogaleids than more terrestrial didelphids.

Morphometric Results

Details of the skeletal samples, measurements, and some of the statistical analyses have been reported previously in Lemelin (1996, 1999). For the purpose of this chapter, the methods and some of the results of these analyses are summarized, in addition to new morphometric results. Five didelphid taxa (*Monodelphis domestica*, *Philander opossum*, *Didelphis virginiana*, *Marmosa robinsoni*, and *Caluromys philander*) and three cheirogaleid taxa (*Microcebus murinus*, *Cheirogaleus medius*, and *Cheirogaleus major*) were considered. The sample sizes are indicated in Figures 3 and 4. The length of each metapodial (M), proximal phalanx (PP), and middle phalanx (MP) was measured on the hands and feet of museum specimens (skeletal specimens or radiographs of pelts). A phalangeal index was computed in order to estimate the degree of prehensility of each ray (I-V) (Napier and Napier, 1967; Napier, 1993). For the first ray of the hand and foot, this index is equal to the length of the PP divided by the corresponding M times 100. For all other rays, the same index is equal to the sum of the PP and MP divided by the corresponding M times 100.

When examining the means and spread of the data within didelphid marsupials, the same basic pattern repeats itself for all rays (with the exception of the hallux): *Marmosa* and *Caluromys*, two fine-branch arborealists, have significantly higher phalangeal indices compared to the more ground-dwelling *Monodelphis*, *Didelphis*, and *Philander* (see Lemelin, 1996, 1999 for details of the statistical results) (Figures 3 and 4). In other words, didelphids that rely on terminal branches or vines when moving and foraging have more prehensile hands and

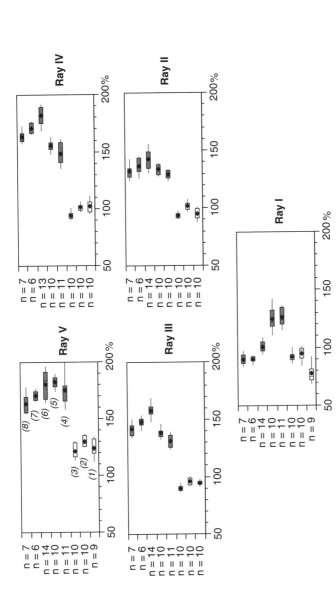

Figure 3. Box-plot graphs of the phalangeal indices for rays I through V of the hand in didelphids and cheirogaleids. The diamonds represent the mean, the vertical line the median, the left and right sides of the rectangle the 25th and the 75th percentiles, and the left and right ends of the horizontal line the 10th and the 90th percentiles. The shaded rectangles represent taxa that move and forage primarily on thin branches and open rectangles represent taxa that move and forage primarily on large arboreal substrates and/or the ground. (1) *Monodelphis brevicaudata*, (2) *Philander opossum*, (3) *Didelphis virginiana*, (4) *Marmosa robinsoni*, (5) *Caluromys philander*, (6) *Microcebus murinus*, (7) *Cheirogaleus medius*, and (8) *Cheirogaleus major*.

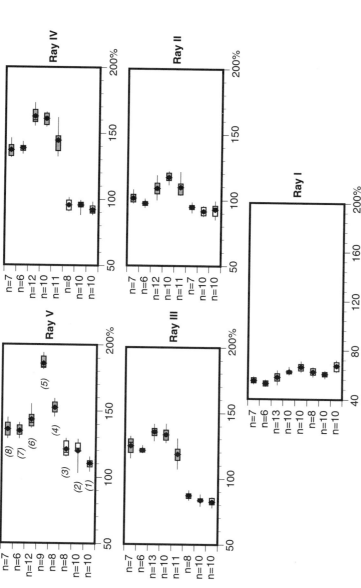

Figure 4. Box-plot graphs of the phalangeal indices for rays I through V of the foot in didelphids and cheirogaleids. The diamonds represent the mean, the vertical line the median, the left and right sides of the rectangle the 25th and 75th percentiles, and the left and right ends of the horizontal line the 10th and the 90th percentiles. The shaded rectangles represent taxa that move and forage primarily on thin branches and open rectangles that move and forage primarily on large arboreal substrates and/or the ground. (1) *Monodelphis brevicaudata*, (2) *Philander oposum*, (3) *Didelphis virginiana*, (4) *Marmosa robinsoni*, (5) *Caluromys philander*, (6) *Microcebus murinus*, (7) *Cheirogaleus medius*, and (8) *Cheirogaleus major*.

feet. The link between arboreal foraging on thin supports and more grasping extremities is reinforced when didelphids are compared to primates. The means and distributions of most phalangeal indices of *Marmosa* and *Caluromys* fall closer to or within the range of those of cheirogaleids (Figures 3 and 4).

Several additional patterns are also worth mentioning. First, it is important to note that the clawless distal phalanx of cheirogaleids (which contributes to about 20% of total thumb length) was not included in the computation of the phalangeal index, explaining why the cheirogaleid distributions of manual ray I fall within those of more terrestrial didelphids (Figure 3). Second, the lack of clear differentiation for the hallux within didelphids, and between didelphids and cheirogaleids (Figure 4) suggests that some climbing abilities have been retained in the foot of more terrestrial didelphids (see following section) and that all didelphids can be derived from an ancestor with a divergent, grasping hallux capable of arboreal locomotion such as climbing (Bensley, 1901a,b; Dollo, 1899; Huxley, 1880; Szalay, 1984, 1994). Finally, the hands and feet of didelphids are characterized by a monotonic increase in the values of the phalangeal indices from ray I through V, whereas those of cheirogaleids show an increase from ray I through IV, then a decrease for ray V (Figures 3 and 4). These proportional differences are not surprising considering the wide gap in phylogenetic heritage between these two distantly related groups, and may underlie two slightly different grasping mechanisms.

Overall similarity of hand and foot morphology among didelphids and cheirogaleids can be summarized using clustering methods (Sneath and Sokal, 1973). Standardized mean values of phalangeal indices were used to compute average taxonomic distances among taxa. The resulting matrix (8×8) was summarized using four different clustering algorithms found in NTSYS v. 2.02 (Exeter Software, Setauket, NY): (a) unweighted pair-group method using arithmetic average (UPGMA), (b) weighted pair-group method using arithmetic average (WPGMA), (c) single-link (nearest neighbor) method, and (4) complete-link (furthest neighbor) method (Manly, 1986; Pimentel, 1979). All four methods pointed to a similar clustering pattern: both groups of fine-branch arborealists cluster together, *Marmosa* and *Caluromys* with cheirogaleids, separately from another group comprising all three more terrestrial didelphid taxa (Figure 5). This similarity in overall proportions of the hands and feet, which are likely to reflect similar grasping abilities between taxa with radically different phylogenetic backgrounds, strengthens the case for functional convergence in response to similar ecological niches.

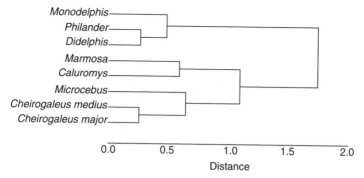

Figure 5. Phenogram of UPGMA clustering (r_{coph} = 0.94) based on average taxonomic distance matrix of all manual and pedal phalangeal indices in didelphids and cheirogaleids. Note the grouping of *Marmosa* and *Caluromys* with cheirogaleids in a cluster separate from more terrestrial didelphids. The cophenetic correlation coefficient (r_{coph}) measures the goodness of fit between the original distance matrix and that resulting from the clustering algorithm.

Nevertheless, it is important to note that the clustering within the fine-branch arborealists also points to a phylogenetic signal (i.e., the two didelphids and three cheirogaleids cluster within their own families), which may be linked to those gradient differences reported above in the values of the phalangeal indices between didelphids and cheirogaleids.

Opossums and primates that spend more time on thin arboreal supports are also characterized by similar intrinsic ray proportions of the hand and foot compared to more terrestrial opossums. The digital portion (i.e., proximal and middle phalanges) of *Marmosa*, *Caluromys*, and all three cheirogaleids always contributes to 50% or more of total ray length. In *Monodelphis*, *Didelphis*, and *Philander*, the metapodials always contribute to 50% or more of total ray length, with the exception of ray V (Figures 6 and 7). For all rays of the hand and foot, the relative contribution of the digit portion in total ray length is always greater in fine-branch arborealists compared to more ground-dwelling taxa. As reported by Hamrick (2001), longer proximal phalanges relative to the metapodials appear very early in life in *Microcebus* and *Caluromys* and are conserved throughout ontogeny. This pattern, found in primates and woolly opossums, is different from that of bats, tree shrews, flying lemurs, as well as plesiadapiforms (Hamrick, 2001).

Short metapodials and long phalanges typical of primates are unusual among mammals (Hamrick, 2001) and appear to characterize only nonprimate

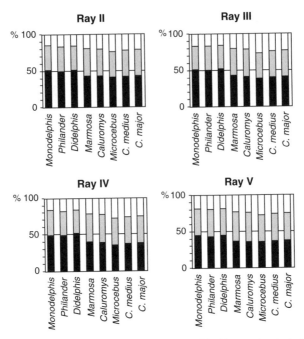

Figure 6. Relative contribution of the metacarpal (black), proximal phalanx (shaded), and middle phalanx (white) in the hand of didelphids and cheirogaleids. The values for rays II through V represent mean percentages of total ray length for each species.

mammals that inhabit fine-branch niches (i.e., *Marmosa* and *Caluromys*). This suggests to us that the invasion of a fine-branch niche by the earliest primates led to the evolution of more prehensile hands and feet (i.e., longer digits relative to the metapodials)—a conclusion also reached by Hamrick (2001). In the following section, we present behavioral observations and tests using the same contrasts among didelphids in order to strengthen the functional link between having more prehensile hands and feet (i.e., longer digits relative to the palm/sole) and improved abilities to negotiate thin branches.

Performance Results

In the previous section, we showed how longer digits relative to the palm/sole are mechanically advantageous to achieve firm grasping of a thin branch and that such cheiridial proportions do indeed characterize mammals, foraging preferentially on thin vines or terminal branches. Behaviorally, several studies

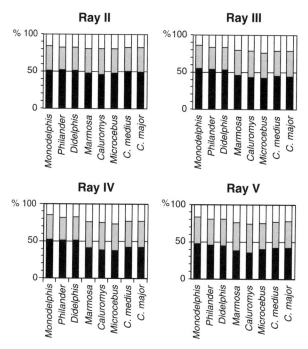

Figure 7. Relative contribution of the metatarsal (black), PP (shaded), and MP (white) in the foot of didelphids and cheirogaleids. The values for rays II through V represent mean percentages of total ray length for each species.

have shown strong similarities between *Marmosa*, *Caluromys*, and cheirogaleids in use of the extremities when grasping thin arboreal substrates or food objects. The comparison of *Marmosa* (Figure 18 in Charles-Dominique et al., 1981) with *Microcebus* (Figure 4 in Martin, 1972) or *Caluromys* with *Mirza* (Figure 2 in Lemelin, 1999) are striking examples of behavioral convergence between these primates and marsupials. What is still missing is a clear demonstration that more terrestrial didelphids with relatively shorter digits "perform" more poorly than *Marmosa* or *Caluromys* during locomotion and posture on thin arboreal supports.

Performance studies provide a direct way to assess how the phenotype of an organism limits its ability to carry out behavioral tasks, and therefore provides crucial links between morphology, behavior, and ecology (Arnold, 1983; Emerson and Arnold, 1989; Wainwright, 1994). In this way, performance also aims to understand causal relationships between morphology and behavior, often employing an experimental approach (see next section for further discussion of this approach with regard to the origins of primate locomotor adaptations). We further tested the link between more prehensile extremities and fine-branch arborealism by examining in the laboratory the performance of two didelphid species moving on arboreal substrates of different size and orientation. *Monodelphis domestica* and *Caluromys philander* were chosen because they are well-differentiated in their substrate preferences (i.e., ground versus thin branches) and cheiridial proportions (i.e., relatively shortest versus relatively longest digits).

Two small trees were fastened onto a wooden platform approximately 1.1 m apart from each other (Figure 8). The side branches coming off the trunks at various angles were 1 cm or less in diameter. One of the side branches of one tree was 30 cm away from two parallel branches of the other tree (Figure 8), so the use of the hands and feet during bridging behavior could be observed. Four

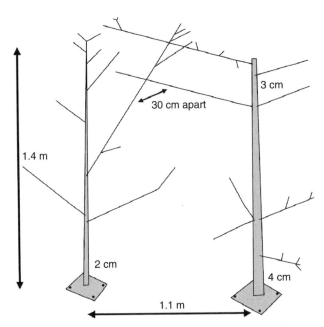

Figure 8. Experimental trees used in the laboratory to examine locomotor and postural performance on thin branches between *Monodelphis domestica* and *Caluromys philander*. The height and diameters of the trunks, and distances between branches of the two trees are given in meters and centimeters. All the side branches are 1 cm in diameter or less.

individuals of *M. domestica* (four males: 115, 122, 150, and 154 g, respectively) and two individuals of *C. philander* (two females: 392 g each) were released by hand one at a time on various portions of the tree setup. A subject was free to follow any path until it came down to the floor where it was captured and released on the setup again. Fruits were impaled at many secondary twigs to promote movement of the animals at the periphery where branches are thinnest and most flexible. Locomotion and posture for each animal was recorded (60 Hz) using a Super VHS Panasonic camera (Matsushita Electric Corp., Secaucus, NJ) resting on a tripod, and two halogen lights were utilized, so the camera could be electronically shuttered (1/1000 s) to avoid motion blur. The behavioral data were analyzed by viewing the videotapes on a JVC videocassette recorder (JVC Professional Products Co., Wayne, NJ) using the frame-by-frame option.

The most stunning result of our performance study was the clear indication that *Monodelphis* had difficulty keeping its balance when moving on thin and flexible branches (less than 1 cm in diameter). In most of the 45 trials for which *Monodelphis* subjects were released on these thin supports, three out of four individuals examined rolled sideways and ended up upside down as soon as they began to move quadrupedally (Figure 9). All the subjects fell at least once from the setup, for a total of nine falls. In most cases, the animals moved back toward the trunk and descended headfirst in a clumsy, deliberate manner until the relative safety of the ground was reached. Only the smallest male (116 g) was able to walk quadrupedally on thin branches, sometimes all the way to the periphery of the tree setup. The same animal showed also more prowess climbing up and down, and performing suspension by the hindlimbs and bridging between trees. However, it is important to note that while moving quadrupedally on thin branches, this individual often rolled sideways and ended up upside down, as observed in the other animals. In this context, Pridmore (1994) reported in his laboratory study of opossum gaits that *Monodelphis* appeared unstable and frequently fell when walking quadrupedally on 6.3 mm poles.

Despite weighing almost three times as much as *Monodelphis*, *Caluromys* had no problem staying on top of thin and flexible branches of the tree setup when moving and collecting pieces of fruit with the hands (Figure 9). Woolly opossums also climbed up and down along the trunk of each tree, explored the periphery of each thin branch for food, and very often crossed from one tree to the other. Woolly opossums never leapt to cross the gap between both trees. Instead, they cantilevered their bodies using the hindlimbs and reached

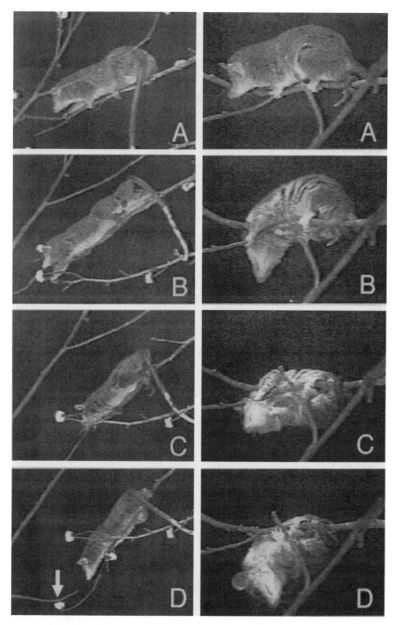

Figure 9. Comparison of the locomotor and grasping abilities of *Caluromys philan-der* (left) and *Monodelphis domestica* (right) while moving quadrupedally on the same thin branch (less than 1 cm in diameter). The sequence on the left shows a woolly opossum able to stay balanced on the top of the pliant support (A, B, and C) as it moves towards the periphery to reach the food treat indicated by the arrow (D). The

a branch with the forelimbs, often in an acrobatic manner. On securing that branch with the forelimbs, the hindlimbs were transferred sequentially to that secured support. Hindlimb suspension (with or without assistance from the tail) and upside-down quadrupedalism were also commonly observed. In essence, the positional behavior and athletic abilities of the woolly opossum observed in the laboratory were very similar to those observed in the wild (Rasmussen, personal communication).

This performance study establishes an important fact with regard to the evolution of grasping hands and feet in primates. A marsupial with relatively longer digits (i.e., more prehensile extremities) like the woolly opossum can negotiate thin and flexible branches more efficiently—in spite of its larger body mass—than the smaller, short-tailed opossum with relatively shorter digits (i.e., less prehensile extremities). These clear differences in locomotor abilities between *Monodelphis* and *Caluromys* strongly suggest that relatively long digits are useful to counteract undesirable torques that have the tendency to roll the animal sideways when walking on relatively slender supports. A causal link can thus be established between more primate-like cheiridia with greater prehensile capabilities and locomotor competence on thin branches. Such causal link between morphology and behavior can only be established with an experimental approach. In the following section, we present the results of a study that test the functional link between locomotor characteristics unique to primates and fine-branch arborealism using an experimental approach.

CONVERGENCE BETWEEN OPOSSUMS AND PRIMATES: EXPERIMENTAL STUDIES

In the previous section, we stressed the importance of testing the causal link between morphology and behavior. More than 25 years ago, Kay and Cartmill (1977) made a similar point, arguing for the need to understand the mechanical relationship between a trait and its function. Kay (1984) cautioned

---

sequence on the right shows a short-tailed opossum losing its balance as it begins to move quadrupedally (A), rolling sideways (B, C), and ending upside-down (D). Despite being three times heavier, woolly opossums are capable of keeping their balance on these relatively small supports compared to short-tailed opossums because of more prehensile hands and feet. Opossums on left and right columns are not pictured on the same scale.

anthropologists, by way of an example, that the fact that all primates share a petrosal bulla and are arboreal does not functionally link this morphological trait with living in the trees. More recently, Lauder (1996) summed up the pitfalls of relying solely on morphology to infer the mechanical function of a trait. He noted that structure and mechanical function are not always tightly matched and pointed out that "... in our desire to draw conclusions about biological design and to support theoretical views of how organisms are built, we have been too willing to make assumptions about the relationship between structure and mechanical function (Lauder, 1996: 56)." Furthermore, Lauder (1996: 56) emphasized that "... we have not often conducted the mechanical and performance tests needed to assess the average quality of organismal design." In other words, in order to better understand the relationship between a morphological trait (or complex of traits) and its mechanical function, we have to rely on experimental methods and techniques.

The importance of an experimentally based approach in biological anthropology was recognized more than 50 years ago by Washburn who called for a "... modern experimental comparative anatomy to take its place among the tools of the student of evolution (Washburn, 1951:67)." Since then, numerous biological anthropologists answered that call with vigor (e.g., Demes et al., 1994; Demes et al., 2001; Fleagle et al., 1981; Hylander, 1979a,b; Jenkins, 1972, 1981; Kimura et al., 1979; Marzke et al., 1998; Schmitt, 1999; Schmitt and Lemelin, 2002; Stern et al., 1977; Tuttle and Basmajian, 1974). Although a wide range of experimental techniques are available to biological anthropologists (Biewener, 2002; Fleagle, 1979), they are all used with the same underlying principle. An experimental analysis involves direct observations, such as muscle activity, movement, external forces, or internal strains of an organism behaving in a certain manner. By "perturbing" the behavior, that is changing the substrate, task, food, or other variable, the experimenter can directly measure the quality of organismal design. The outcome of such an experiment is threefold: in the first case, experimental results support the proposed link between morphology and mechanical function. In the second case, the functional relationship (e.g., large humeral tubercles and terrestrial locomotion) and inferences drawn from that morphology are correct, but the experimental data revise the underlying mechanical role of the morphological trait in question (e.g., supraspinatus muscle is a humeral stabilizer, not protractor; Larson and Stern, 1989). In the last case, experimental data reject the proposed structure-function relationship such as that between long bone cross-sectional shape and loading patterns (Demes et al., 2001; Lieberman et al., 2004).

In the previous section, we assessed the quality of organismal design by testing the link between structure (i.e., relatively longer, more grasping digits) and mechanical function (i.e., stay balanced and moving on thin, flexible branches) in two opossums species. In the following section, we present the results of an experimental analysis that also assesses the quality of organismal design. More specifically, we test the relationship between locomotor characteristics typical of all primates during walking (i.e., DS gaits, protracted arm position at forelimb touchdown, and relatively lower forces on the forelimb) and fine-branch arborealism. Just like grasping extremities, these locomotor features are present in most primates and are likely to have characterized the last common primate ancestor. Therefore, a comparison outside of primates is again needed.

Our comparative strategy is inspired by the phylogenetic bracketing approach (Witmer, 1995), which "serves as a rationale for experimental studies..." (Susman, 1998: 27). However, we take a slightly different approach in that we are not attempting to interpret the anatomy of a fossil species. Instead, we are attempting to understand the functional significance of locomotor features that characterized the earliest primates, using three bracketing taxa. Locomotor mechanics will be compared: (a) between two closely related opossums that have diverged ecologically (*Monodelphis domestica* and *Caluromys philander*) and (b) between two distantly related mammals, a didelphid marsupial and a cheirogaleid primate, that have converged ecologically (*Caluromys philander* and *Cheirogaleus medius*). As stressed earlier, convergence represents one of the strongest sources of evidence to infer adaptation (Brooks and McLennan, 1991; Biewener, 2002; Larson and Losos, 1996; Vogel, 1998).

Materials and Methods

All data were collected in the Animal Locomotion Laboratory at Duke University using methods summarized in Figure 10 and described briefly in Schmitt and Lemelin (2002) and Lemelin and Schmitt (2004). Animals were videotaped while walking quadrupedally on a wooden runway and poles of two different diameters (7 and 28 mm) to simulate terrestrial and slender arboreal substrates. Substrate reaction forces were measured by instrumenting sections of each substrate onto a force platform. Species used in this study included three *Caluromys philander* (two females: 392 g each; one male: 346 g), five

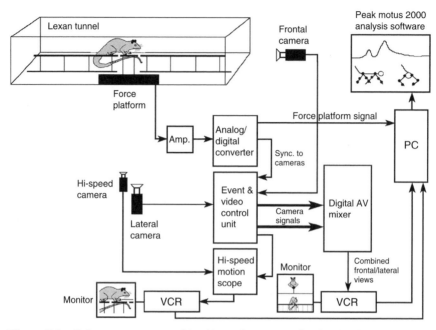

Figure 10. Laboratory setup used in this study. See text for details of the laboratory equipment and connections between various hardware elements.

Monodelphis domestica (five males: 120, 122, 150, 154, and 156 g, respectively), and four *Cheirogaleus medius* (one male: 300 g; three females: 270, 295, and 300 g, respectively).

Prior to each experiment, the fur covering the limbs was shaved to facilitate measurement of arm protraction at forelimb touchdown. Animals were allowed to move freely within a clear Lexan enclosure (3.6-m long). The floor was made of 28-mm thick plywood covered with a coating of polyurethane and sand. In the floor of the runway, a Kistler force platform model 9281B (Kistler Instrument Co., Amherst, NY) was mounted. For locomotor data collected on the runway, a rectangular section of plywood was instrumented onto the surface of the force platform. A plywood mask flush with the floor of the runway and surrounding the instrumented section by 2–3 mm gaps was then positioned to cover the rest of force platform. The simulated arboreal supports consisted of a graphite pole (7 mm in diameter) and wooden pole (28 mm in diameter) painted with a mix of white paint and sand, and coated with polyurethane. Each pole consisted of two 1.2-m-long segments mounted on a series of aluminum struts with adjustable height and fastened onto a wooden board. Each pole section was then attached onto the surface of the runway and positioned on each side of a 5-cm-long central segment with 2–3 mm gaps separating them. This central segment of the pole was attached on a wooden platform by one or two rigid aluminum struts that did not interfere with gripping of the pole during locomotion. This wooden platform was in turn fastened with T-bolts onto the grooves of the surface of the force plate.

We collected data on the ground and 7-mm pole in *Caluromys*, the ground and 28-mm pole in *Cheirogaleus*, and the ground only in *Monodelphis*. Subjects were videotaped at 60 Hz using electronically shuttered (1/1000 s) normal-speed JVC GY-X3 video camera (JVC Professional Products Co., Wayne, NJ) and Panasonic 5100HS video camera (Matsushita Electric Co., Secaucus, NJ) positioned perpendicular and frontal to the moving subject. Both lateral and frontal views were used to assess the footfall pattern of the animals. In addition, a Motion Scope® high-speed digital camera (Redlake MSDA, Inc., San Diego, CA) set at 125 or 250 Hz and positioned lateral to the moving subject was used to obtain precise kinematics of the forelimb. At the same time, substrate reaction forces were recorded with the force platform. The analog signal originating from the force platform was amplified using a Kistler 8-channel charge amplifier model 9865 and then converted into a digital signal using an Analog-Digital Converter designed by Peak Performance Technologies Inc. (Englewood, CO). Views from the normal-speed cameras (lateral and frontal views) were merged into a single, split screen view using a Panasonic Digital AV Mixer WJ-MX50A (Matsushita Electric Co., Secaucus, NJ), and images were recorded on videotape using a Panasonic Video Cassette Recorder AG-7350 (Matsushita Electric Co., Secaucus, NJ). Images from the high-speed camera were also recorded on videotape using the same model VCR.

Video and force plate data were synchronized in the following way. All normal-speed cameras were synchronized using the signal from a single master camera routed through the Event and Video Control Unit (EVCU) and then passed into the gen-lock adapter of other cameras. This assured that all cameras were recording at the same rate and that the fields were synchronized. The force plate was recording data constantly into a buffer until it received a signal from a handheld trigger. At this point, force data were stored in the computer. This signal from the handheld trigger, referred to as the event signal, generated in turn a second signal via the EVCU. This second signal, here called the sync signal, was created as soon as the next available video field passing through the EVCU was encountered. Event and sync signals may occur simultaneously or be off by as much as one field (1/59th s). This sync signal indicated to the computer to store 1 s of data prior and after the triggering of the event signal. The event signal also triggered the high-speed camera to store data before and after its triggering point (1 or 2 s before and after the triggering point at 250 or 125 Hz, respectively). The triggering point of the event signal is designated as 0000 ms on the image of high-speed camera. The EVCU also generated a bar code visible in the right upper hand corner of the videotape image. One line of this code thickened at the triggering of the event signal. This bar code change allowed visual synchronization of all video images, force plate, and high-speed camera. Finally, the signal was digitally encoded onto the videotape. When video was downloaded into the computer, this digital marker was used by the motion analysis software to precisely coordinate video and force data.

For each species, the footfall pattern, arm protraction angle, and peak vertical V_{pk} were analyzed using Peak Motus 2000 movement analysis software (Peak Performance Technologies, Inc.) running on a personal computer. Only strides for which subjects moved steadily across the instrumented section of the force platform (with two strides prior to and after contacting this section) were used for analysis. Video data were imported to the computer and synchronized with the kinetic data from the force platform using the procedure described earlier. For each step examined, footfall pattern was determined using the frame-by-frame analysis and arm protraction (i.e., angle of the arm relative to horizontal body axis) (degree) was calculated from digitized X and Y values at the shoulder, elbow, and wrist joints. After filtering the raw force data (30 Hz Butterworth digital filter), substrate reaction forces ($F_{x,y,z}$) were calculated. Differences in arm protraction between taxa were tested with the nonparametric Mann-Whitney U-test (Sokal and Rohlf, 1995). Mean V_{pk} for the forelimb and hindlimb were calculated. Forelimb/hindlimb peak vertical force ratios were computed and logged (ln), and then compared between taxa.

Gait Patterns

Since Muybridge (1887) first filmed baboons walking quadrupedally, it has been well recognized that primates rely on a different footfall pattern during walking compared to most other mammals (Cartmill et al., 2002; Hildebrand, 1967, 1976, 1985; Larson, 1998; Lemelin et al., 2003; Magne de la Croix, 1936; Rollinson and Martin, 1981; Vilensky and Larson, 1989). During quadrupedal walking, most nonprimate mammals, such as cats, dogs, or horses, use lateral-sequence (LS) walking gaits in which each hind footfall is followed by the *ipsilateral* fore footfall. The resulting footfall sequence is then: right hind, right fore, left hind, left fore (RH RF LH LF). In contrast, primates rely mostly on DS walking gaits in which each hind footfall is followed by the *contralateral* fore footfall. The resulting footfall sequence is then: right hind, left fore, left hind, right fore (RH LF LH RF).

Two main categories of hypotheses have been offered to explain the predominance of DS gaits in primates (Cartmill et al., this volume; Lemelin et al., 2003). Biomechanical hypotheses, such as the more posteriorly placed center of gravity and hindlimb dominance, have been invoked to explain the common usage of DS gaits in primates (Rollinson and Martin, 1981). Neurological hypotheses consider the prevalence of DS gaits in primates as a by-product of greater supraspinal control of forelimb movement involved during grasping, manipulation, and reaching (Larson, 1998; Vilensky and Larson, 1989). Despite disagreeing on the mechanisms involved in the production of DS gaits, both categories of hypotheses make clear that traveling and foraging in a thin-branch habitat was a major impetus in the origin and evolution of such gaits in primates.

Several nonprimate mammals are also reported to use DS walking gaits, including aardvarks, kinkajous, armadillos, and several arboreal marsupials (Cartmill et al., 2002; Goldfinch and Molnar, 1978; Hildebrand, 1967, 1976, 1985; Lemelin, 1996; Lemelin et al., 1999, 2002, 2003; Pridmore, 1994; White, 1990). More terrestrial marsupials like *Dasyurus hallucatus* and *Didelphis virginiana* commonly adopt LS walking gaits (Hildebrand, 1976; White, 1990). At relatively high speeds, *Didelphis* performs DS walks more often, which are also most common in more arboreal taxa such as *Trichosurus vulpecula* and *Dromiciops australis* (Goldfinch and Molnar, 1978; Pridmore, 1994; White, 1990). Results from our lab indicate a similar dichotomy in gait preferences between more terrestrial and more arboreal didelphid marsupials. On both runway and poles of various diameters, LS footfall patterns were the norm in *Monodelphis* during walking (Lemelin et al., 2002, 2003; Schmitt and Lemelin, 2002; Figure 11). In *Caluromys*, DS gaits were most common on the runway, although several steps with a LS footfall pattern were observed as well (6 out 64 gait cycles) (Lemelin et al., 2002, 2003; Schmitt and Lemelin, 2002). On the smallest pole (7 mm in diameter), only DS gaits were observed (43 out of 43 gait cycles) (Lemelin et al., 2002, 2003; Schmitt and Lemelin, 2002). The fact that the woolly opossum displays

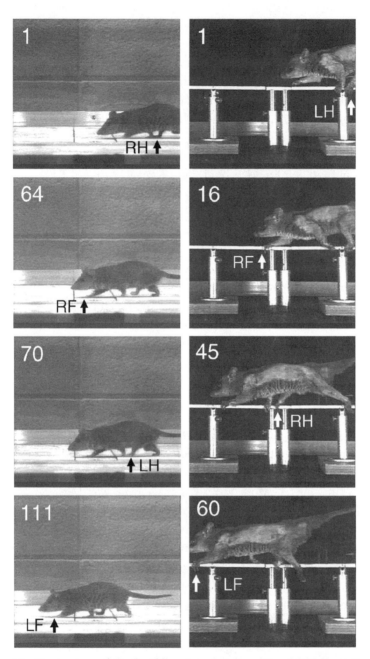

Figure 11. Comparison of the footfall pattern between *Monodelphis domestica* (left) and *Caluromys philander* (right). The left panels show sequentially a lateral-sequence footfall pattern (RH RF LH LF) in *Monodelphis* when walking on the runway. The right panels show sequentially a diagonal-sequence footfall pattern (LH RF RH LF) in *Caluromys* when walking on the 7-mm pole. Each image represents 1/250 s. Opossums on left and right columns are not pictured on the same scale.

DS gaits exclusively when walking on the smallest poles suggests to us that these gaits are more advantageous functionally compared to LS gaits when walking on narrow supports. The functional advantages underlying the origins and evolution of DS gaits in primates and marsupials are discussed in detail in Cartmill et al., (2002, this volume) and Lemelin et al., (2003).

Arm Protraction

Compared to other mammals, primates have relatively long limbs (Alexander et al., 1979; Polk et al., 2000) with more mobile joints (Reynolds, 1985b; Schmitt, 1999). These morphological differences are associated with fundamental kinematic features that distinguish primates from other nonprimate mammals. For example, primates take longer strides when walking quadrupedally (Alexander and Maloiy, 1984; Reynolds, 1987). Larson (1998) and Larson et al., (2000, 2001) reported differences between primates and nonprimate mammals in the degree of arm protraction at forelimb touchdown. Most nonprimate mammals have a retracted arm position at touchdown, that is, the arm lies behind a vertical line going through the glenohumeral joint (i.e., angle between arm and horizontal body axis is less than 90°). In primates, the humerus is more protracted at touchdown, that is the arm is ahead of that same vertical line (i.e., angle between arm and horizontal body axis is greater than 90°). These differences in the degree of arm protraction between primates and small nonprimate mammals have been confirmed for the most part by cineradiographic studies (Fischer et al., 2002; Jouffroy et al., 1983; Schmidt and Fischer, 2000).

This combination of relatively long and more mobile forelimbs, longer strides, and more protracted forelimb postures is believed to have evolved for the need of reaching out when grasping branches and food with the hands (Larson, 1998; Larson et al., 2000; Schmitt, 1998, 1999). Again, the contrasts between more terrestrial and more arboreal opossums (*Monodelphis* and *Caluromys*), as well as between two fine-branch arborealists (*Caluromys* and *Cheirogaleus*) can provide a better understanding of the possible functional link between arm protraction and fine-branch arborealism.

During quadrupedal walking, both species of opossums and the cheirogaleid protracted their arm at forelimb touchdown (Figure 12; Table 1). However, arm protraction angles were higher and significantly different between fine-branch arborealists (woolly opossum and the fat-tailed dwarf lemur) and the more terrestrial short-tailed opossum ($P < 0.0001$). *Caluromys* showed almost identical degrees of arm protraction compared to *Cheirogaleus*

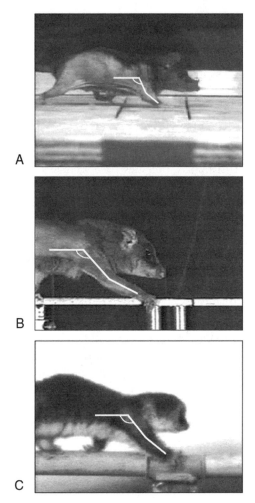

Figure 12. Comparison of arm protraction at touchdown for *Monodelphis domestica* (A), *Caluromys philander* (B), and *Cheirogaleus medius* (C). The arm is more protracted (i.e., higher angle) in the woolly opossum (B), a pattern more similar to that of the fat-tailed dwarf lemur (C) and other primates, compared to the less protracted arm of the short-tailed opossum (A).

(Table 1), and its mean is closer to the primate average of 118° than the marsupial average of 90° reported by Larson et al., (2000). Unlike the changes in gait patterns we reported above for woolly opossums, no differences were observed in the degree of arm protraction at touchdown between the small 7-mm pole and the runway (Table 1).

Table 1.   Arm protraction data for opossums and a cheirogaleid primate

Species (number of individuals)	Mean ± S.D. (°)	Range	Number of steps
Monodelphis domestica (2) (runway)	101.4 ± 6.14	88.1–110.7	13
Caluromys philander (3) (runway)	120.1 ± 8.24	107.6–138.9	23
Caluromys philander (3) (7-mm pole)	119.2 ± 9.16	92.5–132.7	31
Cheirogaleus medius (3) (28-mm pole)	121.9 ± 8.18	106.4–130.4	15

Our arm protraction angle average for *Cheirogaleus* (121.9°) and average arm protraction angle reported by Larson et al., (2000) for cheirogaleids (128°) differ widely from that reported for *Microcebus murinus* (78°) by Fischer et al., (2002). Similarly, our arm protraction angle average for *Monodelphis domestica* (101.4°) differs from the arm protraction angle average reported by Fischer et al., (2002) for the same species (62°). The degree of arm protraction we observed in *Monodelphis* is more similar to that reported by Jenkins and Weijs (1979) in their cineradiography study of *Didelphis*. These differences may be the result of different methodologies involving unrestrained locomotion (Larson et al., 2000, 2001; this study) versus a treadmill for which speed varies constantly in order to keep the animal in front of the X-ray beam (Fischer et al., 2002).

Our results show that animals like *Caluromys* and *Cheirogaleus* that spend more time moving and foraging on thin and flexible branches are characterized by greater ranges of motion in the shoulder region. Such increased ranges of motion may allow greater reaching capabilities when grasping arboreal substrates and collecting food with the hands. In this regard, Argot (2001) and Szalay and Sargis (2001) reported that compared to other more terrestrial didelphids, *Caluromys* has a triangular-shaped scapula with well-developed supra- and infraspinous fossae and a prominent acromion, as well as a more globular humeral head rising above the level of the tubercles. All of these features of the shoulder region are typical of arboreal primates that engage in climbing and other antipronograde behaviors (Ashton and Oxnard, 1964; Ashton et al., 1965; Harrison, 1989; Larson, 1993; Oxnard, 1963, 1967; Roberts, 1974; Rose, 1989). This morphological evidence, in combination with the arm protraction data we presented, strongly suggests these primate-like traits of the forelimb evolved in *Caluromys* to facilitate movement in a discontinuous, fine-branch milieu in which walking with longer strides and reaching with grasping hands were essential behavioral components.

Vertical Force Distribution on Limbs

The evolution of forelimb joints with increased mobility, at the expense of stability in primates is associated with reduced weight-bearing function of the forelimb during locomotion. Two dynamic mechanisms that reduce forces on the forelimbs during locomotion have been identified in primates: active weight shifting on the hindlimbs (Reynolds, 1985a, b) and gait compliance (Schmitt, 1998, 1999). These mechanisms underlie a fundamental kinetic difference that separates most primates from all nonprimates. During walking, the forelimbs of nonprimate mammals experience higher V_{pk} relative to the hindlimbs, whereas the reverse is true for most primates; V_{pk} are lower on the forelimbs relative the hindlimbs (Demes et al., 1994; Kimura et al., 1979; Polk, 2001; Reynolds, 1985b; Schmitt and Hanna, 2004). These differences in vertical force distribution are believed to represent a shift in the functional role of the forelimb that occurred with the origins of primates in an arboreal setting (Kimura et al., 1979; Larson, 1998; Schmitt, 1999, 2003a,b; Schmitt and Lemelin, 2002). This functional shift, from a weight-bearing strut used for propulsion to a grasping organ used for reaching and manipulation, has been a recurrent theme in models of primate origins since Jones (1916). Whether or not a shift in the vertical force distribution of the limbs, and ultimately a shift in the functional role of the forelimb, can be linked to the invasion of a fine-branch niche by the earliest primates can be addressed again with the contrasts between opossums and cheirogaleids.

Unlike any other nonprimate mammal known, *Caluromys* has a primate-like pattern of vertical force distribution (Schmitt and Lemelin, 2002; Figure 13). In a recent study, Schmitt and Lemelin (2002) reported in the woolly opossum V_{pk} forces on the forelimbs that were significantly lower than those of the hindlimbs. When these forces are examined in the form of a ratio (i.e., forelimb/hindlimb peak vertical forces), the woolly opossum clusters with the fat-tailed dwarf lemur, another fine-branch arborealist (Figure 13). The more terrestrial short-tailed opossum shows the opposite pattern (i.e., higher peak vertical forces on the forelimbs), which is typical of nonprimate mammals and a few primates (see Schmitt and Lemelin (2002) for a discussion of these primates). These differences in the pattern of weight distribution between two closely related species of opossums strongly suggest that the dynamics of mammal gaits can change in association with the occupation of different ecological niches. If this is correct, one can argue that the primate-like, vertical force distribution pattern observed in the woolly opossum evolved with the occupation

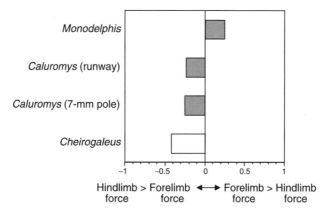

Figure 13. Mean ratios of V_{pk} for the forelimb versus hindlimb during walking in opossums (shaded boxes) and a cheirogaleid primate (white box). The ratios have been ln so that ratios greater or less than zero (i.e., equal forelimb and hindlimb peak vertical substrate reaction forces) are weighted equally and can be directly compared among taxa. Data for woolly opossums have been collected for both runway and 7-mm pole.

of a fine-branch niche. In the context of primate origins, this represents strong evidence that the last common ancestor of all primates was a fine-branch arborealist, very similar to the woolly opossum in behavior and ecology (Schmitt and Lemelin, 2002).

DISCUSSION AND CONCLUSIONS

Using two different comparative strategies—comparisons between more closely related taxa with divergent ecologies and more distantly related taxa with similar ecologies—we showed that mammals inhabiting similar environments, in which locomotion and foraging take place primarily on thin and flexible branches, possess more prehensile cheiridia (i.e., relatively longer digits). We also demonstrated that a mammal with greater prehensile abilities of the extremities can do much better keeping its balance while moving and foraging on thin branches compared to another species with less prehensile cheiridia. Finally, we showed that three locomotor traits typical of most primates during walking (i.e., high frequency of DS gaits, more protracted arm position at forelimb touchdown, and relatively lower peak vertical forces on the forelimbs) also characterize the quadrupedal walking gaits of the woolly opossum (Figure 14). To our knowledge, *Caluromys* is the only nonprimate

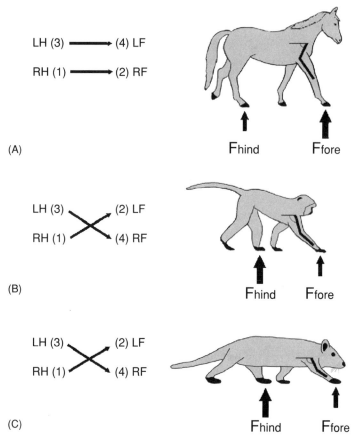

Figure 14. Convergence of locomotor characteristics between primates and the woolly opossum. Compared to horses (A) and most other nonprimate mammals which rely on LS gaits (i.e., hindfoot touchdown is followed by ispilateral touchdown of the forefoot), primates (B) and woolly opossums (C) use DS walking gaits (i.e, hindfoot touchdown is followed by contralateral touchdown of the forefoot). When the forelimb strikes the ground during walking, the arm of horses (A) and most other mammals is more retracted (i.e., arm less than 90° relative to horizontal body axis), whereas that of primates and woolly opossums is more protracted (i.e., arm greater than 90° relative to horizontal body axis). At substrate contact, the forelimbs of horses (A) and other nonprimate mammals are characterized by higher V_{pk} forces relative to those of the hindlimbs. The reverse is true for most primates (B) and woolly opossums (C): the forelimbs are characterized by lower V_{pk} forces relative to those of the hindlimbs. Figure of the horse is adapted from Larson (1998) and that of the vervet monkey from Larson and Stern (1989). Animals are not drawn to scale.

mammal studied to date that shows such close similarity in gait mechanics to primates (Schmitt and Lemelin, 2002).

The significance of these findings for primate origins is profound. Our data support the notion that thin and flexible arboreal substrates were critical for the evolution of prehensile extremities and locomotor specializations in primates. Both the visual predation model (Cartmill, 1972, 1974a,b) and the angiosperm coevolution model (Sussman, 1991, 1995) stressed the importance of vines or terminal branches for the evolution of grasping extremities in primates. This strongly suggests that the origins of grasping extremities and locomotor specializations in primates probably took place in ecological conditions not unlike those encountered by the woolly opossum. From the standpoint of positional adaptations, the last common ancestor of all primates was probably an adept quadruped capable of walking and climbing on thin branches, as well as some running, cantilevering, bridging, and upside-down hanging by the feet. While grasping a thin substrate with the feet, the hands were probably used often to manipulate fruits and capture insects, in a manner similar to *Caluromys* and cheirogaleids (Lemelin, 1996, 1999; Martin, 1972a,b, 1973).

In a recent contribution, Szalay and Sargis (2001: 299) argued "... that didelphid arborealists are drastically unlike a variety of explosive arboreal grasp-leapers encountered among, and probably in the very ancestry, of the Euprimates, particularly Strepsirrhini." We concur with this characterization of didelphids, but disagree with the idea that the earliest primates were "explosive arboreal grasp-leapers". Our study shows that primate-like grasping morphology and locomotor mechanics are found in an arboreal didelphid for which leaping represents a small component of its locomotor repertoire (Charles-Dominique et al., 1981; Rasmussen, 1990). In other words, we believe that leaping did not play a major role in the *origins* of postcranial and locomotor features diagnostic of the order Primates. Although we agree completely with the idea that leaping was an important factor for the later diversification of primates, particularly strepsirrhines as pointed out by Szalay and Sargis (2001) (Szalay and Dagosto, 1988; Szalay et al., 1987), we think that leaping was not a critical component of the locomotor adaptations of the very first primates.

The relevance of opossums as models for primate origins has been questioned recently by Sargis (2001). He argued that arboreal tree shrews (*Tupaia minor and Ptilocercus lowii*) may be better models because they are capable of grasping and are more closely related to primates. Sargis is correct in his assessment of grasping abilities of tree shrews, particularly *Ptilocercus.* However, Lemelin (1996, 1999) found that the hands and feet of woolly opossums are morphologically more similar to those of cheirogaleids than those of other didelphids. In addition, our performance data demonstrate the ability of *Caluromys* to walk and climb on very thin branches, a skill that is not evident in tupaiids whose locomotion has been studied on larger dowels (Jenkins, 1974; Sargis, 2001). Available kinematic data also indicate that *Tupaia glis* is more similar to other nonprimate mammals than primates. When walking, *T. glis* relies on a LS footfall pattern (Jenkins, 1974) and its forelimbs land with a less protracted arm posture (Fischer et al., 2002). Still, it is very possible that more arboreal tree shrews, especially *Ptilocercus*, would display more primate-like locomotor kinematics and kinetics. Sargis (2002a) showed morphological contrasts in the forelimb of more arboreal versus more terrestrial tupaiids very similar to those reported by Argot (2001) for didelphids. Unfortunately, the locomotor behavior of arboreal tupaiids, especially *Ptilocercus*, remains virtually unknown and an effort should be made, both in the field and the laboratory, to gather more data on these small mammals.

Arguing that tree shrews are better models for primate origins solely because they are more closely related to primates negates the use of convergence as a tool to understand adaptation. If this reasoning were to be applied to the problem of whale origins, then an artiodactyl, such as a cow, would be the best model to understand the origins of cetacean swimming (Milinkovitch and Thewissen, 1997). Instead, the study of the kinematics and hydrodynamics of swimming in mammals unrelated to whales has led to a much better understanding of the problem of whale origins (Thewissen and Fish, 1997). The same rationale was used throughout this chapter in addressing the problem of primate origins.

As pointed by Szalay (1984, 1994) and Szalay and Dagosto (1988), it is very unlikely that the arboreality that typifies didelphids and primates evolved from their last common ancestor. Therefore, the extraordinary degree of morphological and behavioral similarities found between *Caluromys* and primates can be best explained as evolutionary events that occurred independently in similar, fine-branch niches (Figure 15). All available ecological, morphological, and locomotor data strongly suggest that the grasping extremities and walking gaits of primates originated in an arboreal quadruped that moved and foraged on thin branches. Whether or not this

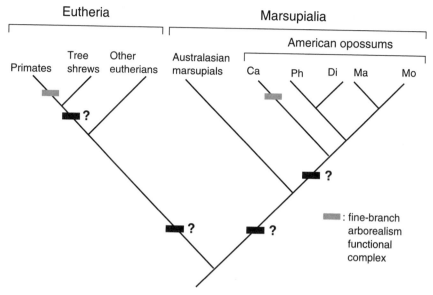

Figure 15. Mapping of a functional complex on a simplified cladogram of mammal relationships. This functional complex for thin-branch arboreal locomotion and foraging (gray rectangle) includes all the following characteristics: (1) the presence of more prehensile hands and feet, (2) predominant use of diagonal-sequence walking gaits, (3) more protracted arm position at forelimb touchdown, and (4) relatively lower V_{pk} forces during quadrupedal walking. The presence of this functional complex has been documented only in woolly opossums and primates, which strongly argues in favor of convergent evolution of these traits for the need of moving and foraging on thin branches in distantly related groups. Nonetheless, the presence or absence of this complex in the last common ancestor of the opossums, marsupials, eutherians, and tree shrews/primates is still equivocal (black rectangles with question marks). The relationships for the opossums are based on work by Kirsch *et al.* (1997) and Jansa and Voss (2000) (Ca: *Caluromys*; Di: *Didelphis*; Ma: *Marmosa*; Mo: *Monodelphis*; Ph: *Philander*). We chose tree shrews as the sister taxon of primates based on the work of Sargis (2002b, this volume).

type of specialized arborealism on thin substrates evolved in both primates and woolly opossums from ancestors that were already adapted for arboreal life remains to be demonstrated (? symbols at various nodes in Figure 15). On the basis of fossil postcranial evidence, it has been argued that basal marsupials and eutherians may have been generalized arborealists (Argot, 2001,

2002, 2003; Ji et al., 2002; Szalay and Sargis, 2001). Behaviors practiced by these early mammals probably involved locomotion and foraging on all kinds of arboreal substrates. In this way, these primitive mammals were already adapted for movements on "uneven, disordered substrates" (Jenkins, 1974: 112). A versatile locomotor repertoire in an animal of the size of a tree shrew or smaller, is just as beneficial in both "arboreal" and "terrestrial" environments as pointed by Jenkins (1974). One can envision such small, generalized arborealist restricted to an environment in which vines or terminal branches are predominant evolving the kind of grasping and locomotor specializations seen in both woolly opossums and primates. It is probably from this initial restriction to a fine-branch niche in the earliest primate ancestors that the order rapidly diversified and various locomotor specializations seen in primates today evolved.

ACKNOWLEDGMENTS

We would like to thank Drs Marian Dagosto and Matt Ravosa for organizing and inviting us to present the results of our research to the "Primate Origins and Adaptations: A Multidisciplinary Perspective" conference. We are grateful to all the attendees, in particular Eric Sargis and Fred Szalay, for stimulating discussions and exchange of ideas. The following curators and staffs gave access to skeletal collections and help: Dr Ross MacPhee, Wolfgang Fuchs, and Bob Randall (American Museum of Natural History, New York), Dr Elwyn Simons and Prithijit Chatrath (Duke University Primate Center, Durham), Dr Bruce Latimer and Lyman Moore Jellema (Cleveland Museum of Natural History, Cleveland), Dr Bruce Patterson and Bill Stanley (Field Museum of Natural History, Chicago), Maria Rutzmoser and Terri McFadden (Museum of Comparative Zoology, Cambridge), Dr Richard Thorington and Linda Gordon (National Museum of Natural History, Washington, DC), Dr Paula Jenkins (British Museum (Natural History, London), Drs Chris Smeenk and René Dekker (Nationaal Natuurhistorisch Museum, Leiden), and Dr Françoise Jouffroy and Francis Renoult (Muséum National d'Histoire Naturelle, Paris). Drs Kenneth Glander and Bill Hylander (Directors, Duke University Primate Center), Kathleen Smith (Duke University), and Martine Atramentowicz (Laboratoire d'Ecologie Générale, CNRS/Unité de Recherche 1183 – ECOTROP) gave access to animals, and Bill Hess, Diane Lewis, Tran Song, Alex Van Nievelt provided invaluable assistance with the animals. Caroline Edwards, Lindsay DeNicola, Laura Gruss, and Jandy Hanna helped with data collection in the laboratory. Finally, we would like to thank Matt Cartmill for countless animated discussions on various aspects of primate origins and life in general. This research was supported by grants from the National Science Foundation (SBR-9318750 and BCS-9904401) and Duke University.

REFERENCES

Alexander, R. McN., Jayes, A. S., Maloiy, G. M. O., and Wathura, E. M., 1979, Allometry of the limb bones of mammals from shrews (*Sorex*) to elephant (*Loxodonta*), *J. Zool., Lond.* **189**: 305–314.

Alexander, R. McN., and Maloiy, G. M. O., 1984, Stride lengths and stride frequencies of primates, *J. Zool., Lond.* **202**: 577–582.

Archer, M., 1984, The Australian marsupial radiation, in: *Vertebrate Zoogeography and Evolution in Australasia*, M. Archer and G. Clayton, eds., Hesperian Press, Perth, pp. 633–808.

Arnold, S. J., 1983, Morphology, performance and fitness, *Amer. Zool.* **23**: 347–361.

Argot, C., 2001, Functional-adaptive anatomy of the forelimb in the Didelphidae, and the paleobiology of the Paleocene marsupials *Mayulestes ferox* and *Pucadelphys andinus*, *J. Morphol.* **247**: 51–79.

Argot, C., 2002, Functional-adaptive anatomy of the hindlimb in the Didelphidae and the paleobiology of the Paleocene marsupials *Mayulestes ferox* and *Pucadelphys andinus*, *J. Morphol.* **253**: 76–108.

Argot, C., 2003, Functional-adaptive anatomy of the axial skeleton in the Didelphidae, and the paleobiology of the Paleocene marsupials *Mayulestes ferox* and *Pucadelphys andinus*, *J. Morphol.* **255**: 279–300.

Ashton, E. H., and Oxnard, C. E., 1964, Functional adaptations of the primate shoulder girdle, *Proc. Zool. Soc. Lond.* **142**: 49–66.

Ashton, E. H., Oxnard, C. E., and Spence, T. F., 1965, Scapular shape and primate classification, *Proc. Zool. Soc. Lond.* **145**: 125–142.

Atramentowicz, M., 1982, Influence du milieu sur l'activité locomotrice et la reproduction de *Caluromys philander* (L.), *Rev. Ecol. (Terre Vie)* **36**: 373–395.

Atramentowicz, M., 1986, Dynamique de population chez trois marsupiaux didelphidés de Guyane, *Biotropica* **18**: 136–149.

Atramentowicz, M., 1988, La frugivorie opportuniste de trois marsupiaux didelphidés de Guyane, *Rev. Ecol. (Terre Vie)* **43**: 47–57.

Bensley, B. A., 1901a, On the question of an arboreal ancestry of the Marsupialia, and the interrelationships of the mammalian subclasses, *Am. Nat.* **35**: 117–138.

Bensley, B. A., 1901b, A theory of the origin and evolution of the Australian Marsupialia, *Am. Nat.* **35**: 245–269.

Biewener, A. A., 2002, Future directions for the analysis of musculoskeletal design and locomotor performance, *J. Morphol.* **252**: 38–51.

Bishop, A., 1964, Use of the hand in lower primates, in: *Evolutionary and Genetic Biology of Primates*, J. Buettner-Janusch, ed., Academic Press, New York, pp. 133–225.

Brooks, D. R., and McLennan, D. A., 1991, *Phylogeny, Ecology, and Behavior*, The University of Chicago Press, Chicago.

Bucher, J. E., and Hoffmann, R. S., 1980, *Caluromys derbianus, Mamm. Species* **140**: 1–4.

Cartmill, M., 1970, *The Orbits of Arboreal Mammals: A Reassessment of the Arboreal Theory of Primate Evolution*, Ph.D. Thesis, University of Chicago, Chicago.

Cartmill, M., 1972, Arboreal adaptations and the origin of the order Primates, in: *The Functional and Evolutionary Biology of Primates*, R. H. Tuttle, ed., Aldine-Atheton, Chicago, pp. 3–35.

Cartmill, M., 1974a, Pads and claws in arboreal locomotion, in: *Primate Locomotion*, F. A. Jenkins, Jr., ed., Academic Press, New York, pp. 45–83.

Cartmill, M., 1974b, Rethinking primate origins, *Science* **184**: 436–443.

Cartmill, M., 1974c, *Daubentonia, Dactylopsila*, woodpeckers and klinorhynchy, in: *Prosimian Biology*, R. D. Martin, G. A. Doyle, and A. C. Walker, eds., Duckworth, London, pp. 655–670.

Cartmill M., 1985, Climbing, in: *Functional Vertebrate Morphology*, M. Hildebrand, D. M. Bramble, K. F. Liem, and D. B. Wake, eds., Harvard University Press, Cambridge, MA, pp. 73–88.

Cartmill, M., 1992, New views on primate origins, *Evol. Anthropol.* **1**: 105–111.

Cartmill, M., Lemelin, P., and Schmitt, D., 2002, Support polygons and symmetrical gaits in mammals, *Zool. J. Linn. Soc.* **136**: 401–420.

Charles-Dominique, P., 1983, Ecological and social adaptations in didelphid marsupials: Comparison with eutherians of similar ecology, in: *Advances in the Study of Mammalian Behavior*, J. F. Eisenberg and D. G. Kleiman, eds., Special Publication No. 7, American Society of Mammalogists, Shippensburg, PA, pp. 395–422.

Charles-Dominique P., and Martin, R. D., 1970, Evolution of lorises and lemurs, *Nature* **227**: 257–260.

Charles-Dominique, P., Atramentowicz, M., Charles-Dominique, M., Gérard, H., Hladik, A., Hladik, C. M., et al., 1981, Les mammifères frugivores arboricoles nocturnes d'une forêt guyanaise: Inter-relations plantes-animaux, *Rev. Ecol. (Terre Vie)* **35**: 341–435.

Collins, L. R., 1973, *Monotremes and Marsupials*, Smithsonian Institution Press, Washington, DC.

Crompton, R. H., 1995, "Visual predation," habitat structure, and the ancestral primate niche, in: *Creatures of the Dark: The Nocturnal Prosimians*, L. Alterman, G. A. Doyle, and M. K. Izard, eds., Plenum Press, New York, pp. 11–30.

Cunha, A. A., and Vieira, M. V., 2002, Support diameter, incline, and vertical movements of four didelphid marsupials in the Atlantic forest of Brazil, *J. Zool., Lond.* **258**: 419–426.

Daegling, D. J., 1992, Mandibular morphology and diet in the genus *Cebus*, *Int. J. Primatol.* **13**: 545–570.

Demes, B., Larson, S. G., Stern, J. T., Jr., Jungers, W. L., Biknevicius, A. R., and Schmitt, D., 1994, The kinetics of primate quadrupedalism: "Hindlimb drive" reconsidered, *J. Hum. Evol.* **26**: 353–374.

Demes, B., Qin, Y-X., Stern, J. T., Jr., Larson, S. G., and Rubin, C. T., 2001, Patterns of strain in the macaque tibia during functional activity, *Am. J. Phys. Anthropol.* **116**: 257–265.

Dollo, L., 1899, Les ancêtres des marsupiaux étaient-ils arboricoles? *Trav. Stat. Zool. Wimereux* 7: 188–203.

Eisenberg, J. F., 1989, Mammals of the neotropics, in: *The Northern Neotropics*, vol. 1, The University of Chicago Press, Panama.

Eisenberg, J. F., and Leyhausen, P., 1972, The phylogeny of predatory behavior in mammals, *Z. Tierpsychol.* **30**: 59–93.

Eisenberg, J. F., and Wilson, D. E., 1981, Relative brain size and demographic strategies in didelphid marsupials, *Am. Nat.* **118**: 1–15.

Emerson, S. B., and Arnold, S. J., 1989, Intra- and interspecific relationships between morphology, performance, and fitness, in: *Complex Organismal Functions: Integration and Evolution in Vertebrates*, D. B. Wake and G. Roth, eds., Wiley, New York, pp. 295–314.

Enders, R. K., 1935, Mammalian life-histories from Barro Colorado Island, Panama, *Bull. Mus. Comp. Zool.* **78**: 385–497.

Erikson, G. E., 1963, Brachiation in New World monkeys and in anthropoid apes, *Symp. Zool. Soc. Lond.* **10**: 135–164.

Etter, H. U. F., 1973, Terrestrial adaptations in the hands of Cercopithecinae, *Folia Primatol.* **20**: 331–350.

Fischer, M. S., Schilling, N., Schmidt, M., Haarhaus, D., and Witte, H., 2002, Basic limb kinematics of small therian mammals, *J. Exp. Biol.* **205**: 1315–1338.

Fleagle, J. G., 1977a, Locomotor behavior and skeletal anatomy of sympatric Malaysian leaf-monkeys (*Presbytis obscura* and *Presbytis melalophos*), *Yrbk. Phys. Anthropol.* **20**: 440–453.

Fleagle, J. G., 1977b, Locomotor behavior and muscular anatomy of sympatric Malaysian leaf-monkeys (*Presbytis obscura* and *Presbytis melalophos*), *Am. J. Phys. Anthropol.* **46**: 297–308.

Fleagle, J. G., 1979, Primate positional behavior and anatomy: Naturalistic and exper-imental approaches, in: *Environment, Behavior, and Morphology: Dynamic Interactions in Primates*, M. E. Morbeck, H. Preuschoft, and N. Gomberg, eds., Gustav Fischer, New York,. pp. 313–325.

Fleagle, J. G., and Meldrum, D. J., 1988, Locomotor behavior and skeletal morphol-ogy of two sympatric pitheciine monkeys, *Pithecia pithecia* and *Chiropotes satanas*, *Am. J. Primatol.* **16**: 227–249.

Fleagle, J. G., Stern, J. T., Jr., Jungers, W. L., Susman, R. L., Vangor, A. K., and Wells, J. P., 1981, Climbing: A biomechanical link with brachiation and with bipedalism, *Symp. Zool. Soc. Lond.* **48**: 429–451.

Gabis, R. V., 1960, Les os des membres des singes cynomorphes, *Mammalia* **24**: 577–607.

Gebo, D. L., 1987, Locomotor diversity in prosimian primates, *Am. J. Primatol.* **13**: 271–281.

Goldfinch, A. J., and Molnar, R. E., 1978, Gait of the brush-tail possum (*Trichosurus vulpecula*), *Aust. Zool.* **19**: 277–289.

Hall, E. R., and Dalquest, W. W., 1963, The mammals of Veracruz, *Univ. Kansas Publ. Mus. Nat. Hist.* **14**: 167–362.

Hamrick, M. W., 2001, Primate origins: Evolutionary change in digital ray patterning and segmentation, *J. Hum. Evol.* **40**: 339–351.

Harrison, T., 1989, New postcranial remains of *Victoriapithecus* from the middle Miocene of Kenya, *J. Hum. Evol.* **18**: 3–54.

Hershkovitz, P., 1969, The evolution of mammals on southern continents. VI. The recent mammals of the neotropical region: A zoogeographic and ecological review, *Q. Rev. Biol.* **44**: 1–70.

Hildebrand, M., 1967, Symmetrical gaits of primates, *Am. J. Phys. Anthropol*, **26**: 119–130.

Hildebrand, M., 1976, Analysis of tetrapod gaits: General considerations and sym-metrical gaits, in: *Neural Control of Locomotion*, R. M. Herman, S. Grillner, P. S. G. Stein, and D. G. Stuart, eds., Plenum Press, New York, pp. 203–236.

Hildebrand, M., 1985, Walking and running, in: *Functional Vertebrate Morphology*, M. Hildebrand, D. M. Bramble, K. F. Liem, and D. B. Wake, eds., Harvard University Press, Cambridge, MA, pp. 38–57.

Hladik, C. M., 1979, Diet and ecology of prosimians, in: *The Study of Prosimian Behavior*, G. A. Doyle and R. D. Martin, eds., Academic Press, New York, pp. 307–358.

Hladik, C. M., Charles-Dominique, P., and Petter, J. J., 1980, Feeding strategies of five nocturnal prosimians in the dry forest of the West Coast of Madagascar, in: *Nocturnal Malagasy Primates: Ecology, Physiology, and Behavior*, P. Charles-Dominique, H. M. Cooper, C. M. Hladik, E. Pages, G. F. Pariente, and A. Petter Rousseaux et al., eds., Academic Press, New York, pp. 41–73.

Howell, A. B., 1944, *Speed in Animals*, The University of Chicago Press, Chicago.

Hunsaker, D. II., 1977, Ecology of New World marsupials, in: *The Biology of Marsupials*, D. Hunsaker, II, ed., Academic Press, New York, pp. 95–156.

Hunsaker, D. II, and Shupe, D., 1977, Behavior of New World marsupials, in: *The Biology of Marsupials*, D. Hunsaker, II, ed., Academic Press, New York, pp. 276–347.

Huxley, T. H., 1880, On the application of the laws of evolution to the arrangement of the Vertebrata, and more particularly of the Mammalia, *Proc. Zool. Soc. Lond.* **1880**: 649–662.

Hylander, W. L., 1979a, Mandibular function in *Galago crassicaudatus* and *Macaca fascicularis:* An in vivo approach to stress analysis of the mandible, *J. Morphol.* **159**: 252–296.

Hylander, W. L., 1979b, The functional significance of primate mandibular form, *J. Morphol.* **160**: 223–239.

Jansa, S. A., and Voss, R. S., 2000, Phylogenetic studies on didelphid marsupials I. Introduction and preliminary results from nuclear IRBP gene sequences, *J. Mamm. Evol.* **7**: 43–77.

Jenkins, F. A., Jr., 1971, Limb posture and locomotion in the Virginia opossum (*Didelphis marsupialis*) and in other non-cursorial mammals, *J. Zool., Lond.* **165**: 303–315.

Jenkins, F. A., Jr., 1972, Chimpanzee bipedalism: Cineradiographic analysis and implications for the evolution of gait, *Science* **178**: 877–879.

Jenkins, F. A., Jr., 1974, Tree shrew locomotion and the origins of primate arborealism, in: *Primate Locomotion*, F. A. Jenkins, Jr., ed., Academic Press, New York, pp. 85–115.

Jenkins, F. A., Jr., 1981, Wrist rotation in primates: A critical adaptation for brachiators, *Symp. Zool. Soc. Lond.* **48**: 429–451.

Jenkins, F. A., Jr., and Weijs, W. A., 1979, The functional anatomy of the shoulder in the Virginia opossum (*Didelphis virginiana*), *J. Zool., Lond.* **188**: 379–410.

Ji, Q., Luo, Z.-X., Yuan, C.-X., Wible, J. R., Zhang, Z.-P., and Georgi, J. A., 2002, The earliest known eutherian mammal, *Nature* **416**: 816–822.

Jones, F. W., 1916, *Arboreal Man*, Edward Arnold, London.

Jones, F. W., 1929, *Man's Place Among the Mammals*, Edward Arnold, London.

Jouffroy, F. K., Renous, S., and Gasc, J.-P., 1983, Étude cinéradiographique des déplacements du membre antérieur du potto de Bosman (*Perodicticus potto*, P.L.S. Müller, 1766) au cours de la marche quadrupède sur une branche horizontale. *Ann. Sci. Nat., Zool.* **13**ème Série, 75–87.

Julien-Laferrière, D., 1995, Use of space by the woolly opossum *Caluromys philander* in French Guiana, *Can. J. Zool.* **73**: 1280–1289.

Julien-Laferrière, D., 1999, Food strategies and food partitioning in the neotropical frugivorous mammals *Caluromys philander* and *Potos flavus*, *J. Zool., Lond.* **247**: 71–80.

Kay, R. F., 1984, On the use of anatomical features to infer foraging behavior in extinct primates, in: *Adaptations for Foraging in Nonhuman Primates*, P. S. Rodman and J. G. H. Cant, eds., Columbia University Press, New York, pp. 21–49.

Kay, R. F., and Cartmill, M., 1977, Cranial morphology and adaptations of *Palaechthon nacimienti* and other Paromomyidae (Plesiadapoidea, ?Primates), with a description of a new genus and species, *J. Hum. Evol.* **6:** 19–35.

Kimura, T., Okada, M., and Ishida, H., 1979, Kinesiological characteristics of primate walking: Its significance in nonhuman walking, in: *Environment, Behavior, and Morphology: Dynamic Interactions in Primates*, M. E. Morbeck, H. Preuschoft, and N. Gomberg, eds., Gustav Fischer, New York, pp. 297–311.

Kirsch, J. A. W., 1977, The comparative serology of Marsupialia, and a classification of Marsupials, *Aust. J. Zool.* **(Suppl. Series)** **52:** 1–152.

Kirsch, J. A. W., Lapointe, F. J., and Springer, M. S., 1997, DNA-hybridisation studies of marsupials and their implications for metatherian classification, *Aust. J. Zool.* **45:** 211–280.

Larney, E., and Larson, S. G., 2004, Compliant gaits in primates: Elbow and knee yield in primates compared to other mammals, *Am J. Phys. Anthropol.* 125: 42–50.

Larson, A., and Losos, J. B., 1996, Phylogenetic systematics of adaptation, in: *Adaptation*, M. R. Rose and G. V. Lauder, eds., Academic Press, San Diego, pp. 187–220.

Larson, S. G., 1993, Functional morphology of the shoulder in primates, in: *Postcranial Adaptations in Nonhuman Primates*, D. L. Gebo, ed., Northern Illinois University Press, DeKalb, pp. 45–69.

Larson, S. G., 1998, Unique aspects of quadrupedal locomotion in nonhuman primates, in: *Primate Locomotion: Recent Advances*, E. Strasser, J. Fleagle, A. Rosenberger, and H. McHenry, eds., Plenum Press, New York, pp. 157–173.

Larson, S. G., Schmitt, D., Lemelin, P., and Hamrick, M. W., 2000, Uniqueness of primate forelimb posture during quadrupedal locomotion, *Am. J. Phys. Anthropol.* **112:** 87–101.

Larson, S. G., Schmitt, D., Lemelin, P., and Hamrick, M. W., 2001, Limb excursion during quadrupedal walking: How do primates compare to other mammals? *J. Zool., Lond.* **255:** 353–365.

Larson, S. G., and Stern, J. T., Jr., 1989, The role of propulsive muscles of the shoulder during quadrupedalism in vervet monkeys (*Cercopithecus aethiops*): Implications for neural control of locomotion in primates, *J. Mot. Behav.* **21:** 457–472.

Lauder, G. V., 1996, The argument from design, in: *Adaptation*, M. R. Rose and G. V. Lauder, eds., Academic Press, San Diego, pp. 55–91.

LeGros Clark, W. E., 1959, *The Antecedents of Man*, Edinburgh University Press, Edinburgh.

Lemelin, P., 1996, *The Evolution of Manual Prehensility in Primates: A Comparative Study of Prosimians and Didelphid Marsupials*, Ph.D. Dissertation, State University of New York at Stony Brook, Stony Brook.

Lemelin, P., 1999, Morphological correlates of substrate use in didelphid marsupials: Implications for primate origins, *J. Zool., Lond.* **247**: 165–175.

Lemelin, P., and Grafton, B. W., 1998, Grasping performance in *Saguinus midas* and the evolution of hand prehensility in primates, in: *Primate Locomotion: Recent Advances*, E. Strasser, J. Fleagle, A. Rosenberger, and H. McHenry, eds., Plenum Press, New York, pp. 131–144.

Lemelin, P., and Schmitt, D., 2004, Seasonal variation in body mass and locomotor kinetics in the fat-tailed dwarf lemur (*Cheirogaleus medius*), *J. Morphol.* **260**: 65–71.

Lemelin, P., Schmitt, D., and Cartmill, M., 1999, Gait patterns and interlimb coordination in woolly opossums: How did ancestral primates move? *Am. J. Phys. Anthropol.* **(Suppl.) 28**: 181–182.

Lemelin, P., Schmitt, D., and Cartmill, M., 2002, The origins of diagonal-sequence walking gaits in primates: An experimental test involving two didelphid marsupials, *Am. J. Phys. Anthropol.* **(Suppl.) 34**: 101.

Lemelin, P., Schmitt, D., and Cartmill, M., 2003, Footfall patterns and interlimb coordination in opossums (Family Didelphidae): Evidence for the evolution of diagonal-sequence gaits in primates, *J. Zool., Lond.* **260**: 423–429.

Lieberman, D. E., Polk, J. D., and Demes, B., 2004, Predicting long bone loading from cross-sectional geometry, *Am. J. Phys. Anthropol.* **123**: 156–171.

Liete, Y. L. R., Costa, L. P., and Stallings, J. R., 1996, Diet and vertical space use of three sympatric opossums in a Brazilian Atlantic forest reserve, *J. Trop. Ecol.* **12**: 435–440.

Magne de la Croix, P., 1936, The evolution of locomotion in mammals, *J. Mammal.* **17**: 51–54.

Malcolm, J. R., 1991, Comparative abundances of neotropical small mammals by trap height, *J. Mammal.* **72**: 188–192.

Manly, B. F. J., 1986, *Multivariate Statistical Methods: A Primer*, Chapman and Hall, London.

Marshall, L. G., Case, J. A., and Woodburne, M. D., 1990, Phylogenetic relationships of the families of marsupials, in: *Current Mammalogy*, H. H. Genoways, ed., vol. 2, Plenum Press, New York, pp. 433–505.

Martin, R. D., 1968, Towards a new definition of primates, *Man* **3**: 377–401.

Martin, R. D., 1972a, A preliminary field-study of the lesser mouse lemur (*Microcebus murinus* JF Miller, 1777), *Z. Comp. Ethol.* **(Suppl.) 9**: 43–89.

Martin, R. D., 1972b, Adaptive radiation and behaviour of the Malagasy lemurs, *Phil. Trans. R. Soc. Lond.* **B264**: 295–352.

Martin, R. D., 1973, A review of the behaviour and ecology of the lesser mouse lemur (*Microcebus murinus* J. F. Miller, 1777), in: *Comparative Ecology and Behaviour of Primates*, R. P. Michael and J. H. Crook, eds., Academic Press, London, pp. 1–68.

Martin, R. D., 1986, Primates: A definition, in: *Major Topics in Primate and Human Evolution*, B. Wood, L. Martin, and P. Andrews, eds., Cambridge University Press, Cambridge, pp. 1–31.

Martin, R. D., 1990, *Primate Origins and Evolution. A Phylogenetic Reconstruction*, Princeton University Press, Princeton.

Marzke, M. W., Toth, N., Schick, K., Reese, S., Steinberg, Hunt, K., Linscheid, R. L., et.al., 1998, EMG study of hand muscle recruitment during hand hammer percussion manufacture of Oldowan tools, *Am. J. Phys. Anthropol.* **105**: 315–332.

McClearn, D., 1992, Locomotion, posture, and feeding behavior of kinkajous, coatis, and raccoons, *J. Mammal.* **73**: 245–261.

McManus, J. J., 1970, Behavior of captive opossums, *Didelphis marsupialis virginiana*, *Am. Midl. Nat.* **84**: 144–169.

McManus, J. J., 1974, *Didelphis virginiana*, *Mamm. Species* **40**: 1–6.

Milinkovitch, M. C., and Thewissen, J. G. M., 1997, Even-toed fingerprints on whale ancestry, *Nature* **388**: 622–624.

Mivart, StG., 1873, On *Lepilemur* and *Cheirogaleus* and on the zoological rank of the Lemuroidea, *Proc. Zool. Soc. Lond.* **1873**: 484–510.

Muybridge, E., 1887, *Animals in Motion* (1957 reprint), Dover, New York.

Napier, J. R., 1961, Prehensility and opposability in the hands of primates, *Symp. Zool. Soc. Lond.* **5**: 115–132.

Napier, J. R., 1967, Evolutionary aspects of primate locomotion, *Am. J. Phys. Anthropol.* **27**: 333–342.

Napier, J. R., 1993, *Hands* (revised edition), Princeton University Press, Princeton.

Napier, J. R., and Napier, P. H., 1967 *A Handbook of Living Primates*, Academic Press, New York.

Nieschalk, U., and Demes, B., 1993, Biomechanical determinants of reduction of the second ray in Lorisinae, in: *Hands of Primates*, H. Preuschoft and D. J. Chivers, eds., Springer-Verlag, Berlin, pp. 225–234.

Nowak, R. M., 1999, *Walker's Mammals of the World*, 6th edn., The Johns Hopkins University Press, Baltimore.

O'Connell, M. A., 1979, Ecology of didelphid marsupials from Northern Venezuela, in: *Vertebrate Ecology in the Northern Neotropics*, J. F. Eisenberg, ed., Smithsonian Institution Press, Washington, DC, pp. 73–87.

O'Connell, M. A., 1983, *Marmosa robinsoni*, *Mamm. Species* **203**: 1–6.

Oxnard, C. E., 1963, Locomotor adaptations of the primate forelimb. *Symp. Zool. Soc. Lond.* **10**: 165–182.

Oxnard, C. E., 1967, The functional morphology of the primate shoulder as revealed by comparative anatomical, osteometric, and discriminant function techniques, *Am. J. Phys. Anthropol.* **26**: 219–240.

Passamani, M., 1995, Vertical stratification of small mammals in Atlantic Hill forest, *Mammalia* **59**: 276–279.

Pimentel, R. A., 1979, *Morphometrics: The Multivariate Analysis of Biological Data*, Kendall/Hunt Publishing Co., Dubuque.

Polk, J. D., 2001, *The Influence of Body Size and Body Proportions on Primate Quadrupedal Locomotion*, Ph.D. Dissertation, State University of New York at Stony Brook, Stony Brook.

Polk, J. D., Demes, B., Jungers, W. L., Biknevicius, A. R., Heinrich, R. E., and Runestad, J. A., 2000, A comparison of primate, carnivoran and rodent limb bone cross-sectional properties: Are primates really unique? *J. Hum. Evol.* **39**: 297–325.

Pridmore, P. A., 1994, Locomotion in *Dromiciops australis* (Marsupialia: Microbiotheriidae), *Aust. J. Zool.* **42**: 679–699.

Preuschoft, H., 1969, The mechanical basis of the morphological differences in the skeleton of apes and man, in: *Recent Advances in Primatology, Proceedings of the 2nd International Congress of Primatology, Atlanta*, H. O. Hofer, ed., vol. 2, Karger, Basel, pp. 160–170.

Preuschoft, H., 1970, Functional anatomy of the lower extremity, in: *The Chimpanzee*, G. H. Bourne, ed., vol. 3, Karger, Basel, pp. 221–294.

Preuschoft, H., Witte, H., and Fischer, M., 1995, Locomotion in nocturnal prosimians, in: *Creatures of the Dark: The Nocturnal Prosimians*, L. Alterman, G. A. Doyle, and M. K. Izard, eds., Plenum Press, New York, pp. 453–472.

Rasmussen, D. T., 1990, Primate origins: Lessons from a neotropical marsupial, *Am. J. Primatol.* **22**: 263–277.

Reig, O. A., Kirsch, J. A. W., and Marshall, L. G., 1987, Systematic relationships of the living and Neocenozoic American opossum-like marsupials (sub-order Didelphimorphia), with comments on the classification of these and of the Cretaceous and Paleogene New World and European metatherians, in: *Possums and Opossums*, M. Archer, ed., vol. 1, pp. 1–90, Surrey Beatty and Sons Pty. Ltd., Chipping Norton (New South Wales).

Reynolds, T. R., 1985a, Mechanics of increased support of weight by the hindlimbs in primates, *Am. J. Phys. Anthropol.* **67**: 335–349.

Reynolds, T. R., 1985b, Stresses on the limbs of quadrupedal primates, *Am. J. Phys. Anthropol.* **67**: 351–362.

Reynolds, T. R., 1987, Stride length and its determinants in humans, early hominids, primates, and mammals, *Am. J. Phys. Anthropol.* **72**: 101–116.

Roberts, D., 1974, Structure and function of the primate scapula, in: *Primate Locomotion*, F. A., Jenkins, Jr., ed., Academic Press, New York, pp. 171–200.

Rodman, P. S., 1979, Skeletal differentiation of *Macaca fascicularis* and *Macaca nemestrina* in relation to arboreal and terrestrial quadrupedalism, *Am. J. Phys. Anthropol.* **51**: 51–62.

Rollinson, J., and Martin, R. D., 1981, Comparative aspects of primate locomotion with special reference to arboreal cercopithecines, *Symp. Zool. Soc. Lond.* **48**: 377–427.

Rose, M. D., 1989, New postcranial specimens of catarrhines from the Middle Miocene Chinji Formation, Pakistan: Descriptions and a discussion of proximal humeral functional morphology in anthropoids, *J. Hum. Evol.* **18**: 131–162.

Sargis, E. J., 2001, The grasping behaviour, locomotion and substrate use of the tree shrews *Tupaia minor* and *T. tana* (Mammalia, Scandentia), *J. Zool., Lond.* **253**: 485–490.

Sargis, E. J., 2002a, Functional morphology of the forelimb of tupaiids (Mammalia, Scandentia) and its phylogenetic implications, *J. Morphol.* **253**: 10–42.

Sargis, E. J., 2002b, The postcranial morphology of *Ptilocercus lowii* (Scandentia, Tupaiidae): An analysis of primatomorphan and volitantian characters, *J. Mamm. Evol.* **9**: 137–160.

Schmitd, M., and Fischer, M. S., 2000, Cineradiographic study of forelimb movements during quadrupedal walking in the brown lemur (*Eulemur fulvus*, Primates: Lemuridae), *Am. J. Phys. Anthropol.* **111**: 245–262.

Schmitt, D., 1998, Forelimb mechanics during arboreal and terrestrial quadrupedalism in Old World monkeys, in: *Primate Locomotion: Recent Advances*, E. Strasser, J. Fleagle, A. Rosenberger, and H. McHenry, eds., Plenum Press, New York, pp. 175–200.

Schmitt, D., 1999, Compliant walking in primates, *J. Zool., Lond.* **248**: 149–160.

Schmitt, D., 2003a, Insights into the evolution of human bipedalism from experimental studies of humans and other primates, *J. Exp. Biol.* **206**: 1437–1448.

Schmitt, D., 2003b, Substrate size and primate forelimb mechanics: Implications for understanding the evolution of primate locomotion, *Int. J. Primatol.* **24**: 1023–1036.

Schmitt, D. and Hanna, J. B., 2004, Substrate alters forelimb to hindlimb peak force ratios in primates, *J. Hum. Evol.* **46**: 237–252.

Schmitt, D., and Lemelin, P., 2002, Origins of primate locomotion: Gait mechanics of the woolly opossum, *Am. J. Phys. Anthropol.* **118**: 231–238.

Schultz, A. H., 1963, Relations between the lengths of the main part of the foot skeleton in primates, *Folia Primatol.* **1**: 150–171.

Smith, G. E., 1912, The evolution of man, *Smithsonian Inst. Ann. Rep.* **1912**: 553–572.

Smith, J. M., and Savage, R. J. G., 1956, Some locomotor adaptations in mammals, *Zool. J. Linn. Soc.* **42**: 603–622.

Smith, R. J., and Jungers, W. L., 1997, Body mass in comparative primatology, *J. Hum. Evol.* **32**: 523–559.

Sneath, P. H. A., and Sokal, R. R., 1973, *Numerical Taxonomy*, W. H. Freeman, San Francisco.

Stern, J. T., Jr., Wells, J. P., Vangor, A. K., and Fleagle, J. G., 1977, Electromyography of some muscles of the upper limb in *Ateles* and *Lagothrix*, *Yrbk. Phys. Anthropol.* **20**: 498–507.

Strasser, E., 1992, Hindlimb proportions, allometry, and biomechanics in Old World monkeys (Primates, Cercopithecidae), *Am. J. Phys. Anthropol.* **87**: 187–213.

Strasser, E., 1994, Relative development of the hallux and pedal digit formulae, *J. Hum. Evol.* **26**: 413–440.

Susman, R L., 1998, Hand function and tool behavior in early hominids, *J. Hum. Evol.* **35**: 23–46.

Sussman, R. W., 1991, Primate origins and the evolution of angiosperms, *Am. J. Primatol.* **23**: 209–223.

Sussman, R. W., 1995, How primates invented the rainforest and vice versa, in: *Creatures of the Dark: The Nocturnal Prosimians*, L. Alterman, G. A. Doyle, and M. K. Izard, eds., Plenum Press, New York, pp. 1–10.

Szalay, F. S., 1984, Arboreality: Is it homologous in metatherians and eutherian mammals? *Evol. Biol.* **18**: 215–258.

Szalay, F. S., 1994, *Evolutionary History of the Marsupials and An Analysis of Osteological Characters*, Cambridge University Press, Cambridge.

Szalay, F. S., and Dagosto, M., 1988, Evolution of hallucial grasping in the primates, *J. Hum. Evol.* **17**: 1–33.

Szalay, F. S., and Sargis, E., 2001, Model-based analysis of postcranial osteology of marsupials from the Palaeocene of Itaboraí (Brazil), and the phylogenetics and biogeography of Metatheria, *Geodiversitas* **23**: 139–302.

Szalay, F. S., Rosenberger, A. L., and Dagosto, M., 1987, Diagnosis and differentiation of the order Primates, *Yrbk. Phys. Anthropol.* **30**: 75–105.

Tattersall, I., 1982, *The Primates of Madagascar*, Columbia University Press, New York.

Thewissen, J. G. M., and Fish, F. E., 1997, Locomotor evolution in the earliest cetaceans: Functional model, modern analogues, and paleontological evidence. *Paleobiology* **23**: 482–490.

Tuttle, R. H., and Basmajian, J. V., 1974, Electromyography of forearm musculature in gorilla and problems related to knuckle-walking, in: *Primate Locomotion*, F. A. Jenkins, Jr., ed., Academic Press, New York, pp. 293–347.

Vilensky, J. A., and Larson, S. G., 1989, Primate locomotion: Utilization and control of symmetrical gaits, *Annu. Rev. Anthropol.* **18**: 17–35.

Vogel, S., 1998, Convergence as an analytical tool in evaluating design, in: *Principles of Animal Design*, E. R. Weibel, C. R. Taylor, and L. Bolis, eds., Cambridge University Press, Cambridge, UK, pp. 13–20.

Wainwright, P. C., 1994, Functional morphology as a tool in ecological research, in: *Ecological Morphology: Integrative Organismal Biology*, P. C. Wainwright and S. M. Reilly, eds., The University of Chicago Press, Chicago, pp. 42–59.

Walker, A., 1979, Prosimian locomotor behavior, in: *The Study of Prosimian Behavior*, G. A. Doyle and R. D. Martin, eds., Academic Press, New York, pp. 543–565.

Ward, S. C., and Sussman, R. W., 1979, Correlates between locomotor anatomy and behavior in two sympatric species of *Lemur*, *Am. J. Phys. Anthropol.* **50:** 575–590.

Washburn, S. L., 1951, The analysis of primate evolution with particular reference to the origin of man, Cold Spring Harbor, *Symp. Quant. Biol.* **15:** 67–77.

White, T., 1990, Gait selection in the brush-tail possum (*Trichosurus vulpecula*), the northern quoll (*Dasyurus hallucatus*), and the Virginia opossum (*Didelphis virginiana*), *J. Mamm.* **71:** 79–84.

Witmer, L. M., 1995, The extant phylogenetic bracket and the importance of reconstructing soft tissues in fossils, in: *Functional Morphology in Vertebrate Paleontology*, J. Thomason, ed., Cambridge University Press, Cambridge, pp. 19–33.

CHAPTER ELEVEN

Evolvability, Limb Morphology, and Primate Origins

Mark W. Hamrick

INTRODUCTION

The diversification of locomotor and postural behaviors among mammals has brought about striking changes in the ancestral pattern of the pentadactyl limb. These evolutionary changes frequently involve modification of either the most marginal digits (e.g., reduction of digits one or five; Morse's law) or the most distal limb segments (e.g., phalangeal elongation or reduction; Shubin et al., 1997). Goslow (1989) and Hinchliffe (1989) suggested that evolutionary change in the proximal elements of the vertebrate limb has been relatively conservative, whereas more marked changes have occurred in the distal limb (autopod) elements. A number of comparative studies demonstrate that the autopod of mammals is quite variable interspecifically, supporting the "proximal stabilization" model. Examples include variability in bat wing phalanx length related to functional variation in the aerodynamics of flight (Norberg, 1994; Norberg and Rayner, 1987), variability in apical pad and terminal phalanx morphology related to variation in arboreal foraging preferences among neotropical primates and marsupials (Hamrick, 1998, 2001a,b, 2003), and variability in claw shape related to the diversification of

Mark W. Hamrick • Department of Cellular Biology & Anatomy, Medical College of Georgia Augusta, GA 30912

foot postures in climbing, perching, and ground-dwelling birds (Feduccia, 1993). Quantitative studies have also revealed an intraspecific trend of increasing variation from the proximal to distal elements of bird and bat wings (Bader and Hall, 1960; Engels, 1938).

Vertebrate limbs consist of hierarchical structural units, undergo temporal transformations in form, and exhibit varying degrees of connectivity with other modules (Raff, 1996). Limbs are therefore important units, or fields, of ontogenetic and phylogenetic change (Gilbert et al., 1996). Recent studies of limb development provide new support for Hinchliffe's model of proximal stabilization and distal variability in limb evolution. These include evidence that the autopodium is developmentally autonomous from the proximal limb, involving different molecular pathways during chondrogenesis and a distinctive third phase of HoxD gene expression regulated by a single "global" enhancer (Chiu and Hamrick, 2002; Hérault et al., 1998). Loss-of-function mutations also reveal that redundancy exists among many of the patterning genes expressed during limb development so that knockout mutants often show alterations in phenotype that are either relatively minor or absent altogether. These experiments demonstrate that developmental perturbations such as changes in cell number and arrangement result in new, but nonlethal, skeletal morphologies.

This chapter has two objectives. First, to present evidence from limb development supporting Hinchliffe's model of evolvability in distal limb structures. Evolvability refers here to a species' capacity to generate heritable phenotypic variation (Kirschner and Gerhart, 1998). Data from experimental genetics indicate that the mammalian distal limb (autopod) fulfills many of the theoretical criteria for "evolvability," suggesting that the origin and radiation of mammalian clades would be expected to include early and rapid changes in autopod morphology. The second objective of this chapter is to present comparative evidence from hindfoot morphology showing that mammalian adaptive radiations frequently involve diversification of digit proportions associated with the evolution of new postural behaviors. These data establish diversification in digit proportions as a repeated pattern in tetrapod evolution, and data are provided documenting the evolution of derived digit proportions in early primates. The fossil evidence for skeletal evolution in early primates and other mammals is therefore consistent with Hinchliffe's (1989, 1991) hypothesis, revealing that the evolution of new digit proportions was a critical first step in the origin and adaptive radiation of the order Primates.

THEORETICAL CRITERIA FOR EVOLVABILITY

The recent synthesis of evolutionary and developmental biology, or EvoDevo (Hall, 1998), has yielded important new insights into the molecular- and cellular-level properties that facilitate morphological change in evolving lineages. Here the focus is on four of these basic properties, reviewed by Conrad (1990), Gerhart and Kirschner (1997), and Kirschner and Gerhart (1998), that increase the potential for evolutionary change in the limb, particularly within the autopodial region. The forelimb autopod includes the carpus, metacarpus, and fingers, whereas the hindlimb autopod includes the tarsus, metatarsus, and toes. The first property of the autopod, which increases its potential for evolutionary change, is modularity or compartmentalization. Modules are defined in different ways by different researchers (Winther, 2001). Modularity can be identified from embryological studies in which a module is an isolatable, transplantable, and well-characterized landmark on the embryo (Gilbert et al., 1996). Such modules are discrete units of development requiring specific selector gene expression within a bounded spatial domain (Gilbert et al., 1996; see also Carroll et al., 2001, for a discussion of selector genes). A module can also be recognized in a group of related organisms as a subset of body plan elements that exhibits adaptive variation more or less autonomously (Von Dassow and Munro, 1999). The most important feature of modularity is autonomy in which one aspect of form may "explore" new structural variants without affecting another (Gerhart and Kirschner, 1997).

The second important prerequisite for evolvability is redundancy. In the case of a module, genetic redundancy protects old functions while at the same time allowing for the acquisition of new ones. In the case of regulatory genes, redundancy and duplication allow different *cis*-regulatory (e.g., enhancer) elements to be acquired over time leading to slight differences in spatiotemporal gene expression (Figure 1; Carroll et al., 2001; Chiu and Hamrick, 2002). Genetic redundancy is, therefore, a means of maintaining a "pool" of evolutionary novelty at the biochemical level (Wagner, 1996). As discussed in a later section, redundancy is best recognized from gene knockout experiments. One example comes from the Msx genes, a group of homeobox genes that are often expressed in overlapping domains during the formation of ectodermal organs such as teeth and hair (Noveen et al., 1995). They are functionally redundant at early stages of organ development and knockouts of either Msx1 or Msx2 still form hair follicles. The expression of these genes differs at later stages of hair morphogenesis, when Msx1 expression is down-regulated and Msx2 expression shifts from the germinal matrix to the root sheath (Satokata et al., 2000). Mice homozygous for the disrupted Msx2 sequence are viable and mice heterozygous for the disrupted Msx2 sequence show no phenotypic differences compared to normal mice. Thus, redundancy appears to reduce the lethality of mutation and may also facilitate divergence of gene function in the form of change in spatiotemporal expression.

The third criterion for evolvability is weak linkage. This refers to the dependence of one stage in a metabolic or transduction pathway on another. The genes of eukaryotes, particularly of metazoans, have large and complex *cis*-regulatory regions that function to activate (e.g., enhancers) or suppress (e.g., repressors) transcription of the gene of interest (Figure 1). Many enhancer proteins are known to bind with relatively low specificity to enhancer sequences (Kirschner and Gerhart, 1998). Moreover, there is evidence to suggest that *cis*-regulatory regions, such as enhancer sequences, can evolve quite rapidly through processes, such as de novo evolution, from previously nonfunctional DNA sequences, duplication, and then divergence from existing regulatory sequences or modification of existing regulatory sequences (Carroll et al., 2001; Chiu and Hamrick, 2002). As Carroll et al. (2001) have noted, the probability of mutational change in enhancer sequences is relatively high; so high, in fact, that the probability of de novo

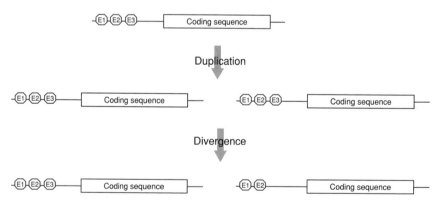

Figure 1. Evolutionary change in the expression of regulatory gene function via gene duplication followed by enhancer sequence (E1, E2, and E3) divergence. Modified from Carroll et al. (2001).

enhancer evolution in *Drosophila* is approximately once per gene. Thus, genes expressed during the patterning of morphogenetic fields may acquire new spatiotemporal patterns of expression via changes in their *cis*-regulatory regions. This flexibility of transcriptional regulation facilitates evolutionary change in morphogenetic pathways via mutation in *cis*-regulatory sequences. These *cis*-regulatory mutations usually have relatively subtle phenotypic effects (Stern, 2000), such as altering the number of bristles occurring on the legs of fruit flies (Stern, 1998a).

The final requirement for evolvability is robusticity, which refers to the ability of a module to withstand mutational changes in cell number, cell arrangement, etc., during development so that these mutations result in nonlethal phenotypes outside the structural norm (Conrad, 1990). The complex cellular arrangements that characterize morphogenetic fields in vertebrates are precisely sculpted by the processes of cell adhesion, cell proliferation, and programmed cell death (apoptosis). Yet, relatively subtle variations in cellular patterning during morphogenesis can yield potentially significant (adaptively) changes in morphology. In the case of tooth morphology, Jernvall's (2000) patterning cascade model of tooth development predicts that minor variations in the diffusion of molecular signals from the primary enamel knot produce slight variations in the locations of secondary enamel knots that later form smaller cusps. The effect is cumulative, where the last cusps to form tend to be the smallest and most variable in terms of size and position. These data explain the high degree of intraspecific variability observed in tooth cusp formation among seals, and may also explain the repeated convergent evolution of small cusps such as the hypocone in early mammalian evolution (Hunter and Jernvall, 1995; Jernvall et al., 1996). Mammalian molar tooth cusp topography is, therefore, relatively robust so that differences in cusp size and number can be variable even within a single seal species (*Phoca hispida*) yielding relatively subtle, nonlethal variation upon which selection may act (Jernvall et al., 2001).

EVIDENCE FOR EVOLVABILITY IN THE AUTOPOD

There is now ample evidence to indicate that the mammalian limb in general, and the autopod in particular, fulfill the criteria enumerated in an earlier section that increase the potential for evolutionary change in morphology. First, experimental embryology demonstrates that the limb itself is a well-defined module. This is supported by the fact that isolated limb buds left to develop in culture form the normal pattern of skeletal elements (Searls, 1968), and entire limbs can be induced from the flank of chick embryos by application of appropriate growth factors (Cohn et al., 1995). The limb is therefore a transplantable, isolatable embryonic landmark that has the potential to develop autonomously, once limb bud formation is initiated and the appropriate selector genes are expressed. The autopodial region also exhibits several features characteristic of a submodule within the limb bud (Richardson, 1999). As mentioned in an earlier section, modules are discrete units of development produced by a hierarchy of genetic interactions within a bounded spatial domain (Gilbert et al., 1996). Autopodial development involves a distinct third phase of Hox gene expression regulated by a single "global" enhancer and the sequence of Hox gene transcription in the autopod is the reverse of that observed in the zeugopodial region (Hérault et al., 1998). Furthermore, chondrogenesis in the autopod involves Activin A expression, whereas this growth factor is not able to induce chondrogenesis in more proximal limb elements (Merino et al., 1999). Thus, the autopod is a unit of development with a distinctive hierarchy of genetic interactions within a clearly defined spatial domain (Chiu and Hamrick, 2002). This compartmentalization allows for changes in one aspect of gene function and cellular patterning without affecting another (Carroll et al., 2001).

Modules may also be recognized in a group of related organisms as anatomical elements that exhibit adaptive variation more or less autonomously (Von Dassow and Munro, 1999). Bock and Miller (1959) showed that the diversification of scansorial, climbing, and perching behaviors in piciform birds (woodpeckers and their relatives) primarily involved diversification in morphology of the hindlimb digits. Feduccia (1993) and Clark et al. (1998) also found that digit morphology alone could be used to infer climbing, perching, and terrestrial habits in *Archaeopteryx*, as well as fossil pterosaurs. Likewise, Howell (1944) illustrated how variability in digit size and number among peramelid marsupials was related to cursorial and fossorial specialization (Figure 2). The author's own data from didelphid marsupials provide further evidence for phylogenetic variation in autopod morphology (Hamrick, 2001a, 2003). Didelphid opossums include terrestrial species (*Philander*), more arboreal taxa (*Marmosa* and *Caluromys*), and even an aquatic forager, the yapok (*Chironectes*; Hamrick, 2003). Comparative data show that the terrestrial, arboreal, and aquatic forms differ from one another

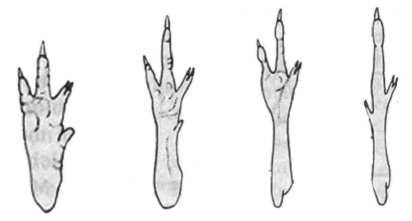

Figure 2. Hindfeet of peramelid marsupials showing variability in autopod morphology. From left to right: *Perameles, Peroryctes, Macrotis,* and *Chaeropus.* Adapted from Howell (1944).

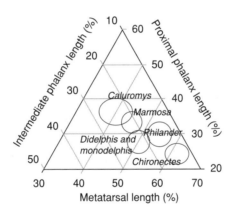

Figure 3. Ternary plot of relative hindfoot segment lengths in didelphid marsupials (n = 6 per genus). Length of the third metatarsal, proximal phalanx, and intermediate phalanx is divided by the total length of the three segments added together multiplied by 100. The ellipses enclose the range of individual values for each species.

in the relative length of their toe segments (Figure 3; see also Lemelin, 1999). Hence, the evidence from extant clades of birds and mammals is consistent with the hypothesis that related taxa diverge from one another in skeletal form and postural behavior primarily by modifying autopod morphology.

The second piece of evidence for evolvability in the autopod is redundancy, demonstrated by Hox gene knockout experiments and analysis of heterozygote crosses. Mice heterozygous for a mutation in the Hoxd13 gene sequence (Hoxd13+/−) show a mild phenotype (36% penetrance) characterized by a sixth digital rudiment and misshapen carpal bones (Davis and Capecchi, 1996). Mice homozygous for the disrupted HoxD13 sequence (Hoxd13−/−) show a more severe autopod phenotype that includes the presence of interdigital webbing, reduced second and fifth digits, and fusion of the first metacarpophalangeal joint (Dollé et al., 1993). Likewise, mice heterozygous for a mutation in the Hoxa13 gene sequence (Hoxa13+/−) show only a reduced first digit compared to normals, whereas mice homozygous for the disrupted Hoxa13 sequence (Hoxa13−/−) usually die in utero and lack all forelimb digits except one (Fromental-Ramain et al., 1996; Mortlock et al., 1996). However, when Hoxd13 heterozygotes (Hoxd13+/−) are crossed with mice heterozygous for the Hoxa13 mutation (Hoxa13+/−), the phenotype is much more severe and resembles the condition seen in the Hoxd13 homozygous (Hoxd13−/−) mutants (Fromental-Ramain et al., 1996). These results indicate that Hox proteins function in a partially redundant manner in which the severity of the mutation is proportional to the total Hox protein "dose" (Favier and Dollé, 1997; Zákány et al., 1997). Results from these and other knockout experiments show that redundancy of Hox function in the autopod does reduce the lethality of mutation, while at the same time providing a potential source for phenotypic variation. The experiments noted in an earlier section also provide evidence for weak linkage, or flexibility of transcriptional regulation, in autopod morphogenesis. Different Hox proteins are quite similar to one another in form and, as shown from many of the knockout experiments referred to in an earlier section, can bind to the same DNA sequences and initiate transcription of the same target genes (Krumlauf, 1994).

Evidence from experimental genetics also confirms that the autopod is relatively robust to mutational changes in the expression of genes involved in skeletal patterning. Growth and differentiation factor-5 (GDF-5) is a signaling molecule that is expressed at various times and locations in the developing limb. In the developing digits, GDF-5 initially has a broad expression domain surrounding the presumptive joint region in which it promotes epiphyseal chondrogenesis, whereas later in limb development, during the process of joint formation, its expression is limited to the joint interzone (Storm and Kingsley, 1999). Mutations in GDF-5 produce a variety of limb defects, referred to as "brachypodism," which include fusion of the proximal and intermediate phalanges in mice (Storm et al., 1994), and severe reduction of the intermediate phalanges in humans (Polinkovsky et al., 1997). More minor alterations include a pollex that is medially divergent compared to that of normals. GDF-5 is not expressed in developing nerves, blood vessels, or muscle precursor cells. These tissues invade the limb normally in brachypod mice, but the muscle attachments differ from those of normals due to the defects in skeletal and tendon patterning. For example, flexor digitorum superficialis sends a tendon to the pollex in brachypod mice but not in normals (Grüneberg and Lee, 1973). This illustrates a key feature of robusticity in developmental patterning of the autopod—mutational changes may be confined to the cartilage rudiments forming future skeletal elements but need not be accompanied by mutations in nervous, vascular, or muscular systems to yield a new, fully viable phenotype (Kirschner and Gerhart, 1998).

PRIMATE ORIGINS: ROLE OF AUTOPOD EVOLUTION

Studies of primate skeletal remains from the Eocene epoch of North America and Europe have brought to light several derived features of the autopod that distinguish primates from their close relatives: the plesiadapiforms, dermopterans, and tree shrews. This discussion of derived autopod features is restricted to the digital rays and does not include a discussion of the carpus and tarsus, as the morphology of these bones in early primates has been dealt with elsewhere (e.g., Dagosto, 1988; Gebo, 1985; Godinot and Beard, 1991; Hamrick, 1996; Hamrick, 1999a). Dagosto (1988) figured a number of terminal phalanges representing the Early Tertiary primate family Omomyidae that showed these early primates had short, broad, nail-bearing digit tips. In contrast, plesiadapiforms, such as *Plesiadapis insignis* (Gingerich, 1976) and *Phenacolemur simonsi* (Beard, 1990) resemble tree shrews and dermopterans in having narrow, compressed, claws. Godinot (1992), however, noted that the terminal phalanges of the adapiforms *Smilodectes* and *Adapis* were somewhat keeled in dorsal view and not as spatulate as those of omomyids. This led Godinot (1992) to suggest that these primates may have had claw-like tegulae rather than broad, flat ungulae.

The structure of mammalian distal limb integumentary appendages (e.g., ungulae, falculae, tegulae) varies in part according to length of the terminal phalanx (Hamrick, 1999b). Growth rate of the nail or claw is correlated with terminal phalanx size, such that the nail on the third (longest) finger of humans grows at a faster rate than the nail on the fifth (shortest) finger; and human fingernails grow up to four times faster than toenails (Williams, 1995). Furthermore, the number of claw layers is proportional to the proximodistal length of the germinal matrix, which is in turn proportional to the length of the terminal phalanx (Hamrick, 1999b). Increased terminal phalanx length relative to body size is, among primates and tree shrews, associated with an increase in claw thickness and number of claw layers (Hamrick, 1999b, 2001b). Comparison of digit proportions among early primates, plesiadapiforms, primitive mammals, such as *Megazostrodon* and *Ptilodus* (Table 1), and modern archontan taxa (Table 2), shows that primates (including adapiforms) have relatively short-terminal phalanges (Figure 4). Thus, reduction in terminal phalanx length associated with claw reduction does appear to be a primitive feature for euprimates likely related to habitual foraging on slender arboreal supports (Cartmill, 1974; Hamrick, 1998, 1999b). Furthermore, it is clear that the various orders of archontan mammals differ considerably from one another in terminal phalanx morphology and proportions (Figure 5), indicating that diversification of autopod morphology was key to the radiation of arboreal behaviors in these mammals.

A second aspect of autopod morphology that is derived from early primates also relates to proportions of the digital ray segments. Arboreal mammals that

Table 1. Fossil specimens included for comparative analysis

Taxon (Specimen number)*	Age (Reference)
Order Triconodonta	
Megazostrodon rudnerae (BMNH M26407)	Triassic (Jenkins and Parrington, 1976)
Order Multituberculata	
Ptilodus kummae (UA 9001)	Paleocene (Krause and Jenkins, 1983)
Order Microchiroptera	
Icaronycteris index (PU 18150)	Eocene (Jepsen, 1966)
Order Plesiadapiformes	
Plesiadapis insignis (MNHN ref. spec.)	Paleocene (Gingerich, 1976)
Plesiadapis cookei (UMMP ref. spec.)	Eocene (Gunnell, pers. com.)
Order Primates	
Europolemur koenigswaldi (SMNK Me-1125A)	Eocene (Franzen and Frey, 1993)
Godinotia neglecta (SMF ref. spec.)	Eocene (Franzen, 2000)
Notharctus tenebrosus (AMNH 11474)	Eocene (Gregory, 1920)

*Abbreviations: BMNH = British Museum of Natural History; UA = University of Alberta; PU = Princeton University; MNHN = Musée National d'Histoire Naturelle; UMMP = University of Michigan Museum of Paleontology; SMNK = Staatliches Museum für Naturkunde Karlsruhe; SMF = Senckenberg Museum Frankfurt; AMNH = American Museum of Natural History.

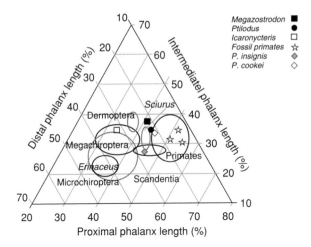

Figure 4. Ternary plot of relative pedal digit segment lengths in living and fossil mammals. Sample sizes are shown in Tables 1 and 2. Length of proximal phalanx, intermediate phalanx, and distal phalanx from the third toe is divided by the total length of the three segments added together multiplied by 100. The ellipses enclose the range of individual values for each group.

prefer to feed among slender vines and branches frequently do so by either sitting atop the branch and grasping with clawless, opposable digits or hanging below the branch and holding on with hook-like fingers and toes (Cartmill, 1985). In each case the digits (phalanges), particularly digits III-V, are long relative to the palm and sole so that they can flex completely around

Table 2. Extant sample included for comparative analysis of digit proportions

Taxon	n	Taxon	n
Order Primates		Order Microchiroptera	
Microcebus murinus	5	*Nycteris* spp.	3
Galago senegalensis	4	*Myotis lucifugus*	3
Tarsius spp.	8	*Phyllostomus discolor*	3
Aotus trivirgatus	5	Order Megachiroptera	
Callicebus moloch	4	*Pteropus hypomelanus*	4
Order Dermoptera		*Rousettus amplexicaudatus*	5
Cynocephalus volans	5	*Nyctimene albiventer*	3
Order Scandentia		Order Rodentia	
Tupaia glis	6	*Sciurus carolinensis*	4
Tupaia tana	3	Order Lipotyphla	
		Erinaceus europaeus	5

Pteropus

Cynocephalus

Otolemur

Tupaia

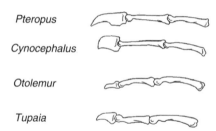

Figure 5. Lateral views of the third toe in a fruit bat (*Pteropus*), colugo (*Cynocephalus*), primate (*Otolemur*), and tree shrew (*Tupaia*) illustrating the relatively long proximal phalanx and reduced terminal phalanx of primates. Not to scale.

the branch in a firm grasp. Primates resemble bats, dermopterans, and *Plesiadapis* in having relatively long toes (Figure 6) and fingers (Hamrick, 2001); however, these mammals have increased the relative length of their fingers and toes in different ways. Dermopterans and bats share relatively long intermediate phalanges, whereas primates and tree shrews have proximal phalanges that are longer than their intermediate phalanges (Figures 4, 5; Hamrick, 2001c; Hamrick et al., 1999). Furthermore, as noted in an earlier section, dermopterans and bats have long, hook-like claws, whereas primates have reduced the length of their claws and terminal phalanges. Thus, primates are derived among these archontan mammals not in having long digits per se, but in having digits that are long due to elongation of the proximal phalanges. It should be noted that this is not true of anthropoids, which have relatively long metatarsals (Dagosto, 1990). The condition shared by adapids, tarsiers, and strepsirhines is inferred to be primitive for euprimates with the anthropoid condition considered a reversal resulting from a shift to more frequent positional behaviors on large-diameter, horizontal supports (Gebo, 1986).

One of the most significant discoveries in recent years concerning the origin of primate limb morphology is the finding that the Early Tertiary plesiadapiform *Carpolestes* had long fingers and toes, like early Euprimates (Bloch and Boyer, 2002). Furthermore, other plesiadapiforms, such as paromomyids and micromomyids, appear to have also had elongate proximal phalanges like *Carpolestes* and euprimates. These plesiadapiforms still retained claws on their fingers and toes, but *Carpolestes* had a divergent hallux with a broad, flat,

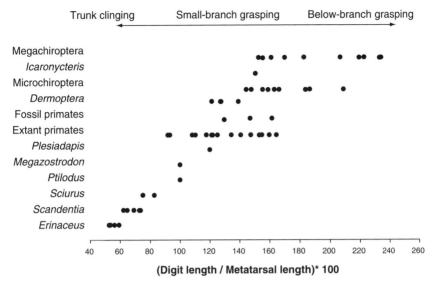

Figure 6. Univariate plot of estimated digit length (proximal phalanx length plus intermediate phalanx length) relative to metatarsal length in several living and fossil mammals. Data are from the third toe and sample sizes are shown in Tables 1 and 2. Data for *Plesiadapis* are from *P. cookei* because complete third metatarsal length is not preserved in the specimen of *P. insignis*.

terminal phalanx indicating that the hallux probably bore a flat nail instead of a claw. These findings suggest that the derived morphology of euprimates may have evolved in a mosaic fashion, with elongate digits appearing prior to the origin of flat nails. These discoveries also demonstrate that the Early Tertiary diversification of angiosperms was accompanied by a corresponding diversification of autopod proportions in small-bodied, arboreal mammals. The evidence discussed in an earlier section pertaining to evolvability in the mammalian autopod provides a mechanistic basis for these patterns observed in the fossil record.

DISCUSSION

Stern (1998b) recently commented that evolutionary biologists have traditionally asked two kinds of questions: What is the evolutionary history of a particular group of organisms revealed by phylogenetic systematics and the fossil record and why, in ecological and adaptive terms, did such evolutionary transformations occur? Previous studies of primate origins have primarily addressed these sorts of questions: What were early primates like in terms of their biology and behavior (e.g., Covert, 1986; Kay, 1984; Szalay, 1981)? What taxa might be the closest relatives of euprimates (e.g., Beard, 1993; Szalay et al., 1987; Wible and Covert, 1987)? What is the functional and adaptive significance of derived euprimate features, such as a postorbital bar, orbital convergence, and grasping extremities (e.g., Cartmill, 1972, 1992; Ravosa et al., 2000)?

Stern (1998b, 2000) has suggested that recent advances in developmental genetics now allow evolutionary biologists to ask a third type of question; namely, how do new morphologies arise via evolutionarily relevant mutations? Thus, one question we may now ask concerning primate origins is how did the derived phenotype that characterized basal primates come into being in the first place? In other words, how did the first primates modify their ancestral developmental program to yield a new skeletal morphology recognizable as distinctive from that of all other placental mammals over 50 million years ago? Experimental observations suggest that the architecture of development facilitates evolutionary change within particular spatial domains or modules. The limb represents one such module, and within the limb field morphology of the autopod (hand and foot) is especially variable interspecifically. Experimental genetics illustrates how redundancy of gene function, robusticity in the face of mutational changes in skeletal patterning, and a degree of developmental autonomy together increase the potential for morphological change within the autopod. These factors contribute toward the production of novel limb phenotypes and are therefore likely to have played an important role in the diversification of skeletal form within modern mammalian clades and among early eutherian mammals as well. The experimental, comparative, and paleontological data presented in this chapter provide evidence that evolutionary change in developmental patterning of the digital rays was key to the origin of the order Primates.

The next challenge for primate evolutionary biology is to identify the specific mutations implicated in the origin of the primate morphotype. As discussed in an earlier section, in the case of the autopod, these evolutionarily relevant mutations do not involve changes in regulatory protein structure. Rather, these mutational changes are more likely to occur in *cis*-regulatory regions, which affect the timing and location of regulatory protein expression. Identifying these mutations is a formidable task, since DNA-binding proteins typically recognize a core 6–9 base pair sequence and these sequences can be nested within long noncoding regions located at variable distances from the particular gene that they influence (Carroll et al., 2001). One strategy that has already proven useful for investigating the evolution of anthropoid globin genes is outlined by Chiu et al. (1999). Briefly, a comparison of DNA sequence alignment among extant primate taxa permits the identification of conserved *cis*-sequence elements that comprise an enhancer region. Protein-binding or transgenic experiments can then be used to investigate the effects of particular enhancer sequences on gene expression and morphogenesis (Chiu and Hamrick, 2002; Chiu et al., 2000). The integration of this experimental research with comparative and paleontological evidence is certain to expand on our present understanding of the mechanisms important in the origin of primate skeletal form.

Data presented here on hindfoot proportions show that primates, including Early Tertiary adapiforms, such as *Europolemur*, *Godinotia*, and *Notharctus*, are derived relative to tree shrews, dermopterans, and *Plesiadapis* in having relatively long-proximal phalanges and short-terminal phalanges. These data provide strong support for the hypothesis that primate origins involved an evolutionary change in digital-ray patterning yielding hands and feet with relatively long digits and reduced claws. These comparative results should, however, be viewed with some caution because the cheiridial morphology of many early primates is still unknown. For example, omomyid primates comprise a family of approximately 80 species, which existed throughout North America, Europe, and Asia during the Eocene epoch, yet, only terminal phalanges and hallucial metatarsals have been described for this group. A diverse haplorhine fauna from the Eocene of China is also now known from dental and tarsal remains (Beard et al., 1994; Gebo et al., 2000a, b), but morphology of the fingers and toes in these animals has not been described. The anatomy of these creatures is critical for understanding the pattern of cheiridial morphology primitive for the order Primates (Euprimate morphotype). Finally, recent molecular studies (Liu et al., 2001; Madsen et al., 2001; Murphy et al., 2001) suggest that tree shrews (Scandentia) and colugos (Dermoptera) are the closest relatives of primates. Unfortunately, the fossil record for these two Orders is very poor. New discoveries of plesiadapiform skeletal remains (Bloch and Boyer, 2002) clearly show that digit proportions among early archontan mammals were much more varied and diverse than previously thought. Our understanding of the temporal and phylogenetic sequence in which the derived skeletal features of euprimates was acquired will remain incomplete until postcranial remains of early tree shrews and dermopterans are found.

ACKNOWLEDGMENTS

The author is grateful to Drs Chi-Hua Chiu and C.O. Lovejoy for insightful discussions on many of the topics covered here and thanks M. Ravosa and M. Dagosto for the opportunity to present this and other contributions directed towards understanding primate evolution. Ms Linda Gordon and Drs J. Mead and A. Gardner provided curatorial assistance at the Smithsonian Institution, Mr B. Stanley at the Field Museum of Natural History, and Mr L. Jellema at the Cleveland Museum of Natural History. Dr G. Gunnell, University of Michigan, generously shared unpublished data on hand and foot proportions in *Plesiadapis cookei* and J. Bloch and D. Boyer provided the opportunity to examine undescribed specimens of *Phenacolemur*.

REFERENCES

Bader, R., and Hall, J., 1960, Osteometric variation and function in bats, *Evol.* **14:** 8–17.

Beard, K. C., 1990, Gliding behaviour and palaeoecology of the alleged primate family Paromomyidae (Mammalia, Dermoptera). *Nature* **345:** 340–341.

Beard, K. C., 1993, Phylogenetic systematics of the Primatomorpha, with special reference to Dermoptera, in: F. Szalay, M. Novacek, and M. McKenna, eds., *Mammal Phylogeny: Placentals*, Springer-Verlag, New York, pp. 129–150.

Beard, K. C., Qi, T., Dawson, M. R., Wang, B., and Li, C., 1994, A diverse new primate fauna from middle Eocene fissure-fillings in southeastern China, *Nature* **368:** 604–609.

Bloch, J. I., and Boyer, D. M., 2002, Grasping primate origins, *Science* **298:** 1606–1610.

Bock, W. J., and Miller, W., 1959, The scansorial foot of the woodpeckers with comments on the evolution of perching and climbing feet in birds, *Am. Mus. Novitates* **1931:** 1–45.

Carroll, S. B., Greiner, J., and Weatherbee, S. D., 2001, *From DNA to Diversity*, Blackwell Science, Massachusetts.

Cartmill, M., 1972, Arboreal adaptations and the origin of the order Primates, in: *The Functional and Evolutionary Biology of Primates*, R. Tuttle, ed., Aldine-Atherton, Chicago, pp. 97–122.

Cartmill, M., 1974, Pads and claws in arboreal locomotion, in: *Primate Locomotion*, F. Jenkins, ed., Plenum, New York, pp. 45–83.

Cartmill, M., 1985, Climbing, in: *Functional Vertebrate Morphology*, M. Hildebrand, ed., Aldine-Atherton, Chicago, pp. 73–88.

Cartmill, M., 1992, New views on primate origins, *Ev. Anth.* **1**: 105–111.

Chiu-C.-H., Amemiya, C., Carr, J., Bhargava, J., Hwang, J., Shashikant, C., et al., 2000, A recombinogenic targeting method to identify large inserts for *cis*-regulatory analysis in transgenic mice: Construction and expression of a 100-kb, zebrafish Hoxa-11b-lacZ reporter gene, *Dev. Genes Evol.* **210**: 105–109.

Chiu-C.-H., Gregoire, L., Gumucio, D., Muniz, J., Lancaster, W., and Goodman, M., 1999, Model for the fetal recruitment of simian g-globin genes based on findings from two New World monkeys *Cebus apella* and *Callithrix jacchus* (Platyrrhini, Primates), *J. Exp. Zool. (Mol. Dev. Evol.)* **284**: 27–40.

Chiu-C.-H., and Hamrick, M. H., 2002, Evolution and the development of the primate limb skeleton, *Ev. Anth.* **11**: 94–.

Clark, J. M., Hopson, J. A., Hernández, R., Fastovsky, D., and Montellano, M., 1998, Foot posture in a primitive pterosaur, *Nature* **391**: 886–889.

Cohn, M. J., Izpisúa-Belmonte, J. C., Abud, H., Heath, J., and Tickle, C., 1995, Fibroblast growth factors induce additional limb development from the flank of chick embryos, *Cell* **80**: 739–746.

Conrad, M., 1990, The geometry of evolution, *Biosystems* **24**: 61–81.

Covert, H. H., 1986, Biology of early Cenozoic primates, in: *Comparative Primate Biology*, D. Swindler and J. Erwin, eds., Alan R. Liss, New York, pp. 335–349.

Dagosto, M., 1988, Implications of postcranial evidence for the origin of Euprimates, *J. Hum. Evol.* **17**: 35–56.

Dagosto, M., 1990, Models for the origin of the anthropoid postcranium, *J. Hum. Evol.* **19**: 121–140.

Davis, A., and Capecchi, M., 1996, A mutational analysis of the 5' HoxD genes: Dissection of genetic interactions during limb development in the mouse, *Development* **122**: 1175–1185.

Dollé, P., Dierich, A., LeMeur, M., Schimmang, T., Schuhbaur, B., Chambon, P., et al., 1993, Disruption of the Hoxd-13 gene induces localized heterochrony leading to mice with neotenic limbs, *Cell* **75**: 431–441.

Engels, W., 1938, Variation in bone length and limb proportions in the coot (*Fulica americana*), *J. Morph.* **62**: 599–607.

Favier, B., and Dollé, P., 1997, Developmental functions of mammalian Hox genes, *Mol. Hum. Reprod.* **3**: 115–131.

Feduccia, A., 1993, Evidence from claw geometry indicating arboreal habits of *Archaeopteryx*, *Science* **259**: 790–793.

Franzen, J. L., 2000, Der sechste Messel-Primate (Mammalia, Primates, Notharctidae, Cercamoniinae), *Senckenbergiana lethaea* **80**: 289–303.

Franzen, J. L., and Frey, E., 1993, *Europolemur* completed, *Kaupia* **3**: 113–130.

Fromental-Ramain, C., Warot, X., Messadecq, N., LeMeur, M., Dollé, P., and Chambon, P., 1996, Hoxa-13 and Hoxd-13 play a crucial role in the patterning of the limb autopod, *Development* **122**: 2997–3011.

Gebo, D. L., 1985, The nature of the primate grasping foot, *Am. J. Phys. Anthropol.* **67**: 269–278.

Gebo, D. L., 1986, Anthropoid origins—The foot evidence, *J. Hum. Evol.* **15**: 421–430.

Gebo, D. L., Dagosto, M., Beard, K. C., and Qi, T., 2000a, The oldest known anthropoid postcranial fossils and the early evolution of higher primates, *Nature* **404**: 276–278.

Gebo, D. L., Dagosto, M., Beard, K. C., and Qi, T., 2000b, The smallest primates, *J. Hum. Evol.* **38**: 585–594.

Gerhart, J., and Kirschner, M., 1997, *Cells, Embryos, and Evolution*, Blackwell Science, Massachusetts.

Gilbert, S. F., Opitz, J. M., and Raff, R., 1996, Resynthesizing evolutionary and developmental biology, *Dev. Biol.* **173**: 357–372.

Gingerich, P. D., 1976, Cranial anatomy and evolution of early Tertiary Plesiadapidae (Mammalia, Primates), *Mus. Paleontol. Univ. Michigan Papers in Paleontology.* **15**: 1–140.

Godinot, M., 1992, Early euprimate hands in evolutionary perspective, *J. Hum. Evol.* **22**: 267–283.

Godinot, M., and Beard, K. C., 1991, Fossil primate hands: A review and an evolutionary inquiry emphasizing early forms, *Hum. Evol.* **6**: 307–354.

Goslow, G. E. J., 1989, How are locomotor systems integrated and how have evolutionary innovations been introduced? in: *Complex Organismal Functions: Integration and Evolution in Vertebrates*, D. B. Wake and G. Roth, eds., Wiley, New York, pp. 205–218.

Gregory, W. K., 1920, On the structure and relations of *Notharctus*: An American Eocene primate, *Mem. Am. Mus. Nat. Hist.* **3**: 51–243.

Grüneberg, H., and Lee, A. J., 1973, The anatomy and development of brachypodism in the mouse, *J. Embyol. Exper. Morph.* **30**: 119–141.

Hall, B. K., 1998, *Evolutionary Developmental Biology*, Kluwer, New York.

Hamrick, M. W., 1996, Locomotor adaptations reflected in the wrist joints of early Tertiary primates (Adapiformes), *Am. J. Phys. Anthropol.* **100**: 585–604.

Hamrick, M. W., 1998, Functional and adaptive significance of primate pads and claws: Evidence from New World anthropoids, *Am. J. Phys. Anthropol.* **106**: 113–127.

Hamrick, M. W., 1999a, First carpals of the Eocene primate family Omomyidae, *Contrib. Museum of Paleontology, University of Michigan* **30**: 191–198.

Hamrick, M. W., 1999b, Pattern and process in the evolution of primate nails and claws, *J. Hum. Evol.* **37**: 293–298.

Hamrick, M. W., 2001a, Morphological diversity in digital skin microstructure of didelphid marsupials, *J. Anat.* **198**: 683–688.

Hamrick, M. W., 2001b, Development and evolution of the mammalian limb: Adaptive diversification of nails, hooves, and claws, *Evol. Dev.* **3**: 355–363.

Hamrick, M. W., 2001c, Primate origins: Evolutionary change in digital ray patterning and segmentation, *J. Hum. Evol.* **40**: 339–351.

Hamrick, M. W., 2003, Evolution and development of mammalian limb integumentary structures, *J. Exp. Zoolog. (Mol. Dev. Evol.)*, **298B**: 152–163.

Hamrick, M. W., Rosenman, B. A., and Brush, J. A., 1999, Phalangeal morphology of the Paromomyidae (?Primates, Plesiadapiformes): The evidence for gliding behavior reconsidered, *Am. J. Phys. Anth.* **109**: 397–413.

Hérault, Y., Beckers, J., Kondo, T., Fraudeau, N., and Duboule, D., 1998, Genetic analysis of a Hoxd-12 regulatory element reveals global versus local modes of controls in the HoxD complex, *Development* **125**: 1669–1677.

Hinchliffe, J. R., 1989, Reconstructing the archetype: Innovation and conservatism in the evolution and development of the pentadactyl limb, in: *Complex Organismal Functions: Integration and Evolution in Vertebrates*, D. B. Wake and G. Roth, eds., Wiley, New York, pp. 171–190.

Hinchliffe, J. R., 1991, Developmental approaches to the problem of transformation of limb structure in evolution, in: *Developmental patterning of the Vertebrate Limb*, J. R. Hinchliffe, J. M. Hurle, and D. Summerbell, eds., Plenum, New York.

Howell, A. B., 1944, *Speed in Animals*, Chicago University Press, Chicago.

Hunter, J., and Jernvall, J., 1995, The hypocone as a key innovation in mammalian evolution, *Proc. Natl. Acad. Sci.* **92**: 10718–10722.

Jenkins, F. A., and Parrington, F. R., 1976, The postcranial skeletons of the Triassic mammals *Eozostrodon*, *Megazostrodon*, and *Erythrotherium*, *Philos. Trans. Zoolog. Soc. Lond.* **273**: 387–431.

Jepsen, G. L., 1966, Early Eocene bat from Wyoming, *Science* **154**: 1333–1339.

Jernvall, J., 2000, Linking development with evolutionary novelty in mammalian teeth, *Proc. Natl. Acad. Sci.* **97**: 2641–2645.

Jernvall, J., Hunter, J., and Fortelius, M., 1996, Molar tooth diversity, disparity, and ecology in Cenozoic ungulate radiations, *Science* **274**: 1489–1492.

Jernvall, J., Këranen, S. V. E., and Thesleff, I., 2001, Evolutionary modification of development in mammalian teeth:quantifying gene expression patterns and topography, *PNAS* **97**: 14444–14448.

Kay, R. F., 1984, On the use of anatomical features to infer foraging behavior inextinct primates, in: *Adaptations for Foraging in Nonhuman Primates*, P. S. Rodman and J. G. H. Cant, eds., Columbia University Press, New York, pp. 21–53.

Kirschner, M., and Gerhart, J., 1998, Evolvability, *Proc. Natl. Acad. Sci.* **95:** 8420–8427.

Krause, D. W., and Jenkins, F. A., 1983, The postcranial skeleton of North American multituberculates, *Bull. Museum of Comp. Zoolog.* **150:** 199–246.

Krumlauf, R., 1994, Hox genes in vertebrate development, *Cell* **78:** 191–201.

Lemelin, P., 1999, Morphological correlates of substrate use in didelphid marsupials: Implications for primate origins, *J. Zool. Lond.* **247:** 165–175.

Liu, F., Miyamoto, M. M., Freire, N., Ong, P., Tennant, M., Young, T., et al., 2001, Molecular and morphological supertrees for eutherian (placental) mammals, *Science* **291:** 1786–1789.

Madsen, O., Scally, M., Douady, C. J., Kao, D. J., Debry, R. W., Adkins, R., et.al., 2001, Parallel adaptive radiations in two major clades of placental mammals, *Nature* **409:** 610–614.

Merino, R., Macias, D., Ganan, Y., Rodriguez, J., Economides, A., Izppisua, J., et al., 1999, Control of digit formation by activin signalling, *Development* **126:** 2161–2170.

Mortlock, D. P., Post, L., and Innis, J., 1996, The molecular basis of hypodactyly (Hd) a deletion in Hoxa13 leads to arrest of digital arch formation, *Nat. Genet.* **13:** 284–289.

Murphy, W. J., Eizirik, E., Johnson, W. E., Zhang, Y. P., Ryder, O. A., and O'Brien, S. J., 2001, Molecular phylogenetics and the origins of placental mammals, *Nature* **401:** 614–618.

Norberg, U., 1994, Wing design, flight performance, and habitat use in bats, in: *Ecological Morphology*, P. C. Wainwright and S. M. Reilly, eds., The University of Chicago Press, Chicago, pp. 205–239.

Norberg, U., and Rayner, R. J., 1987, Ecological morphology and flight in bats (Mammalia, Chiroptera): Wing adaptations, flight performance, and echolocation, *Phil. Trans. Royal Soc. London* **B316:** 335–427.

Noveen, A., Jiang, T.-X., Ting-Berreth, S. A., and Chuong, C.-M., 1995, Homeobox genes Msx-1 and Msx-2 are associated with induction and growth of skin appendages, *J. Invest. Dermatol.* **104:** 711–719.

Polinkovsky, A., Robin, N., Thomas, J., Irons, M., Lynn, A., Goodman, F., et al., 1997, Mutations in CDMP1 cause autosomal dominant brahcydactyly type C, *Nat. Genet.* **17:** 18–19.

Raff, R., 1996, *The Shape of Life*, The University of Chicago Press, Chicago.

Ravosa, M. J., Noble, V. E., Hylander, W. L., Johnson, K. R., and Kowalski, E., 2000, Masticatory stress, orbital orientation, and the evolution of the primate postorbital bar, *J. Hum. Evol.* **38:** 667–693.

Richardson, M. K., 1999, The developmental origins of adult variation, *Bioessays* **21**: 604–613.

Satokata, I., Ma, L., Ohshima, H., Nbei, M., Woo, I., Nishizawa, K., et al., 2000, Msx2 deficiency in mice cause pleiotropic defects in bone growth and ectodermal organ formation, *Nat. Genet.*. **24**: 391–395.

Searls, R., 1968, Development of the embryonic chick limb bud in avascular culture, *Dev. Biol.* **17**: 382–399.

Shubin, N. H., Tabin, C., and Carrol, S. B., 1997, Fossils, genes, and the evolution of animal limbs, *Nature* **388**: 639–648.

Stern, D. L., 1998a, The future of evolutionary biology, *New Scientist* **159**: 1–4.

Stern, D. L., 1998b, A role of *Ultrabiothorax* in morphological differences between *Drosophila* species, *Nature* **396**: 463–466.

Stern, D. L., 2000, Evolutionary developmental biology and the problem of variation, *Evolution* **54**: 1079–1091.

Storm, E., Huynh, T., Copeland, N., Jenkins, N., Kingsley, D., and Lee, S., 1994, Limb alterations in brachypodism mice due to mutations in a new member of the TGFa superfamily, *Nature* **368**: 639–642.

Storm, E., and Kingsley, D. M., 1999, GDF5 coordinates bone and joint formation during digit formation during digit development, *Dev. Biol.* **209**: 11–27.

Szalay, F. S., 1981, Phylogeny and the problem of adaptive significance: The case of the earliest primates, *Folia Primatol.* **36**: 157–182.

Szalay, F. S., Rosenberger, A. L., and Dagosto, M., 1987, Diagnosis and differentiation of the Order Primates, *Ybk. Phys. Anthropol.* **30**: 75–105.

Von Dassow, G., and Munro, E., 1999, Modularity in animal development and evolution: Elements of a conceptual framework for EvoDevo, *J. Exp. Biol.* (*Ml. Dev. Evol.*), **285**: 307–325.

Wagner, A., 1996, Genetic redundancy caused by gene duplication and its evolution in networks of transcriptional regulators, *Biol. Cybern.* **74**: 557–567.

Wible, J. R., and Covert, H. H., 1987, Primates: Cladistic diagnosis and relationships, *J. Hum. Evol.* **16**: 1–20.

Williams, P. L., 1995, *Gray's Anatomy,* Churchill-Livingstone, London.

Winther, R. G., 2001, Varieties of modules: Kinds, levels, origins, and behaviors, *J. Exp. Biol.* (*Mol. Dev. Evol.*), **291**: 116–129.

Zákány, J., Fromental-Ramain, C., Warot, X., and Duboule, D., 1997, Regulation of number and size of digits by posterior Hox genes: A dose-dependent mechanism with potential evolutionary implications, *Proc. Natl. Acad. Sci.* **94**: 13695–13700.

CHAPTER TWELVE

Primate Gaits and Primate Origins

Matt Cartmill, Pierre Lemelin,
and Daniel Schmitt

PECULIARITIES OF PRIMATE GAITS

The order Primates in the strict sense—Euprimates or primates of modern aspect—is defined by a familiar suite of synapomorphies. Some of these may represent adaptively neutral contingencies (for example, the formation of the auditory bulla by an outgrowth from the petrosal, rather than by a separate entotympanic bone). However, others appear to be telling us things about the basal adaptations of the order. Compared to primitive placental mammals, primates have a reduced sense of smell and an enhanced sense of vision. primate eyes point forward and are encircled by a ring of bone. The first toes of primates are stout, divergent grasping organs. All primates have reduced, flattened claws on the first toe, and most of them have them on the other digits as well. The adaptive meaning and origins of some of these morphological synapomorphies of the primate order are discussed in other chapters of this book.

One *behavioral* synapomorphy of primates, which has received less attention in discussions of primate origins, is their distinctive walking gait. When

Matt Cartmill • Department of Biological Anthropology and Anatomy, Duke University Medical Center, Durham, NC 27710 Pierre Lemelin • Division of Anatomy, Faculty of Medicine and Dentistry, University of Alberta, Edmonton, Alberta, Canada, T662H7 Daniel Schmitt • Department of Biological Anthropology and Anatomy, Duke University Medical Center, Durham, NC 27710

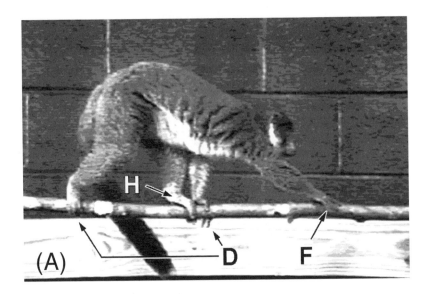

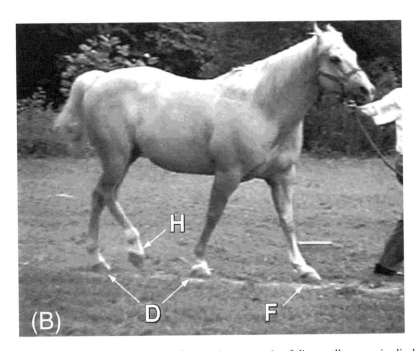

Figure 1. In diagonal-couplets walking gaits, one pair of diagonally opposite limbs (D) supports the body while the other pair is swinging forward. In a typical primate walk (diagonal-couplets in *diagonal* sequence), illustrated here in *Eulemur mongoz* (A), the hindfoot (H) in the forward-swinging pair touches down before the forefoot (F) contacts the support. In a diagonal-couplets walk in *lateral* sequence, illustrated here in *Equus* (B), the forefoot (F) in the forward-swinging pair strikes down before the hindfoot (H). (Stills from experimental videotapes)

quadrupedal primates walk, most of them tend to swing their diagonally opposite legs back and forth more or less together, so that when (say) the left front foot is off the ground and swinging forward, so is the right hindfoot, and so on. This so-called *diagonal-couplets* pattern is equally evident in horses and many other terrestrial mammals (Figure 1). But when horses swing a diagonal pair of limbs forward, the front foot in the pair strikes down first. In a monkey or a lemur, the *hind* foot in the diagonal pair strikes down *before* the forefoot. The horse pattern, in which each hind footfall is followed by the fall of the forefoot on the same side, is called a *lateral-sequence* (LS) walk. Most mammals employ LS gaits in walking. The primate pattern, in which each hind footfall is followed by the fall of the opposite forefoot, is called a *diagonal-sequence* (DS) walk (Hildebrand, 1965, 1966, 1985). DS walks are seldom seen in nonprimates.

The peculiarities of primate gaits were discovered in 1887 by Eadweard Muybridge, but they were not analyzed quantitatively until the 1960s, when Milton Hildebrand devised a way of quantifying the differences between various symmetrical gaits. These are defined as gaits in which the first half of each cycle is the same as the second half, but with the movements of the left and right limbs switched. Symmetrical gaits include the walk, trot, and pace. Hildebrand pointed out that most of the differences among such gaits could be represented by just two numbers, and so any such gait could be specified by a single point on a bivariate plot, the *Hildebrand diagram* (Figure 2; Hildebrand, 1965, 1966, 1985).

The horizontal or x-axis on the Hildebrand diagram, which we will refer to as *duty factor*, represents the percentage of time a foot stays on the ground during one complete gait cycle, from one touchdown of that foot to the next. In general, the higher an animal's speed, the lower the duty factor for each foot (Figure 2A; Demes et al., 1990, 1994; Gatesy and Biewener, 1991; Grillner, 1975; Prost, 1965, 1969, 1970; Vilensky et al., 1988). Hildebrand's y-axis variable, or *diagonality*, is the one that distinguishes primates from most other mammals. This variable expresses the phase difference between the front and the hind end as a percentage of cycle duration (Figure 2B). If the hindfeet are exactly in phase with the forefeet, so that the right fore (RF) and right hind (RH) feet touch down together, followed by the simultaneous touchdown of the left fore (LF) and left hind (LH) feet, diagonality is zero (or 100). Such a gait is called a *pace*. If diagonally opposite feet touch down together (RF + LH, LF + RH), diagonality is 50 and the gait is a *trot*. In a lateral-sequence walk (RF touchdown, then LH, LF, RH), diagonality assumes a value between zero and 50; in a DS walk (RF, RH, LF, LH), it lies between 50 and 100.

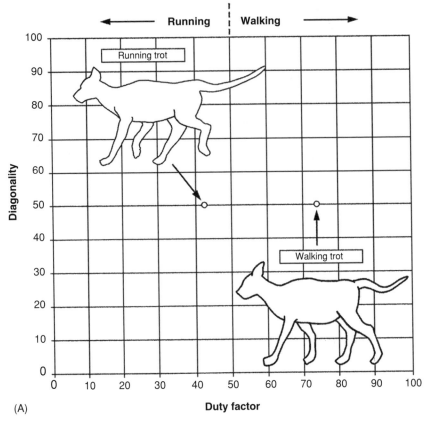

Figure 2. The two axes of the Hildebrand diagram. (A) variation in the x-axis only (trotting gaits, diagonality = 50). The X variable, here called *duty factor*, represents the time that a foot remains on the support as a percentage of one complete gait cycle (i.e., from one touchdown of that foot to the next). In the symmetrical gaits of most mammals, duty factor is approximately the same for all four feet and has a close inverse correlation with speed. When duty factor is less than 50, all four feet are usually off the ground at some point in the cycle (aerial phase), and the gait is a *run*; when it is more than 50, there is no aerial phase, and the gait is a *walk*.

PROBLEMS OF DIAGONAL SEQUENCE WALKS

Most primates preferentially employ the DS footfall sequence in walking (Hildebrand, 1967, 1985; Prost, 1965, 1969, 1970; Rollinson and Martin, 1981; Vilensky, 1989; Vilensky and Larson, 1989). The primate preference for DS walks is problematic, because (as many people have shown) such gaits

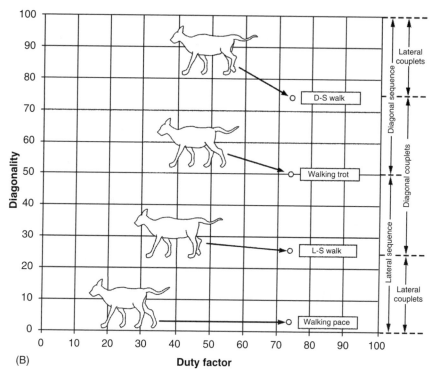

Figure 2. (*Continued*) (B) variation in the y-axis only (walking gaits, duty factor = 74). The Y variable, here called *diagonality*, represents the percentage of time in the gait cycle by which each fore footfall lags behind the fall of the hindfoot on the same side. The four illustrated gaits differ in the phase relationship between the forelimbs (shown in the same position in all four drawings) and the hindlimbs. When both feet on one side are in synchrony in walking (diagonality = 0 or 100), the gait is a *walking pace*. When they are exactly out of phase (diagonality = 50), the gait is a *walking trot*. When the fore footfall lags the ipsilateral hind footfall by more than zero but less than 50% of the cycle period, the gait is a *lateral-sequence walk*; when it lags by more than 50% but less than 100%, the gait is a *DS walk*. The right hindlimb is shaded in the four drawings to emphasize the phase shift.

appear to be inherently less stable at slow speeds than typical LS walks. In the LS walk of a horse or other typical mammalian quadruped, the areas of the tripods of support (gray triangles, Figures 3A, 4–5, 8–9) are maximized, with three support points well spread out along the anteroposterior axis (cf. Figure 1). The vertical line through the animal's center of mass (line of gravity) probably falls inside the triangles during most of the tripedal support phases. In the bipedal phases of the cycle, when the line of gravity necessarily gets off the line of support, the animal tends to roll to the left (Figures 3A) or right (Figures 3A, 7), but the next foot to come down descends in the right place to check the roll (Figure 3A, nos. 4, 8).

The DS walks seen in primates are considerably less stable. Because the hindfoot in such walks strikes down close behind the forefoot on the same side, the unilateral bipod of support (support by only two feet on the same side) is extremely short, and so the tripods or triangles of support are much smaller than they are in an LS walk (Figure 1; Figure 3A, nos. 4, 8, Figure 3B, nos. 6, 8). Depending on the values of certain gait parameters, the animal may have to balance briefly on the small unilateral bipod twice in each gait cycle (Figure 3B, no. 7; cf. Figure 4). The periods of instability in a DS walk do not appear to contribute uniformly to forward movement. In fact, the direction of pitch may be slightly toward the rear when the forefoot comes down (Hildebrand, 1980), so that the animal is periodically on the verge of toppling backward.

It is not clear what advantage DS walks confer that compensate for these apparent disadvantages. Few answers have been suggested. None of them are persuasive, and none of them satisfactorily account for the observed distribution of LS and DS gaits among mammals.

Muybridge (1887) proposed: (1) that the stronger limb on each side always descends immediately before the other (his "Law of the Walk"), and (2) that arboreal climbing has given primates exceptionally strong forelimbs—and hence DS gaits. But as Vilensky and Larson (1989) observe, there is no reason to think that primate forelimbs are stronger, or bear greater stresses, than their hindlimbs. Force-plate studies suggest precisely the reverse (Demes et al., 1994; Kimura, 1985, 1992; Kimura et al., 1979; Lemelin and Schmitt, this volume; Reynolds, 1985; Schmitt and Lemelin, 2002).

Prost (1965) suggested that DS gaits, but not LS gaits, allow a mammal "to use lateral spine bending to increase distance between successive contact points for the same leg" and thereby to increase stride length. At least some primates do in fact increase stride length in this way (Dykyj, 1984; Demes et al., 1990; Shapiro et al., 2001). But so do some nonprimate tetrapods that use LS walking gaits (Carlson et al., 1979; Pridmore, 1992; Ritter, 1995). What a quadruped needs to allow lateral vertebral flexion to enhance stride

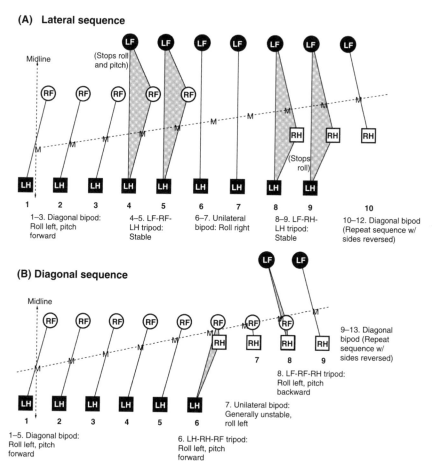

Figure 3. Support polygons in typical LS (A) and DS (B) walking gaits. The diagrams represent foot placement positions during the first half of a gait cycle, beginning with the onset of the LH–RF diagonal bipedal support phase ("bipod"), as viewed from above. Triangles of support are shown in gray. The second half of the cycle, beginning with position 10 in (A) and 9 in (B), would be the mirror image of the first half. In both sequences, time intervals between successive diagrams are roughly constant. The position indicated for the vertical projection through the center of mass (M) is an approximation based on three assumptions: (1) the animal is roughly in balance at the beginning of the diagonal bipod, (2) the center of mass remains in the midline, and (3) the center of mass moves forward at a constant velocity. Footfall timing and spacing are based on Muybridge (1887, p. 28, Pl. 143); graphic convention after Rollinson and Martin (1981) Symbols: black figures, left foot placement (LH, left hind; LF, left fore); white figures, right foot placement (RF, right fore; RH, right hind); squares, hindfeet; circles, forefeet.

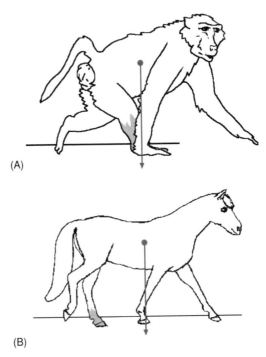

(A)

(B)

Figure 4. Possible adaptive value of diagonal-sequence walking gaits in primates. At the moment of forefoot touchdown, when weight is about to be transferred to a new and untested substrate, the line of gravity (gray arrow: the vertical through the body's center of mass, estimated here as the vertical through the midpoint of an ischium-to-occiput line) will lie much closer to the supporting hindfoot (gray tone) in the D-S walk of the baboon (A) than the L-S walk of the horse (B). In primates or other arboreal animals with marked grasping specializations of the hindfoot, the primate support pattern allows the animal to draw back or regain its balance if the new support breaks or bends precipitously. (Drawings after Muybridge, 1887; from Cartmill et al., 2002)

length is not a DS gait, but a diagonal-couplets gait, in which diagonally opposite limbs swing forward and back as a pair. A diagonal-couplets gait (25< diagonality <75) can be either DS (diagonality >50) or LS (diagonality <50; Figure 2B). Moreover, as Vilensky and Larson (1989) point out, if Prost's (1965) analyses were correct, it is hard to see why other quadrupeds would not adopt DS gaits in order to enhance stride length in walking. This objection can be put more globally: any theory that proposes a benefit accruing to

DS walking gaits needs to explain why most nonprimates have not availed themselves of this benefit.

Prost (1969) subsequently observed that the apparent inferiority of DS gaits (Figure 3) is irrelevant in arboreal locomotion on a narrow branch, since no triangles of support can be formed if all footfalls are collinear. This is an important observation, which refutes some of the supposed adaptive barriers to the adoption of DS walking gaits, but it does not suggest any positive advantage to adopting them. Prost proposed that DS walks are more advantageous than LS walks for an arboreal animal walking on a thin horizontal branch because they allow diagonally opposite limbs to act in concert, and thus reduce rolling and yawing forces during locomotion. Unfortunately, this analysis again confuses diagonal *sequences* (RF footfall follows LH) with *diagonal-couplets* (RF and LH move more or less together). It also fails to explain why many arboreal nonprimates use LS gaits.

Most subsequent analyses of the significance of DS walking gaits have argued that primates have DS gaits because they carry a larger percentage of their weight on the hindlimbs than other mammals do. This idea originated with the work of Tomita (1967). Tomita reasoned that a walking animal swinging the RF and LH limbs forward as a pair (diagonal-couplets) might be expected to put them down in a sequence that depends on the position of the animal's line of gravity: forefoot first (LS walk) to stop forward pitch if the line of gravity passes in front of the line connecting the other two, supporting feet (LF + RH, in this case), and hindfoot first (DS walk) to stop backward pitch if the line of gravity passes behind that line. By heavily loading the hindquarters of dogs, Tomita was able to induce DS walking gaits in a small percentage of trials. He concluded that primates use DS walks because they are tail-heavy compared to nonprimates—which accordingly use LS walks instead.

We believe that Tomita's theory contains a fundamental mistake. In an LS walk, the fore footfall at the end of a diagonal bipedal support period does indeed descend in a position where it effectively checks forward pitch (Figure 3A, nos. 3–4). But the corresponding hind footfall in a typical DS walk descends immediately behind, or alongside of, the forefoot in the diagonal support pair (Figures 3B, nos. 5–6)—and so it is not in a position to arrest backward pitch effectively. Moreover, the direction of pitch in a DS walk at the moment of hindfoot touchdown appears to be forward, not backward (Figures 3B, no. 6; Hildebrand, 1980).

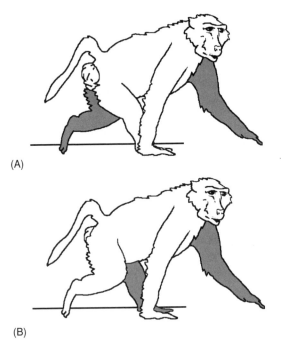

(A)

(B)

Figure 5. Two ways to maximize hindlimb protraction at the moment of forefoot touchdown. (A) the diagonal-sequence, diagonal-couplets pattern, with a diagonality slightly exceeding 50; (B) the lateral-sequence, lateral-couplets pattern, with a diagonality slightly exceeding zero. Baboon A is traced from a photograph by Muybridge (1887, Pl. 143). Baboon B is an artificial construct that has its diagonality lowered by 50—that is, through a 180° phase shift.

Although the symmetry required by Tomita's analysis does not exist, his insights have influenced subsequent thinking about primate locomotion. From his computer simulations of quadrupedal gaits, Yamazaki (1976) reportedly concluded that DS gaits reduce roll in walking—if and only if the hindlimbs bear most of the body weight (Kimura et al., 1979). Unfortunately, Yamazaki's dissertation research has never appeared in print. Kimura and his coworkers, who used force-plates to measure reaction forces on the fore- and hindlimbs of primates and dogs (Kimura et al., 1979), found that vertical forces in primates were greater on the hindlimb than on the forelimb, but that the reverse was the case in dogs. Similar differences have been found in other force-plate studies (Demes et al., 1994; Kimura, 1985, 1992; Reynolds,

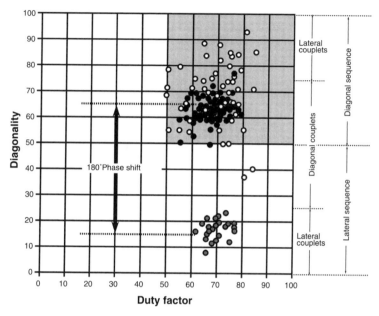

Figure 6. Walking gaits of early and later juvenile *Macaca fuscata*, displayed on the Hildebrand diagram. During ontogeny, an originally wide scatter (white circles) narrows to a more coordinated focus (black circles) in the D-S sector (light gray square) and adds a satellite cluster separated from the main cluster by a phase shift of 180°. This secondary lateral-couplets, lateral-sequence cluster (gray circles) involves a 180° phase shift of the sort illustrated in Figure 5, which represents an alternative but suboptimal way of balancing on a protracted hindfoot at the moment of forefoot touchdown. (Data from Nakano, 1996)

1985; Schmitt and Lemelin, 2002; Lemelin and Schmitt, this volume). On the basis of these facts, Kimura et al., (1979) famously characterized primates as "front steering–rear driving" animals, and conjectured that this accounts for the prevalence of DS gaits among primates. These Japanese studies were summarized and elaborated upon by Rollinson and Martin (1981: 388), who concluded that "the typical diagonal walk sequence found in monkeys is a reflection of the fact that the center of gravity is located further back in the body than in nonprimate mammals."

All these analyses were rebutted by Vilensky and Larson (1989), who offered a radical new approach to the study of primate gaits. Dismissing all the analyses that had seen DS gaits as reflecting some sort of dominance of the hindlimbs in primate locomotion, Vilensky and Larson argued that there is no evidence that primates have a more posterior center of mass than other mammals. While granting that primates adopt postures that actively shift the line of gravity tailward (as demonstrated by the force-plate data), Vilensky and Larson found no evidence that the percentage of weight supported on the hindlimbs is correlated with the frequency of adoption of DS gaits. More fundamentally, they pointed out that many individual lemurs and monkeys occasionally or habitually use LS walks, and questioned whether the use of DS gaits has any adaptive significance at all. "Our hypothesis," they wrote, "is that the choice of which symmetrical gait a particular animal uses is to a large extent arbitrary, at least in the sense that stability is not a factor" (Vilensky and Larson, 1989, p. 28).

Vilensky and Larson suggested that the important difference between primates and other mammals lies in a neurological reorganization that has brought primate locomotion more directly under cerebral control and given primates greater behavioral flexibility in the selection of their gaits. Vilensky and Larson conjectured that the high frequency of DS walking gaits in primates is somehow related to increasing specialization of the forelimb as an organ of manipulation, with major forelimb muscles ceasing to play an active role in propulsion. In their view, correlated neurological changes, from a pattern of "contralateral" to "ipsilateral forehind coordination ... have resulted in a preference for DS gait use in primates. However, the locomotor control system is quite flexible, and slight biases in one complex of neural connections or another result in either DS or LS gaits" (Vilensky and Larson, 1989, 29, 32).

Vilensky and Larson's analysis has had a profound influence on thinking about DS gaits. However, there are three problems with their interpretation. First, no details of the hypothetical neurological mechanisms underlying the preference for DS gaits are provided, so that the proposed explanation in terms of "ipsilateral fore-hind coordination" is really only a different way of saying that primates prefer DS gaits. Second, while the presence of LS gaits in primates attests to their ability to use both gait modes, that ability does not alter the fact that almost all primates preferentially and predominantly use DS walking gaits (Cartmill et al., 2002; Hildebrand, 1967; Rollinson and Martin, 1981), as Vilensky and Larson themselves recognize.

Third and most importantly, DS gaits are also characteristic of arboreal marsupials (Goldfinch and Molnar, 1978; Hildebrand, 1976; Lemelin, 1996; Lemelin and Schmitt, this volume; Lemelin et al., 1999, 2002, 2003; Pridmore, 1994; Schmitt and Lemelin, 2002; White, 1990). Most of these animals are relatively primitive neurologically and poorly encephalized. *Didelphis*, which uses both LS and DS walking gaits (Hildebrand, 1976; White, 1990), retains standard mammalian propulsive functions of major forelimb muscles (Jenkins and Weijs, 1979). The neurological transformation that Vilensky and Larson posit to explain the primate preference for DS gaits is correspondingly unlikely to apply to opossums and phalangers.

Since arboreal marsupials show many detailed resemblances to primates in the functional morphology of their hands and feet (Jones, 1924; Lemelin, 1996, 1999), and have frequently been proposed as ecological and behavioral models for the ancestral primates (Cartmill 1974a, b, 1992; Charles-Dominique, 1983; Henneberg et al., 1998; Lemelin, 1999; Rasmussen, 1990), it seems reasonable to suspect that the DS walking gaits characteristic of both marsupials and primates have a direct adaptive significance, and are not mere epiphenomena of neurological changes having little to do with arboreal locomotion. Some of Vilensky's work reaches similar conclusions (Vilensky and Moore, 1992; Vilensky et al., 1994).

In what follows, we propose a new theory of the adaptive value of DS gaits. This theory, which incorporates insights from the work of Gray, Hildebrand, Tomita, Martin, Vilensky, and others, explains why arboreal marsupials and primates resemble each other in locomotor behavior and differ from typical mammalian quadrupeds.

DS WALKING AND ARBOREAL LOCOMOTION

When the hindfoot strikes down in a typical DS gait, it lands just behind the forefoot on the same side (Figure 3B, no. 6)—which is why the unilateral bipod of support is so small (Figure 3B, no. 7). In a tree that hindlimb will be landing on the same support as the forefoot. The animal has already put weight on its forefoot on this support, and so it knows that the support is safe. If the next foot to descend—namely, the diagonally opposite forefoot (Figure 3B, no. 8)—lands on an insecure support that breaks or shifts, the animal can keep from falling by grasping the proved support with its hindfoot.

This strategy is profitable only for animals that have prehensile specializations of the hindfoot. Most arboreal mammals lack such specializations, and move around mainly on major branches and trunks—which (like the ground) are likely to be just as stable for the footfall ahead as they were for the one behind (Cartmill, 1970, 1972). But primates do much of their foraging out

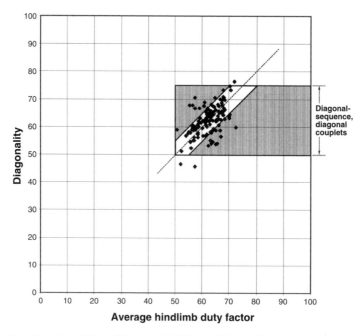

Figure 7. Our data (Cartmill et al., 2002) for 130 walking gaits (hind duty factor >50) for 17 genera of primates (*Microcebus, Mirza, Lemur, Eulemur, Varecia, Hapalemur, Daubentonia, Otolemur, Nycticebus, Loris, Perodicticus, Cebus, Ateles, Erythrocebus, Papio, Macaca, Pan*), plotted on the Hildebrand diagram. The sloping dashed line represents the equation (diagonality = hindlimb duty factor) that a diagonal-couplets walk in DS (boxed gray area: 50 < diagonality < 75) must obey in order to minimize the duration of bipedal support phases in general and of unilateral bipedality in particular (Cartmill et al., 2002). In the gaits that fall above this line, total bipedality is minimal but some percentage of the cycle is spent standing on the unilateral bipod (Figure 3B, no. 7); in those that fall below the line, unilateral bipedality is zero but total bipedality is not minimized. Eighty percent of the primate data lie within an envelope (diagonal white band) of ± 5% deviation from this optimal line.

in terminal branches and twigs, where the next footfall may prove treacherous. In such a situation, it is useful to establish a safe anchor on the last foothold before trusting any weight to the next one. The DS footfall sequences of primates, in which the hindfoot in the diagonal-couplets pair strikes down before the contralateral forefoot in time and just behind the already-planted ipsilateral hindfoot in space (Figure 3B, nos. 6–8), work in combination with their anatomically specialized hindfeet to accomplish that aim.

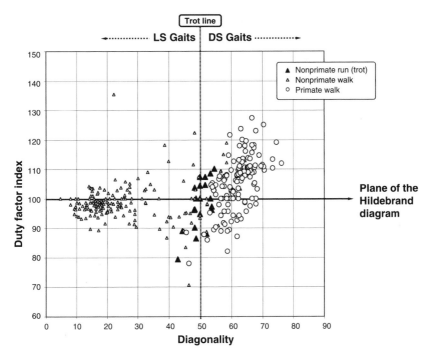

Figure 8. Diagonality plotted against the duty factor index (100 x hind/fore duty factor) for nonprimate gaits and for walking gaits of primates other than *Callithrix* (Cartmill et al., 2002). Nonprimate running trots (black triangles) and some of the fast trotlike LS walks deviate from the plane of the Hildebrand diagram (y = 100) because the stance phases are short in these fast gaits, and so differences of a few frames in the length of one stance phase produce large fluctuations in the index. By contrast, primate walks (open circles) deviate from the Hildebrand plane even at low speeds. If primate DS walks were simply walking trots with enhanced hindlimb duty-factors, their scatter would intersect the plane of the Hildebrand diagram (heavy horizontal line) where diagonality equals 50 (vertical dashed line). Instead, they lie mostly to the right of the scatter of nonprimate trots, even below the plane of the Hildebrand diagram. This distribution shows that the high diagonality of primate walks is not due solely to increase in hindlimb duty factors.

Diagonal-sequence walks are not the only walking gaits in which a hind-foot is already on the support when the forefoot strikes down. *All* quadrupedal walking gaits meet that description, because there is always at least one hindfoot on the ground at every instant in a symmetrical walk cycle. But differences in diagonality affect the placement of the supporting hindfoot at the moment of forefoot touchdown. Figure 4 illustrates limb deployment at that point in the cycle in a primate DS walk and a typical nonprimate LS walk. In the DS walk of the primate, the supporting hindfoot is *protracted* at the moment of forefoot touchdown, placing it beneath or slightly in front of the line of gravity. In the LS walk of the nonprimate, with a diagonality around 25, the supporting hindfoot is stationed far posterior to the line of gravity when the next forefoot comes down. Protraction of the hindfoot when the forefoot comes down is advantageous for primates and other tree-dwelling animals with grasping hindfeet. If the forefoot of such an animal comes down on an insecure support, the animal is still roughly balanced on its grasping hindfoot, and can pull back or right itself more easily. A primate that walked like a horse could not do this.

A tendency to pitch backward at the moment of forefoot touchdown (Figure 3B, no. 8) can be seen in this light as adaptively advantageous. This tendency fits into the generally hindlimb-based pattern of quadrupedal primate locomotion. Unlike most nonprimates, primates in general sustain substantially higher substrate reaction forces on the hindlimb than on the forelimb (Demes et al., 1994; Kimura et al., 1979; Reynolds, 1985; Schmitt and Lemelin, 2002; Lemelin and Schmitt, this volume). This force distribution pattern, in combination with the tendency to neutral or backward pitch at the moment of forelimb touchdown, allows primates to test a support more gently with the forelimb before loading it more vigorously with the hindlimb.

In principle, a primate could balance equally well on a protracted hindlimb at the moment of forefoot touchdown in a lateral-sequence walk if it inverted the phase relationships between its fore and hind limbs. This would lower its diagonality by 50 (that is, through a 180° phase shift), yielding a lateral-couplets, lateral-sequence walk (Figure 5). The supporting hindlimb would be equally protracted in both cases. Such a shift is in fact seen occasionally in some primates—for example, among macaques (Figure 6), where a lateral-sequence walk in lateral-couplets appears as a transitory variant during ontogeny (Hildebrand, 1967; Nakano, 1996). The appearance of this shift corroborates the thesis that primate walking gaits are adaptive because they optimize balance on the protracted hindfoot at the moment of forefoot touchdown.

However, a lateral-couplets walk, in which the two limbs on the same side tend to move as a pair, is in general not desirable for a tree-dwelling animal, because the animal winds up spending most of the cycle standing on two feet on the same side. This unilateral sort of bipedality tends to produce unwanted rolling moments. Camels and some other long-legged mammals prefer lateral-couplets walking and running gaits—perhaps, as Hildebrand (1968) argued, in order to avoid stepping on their forefeet with their hindfeet in running and fast walking, when stride length increases. But such gaits are avoided by most arboreal mammals. For example, *Procyon* and *Nasua* use lateral-couplets gaits; but their monkey-like arboreal relative *Potos* uses strictly diagonal couplets (McClearn, 1992). We infer that the optimal way for a walking primate to balance itself on a securely planted hindlimb when the forefoot comes down is to use diagonal-couplets gaits in diagonal sequence (Figure 5A).

In selecting a walking gait on an arboreal support, a quadrupedal primate faces a complex trade-off between maximizing hindlimb protraction, maximizing the duration of tripedal or quadrupedal support phases (for enhanced stability), and giving the hindfoot enough time to secure a grip on the support before the contralateral fore footfall. As shown above, it is desirable to maximize hindlimb protraction at the moment of forelimb touchdown.[1] Such protraction is maximal if the hind and fore limbs touch down simultaneously—that is, in a walking trot, with a diagonality of 50, or a walking pace, with a diagonality of zero. However, simultaneous diagonal footfalls do not allow the grasping hindfoot any time to establish a grip on the support before the forefoot touches down. (We will refer to this difficulty in a later section as the "grasp-interval problem.") DS gaits approaching the walking trot, with diagonalities between 51 and 59, are seen in many primates at relatively fast walking speeds. The drawback to such low-diagonality DS gaits is that they maximize the percentage of the cycle that the animal spends balancing on only two feet (Cartmill et al., 2002). This may not be a serious disadvantage

---

[1] Our model suggests that primates should maximize hindlimb protraction at the moment of contralateral forefoot touchdown. We might therefore expect primates to exhibit greater maximum overall protraction of the hindlimb (i.e., at the moment of hindfoot touchdown) than typical nonprimates do. This prediction is concordant with the findings of Larson et al. (2001), who report that *angular* protraction of the hindlimb (the angle between the vertical and a line drawn from hip to ankle at the moment of hind footfall) is greater in primates than in other mammals, but is less than seen in marsupials. However, the relevant variable in terms of our theory is not angular protraction, but *linear* protraction—that is, the position of the hindfoot at touchdown relative to the length of the body axis. Since primates have longer limbs and greater stride lengths relative to trunk length than most nonprimate mammals (Alexander and Maloiy, 1984; Alexander et al., 1979; Larson et al., 1999, 2001; Reynolds, 1987), linear protraction of the hindfoot at touchdown should be correspondingly greater in primates than it is in other mammals with a like amount of angular protraction. This expectation remains to be tested.

in fast walking, but it must exacerbate problems of balance for an animal moving more slowly and cautiously (with higher duty factors). Many primates accordingly show a positive correlation between diagonality and duty factor in their walking gaits: the higher the duty factor (and thus the lower the speed), the more the animal deviates from a walking trot (and thus exhibits increased diagonality). In our data, such a correlation is evident for walking gaits of primates other than *Callithrix* (Spearman's ρ = 0.454, p < 0.001). This pattern of covariation between duty factor and diagonality also means that primate walking gaits cluster around the theoretical line on the Hildebrand diagram where both total bipedality (the percentage of the walking cycle spent standing on only two feet) and unilateral bipedality are kept to a minimum (Figure 7; Cartmill et al., 2002).

DUTY-FACTOR RATIOS AND DIAGONALITY

As noted earlier, most primates adopt locomotor postures and gaits that concentrate vertical reaction forces on the hindlimbs. primates also generally have higher duty-factors for their hindlimbs than they do for their forelimbs—that is, their hindlimbs are in contact with the support for a larger percentage of the cycle than their forelimbs are. We believe that this difference in fore- and hindlimb duty-factors has a great deal to do with why primate walks are DS.

In Figure 8, we have plotted the duty factor index (hindlimb duty factor as a percentage of forelimb duty factor) against diagonality for 330 locomotor cycles that we have recorded for a wide variety of mammals, comprising 17 genera of noncallitrichid primates and 21 genera belonging to six other therian orders (Cartmill et al., 2002). The horizontal line along which this index equals 100 represents the plane of the Hildebrand diagram (seen edge-on). The walking gaits of most nonprimate mammals fall quite close to the Hildebrand plane, conforming to Hildebrand's (1966, 1968) observation that fore- and hindlimb duty factors are nearly equal as a rule in symmetrical gaits. However, most primates have hindlimb duty factors that are significantly larger than those for their forelimbs. Moreover, there is a significant positive correlation among primates between the duty factor index and diagonality (ρ = 0.614, p < 0.001). This correlation has a simple mathematical basis. To lengthen the hindfoot's contact period relative to that of the forefoot (and thus increase the value of the duty factor index), the hindfoot has to come down earlier, or be picked up later, or both. If it comes down earlier, diagonality increases by definition. Increasing the duty factor index in this way has the additional advantage of solving the grasp–interval problem inherent in the walking trot: by swinging the hindfoot further forward (and thereby advancing the moment of its touchdown) in a gait that is otherwise similar to a walking trot, a primate can gain enough hindlimb contact time for the foot to establish a secure grip on the support and still be in a trot-like protracted position when the hand comes down on an untested branch.

Does an elevated duty factor index by itself explain why primates have DS walks? To put it another way: are the DS gaits of walking primates simply walking trots in which the hindfoot swings farther forward and thus has an enhanced contact time? If primate DS gaits are merely walking trots with enhanced hindlimb protraction and contact times, then diagonality should be 50 (a walking trot) when hind and fore duty factors are equal. However, all our primate data for which hind- and forelimb duty factors are equal have diagonality values greater than 50 (Figure 8). The scatter of primate walking gaits in Figure 8 lies almost entirely to the right (increased diagonality) of the scatter of nonprimate trots for all values of the duty factor index (100 × hind/fore duty factor). This rightward shift indicates that primate DS walks are not simply walking trots with a prolonged hindlimb stance phase, but rather involve a slight additional shift in the phase relationships of the fore- and hindlimb movement cycles. It is nevertheless clear that the DS walking gaits of primates are in large measure a consequence of an elevated hind- to forelimb duty factor ratio. Just as importantly, the lateral-sequence gaits that occur in some primates are part and parcel of the same pattern, resulting from a reduction of hindlimb contact time relative to that of the forelimb.

DS GAITS: FINE BRANCHES VERSUS FLAT SURFACES

The inferiority of DS gaits lies chiefly in their narrow support polygons. But as Prost noted in 1969, this disadvantage disappears on a narrow support. As footfalls become collinear, the areas of all support polygons shrink and converge on the contact surfaces of the feet, no matter whether the gaits are DS or LS. On a narrow support like a branch or a pole, stability depends mainly on having the line of gravity fall somewhere between the contact point in front and the one in back. A DS gait can meet that criterion well enough if the support is narrow (Figure 3B).

But even if we accept that DS gaits lose most of their disadvantages on sufficiently narrow supports, it remains puzzling that even such highly terrestrial primates as chimpanzees and baboons retain DS gaits on the ground, rather than reverting to the LS walking gaits preferred by other terrestrial mammals. Hildebrand (1967) reported that many primates—lemurs, ceboids, cercopithecoids, and pongids—walking on the ground typically walk somewhat crabwise, especially in faster walks, with the trunk turned at an angle to the direction of travel and both hindlimbs overstriding the ipsilateral forelimbs on the side of the body that the trunk is turned away from. (This overstriding is evident in Figure 4.) Larson and Stern (1987) conjectured that chimpanzees do this to mitigate the risk, common to all diagonal-couplets gaits, of having the hindlimb hit the forelimb on the same side when stride length increases with speed. We conjecture further that this crabwise walking may also address a fundamental shortcoming of DS gaits, by increasing the sizes of the walking animal's support polygons. If the body is turned to one side while the limbs continue to swing back and forth in the direction of movement and stride length is increased, then the contact points of the ipsilateral feet (Figure 3B, no. 7, "unilateral bipod") will move apart from each other, and the areas of the triangles of support will be correspondingly increased (Figure 9). The size of the increase will depend on body proportions, the precise placement of the feet, and the magnitude of the angle between the body axis and the direction of movement (angle α in Figure 9). The extent to which primates walking on the ground actually do enlarge their support polygons in this way remains to be determined experimentally; but in principle, unilateral overstriding would be expected to improve the stability of DS walking gaits on flat surfaces.

LOCOMOTION AND THE ANCESTRAL PRIMATES

Our data give new meaning to the old idea that primate origins involved a differentiation of the functions of the hindlimb and forelimb. This idea dates back almost a century to the work of F. Wood Jones (1916), who wrote about "the emancipation of the forelimb" as a characteristic primate trend, through which the hindlimb becomes the main supportive organ and the forelimb is supposedly freed to reach out, test new supports, and handle things. Whereas Jones and others have seen trends toward the emancipation of the forelimb as a defining primate feature, others have stressed the converse proposition, that

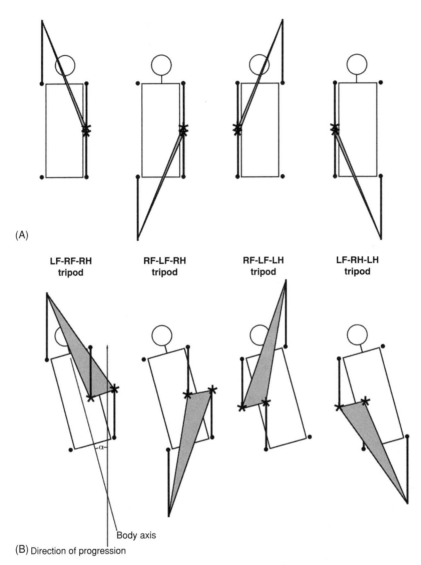

Figure 9. Diagonal-sequence walks on a flat surface: schematic, viewed from above. (A) body axis collinear with direction of progression; (B) body rotated to one side through an angle α (without altering direction of progression or plane of limb movement), and stride length increased slightly, producing unilateral overstriding of the forefeet by the hindfeet. Depending on the size of α and the precise placement of the feet, such overstriding can augment the meager support polygons (hachure) of DS walks by moving the placement points of the two ipsilateral feet (asterisks: "unilateral bipod," Figure 3) further apart during tripedal support phases.

the hindlimb must have had a particular and special importance in the loco-motor behavior of primitive primates. Harking back to Jones in their initial formulations of their theories about vertical clinging and leaping, Napier and Walker coined the phrase "hindlimb-dominated" to describe ancestral primate locomotion (Napier, 1967; Napier and Walker, 1967; Walker, 1967). This phrase has been in use ever since, but its meaning has been shifting. In 1972, Martin gave the phrase a new meaning when he used it to characterize the locomotor behavior of *Microcebus* as a model ancestral primate. Originally, Martin seems to have been thinking about the use of the hindlimb in leaping—among horizontal as well as vertical branches—and in supporting the body in cantilevered postures when reaching for food items. The term "graspleaping," used by Szalay and his coworkers (Szalay and Dagosto, 1988; Szalay and Delson, 1979; Szalay and Sargis, 2001; Szalay et al., 1987), branches off the semantic tree somewhere near this node. In describing primate locomotion as "rear-driven," Kimura et al. (1979) concluded that primates differ from other mammals in using the hindlimb as the main propulsive organ. They suggested that this has somehow resulted in the evo-lution of DS walking gaits, and they followed Wood Jones in relating all this to orthograde posture and the freeing of the forelimb for manipulation. Rollinson and Martin built on Kimura's work in arguing that primates have a posteriorly shifted center of mass, and Martin (1990) subsequently incorpo-rated this conclusion into his earlier characterization of primate locomotion as "hindlimb-dominated."

We think that all these theories are correct in stressing the distinctive importance of the hindlimb in primate locomotion. The grasping modifica-tions of the primate hindfoot provide additional safety in the trees by offer-ing a secure anchor on a support known to be reliable. Primates modify their locomotor behavior to take advantage of this by shifting weight toward the hindlimb and by increasing hindlimb duty factors relative to those of the fore-limb in walking. This increase in hindlimb duty factors enhances the diago-nality of primate walking gaits, so that the body is more or less balanced over the protracted hindlimb when the forelimb comes down on an untested support. These points are demonstrated facts. Our interpretation of these facts, which is subject to debate, is that these primate traits make sense as adaptations to an arboreal locomotor pattern that is more tentative—less headlong—than those of typical terrestrial quadrupeds or most arboreal nonprimates. We suggest that this is part of a set of basal ordinal adaptations for moving and foraging on fine branches.

As noted earlier, DS walking gaits lose many of their comparative disadvantages as support width approaches zero. They seem correspondingly likely to have evolved in animals that moved on supports that were thin relative to the dimensions of the animals' body—in other words, in a fine-branch milieu. But such a milieu implies no intrinsic *advantage* for DS over LS walking. The only such advantages we have been able to identify involve a reliance on grasping hindfeet for security in the trees. Such security is enhanced by increasing hindlimb duty-factors. Increases in hindlimb duty-factors relative to those of the forelimb also automatically increase diagonality. As soon as diagonality surpasses 50, the animal enters the part of the Hildebrand space where it can balance on a protracted hindlimb before putting any weight on an untested forelimb support. This is advantageous for arboreal animals with grasping hindlimbs that are able to grab the support behind if the one ahead fails.

The tempo of the animal's gait is also relevant to the utility of the grasping hindlimb in quadrupedal locomotion. The security of the hindlimb support point is of little consequence to an animal that bounds along branches like a squirrel or a tree shrew, using mainly asymmetrical gaits and moving too rapidly to be able to stop if the branch ahead bends or breaks. It is accordingly not surprising that the infrequent symmetrical arboreal walking gaits of *Tupaia* are diagonal-couplets walks in lateral sequence (Hildebrand, 1967; Jenkins, 1974). To enter the adaptive zone where DS gaits make a difference, an animal has to have grasping specializations of the hindfeet and relatively deliberate locomotor habits. A tendency of the body to pitch backward at the moment of forefoot touchdown may be advantageous in this context, and can best be achieved by using a diagonal-couplets, diagonal-sequence gait.

We conclude that the general preference of primates for DS gaits reflects an ancestral adaptation to careful or controlled locomotion in a fine-branch setting, where the characteristic primate specializations of the hindfoot function to enhance security and stability. There are two kinds of natural experiments that we might look to test this conclusion. First, we can look at walking patterns in primates that have reduced the grasping specializations of the hindfoot and reevolved a claw-based locomotion on large supports. We might expect such animals to revert to LS walking gaits. Recent studies of *Callithrix jacchus* support this contention (Cartmill et al., 2002; Schmitt, 2002, 2003). These marmosets are small-bodied primates with reduced halluces and claws on their other digits. They have the sharpest claws of any callitrichid (Hamrick, 1998), and spend more time clinging to and feeding on large, vertical supports than other callitrichids do (Garber, 1992). Studies of three male *C. jacchus* show that their infrequent walks are exclusively lateral-sequence gaits, plotting with nonprimate walks on the Hildebrand diagram (Cartmill et al., 2002; Schmitt, 2002, 2003). Other callitrichids—*Leontopithecus rosalia, Callimico goeldii,* and *Saguinus midas*—exhibit DS walking gaits (Rosenberger and Stafford, 1994; Schmitt, 2002, 2003). These facts support the association we posit between DS gaits, reliance on pedal grasp, and a preference for narrow supports.

The second test we can run involves looking at nonprimate mammals that resemble the hypothetical ancestral primate in the respects specified by the theory. If they have not moved in evolutionary directions parallel to those of primates, then the theory is defective or incomplete. For this second comparison, the appropriate animals to look at are small marsupials like *Caluromys.* As noted by Lemelin and Schmitt elsewhere in this volume (Chapter 10), woolly opossums have grasping hindfeet similar in proportions and grasping abilities to those of prosimians, and move around in fine branches foraging for fruit and insects. We might therefore predict that woolly opossums too would have DS walking gaits, especially when walking on thin poles. Our data confirm this prediction (Figure 10). By contrast, the terrestrial didelphid *Monodelphis* exhibits mainly LS walking gaits. Like *Callithrix, Monodelphis* has probably reverted to LS gaits secondarily as a result of giving up an arboreal habitus, in which DS gaits have an adaptive advantage.

The interspecific scatter of didelphid walks plotted in Figure 10 passes roughly through the intersection of the trot line (D = 50) with the Hildebrand plane, and does not evince the shift to the right seen in the primate scatter (Figure 8). This is probably not a real difference between primates and didelphids. A more detailed examination of our intraspecific data for *Caluromys* shows that its walking gaits on a 7-mm pole are significantly right-shifted compared to its walking gaits on a flat surface—that is, they have higher diagonalities for a given value of the duty factor index (Lemelin et al., 2003). It remains to be determined whether a similar rightward shift can be detected in gait data for primates walking on arboreal versus terrestrial supports.

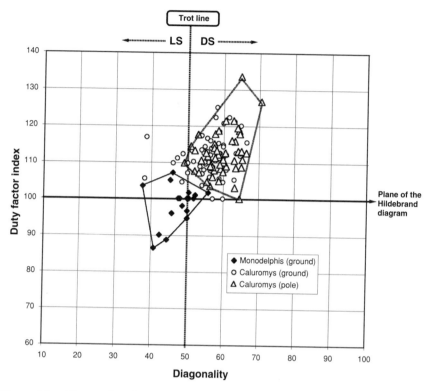

Figure 10. Diagonality plotted against the duty factor index (100 x hind/fore duty-factor) for walking gaits of the didelphid marsupials *Caluromys philander* (open symbols) and *Monodelphis brevicaudata* (black diamonds). Minimal convex polygons are shown for the *Monodelphis* scatter (solid polygon) and for walking gaits of *Caluromys* on a 7-mm-diameter pole (dotted polygon). Data from Lemelin et al. (2003). (cf. Figure 8).

CONCLUSIONS

Our data, analyses, and interpretations support the thesis that the ancestral primates were small animals that moved and foraged "in the dense tangle of small twigs and vines, which characterizes the canopy and forest margins" (Cartmill, 1970: 328)—what Charles-Dominique and Martin (1970) independently identified as "the fine branch and creeper niche." This idea has been accepted by some authors (Allman, 2000; Hamrick, 1998; Larson, 1998; Lemelin, 1999; Rasmussen, 1990; Rollinson and Martin, 1981; Schmitt, 1999; Schmitt

and Lemelin, 2002), but it continues to provoke controversy and misunderstandings. Some of those misunderstandings deserve to be laid to rest. For
example, the proposition that ancestral primates were fine-branch foragers
implies nothing about their diets. That proposition is equally compatible with
the thesis that the primitive primates were predominantly herbivorous
(Sussman, 1991; Szalay, 1972) or with the "visual-predation theory" proposed
by Cartmill (1970, 1972, 1974a,b, 1992). The "fine-branch theory" and the
"visual-predation theory" are not competing alternatives (*contra* Fleagle,
1999: 346). Though the second was originally proposed in conjunction with
the first (Cartmill, 1970), the two are logically independent. The first is a theory about primate locomotor adaptations, whereas the second offers to explain
the primate synapomorphies of the visual system. It is possible that the ancestral (eu)primates originally adapted to the fine-branch milieu to exploit insect
resources. But it is also possible that they developed their distinctive locomotor synapomorphies as adaptations for feeding on plant tissues, and later developed specializations of the visual system to allow them to more effectively stalk
the insects they encountered while foraging among slender terminal branches
(Cartmill, 1975; Rasmussen, 1990).

A similar mosaic sequence of evolutionary changes may account for some
of the apparent conflict between our conclusions and those of Szalay and his
collaborators. They argue that "... rapid, successive, leaping, and landing with
a habitual grasp (i.e., graspleaping)" (Szalay and Dagosto, 1988: 27), represents the ancestral primate locomotor habit, and that the morphotypic peculiarities of the primate locomotor and neural apparatus originated as
adaptations to this sort of fast, jumpy arboreal locomotion (Szalay and
Dagosto, 1988; Szalay and Delson, 1979; Szalay and Sargis, 2001; Szalay
et al., 1987). Szalay and Sargis (2001: 299) accordingly dismiss the relevance
of marsupial analogies in understanding the adaptive significance of primate
traits. "Didelphid arborealists," they aver, "are drastically unlike a variety of
explosive arboreal graspleapers encountered among, and probably in the very
ancestry of, the Euprimates."

These conclusions are hard to reconcile with the manifest and unique
resemblances that we and others have documented between primitive primates and small, fine-branch-haunting marsupials such as *Caluromys*
(Lemelin and Schmitt, this volume). If marsupial arboreality is drastically
unlike that of the ancestral primates, and if the ancestral primate locomotor
repertoire consisted largely of jumping and other fast, irregular, asymmetric

forms of locomotion, then it is difficult to understand why small arboreal marsupials and primates should share a complex of behaviors—diagonal-sequence walking gaits, increased hindfoot duty-factors, and peak vertical forces on the hindlimb that are higher than those on the forelimb—that have not been found as a complex in any other mammals and appear to be functionally related to cautious or deliberate arboreal locomotor habits. If ancestral primates were habitual "explosive grasp-leapers," then why should primates have a distinctive symmetrical slow gait?

It might be argued that increased hindlimb loading in primates appeared originally in connection with an ancestral habit of vertical clinging and leaping, and that DS gaits evolved in various descendants that reverted to quadrupedalism. But the manifest morphological and behavioral convergences between primates and less saltatory arboreal marsupials (Lemelin and Schmitt, this volume; Lemelin et al., 2003; Schmitt and Lemelin, 2002) seem to argue for the opposite conclusion—that ancestral primates evolved these marsupial convergences as adaptations to foraging habits resembling those of *Caluromys*, and that specializations for leaping evolved secondarily in later phases of primate evolution.

It is of course possible that the *Caluromys*-like phase in primate phylogeny predated the last common ancestor of the Euprimates. Bloch and Boyer (2002; Chapter 16, this volume) have suggested that grasping feet are a synapomorphy of a clade that includes *Carpolestes* and Euprimates, but excludes *Plesiadapis* and other "archaic primates" that lacked grasping feet. If this were so, then DS gaits might have been acquired along with grasping feet during a pre-euprimate phase of primate evolution. This account could in principle reconcile our ideas about DS gaits with the reconstruction of the ancestral primates as "graspleapers" (Szalay and Dagosto, 1988; Szalay and Sargis, 2001; Szalay et al., 1987). Unfortunately, the phylogeny of Bloch and Boyer implies that the highly derived dental features that carpolestids share uniquely with other so-called "archaic primates" are convergences (Kirk et al., 2003). We think that it is more likely that grasping extremities arose independently in carpolestids and euprimates. This conclusion is still compatible with the proposition that leaping became an important locomotor mode for many lineages in later phases of primate evolution, or even in the last common ancestor of the living primates. However, the marsupial parallels for the peculiar features of primate gaits favor the conclusion that nonsaltatory, *Caluromys*-like locomotor adaptations were established earlier in the euprimate ancestry.

In some respects, this conclusion is reminiscent of the ideas of Böker (1926, 1932), who thought that the ancestral mammals had been cautious arboreal "clamp-climbers" (*Klammerkletterer*) with lemur-like grasping hands and feet, adapted to a slow "branch walking" locomotor habit (*Schreiten auf den Ästen*), from which their descendants had departed in various ways to increase speed. This is almost certainly not true of mammals as a whole, but something similar may be true of primates. Although the last common ancestors of the living primates were surely not restricted to exaggeratedly slow-motion movements like those of a loris and could no doubt leap or bound when they needed to, they must have been equally well adapted to more deliberate, cautious, or stealthy locomotion when circumstances called for it—for example, in searching for food items in the terminal branches of tropical forests.

ACKNOWLEDGMENTS

We thank Kaye Brown and Peter Klopfer for helping us handle and videotape animals, Chris Vinyard for advice on data analysis, Roshna Wunderlich for loaning us videotapes of chimpanzee locomotion, Kathleen Smith for supplying *Monodelphis* specimens, and Martine Atramentowicz and the Laboratoire d'Ecologie Générale in Brunoy, France (URA 8571 CNRS/MNHN) for *Caluromys* specimens. We are grateful for all the assistance we received from the staffs of the animal facilities where we collected video records. We owe a special debt of thanks to Sitha Bigger (Wake Forest, NC, ZooFauna Zoo), to Lori Widener and Alison Larios (Carnivore Preservation Trust, Pittsboro, NC), and to Ken Glander, David Haring, and Bill Hess (Duke University primate Center). We thank Matt Ravosa and Marian Dagosto for organizing and hosting the 2001 "primate Origins" conference, for inviting us to participate in it, and for bringing it off successfully in the face of grave, unforeseen difficulties. This research was funded by grants SBR-9209004, SBR-9318750, BCS-9904401, and BCS-0137930 from the U. S. National Science Foundation.

REFERENCES

Alexander, R. McN., Jayes, A. S., Maloiy, G. M. O., and Wathura, E. M., 1979, Allometry of the limb bones of mammals from shrews (*Sorex*) to elephant (*Loxodonta*). *J. Zool. Lond.* **189**: 305–314.

Alexander, R. McN., and Maloiy, G. M. O., 1984, Stride lengths and stride frequencies of primates. *J. Zool. Lond.* **202:** 577–582.

Allman, J., 2000, *Evolving Brains*, New York, Scientific American Library.

Bloch, J. L., and Boyer, D. M., 2002, Grasping primate origins, *Science* **298:** 1606–1610.

Böker, H., 1926, Die Entstehung der Wirbeltiertypen und der Ursprung der Extremitäten, *Zeitsch. Morphol. Anthropol.* **26:** 1–58.

Böker, H., 1932, Beobachten und Untersuchungen an Säugetieren während einer biologisch-anatomischen Forschungsreise nach Brasilien im Jahre 1928, *Geg. Morph. Jb.* **70:** 1–66.

Carlson, H., Halbertsma, J., and Zomlefer, M., 1979, Control of the trunk during walking in the cat, *Acta Physiol. Scand.* **105:** 251–253.

Cartmill, M., 1970, *The Orbits of Arboreal Mammals: A Reassessment of the Arboreal Theory of Primate Evolution*, Ph.D. Dissertation, Chicago, University of Chicago.

Cartmill, M., 1972, Arboreal adaptations and the origin of the order Primates, in: *The Functional and Evolutionary Biology of Primates,* R. H. Tuttle, ed., Aldine-Atheton, Chicago, pp. 3–35.

Cartmill, M., 1974a, Rethinking primate origins, *Science* **184:** 436–443.

Cartmill, M., 1974b, Pads and claws in arboreal locomotion, in: *Primate Locomotion,* F. A. Jenkins, Jr., ed., Academic Press, New York, pp. 45–83.

Cartmill, M., 1975, *Primate Origins*, Burgess Publishing Co, Minneapolis.

Cartmill, M., 1992, New views on primate origins, *Evol. Anthropol.* **1:** 105–111.

Cartmill, M., Lemelin, P., and Schmitt, D., 2002, Support polygons and symmetrical gaits in mammals, *Zool. J . Linn. Soc.* **136:** 401–420.

Charles-Dominique, P., 1983, Ecological and social adaptations in didelphid marsupials: Comparison with eutherians of similar ecology, in: *Advances in the Study of Mammalian Behavior,* J. F. Eisenberg and D. G. Kleiman, eds.,. Special Publication No. 7. Shippensburg, PA, American Society of Mammalogists, pp. 395–422.

Charles-Dominique, P., and Martin, R. D., 1970, Evolution of lorises and lemurs, *Nature* **227:** 257–260.

Demes, B., Jungers, W. L., and Nieschalk, U., 1990, Size- and speed-related aspects of quadrupedal walking in slender and slow lorises, in: *Gravity, Posture, and Locomotion in Primates,* F. K. Jouffroy, M. H. Stack, and C. Niemitz, eds., Firenze, Editrice Il Sedicesimo, pp. 175–197.

Demes, B., Larson, S. G., Stern, J. T., Jr., Jungers, W. L., Biknevicius, A. R., and Schmitt, D., 1994, The kinetics of "hind limb" drive reconsidered, *J. Hum. Evol.* **26:** 353–374.

Dykyj, D., 1984, Locomotion of the slow loris in a designed substrate context, *Am. J. Phys. Anthropol.* **52:** 577–586.

Fleagle, J.G., 1999, *Primate Adaptation and Evolution*, 2nd edition, Academic Press, San Diego.

Garber, P. A., 1992, Vertical clinging, small body size, and the evolution of feeding adaptations in the Callitrichinae, *Am. J. Phys. Anthropol.* **88:** 469–482.

Gatesy, S. M., and Biewener, A. A., 1991, Bipedal locomotion: Effects of speed, size and limb posture in birds and humans, *J. Zool. London* **224:** 127–148.

Goldfinch, A. J., and Molnar, R. E., 1978, Gait transition in the brush-tail possum (*Trichosurus vulpecula*), *Austral. Mamm.* **4:** 59–60.

Grillner, S., 1975, Locomotion in vertebrates: Central mechanisms and reflex interaction, *Physiol. Rev.* **55:** 247–304

Hamrick, M. W., 1998, Functional and adaptive significance of primate pads and claws: Evidence from New World anthropoids, *Am. J. Phys. Anthropol.* **106:** 113–127.

Henneberg, M., Lambert, K. M., De Miguel, C., and Haynes, J., 1998, Koalas and primates: What can we learn about primate origins and adaptations by observing koalas, *Am. J. Phys. Anthropol.* (Suppl.) **26:** 92–93.

Hildebrand, M., 1965, Symmetrical gaits of horses, *Science* **150:** 701–708.

Hildebrand, M., 1966, Analysis of the symmetrical gaits of tetrapods, *Folia Biotheor.* **6:** 9–22.

Hildebrand, M., 1967, Symmetrical gaits of primates, *Am. J. Phys. Anthropol.* **26:** 119–130.

Hildebrand, M., 1968, Symmetrical gaits of dogs in relation to body build, *J. Morphol.* **124:** 353–360.

Hildebrand, M., 1976, Analysis of tetrapod gaits: General considerations and symmetrical gaits, in: *Neural Control of Locomotion*, R. M. Herman, S. Grillner, P. S. G. Stein, and D. C. Stuart, eds., Plenum Press, New York, pp. 203–236.

Hildebrand, M., 1980, The adaptive significance of tetrapod gait selection, *Amer. Zool.* **20:** 255–267.

Hildebrand, M., 1985, Walking and running, in: *Functional Vertebrate Morphology*, M. Hildebrand, D. M. Bramble, K. F. Liem, and D. B. Wake, eds., Harvard University Press, Cambridge, MA, pp. 38–57.

Jenkins, F. A., Jr., 1974, Tree shrew locomotion and the origins of p rimate arborealism, in: *Primate Locomotion*, F. A. Jenkins, Jr., ed., Academic Press, New York, pp. 85–115.

Jenkins, F. A., Jr., and Weijs, W. A., 1979, The functional anatomy of the shoulder in the Virginia opossum (*Didelphis virginiana*), *J. Zool. (London)* **188:** 379–410.

Jones, F. W., 1916, *Arboreal Man*, Edward Arnold, London.

Jones, F. W., 1924, *The Mammals of South Australia, Part 1*, British Science Guild, Adelaide.

Kimura, T., 1985, Bipedal and quadrupedal walking of primates: Comparative dynamics, in: *Primate Morphophysiology, Locomotor Analyses and Human Bipedalism*, S. Kondo, ed., University of Tokyo Press, Tokyo, pp. 81–104.

Kimura, T., 1992, Hindlimb dominance during primate high-speed locomotion, *Primates* **33**: 465–476.

Kimura, T., Okada, M., and Ishida, H., 1979, Kinesiological characteristics of primate walking: Its significance in human walking, in: *Environment, Behavior, and Morphology: Dynamic Interactions in Primates*, M. E. Morbeck, H. Preuschoft, and N. Gomberg, eds., Gustav Fischer, New York, pp. 297–311.

Kirk, E. C., Cartmill, M. Kay, R. F., and Lemelin, P., 2003, Comment on "Grasping primate Origins," *Science* **300**: 741.

Larson, S. G., 1998, Unique aspects of quadrupedal locomotion in nonhuman primates, in: *Primate Locomotion: Recent Advances*, E. Strasser, J. Fleagle, A. Rosenberger, and H. McHenry, eds., Plenum Press, New York, pp. 157–173.

Larson, S. G., Schmitt, D., and Sipe, C., 1999, Forelimb and hind limb angular excursions in primates: Which is unique? *Am. J. Phys. Anthropol.* (Suppl. 28): 179.

Larson, S. G., Schmitt, D., Lemelin, P., and Hamrick, M. K., 2001, Limb excursion during quadrupedalism: How do primates compare to other mammals? *J. Zool. London* **255**: 353–365.

Larson, S. G., and Stern, J. T., Jr., 1987, EMG of chimpanzee shoulder muscles during knuckle-walking: Problems of terrestrial locomotion in a suspensory adapted primate, *J. Zool. London* **212**: 629–655.

Lemelin, P., 1996, *The Evolution of Manual Prehensility in Primates: A Comparative Study of Prosimians and Didelphid Marsupials*, Ph.D. Thesis, State University of New York at Stony Brook, Stony Brook.

Lemelin, P., 1999 Morphological correlates of substrate use in didelphid marsupials: Implications for primate origins, *J. Zool. Lond.* **247**: 165–175.

Lemelin, P., Schmitt, D., and Cartmill, M., 1999, Gait patterns and interlimb coordination in woolly opossums: How did ancestral primates move? *Am. J. Phys. Anthropol.* (Suppl. 28): 181–182.

Lemelin, P., Schmitt, D., and Cartmill, M., 2002, The origins of diagonal-sequence walking gaits in primates: An experimental test involving two didelphid marsupials, *Am. J. Phys. Anthropol.* (Suppl. 34): 101.

Lemelin, P., Schmitt, D., and Cartmill, M., 2003, Footfall patterns and interlimb coordination in opossums (Family Didelphidae): Evidence for the evolution of diagonal-sequence gaits in primates, *J. Zool., Lond.*, **260**:423–429.

Martin, R. D., 1972, A preliminary field study of the lesser mouse lemur (*Microcebus murinus* J. F. Miller 1777), *Z. Comp. Ethol.* (Suppl. 9): 43–89.

Martin, R. D., 1990, *Primate Origins and Evolution: A Phylogenetic Reconstruction*, Princeton University Press, Princeton, NJ.

McClearn, D., 1992, Locomotion, posture, and feeding behavior of kinkajous, coatis, and raccoons, *J. Mamm.* **73**: 245–261.

Muybridge, E., 1887, *Animals in Motion*, 1957 reprint, Dover, New York.

Nakano, Y., 1996, Footfall patterns in the early development of the quadrupedal walking of Japanese macaques, *Folia Primatol.* **66**: 113–125.

Napier, J., 1967, Evolutionary aspects of locomotion, *Am. J. Phys. Anthropol.* **27**: 333–342.

Napier, J., and Walker, A. C., 1967, Vertical clinging and leaping: A newly recognized locomotor category of primates, *Folia Primatol.* **6**: 204–219.

Pridmore, P. A., 1992, Trunk movements during locomotion in the marsupial *Monodelphis domestica* (Didelphidae), *J. Morphol.* **211**: 137–146.

Pridmore, P. A., 1994, Locomotion in *Dromiciops australis* (Marsupialia: Microbiotheriidae), *Aust. J. Zool.* **42**: 679–699.

Prost, J. H., 1965, The methodology of gait analysis and the gaits of monkeys, *Am. J. Phys. Anthropol.* **23**: 215–240.

Prost, J. H., 1969, A replication study on monkey gaits, *Am. J. Phys. Anthropol.* **30**: 203–208.

Prost, J. H., 1970, Gaits of monkeys and horses: A methodological critique, *Am. J. Phys. Anthropol.* **32**: 121–127.

Rasmussen, D. T., 1990, Primate origins: Lessons from a neotropical marsupial, *Am. J. Primatol.* **22**: 263–277.

Reynolds, T. R., 1985, Stress on the limbs of quadrupedal primates, *Am. J. Phys. Anthropol.* **67**: 351–362.

Reynolds, T. R., 1987, Stride length and its determinants in humans, early hominids, primates, and mammals, *Am. J. Phys. Anthropol.* **72**: 101–115.

Ritter, D., 1995, Epaxial muscle function during locomotion in a lizard (*Varanus salvator*) and the proposal of a key innovation in the vertebrate axial musculoskeletal system, *J. Exp. Biol.* **198**: 2477–2490.

Rollinson, J., and Martin, R. D., 1981 Comparative aspects of primate locomotion, with special reference to arboreal cercopithecines, *Symp. Zool. Soc. Lond.* **48**: 377–427.

Rosenberger, A. L., and Stafford, B. J., 1994, Locomotion in captive *Leontopithecus* and *Callimico*: A multimedia study, *Am. J. Phys. Anthropol.* **94**: 379–394.

Schmitt, D., 1999, Compliant walking in primates, *J. Zool., London*, **248**: 149–160.

Schmitt, D., 2002, Gait mechanics in the common marmoset: Implications for the origin of primate locomotion, *Am. J. Phys. Anthropol.* (Suppl. 34): 137.

Schmitt, D., 2003, Evolutionary implications of the unusual walking mechanics of the common marmoset (*C. jacchus*). *Am. J. Phys. Anthropol.* **122**: 28–37.

Schmitt, D., and Lemelin, P., 2002, Origins of primate locomotion: Gait mechanics of the woolly opossum, *Am. J. Phys. Anthropol.* **118**: 231–238.

Shapiro, L. J., Demes, B., and Cooper, J., 2001, Lateral bending of the lumbar spine during quadrupedalism in strepsirhines, *J. Hum. Evol.* **40**: 231–259.

Sussman, R. W., 1991, Primate origins and the evolution of angiosperms, *Am. J. Primatol.* **23**: 209–223.

Szalay, F. S., 1972, Paleobiology of the earliest primates, in: *The Functional and Evolutionary Biology of Primates*, R. H. Tuttle, ed., Aldine-Atheton, Chicago, pp. 3–35.

Szalay, F. S., and Dagosto, M., 1988, Evolution of hallucial grasping in the primates, *J. Hum. Evol.* **17**: 1–33.

Szalay, F. S., and Delson, E., 1979, *Evolutionary History of the Primates*, Academic Press, New York.

Szalay, F. S., Rosenberger, A. L., and Dagosto, M., 1987, Diagnosis and differentiation of the order Primates, *Yrbk. Phys. Anthropol.* **30**: 75–105.

Szalay, F. S., and Sargis, E. J., 2001, Model-based analysis of postcranial osteology of marsupials from the Palaeocene of Itaboraí (Brazil) and the phylogenetics and biogeography of Metatheria, *Geodiversitas* **23**: 139–302.

Tomita, M., 1967, A study on the movement pattern of four limbs in walking, (1). Observation and discussion on the two types of the movement order of four limbs seen in mammals while walking, *J. Anthropol. Soc. Nippon* **75**: 126–146.

Vilensky, J. A., 1989, Primate quadrupedalism: How and why does it differ from that of typical quadrupeds? *Brain, Behav. Evol.* **34**: 357–364.

Vilensky, J. A., Gankiewicz, E., and Townsend, D. W., 1988, Effects of size on vervet (*Cercopithecus aethiops*) gait parameters: A cross-sectional approach, *Am. J. phys. Anthropol.* **76**: 463–480.

Vilensky, J. A., and Larson, S. L., 1989, Primate locomotion: Utilization and control of symmetrical gaits, *Ann. Rev. Anthropol.* **18**: 17–35.

Vilensky, J. A., and Moore, A. M., 1992, Utilization of lateral and diagonal sequence gaits at identical speeds by individual vervet monkeys, in: *Topics in Primatology. Evolutionary Biology*, S. Matano, R. H. Tuttle, H. Ishida, and M. Goodman, eds., vol. 3, pp. 129–138, University of Tokyo Press, Tokyo.

Vilensky, J. A., Moore, A. M., and Libii, J. N., 1994, Squirrel monkey locomotion on an inclined treadmill: Implications for the evolution of gaits, *J. Hum. Evol.* **26**: 375–386.

Walker, A. C., 1967, *Locomotor Adaptation in Recent and Fossil Madagascan Lemurs*, Ph. D. Dissertation, University of London, London.

White, T. D., 1990, Gait selection in the brush-tail possum (*Trichosurus vulpecula*), the northern quoll (*Dasyurus hallucatus*), and the Virginia opossum (*Didelphis virginiana*). *J. Mamm.* **71**: 79–84.

Yamazaki, N., 1976, *Keisanki simulation niyoru seibutu no hokou no kenkyu (A Study of Animal Walking by Means of Simulation Methods)*, Ph. D. Dissertation, Gijuku University, Yokohama, Keio.

CHAPTER THIRTEEN

Morphological Correlates of Forelimb Protraction in Quadrupedal Primates

Susan G. Larson

INTRODUCTION

Among the issues that remain unresolved in regard to the origins of primates is the locomotor mode of the ancestral euprimate. In living primates, quadrupedal running and walking is the most common form of locomotion (Rose, 1973). Most analyses of fossil material also suggest that the ancestral primate locomotor type included some amount of quadrupedalism (e.g., Cartmill, 1972, 1974, 1992; Dagosto, 1988; Ford, 1988; Gebo, 1989a; Godinot and Beard, 1991; Godinot and Jouffroy, 1984; Martin, 1972; Szalay and Dagosto, 1980; Szalay and Delson, 1979; however, for a contrary view, see Napier, 1967; Napier and Walker, 1967). That most mammalian species are quadrupedal makes it tempting to view quadrupedalism as simply a primitive retention in primates. However, several lines of research indicate that the form of quadrupedal locomotion displayed by primates differs from that observed in other small to medium mammals (see Larson, 1998; Vilensky, 1987, 1989). These differences include use of a diagonal sequence/diagonal couplets walking gait pattern (Hildebrand, 1967; Howell, 1944; Prost, 1965,

---

Susan G. Larson • Department of Anatomical Sciences, Stony Brook University School of Medicine, Stony Brook, NY 11794-8081

1969; Rollinson and Martin, 1981; Vilensky, 1989; Vilensky and Larson, 1989); infrequent use of a running trot (Hildebrand, 1967); atypical patterns of shoulder muscle recruitment (Larson, 1998; Larson and Stern, 1989a,b; Whitehead and Larson, 1994), greater reliance on hindlimbs for both support and propulsion (Demes et al., 1992, 1994; Kimura, 1985; 1992; Kimura et al., 1979; Reynolds, 1985; however, see Schmitt and Lemelin, 2002); relatively low stride frequencies (Alexander and Maloiy, 1984; Demes et al., 1990), relatively longer stride lengths (Alexander and Maloiy, 1984; Reynolds, 1987; Vilensky, 1980) brought about by relatively long limb bones (Alexander et al., 1979) and large limb angular excursions (Larson, 1998; Larson et al., 2000, 2001; Reynolds, 1987), and use of a compliant gait (Schmitt, 1994, 1998, 1999). The ubiquitous nature of these characteristics among primate quadrupeds suggests a shared origin early in the evolution of primates. However, exactly how these characteristics figure into the debate about the locomotor mode of the ancestral euprimate depends on identifying morphological correlates of these behaviors in fossil forms to determine when and where they might have arisen.

Unfortunately, other than relative limb bone length and functional differentiation between fore- and hindlimbs (by differences in bone cross-sectional properties, e.g., Burr et al., 1989; Demes and Jungers, 1989, 1993; Demes et al., 1991; Kimura, 1991, 1995; Ruff and Runestad, 1992; Schaffler et al., 1985), none of these locomotor characteristics have been related to any known morphological features. The goal of this chapter is to identify morphological correlates of one aspect of the unusual form of primate quadrupedal locomotion, namely, the uniquely protracted posture of the primate forelimb at the beginning of a walking step that helps generate their distinctively large forelimb excursion angles (Larson, 1998; Larson et al., 2000; Schmitt, 1994; Schmitt and Lemelin, 2002). Such morphological correlates can then be used to identify where and when this aspect of the unique form of primate quadrupedal locomotion arose in the course of primate evolution.

METHODS

Beginning a step with a more protracted forelimb posture is likely to entail a glenohumeral joint configuration in which the humerus is more aligned with the scapula, and the cranial margin of the glenoid fossa encroaches upon the anterior edge of the humeral head articular surface and the greater and lesser tubercles. Therefore, the logical places to explore for features related to this distinctive shoulder posture are the proximal humerus and glenoid fossa. Fortunately, the functional morphology of the proximal humerus has been well studied (e.g., Fleagle and Simons, 1982; Godfrey, et al. 1991; Harrison, 1989; Rose, 1989; Larson, 1995, 1996; Ruff and Runestad, 1992; Rafferty and Ruff, 1994), and some distinctive primate characters have already been suggested. In particular, Baba (1988) compared the proximal humeri in five primate species (including humans) to those of three nonprimate mammalian species and described differences in features related to range of motion, stability of the shoulder joint, and power for glenohumeral retraction. Six of his measurements describing the size and shape of the humeral head and proximal humeral epiphysis were included in this study (Figure 1, a–f; Table 1). Additional linear and angular measurements were constructed to attempt to further describe the configuration of the proximal humerus and the orientation of the glenoid fossa (Figure 1, h–i; Figure 2; Table 1).

Linear measurements were taken with a digital caliper, and all angular measurements (except spinoglenoid angle) were taken with the aid of a torsiometer (Krahl, 1944; Larson, 1996). Spinoglenoid angle was measured using a clear plastic goniometer. Holding a scapula up to a light, the base of the scapular spine could be aligned with one limb of the goniometer while the other was rotated until it was in line with the superior and inferior margins of the glenoid.

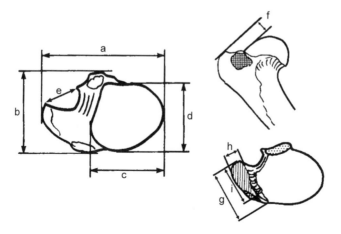

Figure 1. Linear measurements taken on the proximal humeri of the comparative sample (described in Table 1) (redrawn from Baba, 1988). Measurements a–f taken from Baba (1988).

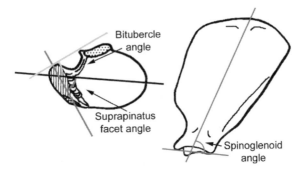

Figure 2. Angular measurements taken on the proximal humeri and scapulae of the comparative sample (described in Table 1).

Table 1. Measurements taken

Sagittal diameter*	Maximum anterio-posterior length of the proximal humeral epiphysis (Figure 1-a)
Transverse diameter*	Maximum medio-lateral width of the proximal humeral epiphysis (Figure 1-b)
Humeral head length*	Maximum anterior-posterior length of the humeral head (Figure 1-c)
Humeral head width*	Maximum medio-lateral width of the humeral head (Figure 1-d)
Intertubercular width*	Width between medial and lateral margins of the intertubercular groove (Figure 1-e)
Greater tubercle height*	Difference between the most superior point of the greater tubercle and the most superior point on the humeral head (Figure 1-f). If tubercle projects above the head, value is positive, if tubercle is below the level of the head, value is negative
Greater tubercle width	Distance between the anteriormost margin and the most posterior point of the greater tubercle (Figure 1-g)
Supraspinatus facet width	Maximum width of supraspinatus insertion facet on the greater tubercle (Figure 1-h)
Supraspinatus facet length	maximum length of supraspinatus insertion facet on the greater tubercle (Figure 1-i)
Supraspinatus facet angle	Angle described by a line passing through the long axis of the supraspinatus insertion facet relative to the axis of the humeral head (Figure 2)
Bitubercle angle	Angle described by a line across the anteriormost points of the greater and lesser tubercles relative to the axis of the humeral head (Figure 2)
Spinoglenoid angle	Angle between a line connecting the superior and inferior margins of the glenoid fossa and the base of the scapular spine (Figure 2)

*From Baba (1988).

The comparative sample consisted of scapulae and humeri of 52 different small- to medium-sized mammalian taxa (derived from the collections of the American Museum of Natural History, and the Stony Brook University Anatomical Sciences Museum), including 9 prosimians, 10 New World monkeys,[1] 11 rodents, 13 carnivores, and 9 marsupials (Table 2). All individuals were adult, as judged from epiphyseal fusion, and each scapula and humerus pair were from the same individual.

Three shape variables and five ratios (using the geometric mean of eight linear measurements on the proximal humerus as a size surrogate) were constructed from the linear measurements to facilitate comparison of species differing in overall size (Table 3). Sample means were derived for each species. Canonical discriminate analysis was used to determine whether taxa would be sorted into their appropriate groups according to the measurements explored here.

Table 2. Comparative sample

Prosimians	n	Rodents	n	Carnivores	n
Lemur catta	3	*Uromys anak*	1	*Vulpes vulpes*	3
Eulemur fulvus	3	*Uromys caudimaculatus*	1	*Genetta genetta*	3
Varecia variegata	3	*Tamiasciurus hudsonius*	2	*Nandinia binotata*	3
Cheirogaleus major	1	*Protoxerus strangeri*	2	*Ailurus fulgens*	3
Loris tardigradus	3	*Sciurus abertmimus*	3	*Herpestes sanguineus*	3
Arctocebus calabarensis	3	*Aplodontia rufa*	3	*Martes americana*	3
Nycticebus coucang	3	*Cratogeomys castanops*	3	*Martes flavigula*	3
Otolemur crassicaudatus	3	*Spermophilus sp.*	1	*Bassariscus astutus*	3
Galago senegalensis	3	*Marmota monax*	1	*Potos flavus*	3
		Cavia porcellus	1	*Nasua nasua*	3
Anthropoids		*Rattus sp.*	1	*Leopardis pardalis*	3
Cacajao calvus	2			*Canis latrans*	3
Pithecia sp.	4	**Marsupials**		*Procyon lotor*	1
Chiropotes satanas	3	*Didelphis virginiana*	3		
Cebus apella	3	*Didelphis albiventis*	3		
Cebus albifrons	3	*Caluromys philander*	3		
Aotus trivirgatus	2	*Philander sp.*	3		
Saimiri sciureus	2	*Trichosurus vulpecula*	2		
Callicebus moloch	3	*Pseudocherius herbertensis*	3		
Callithrix sp.	3	*Phalanger orientalis*	2		
Saguinus fuscicollis	3	*Ailurops ursinus*	2		
		Spilocuscus maculatus	1		

[1] Most estimates suggest that early euprimates were small in body size (e.g., Cartmill, 1974; Dagosto and Terranova, 1992; Martin, 1972; Rose, 1995) and therefore only small taxa were selected for the comparative sample. Most anthropoids of small body size are platyrrhines.

Table 3. Shape variables and ratios

Roundness of proximal humeral epiphysis	(Transverse Diam / Sagittal Diam) × 100
Humeral head shape	(Humeral Head Wd / Humeral Head Lgth) × 100
Humeral head size	(Humeral Head Wd × Humeral Head Lgth)$^{1/2}$
Relative humeral head size	(Humeral Head Size / SIZE) × 100
Supraspinatus facet shape	(Supraspinatus Facet Wd / Supraspinatus Facet Lgth) × 100
Supraspinatus facet size	(Supraspinatus Facet Wd × Supraspinatus Facet Lgth)$^{1/2}$
Relative supraspinatus facet size	(Supraspinatus Facet Size / SIZE) × 100
Relative greater tubercle projection	((Sagittal Diam − Humeral Head Lgth) / Size) × 100
Relative greater tubercle width	(Greater Tubercle Wd / SIZE) × 100
Relative intertubercular width	(Intertubercular Wd / SIZE) × 100
SIZE (Geometric mean of 8 linear measurements)	(Transverse Diam × Sagittal Diam × Humeral Head Wd × Humeral Head Lgth × Supraspinatus Facet Wd × Supraspinatus Facet Lgth × Greater Tubercle Wd × Intertubercular Wd)$^{1/8}$

To investigate the applicability of the resulting analysis to fossil material, three test cases were included. The first was casts of a proximal humerus (USNM 17994#2) and partial scapula (USNM 21815#7) from the Eocene form, *Smilodectes gracilis*. The others were two undescribed possible primate fossil proximal humeri (Figure 3) from Miocene localities in Uganda (MUZM 173 from Moroto I locality dated > 20.6 MYA (Gebo et al., 1997), and BUMP 101 from Napak CC locality thought to be approximately 19 MYA (MacLatchy, pers.com)) loaned to the author by Dr. Laura MacLatchy of the University of Michigan.

RESULTS

There are two main questions that need to be addressed in this attempt to identify morphological correlates of a protracted humeral posture for the purpose of tracing its evolutionary history: (1) are any of the morphological features examined functionally related to humeral protraction, and (2) do these features distinguish primates from other mammalian groups? In answer to the first question, correlations between each of the examined variables and humeral protraction angles from Larson et al., (2000)[2] were analyzed. As summarized in Table 4, protraction angles were significantly correlated with roundness of the proxi-

---

[2] Only a subset of the comparative sample could be used since not all taxa examined here were included in Larson et al. (2000).

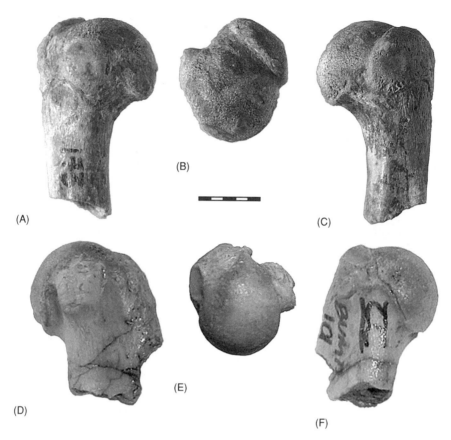

Figure 3. Fossil proximal humeri from Miocene localities in Uganda. Scale bar equals 5 mm. Upper row: MUZM 173 from Moroto 1 locality, (A) medial view, (B) superior view, and (C) lateral view. Lower row: BUMP 101 from Napak CC locality. (D) medial view, (E) superior view, and (F) lateral view.

mal humeral epiphysis, relative humeral head size, relative greater tubercle width, bitubercle angle, and anterior projection of the greater tubercle.

The second question regarding the distinctiveness of these features in primates was addressed by comparing group means for each of the variables (Table 5). Compared to the other mammalian groups, primates possess more obtuse spinoglenoid angles, relatively larger humeral heads, and less anteriorly projecting greater tubercles (not significant for rodents) (Table 5). Primates as a group were most different from carnivores displaying, in addition to those features listed above, larger bitubercle angles, more rounded proximal

Table 4. Correlations between humeral protraction angles at touchdown* during quadrupedal locomotion and study variables (statistically significant correlations in boldface)

Variable name	n[†]	r	Prob r = 0
Spinoglenoid angle	22	0.39	NS
Bitubercle angle	22	**0.62**	**0.002**
Rel Grt Tub Proj	22	**−0.72**	**<0.001**
Rel Grt Tub Wd	22	**−0.44**	**0.04**
Grt Tub Ht[‡]	22	−0.06	NS
Roundness Hum Epi	22	**0.55**	**0.008**
Hum Hd shape	22	0.19	NS
Rel Hum Hd Sz	22	**0.46**	**0.03**
Suprasp facet shape	22	0.30	NS
Suprasp facet angle	22	**−0.42**	**0.05**
Rel supra facet Sz	22	0.05	NS
Rel Intertub Wd	22	−0.05	NS

*Humeral protraction angles taken from Larson et al. (2000)
[†]Humeral protraction angles were available for only 22 out of the 52 species examined here
[‡]Greater tubercle height was not correlated with body size

humeral epiphyses, less long and narrow supraspinatus insertion facets, and shorter and more narrow greater tubercles. However, these latter features did not distinguish primates from either rodents or marsupials. Additional features distinguishing primates from marsupials were humeral head shape and intertubercular groove width, and from rodents was angle of the supraspinatus insertion facet. Figure 4 presents drawings of representative humeri illustrating some of these differences.

Within the set of features examined, therefore, is a subset that is both correlated to humeral protraction angles, and distinguishes primates from (all or at least some) other mammals. Those features are relative humeral head size, relative greater tubercle projection and width, bitubercle angle, and roundness of the proximal humeral epiphysis. In addition, spinoglenoid angle distinguished primates from all three other mammalian groups, although it is not significantly correlated to protraction angle. These six variables were used in a canonical discriminant analysis to explore whether the combination could be used to identify a primate pattern of glenohumeral morphology. Figure 5 displays a bivariate plot of the sampled taxa for canonical discriminant functions 1 and 2, which together explain 91.1% of the variation in the analysis. Function 1 clearly separates primates from rodents and marsupials, and the variable having the highest correlation with this function is spinoglenoid angle. Primates are separated from carnivores along function 2, and variables

Table 5. Group means and comparisons of means (boldface values indicate statistically significant differences between primates and other mammalian groups at p ≤ 0.05)

Variable name	Primates	Rodents	Carnivores	Marsupials	*Smilodectes*	BUMP 101	MUZM 173
Spinoglenoid angle	**81.34**±10.70*	63.61±2.70	73.08±4.58	66.59±2.94	76.00	–	–
Bitubercle angle	**67.50**±5.50	65.88±10.24	47.44±9.98	73.99±10.38	78.67	53.33	60.67
Rel Grt Tub projection	**12.73**±7.61	26.67±14.14	46.66±13.71	**32.51**±7.22	18.48	33.55	27.29
Rel Grt Tub Wd	126.26±8.62	120.25±11.25	**137.67**±6.35	123.15±3.16	131.14	146.66	119.51
Grt Tub Ht	-0.56±1.04	-0.47±0.87	1.19±2.25	-0.19±0.72	-1.81	-0.46	-0.58
Roundness Hum epiphysis	101.26±3.96	102.42±8.17	**85.27**±6.87	103.53±7.67	102.24	90.87	90.05
Hum Hd shape	85.15±4.23	85.08±4.66	83.57±4.76	**91.43**±4.27	83.07	82.31	88.17
Rel Hum Hd size	141.49±8.60	131.20±8.17	125.56±7.34	**130.05**±6.99	141.49	147.89	135.77
Suprasp facet shape	40.08±7.78	41.70±8.42	**27.87**±5.07	40.31±8.12	37.61	29.57	27.48
Suprasp facet angle	32.53±7.36	**41.07**±6.89	38.31±7.28	36.51±6.39	37.33	34.33	38.50
Rel supra facet size	56.91±2.69	58.92±3.57	58.68±4.44	54.80±1.92	50.19	46.43	55.72
Rel intertub Wd	45.08±8.76	49.94±12.39	47.23±5.66	**55.00**±4.69	47.69	41.17	54.98

*One standard deviation.

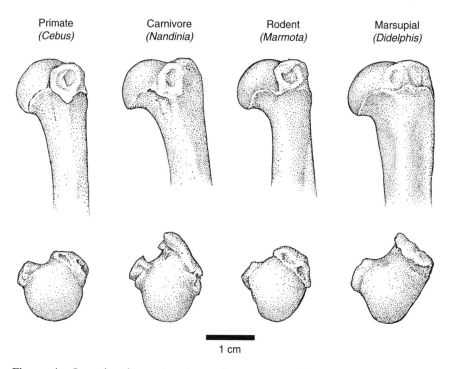

| Primate | Carnivore | Rodent | Marsupial |
| (Cebus) | (Nandinia) | (Marmota) | (Didelphis) |

1 cm

Figure 4. Lateral and superior views of representative humeri from the different mammalian groups examined here. Scale bar represents 1 cm. Primate humeri are characterized by a head that sits more on top of the humeral shaft than those of the other groups.

correlated with this axis are shape of the proximal humeral epiphysis, bitubercle angle, anterior projection of the greater tubercle, and relative greater tubercle width.

Inclusion of the specimen of *Smilodectes gracilis* as a separate group places it within the primate sample between *Eulemur fulvus* and *Otolemur crassicaudatus*. However, *Smilodectes* displayed a higher loading on function 3 (which accounts for an additional 8.7% of the variance) than any of the other primate species, and in this dimension was somewhat more like the Celebes or spotted cuscus (*Ailurops ursinus* or *Spilocuscus maculatus*) (Figure 6).

In order to be able to include the two proximal humeral fragments from the Miocene of Uganda, it was necessary to remove spinoglenoid angle from the analysis. The resulting distribution of taxa along functions 1 and 2, which together explain 89.3% of the variance (Figure 7), shows more overlap between

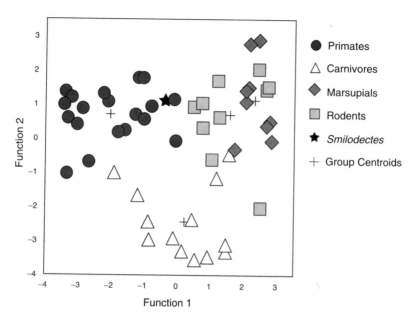

Figure 5. Bivariate plot of functions 1 and 2 of the canonical discriminant analysis based on the six variables that were correlated to humeral protraction angles and/or distinguished primates from some or all other mammalian groups. ●: Primates; △: Carnivores; ◆: Marsupials; ▢: Rodents; ★: *Smilodectes*; +: Group Centroids.

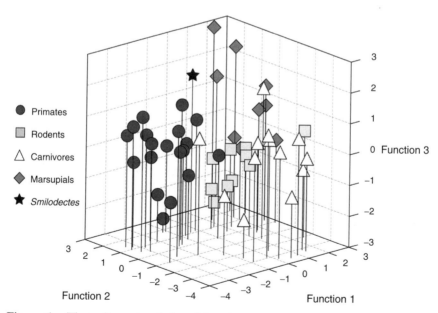

Figure 6. Three dimensional plot of functions 1, 2, and 3 of six variable canonical discriminant analysis. Symbols as in Figure 5. While *Smilodectes* falls among primates along functions 1 and 2, it is nearer to marsupials on function 3.

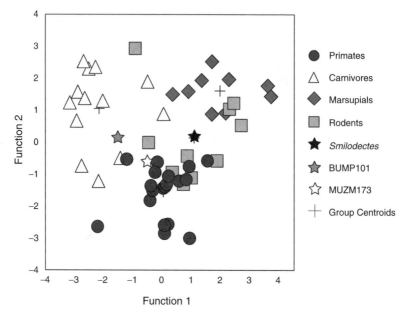

Figure 7. Bivariate plot of functions 1 and 2 of the canonical discriminant analysis with spinoglenoid angle removed from the analysis ★: BUMP 101; ☆: MUZM 173; other symbols as in Figure 5. More overlap between groups is evident than in six variable analysis. *Smilodectes* coincides with rodent group centroid, BUMP 101 falls among the carnivores, but MUZM 173 appears to be a primate.

groups, and predicted group membership places *Saimiri* and *Cebus albifrons* in the rodent group, and the rodent *Uromys sp.* in primates. *Smilodectes* falls within rodents along functions 1 and 2, but is more similar to primates and marsupials on function 3 (accounting for an additional 7.3% of the variance) (Figure 8). BUMP 101 falls within the carnivore group, and MUZM 173 is just within primates, falling nearest to *Lemur catta* on all three discriminate functions.

DISCUSSION

Beginning a step with a protracted glenohumeral joint is but one aspect of the set of behavioral features distinguishing the form of primate quadrupedal locomotion from that of other mammals (Larson, 1998, Larson et al., 2000).

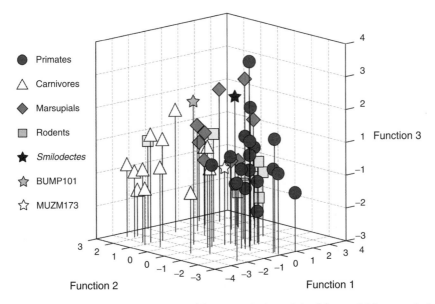

Figure 8. Three dimensional plot of functions 1, 2, and 3 of five variable canonical discriminant analysis. Symbols as in Figures 5 and 7. While *Smilodectes* falls among rodents for functions 1 and 2, along function 3 it is more similar to some marsupials and primates.

Limb protraction has been related to larger total limb excursion (Larson et al., 2000, 2001; Reynolds, 1987), and in turn to relatively long stride lengths in primates (Alexander and Maloiy, 1984; Demes et al., 1990; Reynolds, 1987). Which of these features is cause or effect is less clear. It is also not obvious exactly how they are related to the various other unusual aspects of primate quadrupedal locomotion. Thus attempts to explain how and why any of the features arose in the course of primate evolution at this point are largely speculation (e.g., Larson, 1998).

Of the features that have been examined here, spinoglenoid angle is perhaps most clearly related functionally to degree of humeral protraction. As shown in Figure 9, if a primate and a nonprimate scapula are oriented with their spines parallel, a more obtuse spinoglenoid angle gives the glenoid fossa more of a cranial tilt throughout support phase to articulate with a more protracted humerus. If this is the case, however, why is spinoglenoid angle not more strongly correlated to humeral protraction angle (Table 4)? This lack of correlation may be due in part to the variation in spinoglenoid angles within

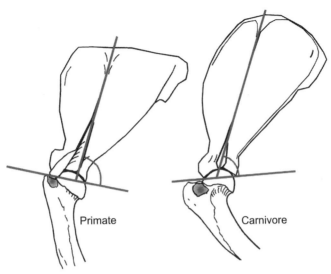

Figure 9. Spinoglenoid angle in a primate and a carnivore. Scapulae have been positioned with parallel scapular spines, and humeri are approximately in a midstance position. A more obtuse spinogleniod angle in primates accommodates articulation with their more protracted humerus.

primates. Anthropoids typically have larger spinoglenoid angles than prosimians, but prosimians tend to display more extreme humeral protraction angles (Larson et al., 2000), the reverse of what the results presented here would predict. However, comparing the cineradiographic observations on shoulder motion in the brown lemur reported by Schmidt and Fischer (2000) to those on the vervet monkey reported by Whitehead and Larson (1994), prosimians appear to begin a walking step with their scapulae in a slightly more oblique position, and thus even though their spinoglenoid angles are more acute, their glenoid fossae have nearly the same orientation as do those of anthropoids at the beginning of a step.

On the primate proximal humerus, the degree to which the greater tubercle projects anteriorly and superiorly has been reduced. Looking at the proximal humerus in side view (Figure 4), it appears that this difference is due to the humeral head being shifted anteriorly so that it encroaches upon the greater tubercle, rather than to the greater tubercle being reduced in size. This places the head more directly in line with the humeral shaft, rather than posterior to it as in other mammals. Therefore, during support phase of a step, the weight of the body passes through the humeral head and along the shaft as the humerus passes from a protracted, to a vertical, and finally to a retracted position. In nonprimate mammals, the humerus begins in a vertical or slightly retracted position and becomes more retracted through a step, and body weight is directed through the more offset head into the humeral neck, which is oblique to the shaft. A possible explanation for the change in humeral head position in primates, therefore, is that it may help to align the humerus and shoulder joint more with the substrate reaction force and thereby reduce joint moments (see Schmitt, 1999).

Regarding the three "test" cases included in the study, the BUMP 101 specimen from the Napak CC Miocene locality in Uganda may be a carnivore, but a more extensive comparative study would be necessary to discern its more specific relationships. The MUZM 173 proximal humerus from the Miocene Moroto I locality in Uganda appears to be a primate, and given its age and geographical location, probably of lorisoid affinities. Previous analyses of Miocene lorisoid postcranial material from East Africa have typically indicated either galago-like leaping abilities or general quadrupedalism (e.g., Gebo 1989b; McCrossin, 1992; Walker, 1970, 1974). Although it is not particularly similar to either galago species in the comparative sample, the fact that the MUZM 173 proximal humerus is close to *Lemur*, *Callithrix* and *Aotus* in the discriminant analysis suggests that it may be from a running, leaping form. Finally, while it is well established that *Smilodectes* is an Eocene primate, judging from the second discriminant analysis without spinoglenoid angle, its proximal humerus appears to be less primate-like than its scapula. In particular, it appears to have a larger bitubercle angle and somewhat more anteriorly projecting greater tubercle than do modern primates. This suggests that this Eocene form may have begun a quadrupedal step with only a modestly protracted humerus.

Obviously, to fully investigate when and where the protracted humeral posture characteristic of primate quadrupedal locomotion evolved it will be necessary to examine the morphology of scapulae and proximal humeri of a much larger sample of early primate taxa. It is also the case that the specimens necessary for such an investigation do not all yet exist. Nonetheless, this study suggests that as new material is discovered, we will have the tools needed to recognize and track the evolution of this particular unique aspect of primate quadrupedal locomotion.

ACKNOWLEDGMENTS

Thanks go to Bob Randall and the American Museum of Natural History for assistance in data collection, and to Bill Jungers for help with the statistical analysis of the data. Special thanks to Laura MacLatchy for the loan of the MUZM 173 and BUMP 101 fossils, and for the photos in Figure 3. Luci Betti-Nash created Figure 4. This research was supported by NSF Grant BCS-0509190.

REFERENCES

Alexander, R. McN., Jayes, A. S., Maloiy, G. M. O., and Wathuta, E. M., 1979, Allometry of the limb bones of mammals from shrews (*Sorex*) to elephant (*Loxodonta*), *J. Zool., Lond.* **189**: 305–314.

Alexander, R. McN., and Maloiy, G. M. O., 1984, Stride lengths and stride frequencies of primates, *J. Zool., Lond.* **202**: 577–582.

Baba, H., 1988, Locomotor adaptation of the structure of proximal humerus in the primates and other mammals, *Bull. Natn. Sci. Mus., Tokyo, Ser. D.* **14**: 9–17.

Burr, D. B., Ruff, C. B., and Johnson, C., 1989, Structural adaptation of the femur and humerus to arboreal and terrestrial environments in three species of macaque, *Am. J. Phys. Anthropol.* **79**: 357–368.

Cartmill, M., 1972, Arboreal adaptations, and the origin of the order Primates, in: *The Functional and Evolutionary Biology of Primates,* R. Tuttle, ed., Aldine, Chicago, pp. 97–122.

Cartmill, M., 1974, Pad and claws in arboreal locomotion, in: *Primate Locomotion.* F. A. Jenkins, Jr., ed., Academic Press, New York, pp. 45–83.

Cartmill, M., 1992, New views on primate origins, *Evol. Anthropol.* **1**: 105–111.

Dagosto M., 1988, Implications of postcranial evidence for the origin of euprimates, *J. Hum. Evol.* **17**: 35–56.

Dagosto, M., and Terranova, C. J., 1992, Estimating the body size of Eocene primates: A comparison of results from dental and postcranial variables, *Int. J. Primatol.* **13**: 307–344.

Demes, B., and Jungers, W. L., 1989, Functional differentiation of long bones in lorises, *Folia Primatol.* **52**: 58–69.

Demes, B., and Jungers, W. L., 1993, Long bone cross-sectional dimensions, locomotor adaptations and body size in prosimian primates, *J. Hum. Evol.* **25**: 57–74.

Demes, B., Jungers, W. L., and Nieschalk, U., 1990, Size- and speed-related aspects of quadrupedal walking in slender and slow lorises, in: *Gravity, Posture and Locomotion in Primates,* F. K. Jouffroy, M. H. Stack, and C. Niemitz, eds., Firenze: Il Sedicesimo, pp. 175–197.

Demes, B., Jungers, W. L., and Selpien, K., 1991, Body size, locomotion, and long bone cross-sectional geometry in Indriid primates, *Am. J. Phys. Anthropol.* **86:** 537–547.

Demes, B., Larson, S. G., Stern, J. T. Jr., and Jungers, W. L., 1992, The hindlimb drive of primates—Theoretical reconsideration and empirical examination of a widely held concept, *Am. J. Phys. Anthropol.,* Suppl. **14:** 69.

Demes, B., Larson, S. G., Stern, J. T. Jr., Jungers, W. L., Biknevicius, A. R., and Schmitt, D., 1994, The kinetics of "hind limb drive" reconsidered, *J. Hum. Evol.* **26:** 353–374.

Fleagle, J. G., and Simons, E. L., 1982, The humerus of *Aegyptopithecus zeuxis:* A primitive anthropoid, *Am. J. Phys. Anthropol.* **59:** 175–193.

Ford, S. M., 1988, Postcranial adaptations of the earliest platyrrhine, *J. Hum. Evol.* **17:** 155–192.

Gebo, D. L., 1989a, Locomotor and phylogenetic considerations in anthropoid evolution, *J. Hum. Evol.* **18:** 201–233.

Gebo, D. L., 1989b, Postcranial adaptation and evolution in Lorisidae, *Primates* **30:** 347–367.

Gebo, D. L., MacLatchy, L., Kityo., Deino, A., Kingston, J., and Pilbeam D., 1997, A hominoid genus from the Early Miocene of Uganda, *Science* **276:** 401–404.

Godfrey, L., Sutherland, M., Boy, D., and Gomberg, N., 1991, Scaling of limb joint surface area in anthropoid primates and other mammals, *J. Zool., Lond.* **223:** 603–625.

Godinot, M., and Beard, K. C., 1991, Fossil primates hands: A review and an evolutionary inquiry emphasizing early forms, *Human Evol.* **6:** 307–354.

Godinot, M., and Jouffroy, F. K., 1984, La main d'*Adapis* (Primates, Adapidae), in: *Actes du symposium paleontologique, G. Cuvier,* E. Buffetaut, J. M. Mazin, and E. Salmion, eds., Montbeliard, pp. 221–242.

Harrison, T., 1989, New postcranial remains of *Victoriapithecus* from the middle Miocene of Kenya, *J. Hum. Evol.* **18:** 3–54.

Hildebrand, M., 1967, Symmetrical gaits of primates, *Am. J. Phys. Anthropol.* **26:** 119–130.

Howell, A. B., 1944, *Speed in Animals,* University of Chicago Press, Chicago.

Kimura, T., 1985, Bipedal and quadrupedal walking of primates: Comparative dynamics, in: *Primate Morphophysiology,* S. Kondo, ed., *Locomotor Analyses and Human Bipedalism,* University of Tokyo Press, Tokyo, pp. 81–104.

Kimura, T., 1991, Long and robust bones of primates, in: *Primatology Today,* A. Ehara, T. Kimura, O. Takenaka, and M. Iwamoto, eds., Elsevier, New York, pp. 495–498.

Kimura, T., 1992, Hind limb dominance during primate high-speed locomotion, *Primates* **33:** 465–476.

Kimura, T., 1995, Long bone characteristics of primates, *Z. Morphol. Anthropol.* **80:** 265–280.

Kimura, T., Okada, M., and Ishida, H., 1979, Kinesiological characteristics of primate walking: Its significance in human walking, in: *Environment, Behavior, and Morphology: Dynamic Interactions in Primates*, M. E. Morbeck, H. Preuschoft, and N. Gomberg, eds., Gustav Fischer, New York, pp. 297–311.

Krahl, V. E., 1944, An apparatus for measuring the torsion angle in long bones, *Science* **99**: 498.

Larson, S. G., 1995, New characters for the functional interpretation of primate scapulae and proximal humeri, *Am. J. Phys. Anthropol.* **98**: 13–35.

Larson, S. G., 1996, Estimating humeral torsion on incomplete fossil anthropoid humeri, *J. Hum. Evol.* **31**: 239–257.

Larson, S. G., 1998, Unique aspects of quadrupedal locomotion in nonhuman primates, in: *Primate Locomotion*, E. Strasser, J. Fleagle, A. Rosenberger, and H. McHenry, eds., Plenum Press, New York, pp.157–173.

Larson, S. G., and Stern, J. T. Jr., 1989a, The role of supraspinatus in the quadrupedal locomotion of vervets (*Cercopithecus aethiops*): Implications for interpretation of humeral morphology, *Am. J. Phys. Anthropol.* **79**: 369–377.

Larson, S. G., and Stern, J. T. Jr., 1989b, The role of propulsive muscles of the shoulder during quadrupedalism in vervet monkeys (*Cercopithecus aethiops*): Implications for neural control of locomotion in primates, *J. Motor Behavior.* **21**: 457–472.

Larson, S. G., Schmitt, D., Lemelin, P., and Hamrick, M., 2000, Uniqueness of primate forelimb posture during quadrupedal locomotion, *Am. J. Phys. Anthropol.* **112**: 87–101.

Larson, S. G., Schmitt, D., Lemelin, P., and Hamrick, M., 2001, Limb excursion during quadrupedal walking: How to primates compare to other mammals? *J. Zool., Lond.* **255**: 353–365.

Martin, R. D., 1972, Adaptive radiation and behavior of the Malagasy lemurs, *Phil. Trans. Royal. Soc. Lond.* **264**: 295–352.

McCrossin, M., 1992, New species of bushbaby from the Middle Miocene of Maboko Island, Kenya, *Am. J. Phys. Anthropol.* **89**: 215–233.

Napier, J. R., 1967, Evolutionary aspects of primate locomotion, *Am. J. Phys. Anthropol.* **27**: 333–342.

Napier, J. R., and Walker, A. C., 1967, Vertical clinging and leaping-A newly recognized category of locomotor behaviour of primates, *Folia Primatol.* **6**: 204–219.

Prost, J. H., 1965, The methodology of gait analysis and gaits of monkeys *Am. J. Phys. Anthropol.* **23**: 215–240.

Prost, J. H., 1969, A replication study on monkey gaits, *Am. J. Phys. Anthropol.* **30**: 203–208.

Rafferty, K. L., and Ruff, C. B., 1994, Articular structure and function in *Hylobates, Colobus,* and *Papio, Am. J. Phys. Anthropol.* **94**: 395–408.

Reynolds, T. R., 1985, Stresses on the limbs of quadrupedal primates, *Am. J. Phys. Anthropol.* **67**: 351–362.

Reynolds, T. R., 1987, Stride length and its determinants in humans, early hominids, primates, and mammals, *Am. J. Phys. Anthropol.* **72**: 101–116.

Rollinson, J., and Martin, R. D., 1981, Comparative aspects of primate locomotion, with special reference to arboreal cercopithecines, in *Vertebrate Locomotion*, M. H. Day, ed., Symposia of the Zoological Society of London, No. 48, Academic Press, London, pp. 377–427.

Rose, K. D., 1995, The earliest primates, *Evol. Anthropol.* **5**: 159–173.

Rose, M. D., 1973, Quadrupedalism in primates, *Primates* **14**: 337–357.

Rose, M. D., 1989, New postcranial specimens of catarrhines from the Middle Miocene Chinji Formation, Pakistan: Descriptions and a discussion of proximal humeral functional morphology in anthropoids, *J. Hum. Evol.* **18**: 131–162.

Ruff, C. B., and Runestad, J. A., 1992, Primate limb bone structural adaptations, *Annu. Rev. Anthropol.* **21**: 407–433.

Schaffler, M. B., Burr, D. B., Jungers, W. L., and Ruff, C. B., 1985, Structural and mechanical indicators of limb specialization in primates, *Folia Primatol.* **45**: 61–75.

Schmidt, M., and Fischer, M. S., 2000, Cineradiographic study of forelimb movements during quadrupedal walking in the brown lemur (*Eulemur fulvus*, Primates: Lemuridae), *Am. J. Phys. Anthropol.* **111**: 245–262.

Schmitt, D., 1994, Forelimb mechanics as a function of substrate type during quadrupedalism in two anthropoid primates, *J. Hum. Evol.* **26**: 441–457.

Schmitt, D., 1998, Forelimb mechanics during arboreal and terrestrial quadrupedalism in Old World monkeys, in: *Primate Locomotion: Recent Advances*, E. Strasser, J. Fleagle, A. Rosenberger, and H. McHenry, eds., Plenum Press, New York, pp. 175–200.

Schmitt, D., 1999, Compliant walking in primates, *J. Zool., Lond.* **248**: 149–160.

Schmitt, D., Larson, S. G., and Stern, J. T. Jr., 1994, Serratus ventralis function in vervet monkeys (*Cercopithecus aethiops*): Are primate quadrupeds unique? *J. Zool., Lond.* **232**: 215–230.

Schmitt, D., and Lemelin, P., 2002, Origins of primate locomotion: Gait mechanics of the woolly opossum, *Am. J. Phys. Anthropol.* **118**: 231–238.

Szalay, F. S., and Dagosto, M., 1980, Locomotor adaptations as reflected on the humerus of Paleogene primates, *Folia Primatol.* **34**: 1–45.

Szalay, F. S., and Delson, E., 1979, *Evolutionary History of the Primates*, Academic Press, New York.

Vilensky, J. A., 1980, Trot-gallop transition in a macaque, *Am. J. Phys. Anthropol.* **53**: 347–348.

Vilensky, J. A., 1987, Locomotor behavior and control in human and nonhuman primates: Comparisons with cats and dogs, *Neurosci. Biobehav. Rev.* **11**: 263–274.

Vilensky, J. A., 1989, Primate quadrupedalism: How and why does it differ from that of typical quadrupeds, *Brain Behav. Evol.* **34**: 357–364.

Vilensky, J. A., and Larson, S. G., 1989, Primate locomotion: Utilization and control of symmetrical gaits, *Annu. Rev. Anthropol.* **18**: 17–35.

Walker, A., 1970, Post-cranial remains of the Miocene Lorisidae of East Africa, *Am. J. Phys. Anthropol.* **33**: 249–262.

Walker, A., 1974, Locomotor adaptations in past and present prosimian primates, in: *Primate Locomotion*, F. A. Jenkins, Jr., ed., Academic Press, New York, pp. 349–381.

Whitehead, P. F., and Larson, S. G., 1994, Shoulder motion during quadrupedal walking in *Cercopithecus aethiops*: Integration of cineradiographic and electromyographic data, *J. Hum. Evol.* **26**: 525–544.

CHAPTER FOURTEEN

Ancestral Locomotor Modes, Placental Mammals, and the Origin of Euprimates: Lessons from History

Frederick S. Szalay

INTRODUCTION

The title of this chapter has a double entendre embedded in it. It is a truism that biological history, in addition to ongoing adaptive demands, is decisive in shaping properties of lineages. But it is also uncontestable that precedent notions, influential contributors, or specific papers, right or wrong, channel and continue to profoundly influence thinking on many issues in science. There is a difference, however, in these two processes of canalization. In the evolutionary dynamic, there is no right or wrong, and the inherited attributes are the initial and boundary conditions that define the avenues open for subsequent phylogenetic/adaptive change. These paths do not only constrain but facilitate as well. At any rate, whales are not fish, so history fundamentally

Frederick S. Szalay • Anthropology, and Ecology and Evolutionary Biology, CUNY, and Department of Anthropology Hunter College, CUNY, 695 Park Avenue, New York 10021

matters in all biological science. Genotype encoded factors which sum up history, beyond the maternal contribution in the egg, guide the change, adaptive or not, which is phylogeny.

Such largely adaptive phylogeny is an ongoing probabilistic outcome of environmental demands, which determine the frequency of individuals that make it through the survival and reproductive bottlenecks of each generation. The necessity to consider this theoretical foundation should be, therefore, neither surprising nor burdensome for natural historians who study morphological attributes or behaviors. Sundry disciplines, in particular research from behavioral ecology, provide fundamentally important plausibility hypotheses for paleobiologists, who seek such questions as this conference set out to do. The task, however, to reconstruct adaptive phylogeny is within the realm of paleontologists and morphologists who must tie the fossil record, through a variety of procedures referred to as modeling (see Szalay and Sargis, 2001), to information and ideas from neontology. This is done by testing specific historical-narrative explanations (i.e., phylogeny or taxon hypotheses; see Figure 1 in Szalay, 2000) against various areas of information. Historical narratives of science are tested against evidence of all sorts (Bock, 1981)—an activity not indulged in by Kipling. So contrary to Popperian thinking, much of science consists of historical-narrative explanations offered within the confines of law-like explanations, in juxtaposition to Cartmill's (1990) opinion that only the law-like statements are scientific. Law-like statements must be part of the context within which the various topics of becoming are explained (Bock, 1981; Szalay and Bock, 1991). But nomological-deductive explanations (law-like statements) alone, obviously, do not suffice in any science where history played a role. It is all those specific and contextual historical "mistakes" in the law-like workings of chemistry that result in consequences for replication, transcription, and translation of nucleic acids where the science of evolutionary change begins and couples with the vicissitudes of the environment.

The aim of this chapter is relatively straightforward, but because of space constraints, it is more of a review of some literature debates and an outline of some issues related to the origin of both the Plesiadapiformes and Euprimates (perhaps best considered as sister orders at present) rather than detailed documentation. To achieve these goals I will (1) examine, in a historical framework, selected examples of hypotheses in which conceptual methods, as well as empirical emphasis or de-emphasis, have had a significant role in the construction of these hypotheses, as well as their consequences, for analyzing the phylogeny of adaptations for plesiadapiforms and euprimates; (2) present my views on modeling in paleobiology and the testing of homology hypotheses; (3) remark on the evidence related to locomotor strategies of Cretaceous and some recent therians that are relevant to the assessment of the ancestral pattern in the Eutheria, and of a clade within that group, the Placentalia; and (4) reassert the importance of the "morphotype locomotor mode" concept as a critical connection between phylogeny estimation and adaptational (functional, in a broad sense) assessment. As an example, I point to some evidence from hard anatomy for the ancestral euprimate locomotor mode. The latter was first referred to as "grasp-leaping" in Szalay and Delson (1979) and subsequently more fully developed in Szalay and Dagosto (1980, 1988). R. H. Crompton (1995) appears to strongly second this view.

The Placentalia, diagnosed elsewhere based on tarsal attributes and four premolars, is the taxon that stems from (back in the Cretaceous) the last common ancestor of the Cenozoic and surviving eutherians. This is not just the living crown group because it includes now extinct orders as well. The corresponding stem group of the Eutheria from which the Placentalia arose is the paraphyletic Eoeutheria that diverged from Metatheria at least 125 MYA.

GLIMPSES OF HISTORY OF RESEARCHES REGARDING ARCHONTAN, PLESIADAPIFORM, AND EUPRIMATE MORPHOTYPE LOCOMOTOR STRATEGIES, AND THEIR INFLUENCE

The customary empirical efforts to study extant forms and fossils often break down into two approaches, the functional (in a broad sense) and the phylogenetic (see Szalay, 2000, for review). As a consequence, the conceptual methods that should guide the analysis of the various facets of a problem become simplified either into functional undertakings or synapomorphy sorting through parsimony analysis, or other phyletic approaches. In addition, it is not unusual at all for many scientists leaning in one or another direction regarding morphological analysis to completely barricade themselves into either of these two, often walled-off, compartments, stating they are not really interested in the "other" questions. This is not an exaggerated rendering of the state of affairs, particularly either for functional anatomy or parsimony cladistics-based studies, with the subsequent distortion of the questions and a loss of the evolutionary explanation that one is interested in.

There have been valiant undertakings to somehow combine function and taxonomic position in one fell swoop, although some of these quantitative efforts have, unfortunately, resulted in such empirical conflation of data that all that followed were strikingly visual "species stamps," rather than any illumination of the role of heritage in the evolution of functional complexes (e.g., Oxnard et al., 1990). While the aim of these studies was laudable, the setting aside of the complex but feasible and complementary interrelationship of functional-adaptive and phylogenetic methods for the analysis of evolutionary origins (problems of transformation from one stage to another in the history of lineages) suffered, or simply was not part of the analysis.

As attested to by this conference and many others before it, primates generate great interest among an inordinately large number of natural historians of all sorts, morphologists among them. This is understandable but it makes for an enormous literature, and extremes of conceptual approaches to problems of adaptation of ancestral conditions and disagreements about the specifics of an ancestral lineage. At this conference (and before), for example, I or Dagosto viewed the ancestral euprimate as the phyletic antecedent of the reasonably well-known Eocene strepsirhines and haplorhines, whereas others considered a cheirogaleid such as *Microcebus* as a stand-in for this ancestor. In spite of such a difference in perception, which is almost never explicitly stated, what is less understandable is how several past contributions on the deep adaptive history of primates were based sometimes on a lack of expertise in evolutionary morphology, on highly selected literature contributions, or on a neglect of the specifics of extant species. Some of these publications were often by primatologists, who have written about bones and fossils with little experience either in the theoretical issues surrounding evolutionary analysis of morphology or the fossil record. This state of affairs, however, has considerably improved recently due to competitive pressures resulting from an upwelling of young talent specializing in these complex and intertwined fields of analysis. But the past has shown its powerful constraint on the collective minds of a whole subfield. Some textbooks and reviews have helped to perpetuate uncorroborated ideas about locomotor inferences regarding protoplacentalians, plesiadapiforms, and the stem euprimates. In the review given in later section, I will comment briefly on some such examples.

Arboreality as a Novel Strategy for the Stem of Archonta

Over and beyond the obvious specifics of arboreal heritage in the morphology of living primates, in the 1960s, the debate over this heritage has entered a new phase with an admittedly confusing framework for considering primates with or without the archaic primarily Paleocene radiation of the Plesiadapiformes. While the evidence now is overwhelming regarding the arboreality of plesiadapiforms (and their close phyletic ties to euprimates without the interference of dermopterans; see Bloch and Boyer, 2001; Bloch et al., 2000, 2001a,b,c 2002; Boyer and Bloch, 2000; Boyer et al., 2001), the history of the literature regarding archontans, tupaiids, and plesiadapiforms is highly instructive.

The initial and widely read impetus (if one was to start somewhere in a quasi-historical assessment such as this) that euprimates owe their particular morphological (and functional in a broad sense) divergence from their ancestry due to a particular locomotor behavior that involved leaping, was the contribution of Napier and Walker (1967)—a study that advocated vertical clinging and leaping as the initial stage of euprimate locomotor evolution. This restatement of previous views on leaping but with greater force and examples were significant because they went beyond the customarily evoked arboreality as an explanation for euprimate attributes. Much of the development of the insight regarding leaping in euprimate ancestry was largely due to the seminal studies of Walker (1967) on the subfossil and extant osteology of the Malagasy strepsirhines. The extended debate about vertical clinging and leaping that ensued is interesting history, but not directly relevant here. The theoretical underpinnings of the ecomorphological assessment of the vertical clinging Malagasy lemurs, galagos, and tarsiers were obviously sound. Much of the following debate focused, correctly, on the applicability of those conclusions to the fossil postcranial morphology, the area of anatomy that should have been logically the most significant for locomotor assessment of the fossil record. But that is not what happened.

Visual Predation as the Strategy for the Stem Lineage of Euprimates

Cartmill (1972) has presented the ambitious "visual-predation hypothesis" based on cranial attributes and grasping hands that was to explain the whole diagnostic structural make up of the protoeuprimate, and, at the same time, came to de-emphasize not arboreal locomotion as such, but the importance of grasping related leaping that shaped this ancestor. The whole argument was an attack on the straw man of "arboreality," without any consideration given to the multitude of ways that adaptations may be required to fulfill various kinds of positional regimes, arboreal or otherwise. The fact that clawed hands work very well in all predatory mammals, from opossums to cats as a tool for prey capture, however detected, was coupled with the need to climb cautiously and grasp tightly on small branches. With an emphasis on grasping hands and the loss of the claws coordinated with stereoscopy and appropriate neurology, Cartmill has relegated the powerful and larger grasping hindfeet (compared to the hands) as a means to allow "...to move cautiously up to insect prey and hold securely onto narrow supports when using both hands to catch the prey" (p. 440). What was largely missing from this overarching hypothesis is the accounting for the skeletal evidence known by then for a number of early euprimate lineages. Cartmill (1975) further developed his views along similar lines. It should be emphasized here that Cartmill's (1972, 1974) views (or those of Hamrick, 1998) regarding the reduction of claws are not supported by the targeted selective loss of the falcula on the hallux in didelphids and descendants. The correlation in extant marsupials appears to be with the powerful grasp of the pes and a postulated selective disadvantage of the sharp falcula on the hallux on smaller branches.

What also complicates matters of historical reconstruction regarding the evolution of various published perspectives on primate morphotype locomotion, and the implicit assumptions that these views rest on, is the apparent inconsistency of some published views. Issues of phylogeny, latent in any adaptive hypothesis, but almost always implicit when they should be explicit, point to some critical inconsistencies in the presentation of the visual predation hypothesis of Cartmill. For example, Cartmill (1974: 74) has given confusing testimony about the historical context of his views on claw "loss" (part of a transformation series of the homologues called digital ungulae) in the protoeuprimate. In fact, Cartmill (1974) and later Hershkovitz (1977) have strongly supported the transformation of falculae (claws) into the tegulae of platyrrhines independently from other euprimates, with Cartmill, in particular, arguing for "greatest parsimony." This is particularly puzzling because Cartmill's "visual-predation" hypothesis launched in 1972, and expanded in 1975 was critically dependent on the assumption of a nailed condition in the euprimate stem. Regarding the loss of claws in euprimate ancestry Cartmill (1974: 74) says that: "The comparative anatomical evidence indicates that the hands and feet of the last common ancestor of the extant primates [i.e., Euprimates] must have resembled those of the opossum; claws have been lost independently in four or five parallel lineages of primates." On the same page further down Cartmill explicitly supports the notion that claw loss can be the result of a "...trend toward increased size in animals inhabiting the higher strata of tropical forest, or from the restriction to the lower strata of a relatively treeless heath or scrub floral community." It is also relevant here that Lewis (1989, based on a series of articles published in 1980) explicitly supported an arboreal ancestry for the last common ancestor of the fossil and living placentalian mammals—a view that Martin (1990) has continued to champion. This appears to be decidedly untrue for the Placentalia, and probably also for the stem of the Early Cretaceous Eutheria as well.

To put it bluntly, contrary to pronouncements, the comparative anatomical evidence never "indicates" anything; one explicitly tests and interprets homology hypotheses, which Cartmill did in an unacceptable way (see detailed discussion of this in Szalay, 1981b: 40–44). But the most striking feature at that time, given the "visual-predation" hypothesis (which one might have thought was based on a homology-based phylogenetic position, i.e., "claw-loss" and postorbital bars) was the concept of parallelisms in Cartmill's theoretical and historical-narrative explanations (i.e., the recurrence of parallel trends in the evolution of euprimates).

Added to this, I believe, was a connection to the "Plesitarsioidea" versus "Anthrolemuroidea" view of primate phylogeny, an interesting historical curiosity (Gingerich, 1974; 1975a,b; see also Krishtalka and Schwartz, 1978; and Schwartz et al., 1978) which is relevant here. This view of primate phylogeny, which posited an unacceptable wedding (then or now) of the plesiadapiforms and one of the early euprimate groups (the Tarsiiformes) as a clade, represented at that time a significant manifestation of primate evolutionary studies in contrast to the strepsirhine–haplorhine dichotomy advocated by others. The disregard for the very accessible postcranial evidence of fossils (Szalay et al., 1975) and the extant postcranial osteology by both the proponents of the "visual-predation" hypothesis and the taxonomic notion of the "Plesitarsiiformes" (this latter derived from, and synonymous with, the "Plesitarsioidea") points out that postcranial attributes (at the level perceived by these authors) were considered (if examined at all) as rife with "parallelisms," hence not very reliable.

But subsequent to Cartmill (1972), Szalay and Decker (1974), and Szalay et al. (1975) have assessed the then known skeletal collections of *Plesiadapis* (the former study emphasizing the tarsus the latter the remainder of the skeleton) and concluded that the only reasonable explanation of the evidence was unquestionable arboreality for the archaic plesiadapiforms. Szalay and Decker (1974), Decker and Szalay (1974), and Szalay et al. (1975) emphasized in particular both the similarities to but also the differences in arboreal adaptations in the tarsus between plesiadapiforms and euprimates in contrast to latest Cretaceous eutherians. These conclusions were dismissed as doubtful by Cartmill (1975: 32), without any indication that he considered the evidence. But then, it appears, that Cartmill was wedded to the notion that arboreality was primitive for the ancestry of living placentalian mammals, and therefore, his attacks on the arboreal theory of primate origins, as he called it, were justified only on the grounds that attributes related to other than some specific arboreal locomotion were necessary to explain the origin of both the Plesiadapiformes and the Euprimates. Martin (1990), in his text also insisted that the plesiadapiforms simply retained arboreal modifications already present in a remote placental ancestor. This unfortunate disregard for the fossil evidence (dubbed as "special problems of the fossil record" by Martin, 1986: 4) was also evident earlier. [Martin's statement (1986: 23) about *Plesiadapis* that its hallux "might have been totally lacking," is particularly revealing in light of the fact that in the same volume Gingerich illustrates and makes a note about the preserved big toe, suggesting a lack of familiarity with the record. Yet, this unfamiliarity with the specifics of fossil evidence did not prevent that author to present high profile discourse about fossil primates elsewhere as well (see for example Martin, 1993)].

There can be little doubt that there was a nearly complete disconnect between phylogenetic thinking and adaptive assessment by Cartmill (1975: 32–33) when one reads that "[if] the characteristic primate traits are the result of progressive adaptation for arboreal visual predation in one line of descent from an early plesiadapoid... thrusting the plesiadapoids...back into the ancestral order Insectivora would make the order Primates more coherent, However, we must not forget...[that]...If, for instance, it turns out that anaptomorphids arose from very early paromomyids, while adapids evolved separately out of the earliest plesiadapids, it might still prove true that the lines leading to the Eocene families went through an adaptive shift to visual predation, in parallel in two different lineages...". It is difficult to see how the more complex areas of the skeleton, particularly the carpus and tarsus, failed to convince these authors both about the unequivocal arboreality in the plesiadapiform ancestry and the unquestioned monophyly of the Euprimates, except if one considers the overwhelming "scenario" bias by Cartmill and a then prevalent dental mindset by Gingerich and associates. After the widely available postcranial evidence had been repeatedly pointed out in the literature in the 1970s and 1980s (Szalay, 1972) the polyphyletic notion of the euprimates was finally abandoned.

It should be added here that parsimony (a useful notion if properly applied to not only relevant "facts" but to all the complex interpretations necessary in the construction of tested hypotheses regarding properties) was much used then as it is now. Such procedures, however, rapidly (and properly) turn into a series of Bayesian considerations. This is an approach not much appreciated by Popperian systematists in primatology who became advocates of a falsificationist approach to cladogeny based on algorithm research, as opposed to an incremental research program leading to phylogeny estimation (e.g., Szalay, 2000). The unfortunate reality has been, however, that either erudite and literary rhetoric about scenarios or unexamined character lists require more than "parsimonious thinking" or scholastic Aristotelian logic (algorithmic or not) for nonmonotonic testing procedures in evolutionary morphology and the testing of historical-narrative explanations. The arguments about plesiadapiform and euprimate relationships and adaptations, and the methods of assessment, continued in the literature.

Kay and Cartmill (1977: 19) in their restudy of a crushed skull of the Torrejonian Paleocene *Palaechthon* concluded that while euprimates were derived from plesiadapiforms, the cranial adaptations of the latter (exemplifying primitive plesiadapiforms) reflect a "...predominantly terrestrial insecteater, guided largely by tactile, auditory, and olfactory sensation in its pursuit of prey." Even more interestingly (and in stark contradiction to Cartmill's views on ancestral placental arboreality), they noted that "Adaptations to living in trees and feeding on plants probably developed in parallel in more than one lineage descended from the ancestral plesiadapoids." It was pointed out subsequently in a critique by Szalay (1981a: 157) that Kay and Cartmill in their analysis of the cranial evidence based their conclusions regarding plesiadapiform adaptations on: (a) nonphylogenetic and static assumptions, (b) misinterpretation of the form and mechanics of the attributes analyzed, and (c) employment of irrelevant characters for the establishment of substrate preference (e.g., infraorbital foramen size). Szalay criticized the general outlines for adaptational analysis espoused by Kay and Cartmill, and the positions taken by these contending parties on the type of character choices and functional interpretations are still the general positions that endure in many debates today. Namely, in dealing with fossils, how should one approach the difficult issue of adaptational assessment (see Szalay, 2000, contra the arguments offered by Anthony and Kay, 1993: 374)?

My arguments in 1981 were in juxtaposition to the practice of indiscriminate use of ancestral characters that could be correlated with some habitat in living animals (e.g., the relative size of the infraorbital foramen in archaic primates used by Kay and Cartmill, 1977, to argue for terrestriality in paromomyid plesiadapiforms). While the persistence of functional correlates of even primitive traits can be useful in framing an adaptational analysis, primitive traits are often revealing of ancestrally acquired adaptations within a different context. The human thorax, shoulder complex, and elbow joint are good examples. These heritage traits, a group's synapomorphies, set the limits for various trajectories of the more derived features. For example, the contact of the fibula with the femur, and also via the parafibula (the fibular fabella), correlates only with some aspect of therian primitiveness in the knee complex, but no ecologically meaningful differentiating function can be associated with it in marsupials that show different habits today. Both the most arboreal and terrestrial marsupials have this as part of the knee complex, although instructively, with different conformation of the proximal fibula. The extreme narrowing of the proximal fibula (and attendant muscular and mechanical correlates) occurs only in highly terrestrial metatherians (see later section). Similarly, the repeated narrowing of the lateral femoral condyle in terrestrial didelphids, bandicoots, basal, and all other kangaroos, as well as in the ancestral placentalian, also closely predicts terrestriality (Szalay and Sargis, 2001). But the narrowing of the proximal fibula that also occurred in protoplacentalians does not rewiden again in Cenozoic and recent arboreal eutherians, nor does the medial femoral condyle changes its proportions. The extant eutherian lineages (and their fossil relatives which postdate the stem of these) are likely all derived from the terrestrially modified eutherian that was the stem of the Placentalia.

The Role of Leaping in the Ancestral Euprimate

By 1979, Szalay and Delson noted that the likely breakthrough *from* an arboreal plesiadapiform ancestry (unequivocally suggested as such by Szalay and Decker, 1974, and corroborated beyond any reasonable expectation by the efforts of Bloch and Boyer) involved "...the establishment of *grasp-leaping* arboreal adaptations ...necessitated by a particular feeding regime" (p. 99) for the stem of the strepsirhines, considered by them to be the best approximation of the euprimate stem. Szalay and Dagosto (1980) in their extended discussion of what they defined as *morphotype locomotor modes* (a concept which incorporated a phylogenetic context into the assessments of locomotor behavior/anatomy) have discussed *claw-climbing* as reflected in the protoplesiadapiform condition. They also emphasized in some detail that the interpretation of skeletal features strongly supports grasp-leaping as a *monophyletic acquisition* of the protoeuprimate. They essentially agreed with Le Gros Clark (1959) that arboreal locomotion (but a particular type) was likely part of the causal nexus of the cranial features one observes in the Eocene primates—a foundation on which modern diversity is based. They disagreed with Cartmill's hypothesis, and stated that "The greater importance and more severe selectional consequences of judging distances by quadrumanous *fast* grasp-leapers would clearly put a greater premium on stereoscopy than just running and walking along branches in an arboreal environment. There is no evidence for uniquely associating quadrumanous primate grasp-leaping with arboreal insectivory-omnivory. The first euprimate grasp-leaper may or may not have been primarily phytophagous, zoophagous, or ominivorous." (p. 35).

In 1992, Cartmill reviewed, with candor, the differences between the grasp-leaping and the visual predation hypothesis as contributing causal factors in the development of the protoeuprimate cranioskeletal complex, although he continued to think of "arboreality" as some monolithic causal agent. He correctly cites my often-stated view (following those of, e.g., Darwin, Gregory, Matthew, and Simpson, and others'), namely that evolutionary transformations are constrained by history in a highly contingent way, and that the new adaptive solutions mirror that heritage, often to a considerable degree. This view, in light of the prevalence of mosaic evolution (bolstered by an understanding of modularity by students of EvoDevo), demands character level, rather than a taxic, analysis of homologies (the former dubbed as null-group comparison; Szalay, 1994; Szalay and Bock, 1991). In order to arrive at reasonable phylogenetic estimates of character complexes (and subsequently taxon phylogeny hypotheses), the development, functional biology, and adaptation of taxonomic properties need to be considered, in contrast to the declared primacy of algorithm-based rooting with taxic outgroups.

Nevertheless, following our debates of the extant and fossil evidence regarding the ancestral stage of euprimate locomotion, Cartmill has come to consider the issue of phyletics of characters and even the notion of (Darwinian, i.e., evolutionary) homology somewhat moot points (see Cartmill, 1994, on the issue of homology hypotheses; and Cartmill, 1990, for his rejection of historical-narrative explanations as science). In arguing against the grasp-leaping euprimate locomotor mode, Cartmill (1992: 107) noted that "...particular evolutionary events cannot in principle be explained except as instances of some more general regularity," and also stated that "...adaptation to a grasp-leaping habit unique to euprimates, explains nothing." He has professed this belief in a variety of ways, in fact arguing against the very practice of historical-narrative explanations in science. I (and others) completely reject such ahistorical theoretical assumptions about the nature of science.

Cartmill (1992: 107) was correct in stating that other arboreal mammals "...do not look much like euprimates." Of course, few other arboreal mammals (with their independent heritage) do the acrobatic antics of those grasp-leaping lemuriforms whose general skeletal anatomy shows the same derived suite of features that can be reasonably attributed to the protoeuprimates as well. And those skeletal attributes appear to be diagnostic of the order based on the Eocene evidence (i.e., they represent a derived suit of features of the stem). But Cartmill's (1992) discussion of the issue of the euprimate morphotype locomotor mode, including his evaluation of the proposals of Sussman (1995) and Rasmussen (1990) were, in my view, deeply flawed. This was so not only on the theoretical grounds regarding his perspective on how one employs living model species to evaluate fossil animals (e.g., Szalay, 1981a,b; Szalay and Sargis, 2001; and later section). But perhaps more importantly than anything else, Cartmill continued to make only casual, if any, use of the highly specific and functionally well-understood aspects of postcranial morphology for interpreting the fossil postcranial evidence when discussing locomotion in the euprimate stem. This is odd enough by itself, but the postulate that (rapid and frequent) leaping and precise landing by grasping small branches has obvious consequences for both the nervous system and vision should not have been ignored. Habitual great leaping ability in the three-dimensional arboreal environment would certainly suggest a causal relationship to enhanced vision and attendant neurology. And to consider the reduction of the snout, olfaction is far less important for the execution of a leap than visually judging distance and points of landing among variable-sized branches for animals, whose size we cannot be certain of. Nevertheless, the issue remains a particular type of arboreal locomotion (grasp-leaping), not just "arbo-reality," and testing of that issue resides primarily in the mechanics of the joints of the skeleton of an inferred common ancestor and their near-fossil relatives.

The general area of modeling ecological morphology and its use for fossil species (see later section) is a lot more complex but also far more applicable than Cartmill's (1992: 107) statement that only parallelisms can be explained adaptively. For example, Szalay (1981a) argued against the thesis presented by Kay and Cartmill (1977) that large infraorbital foramina of the plesiadapiform *Palaechthon* pointed to a terrestrial, hedgehog-like habitus. I pointed out the difficulties of judging habitus (real-time adaptation in a species) based on primitive features because primitive features, while perfectly functional (obvi-ously), do not reflect the most recent shifts in a lineage, unlike their derived attributes. Convergences of complex derived attributes of recognized mechanical consequences, however, are powerful "postdictors" of the habitus of fossil species, and are the most potent tests of historical-narrative explana-tions. I showed that relatively very large infraorbital foramina persist in some very arboreal species. Therefore, such features simply cannot be very useful in interpreting fossils, "parallelism" aside. Rather instructively, the size of vari-ous foramina continues to have a rather checkered history in predicting any-thing, including scenarios pertaining to the hominid realm.

It is exactly the rejection of the analyzed, ordered, and polarized use of character states of homologous features that is missing from the notion of "parallelism" dictated by Cartmill's views on homology. Is one's assessment of parallelism the result of parsimony analysis? Are we considering some con-vergent aspects of features, given distinct phylogenetic/taxonomic contexts? For establishing convergence (a tested, and failed homology hypothesis, with-out the somewhat obfuscating discussion and mixing of levels of organization by Lockwood and Fleagle, 1999), however, one should have some criteria other than the leftover traits expressed as a consequence of "CI" indices of parsimony-derived taxograms. The notion of convergence that Cartmill sub-scribes to in his pledge to taxic analysis as the arbiter of the nature of similar-ities is, *ipso facto*—a residue of a "losing batch of synapomorphies" that one now calls "convergent" (see Szalay, 2000). *But beyond how homology is estab-lished with some probability, there is the key issue of what particular conver-gent/parallel properties one is going to employ to explain a particular facet of adaptational history or a fossil species.*

R. H. Crompton (1995) has presented a detailed analysis of the literature (albeit with some studied omissions) regarding the origin of feeding and loco-motor strategies of the euprimate ancestor. He has paid laudable attention to the connection that must exist between feeding and locomotor strategies. His conflation of the arboreal and scansorial strategy that was suggested for the plesiadapiforms by Szalay (1972) is taken by him as that for the protoeupri-mate, one that is a minor *lapsus* by a primatologist with little practice in sys-tematics or acquaintance with the fossils. What is, however, a recurring pattern in his critique of Cartmill (as well as in Cartmill's own previous con-tribution) is the consistent lack of attention paid to the details of the fossil dental and postcranial evidence. The circular "chop" diagrams of "total adap-tive strategies" of various extant primate species published by Oxnard et al. (1990) are hardly a substitute for the independent assessment of the relevant fossil or even extant evidence. Unfortunately, a remark by Crompton (1995: 19) that a general arboreal form of locomotion "...is typical of many small, primitive mammals...," has less meaning than no statement at all. Within the even conventionally accepted concept of Mammalia, different groups undoubtedly had different primitive locomotor patterns (i.e., morphotype locomotor modes) with their attendant morphological properties that are amenable to specific model-based analysis (see later section).

[One would, in general, hope for the recognition by students of living pri-mate ecology that feeding and locomotor strategies are primarily reflected in the morphologies of the relevant regions of the hard anatomy. Furthermore, it is appropriate to state here that feeding and locomotion can be decoupled not only in terms of morphological mosaic evolution, but also in terms of var-ious solutions for the feeding/locomotion dilemma faced by all lineages. Nothing better exemplifies the mosaic nature of adaptive solutions than the variety of strategies seen within the lorisiforms—a group cited repeatedly by R. H. Crompton.]

Contrary to Crompton's statement (1995: 21), which is relevant here, there is no morphological evidence of any sort that would suggest dwarfism in the ancestry of the living tarsiers, only perhaps if one assumes a large-bodied hap-lorhine ancestry. As I noted earlier, we cannot be certain of the size of either euprimate, strepsirhine, or haplorhine ancestries, even though great antiquity does tend to preserve some aspects of morphology that indicate general func-tional features. Tarsiers are well within the size range of the group—which they are a relict of—the fossil Tarsiiformes of the Eocene. Their enormous eyes are probably a reflection of the compensation required by secondary nocturnality in a probably diurnal lineage that has shed the tapetum lucidum in its ancestry. But in his conclusions, Crompton (1995) seems to agree with the locomotor mode designation that was proposed by Szalay and Dagosto (1980, 1988) as grasp-leaping, and which was specifically tied to ancestral euprimate postcranial morphology and its inferred biological role (Crompton did not use that term, nor did he cite the 1980 article). Crompton's conclusions certainly corroborate those of Szalay and Dagosto (1980, 1988). Crompton, in spite of his strong disagreement with Cartmill, however, goes on to endorse the dietetic component of Cartmill's visual predation hypothesis. Unfortunately, there is no evidence from the dentition of the earliest euprimates, or from the best estimates of the adaptations of the morphotype of euprimates, that insectivory and predation were the preponderant ancestral dietary strategy. The postcranium and inferred leaping is neutral on that important question. The variety of dental pattern is great, however, so inference as to diet is at best a variety of fruit, flower, nectar, gum, and insect feeding, with no clear-cut emphasis in any reconstructed common ancestor.

It must be stressed that early dietary strategies in the protoeuprimate are not as yet understood, in spite of the often-cited deductive argument of Kay (1984) based on the body weight and diet of living primates, asserting that size is a predictor of diet. Body size is also often inferred from fossil teeth themselves, often a poor measure. According to that view, small fossil primates were, *ipso facto*, primarily insectivores—a gross oversimplification even on general grounds restricted to living primates as models. Assertions that because some small living lipotyphlans or primates are primarily insectivorous, all small fossil primates had to be as well, are divorced from morphological analysis. Many small fossil primates (as well as marsupials) with the appropriate dental and cranial attributes were probably oblivious to "Kay's rule" (contra Kay and Covert 1984) when it came to their dietary regimes. Similarly, the extant *Hapalemur* and *Lepilemur*, or even cheirogaleids, do not adhere to such a rule. Morphological and functional patterns, in light of the appropriate models (but not size alone) supply convincing paleobiological explanations. As argued before (Sussman, 1995; Szalay, 1968, 1969, 1972), the dental evidence leaves little doubt that among early plesiadapiforms and euprimates, a mixed feeding strategy, evidenced by relatively low crowned and quasi-bunodont cheek teeth was likely to be both the ancestral and one of the more widespread conditions. One has to look no further than the variety of small rodents who find ample energy and nourishment primarily from seed consumption, ignoring this general "rule." Small fossil primates were not necessarily obligate insectivores unless their morphology corroborates such assessment.

It is also of some importance to note here the relevant point that contrary to Martin (1993), the radiation of a mammalian group is not usually that of an algorithm-based inverted pyramid, and therefore, the living radiation of primates is a poor foundation to model the early story that was driven by ecological context and biogeography, in spite of the putative elegance of such iconography. Given the enormously more extensive favorable habitats for primates in the Paleogene of Holarctica (and probably Africa as well), experimentation of many early lineages among the euprimates probably resulted in a far greater diversity of small omnivorous primates than there is today. An understanding of the fossil record helps in this regard. Massive extinctions with the changing of habitats have resulted in a pattern nearly the opposite of the computer-generated diagram of Martin (1993).

Sussman (1995) and I are in broad agreement on the importance of frugivory early in primate evolution. Regarding the close relationship of habitat and primate strategies, it is perhaps important to note that primates did not "invent the rainforest," although they certainly carry on the roles started by other clades. At least in South America, where primates did not arrive until relatively late in the Tertiary (and certainly never in Australia and New Guinea until humans ventured there), the radiation of arboreal marsupials was well under way since the Latest Cretaceous or Earliest Paleocene in tropical rainforest environments of South America, and sometime later in the antipodes. And even prior to that, a variety of atribosphenic mammals undoubtedly interacted in a number of ways with the tropical forests and angiosperms of that continent. It should be emphasized that the derived suit of postcranial traits of the stem euprimate certainly does not preclude a reliance on fruits, flowers, gums, or seeds, together with insects as the main items of its dietary regime, although such a diet can be attained by a whole variety of ways other than grasp-leaping. The most corroborated explanation for the morphotype skeletal evidence, however, is a regular practice of bounding leaps and landing with a hindfoot/forefoot grasp ("grasp-leaping"). But such interpretation, of course, does not mean that an animal with such morphology cannot slowly climb, walk, shamble, or in any other way get to its food, or stalk insects. But leaping does make a particular combination of *energetic and competitive* sense when particulate and discontinuously distributed clumps of food are sought after by many parties, both intra- and interspecific.

It is gratifying indeed that the general idea of grasp-leaping as the euprimate morphotype locomotor mode is so thoroughly circumscribed and argued for and advocated in all but name (i.e., without reference to the article by Szalay and Dagosto, 1980, where the hypothesis was first explicitly outlined and supported) by R. H. Crompton (1995).

MODELS AND THE LOCOMOTOR STRATEGIES OF EXTINCT TAXA

Central to paleobiological research that aims to explain both aspects of behavioral ecology of extinct forms and patterns of historical factors (these efforts are usually limited to dietary and locomotor strategies) is the analysis of skeletal remains. Living species models with their rigorously analyzed form-function attributes and their ecological causality lay the foundation for not only character analysis in systematics (as opposed to taxic analysis), but also for analyzing, through the use of convergence and matching, the form-function of the fossils as well (Szalay, 2000). Biomechanical generalities, such as occlusal mechanics of teeth or the loading of joints are paramount, but because, due to the uniqueness of lineages, there are no living species that match exactly the habitus of fossil entities.

It is not unusual that a living analog is used to find similarity (a concept fundamentally context- and paradigm-driven) for some sort of fossil morphology without functional, and therefore, causal reasons. The lack of a causal analysis (i.e., ecological, real-time) in the process of modeling does not allow one to conclude that selected matching morphologies indicate adaptive (ecological) similarities between the model and the fossil. Nevertheless, this approach can supply some meaning for paleobiological assessment if a whole skeleton is available for the fossil. Without complete specimens, however, the modularity-based and well-corroborated patterns of mosaic evolution render such assessments problematic for functional units of the skeleton. Such a general similarity evaluation lacks, as its basis, the necessary character analysis that functional-adaptive approaches provide and which test the nature of similarities before these are used either for paleobiological assessment or phylogenetics.

Modeling relies heavily on theoretical perspectives, as well as the experiences of the modelers with the subjects that they are focused on. A far more desirable procedure than mere similarity matching is the construction of mechanically and adaptively meaningful relationships in character complexes in a number of distantly related species that display attributes which are more likely convergent than homologous (e.g., Szalay, 1981a). One may call this a convergence-based "modular-function" approach. It is important to have some strong ecologically compelling evidence that certain recurrent attributes are (given a similar level of basic mechanical organization of the skeletal biology) under strong selectional imperatives for their recurrent development. An understanding of functional-adaptive significance (and consequently the probability of convergence versus homology of properties) is decisive in establishing a list of tested taxonomic properties. This approach has both an inductive component in using the recurrent correlations between morphology and mechanics and the ecological context, as well as a deductive one in applying the correlations to the fossil taxa. *Uncovering consistently convergent, biomechanically significant, features that have strong functional associations with either feeding or locomotor strategies in the skeleton of extant mammals does supply us with powerfully modeled "postdictors" for adaptations in the fossil record.*

Furthermore, if the probability is high that one or more aspects of properties in two or more taxa examined are the result of phyletically independent adaptive responses (rather than ancestral constraints), then, such convergent attributes (not to be considered taxonomic properties at a level higher than species) become excellent indicators of ecologically meaningful aspects of the fossils under study. Once the initial and boundary conditions (both phyletic and adaptive in a morphotype) are established for extant model species, and the fossils can be placed in a particular ecologically meaningful framework, then further analysis of the attributes of these fossil taxa becomes properly constrained for phylogenetically useful character analysis.

Models are particularly significant as they represent results of judiciously chosen surrogate evolutionary processes for a particular set of adaptive transformations. These selected extant models are chosen based on form-functional considerations with the heritage attributes *often necessarily de-emphasized!* These tested models (i.e., whose causal correlations with their various biological roles are well understood), as noted above, like all models, can never be a complete match for extinct organisms, or their aspects, that are subjected to analysis. Nevertheless, when size is controlled for, and functional (mechanical) attributes are correlated with some well-understood adaptations in the living models, many behaviors can be inferred for those fossils that share these features (for detailed examples see Court, 1994, for assessing *Numidotherium*, or Cifelli and Villarroel, 1997, for an interpretation of *Megadolodus*). Such procedures provide a corroborated level of character explanation (to varying degrees), both functional (nomothetic, nomological in essence) and phylogenetic (evolutionary, i.e., unique, idiographic, historical). Szalay and Sargis (2001) have demonstrated this to be the case in their use of selected osteological attributes of four extant model species of metatherians (boosted by numerous other examples examined there in less detail) for interpreting adaptive strategies in fossil marsupials.

In light of the foregoing I should comment here on the use of *Caluromys*, and various concepts of the didelphid ancestry, as models for interpreting the origins of the euprimates or their relatives, the plesiadapiforms. Morphology is the only point of reference that fossils can offer for analysis, and similarly, the assessed morphotype locomotor mode of a group is grounded in osteology. This should be connected with functionally well understood similar, or instructively contrasting, morphology in proper models that represent aspects of extant species, whose biological roles have been well investigated. Explaining fossil morphology should not consist of picking a living species based on some *behavioral criterion*, and stating categorically that its behavioral or physiological state (or another attribute) was probably similar to that in a postulated fossil taxon or an inferred common ancestor. Unfortunately, sometimes this has been done in primatology (not frequently, fortunately) even when the morphology of the designated extant "model" is singularly dissimilar to the inferred fossil condition. This dissimilarity is not only phyletic (as expected) but functional as well. The use of some marsupials is a case in point. For example, Rasmussen (1990) chose the didelphid *Caluromys* as a "model" for the protoeuprimate. Some of the factors he recognized, regarding arboreal adaptations of the euprimates, were no doubt correct, but these are not diagnostic of the stem of that clade. The type of arboreality displayed by arboreal didelphids, however, is a very good approximation of what the emerging evidence suggests for plesiadapiforms. *Caluromys*, therefore, may be a very good model for the origins of arboreality for the archontan or plesiadapiform stem.

Rasmussen (1990) posited that the relatively large brain and eyes, small litters, slow development (meaning postparturition because preparturition development is nearly uniform in all didelphids and fundamentally different from the universally "accelerated" condition of eutherians when these are compared to metatherians), and agile locomotion (compared to clumsier similar-sized arboreal didelphids such as the not infrequently terrestrial and scansorial species of *Didelphis*) represent a suite of attributes that is convergent to the euprimate ancestor. He stated (p. 263) that these "analogous...selection pressures, represent an independent test of the arboreal hypothesis,...the visual predation hypothesis,...and the angiosperm exploitation hypothesis of primate origins." Regrettably, the prehensile-tailed *Caluromys* does not have special similarity in its osteological properties to the diagnostic conditions of early euprimates, and therefore, cannot support the consensus of views envisaged by Rasmussen. Nevertheless, this was a useful analysis in that it resignaled the importance of marsupials for the study of archaic primates. However, among its numerous critical attributes the protoeuprimate, unlike the clawed *Caluromys* (which occasionally indulges in small leaps), had nails (for details see later section) and had a hindleg superbly adapted for leaping. No extant and arboreal marsupial comes close to the level of biomechanical attributes displayed by the Eocene euprimates. There are no osteological attributes of *Caluromys* that parallel euprimate osteological features, and therefore, this genus (or any didelphid) is an inappropriate model for interpreting euprimate ancestry. But a strong case can be made that, osteologically, *Caluromys* probably approximates a good model for the arboreal protodidelphid (but not for the didelphidan or sudameridelphian ancestry)—one that significantly differed in its advanced arboreal abilities from the postcranially more primitive sudameridelphians of the Paleocene (Szalay, 1994; Szalay and Sargis, 2001) whose stem, in a departure from Cretaceous metatherians, may have been more terrestrial. The well-known agility of *Caluromys* (and other didelphids as well) compared to *Didelphis*, which is quite scansorial and is at home on terrestrial substrates, does not provide evidence for the argument that the agile arboreality of *Caluromys* is a derived condition within the Didelphidae. Many smaller species of didelphids are also quite agile and quick in an arboreal environment (see discussion of the Didelphidae in Szalay, 1994). Although a proposed model species like *Caluromys* tells us little about the origins of euprimate skeletal morphology (and therefore the inferred habits from that), it does, however, as noted, may be very useful for comparisons with archontans and plesiadapiforms. The stem euprimate lineage was likely transformed, via a still poorly understood arboreal archontan stage, from an essentially terrestrial placentalian heritage into an ancestor with a relatively well-understood primitive euprimate postcranial state whose obligate leaping behaviors were not unlikely (Dagosto, 1988; Dagosto et al., 1999; Szalay et al., 1987).

In attempting to explain arboreal attributes of the inferred common ancestry of euprimates, Lewis (1989, and references to his previous articles therein) has derived the various primate attributes from an essentially didelphid condition—the latter standing in as a surrogate for a "marsupial stage" prior to eutherian arboreality. Neither the phylogenetically troubling details that primates are eutherians with their own highly taxon specific constraining heritage that circumscribes their morphology, nor the fact that didelphids appear to be a particularly derived arboreal clade among South American Metatheria, have constrained Lewis' explanation. His transformational analysis lacked the necessary and appropriate phylogenetic context. Furthermore, many of the problems with his proposed transformations were also due to a lack of ecomorphologically meaningful assessment of details. The general notion that some aspects of marsupials are probably primitive (e.g., their reproductive or developmental patterns) compared to their eutherian homologues does not mean that there is a functional similarity between eutherian skeletal attributes and those of didelphid marsupials (Szalay, 1984, 1994). Hence, the same applies even more emphatically to any attempt to understand euprimate origins based on didelphids.

Another inappropriate use of various modeled conditions of metatherian and eutherian skeletal adaptations was made by Martin (1990). He provided narratives, based on the contributions of Lewis (summarized in 1989), that were supposed to connect (historically!) metatherian morphology to the Paleocene plesiadapiform evidence, certainly well understood by that time in *Plesiadapis*. The explanations advanced by Martin heavily relied on implicit assumptions about the relevance of didelphid attributes for evaluating fossil eutherians. Martin confused the application of modeled properties in his text. He presented a lengthy, literature-based analysis of selected osteological attributes of euprimates and their possible closest relatives, specifically the plesiadapiforms, colugos, and tupaiids. In writing about the evolution of mammalian locomotion, primate arboreality, and the specifics of the osteological evidence retrieved from the literature, a number of issues that relate to modeling and phylogenetic analysis of the metatherian-eutherian dichotomy framed his account. His views on the alleged homology of arboreality in marsupials and protoplacentalians, on the supposed "primitiveness" of the cheirogaleid primates within the euprimates, and the use of the various didelphid attributes for an arboreal habitat preference have provided confusing examples of modeling. Additionally, gross mistakes were committed when critical morphological details were misperceived or mistakenly reinterpreted from the literature.

It needs to be emphasized how important unexamined assumptions can be in any search for causal explanations of euprimate origins. Martin interpreted morphology in light of his assumption that ancestral placentalians were arboreal—a view which framed his ideas on the origin of the euprimate radiation. Interestingly, one who believed that the stem placentalian was arboreal (and who categorically continued to dismiss the relevance of the Plesiadapiformes) could accept the Archonta in spite of the fact that the modern rebirth of that concept (Szalay and Decker, 1974) was largely based on diagnostic arboreal adaptations (albeit taxon specific ones). Martin's published illustrations do not represent the actual morphology that he used to support his views. He overlooked, and missed the significance of the fact that, unlike the relatively free upper ankle joint adjustments in such primitive living marsupials as didelphids (with their meniscus mediated fibular contact that puts little restraint on the upper ankle joint laterally), the protoplacentalian condition has evolved considerable tibial and fibular restraint for the upper ankle as reflected by the astragalus.

Similar, but taxon specific and independently evolved ankle restriction patterns can be found in obligate terrestrial marsupials like peramelids and macropodids. Martin and others failed to recognize (even though this has been painstakingly detailed in the literature) that the extensive lower ankle joint adjustments of plesiadapiforms, euprimates, and all other obligate arboreal placentalians became constrained by the protoplacentalian adaptation, and that the most extensive adjustments to pedal inversion have invariably occurred in these taxa in the lower ankle joint. As a result, evolution of a morphological complex in the lower ankle joint that facilitates inversion is invariably a derived condition among early placentalians that show such morphotypic attributes, albeit convergently, such as archontans, some lipotyphlans, creodonts, carnivorans, and rodents.

HOMOLOGY IN EVOLUTIONARY MORPHOLOGY

The issue of homology testing cannot be divorced from any discussion of adaptation and phylogeny. So these remarks are very relevant here. It was only in the 1980s that many primatologists and other students of fossil mammals

increasingly accepted the notion that the determination of levels of relatedness between lineages was not tied to any one kind of evidence, such as teeth
or skulls only, but that the whole skeleton (along with other attributes, of
course) was at least as important. What matters in phylogenetic estimation is
the nature of complexity of properties that are being utilized, as well as the
relevance of these to adaptive solutions. The latter assessments aid in the
recognition of heritage features, and the particular stage of evolution designated of a character complex (its polarity), not necessarily in that order.

But what renders discourse sometimes nearly impossible, however, is the
assumptions (both implicit and explicit) of some workers about homology.
Some have stated recently that phylogenetic or Darwinian homology (as
opposed to Owen's views) is "logically" flawed. Such remarks overlook the
fact that a theoretical definition of homology requires specific hypothetical
statements regarding properties in different species, and that these hypotheses are to be operationally and independently tested against specific criteria
relevant to the proposition. Much more cannot be asked of any other science
(contra Cartmill, 1990).

So, impediments to the practice of testing phylogenetic homology are
views that relate to the credo of parsimony cladists, whose assumptions were
explicitly espoused by Cartmill (1994). The roots of such a change are
difficult to trace in anyone's contributions, but the issue of morphological
homology was undoubtedly troubling for Cartmill. In spite of the long and
erudite introduction and his selective use of the literature that led up to his
changed views, what remains is Cartmill's acceptance of algorithmic
analysis as the ultimate arbiter of homology testing. The tone of the bottom
line has the customary declarative "truth component" of theorizing by
parsimony cladists. "The concept can be made intelligible in an evolutionary context only by giving it a cladistic interpretation that makes homology
judgments dependent on the outcome of a phylogenetic analysis. It follows
that such judgments cannot play a role in evaluating conflicting
phylogenetic hypotheses" (p. 115). Clearly, for Cartmill, they cannot, but
they certainly did and continue to do so for the assessment of a large and
growing body of phylogenetic hypotheses, even if many feel the necessity
for an algorithmic, *a posteriori* cloak to legitimize their efforts within a
Kuhnian community.

Similarly, Lieberman's (1999) generally peculiar stance on the "relative
goodness" of homology hypotheses, but particularly Lieberman's (2000: 152)

opinion, misses the theoretical versus operational empirics of homology evaluations. In his deceptively authoritative sounding essay on homology, he overlooks the fundamental requirement for any (Darwinian, hence phylogenetic) homology hypothesis, namely, its phylogenetic (and level specific) context and a rigorous delineation of either the phenotypic or genotypic condition about which a hypothesis is proposed (gene trees, character transformations, taxograms, and phylogenetic trees express different things). Generally the same may be stated regarding the confusion of levels for the equivocating perspective of Lockwood and Fleagle (1999), who analyzed the meaning of homoplasy. Hypothesis and operational testing are (or rather should be) independent from one another. Lieberman (2000: 152), when he states that he agrees that phylogenetic homology concepts are fine "...but it remains true that the concept is *logically* problematic in the absence of *a priori* knowledge of the phylogeny in question" (italics supplied), adds an unwanted level of confusion to the already enormous literature. I note here that Lieberman, like Cartmill, obviously does not believe in the independence of homology testing, and therefore, neither can they *logically* consider testing phylogenies against independently tested and corroborated homologies. So for both Lieberman- and Cartmill-proposed phylogeny hypotheses of taxa should remain just that, vacuous proposals, as they cannot test these against independently corroborated homologies. Lieberman, or anyone else, who holds forth in detail about homology (or homoplasy) without some experience in the procedures of phylogenetic estimations in systematics, and who vaguely cites EvoDevo studies and equivocates on the level-specific meaning of these concepts to somehow support their taxic perspective has a serious problem. These workers have to grapple with the fact that the key conceptual contribution of evolutionary developmental genetics (that character complexes are modular in spite of the phenomenon of epistasis) obviously means that phylogeny estimation of these modules are likely to be independent from those of others in the same species, and therefore, in higher taxa as well. Mosaic evolution is back under the cloak of modularity (contra the opinion expressed by Tattersall, 2000, that it is a "hoary old concept"), showing us that the logical positivism of cladism is incapable of setting the ontological foundations for the theory of descent. Consequently both the choices of characters for analysis and the taxic approach to phylogenetic estimation may have to be seriously reconsidered in the near future.

TRANSITIONS LEADING UP TO THE ARCHONTAN AND EUPRIMATE LOCOMOTOR STRATEGIES AND SUBSTRATE PREFERENCE

I will not belabor the platitude that the postcranial record of Mesozoic eutherian (or other) mammals is still relatively poor, and that such a state of affairs makes for very tentative conclusions regarding locomotor adaptations in the stems of various higher taxa within the Eutheria, Metatheria, and Theria (the latter restricted here to the concept of monophyletically tribosphenic mammals). There is certainly overwhelming evidence that the extant Metatheria had a specifically arboreal ancestry, except perhaps for the Caenolestidae, the stem of the Sudameridelphia (Szalay, 1994; Szalay and Sargis, 2001), and for some early lineages like *Asiatherium* (Szalay and Trofimov, 1996). Similarly, there is little doubt at present that the last common ancestor of extant eutherians (all placentalians), various extinct Cenozoic groups, and lineages related to these extending back to the Cretaceous, were derived of a terrestrially committed stock, the stem of the Placentalia (Szalay, 1984, 1985, 1994; Szalay and Decker, 1974; Szalay and Drawhorn, 1980; Szalay and Lucas, 1993, 1996; Szalay and Schrenk, 1998). Prior to the recent description of some postcranial remains of Cretaceous mammals the same may have been said of the then known Eutheria (Szalay, 1977).

But beginning with the increased recovery of a variety of cladistically unquestioned eutherians from the Cretaceous in the last three decades, it became apparent that the eutherian branch of the Theria probably had a great variety of postcranial properties that cast serious doubt on the wholesale categorization of the stem Eutheria based on the extant forms and Cenozoic fossils. Szalay and Trofimov (1996, Figure 26) made the suggestion that the early, basal, radiation of the Eutheria probably retained a reproductive strategy that could be characterized as "marsupial" in a general way, and from such an undoubtedly many-branched paraphyletic entity (dubbed above as Eoeutheria) arose the last common ancestor of, what I call here, the Placentalia. All of that implies that there is no simple way to characterize the postcranially unknown lineages of 60–70 MY of evolution prior to the Cenozoic. For example Kielan-Jaworowska (1975) reported the presence of epipubics in a clade of early eutherians, and more recently Horovitz (2000) described the tarsus of the asioryctithere *Ukhaatherium*, also from the Cretaceous of Mongolia. The palmate and broad proximal fibula of *Ukhaatherium*, among other features, suggests grasping (as inferred from a well developed peroneus longus that is probably indicated by that type of proximal fibula; Argot, personal communication), and its highly mobile calcaneocuboid joint suggests a marsupial-like mobility of the foot. An ongoing study (Szalay, Sargis, Archibald, and Averianov, in preparation) of mammal postcranials from the Santonian Cretaceous of Uzbekistan (see Archibald et al., 1998) will also help the ongoing assessment of problems regarding early locomotor strategies in the Eutheria. To complicate matters even for the archimetatherian (early) marsupial radiation, the skeleton of *Asiatherium*, from the semi-arid environments of the Late Cretaceous of Mongolia suggests a terrestrial locomotor strategy, very tentatively.

LOCOMOTION AND THE ORIGINS OF EUPRIMATES

I believe that all the known placentalian arboreal adaptations are secondarily derived from a terrestrial stem—a point that has been amply documented before. The previously elaborated explanations that pointed to the derived nature of pedal mobility in the lower ankle joint (Szalay, 1984, 1994) in placentalians are also corroborated from other areas of the skeleton in the known Early Tertiary representative of eutherian orders.

While the issue of Archonta will continue to be debated as new fossils are described, the morphotypic skeletal adaptations unique to the euprimate stem are relatively well established (Dagosto, 1985, 1986, 1988; Decker and Szalay, 1974). Among other attributes, the early euprimates had a flattened ilium to accommodate a musculature hypertrophied for leaping. They had fast, deep, and highly stabilized knee joints superbly constructed for powerful leaping in conjunction with a foot that had an equally speed-adapted upper ankle joint capable of rapid flexion, combined with a highly helical lower ankle joint articulation, totally unlike we see in arboreal didelphids. Although the general condition of the upper ankle joint is a eutherian one, the euprimate condition is highly derived in its astragalar construction for extensive flexion-extension (with its great angular distance of the tibial articular surface) and the attendant speed. While the euprimate feature for obligate inversion was held over in the lower ankle joint from its archontan ancestry (and further evolved for specific regime of locomotion on arboreal substrates), this happened within the highly constrained cruropedal contact that characterizes eutherians (Szalay and Decker, 1974).

Neither merely obligate arboreality, as such, nor visual predation accounts for the postcranial heritage of the euprimates acquired from their last common ancestor. The transformation of claws into nails, and the evolution of hypertrophied feet (compared to smaller hands) and powerful pedal grasping coupled with mechanical solutions of the entire pelvic limb do, however, account for a particular kind of arboreality. These features are related to explosive long jumps, combined with the precise ability to grasp small branches when landing usually with the feet first. Grasp-leaping appears to have been the morphotypic locomotor mode for the stem lineage of the Euprimates.

ACKNOWLEDGMENTS

I am indebted to the organizers Matt Ravosa and Marian Dagosto for inviting me to the conference. I am also grateful to Johanna Warshaw and Eric Sargis for their careful reading of the manuscript, and for useful suggestions.

REFERENCES

Anthony, M. R. L., and Kay, R. F., 1993, Tooth form and diet in ateline and alouattine primates: Reflections on the comparative method, *Am. J. Sci.* **293-A:** 356–382.

Archibald, J. D., Sues, H.-D., Averianov, A. O., King, C., Ward, D. J., Tsaruk, O. A., et al., 1998, Precis of the Cretaceous paleontology, biostratigraphy and sedimentology at Dzharakuduk (Turonian?–Santonian), Kyzylkum Desert, Uzbekistan, *New Mexico Museum of Natural History and Science Bulletin.* **14:** 21–28.

Bloch, J. I., and Boyer, D. M., 2001, Taphonomy of small mammal fossils in freshwater limestones from the Paleocene of the Clarks Fork Basin, Wyoming, in: *Paleocene–Eocene Stratigraphy and Biotic Change in the Bighorn and Clarks Fork Basins, Wyoming,* P. D. Gingerich, ed., University of Michigan Papers on Paleontology, **33:** 185–198.

Bloch, J. I., Boyer D. M., Gunnell, G. F., and Gingerich, P. D., 2000, New primitive paromomyid from the Clarkforkian of Wyoming and dental eruption in Plesiadapiformes, *J. Vert. Paleo.* **20**(3): 30A.

Bloch, J. I., Boyer, D. M., and Gingerich, P. D., 2001a, Small mammals from Paleocene–Eocene freshwater limestones of the Fort Union and Willwood Formations, Clarks Fork Ban, Wyoming: Exceptional preservation in unique depositional environments; Published online at: http://www.paleogene.net/abstract_volume.html

Bloch, J. I., Boyer, D. M., and Gingerich, P. D., 2001b, Positional behavior of late Paleocene *Carpolestes simpsoni* (Mammalia, ?Primates), *J. Vert. Paleo.* **21**(3): 34A.

Bloch, J. I., Boyer, D. M., Gingerich, P. D., and Gunnell, G. F., 2001c, New Paleogene plesiadapiform (Mammalia, ?Primates) skeletons from Wyoming: A diversity of positional behaviors in possible primate ancestors, First International Conference on Primate Origins and Adaptations: A Multidisciplinary Perspective.

Bloch, J. I., Boyer D. M., Gunnell, G. F., and Gingerich, P. D., 2002, Primitive paromomyid from the Clarkforkian of Wyoming and dental eruption in Plesiadapiformes, *J. Vert. Paleo.* **22:** 380–387.

Bock, W. J., 1981, Functional–adaptive analysis in evolutionary classification, *Am. Zool.* **21:** 5–20.

Boyer, D. M., and Bloch, J. I., 2000, Documenting dental–postcranial associations in Paleocene–Eocene freshwater limestones from the Clarks Fork Basin, Wyoming, *J. Vert. Paleo.* **20**(3): 31A.

Boyer, D. M., Bloch, J. I., and Gingerich, P. D., 2001, New skeletons of Paleocene paromomyids (Mammalia, ?Primates): Were they mitten gliders? *J. Vert. Paleo.* **21**(3): 35A.

Cartmill, M., 1972, Rethinking primate origins, *Science* **184:** 436–443.

Cartmill, M., 1974, Pads and claws in arboreal locomotion, in: *Primate Locomotion*, F. A. Jenkins, Jr., ed., Academic Press, New York, pp. 45–83.

Cartmill, M., 1975, *Primate Origins, A Module in Modern Physical Anthropology*, Burgess Publishing Company, Minneapolis, Minnesota.

Cartmill, M., 1990, Human uniqueness and theoretical content in paleoanthropology, *Int. J. Primatol.* **11:** 173–192.

Cartmill, M., 1992, New views on primate origins, *Evol. Anth.* **1**(3): 105–111.

Cartmill, M., 1994, A critique of homology as a morphological concept, *Am. J. Phys. Anthro.* **94:** 115–123.

Cifelli, R. L., and Villarroel, C., 1997, Paleobiology and affinities of *Megadolodus*, in: *Vertebrate Paleontology in the Neotropics: The Miocene Fauna of La Venta, Colombia*, R. F. Kay, R. H. Madden, R. L. Cifelli, and J. J. Flynn, eds., Smithsonian Institution Press, Washington, DC, pp. 265–288.

Court, N., 1994, Limb posture and gait in *Numidotherium koholense*, a primitive proboscidean from the Eocene of Algeria, *Zool. J. Linn. Soc.* **111:** 297–338.

Crompton, R. H., 1995, "Visual predation," habitat structure, and the ancestral primate niche, in: *Creatures of the Dark: the Nocturnal Prosimians*, L. Alterman, G. A. Doyle, and M. K. Izard, eds., Plenum Press, New York, pp. 11–30.

Dagosto, M., 1985, The distal tibia of primates with special reference to the Omomyidae, *Int. J. Primatol.* **6:** 45–75.

Dagosto, M., 1986, *The Joints of the Tarsus in the Strepsirhine Primates*, Ph.D. Dissertation, City University of New York, New York.

Dagosto, M., 1988, Implications of postcranial evidence for the origin of euprimates, *J. Hum. Evol.* **17:** 35–56.

Dagosto, M., Gebo, D. L., and Beard, K. C., 1999, Revision of the Wind River faunas, Early Eocene of central Wyoming. Part 14, Postcranium of *Shoshonius cooperi* (Mammalia: Primates), *Ann. Carnegie Museum* **68**(3): 175–211.

Decker, R. L., and Szalay, F. S., 1974, Origins and function of the pes in the Eocene Adapidae (Lemuriformes, Primates), in: *Primate Locomotion*, F. A. Jenkins, Jr., ed., Academic Press, New York, pp. 261–291.

Gingerich, P. D., 1974, *Cranial Anatomy and Evolution of Early Tertiary Plesiadapidae (Mammalia, Primates)*, Ph.D. Thesis, Yale University.

Gingerich, P. D., 1975a, Systematic position of *Plesiadapis, Nature* **253**: 110–113.

Gingerich, P. D., 1975b, New North American Plesiadapidae (Mammalia, Primates) and a biostratigraphic zonation of the middle and upper Paleocene, *Contr. Mus. Paleont. Univ. Mich.* **24**: 135–48.

Hamrick, M. W., 1998, Functional and adaptive significance of primate pads and claws: Evidence from New World anthropoids, *Am. J. Phys. Anthro.* **106**: 113–127.

Hershkovitz, P., 1977, *New World Monkeys (Platyrrhini)*, vol. 1, University of Chicago Press, Chicago.

Horovitz, I., 2000, The tarsus of *Ukhaatherium nessovi* (Eutheria, Mammalia) from the Late Cretaceous of Mongolia: An appraisal of the evolution of the ankle in basal therians, *J. Vert. Paleo.* **20**(3): 547–560.

Kay, R. F. 1984. On the use of anatomical features to infer foraging behavior in extinct primates, in: *Adaptations for Foraging in Nonhuman Primates: Contributions to an Organismal Biology of Prosimians, Monkeys and Apes*, P. S. Rodman and J. G. H. Cant, eds., Columbia University Press, New York, pp. 21–53.

Kay, R. F., and Cartmill, M., 1977, Cranial morphology and adaptations of *Palaechthon nacimienti* and other Paromomyidae (Plesiadapoidea, ? Primates), with a description of a new genus and species, *J. Hum. Evol.* **6**: 19–54.

Kay, R. F., and Covert, H. H., 1984, Anatomy and behaviour of extinct primates, in: *Food Acquisition and Processing in Primates*, D. J. Chivers, B. A. Wood, and A. Bilsborough, eds., Plenum Press, New York, pp. 467–508.

Kielan-Jaworowska, Z., 1975, Possible occurrence of marsupial bones in Cretaceous eutherian mammals, *Nature* **255**: 698–699.

Krishtalka, L., and Schwartz, J. H., 1978, Phylogenetic relationships of plesiadapiform–tarsiiform primates, *Ann. Carnegie Museum* **47**: 515–540.

Le Gros Clark, W. E., 1959, *The Antecedents of Man*, University Press, Edinburgh.

Lewis, O. J., 1989, *Functional Morphology of the Evolving Hand and Foot*, Oxford University Press, Oxford.

Lieberman, D. E., 1999, Homology and hominid phylogeny: Problems and potential solutions, *Evol. Anth.* 7: 142–151.

Lieberman, D. E., 2000, Reply, *Evol. Anth.* 9:152.

Lockwood, C. A., and Fleagle, J. G., 1999, The recognition and evaluation of homo-plasy in primate and human evolution, *Ybk. Phys. Anthro.* **42**: 189–232.

Martin, R. D., 1986, Primates: A definition, in: *Major Topics in Primate and Human Evolution*, B. Wood, L. Martin, and P. Andrews, eds., Cambridge University Press, Cambridge, pp. 1–31.

Martin, R. D., 1990, *Primate Origins and Evolution*, Princeton University Press, Princeton.

Martin, R. D., 1993, Primate origins: Plugging the gaps, *Nature* **363**: 223–234.

Napier, J. R., and Walker, A. C., 1967, Vertical clinging and leaping—A newly recog-nized category of locomotor behaviour of primates, *Folia Primatol.* **6**: 204–219.

Oxnard, C. E., Crompton, R. H., and Liebermann, S. S., 1990, *Animal Lifestyles and Anatomies*, Washington University Press, Seattle.

Rasmussen D. T., 1990, Primate origins: Lessons from a neotropical marsupial, *Am. J. Primatol.* **22**: 263–277.

Schwartz, J. H., Tattersall, I., and Eldredge, N., 1978, Phylogeny and classification of the primates revisited, *Ybk. Phys. Anthro.* **21**: 95–133.

Sussman, R. W., 1995, How primates invented the rain forest and vice versa, in: *Creatures of the Dark: The Nocturnal Prosimians*, L. Alterman, G. A. Doyle, and M. K. Izard, eds., Plenum Press, New York, pp. 1–10.

Szalay, F. S., 1968, The beginnings of primates, *Evolution* **22**: 19–36.

Szalay, F. S., 1969, Mixodectidae, Microsyopidae, and the insectivore–primate transi-tion, *Bull. Am. Mus. Nat. Hist.* **140**: 193–330.

Szalay, F. S., 1972, Paleobiology of the earliest primates, in: *The Functional and Evolutionary Biology of Primates*, R. Tuttle, ed., Aldine-Atherton, Chicago, pp. 3–35.

Szalay, F. S., 1977, Phylogenetic relationships and a classification of the eutherian Mammalia, in: *Major Patterns in Vertebrate Evolution*, M. K. Hecht, P. C. Goody, and B. M. Hecht, eds., Plenum Press, New York, pp. 315–374.

Szalay, F. S., 1981a, Phylogeny and the problem of adaptive significance: The case of the earliest primates, *Folia Primatol.* **36**: 157–182.

Szalay, F. S., 1981b, Functional analysis and the practice of the phylogenetic method as reflected by some mammalian studies, *Am. Zool.* **21**: 37–45.

Szalay, F. S., 1984, Arboreality: Is it homologous in metatherian and eutherian mam-mals? *Evol. Biol.* **18**: 215–258.

Szalay, F. S., 1985, Rodent and lagomorph morphotype adaptations, origins, and rela-tionships: Some postcranial attributes analyzed, in: *Evolutionary Relationships Among Rodents—A Multidisciplinary Analysis*, W. P. Luckett and J.-L. Hartenberger, eds., (NATO ASI Series, Series A, Life Sciences, vol. 92.), Plenum Press, New York, pp. 83–132.

Szalay, F. S., 1994, *Evolutionary History of the Marsupials and an Analysis of Osteological Characters*, Cambridge University Press, New York.

Szalay, F. S., 2000, Function and adaptation in paleontology and phylogenetics: Why do we omit Darwin? *Palaeontologia Electronica* **3**(2)25 p. 366, KB; http://palaeo–electronica. org/paleo/2000_2/darwin/issue2_00.htm

Szalay, F. S., and Bock, W. J., 1991, Evolutionary theory and systematics: Relationships between process and pattern, *Z. Zool. Syst. Evolut.–forsch.* **29**: 1–39.

Szalay, F. S., and Dagosto, M., 1980, Locomotor adaptations as reflected on the humerus of Paleogene primates, *Folia Primatol.* **34**: 1–45.

Szalay, F. S., and Dagosto, M., 1988, Evolution of hallucial grasping in primates, *J. Hum. Evol.* **17**: 1–33.

Szalay, F. S., and Decker, R. L., 1974, Origin, evolution and function of the tarsus in Late Cretaceous Eutheria and Paleocene primates, in: *Primate Locomotion*, F. A. Jenkins, Jr., ed., Academic Press, New York, pp. 223–259.

Szalay, F. S., and Delson, E., 1979, *Evolutionary History of the Primates*, Academic Press, New York.

Szalay, F. S., and Drawhorn, G., 1980, Evolution and diversification of the Archonta in an arboreal milieu, in: *Comparative Biology and Evolutionary Relationships of Tree Shrews*, W. P. Luckett, ed., Plenum Press, New York, pp. 133–169.

Szalay, F. S., and Lucas, S. G. 1993, Cranioskeletal morphology of archontans, and diagnoses of Chiroptera, Volitantia, and Archonta, in: *Primates and Their Relatives in Phylogenetic Perspective*, R. D. E. MacPhee, ed., Plenum Press, New York, pp. 187–226.

Szalay, F. S., and Lucas, S. G., 1996, Postcranial morphology of Paleocene *Chriacus* and *Mixodectes* and the phylogenetic relationships of archontan mammals, *New Mexico Bulletin of Natural History and Science* **7**: 1–47.

Szalay, F. S., and Sargis, E. J., 2001, Model-based analysis of postcranial osteology of marsupials from the Paleocene of Itaboraí (Brazil) and the phylogenetics and biogeography of Metatheria, *Geodiversitas* **23**(2): 1–166; http://www.mnhn. fr/publication/geodiv/g01n2a1.pdf

Szalay, F. S., and Schrenk, F., 1998, The middle Eocene *Eurotamandua* and a Darwinian phylogenetic analysis of "edentates," *Kaupia* **7**: 97–186.

Szalay, F. S., and Trofimov, B. A., 1996, The Mongolian Late Cretaceous *Asiatherium*, and the early phylogeny and paleobiogeography of Metatheria, *J. Vert. Paleo.* **16**(3): 474–509.

Szalay, F. S., Rosenberger, A. L., and Dagosto, M., 1987, Diagnosis and differentiation of the order Primates, *Ybk. Phys. Anthro.* **30**: 75–105.

Szalay, F. S., Tattersall, I., and Decker, R. L., 1975, Phylogenetic Relationships of Plesiadapis—Postcranial evidence, *Contr. Primatol.* **5**: 136–166.

Tattersall, I., 2000, Paleoanthropology: The last half century, *Evol. Anthro.* **4**(1): 2–16.

Walker, A. C., 1967, *Locomotor Adaptations in Recent and Fossil Madagascan Lemurs*, Ph.D. Dissertation, University of London, London.

CHAPTER FIFTEEN

The Postcranial Morphotype of Primates

Marian Dagosto

INTRODUCTION

The goal of this chapter is to reconstruct aspects of the postcranial morphotype of the order Primates and to assess their significance for the positional behavior of the ancestor. What derived features of the limb skeleton are likely to have distinguished the last common ancestor of primates from more remote ancestors and what implications does this set of features have for the way of life of the ancestor of Primates? In pursuing this goal, the following questions are addressed:

(1) What are the derived characters of the postcranium that characterize the most recent common ancestor of the primates?
(2) What are the functional and biological role attributes of these characters individually?
(3) Do the functional/biological role attributes of the traits as a whole constitute a cohesive story? Can they be explained by a single selective factor or a set of selective factors arising from a particular way of life?
(4) If primate synapomorphies cannot be attributed to a single way of life, does the evidence suggest an order in which characters were added to the morphotype, and thus a plausible functional/behavioral sequence?

Marian Dagosto • Deptartment of Cell and Molecular Biology, Northwestern University, Chicago IL 60611; Research Associate, Division of Mammals, Field Museum of Natural History, Chicago IL

These questions raise many issues, the primary one being, of course, what does one mean by the phrase "Origin of Primates?" This topic is dealt with further at the end of this chapter, but for immediate clarification, the intention is to explain the behavioral significance of the set of features that characterize the Most Recent Common Ancestor (MRCA; Figure 1) of crown group primates. Crown group primates are the anthropoids, lemurs, tarsiers, adapids, and omomyids. There remain few serious challenges to the hypothesis that this group of mammals shared a common ancestor relative to other extant and fossil mammals. The derived features of the postcranium that distinguish this ancestor from the outgroup are given in Table 1. A formal phylogenetic analysis of the features is not given here since it seems fairly certain from both morphological and molecular evidence that primates are part of the Euarchonta (Springer et al., this volume), and for most of the features listed in Table 1, primates differ from any of the most likely outgroups (Scandentia, Dermoptera, Plesiadapiformes, Rodentia, Lagomorpha), as well as from the majority of other mammals. The few exceptions are noted in the text.

Question 2 entails having a philosophy for formulating and evaluating hypotheses about functional and biological role in fossil organisms. This is discussed in the next section. (See also Szalay, this volume.) Questions 3 and 4 ask if the set of traits can be reasonably considered to be a correlated complex—can a single niche, habitus, or way of life explain all or most of them?

The most comprehensive "single niche" model for the Origin of Primates is the "nocturnal visual predation" model (NVP) developed by Cartmill (1972; 1974a; 1974b). It explains the grasping extremities, loss of claws, and optical convergence of primates as being related to a way of life involving visually directed predation in the small branch niche by nocturnal animals.

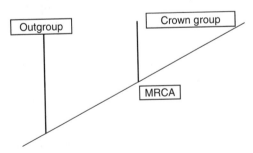

Figure 1. The most recent common ancestor (MRCA), is the common ancestor of the crown group primates, which include the adapids, omomyids, tarsiers, and anthropoids.

Table 1. Derived features that distinguish primates from other Archonta

Trait	Features of the forelimb		
	Suggested mechanical function	Suggested biological role/s	Reference
Forelimb			
Relatively long hands and short forearms	Grasp relatively large branches when landing	Vertical clinging and leaping; seize live prey	(Godinot and Beard, 1993; Jouffroy et al., 1991)
Humerus			
Trochlea elongate and cylindrically shaped with strong lateral edge and deep groove separating the humeroradial and humerolunar joints	Increased support in supinated positions	Grasping	(Szalay and Dagosto, 1980)
Hand			
Paramesaxony; Elongation of third metacarpal			(Jouffroy et al., 1991)
Relatively short carpus/long digits			(Godinot, 1992; Godinot and Beard, 1991; Jouffroy et al., 1991)
Elongate scaphoid tubercle	Deepens radial margin of the carpal tunnel; acts as windlass for the flexor digitorum profundus	Grasping small diameter supports	(Godinot and Beard, 1991; Hamrick, 1997)
Large pisiform, triquetrum quadrangular in dorsal view; hamate without hamulus; capitate short and narrow			(Godinot and Beard, 1991; 1993)
Independent thumb		Grasping	(Altner, 1971)
Claws replaced by nails		?Grasping	

(Continued)

Table 1. Derived features that distinguish primates from other Archonta—(*Continued*)

Trait	Features of the hindlimb		
	Suggested mechanical function	Suggested biological role/s	Reference
Hindlimb			
Elongated relative to body size: femur, tibia, tarsus	Increase distance and/or time of propulsive force	Leaping	(Martin, 1972; Connour, 2000; Polk et al., 2000; Silcox, 2001)
Pelvis			
Flattened, expanded ilium	Hypertrophy of gluteus medius	Leaping	(Szalay et al., 1987)
Long ilium	Increase speed of gluteals for femoral extension; increased power of femoral flexors	Leaping	(Anemone, 1993)
Short ischium	Mechanical advantage of hip extensors	Leaping	(Anemone, 1993; Fleagle and Anapol, 1992; McArdle, 1981)
Femur			
Patellar groove deep and narrow, condyles anteroposteriorly deeper than mediolaterally wide; lateral epicondyle more anteriorly projecting than medial	Increase mechanical advantage of knee extensors	Leaping	(Szalay et al, 1987; Silcox, 2001)
3rd trochanter at same level as lesser	Increased speed of action of hip extensors	Leaping	(Silcox, 2001)
Tibia			
Moderate degree of medial rotation of medial malleolus Prominent, inferiorly long, pyramid shaped medial malleolus with a distally convex surface	Increases abduction of talus relative to tibia during dorsiflexion; allows lamina pedis full range of motion at subtalar and transverse tarsal joints	Grasping/climbing/variable orientations of foot	(Dagosto, 1985; Hafferl, 1932; Lewis, 1980a)

Character	Function	Behavior	Reference
Inferior tibial joint surface as long or longer than wide (same in *Ptilocercus*)	Related to long, narrow talar trochlea		(Dagosto, 1985; Sargis, 2000)
Foot			
Elongation of tarsal elements: talar neck, distal part of calcaneus, navicular, cuboid, cuneiforms	Increase effective length of hindlimb. Elongation of talar neck may be for increased subtalar motion	Leaping Grasping/climbing	(Hall-Craggs, 1965; Jenkins and McClearn, 1984; Martin, 1972; Morton, 1924)
High phalangeal index; proximal phalanges longer than intermediates or terminals		Grasping	(Lemelin, 1999; Hamrick, 1999; Hamrick, 2001)
Relatively short terminal phalanges	Developmentally related to loss of claws		(Hamrick, 1998, 1999)
Reverse alternating foot	Related to tarsal elongation	Leaping	(Lewis, 1980b)
Talus			
Talar body tall	Increase radius of curvature of upper ankle joint	Leaping	(Dagosto, 1986)
Medial and lateral crests of talar trochlea more or less equal in height; sharper edges	Reduce concomitant rotations during flexion/extension at upper ankle joint	Leaping	(Dagosto, 1986)
Lengthen joint facet for tibia on trochlea posteriorly	Increase range of plantarflexion	Leaping	Dagosto, 1986
Posterior trochlear shelf moderately developed	1. Buttress for plantarflexed foot in pushoff 2. Support for lengthened posterior astragalar calcaneal face	Leaping	(Dagosto, 1986)
Talar trochlea is deep	Increases stability of upper ankle joint	Grasping	(Dagosto, 1986; Decker and Szalay, 1974; Szalay and Decker, 1974)
More spherical talonavicular joint; medial and lateral sides are equivalent in size; distal margin of head not indented	Allows axial rotation at talonavicular joint	Leaping Climbing/grasping	(Dagosto, 1986) (Dagosto, 1986; Hooker, 2001)

(*Continued*)

Table 1. Derived features that distinguish primates from other Archonta—(*Continued*)

Trait	Features of the hindlimb		
	Suggested mechanical function	Suggested biological role/s	Reference
Calcaneus			
Sellar shaped calcaneocuboid joint	Reduces translational component of movement at the calcaneocuboid joint; increases axial rotation	Grasping	(Dagosto, 1986; Szalay and Decker, 1974)
Foot			
Medial shift of entocuneiform-first metatarsal joint (also in *Carpolestes*)	Opposability of hallux	Grasping	(Szalay and Dagosto, 1988)
More restrictive Sellar entocuneiform-first metatarsal joint (also in *Carpolestes*)	Stability of hallucal-metatarsal joint		(Szalay and Dagosto, 1988)
Habitual abduction of hallux (also in *Carpolestes*)	Opposability of hallux	Grasping	(Szalay and Dagosto, 1988)
Enlarged peroneal process on first metatarsal	Buttress MT1-entocuneiform joint	Leaping	(Szalay and Dagosto, 1988)
Enlarged hallux		Grasping	(Szalay and Dagosto, 1988)
Nails instead of claws on all pedal digits		?Grasping	

It envisions a slow moving quadrupedal ancestor. In contrast, the "grasp leaping" (GL) hypothesis of Szalay (Szalay and Dagosto, 1980; Szalay and Delson, 1979; this volume) proposes a more agile animal. The term was coined to recognize a unique category of positional behavior that is typical of many primates and thought to be ancestral for Primates. It includes both leaping and grasping as significant elements and thus distinguishes grasp-leapers from other "arboreal quadrupeds." It does *not* claim that behaviors other than leaping (i.e., quadrupedalism, climbing) are not used. It does *not* imply the specialized leaping of galagines, tarsiers, or indriids, which are placed in a separate, more derived locomotor category (vertical clinging and leaping). And, although it hypothesizes that primate limb morphology may be a compromise for the demands of grasping and leaping, it does *not* require tandem coevolution of grasping and leaping. The model presented by Szalay and Dagosto (1988; and later in this chapter) explicitly recognized the presence of a more primitive kind of grasping morphology and behavior in archontans prior to the MRCA.

GL is somewhat more limited than NVP in that it simply seeks to describe a locomotor mode of the ancestor, and is not intended as a complete description of a way of life. It is compatible with, but does not require, the hypothesis of Clark (1959), Collins (1921), and Crompton (1995) that the visual acuity necessary for leaping and landing with the precision characteristic of nocturnal primates may have been an important factor in the development of orbital convergence. GL does not require, nor is it specifically linked to any dietary regime. The positional behavior elements of GL are similar to reconstructions of the ancestral primate offered by Martin (1972; 1990) and Crompton (1995).

RECONSTRUCTING FUNCTION AND BIOLOGICAL ROLES

Some of the disagreement between these two models is the result of differing philosophies regarding the assessment of biological role in fossil organisms. One of Cartmill's (1974b; 1990; 1992) critiques of GL is that it proposes a unique biological role for primate morphology; he holds that evolutionary events can only be explained if they can be understood as examples of more general phenomena. If no analogies or parallelisms exist, no explanation of the trait is possible. His work on primate origins has thus consisted of employing the comparative method to understand the derived features of primates.

There is no doubt that the "comparative method" is a powerful way of elucidating the relationship between "traits" and "behaviors" (or form and biological role in the sense of Bock and von Wahlert, 1965) in both living and fossil organisms. The stronger a correlation between form and biological role (i.e., the more examples of phylogenetically independent co-occurrence), the stronger the inference that the same relationship applied to a taxon or morphotype where the trait is known, but the behavior is not. This is not, however, an infallible line of reasoning since numerous examples of morphological convergence without behavioral convergence are known, as well as behavioral convergence without morphological convergence (Bock, 1977; Rickelfs and Miles, 1994). Because of other factors (allometry, developmental constraints, multiple pathways of adaptation, functional compromise (see Ross et al., 2002 for further discussion), associations rarely exhibit the regularity necessary to deduce biological role from form. The fact that an association between a trait and a behavior exists in living organisms (a population which in itself is a historical accident, not likely to exhibit all the form–function relationships that could possibly exist) is simply no guarantee that the same relationship applied to a fossil or morphotype.

In addition to the inductive nature of a "comparative method" argument, both its proponents and detractors recognize that this sort of analogy is powerless to explain unique traits (Lauder, 1982; Rudwick, 1964; Van Valkenburgh, 1994). Furthermore, if an analogy is based only on a few examples, it really has little statistical strength (Garland and Adolph, 1994). If the comparative method were the only avenue available for investigating form and function, one would have to conclude, like Cartmill, that unique events are "inexplicable." This notion is rejected. Relationships between form, function, and behavior can be (and have been) investigated independently of whether or not there are multiple (or any) extant examples.

The real power of the comparative method is the demonstration of the strength of a relationship and thus the implication that there is a *causal* basis to the relationship (Harvey and Pagel, 1991; Losos and Miles, 1994; Rickelfs and Miles, 1994). The pattern is used to infer process—we make the assumption that similar selective regimes arising from similar biological roles result in similar anatomical solutions (Lauder, 1996). Thus, it is the functional/biomechanical relationship itself, not the number of examples of it that provides the essential link between a trait and a behavior. These causal links are tested through the analysis of performance (Arnold, 1983; Baum and Larson, 1991; Lauder, 1990; 1996; Wainwright, 1994).

In primate biology, both of the explicit strategies for executing the "comparative method" recognize the "functional implication" aspect as an important part of the argument. A hypothesis of biological role is stronger if "All the features specified in the definition of T (traits) have some *functional relationship* to F (function)" (Kay and Cartmill, 1977, p. 20, italics added); "it is based on two or more morphologically and functionally convergent derived features, which are *causally explicable* and exclusively correlated with the same biological role" (Szalay, 1981: 167, italics added). The dual importance of both comparison (i.e., analogy) and functional analysis is also recognized by many other primatologists addressing this topic (e.g., Day, 1979; Fleagle, 1979; Lovejoy, 1979; Preuschoft, 1979; Ross et al., 2002). Both the NVP and GL models make use of the comparative method and functional inference and therefore share the strengths and limitations of both (Ross et al., 2002).

Establishing a plausible functional and therefore causal, relationship between a trait and a behavior provides at least as compelling evidence for a hypothesis of biological role in an extinct animal as does the demonstration of convergent examples of the association. Some have even argued that if such causal relationships are understood, there is in fact no need to rely on analogy at all (Hesse, 1966; Rudwick, 1964). Analogies are useful for suggesting possible biological roles (i.e., they establish prior probabilities (Fisher, 1985)), and narrow the universe of form–function-biological role hypotheses we may wish to subject to further testing. Others have argued that paradigms and biomechanical models are simply another kind of analogy (e.g., Weishampel, 1995). Regardless, with an understanding of a causal relationship between trait and behavior it really does not matter if there are numerous, few, or no examples among living taxa— "...the range of our functional inferences about fossils is limited not by the range of adaptations that happen to be possessed by organisms at present alive, but by the range of our understanding of the problems of engineering" (Rudwick, 1964, p. 33). Thus, unique traits present no problem for this line of reasoning.

Predictions about how variations in morphology will influence performance can be made from theoretical considerations. Arguments will be even stronger when "such predictions are supported by the results of laboratory or field performance experiments" (Wainwright, 1994). Therefore, function, even unique function (the word "function" used here is in the restricted sense of Bock and von Wahlert, 1965), is predictable from form, even in the absence of living examples. What use, if any, an extinct animal (or a morphotype) made of this function i.e., its biological role, is not in principal deducible simply from form (Bock, 1977). Nor is it able to be deduced from analogy or comparison. Nor is the demonstration of a performance benefit in a living model proof that the feature served the same role or gave the same performance benefit in an extinct organism (Koehl, 1996). We can only use these lines of evidence to evaluate the relative probability of competing hypotheses (Fisher, 1985; Ross et al., 2002; Szalay and Sargis, 2001).

In this context, the distinction between "function" and biological role clarified by Bock and von Wahlert (1965) becomes important. In this scheme, function is restricted to "the physical and chemical properties arising from its form" (p. 274). What abilities does a particular morphology enhance or constrain? Biological role is how the form–function complex is actually used by the animal in its natural habitat. A form may have numerous functions; whether we would ever be able to work out all of them and their implications for biological roles is a difficult question. The image quality afforded by frontally directed, convergent orbits may be useful, even crucial, for nocturnal visual predators. This fact, however, does not signify that it is not also useful for other activities, such as visually locating any kind of small, cryptic object, or negotiating a complex three-dimensional environment (Clark, 1959; Collins, 1921; Crompton, 1995; Sussman, 1991, 1995; Szalay and Dagosto, 1980; Szalay and Delson, 1979), even if no living animal uses it for that activity. This confounds function (increased visual acuity) and biological role (increased visual acuity is useful for hunting insects, finding small fruits or flowers, performing acrobatic movement in a complex 3-D environment, etc.). Thus, when evaluating the biological role of fossil organisms (or morphotypes) we are usually left in the position of having several reasonable unfalsified alternate hypotheses. How can we judge between them?

Like comparisons, functional implications of form can be used to provide relative levels of support for competing hypotheses of biological role. How good is the "fit" between the performance attributes of the morphology (either inferred or actually tested) and the mechanical demands of the biological role? A serious drawback for the particular problem addressed here is that virtually all the connections between form, function, and biological role for the traits in Table 1 are only inferred from form or by analogy; their performance attributes have not been experimentally tested.

Other criteria need to be brought in. Suggestions for these other criteria include:

(1) Comprehensivity: Does the proposed biological role account for other features of the organism (Rudwick, 1964; Van Valkenburgh, 1994)?

(2) Phylogenetic hierarchy: Szalay (1981) has argued that more recently acquired traits will be more reliable predictors of biological role than either retained primitive features or a general assessment of overall morphology.

(3) Extant phylogenetic bracket: Witmer (1995) has developed an explicitly phylogenetic model wherein the first two extant outgroups of the fossil taxon of interest can provide constraints regarding the interpretation of the fossil taxon.

NVP is corroborated not only by the strength of analogy, since most nocturnal visual predators have greater orbital convergence than their ancestors, but also by its comprehensivity, since it presents a reasonable explanation not only for orbital convergence but also the grasping extremities of primates. The dental evidence is somewhat problematic since it indicates a variety of dietary strategies in the earliest primates (Szalay, 1968; 1972; 1981; Szalay and Delson, 1979). NVP is attempting to explain derived characteristics of primates, but the EPB approach is inconclusive because there are so few extant primates that are nocturnal visual predators, unless this niche is very broadly defined to include any primates for which insects provide an important resource.

GL also attempts to explain derived primate morphology. Since GL is a locomotor mode unique to primates, the comparative approach is of limited value. However, the separate elements of the mode, grasping and leaping, do occur in other mammals, and can be evaluated with both the comparative method and functional inference, obviating one of Cartmill's prime objections to the hypothesis. GL is not as comprehensive as NVP; it may explain orbital convergence, but it does not imply any specific dietary regime. Most extant primates leap, but members of the most likely outgroups do not, so the phylogenetic bracket approach yields only equivocal results. None of these other lines of evidence resolve the differences between NVP and GL in their reconstruction of the locomotor mode of the ancestral primate.

DERIVED FEATURES OF THE MRCA

There are a number of characteristics of the postcranium that are hypothesized to distinguish the MRCA of primates from other mammals (even though most have been modified in some groups of primates), and most importantly from the most likely sister taxon of Primates, whether that be some member of the Plesiadapiformes, the Scandentia, or the Dermoptera. A partial list of these characters, their presumed mechanical implications and probable biological roles is given in Table 1.

Features Related to Grasping

As noted by many others, a significant number of features in the list are attributed to the presence of grasping extremities, particularly the foot. Among these are—an opposable hallux (which is a shorthand way of referring to numerous morphological modifications, including changes in the shape of the entocuneiform-first metatarsal joint (Szalay and Dagosto, 1988)), replacement of the hallucal claw with a nail, and changes in the shapes and orientations of tarsal bones and joint facets on crural and tarsal bones to accommodate an inverted foot, most of which serve to increase the range of inversion and eversion of the lamina pedis. Inversion is necessary to advantageously position the foot for use of an opposable hallux.

The modifications to the primate tibiotalar articulation have been described and discussed by several authors (Dagosto, 1985; Hafferl, 1932; Lewis, 1980a). The rotation of the joint surface of the medial malleolus, the convexity of the malleolus and the coordinated concavity of the malleolar cup on the talus, and the medial curve of the tibial facet on the talus dictate that as the talus moves from a plantarflexed to a dorsiflexed position, it also abducts relative to the tibia. The resulting close-packed dorsiflexed-abducted-pronated position of the talus allows for full range of motion of the lamina pedis at the subtalar and transverse tarsal joints with a stable talus that is functionally part of the leg. Lewis (1980a) proposed that this suite of features is related to grasping, and there is no doubt that the lamina pedis would be advantageously positioned to invert and grasp. However, there may be a more general relationship to the requirements of an arboreal milieu. The freer motion at the subtalar joint (STJ) and transverse tarsal joint (TTJ) that the upper ankle joint (UAJ) morphology allows simply permits the foot to attain variable positions.

The subtalar joint of primates is not greatly modified relative to any archontan ancestor. The posterior talocalcaneal facets are slightly more elongated (reflected in the posterior trochlear shelf). The anterior talocalcaneal facet is longer than in plesiadapids, but not much longer than in *Ptilocercus*, most likely as the result of its longer talar neck. Elongation of the talar neck may be related to the general lengthening of the tarsus typical of leapers. However, other arboreal nonleaping mammals (*Potos, Felis wiedii, Ptilocercus*) exhibit long talar necks, and it may simply permit greater subtalar motion for inversion–eversion (Jenkins and McClearn, 1984).

The transverse tarsal joint of primates exhibits several derived characters. The talonavicular joint is more spherical (mediolateral and dorsoplantar dimensions of the talar head approximately equal) compared to the condition seen in plesiadapids, tupaiids, or dermopterans, and mammals generally, where the mediolateral dimension exceeds the dorsoplantar (Dagosto, 1986, 1988; Hooker, 2001). Probably correlated with this, the dorsal margin of the talar head is not indented as in the other taxa (Hooker, 2001). In other mammals motion at the TTJ is a combination of pronation–supination and mediolateral translation (Jenkins and McClearn, 1984). The unmodified ovoid shape of the primate talonavicular joint dictates that the pronation–supination component is enhanced and the mediolateral translation is reduced, producing a more purely axial rotation. This is possibly related to the enlargement and functional independence of the hallux, and thus to grasping abilities (Dagosto, 1986).

There is some disagreement concerning the morphology, function, and polarity of the calcaneocuboid joint. Decker and Szalay (1974) contrasted the form of this joint in primates where a subconical projection on the proximoplantar edge of the cuboid fits into a depression on the distoplantar surface of the calcaneus, with that of Plesiadapiformes and other mammals where the more evenly rounded convex cuboid surface articulates with a more evenly concave calcaneocuboid facet. Stressing the shape differences, they emphasized the pivotal nature of the joint. Lewis (1980b) criticized their interpretation, noting that pivotal action at the joint does not require a physical pivot (i.e., protuberance). Both kinds of joints function as "pivots" (i.e., they allow some rotation between the calcaneus and cuboid), but in different ways. In terms of joint geometry, the nonprimate type of joint is a modified ovoid, and the primate joint is an unmodified sellar (nomenclature of joint shape is from MacConaill, 1973). The modified ovoid allows a limited degree of conjunct rotation (inversion–eversion) accompanying medial and lateral translation (adduction–abduction), while the sellar shape of the primate joint, plus the central location of the projection, allows an arcuate slide (inversion–eversion) but limits the accompanying degree of mediolateral translation (Hafferl, 1929).

The form of the primate calcaneocuboid joint, like that of the talonavicular joint, allows increased inversion–eversion at the transverse tarsal joint without appreciable mediolateral sliding of the cuboid and navicular against the talus and calcaneus. This is also likely a mechanism for repositioning the foot into a grasping attitude, while increasing stability at the transverse tarsal joint.

A more serious problem for recognizing this feature as a primate apomorphy is that similar morphologies occur in other mammals including *Cynocephalus*, *Tupaia*, and some plesiadapiforms (Beard, 1993; Lewis, 1980b). Although *Cynocephalus* clearly has a primatelike protuberance, neither *Ptilocercus* nor *Dendrogale* have a protuberance, and *Tupaia* and *Urogale* have only the smallest hint of one. The presence of a protuberance in Plesiadapiformes is variable (Beard, 1989). In contrast to Beard, who favorably compared the function and morphology of the dermopteran and primate form of the joint, Hooker (2001) questions the homology between them.

Another feature of the primate foot that has been associated with grasping small supports is the high phalangeal index (digits are long relative to metatarsus) (Lemelin, 1999). Hamrick (this volume) demonstrates that this is accomplished through elongation of the proximal, rather than intermediate phalanges.

Some features of the primate forelimb skeleton have also been thought to be related to grasping. Godinot (Godinot, 1992; Godinot and Beard, 1991) cites the relatively short carpus, long digits, and elongated scaphoid tubercle (see also Hamrick, 1997). The thumb shows some independence and grasping ability (Altner, 1971), and the phalangeal index is high (Lemelin, 1999).

There seems to be relatively little disagreement about either the functional implication of these features (e.g., they enhance the ability of the extremity to grasp), or that the biological role of grasping is to allow/facilitate movement on supports that are small relative to the size of the animal. "Small branches," however, is a rather inexact term that needs more clarification in order to assess selective forces and performance attributes of differing morphologies (Crompton, 1995). The terms "small branch," "relatively small branch," "terminal branches," "fine branches," "canopy," and "shrub layer" are used almost interchangeably, but may have different implications for morphology.

An opposable hallux gives a performance advantage when an animal is balancing or moving on top of a branch small enough so that the narrow triangle of support induces high moments of pitching and rolling (Napier, 1967). There are also performance advantages in other situations, e.g., climbing on or clinging to an oblique or vertical support (Napier, 1967). The ability to use a power grip on a single support is, however, compromised as support size decreases relative to the size of the animal, or more precisely, the size of the grasping organ (Napier, 1980). Studies of human grasping have shown that there is a rather narrow range of cylinder size that provides optimal perform- ance; larger and smaller diameter tool handles require greater muscle activity, and increase the rate of fatigue (Ayoub and LoPresti, 1971; Pheasant and O'Neill, 1975). Bishop (1964) found that on fine branches (<1.2 cm), G. sene- galensis does not use a power grasp; rather it balances with the support across its palm. When challenged with such tiny supports, lemurs, indriids, and even tarsiers either grasp several at once or sit, stand, hang, or cling from larger branches, rather than attempt to support themselves or walk on top of single, very small twigs. For any particular body size, there is an optimal substrate size range on which a power grasp can be applied. This optimum should be deter- minable from hand or foot size, span of grasp (which depends on the oppos- ability of the hallux and the size of the palm or sole), and hand-phalangeal proportions. Bishop (1964) for example, has measured the "effective grasp" of the hand in primates. Her data indicates that *Tupaia glis* has a smaller effective grasp (both absolutely and relative to body size) than in prosimians. This sug- gests that, at least at the upper end of the "optimal substrate size range" prosimian hands are designed to grasp branches that are large relative to their body size. Although no similar studies have been done on primate feet, the opposable hallux suggests that they are also designed to allow the use of a power grasp on a range of support sizes (and types, see below), including sup- ports that are larger than most terminal branches. Unless one also proposes that the ancestral euprimate was small enough that the largest support it could reasonably grasp was a few millimeters or less in diameter (Gebo, 2004) there is no reason to suppose that an opposable hallux evolved in an animal restricted to walking along single supports in the terminal branch area. Depending on the size of the animal, an opposable hallux is also useful in areas of the arboreal environment other than "terminal branches" and in situations other than walking slowly quadrupedally on single branches.

Beyond Grasping and "Small Branches"

The grasping foot, and its implication for small branch use, is a key feature of the nocturnal visual predation (NVP) model. It was in fact the only postcranial system considered in the original formation of the model. There are, however, aspects of both the grasping mechanism and other features of the primate postcranial morphotype that differ from other mammalian small branch users. This suggests that, for the primate skeleton, something more than simply grasping small branches is involved.

The most obvious of these differences is the presence of nails on the lateral digits. There is no extant mammalian analog for this, making it difficult to interpret. The suggestions are: (1) Nails are superior to claws when moving within the small branch milieu either because claws interfere with digital flexion when grasping very small supports, or because the expansion of the digital pads, which increases stability on small supports, is correlated with reduction of terminal phalanx length and claw length (Cartmill, 1974a; 1974b; Clark, 1959; Hamrick, 1998; Hershkovitz, 1970; Lemelin and Grafton, 1998; Napier, 1980); and (2) Claws interfere with landings after leaps (Szalay, 1972). In support of the first hypothesis, Hamrick (1998) has demonstrated a morphocline in apical pad size, terminal phalanx breadth, and small branch use in platyrrhines. Cartmill cites the examples of *Cercartetus* and *Burramys*, small marsupials that have reduced claws on lateral digits. *Burramys*, however, does not seem to exploit a small branch environment (see Rasmussen and Sussman, this volume), and *Caluromys*, a marsupial which does, retains claws, as do most other arboreal didelphids and phalangerids with grasping extremities. The difficulty with Szalay's hypothesis, according to Cartmill (1974a), is that other mammals that jump from trees retain claws. It is argued below, however, that the frequency of leaping in primates distinguishes them from other mammals, and thus may have been a more significant factor in the evolution of their postcranium.

It is also possible that in this case, we have misidentified the target of selection. The developmental relationship between terminal phalanx length and claw size (Hamrick, 1998; 1999) suggests that if short terminal phalanges are biomechanically advantageous for some aspect of behavior (leaping, climbing, or grasping) claw reduction may simply be a passive result.

The anatomy of the primate pedal grasping mechanism itself is quite different from that of tree shrews, plesiadapids, or dermopterans (Szalay and Dagosto, 1988). Features of the ancestral primate that distinguish it from these animals include:

(1) Strong medial shift of the facet for the first metatarsal on the entocuneiform which increases the arc of the joint and therefore the range of movement, and also puts the hallux into a permanently abducted position.

(2) Change from a shallow sellar joint to a deeper sellar joint, which increases the stability of the articulation (matched only by *Carpolestes* (Bloch and Boyer, 2002)).

(3) Enlarged peroneal process on the first metatarsal.

The NVP proposes that the grasping extremities of primates evolved to facilitate "cautious and controlled movements in pursuit of prey" (Cartmill, 1974b, p. 76) or "prolonged and stealthy locomotion on slender terminal branches" (Cartmill, 1974a, p. 442). Therefore, one possible explanation of these traits is that they contribute to better control in slow movement or maintaining stationary postures on small supports. Lorises, however, which fit this behavioral profile, do not have a deeper joint or a larger peroneal process than other primates, in fact the opposite is true (Szalay and Dagosto, 1988). Likewise, arboreal didelphids do not exhibit an appreciably deeper joint nor a larger peroneal process than their more terrestrial relatives.

These differences could also possibly be attributed to more frequent use of small supports or use of differently sized supports. In the absence of good support use data for small marsupials, tree shrews, and even primates, this is hard to evaluate. This is a critical question for determining the reason for the unique foot morphology of primates—is the minimal optimal power grasp substrate size of primates larger, smaller, or the same as in other small branch users? *Caluromys*, for example, seems to be able to move in "fine branches" without any particular disadvantage without having nails on lateral digits or any of the primate specializations of the hallucal-metatarsal joint (Charles-Dominique et al., 1981; Rasmussen, 1990). If support size characteristics of primates are not different from those of other "small branch users" there needs to be an alternate explanation for the unique features of the primate foot. More frequent or critical use of oblique and/or vertical branches is one possibility; the mechanical consequences of using supports of various orientations needs to be further investigated. The influence of aspects of positional behavior other than quadrupedalism also needs to be considered. Climbing on or clinging to relatively small supports may prove to be an important factor in the evolution of the unique primate grasping mechanism. Alternatively, Szalay and Dagosto (1988) have hypothesized that the distinctive features of the primate entocuneiform-first metatarsal joint are adaptations to buttress the joint during landings after leaps.

Features Related to Leaping

In addition to the features associated with grasping, there is another set of features present in the primate morphotype that seem to be related mechanically to facilitate leaping. The importance of leaping in the behavior of early primates, for the structure of the primate skeleton, and for the origin of primates has been recognized by numerous authors (Clark, 1959; Conroy and Rose, 1983; Crompton, 1995; Dagosto, 1988; Gregory, 1920; Martin, 1972; Morton, 1924; Napier and Walker, 1967a, 1967b; Szalay and Dagosto, 1980; Szalay and Delson, 1979). Demes and colleagues (Demes et al., 1995, 1996, 1999) have demonstrated that the forces limbs are exposed to during leaping are larger than those of other locomotor behaviors and are especially large in small animals, so we should expect this behavior to leave its particular mark on primate limbs.

Like "small branches," leaping is an imprecise term (Oxnard et al., 1990). Virtually all mammals are capable of propelling themselves into the air using their limbs. Proponents of GL are proposing something more—that leaping constitutes a significant means of displacement. "Significant" means that leaping is used frequently, regularly, and contributes appreciably to a meter of travel. It does not, however, require that leaping is the only method employed in moving from one location to another, nor does it imply the inability to walk slowly on small branches. Most primates are capable of performing both these behaviors. The GL argument is that leaping is more consequential to the MRCA than it is to other archontans, plesiadapids, or other arboreal mammals that are described as "agile," "acrobatic," or occasionally leaping (e.g., squirrels, marsupials, carnivores, tree shrews). It is clear that leaping is an important part of the locomotor behavior of most primates (Oxnard et al., 1990; Walker, 1974). For the majority of prosimians and small anthropoids in the GL category leaping makes up at least 20%, and usually significantly more, of locomotor events. For the few nonprimate arboreal mammals on which there is data, leaping is never more than 30% of events, and is usually much less (Garber and Sussman, 1984; Sargis, 2001; Stafford et al., 2003; Youlatos, 1999).

Leaping is also a multifaceted behavior. Factors that need to be considered in any analysis are: (a) frequency of leaps relative to other locomotor behaviors; (b) number of leaps per unit time; (c) length of leaps relative to body size; (d) size, orientation, and compliance of landing and takeoff supports; (e) body size; (f) the specific kinematic properties of leaping performed; and (g) the histochemical properties of muscles (Crompton, 1993; Demes et al.,

1996; Emerson, 1985; Warren and Crompton, 1998). Other than the first, these aspects of leaping are, unfortunately, rarely documented.

There are several features of the primate postcranium that appear to be morphological compromises to permit/facilitate leaping in a hindlimb otherwise adapted for the mobility necessary for grasping. Szalay and Dagosto (1988) interpreted the shape of the primate entocuneiform-first metatarsal joint as a mechanism to resist forces encountered during takeoff and landing while still allowing the rotations necessary for grasping. Similarly, a morphological compromise to preserve the grasping ability of the foot, while remodeling it to facilitate leaping, is the hypothesis proposed by Morton (1924) to explain the lengthening of the tarsal, rather than metatarsal region of the foot in small arboreal primates, and by Preuschoft et al. (1995) to explain the lengthening of the proximal rather than distal limb segments. Likewise, some of the upper ankle joint characters discussed above (e.g., rotation of medial malleolus, etc.) are more exaggerated in taxa that de-emphasize leaping and least expressed in those that leap most frequently (Dagosto, 1985). Both notharctines and omomyids show only moderate expression of the features, suggesting that they were frequent leapers, moderate expression is primitive for primates, and/or that primitive primates were frequent leapers.

In addition to these "compromise" features, the primate morphotype is also characterized by others that are best explained (by both mechanical implications of the features and distribution within both mammals and primates), as functionally related to leaping. Among these features are an elongated ilium and short ischium, elongation of hindlimbs and elements of the hindlimb (femur, tibia, tarsus), a proximally placed third trochanter, an antero-posteriorly deepened knee, and a relatively tall talar body.

Within primates, specialized leapers have especially short ischia and somewhat longer ilia (Anemone, 1993; Anemone and Covert, 2000; McArdle, 1981; Napier and Walker, 1967a; Walker, 1974). The relationship between these features and leaping in mammals as a whole is more complex—other mammalian leapers have a long ischium and long ilia are found among cursors as well as leapers (Howell, 1944; Smith and Savage, 1956). Compared to other archontans, extant prosimian primates are characterized by a relatively long ilium and short ischium (Anemone, 1993; Anemone and Covert, 2000). *Notharctus* and *Omomys*, however, have relatively longer ischia and short ilia compared to most extant prosimians, exhibiting values more similar to those of tupaiids (Anemone and Covert, 2000; Gregory, 1920). On the other hand, they do exhibit the dorsal projection of the ischium typical of leaping primates (Fleagle and Anapol, 1992).

Several contributions have demonstrated that primates have longer hindlimbs relative to forelimb length, trunk length, or body mass than other mammals including arboreal members of the Carnivora and Rodentia (Alexander et al., 1979; Anemone, 1990, 1993; Polk et al., 2000). Anemone (1993) provided the most phylogenetically informative comparison demonstrating that tupaiids have a lower hindlimb index (femur + tibia relative to trunk length) than prosimian primates. A potential problem with using trunk length as a body mass surrogate is that leapers often have short trunks as well as long limbs (Emerson, 1985). A comparison of hindlimb length (femur + tibia) with body mass is presented in Figure 2 (statistics are given in Table 3). It illustrates that primates do indeed have hindlimbs that are longer than tree shrews and those arboreal marsupials that have been proposed as models for primitive primates. The exceptions to this generalization are the cheirogaleids, which have the shortest hindlimbs relative to body mass of any

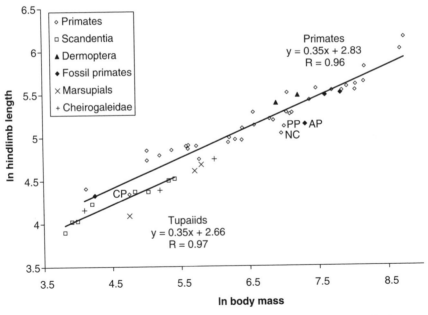

Figure 2. Regression (OLS) of ln hindlimb length (femur + tibia) on ln body mass. AP = *Adapis parisiensis*; CP = *Cebuella pygmaea*; NC = *Nycticebus coucang*; PP = *Perodicticus potto*. See Table 2 for samples and Table 3 for additional information.

primate (Jungers, 1985), and *Cebuella*, which has the shortest hindlmb of any platyrrhine (Davis, 2002). *Notharctus, Smilodectes, Shoshonius,* and *Adapis* are the only fossil primates for which this value can be estimated, and with the exception of *Adapis* (and possibly *Europolemur* (Franzen, 2000)), they too have relatively long hindlimbs. Dermopterans, however, also have long hindlimbs like primates although presumably for different reasons (Runestad and Ruff, 1995).

Simple ballistic formulas dictate the relationship between limb length and leaping ability (Alexander, 1968, 1995; Emerson, 1985). Other things being equal, a long limb permits the propulsive force to act over a longer distance and therefore for a longer time, enabling a longer (or higher) leap for a given amount of muscle mass. A long hindlimb has the additional advantage of being able to be used as a brake to absorb landing forces (Preuschoft et al., 1995). Longer hindlimbs (relative to body mass) are found not only in primates, but in kangaroos, jumping rodents, and frogs (Emerson, 1985; Marsh, 1994). Experimental work has demonstrated that an increase in hindlimb length increases jumping performance (as measured by leap distance) in inter-specific comparisons of *Anolis* lizards and frogs (Emerson, 1985; Losos, 1990; Marsh, 1994).

Each element of the hindlimb is longer in primates (including notharctines and omomyids) than in tree shrews (or plesiadapids). The femur and tibia show essentially the same pattern as the total hindlimb (Figure 3A and B), with the femur being somewhat more elongated than the tibia. In primates, the femur is often longer than the tibia, while in tree shrews (and most other mammals), the tibia is longer than the femur. Among mammalian leapers, primates appear to be unique in lengthening the femur rather than the metatarsus and tibia. This might be taken as evidence that femoral elongation is not associated with leaping; however Morton (1924) and Preuschoft et al. (1995) have argued that the grasping foot of primates, and its associated musculature, acts as a constraint against elongation of the distal elements of the limb. Among primates, special-ized leapers have longer femora than more quadrupedal forms (Anemone, 1990, 1993; Connour, 2000; Jouffroy and Lessertisseur, 1979; Lessertisseur and Jouffroy, 1973).

The elongated tarsus of primates, which results in a "reverse alternating" foot (Lewis, 1980b) has also been thought to be functionally related to leap-ing. Alexander (1995) has shown with modeling that a three-segment limb produces longer jumps than a two segment limb, and this is the likely reason

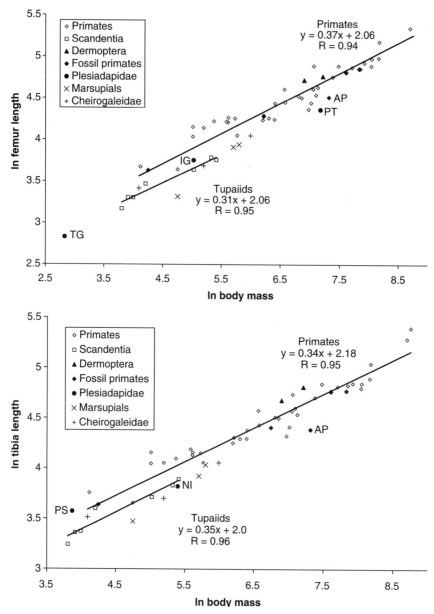

Figure 3. (A) Regression (OLS) of ln femur length on ln body mass. Conventions as in Figure 2. IG = *Ignacius graybullianus;* PT = *Plesiadapis tricuspidens;* TG = *Tinimomys graybulliensis.* (B) Regression (OLS) of ln tibia length on ln body mass. NI = *Nannodectes intermedius;* PS = *Phenacolemur simonsi.*

Table 2. List of taxa, sample sizes, body weights, and sources of information for variables (Analyzed in Table 3 and illustrated in Figures 2–7)

Taxon	N for long bones	N for foot bones	Source for body weight	Additional sources and notes
Prosimians				
Microcebus murinus	13–18	14	1	
Cheirogaleus medius	6	6	1	
Cheirogaleus major	8	7	1	
Daubentonia madagascariensis	5	5	1	
Galago senegalensis	6	6	1	
Galago moholi	9	9	1	
Galago crassicaudatus	10	7	1	
Galago alleni	5	5	1	
Galagoides demidovii	15	13	1	
Euoticus elegantulus	12	16	1	
Avahi laniger	8	11	1	
Propithecus verreauxi	9	11	1	
Propithecus diadema	3	11	1	
Indri indri	8	12	1	
Eulemur fulvus	15–19	25	1	
Eulemur mongoz	5	7	1	
Lemur catta	11	13	1	
Varecia variegata	11–14	14	1	
Hapalemur griseus	9–13	12	1	
Lepilemur mustelinus	17	17	1	
Perodicticus potto	10–15	15	1	
Arctocebus calabarensis	10	11	1	
Nycticebus coucang	12	13	1	
Loris tardigradus	8	7	1	
Tarsius syrichta	10–15	12	1	
Tarsius bancanus	9	6	1	
Anthropoids				
Aotus azarae	5	5	1	
Aotus lemurinus	6	6	1	

(Continued)

Table 2. List of taxa, sample sizes, body weights, and sources of information for variables (Analyzed in Table 3 and illustrated in Figures 2–7)—(Continued)

Taxon	N for long bones	N for foot bones	Source for body weight	Additional sources and notes
Saimiri sciureus	10	15	1	
Callicebus torquatus	1	2	1	
Callicebus donacophilus	5	3	1	
Cebus apella	6	6	1	
Cebus capucinus	7	7	1	
Chiropotes satanas	4	4	1	
Pithecia pithecia	6	6	1	
Callimico goeldii	3	3	1	
Leontopithecus rosalia	4	4	1	
Saguinus leucopus	4	3	1	
Saguinus midas	4	4	1	
Callithrix jacchus	5	4	1	
Cebuella pygmaea	5	5	1	
Tupaiids	Sargis, 2000			
Ptilocercus lowii	1	1	2	
Dendrogale murinus	1	1	3	
Tupaia glis	3	3	2	
T. gracilis	1	1	2	
T. minor	1	1	4	
T. tana	2	2	2	
Urogale everetti	3	3	3	Sargis, 2000
T. montana		0	2	
Dermopterans				
Cynocephalus variegatus	5	2	5	
Cynocephalus volans	6	3	5	
Fossil primates				
Notharctus tenebrosus	1–2	4	6	(Gebo et al., 1991)
Smilodectes gracilis	1–2	7	6	(Gebo et al., 1991)
Cantius mckennai	0	0–5	6	(Gebo et al., 1991)

				Sources
Cantius trigonodus	0	4–6	6	(Gebo et al., 1991)
Cantius abditus	0	2–8	6	(Gebo et al., 1991)
Adapis parisiensis	1–4	2(a), 19(c)	6	(Dagosto, 1983)
Shoshonius cooperi	1–2	2–3		(Dagosto et al., 1999)
Bridger B ?Omomys	0	8 (a), 14 (c)	6	(Dagosto et al., 1999)
Bridger C ?Hemiacodon	0	10 (a), 5 (c)	6	(Dagosto et al., 1999)
Tetonius homunculus	0	2–3	6	(Dagosto et al., 1999)
Washakius insignis	0	1–2	6	(Dagosto et al., 1999)
Plesiadapids				
Plesiadapis tricuspidens	1–3	1–2	7	Beard, pers comm.
Phenacolemur simonsi	1	0	7	Beard, pers. comm
Nannodectes intermedius	1	0	8	Beard, pers. comm.
Ignacius graybullianus	1	1	7	Beard, pers. comm.
Tinimomys graybulliensis	1	0	8	
Nannodectes gidleyi	0	1	7	Est from figure in (Simpson, 1940)
Marsupials				
Caluromys derbianus	7	0	9	
Philander opossum	3	0	10	
Marmosa robinsoni	3	0	10	

All measurements are from the author's own data unless otherwise noted. In the "N for foot bones" column, a = number of tali; c = number of calcanei. Sources for body weights: 1. (Smith and Jungers, 1997), male and female values averaged; field weights preferred to lab weights; 2. (Emmons, 2000); 3. (Sargis, 2000); 4. (Sargis, 2001); 5. (Runestad and Ruff, 1995); 6. (Dagosto and Terranova, 1992); 7. (Conroy, 1987) prosimian equation; 8. (Fleagle, 1999); 9. (Rasmussen, 1990); and 10. (Lemelin, 1999).

Table 3. Results of tests of slopes and intercepts of limb variables regressed against body mass

	Tupaiids versus primates		Tupaiids versus prosimians		Tupaiids versus Anthropoids	
	OLS	II	OLS	II	OLS	II
Hindlimb length	0.001	0.001	0.001	0.005	0.050/0.024	0.009/0.007
Femur length	0.001	0.000	0.000	0.000	0.029/0.026	0.003/0.001
Tibia length	0.004	0.005	0.005	0.005	0.133/0.043	0.056/0.003
Femoral condyle depth	0.020	*	0.018	0.07	*/*	*/*
Talar height	0.004	0.004	0.001	0.002	0.890/0.830	0.976/0.659
Tarsus length	0.000	*	0.000	*	0.292/0.250	0.043/0.008
Humerus length	0.268	0.096	0.541	0.401	*/*	*/*
Forelimb length	0.097	*	0.140	0.102	*/*	*/*

Results are presented for least squares regression (OLS) and Model II (II). p values in the Table are for a test of elevation differences after slopes were found not to differ; *, the slopes differed. Results for the "Tupaiid versus Anthropoids" comparison are given first including, and then excluding *Cebuella*, which is an outlier and has a strong influence on the results. ANOVA and ANCOVA were used to test slopes and intercepts for OLS regressions. (S)MATR (Falster et al., 2003) was used to test for slope and intercept differences for reduced major axis regression. See also Figures 2–7 and their legends.

for a long tarsus in primates and the extremely long tarsus of specialized leapers like galagos and tarsiers. Analysis of the tarsus (calcaneus + cuboid) is complicated by the extreme elongation of this element in tarsiers and galagos, and the different scaling among subgroups of primates (Figure 4). Tree shrews, dermopterans, and *Plesiadapis tricuspidens* have a relatively shorter tarsus than similarly sized nonlorisine prosimians, but anthropoids do not differ from tree shrews in this regard.

In addition to elongation of the hindlimb, each joint of the primate hindlimb evidences the role of leaping. Compared to tree shrews, dermopterans, or plesiadapids, primates have more proximally placed femoral third trochanters that are at approximately the same level as the lesser trochanter, rather than being distal to it (Anemone, 1993; Bloch and Boyer, this volume ; Sargis, 2000; Silcox, 2001). This has the likely effect of decreasing the mechanical advantage of the hip extensors, but increasing their speed of action, and is a morphology typical of leaping primates (Anemone, 1993; McArdle, 1981). The same morphology is seen in all adapids and omomyids for which femora are known.

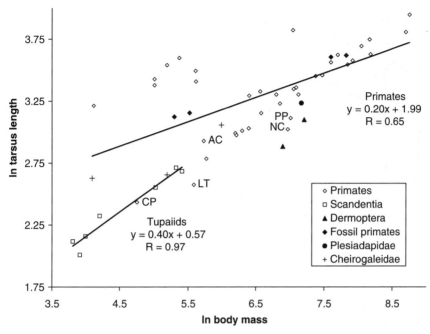

Figure 4. Regression (OLS) of ln tarsus length (calcaneus + cuboid) on ln body mass. Conventions as in Figure 2. AC = *Arctocebus calabarensis;* LT = *Loris tardigradus.*

Primates (including adapids and omomyids) have femoral condyles that are antero-posteriorly deep relative to their medio-lateral width, the consequence of which is to increase the mechanical advantage of the knee extensors (Anemone, 1993; Napier and Walker, 1967a, 1967b; Tardieu, 1983; Walker, 1974). Among primates, the depth of the condyles is greater in more frequently leaping primates, and shallowest in the slow climbing lorises. Figure 5 illustrates that the anterior–posterior dimension of the lateral condyle is larger in most primates than in tree shrews, dermopterans, or plesiadapids. Neither tree shrews nor other Archonta share the particular development of the ridge of the lateral condyle, a feature which likely prevents patellar dislocation by an enlarged vastus lateralis in leaping primates (Anemone, 1993; Bloch and Boyer, this volume; Sargis, 2000; Silcox, 2001).

The upper ankle joint (UAJ) of primates differs from that of plesiadapids, dermopterans, and tupaiids in several ways. Most strikingly, the body of the talus is very high relative to talar length or body mass (Figure 6). The medial

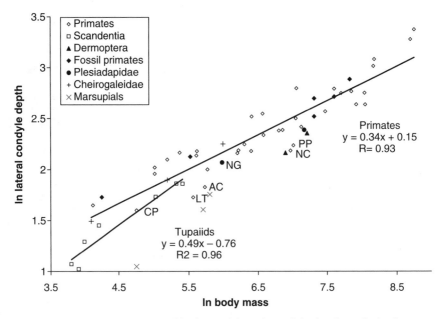

Figure 5. Regression (OLS) of ln femoral lateral condyle depth on ln body mass. Conventions as in Figure 2.

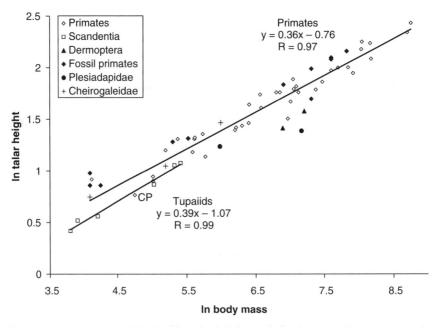

Figure 6. Regression (OLS) of ln talar height on ln body mass. Conventions as in Figure 2.

and lateral crests are approximately equal in height (the lateral is slightly higher in *Plesiadapis, Nannodectes,* tupaiids, *Ptilocercus,* and *Cynocephalus*). In primates, the medial and lateral edges of the trochlea are sharper and more defined than the more rounded edges typical of plesiadapids and dermopterans, although there is less of a difference between primates and tupaiids. The trochlea of primates is deeper than in Dermoptera or Plesiadapiforms, but not so different from tupaiids.

Within lemuriform primates it is the more frequently leaping species that have taller talar bodies, increased radii of curvature, and longer arclengths at the UAJ (Dagosto, 1986; Ward and Sussman, 1979), suggesting a benefit of this morphology to leaping primates. Other things being equal (i.e., the degree of overlap of the tibia on the talus), the greater height of the talar body increases the arc length of the upper ankle joint and therefore the range of plantarflexion–dorsiflexion. The greater depth of the talar trochlea and the sharper talar borders increase medio-lateral stability at this joint. Tarsiers are a notable exception to this characterization, having relatively low talar bodies (Godinot and Dagosto, 1983).

In sum, numerous features of the hindlimb support the hypothesis that leaping was an important factor in the evolution of the primate postcranial skeleton.

DO THE FUNCTIONAL/BIOLOGICAL ROLE ATTRIBUTES OF THE TRAITS AS A WHOLE CONSTITUTE A COHESIVE STORY?

Can this constellation of features be explained by a single selective factor or a set of selective factors arising from a way of life? The answer is possibly not. There seem to be at least two separate selective regimes influencing the primate postcranial morphotype. One is a "substrate size" component that is reflected in the grasping extremities, and the other is a "mode of displacement" component, which is reflected in the leaping features. Of course, these two components may be related (e.g., they evolved concurrently as part of a single mode of positional behavior) but there is at least the possibility that they may have independent histories (i.e., one preceded another in the "origin" of Primates).

The locomotor part of the NVP hypothesis does not recognize the role of leaping in ancestral primates. If the importance of leaping to the MRCA is

in this otherwise comprehensive model? On one hand, Fleagle and Mittermeier (1980) have associated leaping with both small body size and use of the more discontinuous understory, both of which are elements of the NVP model. Cartmill (1974a) cites *Microcebus* as a potential model for an ancestral primate. Although it may employ deliberate quadrupedalism when foraging for insects, its dominant locomotor mode is described as rapid scurrying and leaping (Gebo, 1987; Walker, 1974). As the example of *Tarsius* shows, NVP as an ecological strategy is certainly not incompatible with significant use of leaping or vertical supports.

On the other hand, leaping was certainly not stressed in the original formulation of the hypothesis, and continues to be criticized as an explanatory factor for primate origins (Cartmill, 1982; 1992; Schmitt and Lemelin, 2002, articles in this volume). Indeed, part of Cartmill's purpose was to critique hypotheses that invoke leaping to explain the presence of nails on lateral digits and the evolution of orbital convergence, and his model requires nothing more than a deliberate, slow moving quadrupedal animal (Cartmill, 1974a). If slow, deliberate quadrupedalism is meant as the predominant mode of displacement of the ancestral primate, and not just a locomotor mode employed during insect foraging, then this model and GL do seem contradictory. Are there other factors that might explain the incompatibilities between this aspect of the NVP model and GL or that may allow a reconciliation of the two models?

Are the Biological Roles of Features Misidentified?

It is of course possible that the function or biological role of the features supporting GL (or NVP) are misidentified; perhaps the features cited above are not related to leaping. As noted above, forms have numerous functions; are there other possible explanations for the primate features? For example, long hindlimbs (and elements of the limb) are useful for:

(1) Leaping, as argued in an earlier section;
(2) Increasing stride length, and therefore speed of quadrupedal locomotion. Primates do generally have longer stride lengths than other mammals (Larson, 1998). Demes et al. (1990) explain the long limbs of lorises as an adaptation to increase speed without the need to increase stride frequency, which would decrease stability on small branches. On

the other hand, most cursorial mammals lengthen the distal elements of the limb, especially the metatarsus, unlike primates where the femur, tibia, or tarsus are elongated (Garland and Janis, 1993; Hildebrand, 1982), (Some lizards do lengthen proximal elements (Garland and Losos, 1994)). Alexander (1995) has shown that for leaping, the distribution of mass in the limbs is less important than total limb mass, so that unlike running, mass does not have to be concentrated in the short proximal element. Primates, of course, do not exhibit any of the other postcranial characteristics of cursors. From a mechanical and comparative perspective, the hypothesis that hindlimbs elongated to enhance speed does not seem to be a better alternative than that they elongated to enhance leaping, and in any case speedy locomotion itself is inconsistent with the NVP slow quadruped model;

(3) Mammals that glide have longer femora (and humeri) than related nongliders (Runestad and Ruff, 1995) because long limbs are biomechanically important for maintaining aspect ratio and patagial loading. Since there are no other indications of gliding in primates, this also seems an unlikely explanation for their long femora and tibiae; and

(4) A general indication of "arboreality." Polk et al. (2000) have shown that arboreal members of the Carnivora and Rodentia have longer femora (and humeri) than terrestrial members of the same lineage. Therefore, long limbs may be an indication of some other aspect of arboreality. Three (nonmutually exclusive) possibilities here are: (a) climbing (Preuschoft and Witte, 1991; Preuschoft et al., 1995); (b) bridging (Polk et al. (2000); and (c) long stride length contributing to the compliant gait that facilitates movement on small supports (Larson et al., 2000, 2001; Schmitt, 1998; 1999).

None of these hypotheses, alone or in concert, seems adequate to explain the extreme degree of lengthening of the primate hindlimb compared to other arboreal mammals, including *Caluromys*, which does not appear to have hindlimbs any longer than its terrestrial relatives (Figures 2, 3A). Although humeri are also longer in primates than in arboreal carnivores or rodents (Polk et al., 2000), the difference in femur length is more marked. Primates do differ significantly from tupaiids and plesiadapids in hindlimb length, but not in humeral or forelimb length; in fact plesiadapids appear to have humeri that are as long or longer than primates (Figure 7, Table 3).

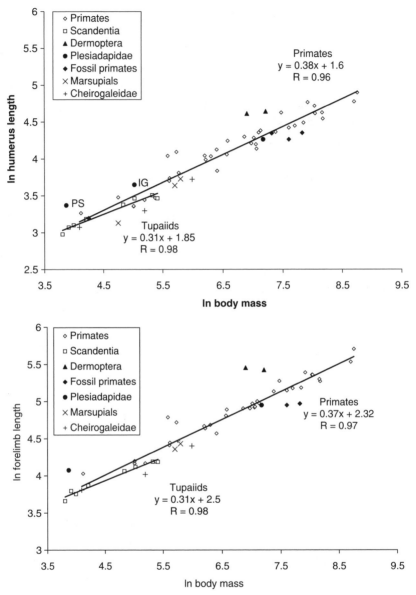

Figure 7. (A) Regression (OLS) of ln humerus length on ln body mass. Conventions as in Figure 2. (B) Regression (OLS) of ln forelimb length (humerus + radius) on ln body mass. Conventions as in Figure 2.

If primates were shown to climb vertical supports or bridge more often or more efficiently than other arboreal mammals there could be some validity to these hypotheses. Among primates, however, the lorises, likely the most frequent climbers and bridgers, do not have longer hindlimbs than other primates, while leaping primates do (Connour, 2000; Runestad, 1997). Prosimians that use vertical supports more frequently (indriids, tarsiers) also have relatively longer femora, but these taxa are also leapers, complicating attempts to separate these factors.

Although it is most certainly naïve to think that a single selective factor is involved in the origin or evolution of any trait, the combined evidence of hindlimb morphology (both limb elongation and joint shape) is strongly supportive that the mechanical requirements of leaping have driven, at least in part, the evolution of the primate hindlimb.

Are these Features Part of the Primate Morphotype?

It is also possible leaping features were not part of the primate morphotype, but evolved after the MRCA only in a clade of prosimian primates, or independently in several different groups of primates. Many of the features listed as primate apomorphies are more strongly expressed in notharctines, omomyids, and lemurs and less so in anthropoids. Anthropoids, for example, do not differ greatly from tupaiids in talar height (Table 3). If we accept the traditional phylogeny where anthropoids evolved from some sort of prosimian, the hypothesis that leaping features were part of the euprimate morphotype is strongly supported. However, if a phylogeny in which prosimians are the sistergroup of anthropoids and in which anthropoids exhibit primitive primate postcranial morphology is accepted (Ford, 1988, 1994; Godinot, 1992; Godinot and Beard, 1991; Godinot and Jouffroy, 1984), the hypothesis is less certain (Figure 8).

These propositions are problematic on several levels. Anthropoids, although not as specialized as most prosimians, still differ significantly from tupaiids or plesiadapids in several critical features, particularly hindlimb length, suggesting that this was a feature present in the MRCA of all primates. The convergence hypothesis also requires the independent development of leaping and the same leaping features in several groups of primates, an idea made unlikely by the presence of these features in adapids, omomyids, and eosimiids (Gebo et al., 2000), the likely stem groups for the living clades. Furthermore, if the grasping complex, which differs significantly in morphological detail between anthropoids and prosimians (Szalay and Dagosto, 1988), is considered to be

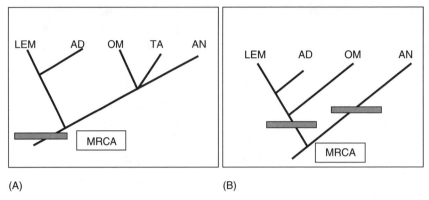

(A)   (B)

Figure 8. Two versions of primate phylogenetic relationships. In (A) anthropoids are derived from an ancestor with prosimian-like ancestry. Leaping features (gray box) are characteristic of the primate MRCA. In (B) anthropoids are the sistergroup of prosimians, leaping features are not characteristic of the MRCA, but of the prosimian ancestor.

a highly corroborated synapomorphy of primates, then isn't it inconsistent to simultaneously argue that the morphology of the knee joint, hip joint, or ankle joint, which differ less in detail, are convergences?

Origin of Primates as a Process Not an Event

The most likely explanation for the apparent discrepancy between the locomotor component of NVP and GL is that any effort to relate **all** derived primate characters to a single way of life is oversimplified. Surely the "Origin of Primates," like that of any other group, was not a single event (i.e., a single speciation), but a series of events occurring over a long period of time (Figure 9; see also Rasmussen and Sussman, this volume, and the distinction between "origin" of a group and the "last common ancestor" in Soligo et al. (this volume)). The presence of a long list of synapomorphies for the MRCA is an illusion attributable to the lack of a good fossil record or the continued existence of intermediate forms (Gebo and Dagosto, 2004). The MRCA had a way of life, but all of its features may not be related to that; some were acquired by ancestors along the stem as a result of their ways of life. The real problem is to identify the sequence of "ways of life," to determine how earlier ones may have channeled later ones or constrained the morphological solutions available to descendants (Cartmill, 1982; Rasmussen and Sussman, this volume; Szalay and Sargis, 2001).

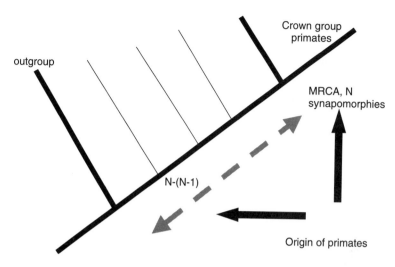

Figure 9. The "Origin of Primates" can be interpreted as either the MRCA with N synapomorphies, or as the stem lineage (dashed arrow) with numerous stages, the earliest of which has only one or few of the "primate" synapomorphies. Each of the stages along the stem may have very different ways of life which constrain paths of evolution of future stages, or constrain the morphological response to new behaviors. Modified from Gebo and Dagosto (2004).

Given the current evidence, the most likely scenario is that grasping either preceded or evolved concurrently with leaping (Figure 10). Both tree shrews, especially *Ptilocercus*, and plesiadapiforms exhibit features of the tarsus and hallucal-metatarsal joint which indicate some independence of the hallux, incipient grasping abilities, and ability to invert the foot greater than in the primitive eutherian state (Beard, 1991; Bloch and Boyer, 2002; Sargis, 2001; Szalay and Dagosto, 1988; Szalay and Drawhorn, 1980). Some rudimentary form of grasping was already present in the ancestral archontan. The elaboration of the primate grasping appendage with its increased hallucal independence, habitual abducted stance, and replacement of hallucal claw by nail indicates more frequent or critical use of relatively small branches (Cartmill, 1974a). Didelphids and phalangerids provide interesting parallels for this phase of postcranial evolution (Rasmussen and Sussman, this volume). The elongated hindlimbs and the stabilized knee, ankle, and hallucal-metatarsal joints evolved as leaping became an integral and frequent part of the positional behavior of the ancestral primate.

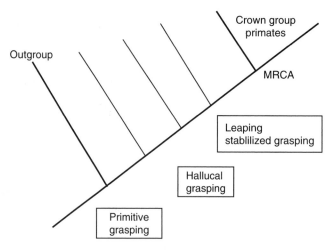

Figure 10. Phases in the evolution of the primate postcranium. A primitive form of grasping, present in the archontan ancestor is replaced by more derived hallucal grasping (opposable hallux, hallucal nail). Leaping either follows this, or evolves concurrently with it.

The primary difference between the Szalay-Dagosto point of view and that of Cartmill, Lemelin, and Schmitt (Cartmill, 1974a; Schmitt and Lemelin, 2002, articles in this volume), is that we believe the evidence clearly shows that the adoption of leaping and its morphological correlates is a shared-derived feature of all crown-group primates and therefore must have occurred **before** or in the MRCA, while the NVPers believe their evidence indicates that some primate locomotor apomorphies are most likely to have evolved in a more slowly moving quadruped and therefore that leaping and morphological changes associated with it must evolve convergently after the appearance of the MRCA. The two models could be reconciled if the NVP hypothesis does not apply to the MRCA, but is emphasizing events earlier in the transition from archontan to Primate; the "leaping" part of GL, later events.

The primary point to take away from this essay is that there are derived features of the primate skeleton that are most parsimoniously interpreted as being present in the MRCA, and are not easily explained simply by grasping small supports, by deliberate quadrupedalism on horizontal branches, or even by rapid scurrying and bounding by clawless animals on horizontal supports. Leaping remains the best explanation for the biological role of these features.

Marsupials such as *Caluromys* mimic *some* aspects of early stages of primate evolution, but do not provide a comprehensive model for the postcranium of the MRCA of Primates, since they lack all of the key leaping-related features (see also Szalay, this volume). Models for the Origin of Primates are incomplete if they do not account for or incorrect if they cannot accommodate the leaping component of the primate morphotype.

ACKNOWLEDGMENTS

I thank the National Science Foundation (BCS-0129349), the Wenner-Gren Foundation for Anthropological Research, the Field Museum of Natural History, and Northwestern University for financial support of the Primate Origins conference. Dan Gebo and another reviewer gave helpful criticism, and Chris Beard provided some unpublished data. I also thank all the museums and personnel who have granted me access to their collections of living and fossil mammals over the years.

REFERENCES

Alexander, R. M., 1968, *Animal Mechanics*, University of Washington Press, Seattle.

Alexander, R. M., 1995, Leg design and jumping techniques for humans, other vertebrates and insects, *Phil. Trans. R. Soc. London* **347**: 235–248.

Alexander, R. M., Jayes, A. S., Malohy, M. O., and Wathuta, E. M., 1979, Allometry of the limb bones of mammals from shrews (*Sorex*) to elephant (*Loxodonta*), *J. Zool., Lond.* **189**: 305–314.

Altner, G., 1971, Histologische und vergleichend-anatomische Untersuchungen zur Ontogenie und Phylogenie des Handskeletts von *Tupaia glis* (Diard, 1820) und *Microcebus murinus* (J. F. Miller, 1777), *Folia Primatol.* **14**: 1–106.

Anemone, R. L., 1990, The VCL hypothesis revisited: Patterns of femoral morphology among quadrupedal and saltatorial prosimian primates, *Am. J. Phys. Anthropol.* **83**: 373–393.

Anemone, R. L., 1993, The functional anatomy of the hip and thigh in primates, in: *Postcranial Adaptation in Nonhuman Primates*, D. L. Gebo, ed., Northern Illinois University Press, DeKalb, IL, pp. 150–174.

Anemone, R. L., and Covert, H. H., 2000, New skeletal remains of *Omomys* (Primates, Omomyidae): Functional morphology of the hindlimb and locomotor behavior of a Middle Eocene primate, *J. Hum. Evol.* **38**: 607–634.

Arnold, S. J., 1983, Morphology, performance, fitness, *Am. Zool.* **23**: 347–361.

Ayoub, M. M., and LoPresti, P., 1971, The determination of an optimum size cylindrical handle by use of electromyography, *Ergonomics* **14**: 509–518.

Baum, D. A., and Larson, A., 1991, Adaptation reviewed: A phylogenetic methodology for studying character evolution, *Syst. Zool.* **40**: 1–18.

Beard, K. C., 1989, Postcranial anatomy, locomotor adaptations, and paleoecology of early Cenozoic Plesiadapidae, Paromomyidae, and Micromomyidae (Eutheria, Dermoptera). Ph.D., Johns Hopkins University, Baltimore.

Beard, K. C., 1991, Vertical postures and climbing in the morphotype of primatomorpha: Implications for locomotor evolution in primate history, in: *Origine(s) de la bipédie chez les hominids*, Y. Coppens and B. Senut, eds., CNRS, Paris, pp. 79–87.

Beard, K. C., 1993, Phylogenetic systematics of the Primatomorpha, with special reference to Dermoptera, in: *Mammal Phylogeny: Placentals*, F. Szalay, M. Novacek, and M. McKenna, eds., Springer-Verlag, New York, pp. 129–150.

Bishop, A., 1964, Use of the hand in lower primates, in: *Evolutionary and Genetic Biology of Primates*, J. Beuttner-Janusch, ed., Academic Press, New York, pp. 133–225.

Bloch, J. I., and Boyer, D. M., 2002, Grasping Primate Origins, *Science* **298**: 1606–1610.

Bock, W. J., 1977, Adaptation and the comparative method, in: *Major Patterns in Vertebrate Evolution*, M. Hecht, P. Goody, and B. Hecht, eds., Plenum Press, New York, pp. 57–82.

Bock, W. J., and von Wahlert, G., 1965, Adaptation and the form-function complex, *Evolution* **19**: 269–299.

Cartmill, M., 1972, Arboreal adaptations and the origin of the Order Primates, in: *The Functional and Evolutionary Biology of Primates*, R. Tuttle, ed., Aldine-Atherton, Chicago, pp. 97–122.

Cartmill, M., 1974a, Pads and claws in arboreal locomotion, in: *Primate Locomotion*, F. Jenkins, ed., Plenum, New York, pp. 45–83.

Cartmill, M., 1974b, Rethinking primate origins, *Science* **184**: 436–443.

Cartmill, M., 1982, Basic Primatology and Prosimian Evolution, in: *A History of American Physical Anthropology, 1930–1980*, F. Spencer, ed., Academic Press, New York, pp. 147–186.

Cartmill, M., 1990, Human uniqueness and theoretical content in paleoanthropology, *Int. J. Primatol.* **11**: 173–192.

Cartmill, M., 1992, New views on primate origins, *Ev. Anth.* **1**: 105–111.

Charles-Dominique, P., Atramentowicz, M., Charles-Dominique, M., Gérard, H., Hladik, A., Hladik, C. M., et al., 1981, Les mammifères frugivores arboricoles nocturnes d'une forêt guyanaise: Inter-relations plantes-animaux, *Rev. Ecol. (Terre Vie)* **35**: 341–435.

Clark, W.E.L.G., 1959, *The Antecedents of Man*, Quadrangle Books, Chicago.

Collins, E. T., 1921, Changes in the visual organs correlated with the adoption of arboreal life and with the assumption of erect posture, *Trans. Opth. Soc., UK* **41**: 10–90.

Connour, J. R., 2000, Postcranial adaptations for leaping in primates, *J. Zool., Lond.* **251:** 79–103.

Conroy, G. C., 1987, Problems of body-weight estimation in fossil primates, *Int. J. Primatol.* **8:** 115–137.

Conroy, G. C., and Rose, M. D., 1983, The evolution of the primate foot from the earliest primates to the Miocene hominoids, *Foot and Ankle* **3:** 342–364.

Crompton, R. H., 1993, Energetic efficiency and ecology as selective forces in the saltatory adaptation of prosimian primates, *Proc. R. Soc. Lond. B* **254:** 41–45.

Crompton, R. H., 1995, "Visual predation," habitat structure, and the ancestral primate niche, in: *Creatures of the Dark: The Nocturnal Prosimians*, L. Alterman, G. Doyle, and M. Izard, eds., Plenum Press, New York, pp. 11–30.

Dagosto, M., 1983, Postcranium of *Adapis parisiensis* and *Leptadapis magnus* (Adapiformes, Primates), *Folia Primatol.* **41:** 49–101.

Dagosto, M., 1985, The distal tibia of primates with special reference to the Omomyidae, *Int. J. Primatol.* **6:** 45–75.

Dagosto, M., 1986, The Joints of the Tarsus in the Strepsirhine Primates, Ph.D., City University of New York.

Dagosto, M., 1988, Implications of postcranial evidence for the origin of Euprimates, *J. Hum. Evol.* **17:** 35–56.

Dagosto, M., Gebo, D. L., and Beard, K. C., 1999, Revision of the Wind River Faunas, Early Eocene of Central Wyoming, Part 14, Postcranium of *Shoshonius cooperi* (Mammalia, Primates), *Ann. Carn. Mus.* **68:** 175–211.

Dagosto, M., and Terranova, C. J., 1992, Estimating the body size of Eocene primates: A comparison of results from dental and postcranial variables, *Int. J. Primatol.* **13:** 307–344.

Davis, L. C., 2002, Functional Morphology of the Forelimb and Long Bones in the Callitrichidae (Platyrrhini, Primates). Ph.D., Southern Illinois University, Carbondale, Illinois.

Day, M. H., 1979, The locomotor interpretation of fossil primate postcranial bones, in: *Environment, Behavior, Morphology: Dynamic Interactions in Primates*, M. Morbeck, H. Preuschoft, and N. Gomberg, eds., Gustav Fischer, New York, pp. 245–258.

Decker, R. L., and Szalay, F. S., 1974, Origin and function of the pes in the Eocene adapids (Lemuriformes, Primates), in: *Primate Locomotion*, F. Jenkins, ed., Academic Press, New York, pp. 261–291.

Demes, B., Fleagle, J. G., and Jungers, W. L., 1999, Takeoff and landing forces of leaping strepsirhine primates, *J. Hum. Evol.* **37:** 279–292.

Demes, B., Jungers, W., Fleagle, J., Wunderlich, R., Richmond, B., and Lemelin, P., 1996, Body size and leaping kinematics in Malagasy vertical clingers and leapers, *J. Hum. Evol.* **31:** 366–388.

Demes, B., Jungers, W. L., Gross, T. S., and Fleagle, J. G., 1995, Kinetics of leaping primates: Influence of substrate orientation and compliance, *Am. J. Phys. Anthropol.* **96:** 419–430.

Demes, B., Jungers, W. L., and Nieschalk, U., 1990, Size and Speed related aspects of quadrupedal walking in slender and slow lorises, in: *Gravity, Posture, and Locomotion in Primates,* F. Jouffroy, Stack, M. H. and C. Niemitz, eds., Il Sedicesimo, Firenze, pp. 176–197.

Emerson, S. B., 1985, Jumping and leaping, in: *Functional Vertebrate Morphology,* M. Hildebrand, D. M. Bramble, K. F. Liem, and D. B. Wake, eds., Harvard University Press, Cambridge, pp. 58–72.

Emmons, L. H., 2000, *Tupai: A field Study of Bornean Treeshrews,* University of California Press, Berkeley.

Falster, D. S., Warton, D. I., and Wright, I. J., 2003, (S)MATR: Standardised major axis tests and routines: http://www.bio.mq.au/ecology/SMATR

Fisher, D. C., 1985, Evolutionary morphology: Beyond the analogous, the anecdotal, and the ad hoc, *Paleobiology* **11:** 120–138.

Fleagle, J. G., 1979, Primate positional behavior and anatomy: Naturalistic and experimental approaches, in: *Environment, Behavior and Morphology: Dynamic Interactions in Primates,* M. Morbeck, H. Preuschoft, and N. Gomberg, eds., G. Fischer, New York, pp. 313–325.

Fleagle, J. G., 1999, *Primate Adaptation and Evolution,* Academic Press, New York.

Fleagle, J. G., and Anapol, F. C., 1992, The indriid ischium and the hominid hip, *J. Hum. Evol.* **22:** 285–306.

Fleagle, J. G., and Mittermeier, R. A., 1980, Locomotor behavior, body size, and comparative ecology of seven Surinam monkeys, *Am. J. Phys. Anthropol.* **52:** 301–314.

Ford, S. M., 1988, Postcranial adaptations of the earliest platyrrhine, *J. Hum. Evol.* **17:** 155–192.

Ford, S. M., 1994, Primitive platyrrhines? Perspectives on Anthropoid origins from Platyrrhine, Parapithecid, and Preanthropoid Postcrania, in: *Anthropoid Origins,* J. G. Fleagle and R. F. Kay, eds., Plenum Press, New York, pp. 595–673.

Franzen, J. L., 2000, *Europolemur kelleri* n.sp. von Messel und ein Nachtrag zu *Europolemur koenigswaldi* (Mammalia, Primates, Notharctidae, Cercamoniinae), *Senckenbergiana lethaea* **80:** 275–287.

Garber, P. A., and Sussman, R., 1984, Ecological distinctions between sympatric species of *Saguinus* and *Sciurus, Am. J. Phys. Anthropol.* **65:** 135–146.

Garland, T., and Adolph, S. C., 1994, Why not to do two-species comparative studies: Limitations on inferring adaptation, *Phys. Zool.* **67:** 797–828.

Garland, T., and Janis, C. M., 1993, Does metatarsal/femur ratio predict maximal running speed in cursorial mammals? *J. Zool., Lond.* **229:** 133–151.

Garland, T., and Losos, J. B., 1994, Ecological morphology of locomotor performance in squamate reptiles, in: *Ecological Morphology*, P. C. Wainwright and S. M.
Reilly, eds., University of Chicago Press, Chicago, pp. 240–302.

Gebo, D. L., 1987, Locomotor diversity in prosimian primates, *Am. J. Primatol.* **13:**
271–281.

Gebo, D. L., 2004, A shrew sized origin for primates, *Yearb. Phys. Anthropol.* **47:**
40–62.

Gebo, D. L., and Dagosto, M., 2004, Anthropoid origins: Postcranial evidence from
the Eocene of Asia, in: *Anthropoid Origins: New Visions*, C. F. Ross and R. F. Kay,
eds., Kluwer Academic/Plenum Publishers, New York, pp. 369–380.

Gebo, D. L., Dagosto, M., Beard, K. C., and Qi, T., 2000, The oldest known anthropoid postcranial fossils and the early evolution of higher primates, *Nature* **404:**
276–278.

Gebo, D. L., Dagosto, M., and Rose, K. D., 1991, Foot morphology and evolution
in early Eocene *Cantius*, *Am. J. Phys. Anthropol.* **86:** 51–73.

Godinot, M., 1992, Early euprimate hands in evolutionary perspective, *J. Hum. Evol.*
22: 267–283.

Godinot, M., and Beard, K. C., 1991, Fossil primate hands: A review and an evolutionary inquiry emphasizing early forms, *Human Evolution* **6:** 307–354.

Godinot, M., and Beard, K. C., 1993, A survey of fossil primate hands, in: *Hands of
Primates*, H. Preuschoft and D. Chivers, eds., Springer-Verlag, New York,
pp. 335–378.

Godinot, M., and Dagosto, M., 1983, The astragalus of *Necrolemur* (Primates,
Microchoerinae), *J. Paleo.* **57:** 1321–1324.

Godinot, M., and Jouffroy, F.-K., 1984, La main d'*Adapis* (Primate, Adapidé), in:
Actes du symposium paléontologique G. Cuvier, J. Mazin and E. Salmion, eds.,
Montbéliard, France, pp. 221–242.

Gregory, W. K., 1920, On the structure and relations of *Notharctus*: An American
Eocene primate, *Mem. Am. Mus. Nat. Hist.* **3:** 51–243.

Hafferl, A., 1929, Bau und funktion des Affenfusses. Ein beitrag zur gelenk und
Muskelmechanic. Die prosimier, *Z. Anat. Entwick. Gesell.* **90:** 46–51.

Hafferl, A., 1932, Bau und funktion des Affenfusses, Ein beitrag zur Gelenk und
Muskelmechanic, II. Die prosimier, *Z. Anat. Entwick. Gesell.* **99:** 63–112.

Hall-Craggs, E. C. B., 1965, An osteometric study of the hindlimb of the Galagidae,
J. Anat. **99:** 119–126.

Hamrick, M. W., 1997, Functional osteology of the primate carpus with special reference to Strepsirhini, *Am. J. Phys. Anthropol.* **104:** 105–116.

Hamrick, M. W., 1998, Functional and adaptive significance of primate pads and claws:
Evidence from New World anthropoids, *Am. J. Phys. Anthropol.* **106:** 113–127.

Hamrick, M. W., 1999, Pattern and process in the evolution of primate nails and claws, *J. Hum. Evol.* **37**: 293–298.

Hamrick, M. W., 2001, Primate origins: Evolutionary change in digital ray patterning and segmentation, *J. Hum. Evol.* **40**: 339–351.

Harvey, P., and Pagel, M. D., 1991, *The comparative method in evolutionary biology*, Oxford University Press, New York.

Hershkovitz, P., 1970, Notes on Tertiary platyrrhine monkeys and description of a new genus from the Late Miocene of Colombia, *Folia Primatol.* **12**: 1–37.

Hesse, M., 1966, *Models and analogies in science*, University of Notre Dame Press, Notre Dame.

Hildebrand, M., 1982, *Analysis of Vertebrate Structure*, John Wiley and Sons, New York.

Hooker, J. J., 2001, Tarsals of the extinct insectivoran family Nyctitheriidae (Mammalia): Evidence for archontan relationships, *Zool. J. Linn. Soc.* **132**: 501–529.

Howell, A. B., 1944, *Speed in Animals*, Chicago University Press, Chicago.

Jenkins, F. A., and McClearn, D., 1984, Mechanisms of hind foot reversal in climbing mammals, *J. Morphol.* **182**: 197–219.

Jouffroy, F. K., Godinot, M., and Nakano, Y., 1991, Biometrical characteristics of primate hands, *Human Evolution* **6**: 269–306.

Jouffroy, F. K., and Lessertisseur, J., 1979, Relationships between limb morphology and locomotor adaptations among prosimians: An osteometric study, in: *Environment, Behavior, and Morphology: Dynamic Interactions in Primates*, M. Morbeck, H. Preuschoft, and N. Gomberg, eds., G. Fischer, New York, pp. 143–182.

Jungers, W. L., 1985, Body size and scaling of limb proportions in primates, in: *Size and Scaling in Primate Biology*, W. L. Jungers, ed., Plenum Press, New York, pp. 345–381.

Kay, R. F., and Cartmill, M., 1977, Cranial morphology and adaptations of *Palaechthon nacimienti* and other Paromomyidae (Plesiadapoidea, ?Primates) with a description of a new genus and species, *J. Hum. Evol.* **6**: 19–54.

Koehl, M. A. R., 1996, When does morphology matter? *Ann. Rev. Ecol. Syst.* **27**: 501–542.

Larson, S. G., 1998, Unique aspects of quadrupedal locomotion in nonhuman Primates, in: *Primate Locomotion: Recent Advances*, E. Strasser, J. G. Fleagle, A. L. Rosenberger, and H. M. McHenry, eds., Plenum, New York, pp. 157–173.

Larson, S. G., Schmitt, D., Lemelin, P., and Hamrick, M. H., 2000, Uniqueness of primate forelimb posture during quadrupedal locomotion, *Am. J. Phys. Anthropol.* **112**: 87–101.

Larson, S. G., Schmitt, D., Lemelin, P., and Hamrick, M. H., 2001, Limb excursion during quadrupedal walking: how do primates compare to other mammals? *J. Zool., Lond.* **255**: 353–365.

Lauder, G. V., 1982, Historical biology and the problem of design, *J. Theor.l Biol.* **97**: 57–67.

Lauder, G. V., 1990, Functional morphology and systematics: Studying functional patterns in an historical context, *Ann. Rev. Ecol. Syst.* **21**: 317–340.

Lauder, G. V., 1996, The argument from design, in: *Adaptation*, M. R. Rose and G. V. Lauder, eds., Academic Press, New York, pp. 55–91.

Lemelin, P., 1999, Morphological correlates of substrate use in didelphid marsupials: Implications for primate origins, *J. Zool., Lond.* **247**: 165–175.

Lemelin, P., and Grafton, B. W., 1998, Grasping performance in *Saguinus midas* and the evolution of hand prehensility in primates, in: *Primate Locomotion: Recent Advances*, E. Strasser, J. G. Fleagle, A. L. Rosenberger, and H. M. McHenry, eds., Plenum, New York, pp. 131–144.

Lessertisseur, J., and Jouffroy, F. K., 1973, Tendances locomotrices des primates traduites par les proportions du pied, *Folia Primatol.* **20**: 125–160.

Lewis, O. J., 1980a, The joints of the evolving foot, Part I. The ankle joint, *J. Anat.* **130**: 527–543.

Lewis, O. J., 1980b, The joints of the evolving foot, Part III. The fossil evidence, *J. Anat.* **131**: 275–298.

Losos, J. B., 1990, The evolution of form and function: Morphology and locomotor performance in West Indian Anolis lizards, *Evolution* **44**: 1189–1203.

Losos, J. B., and Miles, D. B., 1994, Adaptation, constraint, and the comparative method: Phylogenetic issues and methods, in: *Ecological Morphology*, P. Wainwright and S. Reilly, eds., University of Chicago Press, Chicago, pp. 60–98.

Lovejoy, O., 1979, Contemporary methodological approaches to individual primate fossil analysis, in: *Environment, Behavior, Morphology: Dynamic Interactions in Primates*, M. Morbeck, H. Preuschoft, and N. Gomberg, eds., Gustav Fischer, New York, pp. 229–244.

MacConaill, M. A., 1973, A structurofunctional classification of synovial articular units, *Irish J. Med. Sci.* **142**: 19–26.

Marsh, R. L., 1994, Jumping ability of anuran amphibians, *Adv. Vet. Sci. Comp. Med.* **38B**: 51–111.

Martin, R. D., 1972, Adaptive radiation and behavior of the Malagasy lemurs, *Phil. Trans. Roy. Soc. London* **264**: 295–352.

Martin, R. D., 1990, *Primate Origins and Evolution -a phylogenetic reconstruction*, Princeton University Press, Princeton.

McArdle, J. E., 1981, Functional morphology of the hip and thigh of the lorisiformes, *Contrib. Primatol.* **17**: 1–132.

Morton, D., 1924, Evolution of the human foot, *Am. J. Phys. Anthropol.* **7**: 1–52.

Napier, J. R., 1967, Evolutionary aspects of primate locomotion, *Am. J. Phys. Anthropol.* **27**: 333–342.

Napier, J. R., 1980, *Hands*, Pantheon Press, New York.

Napier, J. R., and Walker, A., 1967a, Vertical clinging and leaping–a newly recognised category of locomotor behaviour of primates, *Folia Primatol.* **6**: 204–219.

Napier, J. R., and Walker, A. C., 1967b, Vertical clinging and leaping in living and fossil primates, in: *Neue Ergebnisse der Primatologie (Progress in Primatology)*, D. Starck, R. Schneider, and H. Kuhn, eds., Fischer, Stuttgart, pp. 66–69.

Oxnard, C., Crompton, R. H., and Lieberman, S. S., 1990, *Animal Lifestyles and Anatomies*, University of Washington Press, Seattle.

Pheasant, S., and O'Neill, D., 1975, Performance in gripping and turning, *Appl. Ergon.* **6**: 205–208.

Polk, J. D., Demes, B., Jungers, W. L., Biknevicius, A. R., Heinrich, R. E., and Runestad, J. A., 2000, A comparison of primate, carnivoran, and rodent limb bone cross-sectional properties: Are primates really unique? *J. Hum. Evol.* **39**: 297–326.

Preuschoft, H., 1979, Motor behavior and shape of the locomotor apparatus, in: *Environment, Behavior, and Morphology: Dynamic Interactions in Primates*, M. E. Morbeck, H. Preuschoft, and N. Gomberg, eds., Gustav Fischer, New York, pp. 263–275.

Preuschoft, H., and Witte, H., 1991, Biomechanical reasons for the evolution of hominid body shape, in: *Origine(s) de la Bipédie chez les hominidés*, Y. Coppens and B. Senut, eds., CNRS, Paris, pp. 59–77.

Preuschoft, H., Witte, H., and Fischer, M., 1995, Locomotion in nocturnal prosimians, in: *Creatures of the Dark*, L. Alterman, G. Doyle, and M. Izard, eds., Plenum Press, New York, pp. 453–472.

Rasmussen, D. T., 1990, Primate Origins: Lessons from a Neotropical marsupial, *Am. J. Primatol.* **22**: 263–277.

Rickelfs, R. E., and Miles, D. E., 1994, Ecological and evolutionary inferences from morphology: An ecological perspective, in: *Ecological Morphology*, P. Wainwright and S. Reilly, eds., University of Chicago Press, Chicago, pp. 13–41.

Ross, C. F., Lockwood, C. A., Fleagle, J. G., and Jungers, W. L., 2002, Adaptation and behavior in the Primate Fossil Record, in: *Reconstructing Behavior in the Primate Fossil Record*, J. M. Plavcan, R. F. Kay, W. L. Jungers, and C. P. van Schaik, eds., Kluwer Academic/Plenum Publishers, New York, pp. 1–41.

Rudwick, M. J. S., 1964, Inference of function from structure in fossils, *Brit. J. Phil. Sci.* **15**: 27–40.

Runestad, J. A., 1997, Postcranial adaptations for climbing in Loridae (Primates), *J. Zool., Lond.* **242**: 261–290.

Runestad, J. A., and Ruff, C. B., 1995, Structural adaptations for gliding in mammals with implications for locomotor behavior in paromomyids, *Am. J. Phys. Anthropol.* **98**: 101–119.

Sargis, E. J., 2000, The functional morphology of the postcranium of *Ptilocercus* and Tupaiines (Scandentia, Tupaiidae): Implications for the relationships of Primates and other archontan mammals, Ph.D., City University of New York, New York.

Sargis, E. J., 2001, The grasping behaviour, locomotion, and substrate use of the tree shrews *Tupaia minor* and *T. tana* (Mammalia, Scandentia), *J. Zool., Lond.* **253:** 485–490.

Schmitt, D., 1998, Forelimb mechanics during arboreal and terrestrial quadrupedalism in Old World monkeys, in: *Primate Locomotion: Recent Advances,* E. Strasser, J. G. Fleagle, A. L. Rosenberger, and H. M. McHenry, eds., Plenum, New York, pp. 175–200.

Schmitt, D., 1999, Compliant walking in primates, *J. Zool., Lond.* **248:** 149–160.

Schmitt, D., and Lemelin, P., 2002, Origins of primate locomotion: Gait mechanics of the Woolly Possum, *Am. J. Phys. Anthropol.* **118:** 231–238.

Silcox, M. T., 2001, A Phylogenetic analysis of Plesiadapiformes and their relationship to Euprimates and other archontans, Ph.D., Johns Hopkins University, Baltimore.

Simpson, G. G., 1940, Studies on the earliest primates, *Bull. Am. Mus. Nat. Hist.* **77:** 185–212.

Smith, J. M., and Savage, R. J. G., 1956, Some locomotory adaptations in mammals, *J. Linn. Soc. London* **42:** 603–622.

Smith, R. J., and Jungers, W. L., 1997, Body mass in comparative primatology, *J. Hum. Evol.* **32:** 523–559.

Stafford, B. J., Thorington, R. W., and Kawamichi, T., 2003, Positional behavior of Japanese Giant Flying Squirrels (*Petaurista leucogenys*), *J. Mamm.* **84:** 263–271.

Sussman, R. W., 1991, Primate origins and the evolution of angiosperms, *Am. J. Primatol.* **23:** 209–224.

Sussman, W., 1995, How primates invented the rainforest and vice versa, in: *Creatures of the Dark: The Nocturnal Prosimians,* L. Alterman, G. Doyle, and M. Izard, eds., Plenum Press, New York, pp. 1–10.

Szalay, F. S., 1968, The beginnings of Primates, *Evolution* **22:** 19–36.

Szalay, F. S., 1972, Paleobiology of the earliest primates. in: *The Functional and Evolutionary Biology of the Primates,* R. Tuttle, ed., Aldine-Atherton, Chicago, pp. 3–35.

Szalay, F. S., 1981, Phylogeny and the problem of adaptive significance: The case of the earliest primates, *Folia Primatol.* **36:** 157–182.

Szalay, F. S., and Dagosto, M., 1980, Locomotor adaptations as reflected on the humerus of Paleogene Primates, *Folia Primatol.* **34:** 1–45.

Szalay, F. S., and Dagosto, M., 1988, Evolution of hallucal grasping in the primates, *J. Hum. Evol.* **17:** 1–33.

Szalay, F. S., and Decker, R. L., 1974, Origins, evolution, and function of the tarsus in Late Cretaceous Eutheria and Paleocene Primates, in: *Primate Locomotion*, F. Jenkins, ed., Academic Press, New York, pp. 223–259.

Szalay, F. S., and Delson, E., 1979, *Evolutionary History of the Primates*, Academic Press, New York.

Szalay, F. S., and Drawhorn, G., 1980, Evolution and diversification of the Archonta in an arboreal milieu, in: *Comparative Biology and Evolutionary Relationships of Tree Shrews*, W. Luckett, ed., Plenum Press, New York, pp. 133–169.

Szalay, F. S., Rosenberger, A. L., and Dagosto, M., 1987, Diagnosis and differentiation of the Order Primates, *Yearb. Phys. Anthropol.* **30:** 75–105.

Szalay, F. S., and Sargis, E. J., 2001, Model based analysis of postcranial osteology of marsupials from the Paleocene of Itaborai (Brazil) and the phylogenetics and biogeography of Metatheria, *Geodiversitas* **23:** 139–302.

Tardieu, C., 1983, L'articulation du genou: Anlayse morpho-functionelle chez les primates et les hominides fossils, *Cahiers de Paleoanthropologie* 1–108.

Van Valkenburgh, B., 1994, Ecomorphological analysis of fossil vertebrates, in: *Ecological Morphology*, P. C. Wainwright and S. M. Reilly, eds., University of Chicago Press, Chicago.

Wainwright, P. C., 1994, Functional morphology as a tool in ecological research, in: *Ecological Morphology*, P. C. Wainwright and S. M. Reilly, eds., University of Chicago Press, Chicago, pp. 42–59.

Walker, A. C., 1974, Locomotor adaptations in past and present prosimian primates, in: *Primate Locomotion*, F. Jenkins, ed., Academic Press, New York, pp. 349–381.

Ward, S. C., and Sussman, R. W., 1979, Correlates between locomotor anatomy and behavior in two sympatric species of *Lemur*, *Am. J. Phys. Anthropol.* **50:** 575–590.

Warren, R. D., and Crompton, R. H., 1998, Diet, Body size and the energy costs of locomotion in saltatory primates, *Folia Primatol.* **69** (Suppl 1): 86–100.

Weishampel, D. B., 1995, Fossils, function, and phylogeny, in: *Functional Morphology in Vertebrate Paleontology*, J. Thomason, ed., Cambridge University Press, Cambridge, pp. 34–54.

Witmer, L. M., 1995, The extant phylogenetic bracket and the importance of reconstructing soft tissues in fossils, in: *Functional Morphology in Vertebrate Paleontology*, J. Thomason, ed., Cambridge University Press, Cambridge, pp. 19–33.

Youlatos, D., 1999, Locomotor and postural behavior of *Sciurus igniventris* and *Microsciurus flaviventer* (Rodentia, Sciuridae) in eastern Ecuador, *Mammalia* **63:** 405–416.

CHAPTER SIXTEEN

New Skeletons of Paleocene–Eocene Plesiadapiformes: A Diversity of Arboreal Positional Behaviors in Early Primates

Jonathan I. Bloch and Doug M. Boyer

INTRODUCTION

Knowledge of plesiadapiform skeletal morphology and inferred ecological roles are critical for establishing the evolutionary context that led to the appearance and diversification of Euprimates (see Silcox, this volume). Plesiadapiform dentitions are morphologically diverse, representing over 120 species usually classified in 11 families from the Paleocene and Eocene of North America, Europe, and Asia (Hooker et al., 1999; Silcox, 2001; Silcox and Gunnell, in press). Despite this documented diversity in dentitions,

Jonathan I. Bloch • Florida Museum of Natural History, University of Florida, P. O. Box 117800, Gainesville, FL 32611-7800 **Doug M. Boyer** • Department of Anatomical Science, Stony Brook University, Stony Brook, NY 11794-8081

implying correlated diversities in diets and positional behaviors, very little is known about postcranial morphology among plesiadapiforms. What is known has been largely inferred from a limited number of plesiadapid specimens, representing only a small sample of the known taxonomic diversity from North America and Europe (Beard, 1989; Gingerich, 1976; Russell, 1964; Simpson, 1935a; Szalay et al., 1975). While it has been suggested that plesiadapids may have been terrestrial, similar to extant *Marmota* (Gingerich, 1976), the consensus in the literature is that they were arboreal (Beard, 1989; Godinot and Beard, 1991; Rose et al., 1994; Russell, 1964; Szalay and Dagosto, 1980; Szalay and Decker, 1974; Szalay and Drawhorn, 1980; Szalay et al., 1975). While it has been further suggested that plesiadapids might have been gliders (Russell, 1964; Walker, 1974) or arboreal quadrupeds (Napier and Walker, 1967), they are now thought to have been more generalized arborealists with some specializations for vertical postures (Beard, 1989; Godinot and Beard, 1991; Gunnell and Gingerich, 1987; Silcox, 2001). Commenting on the need for a taxonomically broader sample of plesiadapiform postcranial skeletons, F. S. Szalay wrote: "It may be that once postcranial elements of the Paleocene primate radiation become more common, *Plesiadapis* might become recognized as a relatively more aberrant form than the majority of early primates" (Szalay, 1972: 18). In fact, this prediction has been validated in the course of the last 15 years of paleontological field and laboratory research.

Since the early 1980's, field crews and fossil preparation labs of the University of Michigan Museum of Paleontology (UM), New Mexico State University (fossils housed at the U.S. National Museum of Natural History, USNM), and John Hopkins University (fossils also in the USNM) have recovered a number of plesiadapiform skeletons representing groups other than the Plesiadapidae. Several of these specimens with associated dentition and postcrania were collected from mudstones in the Bighorn Basin (Beard, 1989, 1990; Rose, 2001); however, the most complete specimens, including semi- to fully-articulated individuals, are derived from fossiliferous limestones in the Clarks Fork Basin (Bloch, 2001; Bloch and Boyer, 2001; 2002a,b; Bloch et al., 2001, 2003; Boyer and Bloch, 2000, 2002a,b; Boyer et al., 2001).

Beard (1989, 1990, 1993a,b) studied postcranial specimens attributed to paromomyid and micromomyid plesiadapiforms and concluded that these taxa were very different from known plesiadapids in their locomotor repertoire. Specifically, Beard proposed that micromomyids and paromomyids

were mitten-gliders and shared a sister-group relationship with extant dermopterans (=Eudermoptera of Beard, 1993a). Both the mitten-gliding
hypothesis and the character support for Eudermoptera have since been
questioned both with respect to the original evidence (Hamrick et al., 1999;
Krause, 1991; Runestad and Ruff, 1995; Silcox, 2001, 2003; Stafford and
Thorington, 1998; Szalay and Lucas, 1993, 1996) and based on new lime-
stone-derived specimens that are far more complete and have more carefully
documented dental-postcranial associations (Bloch, 2001; Bloch and Boyer,
2001; 2002a,b; Bloch and Silcox, 2001; Bloch et al., 2001, 2003; Boyer and
Bloch, 2000; 2002a,b; Boyer et al., 2001). Despite doubt regarding Beard's
original arguments for gliding and a close relationship to Dermoptera, the
observation that micromomyids and paromomyids are postcranially distinct
from the better known plesiadapids is not disputed. Furthermore, a recent
study of a carpolestid plesiadapiform skeleton (Bloch and Boyer, 2002b)
indicates that these animals were different from plesiadapids, paromomyids
and micromomyids in exhibiting capabilities for strong pedal grasping in a
manner similar to euprimates (Bloch and Boyer, 2002a). Overall, these
skeletons confirm the implications of the diverse dental remains by suggesting a commensurate diversity in positional behaviors among plesiadapiforms.

This chapter includes: (1) a review of the methods for documenting post-
cranial-dental associations in freshwater limestone deposits from which most
of the new significant plesiadapiform material is derived, (2) a summary of the
postcranial anatomy and inferred positional behaviors of plesiadapiforms
based on these new specimens, and (3) a discussion of the implications of the
newly discovered postcranial anatomy for phylogenetic reconstructions and
understanding primate origins and evolution.

CLARKS FORK BASIN FOSSILIFEROUS FRESHWATER LIMESTONES

Despite the high diversity of mammals known from the Paleocene and
Eocene of North America, most species are known only from isolated teeth
and jaws. Associations of teeth to postcrania, for many taxa, are unknown
(Bown and Beard, 1990; Rose, 2001; Winkler, 1983). This lack of skeletal
association, coupled with the fact that most traditional collecting methods are
biased against recovery of skeletons of mammal less than 1 kg, partly explains

why an understanding of positional behaviors of most Paleocene–Eocene small mammals has been elusive.

Fossiliferous freshwater limestones are known throughout the Fort Union (Paleocene) and Willwood (Late Paleocene and Early Eocene) formations of the Clarks Fork and Crazy Mountains Basins of Wyoming and Montana (Bloch and Bowen, 2001; Bloch and Boyer, 2001; Bowen and Bloch, 2002; Gingerich et al., 1983; Gunnell and Gingerich, 1987). Through careful application of acid preparation techniques, limestones have yielded many exceptionally preserved skulls and skeletons of Late Paleocene and Early Eocene vertebrates (Beard, 1989, 1990, 1993a,b; Bloch, 2001; Bloch and Boyer, 2001, 2002a,b; Bloch and Gingerich, 1998; Bloch and Silcox, 2001, 2006; Bloch et al., 2001; Boyer and Bloch, 2000, 2003; Boyer et al., 2001; Gunnell and Gingerich, 1987; Houde, 1986, 1988; Kay et al., 1990, 1992).

Fossiliferous freshwater limestones record a complex depositional and diagenetic history, with precipitation of micritic low-Mg calcite and accumulation of bone probably having occurred in low-energy, ponded water (Bloch and Bowen, 2001; Bowen and Bloch, 2002). The fossil assemblages contained within the limestones likely represent faunas derived from rarely sampled floodplain microenvironments (Bloch, 2001; Bloch and Bowen, 2001; Bloch and Boyer, 2001; Bowen and Bloch, 2002). Skeletal element frequencies and occasional preservation of articulated skeletons indicate that mammals likely entered the limestone assemblage as complete skeletons that were subsequently partially disarticulated by bioturbation. It is likely that predation and scavenging, pit-trapping, and normal attritional processes all contributed to the concentration of bone (Bloch, 2001; Bloch and Boyer, 2001).

Documenting Postcranial-Dental Associations

The following is a summary of the method we use for preparation of matrix and documenting association and articulation of skeletons in fossiliferous freshwater limestones (from Bloch and Boyer, 2001). Limestones are usually chosen for study based upon surficial visibility of fossil vertebrates. Once a limestone has been selected, exposed bone is coated with polyvinylacetate (PVA) to protect the bone against etching and breaking. Limestones are dissolved with 7% formic acid buffered with calcium phosphate tribasic. Each acid reduction run lasts from 1 to 3 h, and is followed by a rinse period in running water of 2–6 h.

Documentation of skeletal association is accomplished by careful mapping of bone distributions and, in some cases, through preservation of articulation. When bones are articulated, we try to preserve the articulation by gluing adjacent surfaces together as the bones are exposed. Using this method for preserving articulations for as long as possible during the etching process reveals patterns in the distributions of skeletons that would have otherwise been lost. In order to further illustrate this process, we provide an example of this type of documentation in the following section.

Micromomyid Skeleton: An Example from a Late Paleocene Limestone

We are in the process of preparing a block, originally 20 kg in mass, of fossiliferous limestone from the last zone of the Clarkforkian land-mammal age (Cf-3, locality SC-327; see Bloch and Boyer, 2001 for locality information). One amazing aspect of this rather large block is that all of the exposed skeletons, representing at least 11 individuals, are articulated (80–100% complete; see Bloch and Boyer, 2001, Figure 5). At least one of the individuals is a new genus and species of micromomyid plesidapiform (Figure 1A). Bone orientations and positions within the block were documented in detail during preparation of the specimen by frequently taking digital photographs of exposed bones and by making drawings that summarized the information in separate photographs with precision on the order of 1 mm or less. The micromomyid skeleton was isolated and not likely to be mixed up with any adjacent skeletons. Our main concern was documenting associations of phalanges to hands or feet, and between individual metacarpals, as persistent functional and phylogenetic questions have gone unanswered simply because cheiridial elements could not be confirmed as belonging to either the hands or feet (Hamrick et al., 1999; Krause, 1991). After extraction, bones were stored with numbers that correspond to the spatial documentation. When dissolution was complete, the photographs and sketches were compiled to produce a map of how the bones were distributed in the limestone (Figures 1B, 2A). The result was recovery of the most complete and clearly dentally associated skeleton of a micromomyid plesiadapiform yet known.

In this specimen, the metacarpals from the left hand (Figure 2A; bone numbers 30, 103–106) were almost perfectly articulated with each other and also closely associated with proximal ends of proximal phalanges 35–38. Furthermore, proximal phalanges 35 and 36, at least, had their distal ends

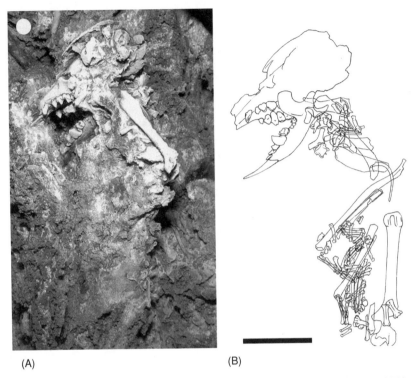

(A)   (B)

Figure 1. (A) Micromomyid plesiadapiform skull and skeleton (UM 41870) partially prepared from fossiliferous limestone, University of Michigan Locality SC-327, late Clarkforkian (Cf-3) North American Land Mammal Age. (B) Composite map of the bones recovered. Scale = 1 cm.

closely associated with the proximal ends of intermediate phalanges 15 and 16, suggesting that they belong to the same hand. The positional relationships described above make interpretation of metacarpal position relatively certain, and allow for confident attribution of proximal and intermediate phalanges to the left hand. In the foot, metatarsals 72, 74–76 were almost perfectly articulated. The distal ends of metatarsals 74 and 75 were articulated with proximal phalanges 63 and 64. Metatarsal 72 is closely associated with the proximal end of 40. In turn, 40, 63, and 64 had their distal ends associated with the proximal ends of intermediate phalanges 80–83. Based on these associations, we are confident that all these bones belong to the same foot.

None of the ungual phalanges recovered are attributed to the foot. It is possible that some, which were not closely associated with a particular manual or pedal intermediate phalanx, were wrongly attributed to the hand (i.e., 102, 41, 42, and 44). However, we are prohibited from attributing any to the feet by two factors: (1) no unguals were recovered posterior to the "knuckles" of the flexed toes, instead, all were clustered around the hand and wrist elements; and (2) there are no consistently diagnosable differences between any of the unguals (due at least partly to their small size and variable preservation quality) that could be used to partition them between hand and foot when clear associations were lacking.

Articulations and associations allowed for identification and subsequent morphological differentiation of manual and pedal intermediate and proximal phalanges in this specimen (Figure 2B). Pedal proximal phalanges are longer and have better developed flexor sheath ridges than those of the hand. Pedal intermediate phalanges differ from those of the hand in: (1) being absolutely longer with mediolaterally relatively narrower shafts, (2) having tubercles for the annular ligament of the flexor digitorum profundus and superficialis muscles with relatively more prominent ventral projections, and (3) having a distal trochlea that is dorsoventrally relatively deeper, with a greater proximal extension of the dorsal margin. Such distinctions allowed attribution of other, more ambiguously positioned cheiridial elements to either hand or foot. These associations of manual and pedal phalanges allow functional interpretations that are more valid than those based on phalanges associated through assumptions about what morphological differences between hand and foot are expected to be (e.g., Beard, 1990, 1993).

Newly Discovered Plesiadapiform Skeletons

Using similar techniques to those outlined above, four other fairly complete plesiadapiform skeletons have been recovered from Paleocene limestones (Figure 3). These include the most complete paromomyid and plesiadapid skeletons yet discovered (Bloch and Boyer, 2001; Boyer et al., 2001; Gunnell and Gingerich, 1987) and the only skeleton of a carpolestid yet known (Bloch and Boyer, 2001, 2002a).

POSTCRANIAL MORPHOLOGY AND INFERRED POSITIONAL BEHAVIORS

Plesiadapiforms as Claw-Climbing Arborealists

Plesiadapiform taxa included in the families Carpolestidae, Micromomyidae, Paromomyidae, and Plesiadapidae are similar to each other in many postcranial

(A)

Figure 2. (A) Composite drawing of micromomyid plesiadapiform skull and skeleton (UM 41870) with numbers on bones corresponding to those of anatomical layout. Scale = 1 cm.

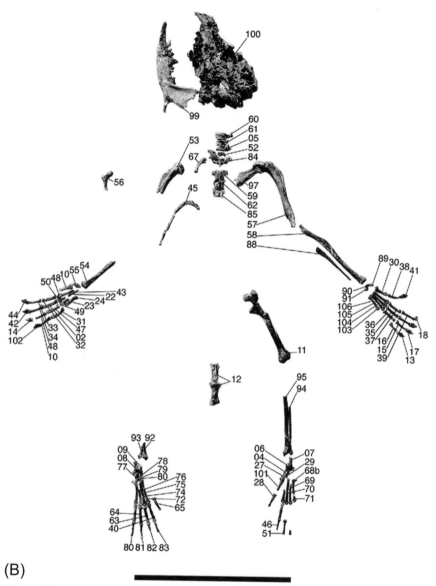

(B)

Figure 2. *Continued* (B) Skeleton of micromomyid (UM 41870) laid out in anatomical position with bones attributed to regions based on positional information. Scale = 3 cm. Note that Figure 2B was made before all of the bones were prepared from the rock. As such, not all bones depicted in Figure 2A are laid out in Figure 2B. Furthermore, a few bones attributed to the skeleton are not depicted in either A or B (see Figure 7).

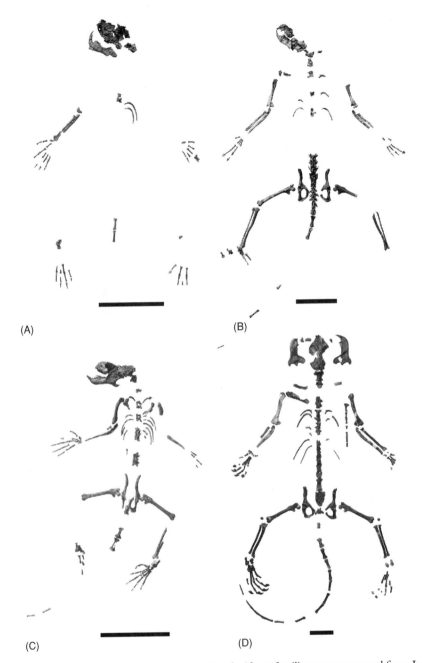

(A)

(B)

(C)

(D)

Figure 3. Skeletons representing three plesiadapiform families were recovered from Late Paleocene limestones. Paromomyidae is represented by (A) *Acidomomys hebeticus* (UM 108207) and (B) *Ignacius* cf. *I. graybullianus* (UM 108210 and UM 82606). Carpolestidae is represented by (C) *Carpolestes simpsoni* (UM 101963; figure from Bloch and Boyer, 2002a, Figure 2A). Plesiadapidae is represented by (D) *Plesiadapis cookei* (UM 87990). Scales = 5 cm.

characteristics that are indicative of arboreality. Specifically, plesiadapiforms are inferred to have been capable of clinging and claw climbing on large diameter vertical tree trunks, as well as grasping smaller branches with their hands and feet (Godinot and Beard, 1991; Sargis, 2001a, 2002b,c,d; Szalay and Dagosto, 1988; Szalay et al., 1975, 1987). Callitrichine primates (Bloch and Boyer, 2002a,b; Bloch et al., 2001; Boyer and Bloch, 2002b; Boyer et al., 2001), arboreal phalangerids (Bloch and Boyer, 2002a), and ptilocercine tree shrews (Sargis, 2001a,b, 2002a,b,c,d) have all been cited as close structural analogues to plesiadapiforms. We draw primarily on studies of behavior [Garber, 1992; Kinzey et al., 1975; Sargis, 2001a (see references therein); Sussman and Kinzey, 1984 (see references therein); Youlatos, 1999] and in some cases on understandings of form-function relationships in extant taxa (Godinot and Beard, 1991; Hamrick, 1998, 2001; Sargis, 2001a,b, 2002a,b) to interpret the functional significance of the features shared by all plesiadapiforms that we have studied.

Morphological correlates of vertical clinging and climbing are numerous and easily observed in the appendicular skeleton, as this region is relatively frequently preserved in fossil taxa and thus has been the focus of many studies. Conversely, the vertebral column of plesiadapiforms has received little attention due to the scarce occurrence of skeletons with associated material from this region. However, plesiadapiform vertebral columns are both diagnosably distinctive in their morphology and functionally informative. Distinctive features include: (1) vertebral bodies that increase markedly in size from the cranial to caudal end of the trunk, (2) vertebral bodies of the cervical and lumbar vertebrae that are dorsoventrally shallower than mediolaterally broad, (3) spinous process of the axis caudally oriented, (4) spinous processes of postdiaphragmatic thoracic and lumbar vertebrae cranially oriented, (5) transverse processes of the lumbar vertebrae arise from the pedicle where it contacts the body, (6) postzygapophyses of the postdiaphragmatic thoracic and lumbar vertebrae mediolaterally broadly-spaced, with facets that are craniocaudally short, have a rectangular (rather than elliptical) margin, and face ventrolaterally, (7) a sacrum that has three vertebrae and in which the long axis of the auricular facet is oriented craniocaudally, and (8) a tail that is relatively long.

While the vertebral column varies in functionally significant ways among taxa, our preliminary study suggests that the center of gravity of the plesiadapiform neck and trunk vertebrae was not midway between the pectoral and pelvic girdles as in suspensory taxa and terrestrial cursors, but was caudally shifted, and had more sagittal than lateral flexibility. Morphology of the vertebral column, viewed as an integrated unit, indicates that plesiadapiforms were capable of bound-galloping in which the brunt of the weight is born on the hindlimbs and flexion and extension of the back contributes to the stride (Gambaryan, 1974). Furthermore, features shared with vertically clinging callitrichine primates (e.g., widely spaced postzygapophyses of lumbar vertebrae that face ventrolaterally), also suggest orthograde postures in plesiadapiforms.

Other plesiadapiform traits that suggest claw climbing on large diameter supports are found mainly in the appendicular skeleton. Many previous studies document and discuss the functional significance of the limb elements in plesiadapiforms (Beard, 1989, 1990, 1991a, 1993a,b; Godinot and Beard, 1991; Sargis, 2002b; Szalay and Dagosto, 1980, 1988; Szalay et al., 1975, 1987). The humerus of plesiadapiforms indicates a mobile forelimb with capabilities for powerful and sustained extension and flexion at the shoulder and elbow joints respectively, as required in vertical clinging postures (Szalay and Dagosto, 1980). The humeral head is spherical and extends superiorly beyond the greater and lesser tuberosities, allowing mobility at the glenohumeral joint (Sargis, 2002a, and references therein) and possibly some stability by providing more room for attachment of the rotator cuff muscles on these tuberosities (Fleagle and Simons, 1982; Grand, 1968; Harrison, 1989). The lesser tuberosity flares medially providing a large insertion site for the subscapularis muscle that extends, adducts, and medially rotates the humerus, making it important during vertical clinging postures and the support phase of vertical climbing (Beard, 1989; Larson, 1993; Sargis, 2002a). The distal humerus has a posterolaterally flaring supinator crest, indicating that plesiadapiforms had a high degree of powerful flexion at the elbow (Dagosto et al., 1999; Gregory, 1920; Szalay and Dagosto, 1980). Presence of a shallow olecranon fossa on the humerus suggests limited extension of the forearm. An extended entepicondyle of the humerus would have provided room for origination of strong flexor muscles of the wrist and fingers, such as flexor carpi radialis and flexor digitorum superficialis muscles (Sargis, 2002a). The capitulum and ulnar trochlea are separated by a deep zona conoidea indicating that the radius and ulna were not highly integrated in their functions (Sargis, 2002a). Instead, the spherical to slightly elliptical capitulum allowed the radius to rotate freely about the ulna (Sargis, 2002a; Szalay and Dagosto, 1980). Therefore, in many regards, the humerus suggests locomotion in an arboreal setting on large diameter supports.

The ulna of plesiadapiforms has a shallow trochlear notch and long, anteriorly inflected olecranon process, indicating habitual flexion as would be used in orthograde clinging and pronograde bounding (Rose, 1987). A flat to slightly convex proximal articulation with the radius is consistent with independent function of that element in axial rotation. The shaft is medially bowed in cross-section, such that a strong lateral groove, which begins proximally as a deep fossa on the olecranon process, runs along the length of at least its proximal two-thirds. Such a groove expands the area for the origin of extensor muscles of the fingers. The shaft is typically slender without marked expression of an interosseous crest.

The proximal radius of plesiadapiforms typically has a spherical fossa and broad lateral lip that matches the spherical capitulum of the humerus (Beard, 1993a), allowing for a large degree of axial mobility (MacLeod and Rose, 1993; Sargis, 2002a; Szalay and Dagosto, 1980). The bicipital tuberosity of the radius is large and proximally located, indicating the presence of a strong biceps brachii muscle. The shaft of the radius is generally mediolaterally wide and flattens distally in its dorsopalmar aspect. Medial and lateral longitudinal ridges for the deep digital flexor muscles often mark the palmar aspect of the radial shaft. The distal end of the radius supports most of the carpus while the ulna is typically reduced. Because the wrist joint in plesiadapiforms is almost entirely formed by the radius, rotation of this element about the ulna does not compromise stability of the wrist. The distal articular surface of the radius is canted palmarly indicating habitual palmar-flexion of the proximal carpal row.

Though the hand is specialized differently in the four plesiadapiform families considered here, there are several features that are shared and suggest similar functions in an arboreal environment. A divergent pollical metacarpal in all plesiadapiforms indicates that they were capable of effective grasping of small diameter supports. Long proximal phalanges relative to the metacarpals (=prehensile proportions; Bloch and Boyer 2002a; Hamrick, 2001; Lemelin and Grafton, 1998) in non-plesiadapid plesiadapiforms also indicate specialized grasping abilities. The proximal phalanges of all plesiadapiforms have strong ridges for annular ligaments that prevent bowstringing of tendons of the flexor digitorum profundus and superficialis muscles during strong grasping in which the intermediate-proximal phalangeal joint is flexed at a highly acute angle. The distal articular surface of the proximal phalanx is not smooth, but has raised medial and lateral margins that create a broad, central groove into which the grooved proximal articular surface of the intermediate phalanx fits tightly (see description in Godinot and Beard, 1991: 311). Such a grooved surface prevents torsion and mediolateral deviation at this joint. The distal phalanx of plesiadapiforms, like that of *Ptilocercus* and other arboreal mammals, is dorsopalmarly deep and mediolaterally narrow providing better resistance against sagittal bending loads incurred during vertical claw clinging and climbing (Beard, 1989; Hamrick et al., 1999; Sargis, 2002a). The distal phalanx is usually characterized by an articular surface that is ventrally canted, indicating habitual palmar-flexion during clinging. It also has a large flexor tubercle that supported a robust tendon for a powerful flexor digitorum profundus muscle, allowing frequent and sustained use of such claw-clinging postures.

While the innominate of plesiadapiforms varies in functionally significant respects among the taxa considered here, all seem to share characteristics that reflect functions and postures associated with vertical clinging behaviors (Beard, 1991a). In contrast to the acetabulum of cursorial animals and more terrestrial scansorialists (e.g., tupaiine tree shrews), plesiadapiforms have a more elliptical acetabulum, the major axis of which is craniocaudally oriented (e.g., Silcox et al., 2005, Figure 9.5A). This indicates a limited range of sagittal flexion and extension during which the joint surfaces of the femur and acetabulum fit tightly together, and maintain maximal stability. Such morphology suggests a joint that has a large range of stable adduction and abduction (Beard, 1991a). When the femur is articulated with the acetabulum the joint surfaces conform most closely, (fovea capitis femoris aligned with the center of the acetabular fossa), when the femur is flexed and the shaft is abducted by about 45° from the sagittal plane. We infer that this orientation of the femur relative to the innominate represents a component of habitual posture. The acetabulum is cranially buttressed and its axis is dorsally rotated in plesiadapiforms, indicating that this joint was probably subject to caudally directed forces experienced during orthograde postures (Beard, 1991a). The ilium is generally slender and triangular in cross section, much the same as in extant *Ptilocercus* (Sargis, 2002b,c). This is in contrast to the condition in euprimates (including extant callitrichines), which are characterized by a hugely expanded dorsolateral face of the ilium, reflecting the origination of hypertrophied gluteal muscles for powerfully extending the femur during leaping or quadrupedal bounding (see Anemone, 1993; Sargis, 2002b; Taylor, 1976).

The femur of plesiadapiforms has been figured for paromomyids, plesiadapids and micromomyids, with its morphology and functional significance discussed many times (Beard, 1991a, 1993b; Sargis, 2002c; Simpson, 1935a; Szalay et al., 1975, 1987). In all plesiadapiforms for which it is known, the posterior margin of the femoral head extends onto a short neck, and farther onto the medial margin of the greater trochanter. This extension gives the articular surface an elliptical or cylindrical form that, in conjunction with a distinct, dorsoposteriorly positioned fovea capitis femoris, indicates abducted limb postures (Beard, 1989;

Sargis, 2002b; Szalay and Sargis, 2001). The femoral neck is typically oriented at a high angle to the femoral shaft and the greater trochanter does not extend beyond the superior margin of the femoral head. Such a configuration allows for mobility at this joint, especially in abduction (Sargis, 2002b), in contrast to taxa that bound-gallop or run using pronograde postures frequently (Gebo and Sargis, 1994; Harrison, 1989; Sargis, 2002b; Szalay and Sargis, 2001). Although the greater trochanter is relatively short, the trochanteric fossa is typically deep and proximodistally oriented, providing ample room for insertion of internal and external obturator and gemelli muscles that serve to abduct or laterally rotate the thigh depending on orientation (Szalay and Schrenk, 2001). In contrast to the condition in cursorial and bounding taxa, the lesser trochanter is medially extended beyond the head, distally positioned on the shaft, and has a large area of attachment for the iliopsoas muscle. This configuration allows the femur to remain somewhat abducted even when the iliopsoas is fully contracted, and hence, when the femur is fully flexed. Furthermore, the distal position of the trochanter gives the iliopsoas muscle a long moment arm for powerful hip flexion (Anemone, 1993) and would have reduced the effort for holding the leg in flexed positions during vertical climbing (Rose, 1987). The third trochanter is relatively small and flares laterally immediately distal to the maximum peak of the lesser trochanter. This is in contrast to the condition in more active, terrestrial tree shrews in which this process flares prominently and is positioned farther distally, allowing the inserting gluteus superficialis muscle to more powerfully extend the thigh (Sargis, 2002b). The femoral shaft is either equal in mediolateral and anteroposterior dimensions or is slightly anteroposteriorly flattened. The distal end is rotated laterally relative to the proximal end, effectively orienting the plane of flexion of the knee mediolaterally. This orientation of the distal femur in plesiadapiforms is similar to that of callitrichines, and differs from leapers and bounders (e.g., *Saimiri* and tupaiids) in which the knees flex in the sagittal plane to accommodate the frequent use of small diameter supports instead of large ones. The medial margin of the patellar groove is buttressed relative to its lateral margin such that, despite the lateral rotation of the distal end, the anterior aspect of the patellar groove lies parallel to the plane defined by the shaft and a transect between the fovea capitis femoris and the tip of the greater trochanter. In extant leaping euprimates and saltatorial lagomorphs, frequent and strong full-extension of the knee is reflected in the distal femur by a deep patellar groove that prevents mediolateral deviation of the patella (Anemone, 1993), a raised patellar groove that increases the moment arm of the quadriceps muscles (Anemone,

1993) and a proximally extended groove that allows the patella to shift high on the thigh during extreme contraction of the quadriceps muscles. In contrast, the patellar groove of plesiadapiforms is shallow, not raised anteriorly above the level of the anterior surface of the shaft, and not extended proximally on the shaft. This form suggests infrequent forceful full-extension of the knee. Notably, the patellar grooves of both *Ptilocercus* and marmosets are nearly identical to those of plesiadapiforms in these respects (Sargis, 2002b,c). Posteriorly, the distal intercondylar area is angled ~10° lateral to the shaft, suggesting that the tibia would have rotated laterally, contributing to pedal inversion, during extension of the knee. Medial and lateral margins of the condyles slope away from each other proximally at an angle of ~45°. This results in the posteroproximal part of the condyles being broader and more robust, again indicating that flexion was the habitual posture with loads being sustained on extended limbs only infrequently. The lateral condyle is generally ~50% wider than the medial condyle.

The morphology of the proximal tibia of plesiadapiforms reflects similar positional behaviors as that of the distal femur. The proximal tibia is antero-posteriorly compressed, unlike that of leapers and runners such as tupaiines (Sargis, 2002b), lagomorphs, tarsiers, and felids. The medial facet is usually smaller than the lateral facet, concave, oriented somewhat posteriorly and sunk below the level of the lateral facet, which is flat to convex and extends higher proximally. Both facets face slightly laterally, rotating the tibial shaft out of the plane of flexion with the knee. The proximal half of the shaft is compressed mediolaterally and triangular in cross section. The posterior and lateral surfaces of the proximal shaft of the tibia are concave in cross section, providing ample room for strong pedal plantar-flexors (soleus and tibialis posterior), and digital flexors (flexor digitorum tibialis), respectively. The distal part of the shaft of the tibia is strongly bowed both in the medial and anterior directions. This makes the foot of plesiadapiforms permanently somewhat inverted when flexed. The medial malleolus on the distal tibia is weaker and more distally restricted than that of other arboreal placentals such as *Ptilocercus* and primitive euprimates. The astragalar facet on the distal tibia is ungrooved, square, angled somewhat posterolaterally, and forms an obtuse angle with its extension on the medial malleolus. With regard to all of these features, the distal tibia of plesiadapiforms is most comparable to that of phalangerid marsupials (e.g., *Petaurus* and *Trichosurus*) in which the distal tibia and fibula have a flexible articulation with each other and the tarsals, allowing for a greater degree of mobility between the tibia and astragalus (the upper ankle joint = UAJ) than is typical for placentals.

Among plesiadapiforms, the proximal articulation of the tibia and fibula is only known in paromomyids and micromomyids. It is transversely oriented and may have been synovial. The distal articulation is known in all four groups discussed here and, except in carpolestids (see the section on Carpolestidae below), appears to have been a flexible syndesmosis (Beard, 1989, 1991a). There are grooves on the posterior surface of the tibia and fibula. On the posterior tibia, such grooves mark the course of the tendons of the tibialis posterior, flexor fibularis, and flexor digitorum tibialis muscles, while on the posterior fibula they mark the course of the tendon of the peroneus brevis muscle. These muscles would have served to resist mediolateral forces at the UAJ, thereby compensating for the stability given up for mobility between joint surfaces, and facilitate inversion and eversion at the lower ankle joint, as they do in arboreal marsupials and some rodents (Gunnell, 1989; Jenkins and McClearn, 1984).

The astragalus, calcaneum, and cuboid of plesiadapiforms have been discussed extensively in terms of their diagnostic and functional features (Beard, 1989, 1993b; Dagosto, 1983; Decker and Szalay, 1974; Gebo, 1988; Gunnell, 1989; Szalay and Decker, 1974; Szalay and Drawhorn, 1980). Plesiadapiforms are limited in the degree of plantar-flexion that can be accomplished at the UAJ by the small arc formed by the tibial facet of the astragalus. A slight amount of pedal inversion, limited by malleoli bracketing the astragalus, results from plantar-flexion at the UAJ (Beard, 1989). The lower ankle joint is axially mobile, the calcaneum being capable of rotating medially and shifting distally to invert the foot (Szalay and Decker, 1974). At the transverse tarsal joint, such rotation is not limited by the calcaneo-cuboid articulation, which is transverse (Beard, 1989; Jenkins and McClearn, 1984). On the cuboid, the orientation of the groove for the tendon of the peroneus longus muscle is transverse, facilitating eversion by this muscle.

Because the distal tarsal rows have rarely been preserved in association, little has been said about them. However, new specimens of micromomyids, paromomyids, and plesiadapids—all show a similar configuration in which the tarsometatarsal articulation faces slightly laterally, causing the foot to be abducted relative to the upper ankle, when dorsiflexed. The mesocuneiform is shorter than the entocuneiform and ectocuneiform such that metatarsal II articulates out of plane with the rest of the metatarsals and is dove-tailed within the distal tarsal row, creating a rigid tarsometatarsal articulation.

The entocuneiform and first metatarsal have been discussed extensively for plesiadapids and paromomyids (Beard 1989, 1993a; Sargis, 2002b,c,d; Szalay and Dagosto, 1988). These elements are nearly identical among the plesi-adapiforms considered here with the exception of those in carpolestids. In plesiadapids, paromomyids, and micromomyids the robust plantar process on the entocuneiform reflects frequent pedal inversion and possibly the presence of powerful pedal and digital flexors (contra Beard, 1993a; but see Sargis, 2002b; Szalay and Dagosto 1988). The hallux is strikingly similar to that of *Ptilocercus* in that the articulation for the hallucal metatarsal on the ento-cuneiform is dorsally broad and saddle-shaped (Sargis, 2002b,c,d; Szalay and Dagosto, 1988), and the hallucal metatarsal is robust, divergent from the other metatarsals, exhibits slight torsion of the distal end, and has peroneal and medial processes that are reduced such that the entocuneiform joint is open and mobile. These features indicate that the hallux was used in grasping (Sargis, 2002b,c,d; Szalay and Dagosto, 1988), as convincingly demonstrated for that of *Ptilocercus* through behavioral observations (Gould, 1978; Sargis, 2001a and references therein) as well as by myological and osteological stud-ies (e.g., Gregory, 1913, Le Gros Clark, 1926, 1927; Sargis 2002b,c; Szalay and Dagosto, 1988). The hallux seems to have been used primarily as a load-bearing "hook" while the rest of the foot served as a lateral brace during loco-motion on subhorizontal supports with relatively small diameters (Sargis, 2002b). Sargis (2001a) considered such grasping to potentially represent the antecedent condition to the powerful grasping of euprimates, as well as the primitive condition in the ancestral archontan or euarchontan (see also Sargis, 2002b,c,d; Szalay and Dagosto, 1988).

Except in carpolestids, the metatarsal/phalangeal proportions of plesiadapi-form feet are not as extreme as those of the hands (i.e., the feet do not exhibit prehensile proportions), thus, the feet have proportions unlike those in the feet of specialized slow-moving graspers, and more like those of generalized arbo-realists that use a bounding gait. This is because the metatarsals are generally relatively long. The long metatarsals that rigidly articulate with the tarsals of plesiadapiforms are similar to those of callitrichine primates, *Ptilocercus*, and many other arboreal and scansorial mammals. These features are indicative of a bounding gait, similar to that usually used on horizontal substrates most fre-quently by more terrestrial scansorialists such as tupaiine tree shrews (Jenkins, 1974; Sargis, 2002b) and sciurid rodents. Long metatarsals increase the dis-tance that can be covered with each "bound." The toes are longer than the fin-gers and thus may have been relied on to support the body weight in clinging and hanging positions more frequently than the fingers.

Overall, the morphology of the appendicular skeleton of plesiadapiforms indicates a mobile forelimb capable of strong flexion at the shoulder and elbow joints that helped keep the body close to the substrate during vertical postures and that assisted the hindlimbs in scrambling up a vertical substrate. The hands could be supinated and pronated freely and were effective at grasping, allowing these animals to move easily through broken substrates on smaller supports. The hindlimbs indicate habitual flexion with a broad foot stance, consistent with habitual use of large diameter supports. Furthermore, the plane of flexion of the hindlimbs is not sagittal, as in terrestrial bounders and runners, but is rotated significantly laterally. Thus, instead of pushing away from a vertical substrate during upward propulsion, the hindlimbs extended more parallel to the substrate. The ankle exhibits axial flexibility and the capability to invert the foot. Such mobility would have allowed the substrate to be grasped from a plantar-flexed position, thereby facilitating head-first descent of tree trunks, as well as moving on small horizontal branches. Although the feet are generally not as committed to grasping as the hands, they too would have been effective on small supports and discontinuous substrates owing to a somewhat divergent, prehensile hallux, like that of *Ptilocercus* and callitrichines.

Plesiadapiform Specializations: A Diversity of Arboreal Behaviors

Despite the large amount of similarity between the four groups of plesiadapiforms considered here, each is also unique in its own way. In some cases, morphological differences are probably engendered by size extremes that change the nature of the arboreal milieu experienced by a given taxon. In other cases, these features truly represent specialized behaviors beyond clinging and claw climbing on large diameter vertical supports and the ability to grasp smaller supports with the hands and feet.

Paromomyidae

New skeletons of Late Paleocene paromomyids *Acidomomys hebeticus* (Bloch et al., 2002a) and a new species of *Ignacius* (Bloch et al., in review) are the most complete known and have clear dental-postcranial associations, allowing for a more refined and better supported understanding of postcranial anatomy and inferred positional behaviors for the group. Elements of the hands and feet have been recovered for *Acidomomys* (Figure 3A), and nearly the whole skeleton has been recovered for *Ignacius* (albeit a composite of two individuals; Figure 3B). Both paromomyids conform to the general plesiadapiform body plan (described in section on Plesiadapiforms as Claw-Climbing Arborealists above) in most respects. Of all plesiadapiform families considered here, paromomyids are most appropriately described as "callitrichine-like" because of their similar body size of 100–500 g (Fleagle, 1999; Garber, 1992), inferred diet of exudates (Gingerich, 1974; Vinyard et al., 2003) and specific locomotor repertoire that likely included bound-galloping, as well as grasping and foraging on small diameter supports in addition to a large amount of clinging and foraging on large diameter supports. This interpretation is contrary to a previous hypothesis that paromomyids were capable of dermopteran-like mitten-gliding (Beard, 1989, 1991a, 1993b) based on fragmentary, composite specimens with undocumented associations that were proposed to have the hallmark osteological feature of mitten-gliding: elongate intermediate phalanges of the hand. The mitten gliding hypothesis has since been questioned (Hamrick et al., 1999; Krause, 1991; Runestad and Ruff, 1995). Furthermore, the new specimens discussed here do not support the mitten-gliding hypothesis because the intermediate phalanges of the hands in paromomyids are not longer than their proximal phalanges (Figure 4A). Even in the face of such evidence against "mitten-gliding," one might argue that it is still possible that more general gliding behaviors (e.g., Petauristinae; Thorington and Heaney, 1981) could have been an aspect of the locomotor repertoire of paromomyids. Similar gliding behaviors, with correspondingly similar specialized anatomical structures (but not homologous), have evolved at least four times in mammals (Petauristinae, Anomaluridae, Phalangeridae, Cynocephalidae). Each of these experiments in gliding is also associated with unique aspects of anatomy reflecting very specific differences in behavior and evolutionary history (Essner and Scheibe, 2000; Scheibe and Essner, 2000; Thorington et al., 2005; Thorington and Heaney, 1981). While such differences can be subtle, this is distinctly not the case for dermopterans, which have unique anatomy among gliding animals reflective of the presence of an interdigital patagium and quadrupedal suspensory behaviors (Simmons, 1995; Simmons and Quinn, 1994; Stafford, 1999). Thus, even if evidence for more generalized gliding behaviors were to be found, it would still be inconsistent with Beard's (1993a,b) hypothesis that paromomyids were mitten-gliding. In fact, paromomyids lack any trace of the osteological correlates for gliding behavior.

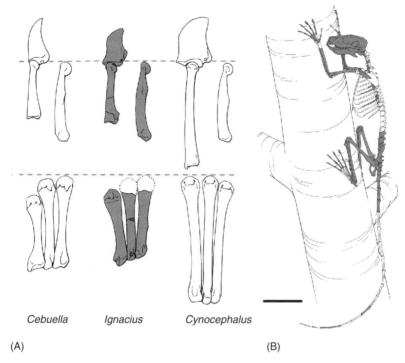

Cebuella   Ignacius   Cynocephalus

(A)   (B)

Figure 4. (A) Manual digit rays (top) and metacarpals (bottom) of paromomyid, *Ignacius* cf. *I. graybullianus*, the dermopteran, *Cynocephalus volans*, and a callitrichine primate, *Cebuella pygmaea*. Phalanges are in lateral view with distal and intermediate phalanges articulated on the left and proximal phalanx on the right. Below phalanges, from left to right, metacarpals V-III are depicted in palmar view. Note that *Ignacius* lacks the elongate intermediate phalanges and metacarpals of *Cynocephalus* and instead has overall proportions comparable to those of *Cebuella*. In this way, *Ignacius* is similar to euprimates that use their relatively long fingers for grasping (prehensile phalangeal proportions of Hamrick, 2001). (B) Reconstruction of *Ignacius* cf. *I. graybullianus* in a habitual resting or foraging posture on a large diameter trunk. The proportions and morphology of both limbs and vertebrae indicate that it was probably more adept at pronograde bounding than some other plesiadapiforms. Gray areas depict bones present in UM 108210 and another individual (UM 82606) from a different region within the same limestone block. Scale = 5 cm.

Comparative functional studies show that there is a suite of osteological features shared by flying squirrels and dermopterans that appear to be gliding adaptations (e.g., Thorington and Heaney, 1981; Runestad and Ruff, 1995; Stafford, 1999), which are apparently lacking in Paleocene paromomyids (Boyer et al., 2001; Boyer and Bloch, 2002b; Bloch et al., in review). Instead, features uniquely exhibited by paromomyids indicate agile arboreality that involved more frequent use of pronograde bounding and scampering than inferred for plesiadapiforms generally. These tendencies are reflected in the limb proportions, the limb to trunk proportions, and the morphology of the vertebral column, sacrum, and innominate.

Ignacius has an intermembral index of ~80, which is comparable to that of most callitrichine primates except the pygmy marmoset, *Cebuella*, in which it is 82–84 (Fleagle, 1999). Other plesiadapiforms have intermembral indices ranging between ~84 for *Carpolestes* and 89 for *Plesiadapis cookei*. Among clawed agile arborealists, including taxa classified in Rodentia, Scandentia (Sargis, 2002a), and Callitrichinae (Fleagle, 1999), higher intermembral indices may correspond to more frequent and sustained use of vertical clinging postures since the arms take a more active role in supporting and lifting the body, instead of acting as struts that must withstand impacts after propulsion by the hindlimbs, as they do in pronograde bounders (Gambaryan, 1974). Just as relative lengths of hindlimbs and forelimbs are behaviorally indicative, so is overall length of the limbs, relative to the trunk, which increases with frequency of use of vertical clinging postures (Boyer and Bloch, in review). Although trunk length estimates are not yet available for micromomyids or plesiadapids, comparison of the limb to trunk proportions of *Ignacius* with callitrichines shows *Ignacius* to be similar to tamarins, such as *Saguinus*, which have substantially shorter limbs than the more arboreally committed *Cebuella*.

Vertebral morphology and proportions in *Ignacius* suggest agility and an emphasis on the hindlimb in forward propulsion. It is comparable to other squirrel-like and primate-like taxa in having a narrow atlas and a short neck relative to the trunk. The posterior lumbar vertebrae are larger and more elongate than the thoracic vertebrae. The sacrum is robust and the tail is long and robust. Such a configuration results in a posteriorly shifted center of gravity (COG) of the vertebral column relative to that of a non-bounder (Shapiro and Simons, 2002) or a quadrupedal runner (Emerson, 1985), thereby reducing the offset between the COG and the pelvic girdle, where the main propulsive force is applied by the hindlimbs. The morphology of the lumbar vertebrae also indicates the ability to powerfully flex and extend the trunk. These vertebrae have narrow, cranially angled spinous processes and long cranioventrally oriented transverse processes. They are qualitatively similar to those of scansorialists that use a bounding gait and strepsirrhines that leap (Shapiro and Simons, 2002). Furthermore, in bounding taxa, the relationships of the dimensions of these aspects of vertebral morphology to overall body mass are significantly different from those of non-bounders (Boyer and Bloch, 2002b; in review), with bounding taxa having lumbar spinous processes that are narrower craniocaudally and transverse processes that extend farther ventrally than those of non-bounders. *Ignacius* fits the scaling relationship characterizing extant bounders.

The sacrum of paromomyids has a reduced spinous process on its first vertebra and tall, narrow, caudally oriented ones on the second and third vertebrae. Such a configuration is similar to that of hindlimb-propelled taxa in which a large degree of flexibility at the lumbosacral joint is required. Not only does the spinous process of the first sacral vertebra not impede extension, but the supraspinous ligament, which might have spanned two vertebrae instead of one (Gambaryan, 1974), would have allowed a greater range of mobility for the same elastic strain than it would separated into two segments.

The innominate of paromomyids differs from that of other plesiadapiforms in having an ilium with a relatively broader dorsolateral surface, an ischium that is relatively longer and more expanded, and an ischiopubic symphysis that is longer and more cranially positioned (relative to the acetabulum). These features indicate a sturdy pelvic girdle with ample room for origination of the hip extensor muscles. Such a pelvis would be capable of withstanding impacts of a bounding gait and would allow room for the attachment of powerful muscles adequate for the effective use of such an active locomotor style.

In summary, the skeleton of *Ignacius* indicates a versatile locomotor repertoire with no specific features detracting from its ability to use vertical postures (Figure 4B), but with additional features that allowed it to effectively exploit horizontal substrates using above branch postures.

Carpolestidae

Insights into the behavior of the Carpolestidae are derived primarily from a single specimen of the Late Paleocene taxon, *Carpolestes simpsoni* (Bloch and

Gingerich, 1998). The specimen is fairly complete (Figure 3C) and the dental associations are well documented (Bloch and Boyer, 2001, 2002b).

Carpolestes is unique among plesiadapiforms in having a foot that is better adapted for powerfully and precisely grasping small diameter supports, a UAJ that reflects even more freedom of motion, a humerus that suggests relatively stronger grasping, and a vertebral column that indicates only infrequent use of a bounding gait on either vertical or horizontal substrates. In terms of behavior, these features suggest that *Carpolestes* spent relatively little time on large diameter supports, and instead most frequently occupied a small branch niche where grasping is more useful than claw-clinging and bridging is more effective than bounding.

In contrast to the condition in other plesiadapiforms, the feet of *Carpolestes* are similar to the hands in having prehensile proportions, a result of unusually short metatarsals and long toes (Bloch and Boyer, 2002b; Figure 5A). In both the fingers and toes, the proximal phalanges are more curved than those of other plesiadapiforms. The intermediate phalanges are not mediolaterally compressed, but have a more spherical cross section than those of other plesiadapiforms (Bloch and Boyer 2002a). The unguals are relatively smaller and slightly broader than in other plesiadapiforms. The articular surface for the intermediate phalanx has a slight dorsal orientation such that when articulated, it is canted dorsally rather than palmarly on the hands and feet. Furthermore, on the ventral surface of the shaft, distal to the flexor tubercle, there is an expanded area that may reflect the presence of an expanded dermal pad in life, as a similar structure seems to do in the unguals of *Petaurus* as well as in the grooming claws of the euprimate, *Tarsius*. These features, taken together, suggest less frequent use of the hands and feet for claw-clinging and more habitual grasping of small diameter substrates (Figure 5B).

Prehensile proportions and phalangeal morphology in *Carpolestes* are subtle expressions of grasping behavior compared to the condition of the hallux (Figure 6). The structure of the joint between the entocuneiform and the hallucal metatarsal, as well as the structure of both this metatarsal and the hallucal distal phalanx are strikingly similar to those of euprimates, indicating specialized, powerful grasping, beyond that inferred from the usual plesiadapiform condition. The entocuneiform is short with a huge plantar process that would have buttressed hypertrophied pedal flexors and on to which may have inserted the tendon of tibialis anterior, a powerful pedal inverter (Sargis, 2002b;

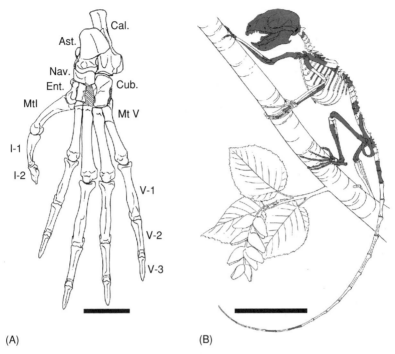

Figure 5. (A) Reconstructed left foot and ankle of *Carpolestes simpsoni* (figure from Bloch and Boyer, 2002a; fig. 4a). Note that the hallux is divergent from and in opposition to the other digits, the metatarsals are short, and the nonhallucal digits are relatively long. All of these features indicate euprimate-like grasping. The foot is unlike that of euprimates, however, in having short tarsals and a diminutive peroneal process on the proximal hallucal metatarsal. Long tarsals and a prominent peroneal process in euprimates are thought to facilitate powerful leaping with stable landings (Szalay and Dagosto, 1988). Abbreviations: *Ast.*, astragalus; *Cal.*, calcaneum; *Cub.*, cuboid; *Ent.*, entocuneiform; *Mt.*, metatarsal; *Nav.*, navicular; *I-1*, proximal phalanx, first digit; *I-2*, distal phalanx, first digit; *V-1*, proximal phalanx, fifth digit; *V-2*, middle phalanx, fifth digit; *V-3*, distal phalanx, fifth digit. (B) Reconstruction of *Carpolestes simpsoni* (figure from Bloch and Boyer, 2002a; fig. 2b). Locomotion on small diameter supports, depicted here, is inferred from the specialized grasping hands and feet; strong, mobile elbow; robust fibula; mobile ankle joints; mobile vertebral column; gracile pelvis; and specialized dentition (Bloch and Boyer, 2002a). Gray areas in B represent bones present in UM 101963. Scale in A = 5 mm. Scale in B = 5 cm.

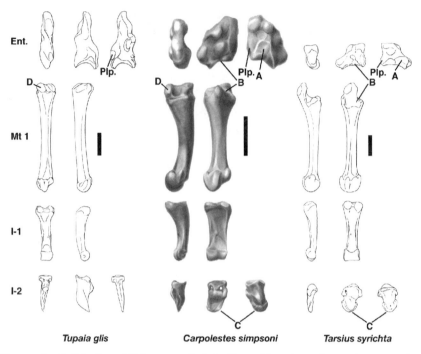

Figure 6. Left hallux of Paleocene plesiadapiform *Carpolestes simpsoni* compared to those of extant euprimate *Tarsius syrichta* and extant tree shrew *Tupaia glis* (figure from Bloch and Boyer, 2002a; fig. 3). The entocuneiform (from left to right) is in ventral, lateral, and medial views, the metatarsal and proximal phalanx are in ventral and lateral views, and the distal phalanx is in ventral, lateral, and medial views. Euprimate traits present in the hallux of *C. simpsoni* include a medial expansion of the distal facet on the entocuneiform (A) for articulation with the first metatarsal that forms a saddle-shaped, or sellar joint (B), and a distal phalanx that supported a nail instead of a claw (C). Primitive traits, also seen in the tree shrew, include a first metatarsal with a peroneal process that is not enlarged (D). Note that the distal, relative to the proximal, end of the hallucal metatarsal of *C. simpsoni* is laterally rotated about 90° compared to the condition in that of tupaiids. Similarities to euprimates are reflective of *C. simpsoni* having a divergent and opposable hallux, while the similarities to tree shrews (and not to euprimates) are reflective of *C. simpsoni* not being a specialized leaper. Size of hallux normalized to the length of the metatarsal. Abbreviations: *Ent.*, entocuneiform; *Plp*, plantar process of entocuneiform; *Mt 1*, metatarsal, first digit; *I-1*, proximal phalanx, first digit; *I-2*, distal phalanx. Scale = 2 mm.

Szalay and Dagosto 1988). Furthermore, the distal articular surface is saddle-shaped, narrow and cylindrical on its ventral margin, and expanded proximally on its medial side. The articulating metatarsal can rotate medially from its most adducted position by ~60°, at which point a blunt medial process on the metatarsal meets a correspondingly spherical depression on the medial side of the entocuneiform. Once these surfaces are in contact, there is an increase in the axial mobility of the metatarsal that allows the abducted hallux to form a more stable grip on the substrate than it might be able to achieve otherwise. The metatarsal itself is no more robust than in other plesiadapiforms, but it shows a greater degree of torsion, which makes the hallux more completely oppose the rest of the digits. The proximal hallucal phalanx is flattened with a mediolaterally broad, but proximodistally short distal articular surface that is almost completely plantar-facing. This wide, shallowly dished surface accommodates the distal phalanx that is dorsoplantarly shallow and mediolaterally expanded, distinctly unlike the nonhallucal unguals of this animal and more consistent with morphology that reflects the presence of a large dermal pad and nail in most extant euprimates, as well as some marsupials and rodents.

The ankle of *Carpolestes* differs from other plesiadapiforms in that: (1) the fibula is relatively larger; (2) the groove for the tendon of the tibialis posterior muscle and the groove for the tendon of the peroneous brevis muscle, on the tibia and fibula respectively, are deeper; and (3) the opposing articular facets on both the tibia and fibula are convex indicating increased axial mobility and possibly a synovial articulation. As might be expected, the astragalus reflects this added mobility in lacking the distinct, often acute, ridge marking the boundary between the tibial and fibular facets on the astragalar body, which restricts the UAJ to plantar and dorsiflexion in other plesiadapiforms.

The greater emphasis on grasping behavior in *Carpolestes* is reflected in the forelimb primarily by a relatively large and medially extended entepicondyle. Such medial extension provides relatively more room for the attachment of the flexor muscles of the wrist and digits. Furthermore, the distal articular surface of the humerus suggests even more complete segregation in the function of the radius and ulna. The zona conoidea is so deep that it creates both a lateral keel on the trochlea for articulation with the ulna (on the medial margin of the zona conoidea) and a lateral ridge medial to the capitulum. This condition is otherwise unique to euprimates and microsyopid plesiadapiforms (Beard, 1991b; Silcox, 2001).

Finally, the vertebral column of *Carpolestes* has an anticlinal vertebra positioned within the thoracic region and narrow spinous processes on the lumbar and posterior thoracic vertebrae, indicating that it was capable of sagittal flexion. However, the vertebral column is not particularly suited for the powerful sagittal flexion and extension required in a bounding gait, such as that inferred for paromomyids (Boyer and Bloch, 2002a,b). Such a de-emphasis on features reflective of a bounding gait is expected for an animal that spends the majority of its time in a small branch niche where bridging is safer and more effective than bounding (Sargis, 2001b).

Based on this specimen, carpolestids appear to diverge from the general plesiadapiform body type more than any other group we have studied. Interestingly, many of the deviant aspects in carpolestid morphology and inferred behavior are similar to those observed and/or inferred for early euprimates.

Plesiadapidae

Plesiadapid postcrania are currently known from a wider geographic and temporal range and from a greater diversity of species than are those of any other plesiadapiform group. Not surprisingly, they exhibit greater morphological diversity than seen in any of the other three families considered here. A skeleton of *Plesiadapis cookei* (Figure 3D; Gunnell and Gingerich, 1987; Gunnell, 1989; Gingerich and Gunnell, 1992, 2005; Hamrick, 2001), a large species known exclusively from western North America, is in the process of being described (Boyer, in preparation). Plesiadapids obtain a large size rather early in their evolutionary history, and this may explain many of their characteristic features (Gingerich, 1976).

Clinging and climbing on large diameter substrates appears to be a major feature of the locomotor repertoire of *Plesiadapis* (Gingerich and Gunnell, 1992; Gunnell, 1989). The ability to grasp small-diameter supports with the hands and feet is reduced, and agile pronograde bounding would probably have been infrequent (Gunnell, 1989).

The unguals of plesiadapids differ from those of other plesiadapiforms in having a shaft that is relatively long, an extensor tubercle that is reduced and proximally extended such that the articular surface is more plantarly oriented, and a flexor tubercle that faces plantarly instead of proximally. As a consequence of reduction in the extensor tubercle, the dorsal margin of the ungual shaft is generally convex for its entire length. The digit ray as a whole is most comparable to that of semi-arboreal new world porcupines such as *Erethizon* and *Sphiggurus* and thus, consistent with the hypothesis of clinging and climbing on large diameter vertical supports. The proximal ends of the unguals of at least *Plesiadapis cookei* are, however, strikingly similar to those of sloths and the pedal unguals of *Pteropus* (Megachiroptera: Pteropodidae), possibly indicating some suspensory behaviors. Godinot and Beard (1991) illustrate the digit ray for *Plesiadapis tricuspidens* showing it to not have this suspensory feature. Furthermore, Beard (1989) illustrated the phalanges of another plesiadapid, *Nannodectes intermedius* demonstrating that it is more like *P. tricuspidens* in this regard.

Although the pollex of plesiadapids is divergent and probably fairly mobile (Beard, 1989, 1990), they have been described as lacking prehensile pha-langeal proportions (Beard 1990; Hamrick, 2001), suggesting a reduction in their ability to grasp small diameter supports in a euprimate-like way (Beard, 1990; Boyer and Bloch, 2002; Hamrick, 2001). However, Godinot and Beard's (1991) reconstruction of the *P. tricuspidens* ray shows it to have a short metacarpal, making it more similar to other plesiadapiforms (e.g., diff-erent from *P. cookei*) in this respect.

The humerus of *Plesiadapis cookei* suggests less emphasis on euprimate-like grasping (Gunnell and Gingerich, 1987) and might be more similar to that of sloths and dermopterans in features that represent suspensory tendencies. This is distinctly *not* the case for *Nannodectes intermedius* in which the humerus is more like that of other plesiadapiforms (Beard, 1989).

The close similarity of some plesiadapid unguals to those of sloths and bats, the similarity of at least some plesiadapid humeri to sloths and dermopterans, and the lack of prehensile phalangeal proportions, indicate more frequent use of underbranch clinging. Whereas smaller-bodied plesiadapiforms could nav-igate small branches using strong grasping (similar to extant *Ptilocercus* and callitrichines) and pronograde postures, plesiadapids were also likely capable of some grasping, but may have also relied more on suspensory behaviors to distribute their weight and avoid torques when moving on small branches, as large-bodied platyrrhine and hominoid primates do today.

Micromomyidae

Micromomyids are by far the smallest (30–40 g) plesiadapiforms for which postcrania are known. Postcranial specimens are late Clarkforkian to middle

Wasatchian in age. They represent three genera: *Chalicomomys*, a *Chalicomomys*-like new genus, and *Tinimomys*. Taking all of these specimens into consideration reveals the morphology and inter-element proportions of almost the entire skeleton (Figure 7). While no specific features seem to detract from the ability of these animals to use vertical postures in the man-

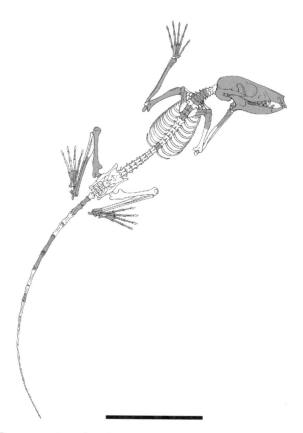

Figure 7. Reconstruction of a micromomyid on a large-diameter support. Features it shares with other plesiadapiforms described here support such a posture. Gray areas depict bones present in one specimen (UM 41870). Note that the posterior three vertebrae and a proximal ulna are not depicted in either Figure 2A or B. These were recently recovered from a block discovered to have broken off from the main limestone (block 821419 from SC-327) early in the preparation process. The association was confirmed by a connection between the broken ulnar shaft and the proximal end of the left ulna. Scale = 3 cm.

ner suggested by the general plesiadapiform morphology, they also exhibit a suite of unique features indicative of some behavioral peculiarities reflected in the morphology and relative length of the radius, the morphology of the innominate, the morphology and relative length of the tibia and fibula, and the morphology of the astragalus.

The radius is unique in having a large, raised area for the origination of the pronator teres muscle on the lateral aspect of the shaft just proximal to its midpoint; a shaft that is mediolaterally expanded starting at the level of the pronator teres muscle (a tuberosity) and continuing to the distal end (providing room for origination of powerful digital flexors and the pronator quadratus muscle, respectively); and a distal articular surface that is deeply cupped, elliptical, and has a distinct dorsal ridge that causes this surface to face palmarly. The form of the distal end is most comparable to that of sloths, dermopterans, and *Ptilocercus* in which it presumably reflects use of suspensory postures wherein the palmar-flexed hand is "hooked" over relatively small-diameter, sub-horizontal supports. Bats have a similar dorsal ridge and palmar-facing articulation, but the shape of the articular surface itself is much different in being almost sloth-like in micromomyids. Taken together, the morphology of the radius seems to indicate sustained use of vertical and underbranch clinging. During underbranch clinging, a strong pronator teres muscle would resist supinatory torque (Miller et al., 1964) produced by gravity, tending to pull the hands out of plane with and away from the substrate.

The fibula and UAJ in micromomyids are substantially different from those of other plesiadapiforms. At such a small body size, micromomyids experienced an arboreal milieu presenting relatively larger diameter supports, and in part these morphological differences seem to reflect that. More specifically, micromomyids appear to have been capable of stronger flexion of the digits and foot, and stronger resistance to pedal inversion than other plesiadapiforms. These inferred functional implications of the ankle morphology are similar to and consistent with those from the forelimb morphology. The proximal end of the fibula flares anteroposteriorly and is blade-like, unlike that known for any other plesiadapiform. The shaft then gradually tapers distally until it obtains a circular cross section. This proximal, blade-like expansion of the shaft provides a large area for origination of pedal plantar-flexor muscles and the pedal evertor muscle, peroneus longus.

The astragalus of micromomyids differs from that of other plesiadapiforms in having, on the body, a relatively high medial ridge that reduces the degree of natural inversion of the foot; on the tibial facet, a deeper groove that limits the UAJ to sagittal flexibility; and an enormous groove for the tendons of the pedal plantar-flexor muscles (flexor tibialis and fibularis) on its plantar aspect as would be expected from the large area for origination of these muscles on the tibia and fibula. The calcaneum of micromomyids differs from that of other plesiadapiforms [except *Phenacolemur praecox* and some other paromomyids (Beard, 1989; Szalay and Drawhorn, 1980)] in having a longer tuberosity, giving the gastrocnemius and soleus muscles more leverage, and in having a more distally and laterally extended peroneal tuberosity, giving the tendon of peroneous longus an even more transverse line of action and making it a more devoted pedal evertor.

Although leaping between vertical supports is not out of the realm of possibility for micromomyids, given the idiosyncrasies described thus far, pronograde postures and any sort of bounding were probably infrequent. Such obligate arboreality is also probably reflected in the innominate. Unlike in *Ignacius* and bounding taxa generally, the ilium is extremely long and rod-like (Sargis, 2002c), the ischium is relatively short and rod-like; and the ischiopubic symphysis is short and caudally shifted relative to the acetabulum, similar to that of dermopterans (Sargis, 2002c) and lorises, both of which often use suspensory postures and neither of which use pronograde bounding. We note that such features also characterize *Ptilocercus* (Sargis, 2002b,c) and primitive eutherians such as *Ukhaatherium* (Horovitz, 2000, 2003), and may be more reflective of the primitive condition (see Szalay et al., 1975) than a behavioral specialization.

The major differences between micromomyids and other plesiadapiforms reflect the ability of micromomyids to more powerfully flex the digits and manus, to plantarflex the pes, to resist supination and inversion, and to less effectively use pronograde postures. Such adaptations suggest more time spent on the undersides of branches (Bloch et al., 2003), where they would be out of sight of aerial predators.

PHYLOGENETIC IMPLICATIONS: PRIMATE ORIGINS AND ADAPTATIONS

Plesiadapiformes have long been considered an archaic radiation of primates (Gidley, 1923; Gingerich, 1975, 1976; Russell, 1959; Simons, 1972; Simpson, 1935b,c; Szalay, 1968, 1973, 1975; Szalay et al., 1975, 1987). In the last 30+ years, many researchers have questioned a plesiadapiform–euprimate link and have suggested removing plesiadapiforms from the primate order (Beard, 1989, 1990, 1993a,b; Cartmill, 1972; Gingerich and Gunnell, 2005; Gunnell, 1989; Kay et al., 1990, 1992; Martin, 1972; Wible and Covert, 1987). Discovery of a paromomyid plesiadapiform skull (Kay et al., 1990, 1992) and independent analysis of postcrania referred to Paromomyidae (Beard, 1989, 1990, 1991, 1993a,b) have led some investigators to conclude that micromomyid and paromomyid plesiadapiforms were mitten-gliders (Beard, 1993b) and shared a closer relationship to extant flying lemurs (classified together in Eudermoptera; Beard, 1993a,b) than Euprimates (Beard, 1989, 1993a,b; Kay et al., 1990, 1992). Despite the fact that this "mitten-gliding hypothesis," as well as the character support for Eudermoptera, have been strongly challenged in the past 15 years (Bloch and Boyer 2002a,b; Bloch and Silcox, 2001, 2006; Bloch et al., 2001, 2002b; Boyer and Bloch, 2002a,b; Boyer et al., 2001; Hamrick et al., 1999; Krause, 1991; Runestad and Ruff, 1995; Sargis, 2002c; Silcox, 2001, 2003; Stafford and Thorington, 1998; Szalay and Lucas, 1993, 1996), a plesiadapiform–dermopteran relationship has gained currency (e.g., McKenna and Bell, 1997). In contrast, based on a wealth of new postcranial data, we have demonstrated that: (1) no plesiadapiform yet studied shows morphological characteristics reflective of dermopteran-like mitten-gliding (Bloch and Boyer, 2002a,b; Bloch et al., in review; Boyer et al., 2001); (2) many aspects of the generalized plesiadapiform postcranium indicate committed arboreality possibly homologous to that of *Ptilocercus*, which suggests that features related to such a lifestyle, previously thought to uniquely link flying lemurs and paromomyids are, instead, reflective of the primitive condition for Euarchonta (Bloch and Boyer, 2002a,b; Bloch et al., 2001, 2002b, 2003, in review; Boyer and Bloch, 2002a,b; Boyer et al., 2001; Sargis, 2001a,b, 2002a,b,c; Szalay and Lucas, 1993, 1996); and (3) cladistic analyses suggest that some of the more specialized arboreal adaptations of certain plesiadapoid plesiadapiforms (specifically Carpolestidae) are uniquely shared with Euprimates, indicating a closer relationship between these two groups than previously supposed (Bloch and Boyer, 2002a, 2003; Bloch et al., 2002b, in review).

Recent cladistic analyses, drawing on different classes of osteological data and including different groups of taxa, support a monophyletic relationship between Plesiadapiformes and Euprimates (Primates, sensu lato; Figure 8). Silcox (2001; also this volume) included dental, cranial, and postcranial data for a large sample of plesiadapiforms, euprimates, scandentians, dermopterans and chiropterans. Her study concluded that plesiadapiforms are the sister

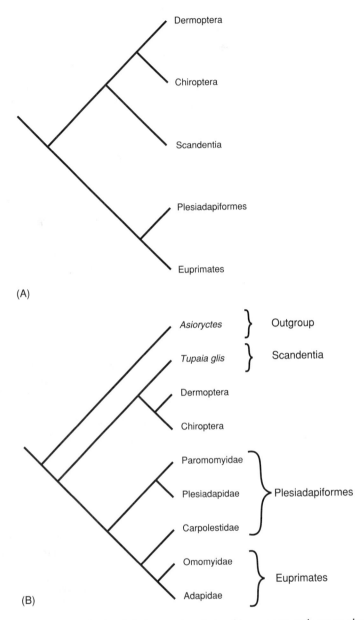

Figure 8. (A) Hypothesis of phylogenetic relationships among archontans that is well supported by dental, cranial, and postcranial evidence presented elsewhere (Silcox, 2001). (B) Hypothesis of phylogenetic relationships among select archontans illustrating phylogenetic position of Carpolestidae based on cladistic analysis of 65 postcranial characters (figure from Bloch and Boyer, 2002a; Figure 1). Note that both topologies support a plesiadapiform-euprimate link, while the cladogram based on new postcranial data presented in Bloch and Boyer (2002a) specifically allies Carpolestidae with Euprimates (Omomyidae + Adapidae).

group to Euprimates to the exclusion of all other included mammals (Figure 8A). Bloch and Boyer (2002a) presented a cladistic analysis of the new post-cranial data discussed here. The results of their postcranial analysis are consistent with those of Silcox (2001) in supporting a plesiadapiform-euprimate relationship but, unlike those of Silcox (2001), they suggest that Carpolestidae falls out with Euprimates to the exclusion of other plesiadapiforms (Figure 8B). Analyses that combine new dental, cranial, and postcranial data from these two analyses, as well as that from the work of Sargis (2001b, 2002a,b,c, also this volume), are underway (Bloch et al., in review; but see Bloch and Boyer, 2003; Bloch et al., 2002b). Preliminary results of this project indicate that plesiadapoid plesiadapiforms (including Carpolestidae, Plesiadapidae, Saxonellidae, and Asian *Chronolestes simul*; Silcox, 2001) form a mono-phyletic clade that is the sister group to Euprimates to the exclusion of all other fossil and living euarchontan mammals (Bloch et al., in review).

This hypothesis of relationships, coupled with new functional interpretations of plesiadapiform skeletons, provides a more resolved picture of the sequence of character acquisitions in early primate evolution than was previously possible through analyses of fragmentary postcrania (e.g., Beard, 1991a, 1993a,b; Szalay and Dagosto, 1980) or through indirect means, such as comparative studies of extant mammals (Cartmill, 1972, 1974; Rasmussen, 1990).

Both arboreal tree shrews (Sargis, 2001a) and didelphid marsupials (Lemelin, 1999) have been presented as living ecological models for plesi-adapiforms and the ancestral euprimate, respectively. It is plausible that the earliest primates were capable of grasping in a manner similar to living arbo-real tree shrews like *Ptilocercus* (Sargis, 2001a, 2002b,c; Szalay and Dagosto, 1988), and in that regard are perhaps best represented in the known postcra-nial fossil record by micromomyids and paromomyids. The specialized eupri-mate foot, which includes a divergent and opposable hallux with a nail (see Dagosto, 1988), likely evolved next in a form similar to that of *Carpolestes*, independent of leaping or orbital convergence. This stage of primate evolu-tion might be best modeled by arboreal delphids like *Caluromys* among living mammals (Lemelin, 1999; Rasmussen, 1990).

We acknowledge that plesiadapiform taxa currently known from post-cranial material are dentally relatively derived (see Kirk et al., 2003) and are unlikely to represent direct ancestors along a lineage leading to the first eupri-mates. However, this type of evidence is usually lacking in the fossil record. If paleontologists were to restrict themselves to studying only those species that were plausibly *directly ancestral* in their studies of the stem lineages of major clades, then we would know very little about the early evolution of, for example, either Hominini (i.e., australopiths) or Cetacea (i.e., archaeocetes). As is the case for these stem taxa and the origin of humans and whales, respectively, we are confident that analyses of plesiadapiform primate skeletons provide useful data in evaluating the competing adaptive scenarios of euprimate origins (Bloch and Boyer, 2003).

At least three possibilities exist concerning the nature of the postcranial similarities between plesiadapiforms and euprimates: (1) plesiadapiforms and euprimates do not share a recent common ancestry, and all of their uniquely shared postcranial similarities are the result of convergence; (2) plesiadapiforms and euprimates do share a recent common ancestry, but all of their uniquely shared postcranial similarities are the result of parallel evolution; or (3) some, or all, of the uniquely shared postcranial similarities are synapomorphies of a clade that either includes carpolestids and euprimates (as suggested by cladistic analysis of only postcranial data; Bloch and Boyer, 2002a), or all plesiadapoid plesiadapiforms (including carpolestids) and euprimates (as suggested by cladistic analysis of dental, cranial, and postcranial data; Bloch and Boyer, 2003; Bloch et al., 2002b, in review). Evidence for and against each of these explanations is outlined below.

It has been suggested that any unique characteristics shared by plesiadapiforms and euprimates must be the result of convergence because the two groups do not share a recent common ancestry (Kay and Cartmill, 1977; Martin, 1990). Evidence for (or against) this interpretation stems from phylogenetic analyses that do not (or do, respectively) support a monophyletic plesiadapiform-euprimate clade. Results of recent phylogenetic analyses unambiguously support a monophyletic plesiadapiform-euprimate clade, based on a larger sample of taxa with more complete morphologic data than ever before analyzed (e.g., Bloch and Boyer, 2003; Bloch et al., 2002b, in review; Silcox, 2001), although these results are not without controversy (Bloch et al., 2003; Kirk et al., 2003). Regardless, there is at least consensus in the literature that plesiadapiforms are euarchontans, and as such, are closer to the origin of euprimates in phylogenetic space and time than are didelphid marsupials (Lemelin, 1999) and arboreal rodents (Kirk et al., 2003) and would be better ecological models and have at least as much, if not more, bearing on the competing adaptive scenarios for euprimate origins as these groups do.

In a similar but not equivalent argument, it is possible that unique similarities between plesiadapiforms and euprimates could have been acquired in parallel from a relatively recent common ancestor (Bloch and Boyer, 2002a, 2003; Kirk et al., 2003). We emphasize that it is implicitly acknowledged in this explanation that plesiadapiforms share a relatively recent common ancestry with euprimates and is thus in broad agreement with recently published phylogenetic hypotheses (Bloch and Boyer, 2003; Bloch et al., 2002b, in review; Silcox, 2001). The most convincing evidence for a "parallel evolution" explanation is that large-bodied plesiadapids, which might share a sister-relationship with carpolestids, lack some of the unique euprimate characters. If *Plesiadapis* represents the primitive condition for Plesiadapoidea, then these characters (e.g., specialized opposable hallux with a nail) would have evolved in parallel. Alternatively, it could be argued that large-bodied species of *Plesiadapis* are derived, and that more primitive, and therefore more phylogenetically relevant, plesiadapids, such as *Nannodectes* (Beard, 1989, 1990), might share more in common with carpolestids than previously recognized. Thus, it is plausible that the primitive plesiadapoid condition is more closely represented by *Carpolestes* (Bloch and Boyer, 2002a) than by *Plesiadapis* (Kirk et al., 2003). However, even if grasping did evolve in parallel from the common ancestor of plesiadapoids and Euprimates, it would represent an example of the parallel evolution of a strikingly euprimate-like mammal from the same arboreal ancestor in potentially identical ecological conditions, and would still be very relevant for assessing hypotheses of euprimate origins (Bloch and Boyer, 2003). Such a scenario would require the common ancestor of euprimates and plesiadapoids to have differed from other euarchontans in having more bunodont teeth and better grasping capabilities. Both of these features are consistent with increased frugivory (Szalay, 1968) and locomotion in terminal branches. In the subsequent hypothetical parallel radiations of euprimates and plesiadapiforms, both could plausibly have evolved more specialized grasping independently, but in similar ways for similar reasons. It is also plausible, although no *direct* evidence supports it yet, that the first euprimates could have then co-opted this initial adaptation to terminal branch frugivory for visually directed predation (Bloch and Boyer, 2003; Ravosa and Savakova, 2004). On the other hand, direct fossil evidence does support the hypothesis that this initial adaptation was co-opted for grasp leaping locomotion in the earliest euprimates (Szalay and Lucas, 1996).

The last argument, and the one preferred here, is that the uniquely shared characteristics of plesiadapoids and euprimates were inherited from a relatively recent common ancestor (Bloch and Boyer, 2003). Arguments against this interpretation are the same as those listed as evidence supporting the convergent and parallel evolution hypothesis outlined above. Furthermore, evidence for this interpretation is the same as that used in the arguments against these two hypotheses: phylogenetic hypotheses that entertain a monophyletic plesiadapoid-euprimate clade (e.g., Bloch and Boyer, 2003; Bloch et al., in review) have greater explanatory power in the face of all of the known dental, cranial, and postcranial data than those based on partitioned data sets (Beard, 1993; Bloch and Boyer, 2002a; Bloch and Silcox, 2006; Kay et al., 1992). If one accepts the hypothesis that Plesiadapoidea are the sister clade to Euprimates, then Euprimates must have originated by around 64 MYA as indicated by the earliest occurrence of a plesiadapoid plesiadapiform (*Pandemonium*; Van Valen, 1994). In this case the first 9 MY of euprimate evolution remains unknown. In this scenario, acquisition of specialized grasping features for terminal branch locomotion would have preceded the evolution of visual specializations in stem-primates and would thus not be considered a specific adaptation for nocturnal, visual predation.

In the words of M. Cartmill (1992: 111) "[w]e can only hope that new fossil finds will help us to tease apart the various strands of the primate story, giving us clearer insights into the evolutionary causes behind the origin of the primate order to which we belong." Older and more primitive skeletons of plesiadapiforms are needed to test our ideas about the evolution of euprimate-like grasping (Bloch and Boyer, 2002a). Likewise, more complete postcranial fossils of the earliest euprimates, and a better sampling of the Paleocene fossil record of Africa, Asia, and the Indian subcontinent, are needed to address how euprimate-like leaping and forward facing orbits might have evolved from a terminal branch-foraging ancestor.

ACKNOWLEDGMENTS

We thank C. Ross, D. Krause, M. Silcox, and E. Sargis for their helpful comments and for reviewing the manuscript. We also thank K. C. Beard, D. Fisher, D. Fox, E. Kowalski, K. Rose, W. Sanders, R. Secord, F. Szalay, J. Trapani, P. Houde, J. Wilson, I. Zalmout, and the participants of the Primate Origins Conference for many helpful conversations. We especially thank G. Gunnell and P. Gingerich for their tremendous help and support during this ongoing project and for access to the unpublished skeleton of *Plesiadapis cookei*. We thank M. Dagosto and M. Ravosa for inviting us to be part of this symposium. Research was supported by grants from the National Science Foundation: BCS-0129601 and EAR-0308902.

REFERENCES

Anemone, R. L., 1993, Functional anatomy of the hip and thigh among Primates, in: *Postcranial Adaptation in Nonhuman Primates*, D. Gebo, ed., Northern Illinois University Press, pp. 150–174.

Beard, K. C., 1989, *Postcranial Anatomy, Locomotor Adaptations, and Paleoecology of Early Cenozoic Plesiadapidae, Paromomyidae, and Micromomyidae (Eutheria, Dermoptera)*, Ph.D. dissertation, Johns Hopkins University School of Medicine, Baltimore.

Beard, K. C., 1990, Gliding behavior and paleoecology of the alleged primate family Paromomyidae (Mammalia, Dermoptera), *Nature* **345:** 340–341.

Beard, K. C., 1991a, Vertical Postures and climbing in the morphotype of Primatomorpha: Implications for locomotor evolution in primate history, in: *Origines de la Bipedie chez les Hominides*, Y. Coppens and B. Senut, eds., Editions du CNRS (Cahiers de Paleoanthropologie), Paris, pp. 79–87.

Beard, K. C., 1991b. Postcranial fossils of the archaic primate family Microsyopidae, *Am. J. Phys. Anthropol.* (Suppl.) **12:** 48–49.

Beard, K. C., 1993a, Phylogenetic systematics of the Primatomorpha, with special reference to Dermoptera, in: *Mammal Phylogeny: Placentals*, F. S. Szalay, M. J. Novacek, and M. C. McKenna, eds., Springer-Verlag, New York, pp. 129–150.

Beard, K. C., 1993b, Origin and evolution of gliding in early Cenozoic Dermoptera (Mammalia, Primatomorpha), in: *Primates and their Relatives in Phylogenetic Perspective*, R. D. E. MacPhee, ed., Plenum Press, New York, pp. 63–90.

Bloch, J. I., 2001, *Mammalian Paleontology of freshwater limestones from the Paleocene-Eocene of the Clarks Fork Basin, Wyoming*, Ph.D., dissertation, University of Michigan, Department of Geological Sciences, Ann Arbor.

Bloch, J. I., and Bowen, G. J., 2001, Paleocene-Eocene microvertebrates in freshwater limestones of the Clarks Fork Basin, Wyoming, in: *Eocene Biodiversity: Unusual Occurrences and Rarely Sampled Habitats*, G. F. Gunnell, ed., Plenum Publishing Corporation, New York, pp. 94–129.

Bloch, J. I., and Boyer, D. M., 2001, Taphonomy of small mammals in freshwater limestones from the Paleocene of the Clarks Fork Basin, Wyoming, in: *Paleocene-Eocene Stratigraphy and Biotic Change in the Bighorn and Clarks Fork Basins, Wyoming*, vol. **33,** P. D. Gingerich, ed., University of Michigan Papers on Paleontology, pp. 185–198.

Bloch, J. I., and Boyer, D. M., 2002a, Grasping primate origins, *Science* **298**: 1606–1610.

Bloch, J. I., and Boyer, D. M., 2002b, Phalangeal morphology of Paleocene Plesiadapiformes (Mammalia: ?Primates): Evaluation of the gliding hypothesis and the first evidence for grasping in a possible primate ancestor, *Geological Society of America Abstracts with Programs* **34**: 13A.

Bloch, J. I., and Boyer, D. M., 2003, Response to comment on "Grasping Primate Origins," *Science* **300**: 741c.

Bloch, J. I., Boyer, D. M., and Gingerich, P. D., 2001, Positional behavior of late Paleocene *Carpolestes simpsoni* (Mammalia ?Primates), *J. Vertebr. Paleontol.* **21**: 34A.

Bloch, J. I., Boyer, D. M., Gingerich, P. D., and Gunnell, G. F., 2002a, New primitive paromomyid from the Clarkforkian of Wyoming and dental eruption in Plesiadapiformes, *J. Vertebr. Paleontol.* **22**: 366–379.

Bloch, J. I., Boyer, D. M., and Houde, P., 2003, New skeletons of Paleocene-Eocene micromomyids (Mammalia, Primates): Functional morphology and implications for euarchontan relationships, *J. Vertebr. Paleontol.* **23**: 35A.

Bloch, J. I., and Gingerich, P. D., 1998, *Carpolestes simpsoni*, new species (Mammalia, Proprimates) from the late Paleocene of the Clarks Fork Basin, Wyoming, *Contributions from the museum of Paleontology, University of Michigan* **30**: 131–162.

Bloch, J. I., and Silcox, M. T., 2001, New basicrania of Paleocene-Eocene *Ignacius*: Re-evaluation of the Plesiadapiform-Dermopteran link, *Am. J. Phys. Anthropol.* **116**: 184–198.

Bloch, J. I., and Silcox, M. T., 2006, Cranial anatomy of the Paleocene plesiadapiform *Carpolestes simpsoni* (Mammalia, Primates) using ultra high-resolution X-ray computed tomography, and the relationships of plesiadapiforms to Euprimates, *J. Hum. Evol.* **50**: 1–35.

Bloch, J. I., Silcox, M. T., Boyer, D. M., and Sargis, E. J., in review, New Paleocene skeletons and the relationship of plesiadapiforms to crown-clade primates.

Bloch, J. I., Silcox, M. T., and Sargis, E. J., 2002b, Origin and relationships of Archonta (Mammalia, Eutheria): Re-evaluation of Eudermoptera and Primatomorpha, *J. Vertebr. Paleontol.* **22**: 37A.

Bowen, G. J., and Bloch, J. I., 2002, Petrography and geochemistry of floodplain limestones from the Clarks Fork Basin, Wyoming, USA: Carbonate deposition and fossil accumulation on a Paleocene-Eocene floodplain, *J. Sediment. Res.* **72**: 50–62.

Bown, T. M., and Beard, K. C., 1990, Systematic lateral variation in the distribution of fossil mammals in alluvial paleosols, lower Eocene Willwood Formation, Wyoming, in: *Dawn of the Age of Mammals in the Northern Part of the Rocky Mountain Interior*, vol. **243**, T. M. Bown and K. D. Rose, eds., Geological Society of America Special Paper, pp. 135–151.

Boyer, D. M., and Bloch, J. I., 2000, Documenting dental-postcranial association in Paleocene-Eocene freshwater limestones from the Clarks Fork Basin, Wyoming, *J. Vertebr. Paleontol.* **20**: 31A.

Boyer, D. M., and Bloch, J. I., 2002a, Bootstrap comparisons of vertebral morphology of Paleocene plesiadapiforms (Mammalia ?Primates): Functional implications, *Geological Society of America Abstracts with Programs* **34**: 13A.

Boyer, D. M., and Bloch, J. I., 2002b, Structural correlates of positional behavior in vertebral columns of Paleocene small mammals, *J. Vertebr. Paleontol.* **22**: 38A.

Boyer, D. M., Bloch, J. I., 2003, Comparative cranial anatomy of the pentacodontid *Aphronorus orieli* (Mammalia: Pantolesta) from the Paleocene of the Western Crazy Mountains Basin, Montana, *J. Vertebr. Paleontol.* **23**: 36A.

Boyer, D. M., and Bloch, J. I., in review, Evaluating the mitten-gliding hypothesis for Paromomyidae and Micromomyidae (Mammalia, "Plesiadapiformes") using comparative functional morphology of new Paleogene skeletons, in: *Mammalian Evolutionary Morphology*, E. J. Sargis and M. Dagosto, eds., Springer: Dordrecht, The Netherlands.

Boyer, D. M., Bloch, J. I., and Gingerich, P. D., 2001, New skeletons of Paleocene paromomyids (Mammalia, ?Primates): Were they mitten gliders? *J. Vertebr. Paleontol.* **21**: 35A.

Cartmill, M., 1972, Arboreal adaptations and the origin of the order Primates, in: *The Functional and Evolutionary Biology of Primates*, R. Tuttle, ed., Aldine-Atherton, Chicago, pp. 97–122.

Cartmill, M., 1974, Rethinking primate origins, *Science* **184**: 436–443.

Cartmill, M., 1992, New views on primate origins, *Evol. Anthropol.* **1**: 105–111.

Dagosto, M., 1983, Postcranium of *Adapis parisiensis* and *Leptadapis magnus* (Adapiformes, Primates), *Folia Primatol.* **41**: 49–101.

Dagosto, M., 1988, Implications of postcranial evidence for the origin of euprimates, *J. Hum. Evol.* **17**: 35–56.

Dagosto, M., Gebo, D. L., and Beard, K. C., 1999, Revision of the Wind River faunas, early Eocene of central Wyoming, Part 14, Postcranium of *Shoshonius cooperi* (Mammalia, Primates), *Ann. Carnegie Mus.* **61**: 1–3.

Decker, R. L., and Szalay, F. S., 1974, Origins and function of the pes in the Eocene Adapidae (Lemuriformes, Primates), in: *Primate Locomotion*, F. A. Jenkins, Jr., ed., Academic Press, New York, pp. 261–291.

Emerson, S., 1985, Jumping and leaping, in: *Functional Vertebrate Morphology*, M. Hildebrand, D. Bramble, K. Liem, and D. Wake, eds., Harvard University Press, Cambridge, pp. 58–72.

Essner, R. L., and Scheibe, J. S., 2000, A comparison of scapular shape in flying squirrels using relative warp analysis (Rodentia: Sciuridae), in: *The Biology of*

Gliding Mammals, R. Goldingay and J. S. Scheibe, eds., Filander Press, Germany, pp. 213–228.

Fleagle, J. G., 1999, *Primate Adaptation and Evolution*, Academic Press, San Diego.

Fleagle, J. G., and Simons, E. L., 1982, The humerus of *Aegyptopithecus zeuxis*—A primitive anthropoid, *Am. J. Phys. Anthropol.* **59**: 175–193.

Gambaryan, P. P., 1974, *How Mammals Run*, Wiley, New York.

Garber, P. A., 1992, Vertical Clinging, small body size and the evolution of feeding adaptations in the Callitrichinae, *Am. J. Phys. Anthropol.* **88**: 469–482.

Gebo, D. L., 1988, Foot morphology and locomotor adaptation in Eocene primates, *Folia Primatol.*, **50**: 3–41.

Gebo, D. L., and Sargis, E. J., 1994, Terrestrial adaptations in the postcranial skeletons of guenons, *Am. J. Phys. Anthropol.* **93**: 341–371.

Gidley, J. W., 1923, Paleocene Primates of the Fort Union Formation, with discussion of relationships of Eocene primates, *Proc. U.S. Natl. Museum* **63**: 1–38.

Gingerich, P. D., 1974, Function of pointed premolars in *Phenacolemur* and other mammals, *J. Dental Res.* **53**: 497.

Gingerich, P. D., 1975, New North American Plesiadapidae (Mammalia, Primates) and a biostratigraphic zonation of the middle and upper Paleocene, *Contributions from the Museum of Paleontology, University of Michigan* **24**: 135–148.

Gingerich, P. D., 1976, Cranial anatomy and evolution of early Tertiary Plesiadapidae (Mammalia, Primates), *University of Michigan Papers on Paleontology* **15**: 1–141.

Gingerich, P. D., and Gunnell, G. F., 1992, A new skeleton of *Plesiadapis cookei*, *The Display Case: A Quarterly Newsletter of The University of Michigan Exhibit Museum* **6**: 1–3.

Gingerich, P. D., and Gunnell, G. F., 2005, Brain of *Plesiadapis cookei* (Mammalia, Proprimates): Surface morphology and encephalization compared to those of Primates and Dermoptera, *Contributions from the Museum of Paleontology, University of Michigan* **31**: 185–195.

Gingerich, P. D., Houde, P., and Krause, D. W., 1983, A new earliest Tiffanian (Late Paleocene) mammalian fauna from Bangtail Plateau, Western Crazy Mountain Basin, Montana, *J. Paleontol.* **57**: 957–970.

Godinot, M., and Beard, K. C., 1991, Fossil primate hands: A review and an evolutionary inquiry emphasizing early forms, *J. Hum. Evol.* **6**: 307–354.

Gould, E., 1978, The behavior of the moonrat *Echinosorex gymnurus* (Erinaceidae) and the pentailed tree shrew, *Ptilocercus lowii* (Tupaiidae) with comments on the behavior of other Insectivora, *Z. Tierpsychol* **48**: 1–27.

Grand, T. I., 1968, Functional anatomy of the upper limb, in: *Biology of the Howler Monkey (Alouatta caraya)*, M. R. Malinow, ed., Karger, Basel, pp. 104–125.

Gregory, W. K., 1913, Relationship of the Tupaiidae and of Eocene Lemurs, especially *Notharctus, Bull. Geol. Soc. Am.* **24**: 247–252.

Gregory, W. K., 1920, On the structure and relations of *Notharctus*, an American Eocene primate, *Memoirs of the American Museum of Natural History* **3**: 51–243.

Gunnell, G. F., 1989, Evolutionary history of Microsyopoidea (Mammalia, ?Primates) and the relationship between Plesiadapiformes and Primates, *University of Michigan Papers on Paleontology* **27**: 1–157.

Gunnell, G. F., and Gingerich, P. D., 1987, Skull and partial skeleton of *Plesiadapis cookei* from the Clarks Fork Basin, Wyoming, *Am. J. Phys. Anthropol.* **72**: 206A.

Hamrick, M. W., 1998, Functional adaptive significance of primate pads and claws: Evidence from new world anthropoids, *Am. J. Phys. Anthropol.* **106**: 113–127.

Hamrick, M. W., 2001, Primate origins: Evolutionary change in digit ray patterning and segmentation, *J. Hum. Evol.* **40**: 339–351.

Hamrick, M. W., Rosenman, B. A., and Brush, J. A., 1999, Phalangeal morphology of the Paromomyidae (?Primates, Plesiadapiformes): The evidence for gliding behavior reconsidered, *Am. J. Phys. Anthropol.* **109**: 397–413.

Harrison T., 1989, New postcranial remain of *Victoriapithecus* from the middle Miocene of Kenya, *J. Hum. Evol.* **18**: 3–54.

Hooker, J. J., Russell, D. E., and Phélizon, A., 1999, A new family of Plesiadapiformes (Mammalia) from the old world lower Paleogene, *Palaeontology* **42**: 377–407.

Horovitz, I., 2000, The tarsus of *Ukhaatherium nessovi* (Eutheria, Mammalia) from the late Cretaceous of Mongolia: An appraisal of the evolution of the ankle in basal eutherians, *J. Vertebr. Paleontol.* **20**: 547–560.

Horovitz, I., 2003, Postcranial skeleton of *Ukhaatherium nessovi* (Eutheria, Mammalia) from the late Cretaceous of Mongolia, *J. Vertebr. Paleontol.* **23**: 857–868.

Houde, P., 1986, Ostrich ancestors found in the northern hemisphere suggest new hypothesis of ratite origins, *Nature* **324**: 563–565.

Houde, P., 1988, Paleognathous birds from the early Tertiary of the northern hemisphere, *Publications of the Nuttall Ornithological Club* **22**: 1–148.

Jenkins, F. A., Jr., 1974, Tree shrew locomotion and the origins of primate arborealism, in: *Primate Locomotion*, F. A. Jenkins, Jr., ed., Academic Press, New York, pp. 85–116.

Jenkins, F. A., and McClearn, D., 1984, Mechanisms of hind foot reversal in climbing mammals, *J. Morphol.* **182**: 197–219.

Kay, R. F., and Cartmill, M., 1977, Cranial morphology and adaptations of *Palaechthon nacimienti* and other Paromomyidae (Plesiadapoidea, Primates), with a description of a new genus and species, *J. Hum. Evol.* **6**: 19–53.

Kay, R. F., Thewissen, J. G. M., and Yoder, A. D., 1992, Cranial anatomy of *Ignacius graybullianus* and the affinities of the Plesiadapiformes, *Am. J. Phys. Anthropol.* **89**: 477–498.

Kay, R. F., Thorington, R. W. Jr., and Houde, P., 1990, Eocene plesiadapiform shows affinities with flying lemurs not primates, *Nature* **345**: 342–344.

Kinzey W. G., Rosenberger, A. L., and Ramirez, M., 1975, Vertical clinging and leaping in a neotropical anthropoid, *Nature* **255:** 327–328.

Kirk, E. C., Cartmill, M., Kay, R. F., and Lemelin, P., 2003, Comment on "Grasping Primate Origins," *Science* **300:** 741.

Krause, D. W., 1991, Were paromomyids gliders? Maybe, maybe not, *J. Hum. Evol.* **21:** 177–188.

Larson, S. G., 1993, Functional morphology of the shoulder in Primates, in: *Postcranial Anatomy of Nonhuman Primates*, D. L. Gebo, ed., Northern Illinois University Press DeKalb, pp. 45–69.

Le Gros Clark, W. E., 1926, On the anatomy of the pen-tailed tree shrew (*Ptilocercus lowii*), *Proc. Zoolog. Soc. Lond.* **1926:** 1179–1309.

Le Gros Clark, W. E., 1927, Exhibition of photographs of the tree shrew (*Tupaia minor*), Remarks on the tree shrew *Tupaia minor*, with photographs, *Proc. Zoolog. Soc. Lond.* **1927:** 254–256.

Lemelin, P., 1999, Morphological correlates of substrate use in didelphid marsupials: Implications for primate origins, *J. Zoolog.* **247:** 165–175.

Lemelin, P., and Grafton, B., 1998, Grasping performance in *Saguinus midas* and the evolution of hand prehensility in primates, in: *Primate Locomotion*, E. Strasser and J. Fleagle, eds., Plenum Press, New York, pp. 131–144.

Macleod, N., and Rose, K. D., 1993, Inferring locomotor behavior in Paleogene mammals via eigenshape analysis, *Am. J. Sci.* **293A:** 300–355.

Martin, R. D., 1972, Adaptive radiation and behavior of the Malagasy lemurs, *Philos. Trans. R. Soc. Lond.* **264:** 295–352.

Martin, R. D., 1990, *Primate Origins and Evolution: A Phylogenetic Reconstruction*, Princeton University Press, Princeton.

McKenna, M. C., and Bell, S. K., 1997, *Classification of Mammals Above the Species Level*, Columbia University Press, New York.

Miller, M. E., Christensen, G. C., and Evans, H. E., 1964, *Anatomy of the Dog*, W. B. Saunders Company, Philadelphia.

Napier, J. R., and Walker, A. C., 1967, Vertical clinging and leaping—A newly recognized category of locomotor behavior of primates, *Folia Primatolog.* **6:** 204–219.

Rasmussen, D. T., 1990, Primate Origins: Lessons from a neotropical marsupial, *Am. J. Primatol.* **22:** 263–277.

Ravosa, M. J., and Savakova, D. G., 2004, Euprimate origins: The eyes have it, *J. Hum. Evol.* **46:** 355–362.

Rose, K. D., 1987, Climbing adaptations in the early Eocene mammal *Chriacus* and the origin of Artiodactyla, *Science* **236:** 314–316.

Rose, K. D., 2001, Compendium of Wasatchian mammal postcrania from the Willwood Formation of the Bighorn Basin, in: *Paleocene-Eocene Stratigraphy and*

Biotic Change in the Bighorn and Clarks Fork Basins, Wyoming, Vol. **33**, P. D. Gingerich, ed., University of Michigan Papers on Paleontology, pp. 157–183.

Rose, K. D., Godinot, M., and Bown, T. M., 1994, The early radiation of euprimates and the initial diversification of Omomyidae, in: *Anthropoid Origins,* J. G. Fleagle and R. F. Kay, eds., Plenum Press, New York, pp. 1–27.

Runestad, J. A., and Ruff, C. B., 1995, Structural adaptations for gliding in mammals with implications for locomotor behavior in paromomyids, *Am. J. Phys. Anthropol.* **98**: 101–119.

Russell, D. E., 1959, Le crane de *Plesiadapis. Bulletin de Societé Gèologique de France* **4**: 312–314.

Russell, D. E., 1964, Les Mammifères Paléocène d'Europe. *Mémoires du Muséum National d'Histoire Naturelle, nouvelle série* **13**: 1–324.

Sargis, E. J., 2001a, Grasping behaviour, locomotion and substrate use of the tree shrews *Tupaia minor* and *T. tana* (Mammalia, Scandentia), *J. Zoolog.* **253**: 485–490.

Sargis, E. J., 2001b, A preliminary qualitative analysis of the axial skeleton of tupaiids (Mammalia, Scandentia): Functional morphology and phylogenetic implications, *J. Zoolog.* **253**: 473–483.

Sargis, E. J., 2002a, Functional morphology of the forelimb of tupaiids (Mammalia, Scandentia) and its phylogenetic implications, *J. Morphol.* **253**: 485–490.

Sargis, E. J., 2002b, Functional morphology of the hind limb of tupaiids (Mammalia, Scandentia) and its phylogenetic implications, *J. Morphol.* **254**: 149–185.

Sargis, E. J., 2002c, The postcranial morphology of *Ptilocercus lowii* (Scandentia, Tupaiidae): An analysis of primatomorphan and volitantian characters, *J. Mamm. Evol.* **9**: 137–160.

Sargis, E. J., 2002d, Primate origins nailed, *Science* **298**: 1564–1565.

Scheibe, J. S., and Essner, R. L., 2000, Pelvic shape in gliding rodents: Implications for the launch, in: *The Biology of Gliding Mammals,* R. Goldingay and J. S. Scheibe, eds., Filander Press, Germany, pp. 167–184.

Shapiro, L. J., and Simons, C. V. M., 2002, Functional aspects of strepsirrhine lumbar vertebral bodies and spinous processes, *J. Hum. Evol.* **42**: 753–783.

Silcox, M. T., 2001, *A Phylogenetic Analysis of Plesiadapiformes and Their Relationships to Euprimates and Other Archontans,* Ph.D. dissertation, Johns Hopkins University School of Medicine, Baltimore.

Silcox, M. T., 2003, New discoveries on the middle ear anatomy of *Ignacius graybullianus* (Paromomyidae, Primates) from ultra high resolution X-ray computed tomography, *J. Hum. Evol.* **44**: 73–86.

Silcox, M. T., Bloch, J. I., Sargis, E. J., and Boyer, D. M., 2005, Euarchonta (Dermoptera, Scandentia, Primates), in: *The Rise of Placental Mammals,* K. D. Rose and J. D. Archibald, eds., The Johns Hopkins University Press, pp. 127–144.

Silcox, M. T., and Gunnell, G. F. (in press). Plesiadapiformes, in: *Evolution of Tertiary Mammals of North America*, **vol. 2**, C. M. Janis, G. F. Gunnell, and M. D. Uhen, eds., Cambridge University Press, Cambridge.

Simmons, N. B., 1995, Bat relationships and the origin of flight, *Symp. Zoolog. Soc. Lond.* **67**: 27–43.

Simmons, N. B., and Quinn, T. H., 1994, Evolution of the digital tendon locking mechanism in bats and dermopterans: A phylogenetic perspective, *J. Mamm. Evol.* **2**: 231–254.

Simons, E. L., 1972, *Primate Evolution: An Introduction to Man's Place in Nature*, The Macmillan Company, New York.

Simpson, G. G., 1935a, The Tiffany fauna, Upper Paleocene. II. Structure and relationships of *Plesiadapis*, *American Museum Novitates* **816**: 1–30.

Simpson, G. G., 1935b, The Tiffany fauna, Upper Paleocene. III. Primates, Carnivora, Condylarthra, and Ambylpoda, *American Museum Novitates* **817**: 1–28.

Simpson, G. G., 1935c, New Paleocene mammals from the Fort Union of Montana, *Proc. U. S. Natl. Mus.* **83**: 221–244.

Stafford, B. J., 1999, *Taxonomy and Ecological Morphology of the Flying Lemurs (Dermoptera, Cynocephalidae)*, Ph.D. dissertation, City University of New York, New York.

Stafford, B. J., and Thorington, R. W., 1998, Carpal development and morphology in archontan mammals, *J. Morphol.* **235**: 135–155.

Sussman, R. W., and Kinzey, W. G., 1984, The ecological role of the Callitrichidae: A review, *Am. J. Phys. Anthropol.* **64**: 419–449.

Szalay, F. S., 1968, The beginnings of primates, *Evolution* **22**: 19–36.

Szalay, F. S., 1972, Paleobiology of the earliest primates, in: *The Functional and Evolutionary Biology of Primates*, R. H. Tuttle, ed., Aldine-Atherton, Chicago, pp. 3–35.

Szalay, F. S., 1973, New Paleocene primates and a diagnosis of the new suborder Paromomyiformes, *Folia Primatolog.* **22**: 243–250.

Szalay, F. S., 1975, Where to draw the nonprimate-primate taxonomic boundary, *Folia Primatolog.* **23**: 158–163.

Szalay, F. S., and Dagosto, M., 1980, Locomotor adaptations as reflected on the humerus of Paleogene Primates, *Folia Primatolog.* **34**: 1–45.

Szalay, F. S., and Dagosto, M., 1988, Evolution of hallucial grasping in primates, *J. Hum. Evol.* **17**: 1–33.

Szalay, F. S., and Decker, R. L., 1974, Origins, evolution, and function of the tarsus in late Cretaceous Eutheria and Paleocene Primates, in: *Primate Locomotion*, F. A. Jenkins, ed., Academic Press, New York, pp. 223–359.

Szalay, F. S., and Drawhorn, G., 1980, Evolution and diversification of the Archonta in an arboreal milieu, in: *Comparative Biology and Evolutionary Relationships of Tree Shrews*, W. P. Luckett, ed., Plenum Press, New York, pp. 133–169.

Szalay, F. S., and Lucas, S. G., 1993, Cranioskeletal morphology of Archontans, and diagnoses of Chiroptera, Volitantia, and Archonta, in: *Primates and Their Relatives in a Phylogenetic Perspective*, R. D. E. MacPhee, ed., Plenum Press, New York, pp. 187–226.

Szalay, F. S., and Lucas, S. G., 1996, The postcranial morphology of Paleocene *Chriacus* and *Mixodectes* and the phylogenetic relationships of archontan mammals, *New Mexico Museum of Natural History & Science Bulletin* 7: 1–47.

Szalay, F. S., Rosenberger, A. L., and Dagosto, M., 1987, Diagnosis and differentiation of the order Primates, *Am. J. Phys. Anthropol.* 30: 75–105.

Szalay F. S., and Sargis, E. J., 2001, Model-based analysis of postcranial osteology of marsupials from the Paleocene of Itaborai (Brazil) and the phylogenetics and biogeography of Metatheria, *Geodiversitas* 23: 139–302.

Szalay, F. S., and Schrenk, F., 2001, An enigmatic new mammal (Dermoptera?) from the Messel middle Eocene, Germany, *Darmstadter Beitrage Zur Naturgeschichte* 11: 153–164.

Szalay, F. S., Tattersall, I., and Decker, R. L., 1975, Phylogenetic relationships of *Plesiadapis*—Postcranial evidence, in: *Approaches to Primate Paleobiology*, F. S. Szalay, ed., Karger, Basel, pp. 136–166.

Taylor, M. E., 1976, The functional anatomy of the hindlimb of some African Viverridae (Carnivora). *J. Morphol.* 148: 227–254.

Thorington, R. W., and Heaney, L. R., 1981, Body proportions and gliding adaptations of flying squirrels (Petauristinae), *J. Mamm.* 62: 101–114.

Thorington, R. W., Schennum, C.E., Pappas, L. A., and Pitassy, D., 2005, The difficulties of identifying flying squirrels (Scuiridae: Pteromyini) in the fossil record, *J. Vertebr. Paleontol.* 25: 950–961.

Van Valen, L. M., 1994, The origin of the plesiadapid primates and the nature of *Purgatorius, Evolutionary Monographs* 15: 1–79.

Vinyard, C. J., Wall, C. E., Williams, S. H., and Hylander, W. L., 2003, Comparative functional analyses of skull morphology of tree-gouging primates, *Am. J. Phys. Anthropol.* 120: 153–170.

Walker, A. C., 1974, Locomotor adaptations in past and present prosimian primates, in: *Primate locomotion*, F. A. Jenkins, Jr., ed., Academic Press, New York, pp. 349–381.

Wible, J. R., and Covert, H. H., 1987, Primates – cladistic diagnosis and relationships, *J. Hum. Evol.* 16: 1–22.

Winkler, D. A., 1983, Paleoecology of an early Eocene mammalian fauna from paleosols in the Clarks Fork Basin, northwestern Wyoming (USA), *Palaeogeography, Palaeoclimatology, Palaeoecology*, 43: 261–298.

Youlatos, D., 1999, Positional behavior of *Cebuella pygmaea* in Yasuni national National Park, Ecuador, *Primates* 40: 543–550.

CHAPTER SEVENTEEN

Start Small and Live Slow: Encephalization, Body Size, and Life History Strategies in Primate Origins and Evolution

Brian T. Shea

INTRODUCTION

Key adaptations of living and extinct primates are central to any understanding of the origins and early evolution of our order. One such attribute of the primate order is the tendency toward relatively high levels of encephalization (e.g., Gould, 1975a,b; Jerison, 1973, 1979; LeGros Clark, 1971). Although there is of course considerable overlap in relative brain size with other mammalian species and groups, primates represent the most diverse mammalian order generally characterized by such high levels of encephalization. Hominoids as a group—and our own excessively encephalized species in particular—may garner much of the focus in discussions of primate brain evolution, but I argue here that it is an understanding of early primate brain evolution, which

Brian T. Shea ● Department of Cell and Molecular Biology, Northwestern University, 303 E. Chicago Avenue, Chicago, IL 60611

is perhaps most central to explicating subsequent patterns of encephalization in our order. Jerison (1979) was among the first to stress that large relative brain size was probably a characteristic of the early primates, compiling data that could be reliably assembled for fragmentary early fossil primates and other contemporary mammals. Gould (1975b) explicitly stressed that any true understanding of anthropoid and human encephalization required an explanation for the large relative brain size in early primates. He stated (Gould, 1975b: 26):

> Primates have been ahead right from the start; our large brain is only an exaggeration of a pattern set at the beginning of the age of mammals. But why did such a large brain evolve in a group of small, primitive, tree-dwelling mammals, more similar to rats and shrews than to mammals conventionally judged as more advanced? . . . we simply do not know the answer to one of the most important questions we can ask.

Here I will address this central and unanswered question, and consider some ramifications for the subsequent evolution of brain size in primates. My focus will be on brain size within the context of body size, life history adaptations and reproductive strategies. In line with a theoretical emphasis on how an understanding of development and ontogeny can elucidate primate evolutionary morphology (e.g., Shea, 1988, 1990, 1992a), this paper will complement traditional foci on adult morphology and selective scenarios by incorporating discussion of neonatal and subadult adaptations in early and subsequent primate evolution.

I will argue that primatologists have not previously appreciated how the consequences of body size variations are key to explicating the evolution of relative brain size in both early primates, and their subsequent and long-term adaptive diversifications. The input of body size here is not predominantly viewed in the direct causal or allometric (Gould, 1966) sense; the orientation is rather one in which size is but a single component within a complex and synergistic adaptive network of features. Aboreality and life history features, such as reproductive strategy, are other key elements. The central argument may be abstracted as two sequential elements. First, the high encephalization of early primates relative to their mammalian contemporaries is seen as linked to the evolution of strongly precocial reproductive strategies at relatively small body sizes. In originating at small body size and "living slowly," the early primates were quite unusual mammals since most small species "live fast and die young" (Eisenberg, 1981; Promislow and Harvey, 1990; Read and Harvey,

1989). Second, the characteristic ordinal feature of high relative brain size is viewed as, in large measure, a result of subsequent diversification in both body size and other adaptive strategies among the major groups of primates. In essence, the initial primate head-start is "translated up" to subsequent larger body sizes and adaptive configurations.

Primates are the only mammalian order characterized by this combination of marked size and adaptive diversification from a foundation of small size, precociality, and high relative brain size. I argue that generally high levels of encephalization perhaps comparable to what is seen in modern primates would likely have been evolved if other small and precocial mammals—such as bats, dermopterans, and elephant shrews—had undergone comparable evolutionary diversification in size and adaptive strategy.

RELATIVE BRAIN SIZE IN PRIMATES

There are several key issues that underlie determinations of relative brain size in primate evolution. The first is allometric correction—long recognized as necessary due to general patterns of negative allometry or progressive relative diminution of brain size compared to body size in mammals and other vertebrates. Some of the earliest studies of allometry dealt with ontogenetic, interspecific, and phylogenetic scaling of brain size (see Gould, 1966: Huxley, 1932: for references). Any number of excellent review papers may be consulted for discussion of various empirical and theoretical issues involved in recognizing and explaining general patterns of brain–body scaling, as well as utilizing these scaling baselines for computation of individual species' encephalization quotients (EQ) or other such residualized determinations (e.g., Gould, 1966, 1975a; Jerison, 1979; Martin, 1989). Here I follow traditional broad interspecific scaling analyses, and the use of these baselines to determine residuals for individual specimens or species. These approaches work reasonably well for addressing general questions, such as whether extant primates tend to cluster above the size-corrected norm for all mammals in static comparisons of extant or fossil species (Figure 1). However, it is vital to keep in mind that such static patterns mask true phylogenetic contexts and comparisons. Evolution consists of transformations of antecedent states, and only hypothetical ancestral values can provide reliable baselines for such assessments—no species ever evolved from the predicted value of a regression line best-fit to a static scatter of diverse species! Moreover, the expectation of

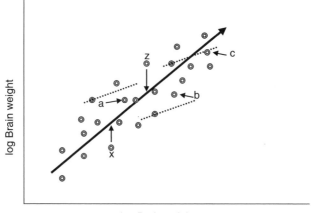

Figure 1. A schematized plot of brain-body scaling in an interspecific series of mammals. Relative brain size (EQ) is traditionally determined by comparing observed values (here for species X and Z as representative examples) against regression-adjusted predicted values for species of that body size, defined as the solid line with arrowhead. These residualized values are shown here as arrows from the predicted values for species X (negative residual, smaller-than-average brain size) and species Z (positive residual, relatively large brain size).

Residuals may also be determined within a more phylogenetically controlled series, either by using genus-level regressions or fitting a line of slope 0.33 through a smaller sister species (see text and Lande, 1979). Three cases of the latter are illustrated here, with the dashed lines fit to the selected values as shown. The species indicated by the "a," "b," and "c" are here taken as phyletically-enlarged descendants of these three species. The disparities between residualized values determined relative to these intrageneric lines, as opposed to the overall line of best fit (solid line), are significant. For species a, encephalization has decreased relative to the expectations based on the intrageneric fit, but the species still has a positive residual value relative to the overall trend. For species b, the opposite pattern holds, i.e., an increase in relative brain size is observed in comparison to the intrageneric prediction, but species b still has a negative residual relative to the broad sample regression. In the case of species c, the observed brain size is exactly as predicted given intrageneric allometric scaling, but the ancestral species has a positive residual and the descendant species a negative residual relative to the line of overall best-fit. The key point here is that the EQ residual values in these cases are coded quite differently, and thus would pattern differently in comparative studies associating relative brain size with factors such as variance in diet, habitat, social structure, etc. See text and Shea (1983), and Williams (2002) for additional discussion.

brain size change correlated with short-term selection on body size yields a different, and substantially lower, scaling exponent than the static broad inter-specific value of 0.66–0.75 (Gould, 1975a; Lande, 1979, 1985; Riska and Atchley, 1985; Shea, 1983, 1987, 2005; Williams, 2002). I have previously argued that such lowered scaling exponents result because the body size diversification is generated by differential selection on postnatal growth rates, which exhibit reduced correlations with early (prenatal) growth periods, when brain size is increasing differentially (Shea, 1983, 1992a,b). This principle is also illustrated in Figure 1, where lower-than-average residualized values relative to the all-mammal plot may actually be coincident with a derived *increase* in encephalization, once the proper baseline is recognized. Other complex and counterintuitive arrangements of the standard EQ values are also possible (see Figure 1 and accompanying legend).

An important issue related to the use of interspecific scaling baselines for calculation of EQ residuals is the specific slope value of the broad interspecific trend. Jerison's (1973, 1979) EQ values utilized slopes of 0.66, originally derived from empirical scatters of extant mammal, and subsequently fit to the grand means of various brain–body assemblages. Following Jerison's early study, various researchers (e.g., Armstrong, 1983; Eisenberg, 1981; Hofman, 1983; Martin, 1981) concluded that a slope of 0.75 provided a better fit to the extant mammal scatter. This will of course alter the calculation of residuals, such that smaller species will on average exhibit higher EQ values in the new calculation, as the line of best-fit through the grand mean is repositioned lower within the scatter at small values of body size, and larger species will on average exhibit lower EQ values as the line is repositioned higher within the scatter at large values of body size. Nevertheless, I utilized Gould's (1975a), Jerison's (1973, 1979), and Radinsky's (1970, 1975) original values here for several reasons. First, these are *relative* assessments, and thus the ordering and comparability is robust to such changes in the overall slope value. In this chapter, I am not focused on either the absolute values of these EQ's, or the underlying slope from which they are derived. In any case, the many discussions of the "new" (circum 0.75) versus "old" (circum 0.66) slope values are in all probability off the mark, since the best-sampled scatters evidence considerable curvilinearity (see Martin, 1981, Figure 1), and the overall slope is unlikely to be linked to any single underlying factor, be it metabolic rate or some other input. A convincing criterion of "functional equivalence" based on neural capacities or brain functions has never been advanced for the 0.66 or 0.75 coefficients.

In fact, arguments linking measured cognitive performance to size-corrected EQ values in anthropoids and hominoids of divergent body size support a criterion of subtraction value in the 0.2–0.4 range (Williams, 2002). Much additional work will be needed to address these complex issues in the comparative study of brain size and scaling.

Jerison's (1973, 1979) computation of EQ in living and fossil mammals represented landmark advances in our understanding of the evolution of gross brain size. However, various other difficulties with these and other such studies must be acknowledged. One is the question of the biological meaning of total brain size and the myriad empirical and theoretical issues related to gross brain size, the brain's internal allometries, and the evolvability of its components. Here I follow Jerison (1979) in acknowledging that gross size of the brain and/or its components serve only as surrogate measures and correlates of more salient aspects of organismal performance and fitness (Arnold, 1983). While brain size scaling regularities—externally with body size and internally among its components—are indisputable, there is obviously also considerable residual variance attesting to the evolvability of total brain size and its localized regions. Much of this variance is related to particular sensory and cognitive specializations of various mammalian species.

Encephalization in Extant Primates

Are the living primates truly the most highly encephalized order among extant mammals? This is partly a semantic issue, revolving around how one defines "most highly encephalized." Stephan (1972) maintained that while high encephalization was obviously characteristic of *Homo*, it was just as clearly not uniformly true of the entire order. Martin (1973), Radinsky (1970, 1975), and others have also shown that primates are not unique in their high level of encephalization. There is considerable overlap with other mammals, particularly those highly encephalized cetaceans (Jerison, 1973; Marino, 1996, 1997), but also with selected carnivores, rodents, bats, and other taxa. Yet these assessments deal with issues of overall variance, range, and overlap. Of course *not all primates exceed all other mammals* in encephalization, and various extant primate species (particularly among the strepsirhines) exhibit only average levels of relative brain size. Nevertheless, *as a group*, the primates exhibit the highest *average level* of encephalization among mammals.

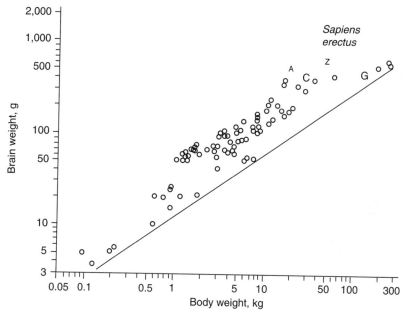

Figure 2. Brain-body scaling in mammals (solid line determined from Jerison's 1979 dataset), illustrating the general high relative brain size of primates (open circles) in comparison to mammalian averages. Additional points labeled: *H. sapiens*; *H. erectus*; Z = *Zinjanthropus* (*Australopithecus boisei*); A = *Australopithecus africanus*; C = *Pan troglodytes*; G = *Gorilla gorilla* (from Gould, 1975a).

Various lines of evidence support the generally high level of encephalization seen in extant primates. Figure 2 illustrates a plot of selected primate species relative to Jerison's (1973) extant mammalian baseline. While some primate species approximate the line for average brain size at a given body size, many are well above the central trend. Martin's (1981) plot of brain and body sizes in extant mammals also demonstrates the generally high level of encephalization in the primates. Austad and Fischer's (1992) tabular summary of EQ values for modern mammals, computed from Eisenberg's (1981) database relative to a line of best-fit with a slope near 0.75, demonstrates the high average level of encephalization for primates (Table 1). A plot of average ordinal values also demonstrates primates as a group to have the greatest positive deviation from the line of best-fit (Figure 3).

Table 1. Average encephalization quotient (EQ) values for selected orders of mammals, illustrating the high mean value for primates

Order	Number of species	EQ
Chiroptera (bats)	50	0.94
Primates (primates)	77	2.54*
Dermoptera (colugos)	1	–
Edentata (edentates)	9	0.95
Scandentia (tree shrews)	2	1.34
Proboscidea (elephants)	2	1.59
Carnivora (carnivores)	109	1.22
Artiodactyla (even-toed ungulates)	92	0.84
Perissodactyla (odd-toed ungulates)	12	0.92
Rodentia (rodents)	131	0.99
Hyracoidea (hyraxes)	3	0.90
Lagomorpha (rabbits, hares)	7	0.62
Insectivora (insectivores)	14	0.55
Macroscelidae (elephant shrews)	6	1.14
Marsupialia (marsupials)	67	0.61
Monotremata (monotremes)	3	0.83

*Data from Harvey et al., 1987.
Data from Eisenberg (1981), except for primates, as noted; modified from Austad and Fischer, 1992.

Encephalization in Extinct Primates

Are the primates of the Paleogene—the period of the Early Tertiary comprising Paleocene, Eocene, and Oligocene epochs—also generally highly encephalized? This question must be addressed in relative fashion as well. Almost all mammalian lineages with substantial fossil records show varying degrees of progressive increase in encephalization throughout the Tertiary (Gould, 1975a; Jerison, 1973); therefore, any determination of relative brain size in extinct early primates must of necessity be rooted in a contemporary mammal-wide context. For example, studies by Gurche (1982), Jerison (1973, 1979), and Radinsky (1970, 1975, 1977, 1979) have revealed that Paleogene primate endocasts yield estimated EQ values which in general are lower than seen in average extant prosimians or strepsirhines (Figures 4 and 5). The absolute values given by Jerison (1979) are summarized in Table 2 and demonstrate levels generally lower than the value of 1.0 for average living mammals (as defined by Jerison's 1973 sample and best-fit). However, Jerison (1973: 373) calculated an average value of only 0.20 (or roughly one-fifth the size of an average living mammal) for archaic mammals of the Early Tertiary, and therefore, the various taxa of fossil Eocene euprimates do indeed exhibit generally high levels of encephalization than their contemporaries. Some

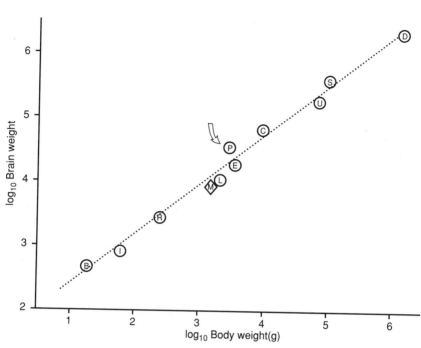

Figure 3. Group averages for brain and body weight in mammals. Note the positive deviation for the primates. Symbols: B = bats; I = insectivores; R = rodents; M = marsupials; L = lagomorphs; E = edentates; P = primates; C = carnivores; U = ungulate; S = pinnipeds; D = cetaceans (modified from Martin, 1983).

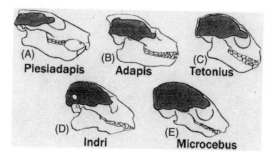

Figure 4. Figures of endocranial size and skull morphology in three extinct primates compared to extant *Indri* and *Microcebus*. All reconstructions drawn to same approximate size (modified from Radinsky, 1979).

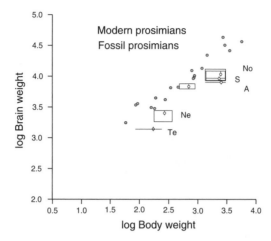

Figure 5. A plot of brain-body values for extant prosimian primates, with reconstructed estimates for selected taxa of fossil primates. Note the smaller average relative brain size in the fossil taxa. (No = *Notharctus*, S = *Smilodectes*, A = *Adapis*, Ne = *Necrolemur*, Te = *Tetonius*) (from Gurche, 1982, with the kind permission of Springer Science and Business Media).

Table 2. Estimates of relative brain size in selected fossil primates

Species	Source of estimate	Mean EQ	Minimum EQ	Maximum EQ
Plesiadapis tricuspidens	HJ	0.50	0.40	0.60
Smilodectes gracilis	JG	0.47	0.35	0.86
	JG	0.44	0.36	0.64
	LR	0.41		
	HJ	0.53		
Notharctus tenebrosus	JG	0.49	0.36	0.92
Tetonius homunculus	JG	0.43	0.33	0.67
	LR	0.42		
	HJ	0.71		
Necrolemur antiquus	JG	0.56	0.35	0.76
	LR	0.79		
	HJ	0.94		
Rooneyia viejaensis	JG	0.81	0.60	1.07
	LR	0.97		
	HJ	1.23		

Adapted from Conroy, 1990 and Gurche, 1982; as determined by Gurche (JG), Radinsky (LR), and Jerison (HJ). Conroy (1990) provides a range of 0.67–1.89, with a mean of 1.09, for extant prosimians (based on the data of Stephan et al., 1970). Note that Jerison (1979) calculated a mean EQ value of 0.20 for archaic mammals of the Early Tertiary.

estimates of relative brain size from the crushed skull of *Plesiadapis tricuspidens* also yield values higher than average contemporary mammals (Jerison, 1979). The traditional characterization of early primates as having relatively large brain sizes (e.g., LeGros Clark, 1971; Martin, 1973) appears quite valid, although this pattern requires additional corroboration through the retrieval of reasonably well-preserved fossil skulls from the earliest Tertiary deposits.

PRECOCIALITY AND ENCEPHALIZATION IN PRIMATE EVOLUTION

Reproductive Strategies and Primate Evolution

Research has indicated that one primary input to the encephalization levels observed in primates and other mammals relates to reproductive and life history strategies. It is necessary to review our general understanding of these mammalian and primate reproductive strategies prior to assessing these associations with encephalization, however. Building directly on previous work by Adolf Portmann (1939, 1965), Martin (1972, 1973, 1975a,b) noted that mammalian reproductive strategies could be organized into two alternative modes: altricial and precocial. The salient aspects of these alternative reproductive strategies are given in Table 3. In essence, these strategies correspond to the broader life history "fast–slow continuum" among mammals (Charnov, 1993; Charnov and Berrigan, 1993; Jones and MacLarnon, 2001; Promislow and Harvey, 1990; Read and Harvey, 1989), derived from the r–K continuum (MacArthur and Wilson, 1967) and density-dependent versus density-independent mortality rates. The altricial mode of reproductive strategy has been characterized as "living fast and dying young" and is linked to relatively high levels of natural mortality (Promislow and Harvey, 1990) and intrinsic rates of natural increase (Hennemann, 1983). These species tend to produce large litters of relatively undeveloped young after short gestation periods. Altricial species are typically small in body size and relative brain size. The tailless tenrec, *Tenrec ecaudatus*, from Madagascar provides one example of a highly altricial mammal (Figure 6A). This tenrec may produce litters of over 30 offspring. Precocial species, by contrast, develop more slowly and take longer to mature. They are characterized by lower levels of mortality and the production of small litters of well-developed offspring following extended gestations. Intrinsic rates of natural increase are lower in such groups (Hennemann, 1983). Precocial mammals tend to be of moderate-to-large body size, and relatively highly encephalized. The elephants provide a good example of a precocial mammal (Figure 6B).

Table 3. Alternative states for development of offspring at birth in mammals

Altricial type	Precocial type
1. Adults usually construct nests, at least when dependent offspring are present. ***Adult body size typically small***	1. Adults do not normally construct nests at any stage. ***Adult body size tends to be medium or large***
2. Infants are born naked, and the ears and eyes are closed by a membrane for some time after birth. Initially, the young usually exhibits imperfect homeothermy compared to adults	2. Infants are born with at least a moderate covering of hair, and the ears and eyes born with atleast a moderate are open at birth or soon afterwards. Homeothermy typically well developed at birth compared to the adult condition
3. The lower jaw is incompletely developed at birth and the middle ear is hence at an early stage of development. The teeth erupt quite late in postnatal development	3. The lower jaw is well developed at birth and the middle ear is fairly well developed. The teeth quite soon after birth, at least in small-bodied forms
4. The gestation period is relatively short; litter size and teat count are large	4. The gestation period is relatively long; litter size and teat count are very small
5. Infants typically have low mobility at birth	5. Infants typically have high mobility at birth
6. ***The relative brain size of the neonate and of the adult is small,*** and the brain usually grows considerably after birth	6. ***The relative brain size of the neonate and of the adult is large,*** and the brain grows only moderately after birth
7. The adults are generally nocturnal in habits	7. The adult tends to be diurnal in habits, though a fair number of smaller species are nocturnal

From Martin, 1975b, following Portmann, 1939, 1965; italics added; Note particularly the entries on relative brain size (# 6) and overall body size (#1).

Mammalian orders are often characterized as being either precocial or altricial, although "mixed" orders such as the carnivores and rodents are acknowledged (Martin, 1975a,b). However, it is perhaps not surprising that with multiple inputs into these reproductive and life history strategies, it is sometimes difficult to view mammalian species as simply one mode or the other. Derrickson (1992) assessed four criteria (thermoregulatory, sensory, locomotory, and nutritional) of offspring development and noted various intermediate or mixed groupings (see Table 4). Studies by Case (1978) and Eisenberg (1981) utilized somewhat different developmental criteria, and again uncovered taxa not easily classified as one mode or the other. Martin and MacLarnon (1985) characterized various species as "intermediate" between the precocial and altricial modes in their study of gestation length scaling. It

(A)

(B)

Figure 6.   These two taxa offer a striking contrast in mammalian reproductive strategies. A. The highly altricial tailless tenrec (*Tenrec ecaudatus*) has very large litter sizes, in rare cases as many as 30 or more neonates (Louwman, 1973) (with the kind permission of Blackwell Publishers); B. The strongly precocial elephant, typically bearing a single well-developed offspring (with the kind permission of David Pride).

Table 4. Neonatal development in 16 orders of mammals

Number of genera (families) within developmental category

Order	0	1	2	3	Total
Insectivora	9(2)	1(1)	–	–	10(3)
Macroscelidae	–	–	–	1(1)	1(1)
Chiroptera	7(2)	1(1)	3(2)	–	10(3)
Scandentia	–	1(1)	–	–	1(1)
Primates	–	2(2)	18(9)	–	20(9)
Edentata	1(1)	–	2(1)	1(1)	4(3)
Pholidota	1(1)	–	–	–	1(1)
Lagomorpha	2(1)	2(2)	–	1(1)	4(2)
Rodentia	58(9)	5(3)	4(4)	17(12)	83(19)
Cetacea	–	–	–	2(1)	2(1)
Carnivora	12(5)	14(5)	2(2)	1(1)	27(7)
Pinnipedia	–	–	–	4(3)	4(3)
Tubulidentata	–	–	1(1)	–	1(1)
Proboscidea	–	–	–	1(1)	1(1)
Hyracoidea	–	–	–	1(1)	1(1)
Artiodactyla	–	–	1(1)	12(5)	13(6)
Total	90(21)	26(15)	31(20)	41(27)	183(62)

From Derrickson, 1992; Developmental categories represent a scale from 0 (highly altricial) to 3 (highly precocial), quantified as a composite value based on neonatal independence in four key areas: *thermoregulatory, sensory, locomotory, and nutritional.* Note that total numbers of genera and families may differ from the sum of the developmental categories, due to the fact that some taxa may be represented in more than one category.

is clear that continuous variation exists for mammal species in some of the factors contributing to the dichotomous precocial and altricial reproductive strategies (Zeveloff and Boyce, 1986), and much additional work remains to be undertaken in order to clarify such variation and its bases.

Primates as an order are classified as strongly precocial (Derrickson, 1992; Eisenberg, 1981; Martin, 1975a,b). The great majority of species give birth to a single, well-developed offspring with relatively large neonatal brain size, following a prolonged gestation period (Martin and MacLarnon, 1985, 1988; Pagel and Harvey, 1988). Significant variation among primate (predominantly strepsirhine) species does exist in the number of offspring, teat number, mother–infant relations, nest-using behavior, and related attributes (Martin, 1975a,b). Kappeler's (1995, 1996, 1998) studies have contributed considerably to the documentation and phylogenetic analysis of this variation. He notes that primates range from those producing the most precocial young (such as *Eulemur*, *Lemur catta*, indris, lorises, tarsiers, and the anthropoids), where neonates are very well developed and capable of grasping the mother's fur or other supports, to certain species characterized by less precocial young (e.g., *Varecia* and many cheirogaleids), where infants may have eyes closed at birth, exhibit difficulty with coordinated movement, and are born in litters. Some *Microcebus* species produce up to two litters of 2–3 rapidly growing offspring in their first year of life, and *Varecia variegata*, the variegated lemur, builds nests for litters of 2–3 infants. Many galago and cheirogaleid offspring are also left in nests and tree holes for some time after birth. Various primates, such as lorises, tarsiers, lepilemurs, and the wooly lemur (*Avahi laniger*), carry their young from birth onward, in certain cases "parking" them for brief periods while the mother forages. On a higher taxonomic level, it is a fair generalization to note that lorisids are quite precocial and slow-growing, cheirogaleids are generally less precocial with more rapid development, galagids are somewhat mixed and intermediate, with the lemurids being predominantly quite precocial (Kappeler, 1996). The data compilations and phylogenetic reconstructions for features including nest building, tree-hole use, infant oral transport, infant carrying, teat number, litter size, activity pattern, and social organization should be consulted for further information (Kappeler, 1998).

This diversity raises issues of both reconstructed ancestral states for primates, and the ecological factors underlying this observed variation. Martin (1975a,b) argued that the strong precociality of extant primates is primitive for the order, and linked to evolution in stable environments with relatively predictable resources and competition, along with low levels of neonatal, juvenile, and adult mortality. He further suggested that those primates characterized by somewhat higher rates of reproductive turnover (such as some of the cheirogaleids and *Varecia variegata*) had secondarily evolved these patterns in response to more unpredictable seasonal environments. Kappeler's (1998) phylogenetic reconstructions have argued that the earliest primates were nocturnal and solitary, with three pairs of teats, and a single offspring which was initially kept in a shelter (tree hole or nest) and subsequently mouth-carried to a parking place where they grasp for periods of time while the mother forages for food. This reconstructed state is similar to that seen in many extant galagids. Kappeler (1995) further stresses that although certain nocturnal strepsirhines do indeed exhibit several ancestral mammalian reproductive traits, such as multiple offspring, nest building, and infant parking, their overall life history strategies are essentially primate-like and reflect a clear evolutionary shift toward the production of quite precocial young. In sum, early primates were probably

Table 5. Life history and encephalization data for selected small, precocial mammals

Species Order; Family	Body weight (g)	Brain weight (g)	Litter size	Gestation length (days)	IP and/or EQ*
Elephantulus fuscipes Macroscelidea; Macroscelididae	57[a]	1.33[a]	1–2[b]	60[b]	287[a]; 0.75[c]
Elephantulus intufi Macroscelidea; Macroscelididae	49[d]	1.14[d]	1–2[d]	51[d]	—; 0.70[e]
Elephantulus myurus Macroscelidea; Macroscelididae	64[d]	1.37[d]	1–2[d]	46[d]	—; 0.70[e]
Cynopterus horsfieldii Chiroptera; Pteropodidae	53[f]	1.24[f]	1–2[b]	115–125[g]	236[f]; 0.72[e]
Microcebus murinus Primates; Cheirogaleidae	54[a]	1.78[a]	1–3[b]	59–62[g]	334[a]; 0.86[c]
Tupaia javanica Scandentia; Tupaiidae	105[a]	2.55[a]	2[g]	—	315[a]; 0.94[e]
Tupaia glis Scandentia; Tupaiidae	150[a]	3.15[a]	2–3[g]	46–50[g]	310[a]; 0.92[e]
Urogale everetti Scandentia; Tupaiidae	275[a]	4.28[a]	1–2[g]	54–56[b]	287[a]; 0.83[e]
Cheirogaleus medius Primates; Cheirogaleidae	177[a]	3.14[a]	2–3[b]	70[b]	279[a]; 0.82[e]
Pteropus edwardsi (Wirz) Chiroptera; Pteropodidae	287[f]	8.00[f]	—	—	527[f]; 1.50[e]
Pteropus edwardsi (Warncke) Chiroptera; Pteropodidae	375[f]	6.85[f]	—	—	382[f]; 1.08[e]
Loris tardigradus Primates; Lorisidae	322[a]	6.60[a]	1[b]	180[b]	402[a]; 1.15[e]
Cheirogaleus major Primates; Cheirogaleidae	450[a]	6.80[a]	2–3[b]	70[b]	336[a]; 0.95[e]
Chinchilla laniger Rodentia; Chinchillidae	432[d]	5.25[d]	2[d]	110[d]	—; 0.75[e]

Species					
Rhynchocyon stuhlmanni Macroscelidea; Macroscelididae	432[d]	5.25[d]	2[d]	110[d]	—; 0.75[c]
Myoprocta pratti Rodentia; Dasyproctidae	780[d]	9.90[d]	1.2[d]	98[d]	—; 0.95[c]
Otolemur crassicaudatus Primates; Galagidae	850[a]	10.30[a]	1[d]	135[d]	341[a]; 0.94[c]
Avahi laniger occidentalis Primates; Indriidae	860[a]	9.67[a]	1[b]	120–150[b]	317[a]; 0.87[c]
Lepilemur ruficaudatus Primates; Lepilemuridae	915[a]	7.60[a]	1[b]	120[g]	240[a]; 0.66[c]
Avahi laniger laniger Primates; Indriidae	1270[a]	11.45[a]	1[b]	120–150[b]	294[a]; 0.79[c]
Hapalemur simus Primates; Lemuridae	1300[a]	9.53[a]	1[g]	160[g]	—; 0.65[c]
Pteropus vampyrus Chiroptera; Pteropodidae	1220[a]	10.20[f]	1[b]	180[b]	271[f]; 0.73[c]
Potos flavus Carnivora; Procyonidae	1970[d]	31.20[d]	1.1[d]	77[d]	—; 1.61[c]
Lemur catta Primates; Lemuridae	2100[d]	22.00[d]	1[d]	135[d]	—; 1.09[c]
Dasyprocta aguti Rodentia; Dasyproctidae	2800[d]	20.30[d]	1.3[d]	104[d]	—; 0.83[c]

*Derived using the formulas or data of Bauchot and Stephan (1966) for the "index of progression" (IP) or Jerison (1973) for the "encephalization quotient" (EQ).

[a]Bauchot and Stephan (1966).

[b]Walker et al. (1975).

[c]Jerison (1973).

[d]Sacher and Staffeldt (1974).

[e]computed by this author.

[f]Pirlot and Stephan (1970).

[g]Asdell (1964).

Source: Data arranged in approximate groupings of increasing size.

strongly precocial mammals, and thus this has been a defining feature of our order from its very origins.

Encephalization and Precociality

The summaries presented in Table 3 indicate that one key component of the precocial reproductive strategy within mammals is relatively high encephalization (Martin, 1975a,b; 1982; Martin and MacLarnon, 1985). This makes sense in developmental, functional, and ecological terms. Precocial mammals are characterized by elongated gestation periods (Martin and MacLarnon, 1985), with their typically high rates of intrauterine brain growth (Sacher and Staffeldt, 1974). Large neonatal and adult relative brain size is a key component of an overall adaptive strategy based on low mortality, complex social organization, and high learning requirements (Dunbar, 1998). As noted, this adaptive configuration is likely linked to evolution in relatively stable tropical environments with predictable resource distributions, competition levels, and mortality schedules.

Martin (1981), and Martin and MacLarnon (1985) quantified the association of precociality in mammals and size-corrected estimates of brain size. There is a significant upward transposition or "grade shift" for precocial mammals versus altricial mammals (Martin, 1981). Figure 7 illustrates this schematically and gives Martin's (1981) regressions for the two groups of mammals. It seems that having relatively large brain size is a key component of reproductive and life history strategy in the precocial mammals. This is undoubtedly linked at least in part to the role of learning, and various cognitive and memory skills in species with prolonged growth, low reproductive turnover, and, in many cases, complex social organization (Dunbar, 1998; Kudo and Dunbar, 2001). There are of course other inputs to relative brain size that are likely operative in the case of particular groups that are precocial. Considering only two examples, the high encephalization of dolphins and some other cetaceans appears related to some extent to diving patterns and/or social complexity (see Marino 1996, 1997), while a significant component of primate encephalization may be due to cortical enlargement linked to visual processing (Preuss, this volume).

The association between precociality and high adult relative brain size is not absolute by any means. Various altricial mammals—particularly some of the carnivores—are also quite highly encephalized. In such species the brain grows considerably after birth. This pattern also characterizes our own

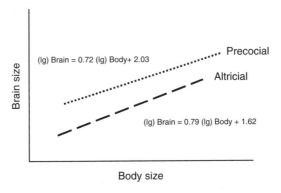

Figure 7. A schematic representation of brain-body scaling in precocial versus altricial adult mammals, with regression values given for each group. Precocial species exhibit a significant upward transposition or "grade shift" relative to altricial species, as indicated by the higher value for the y-intercept parameter. Note that this effect is most strongly marked in smaller size ranges. See text for additional discussion (after Shea, 1987, and based on data from Martin and MacLarnon, 1985, 1988).

species, often described as "secondarily altricial" (Gould, 1976), although this aspect of human development obviously overlays typically strong anthropoid precociality. In birds, altricial species are generally more encephalized as adults than are precocial species (Ricklefs and Starck, 1998). These altricial birds merely exhibit higher duration and rates of postnatal brain growth.

Following Martin's (1981) claims regarding relative brain size and reproductive strategies in mammals, Bennet and Harvey (1985) stressed that these results emerged essentially from phylogenetic bias (i.e., having too many primates in the overall sample). They suggested that there is in fact no statistical difference in EQ values for precocial versus altricial mammals once such phylogenetic factors are controlled. In order to more fully address this issue, we need to consider an additional key factor in these analyses. That factor is the role that body size may play in the broader context of reproductive strategies and encephalization.

PRECOCIALITY, ENCEPHALIZATION, AND SMALL BODY SIZE IN EARLY PRIMATE EVOLUTION

I previously have emphasized that an association between precociality and encephalization in mammals has special implications for early primate evolution (Shea, 1987). The summary in Table 3 reveals that small-bodied mammals are typically altricial, while moderate-to-large sized mammals are generally precocial. This is related to the fact that most small mammals are characterized by the "live fast and die young" strategy of rapid reproductive turnover and high mortality (e.g., Promislow and Harvey, 1990; Read and Harvey, 1989), while large body size with its positively associated increases in gestation lengths and longevity are generally linked to precocial adaptations. But most strepsirhines and many Eocene forms, which might serve as models for early euprimate evolution, are quite diminutive. This means that the small early primates were unusual mammals in adopting such strongly precocial reproductive and life history strategies at quite small body sizes (Shea, 1987).

Reconstruction of early primate body size is somewhat problematic due to the sparse fossil record and other factors. However, most authorities (Cartmill, 1974; Fleagle, 1978, 1988; Martin, 1975b) favor a body size of perhaps several hundred grams, in the range of various cheirogaleids and galagids. Suggestions by Soligo and Muller (1999), based on nail and claw morphology, that the first primates were at least a kilogram in size, and that all primates smaller than this rubicon are therefore "phyletic dwarfs," remain unconvincing (Hamrick, 1999, 2001). The finding of extremely small forms interpreted by Gebo et al. (2000) to be early (Middle Eocene) haplorhines raises at least the possibility that the initial diversification of anthropoids occurred at body sizes considerably smaller than observed in any living primates. Additional fossils and analyses will further clarify these issues. Our current best reconstruction of body size in the first euprimates would therefore range from about 100 to 500 g at most. Of key relevance to this chapter, however, is the central fact that any reasonable reconstruction of ancestral body size in early euprimates puts them toward the smaller end of mammalian and current primate body size ranges.

There are several reasons why the combination of small body size and marked precociality might be expected to be associated with high relative brain size. First, the general negative allometry of brain–body scaling means that smaller mammals within the general ellipsoid scatter have relatively large brains compared to mammals of greater body size (Gould, 1966, 1975a). Additionally, precocial forms exhibit relatively long gestation periods, characterized by high rates of brain growth, as noted earlier. These factors fit nicely with a further elaboration of Martin's (1981) findings summarized in Figure 7. Significantly, it is in the smaller body size ranges where the association of precociality and encephalization is most marked. This at least partially addresses

Bennett and Harvey's (1985) critique of Martin's (1981) linkage between precociality and encephalization. It may eventuate that the association holds in smaller size ranges, but not in larger ranges. Larger mammals in general tend to be highly precocial, with the exception of some carnivoran and other species, so such comparisons are skewed phylogenetically.

A comparison of selected primate taxa with comparably sized mammals exhibiting small litter sizes and a significant degree of precociality allows an indirect test of these associations. Various taxa for which brain and body sizes are available have been collected in Table 4. The fact that direct comparisons are hard to come by within the mammals is in itself a statement that the combination of small body size and marked precociality is an unusual adaptive strategy. Small nonprimate mammals exhibiting precocial adaptations include some of the rodents (e.g., chinchillas, agoutis), elephant shrews, hyraxes, and some bats and tree shrews (though these are more appropriately characterized as having small litter sizes rather than particularly precocial young). Other species, such as the moderately precocial spiny mouse (*Acomys cahirinus*), would provide important contrasts with altricial related species (Brunjes, 1990), but data on gross and relative brain sizes are unfortunately not available.

The comparisons in Table 4, grouped in three size classes ranging up to 2 kg, indicate that other precocial mammals comparable in size to the selected primate species have similar estimates of relative brain size. In the smallest size range, the elephant shrews and the pteropid bat compare favorably with *Microcebus murinus*. The middle size range representatives include several tree shrew species and two pteropids compared to a cheirogaleid and lorisid. Again, encephalization levels in these nonprimate precocial forms are comparable to the primate examples. This is also the case in the selected species clustering around 2 kg in Table 4. The kinkajou has the highest relative brain size here, and it is among the most precocial of the carnivores.

The hypothesis that the combination of small size and marked precociality is generally associated with higher-than-average relative brain size in mammals deserves further scrutiny. The rodents are probably the best group in which to assess these relationships within a controlled phylogenetic context, since they provide examples of both altricial and precocial groups which overlap in body size. Other key mammalian groups include the elephant shrews, which are small and quite precocial forms; the bats (Jones and MacLarnon, 2001); and the hyraxes, which, though currently moderately small, were represented by much larger forms in the past.

SELECTION FOR PRECOCIALITY IN EARLY PRIMATES

An important related question is *why* the early primates were precocial. What are the selective contexts in which precociality seems to be favored in mammal evolution? A key here appears to be some component or components of the arboreal environment. It has been noted previously that arboreal, gliding, and flying mammals generally have small litter sizes of relatively well-developed young (Luckett, 1980; Martin, 1969). Meier (1983) demonstrated that arboreality was correlated with longer gestations, slower growth rates, smaller litter sizes, and increased encephalization among 33 species of sciurid rodents in North America. Eisenberg (1975) emphasized that the arboreal anteaters *Tamandua* and *Cyclopes* exhibit reduced litter sizes, increased longevity, and higher EQs compared to their terrestrial relatives. Eisenberg and Wilson (1981) have demonstrated a similar trend within didelphid marsupials. Further research by Rasmussen (1990; Rasmussen and Sussman, this volume) on the didelphid *Caluromys* explicates this pattern. In addition to various cranial and postcranial morphological adaptations possibly associated with their arboreal habitus, species of *Caluromys* are known to be characterized by relatively small litters, low reproductive rates, extended longevity, and relatively high EQ compared to other didelphids. Jones and MacLarnon (2001) link arboreality to reproductive strategy and encephalization in chiropterans.

Precise relationships among the variables in this complex web of features are far from clearly established. In Meier's (1983) study of sciurids noted earlier, higher encephalization was correlated with arboreal habitus, but the arboreality was associated with not only the life history features, but also divergent locomotor and dietary habits (storage of nuts and seeds versus a terrestrial diet of grasses, etc.). This raises the issue of which of these (or other correlated but uninvestigated) variables is directly and causally associated with the relative brain size. Lemen (1980) has suggested that climbing ability and the need to navigate structurally complex environments is the key input to higher relative brain size in arboreal rodents (though he did not examine influence of or control for reproductive strategy variation). Moreover, we cannot simply assume brain size is the primary "independent" variable of selective and fitness significance. Meier (1983) in fact argues that the small relative brain size of terrestrial sciurids is possibly a correlated artifact of the rapid postnatal growth rates and larger adult body size of these species, rather than any indication for decreased levels of cognitive capacity or behavioral complexity. This accords with observations that even though large-bodied ground squirrels exhibit many life history traits of a K-selected strategy (increased longevity, smaller litters, and high sociality), they have relatively smaller brains (Armitage, 1981). The reasons for this most likely have to do with the large body size of these species and the evolution of this size differentiation via postnatal growth rates among closely related species, as I have previously discussed in a comparison of chimpanzees to gorillas and generalized to other primate and mammal groups (Shea, 1983; see also O'Shea and Reep, 1990; Riska and Atchley, 1985). I believe that the developmental basis of rapid size change may be a factor in many analyses which note a correlation between relative brain size and diet (or other ecological factors). In such cases, unusually high or low EQ values determined relative to the general class or ordinal interspecific trends would reflect spurious correlations and not specific selection on brain size per se.

It is also necessary to stress that while certain aspects of the arboreal environment may indeed select for the complex of reduced litter size, precociality, and encephalization, this pattern is of course not restricted to arboreal versus terrestrial comparisons. For example, Eisenberg (1975) has shown that the same pattern holds among the many species of tenrecs, which exhibit litter size variation from over 30 in *Tenrec ecaudatus* to only 1 or 2 in *Microgale talazaci*. The latter also exhibits the greatest longevity, longest developmental time, and highest EQ of the tenrecs. The macroscelids, as noted above, are also relatively highly precocial and encephalized, yet clearly terrestrial. So what is it, specifically, about the arboreal environment that might also be found in certain other environments and select for the production of highly precocial offspring? Relatively stable and predictable levels of resource availability and competition, along with low levels of neonatal, juvenile, and adult mortality, appear to be key elements (Eisenberg, 1975; Martin, 1975b). Kappeler et al. (2003); Purvis et al. (2003), and Van Shaik and Deaner (2002) have also commented on the possible links between arboreality, reproductive strategy, and relative brain size in their discussions of primate life history evolution. Purvis et al. (2003: 38) further stress that "body size seems to be an adaptive response to life history strategies, rather than the other way around." This view may have great relevance for our understanding of early primate evolution.

Particular aspects of stable tropical rain forest arboreal environments may have selected for small litter size and generalized neonatal precociality in the early primates, but there are additional key primate adaptations which deserve renewed scrutiny in this light. The grasping hallux and appendages of primates have long been recognized as a hallmark and key adaptation in primate evolution (Cartmill, 1972, 1974; LeGros Clark, 1971; Martin, 1990; Szalay and Dagosto, 1988). Studies have provided new evidence that the plesiadapiform primates were well adapted to arboreal climbing and grasping movements, extending this basal adaptation deeper into the phylogeny and perhaps to the entire archontan or euarchontan assemblage (Bloch and Boyer, 2002; Sargis, 2002; Silcox, 2002; Youlatos and Godinot, 2004). These features have traditionally been discussed solely in the selective contexts of mobile and independent animals—typically adults—engaged in visual predation in the terminal branch niche for insects (e.g., Cartmill 1974), in foraging for small angiosperm components in terminal branches (e.g., Sussman, 1991; Rasmussen, 1990; Rasmussen and Sussman, this volume), in generalized terminal branch grasping, and feeding (Bloch and Boyer, 2002), or in "grasp-leaping" across canopy gaps, and among moderate and small branches (Szalay and Dagosto, 1988). These last researchers have also maintained that "the increased speed of locomotion and the precariousness of landings dictated the selective forces which adaptively constrained the euprimates towards reducing the number of young per pregnancy" (Szalay and Dagosto, 1988: 27). These scenarios may all be completely correct as stated, but they ignore an important function and biological role of the grasping appendages in small arboreal primates, which occurs earlier in life history than such independent foraging for plant parts and insects. Primate infants themselves must be able to effectively grasp the mother's fur as she moves throughout the arboreal environment (Figure 8A), or cling precariously to slender branches while they are "parked" during maternal foraging trips (Figure 8B). This function of the mother as a "moveable nest" has been previously stressed (e.g., Eisenberg, 1975), and it seems likely that there is rather strong selection and significant fitness consequences for infants not sufficiently developed and coordinated enough to effectively cling to the mother as she moves about the forest canopy. Eisenberg (1975: 64) summarized this key element of primate life history adaptation and the grasping adaptations of the precocial young:

> Bearing young in an arboreal environment poses special problems. The young must have sufficient muscular coordination so that they will not fall before they leave the nest area. If a trend toward precocial young

Figure 8. A female ring-tailed lemur (*Lemur catta*) is shown on the left with two offspring clinging to her ventrum and dorsum (8a with the kind permission of the Cologne Zoo and Werner Kaumannus). On the right, a newborn lesser bush baby (*Galago moholi*) grasps a thin branch, having been "parked" by its mother while she forages for food. (8b with the kind permission of R.D. Martin)

should be selected for, then the transition phase must be bridged from, at one extreme, a young that is helpless in a nest. In general, this has been done either by retaining a nest phase in the rearing cycle of the young or eliminating it by producing a small number (e.g., 1–2) of very precocial young which can cling to the mother. The mother thus assumes the function of a "moveable nest."

Even in primates that nest or park their infants, there is always considerable maternal carriage of the offspring (Kappeler 1996, 1998). There is also frequent transport of the grasping infant on the belly or back in those species which also occasionally "mouth-carry" infants to and from nests and parking places. Clinging to thin branches during these "parked intervals" when they are entirely on their own may be as significant a challenge to the grasping appendages of infants as when they are holding onto the moving mother's fur (Figure 8B). As infants age and begin to move about independently, the advanced degree of physical and motor development characteristic of precocial mammals also provides a key advantage in the spatially complex arboreal environment (e.g., Pagel and Harvey, 1988).

Use of the grasping hallux by the infant to cling to the moving primate mother and small branches needs to be more fully incorporated into current evolutionary scenarios focused on these key morphological adaptations in locomoting adults. While the selective challenges to infant competence in grasping and early independent movement by no means *supercede* those of locomotor behaviors in adults during foraging and other movements, they most certainly do *antecede* these. Renewed emphases on the selective contexts of infant grasping and maternal carriage also mesh with a balanced focus on postural behaviors in addition to locomotor behaviors in formulating a complete and integrated understanding of anatomical structures and adaptations (e.g., Ripley, 1967). This more balanced approach accords well with many attempts in evolutionary biology to investigate the actions of natural selection throughout the entire life history, and not merely at the adult stage.

SCALING PRINCIPLES AND SUBSEQUENT PRIMATE EVOLUTION

I have argued above that early primates were unusual mammals in combining small body size and a highly precocial reproductive strategy, and that these factors were related to their evolution of large relative brain size. But this was merely the foundation—or initial "grade-shift"—on which further brain size evolution occurred. It is fundamental to our understanding of primate brain evolution to emphasize that this characteristic of the early primates was "translated up" and elaborated at the larger body size ranges characteristic of more recent primate groups.

A consideration of general brain–body scaling principles in broad series of mammals supports this conclusion. The regularities of brain–body scaling and their bases in development, physiology and evolution, have been most cogently discussed by Gould (1975a). There are two primary scaling criteria relevant to any assessment of early-to-recent primate evolution. These are: (1) static mammalian scaling, and (2) broad phylogenetic scaling (Gould, 1975a; Jerison, 1973). Static mammalian scaling is illustrated in Figure 9 and utilizes a scaling coefficient in the neighborhood of 0.66–0.75. These coefficients are derived from many studies of brain–body scaling in static series of living mammals. They provide our best—if still far from completely understood—criteria

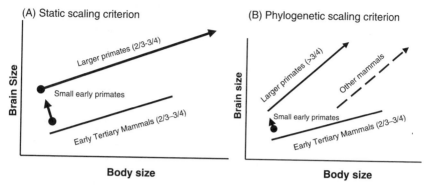

Figure 9. Two schematic frameworks, one static (A) and one phylogenetic (B), each illustrating both the early increase in relative brain size observed in small early primates (short, thick arrows vertically directed), and the subsequent scaling or translation up into larger size ranges (longer, obliquely-directed arrows). In each case, early Tertiary mammals are represented by a trajectory with a slope typical of static mammalian series (2/3 – 3/4). In the case of a static grade shift (A), early to more recent primates scale at a rate typical of static extant mammals (2/3 – 3/4). With phylogenetic scaling (B), primates and other groups of mammals exhibit progressive increase in relative brain size throughout the Tertiary, but the effect is exaggerated in primates due to the upward transposition at small body size in the early primates. See text for additional discussion.

for scaling or predicting brain size changes associated with body size differentiation across broad ranges in living mammals. Using this scaling criterion, the shift in relative brain size characterizing the early primates in comparison to other Paleogene mammals would result in later primates as a group also being transposed above their own contemporaries, yielding the pattern observed for extant mammals illustrated in Figure 9. This relationship expressed across broad size ranges is an example of the transpositions discussed by Gould (1966, 1971, 1975a; White and Gould, 1965) and later termed a "grade shift" by various authors (e.g., Martin, 1989). These grade shifts are merely shorthand ways of expressing higher or lower baselines of scaling at roughly comparable coefficients across body size ranges, and they were one of the earliest patterns recognized in broad studies of static brain–body allometry (e.g., Jerison, 1973).

Static scaling principles do not define phylogenetic trajectories of brain–body scaling, however, as Gould (1975a) and Jerison (1973) have argued in some detail. Jerison (1973), Martin (1973), and others have noted that almost all eutherian lineages have evidenced progressive increase in relative brain size during the Tertiary. The causes for this relative increase in encephalization are not fully known, and the rates and bases of change vary considerably from group to group, as would be expected in any nonorthogenetic evolutionary framework. Certain progressive increases may actually be linked across groups (such as Jerison, 1973, suggested for carnivores and ungulates—but see Radinsky, 1978), while others are probably related to specific ecological contexts and other factors (e.g., Barton and Harvey, 2000; Clark et al. 2001; de Winter and Oxnard, 2001; Eisenberg, 1981; Jerison, 1973; Rilling and Insel, 1998). For our purposes here it is most important to stress that such phylogenetic scaling criteria would be expected to generally exceed the 0.66–0.75 range for broad static scaling coefficients. As illustrated in Figure 9, this "phylogenetic transposition" for primates would combine with their early high relative brain size to yield the generally high levels of encephalization observed in the order relative to other mammals. Thus, either in the case of static or phylogenetic scaling, it was the large relative brain size of the early primates compared to their Early Tertiary mammalian relatives that establishes a baseline elevation which is generally preserved and translated up to larger sizes as the primates and other mammalian lineages underwent adaptive diversification and increases in body size. This represents a fundamentally central "grade shift" in the evolution of primate encephalization and body size, but here we emphasize not merely the general tranposition, but rather the key adaptive shift in the early small-bodied primates.

It is extremely important to stress that the above comments in no way are meant to imply that all brain size diversification within primate evolution is allometric, or causally related to body size changes. This would surely be an absurd claim. We cannot even say with any certainty that static interspecific scaling coefficients around the 2/3–3/4 value in mammals do not incorporate substantial adaptive increases in relative brain size which are merely correlated with, but not directly related to, increased body size. In other words, there may be considerable adaptive change in function and cognitive complexity *along* the central trend, and not merely away from it, as has often been assumed. We still have no corroborated criterion of "functional equivalence" which might convincingly account for observed exponents. Moreover, the phylogenetic scaling criterion depicted in Figure 9 exceeds the 0.66 − 0.75 allometric coefficient for static interspecific comparisons, and therefore, we must assume that the phylogenetic vector incorporates brain size changes associated with novel adaptive diversification. In fact, we know from quantitative genetic theory that any brain–body coefficient exceeding 0.33 in mammals provides evidence that selection has acted on brain size independent of its covariant developmental and genetic relationship with body size (Lande, 1979, 1985; Riska and Atchley, 1985; Shea, 1983, 2005)—and no slope exceeding the 0.66–0.75 range in a broad static series can be viewed as mere interspecific functional equivalence (Pilbeam and Gould, 1974; Shea, 1983, 2005). The fact that in many Tertiary mammalian lineages the adaptation and diversification of relative brain size and its components occurs *in concert with* adaptive shifts in body size is best viewed as a classic example of potentially misleading spurious correlation (see Fleagle, 1985, for discussion). Jerison (1973) details many of the possible extrinsic factors influencing increase in relative brain size for various mammalian orders throughout the Tertiary. Within primates, the evolution of relatively larger brains in haplorhines compared to early primates and strepsirhines may also be correlated with the evolution of larger average body sizes. But these are general associations and not direct allometric influences. Factors such as complex sociality, extended learning periods, delayed maturity, and increased longevity, themselves perhaps linked to adaptations to new (including terrestrial) habitats, are most likely at least in part causally related to the changes in relative size of the brain and its components (Dunbar, 1998; Kudo and Dunbar, 2001).

PRIMATES AND OTHER MAMMALS

If primates indeed exhibit generally high levels of encephalization because their early relative brain size was translated up to larger sizes during subsequent size and adaptive diversification, do we see comparable trends in other groups of mammals? It is very significant that no other mammalian groups seem to qualify for direct comparison. None of the other candidate groups of small, precocial mammals—such as elephant shrews or hyraxes or selected precocial rodents—ever gave rise to broad and long-term radiations characterized by size increase and adaptive diversification. Other mammalian groups where small body size, moderately large brain size, and reduced litter size (if not full-blown precociality—Shea, 1987) combined are the tree shrews, elephant shrews, bats, and flying lemurs. These specialized lineages have enjoyed variable evolutionary success, but none have further diversified through broad ranges of body size and adaptive zones. *The primates stand alone as the only eutherian mammals to fit this pattern—no other mammals started small, lived slowly, and diversified broadly in terms of body size and adaptive strategies.*

This may be a fundamental key to comparative studies of mammalian brain evolution. While many groups have produced one or multiple taxa which fall with primates in the higher range of relative brain size for mammals, these are generally somewhat atypical for the general pattern of their ordinal relatives. Examples include the seals and otters among the carnivores and dolphins and other taxa among the cetaceans. The fact that we find no other groups of mammals as uniformly above-average in relative brain size as the primates relates to these basal, ordinal adaptations involving the synergistic relationship between precociality and relative brain size in early primates. If we were to engage in a type of theoretical life history analysis and hypothetically select another mammalian order in which comparable potential for generally high encephalization resides, the best candidate would probably be the Macroscelidea. The bizarre and generally nocturnal elephant shrews might not be the first group which springs to mind when considering high encephalization in the mammals, but the evidence of primate evolution suggests there may be tremendous evolutionary potential for such development resident within these small, precocial creatures, which scurry through leaf litter searching for insects and other foods in dry African environments. The chiropterans represent another possibility (Jones and MacLarnon, 2001), though flight has obviously set an adaptive constraint on body size evolution in this order. The earliest cetaceans were likely precocial, and their generally high levels of encephalization are probably related to this in part (Pagel and Harvey, 1988; Shea, 1987), but the order did not originate at small body sizes, and the evolution of extreme size has had a depressing effect on EQ values in this group (Jerison, 1973) and other aquatic mammals (O'Shea and Reep, 1990).

Superordinal relationships within Mammalia are of great interest to many, and attempts to reconstruct these are currently in a state of high activity. I stressed in a previous discussion (Shea, 1987) of encephalization and reproductive strategy in primate evolution that the Archonta of Szalay and Drawhorn (1980), combining primates, bats, tree shrews, and dermopterans, might be linked by some degree of precociality, arboreality, and encephalization as an ancestral adaptive suite of features. Accumulating molecular data indicate that these similarities in reproductive strategy are likely not homologous for the Archonta as thus constituted, since chiropterans consistently root distantly (e.g., Madsen et al., 2001; Murphy et al., 2001). Their life history similarities to primates (Jones and MacLarnon, 2001) would therefore likely not be homologous. However, the grouping of primates with tree shrews and flying lemurs in these and other investigations does suggest the possibility that the common ancestor of these three orders of mammals was characterized by reduced litter size, relatively small adult body size, moderate encephalization, and an arboreal habitus. This is not to blur the real distinctions between primate reproductive strategies and those of the tree shrews (e.g., Martin, 1968, 1969, 1975a,b) and little-known dermopterans (e.g., Wharton, 1950), which obviously manifest their own specializations. But the selective pressures favoring production of a few relatively well-developed offspring in the arboreal environment were possibly operating not only on primates but also on their closest relatives.

The diversity and polarity change of reproductive strategies observed within and across mammalian groups suggests that this character is subject to considerable homoplasy, as might be expected. Nevertheless, it is of some interest to examine what independent and molecular analyses of mammalian higher-level relationships might tell us about the evolution of such life history complexes. The phylogeny of living placental mammals suggested by Madsen et al. (2001) and Murphy et al. (2001) corroborates a close relationship between primates, tree shrews, and flying lemurs. This fits well with a reconstruction of a basal adaptive complex for this triad characterized by a significant degree of precociality, relatively small body size, and moderately high encephalization within an arboreal environment, as noted. Early primates likely further specialized along this trajectory, increasing the precociality and high brain size at small body size.

These molecular phylogenies also support the higher grouping known as Afrotheria, composed of Afrosoricids (tenrecs and African golden moles), macroscelids (elephant shrews), aardvarks (Tubulidentates), and a subgroup comprising sirenians, hyracoids, and proboscideans (Madsen et al., 2001; Murphy et al., 2001). This phylogenetic grouping raises the possibility that the ancestral life history adaptation in the Afrosoricids was one of significant precociality and small litter size (along with moderate-to-high degrees of encephalization), with an important reversal seen within the tenrec (plus golden mole?) component of the clade. It is of interest here that the tenrec radiation runs the gamut from relatively precocial forms (*Microgale talazaci*, with a litter size of 1–2) to highly altricial ones (*Tenrec ecaudatus* with a maximum reported litter size of over 30; see Louwman, 1973). African golden moles generally have small litter sizes (1–2), though data on developmental state of the neonates are unknown to me. Additional evidence is clearly needed, but the generally shared precocial reproductive strategy of most of the Afrotherian species raises interesting issues regarding the evolution of life history strategies in these and other mammalian assemblages. Recent published analyses by Symonds (2005) may greatly clarify these issues.

SUMMARY

These preceding discussions may be summarized through the development of a four-input framework, which accounts for the key elements and variance in relative brain size within the primates, and between primates as a whole and other mammalian groups. This framework is summarized schematically in Figure 10. First, there is the "grade shift" linked to strong precociality and small body size in early primate evolution (Figure 10, arrow #1). Second is the primate-wide phylogenetic trajectory, exceeding typical static interspecific coefficients (Figure 10, line 2). This component of primate brain–body scaling "preserves" the initial grade shift of early primates and transfers it through larger body size ranges during primate evolution. Because the slope of this general phylogenetic trajectory exceeds any reasonable estimate of "functional equivalence," selection on relative brain size is an integral component of this pattern, whatever the specific selective and ecological bases. Among other key developments in primate evolution, features evolved by the anthropoids and associated with diet, habitat, and social structure/complexity were undoubtedly central to this phylogenetic patterning of brain–body scaling.

The third input involves changes in relative brain size that reflect allometric correlates of simple increase or decrease in body size. This case is represented by line #3 in Figure 10. Such cases are typically relevant to more restricted comparisons of phylogenetically linked species, usually on the genus level. Slope values for these groupings are approximately 0.2–0.4, as predicted by quantitative genetic theory (e.g., Lande, 1979) and the observation that such size-variant series of closely related forms typically exhibit interspecific coefficients resembling values for adult intraspecific variation (Gould, 1975a; Pilbeam and Gould, 1974; Shea, 1983). These phylogenetically restricted comparisons result in marked changes in relative brain size (or EQ's)

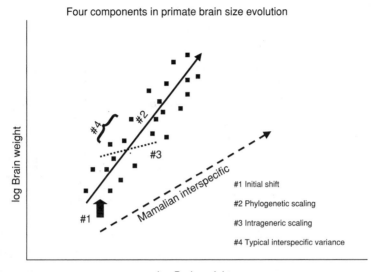

Figure 10. A schematized summary of four primary inputs to the patterning of brain-body values in primate evolution. The initial increase (arrow #1) in relative brain size in early primates is translated up to larger sizes via phylogenetic scaling (solid line, trajectory #2). Intrageneric scaling (dashed line, trajectory #3) characterizes simple allometric transformations among closely related species differing in body size. The substantial residual variance about all scaling trajectories is represented here by #4 and the individual data points. See text for additional discussion.

determined in comparison to broad interspecific patterns (i.e., 0.66–0.75 best-fit trajectories). Specific examples within the primates include the elevated EQ for talapoin monkeys, argued by many to represent a dwarfed *Cercopithecus* (Bauchot and Stephan, 1969; Gould, 1975a; Shea, 1983, 1992b), and the depressed EQ for *Gorilla*, viewed by most as a phyletically enlarged African ape (Gould, 1975a; Pilbeam and Gould, 1974; Shea, 1983). Attempts to establish increased or decreased brain size relative to the expectations of size diversification alone, or efforts to reconstruct relative brain size in specific fossil taxa, should utilize the 0.2–0.4 baseline criterion and the phylogenetically most relevant sample of near relatives (Shea, 2005). Strong supporting evidence for this claim has been provided by Williams (2002), who demonstrated a much more precise fit between literature-based assessments of learning ability and EQ determined relative to a log–log coefficient of 0.28 for anthropoids, than between the cognitive assessments and EQ values determined using 0.67 or 0.75 slopes.

Finally, the fourth input in this framework is the key variance about any broad interspecific lines of best-fit, whether these are static or phylogenetic scaling trends. These real increases or decreases in total relative brain size or its component parts—as determined against broad slopes of best-fit (Figure 10, lines 4a) or within specific ancestor–descendant comparisons (Figure 10, lines 4b)—reflect direct selection for altered relative brain size and function unrelated to allometric changes associated with body size. Likely bases include changes in diet, habitat use, social organization, etc., as discussed in various important papers on primate evolution too numerous to list here. The fact that the present chapter differentially focuses on grade shifts and scaling patterns in no way denies recognition of this key variance, or the selective and ecological contexts for such substantial changes in relative brain size and its component parts. For many primatologists, these changes are the most interesting in studies of brain–body relationships; indeed, the initial shift argued here to be related to precociality and small body size provides but one important example. Unfortunately, even when such acknowledgments are explicitly offered, some critics may insist on caricaturing allometric studies and chastising researchers for ignoring the existence and biological significance of variance about general scaling trajectories. One such example is the reaction to the papers by Finlay and colleagues (Finlay and Darlington, 1995; Finlay et al., 2001). In spite of the fact that the stated primary emphasis of these workers was on the internal scaling generalities and developmental sequencing of brain size, and proportions during mammalian ontogeny and phylogeny, they did very clearly acknowledge the existence of substantial important residual variance which many previous studies had cogently shown to be correlated with key taxonomic, ecological, and functional inputs in primates and other mammals. Nevertheless, several papers (e.g., Barton and Harvey, 2000; Clark et al. 2001; de Winter and Oxnard, 2001; Rilling and Insel, 1998) have erroneously argued or implied that Finlay and colleagues in fact viewed all meaningful brain size and proportion variance as rigidly linked solely to developmental and allometric constraints. In light of these critiques, we must redundantly stress here that such nonallometric variance is both pervasive and central to any general explanations of relative brain size in primates and other mammals. In the framework presented here, it is represented by the fourth primary input into overall variance in primate brain size. The goals of accounting for *both* general scaling trends of brain–body associations, as well as key residual variance in brain size from these broad patterns, are in no way contradictory or substantially at odds. This reasonable and synthetic perspective has been emphasized by Kass and Collins (2001) in regard to the flurry of interest and commentary spurred by the Finlay and Darlington's (1995) paper.

In conclusion, we return to Gould's prescient insight quoted at the beginning of this chapter. A central component of the generally high levels of encephalization observed in our order was indeed the high relative brain size of the early primates. This high encephalization is seen here as significantly related to the highly precocial reproductive strategy, and well-developed neonates and young of the early primates. Their adoption of a life history strategy characterized by marked precociality at small body size proved to be a highly successful, if unusual, evolutionary development for mammals. It also served as a key basis for continued selection for high encephalization and complex sociality in subsequent primate evolution, as Vaughan et al. (2000: 356) stress when they note that "the evolution of an early-primate reproductive pattern involving long gestation and precocial young may have been critical in setting the stage for the highly social lives of higher primates."

ACKNOWLEDGMENTS

I would like to thank Matt Ravosa and Marian Dagosto for their invitation to participate in the primate origins—symposium and volume. The ideas expressed in this paper were considerably influenced by comments offered by Stephen Gould, Bob Martin and John Eisenberg, and I am grateful to have had access to their input as well as their publications.

REFERENCES

Armitage, K. B., 1981, Sociality as a life-history tactic of ground squirrels, *Oecologia* **48**: 36–49.

Armstrong, E., 1983, Relative brain size and metabolism in mammals, *Science* **220**: 1302–1304.

Arnold, S. J., 1983, Morphology, performance, and fitness, *Am. Zool.* **23**: 347–361.

Asdell, S. A., 1964, *Patterns of Mammalian Reproduction*, 2nd Ed., Cornell University Press, New York.

Austad, S. N., and Fischer, K. E., 1992, Primate longevity: Its place in the mammalian scheme, *Am. J. Primatol.* **28**: 251–261.

Barton, R. A., and Harvey, P. H., 2000, Mosaic evolution of brain structure in mammals, *Nature* **405**: 1055–1058.

Bauchot, R., and Stephan, H., 1966, Donnees nouvelles sur l'encephalisation des insectivores et des prosimiens, *Mammalia* **30**: 160–196.

Bauchot, R., and Stephan, H., 1969, Encephalisation et niveau evolutif chez les simiens, *Mammalia* **32**: 228–275.

Bennett, P. M., and Harvey, P. H., 1985, Brain size, development and metabolism in birds and mammals, *J. Zool., Lond.* **207**: 491–521.

Bloch, J. I., and Boyer, D. M., 2002, Grasping primate origins, *Science* **298**: 1606–1610.

Brunjes, P. C., 1990, The precocial mouse *Acomys cahirinus, Psychobiology* **18**: 339–350.

Cartmill, M., 1972, Arboreal adaptations and the origin of the order Primates, in: *The Functional and Evolutionary Biology of the Primates,* R. Tuttle, ed., Aldine, Chicago, pp. 97–122.

Cartmill, M., 1974 Rethinking primate origins, *Science* **184**: 436–443.

Case, T. J., 1978, On the evolution and adaptive significance of postnatal growth rates in terrestrial vertebrates, *Q. Rev. Biol.* **53**: 243–282.

Charnov, E. L., 1993, *Life History Invariants: Some Explorations of Symmetry in Evolutionary Ecology,* Oxford, Oxford University Press.

Charnov, E. L., and Berrigan, D., 1993, Why do female primates have such long life spans and so few babies? or Life in the slow lane, *Evol. Anthropol.* **2**: 191–194.

Clark, D. A., Mitra, P. P., and Wang, S. S.-H., 2001, Scalable architecture in mammalian brains, *Nature* **411**: 189–193.

Conroy, G. C., 1990, *Primate Evolution,* Norton, New York.

Derrickson, E. M., 1992, Comparative reproductive strategies of altricial and precocial eutherian mammals, *Funct. Ecol.* **6**: 57–65.

de Winter, W., and Oxnard, C. E., 2001, Evolutionary radiations and convergences in the structural organization of mammalian brains, *Nature* **409**: 710–714.

Dunbar, R. I. M., 1998, The social brain hypothesis, *Evol. Anthropol.* **6**: 178–190.

Eisenberg, J. F., 1975, Phylogeny, behavior and ecology in the Mammalia, in: *Phylogeny of the Primates,* W. P. Luckett and F. S. Szalay, eds., Plenum Press, New York, pp. 47–68.

Eisenberg, J. F., 1981, *The Mammalian Radiations,* University of Chicago Press, Chicago.

Eisenberg, J. F., and Wilson, D. E., 1981, Relative brain size and demographic strategies in didelphid marsupials, *Am. Nat.* **118**: 1–15.

Finlay, B. L., and Darlington, R. B., 1995, Linked regularities in the development and evolution of mammalian brains, *Science* **268**: 1578–1584.

Finlay, B. L., Darlington, R. B., and Nicastro, N., 2001, Developmental structure in brain evolution, *Behav. Brain Sci.* **24**: 263–278.

Fleagle, J. F., 1978, Size distributions of living and fossil primate faunas, *Paleobiology* **4**: 67–76.

Fleagle, J. F., 1985, Size and adaptation in primates, in: *Size and Scaling in Primate Biology,* W. L. Jungers, ed., Plenum Press, New York, pp. 1–20.

Fleagle, J. F., 1988, *Primate Adaptation and Evolution,* Academic Press, San Diego, CA.

Gebo, D. L., Dagosto, M., Beard, K. C., and Qi, T., 2000, The smallest primates, *J. Hum. Evol.* **38**: 585–594.

Gould, S. J., 1966, Allometry and size in ontogeny and phylogeny, *Biol. Rev.* **41**: 587–640.

Gould, S. J., 1971, Geometric similarity in allometric growth: A contribution to the problem of scaling in the evolution of size, *Am. Nat.* **105**: 113–136.

Gould, S. J., 1975a, Allometry in primates, with emphasis on scaling and evolution of the brain, in: *Approaches to Primate Paleobiology (Contr. Primatol. 5),* F. S. Szalay, ed., Karger, Basel, pp. 244–292.

Gould, S. J., 1975b, Evolution and the brain, *Nat. Hist.* **84**(1): 24–26.

Gould, S. J., 1976, Human babies as embryos, *Nat. Hist.* **85**(2): 22–26.

Gurche, J. A., 1982, Early primate brain evolution, in: *Primate Brain Evolution,* E. Armstrong and D. Falk, eds., Plenum Press, New York, pp. 227–246.

Hamrick, M. W., 1999, Pattern and process in the evolution of primate nails and claws, *J. Hum. Evol.* **37**: 293–297.

Hamrick, M. W., 2001, Primate origins: Evolutionary change in digital ray patterning and segmentation, *J. Hum. Evol.* **40**: 339–351.

Hennemann, W. W., 1983, Relationship among body mass, metabolic rate, and the intrinsic rate of natural increase in mammals, *Oecologia* **56**: 104–108.

Hofman, M., 1983, Evolution of brain size in neonatal and adult placental mammals: A theoretical approach, *J. Theor. Biol.* **105**: 317–332.

Huxley, J. S., 1932, *Problems of Relative Growth,* Methuen, London.

Jerison, H. J., 1973, *Evolution of the Brain and Intelligence,* Academic Press, New York.

Jerison, H. J., 1979, Brain, body and encephalization in early primates, *J. Hum. Evol.* **8**: 615–635.

Jones, K. E., and MacLarnon, A., 2001, Bat life histories: Testing models of mammalian life history evolution, *Evol. Ecol. Res.* **3**: 465–476.

Kappeler, P. M., 1995, Life history variation among nocturnal prosimians, in: *Creatures of the Dark. The Nocturnal Prosimians,* L. Alterman, G. A. Doyle, and M. K. Izard, eds., Plenum Press, New York, pp.75–92.

Kappeler, P. M., 1996, Causes and consequences of life-history variation among strepsirhine primates, *Am. Nat.* **148**: 868–891.

Kappeler, P. M., 1998, Nests, tree holes, and the evolution of primate life histories, *Am. J. Primatol.* **46**: 7–33.

Kappeler, P. M., Pereira, M. E., and van Shaik, C. P., 2003, Primate life histories and socioecology, in: *Primate Life Histories and Socioecology*, P. M. Kappeler and M. E. Pereira, eds., University of Chicago Press, Chicago, pp. 1–24.

Kass, J. H., and Collins, C. E., 2001, Evolving ideas of brain evolution, *Nature* **411:** 141–142.

Kudo, H., and Dunbar, R. I. M., 2001, Neocortex size and social network size in primates, *Anim. Behav.* **62:** 711–722.

Lande, R., 1979, Quantitative genetic analysis of multivariate evolution, applied to brain:body size allometry, *Evolution* **33:** 402–416.

Lande, R., 1985, Genetic and evolutionary aspects of allometry, in: *Size and Scaling in Primate Biology*, W. L. Jungers, ed., Plenum Press, New York, pp. 21–32.

LeGros Clark, W. E., 1971, *The Antecedents of Man*, 3rd Ed., Quadrangle, Chicago.

Lemen, C., 1980, Relationship between relative brain size and climbing ability in *Peromyscus, J. Mammal.* **61:** 360–364.

Louwman, J. W. W., 1973, Breeding of the tailless tenrec (*Tenrec ecaudatus*) at Wassenaar Zoo. *Intl. Zoo Yrbk.* **13:** 125–126.

Luckett, W. P., 1980, The use of reproductive and developmental features in assessing tupaiid affinities, in: *Comparative Biology and Evolutionary Relationships of Tree Shrews*, W. P. Luckett, ed., Plenum Press, New York, pp. 245–268.

MacArthur, R. H., and Wilson, E. O., 1967, *The Theory of Island Biogeography*, Princeton University Press, Princeton, NJ.

Madsen, O., Scally, M., Douady, C. J., Kao, D. J., DeBry, R. W., Adkins, R. et al., 2001, Parallel adaptive radiations in two major clades of placental mammals, *Nature* **409:** 610–614.

Marino, L., 1996, What can dolphins tell us about primate evolution? *Evol. Anthropol.* **5:** 81–86.

Marino, L., 1997, The relationship between gestation length, encephalization and body weight in odontocetes, *Mar. Mammal. Sci.* **14:** 143–148.

Martin, R. D., 1968, Reproduction and ontogeny in tree-shrews (*Tupaia belangeri*) with reference to their general behavior and taxonomic relationships, *Z. Tierpsychol.* **25:** 409–532.

Martin, R. D., 1969, The evolution of reproductive mechanisms in primates, *J. Reprod. Fertil.* **6**(Suppl.): 49–66.

Martin, R. D., 1972, Adaptive radiation and behavior of the Malagasy lemurs, *Phil. Trans. R. Soc. Lond., B* **264:** 295–352.

Martin, R. D., 1973, Comparative anatomy and primate systematics, *Symp. Zool. Soc. London* **33:** 301–377.

Martin, R. D., 1975a, Strategies of reproduction, *Nat. Hist.* **84**(11): 48–57.

Martin, R. D., 1975b, The bearing of reproductive behavior and ontogeny on strepsirhine phylogeny, in: *Phylogeny of the Primates*, W. P. Luckett and F. S. Szalay, eds., Plenum Press, New York, pp. 47–68.

Martin, R. D., 1981, Relative brain size and basal metabolic rate in terrestrial vertebrates, *Nature* **293:** 57–60.

Martin, R. D., 1983, Human brain evolution in an ecological context, *Fifty-second James Arthur Lecture on the Evolution of the Human Brain,* American Museum of Natural History, New York.

Martin, R. D., 1989, Size, shape and evolution, in: *Evolutionary Studies. A Centenary Celebration of the Life of Julian Huxley,* M. Keynes and G. A. Harrison, eds., *Proc. 24th Ann. Symp. Eugenics Soc. Lond., 1987,* MacMillan, London.

Martin, R. D., 1990, *Primate Origins and Evolution,* Princeton University Press, Princeton, NJ.

Martin, R. D., and MacLarnon, A. M., 1985, Gestation period, neonatal size and maternal investment in placental mammals, *Nature* **313:** 220–223.

Martin, R. D., and MacLarnon, A. M., 1988, Comparative quantitative studies of growth and reproduction, *Symp. Zool. Soc. Lond.* **60:** 39–80.

Meier, P. T., 1983, Relative brain size within the North American Sciuridae, *J. Mammal.* **64:** 642–647.

Murphy, W. J., Eizirik, E., O'Brien, S. J., Madsen, O., Scally, M., Douady, C. J., Teeling, E., Ryder, O. A., Stanhope, M. J., De Jong, W. W., and Springer, M. S., 2001, Resolution of the early placental mammal radiation using Bayesian phylogenetics. *Science* **294:** 2348–2351.

O'Shea, T. J., and Reep, R. L., 1990, Encephalization quotients and life-history traits in the Sirenia, *J. Mammal.* **71:** 534–543.

Pagel, M. D., and Harvey, P. H., 1988, How mammals produce large-brained offspring, *Evolution* **42:** 948–957.

Pilbeam, D., and Gould, S. J., 1974, Size and scaling in human evolution, *Science* **186:** 892–901.

Pirlot, P., and Stephan, H., 1970, Encephalization in *Chiroptera, Can. J. Zool.* **48:** 433–444.

Portmann, A., 1939, Die Ontogenese er Saugetiere als Evolutionsproblem, *Biomorphol.* **1:** 109–126.

Portmann, A., 1965, Uber die Evolution der Tragzeit bei Saugetieren, *Rev. Suisse Zool.* **72:** 658–666.

Promislow, D. E. L., and Harvey, P. H., 1990, Living fast and dying young: A comparative analysis of life-history variation among mammals, *J. Zool., Lond.* **220:** 417–437.

Purvis, A., Webster, A. J., Agapow, P.-M., Jones, K. E., and Isaac, N. J. B., 2003, Primate life history and phylogenies, in: *Primate Life Histories and Socioecology,* P. M. Kappeler and M. E. Pereira, eds., University of Chicago, Chicago, pp. 25–40.

Radinsky, L. B., 1970, The fossil evidence of prosimian brain evolution, in: *The Primate Brain,* C. R. Noback and W. Montagna, eds., Appleton, New York, pp. 209–224.

Radinsky, L. B., 1975, Primate brain evolution, *Am. Sci.* **63**: 656–663.

Radinsky, L. B., 1977, Early primate brains: Facts and fiction, *J. Hum. Evol.* **6**: 79–86.

Radinsky, L. B., 1978, Evolution of brain size in carnivores and ungulates, *Am. Nat.* **112**: 815–831.

Radinsky, L. B., 1979, The fossil record of primate brain evolution, *Forty-Ninth James Arthur Lecture on the Evolution of the Human Brain,* American Museum of Natural History, New York.

Rasmussen, D. T., 1990, Primate origins: Lessons from a neotropical marsupial, *Am. J. Primatol.* **22**: 263–277.

Read, A. F., and Harvey, P. H., 1989, Life history differences among the eutherian radiations, *J. Zool., Lond.* **219**: 329–353.

Ricklefs, R. E., and Starck, J. M., 1998, Evolution of developmental modes in birds, in: *Avian Growth and Development, Evolution Within the Altricial-Precocial Spectrum,* J. M. Starck and R. E. Ricklefs, eds., Oxford University Press, New York, pp. 366–380.

Rilling, J. K., and Insel, T. R., 1998, Evolution of the cerebellum in primates: Differences in relative volume among monkeys, apes and humans, *Brain Behav. Evol.* **52**: 308–314.

Ripley, S., 1967, The leaping of langurs: A problem in the study of locomotor adaptation, *Am. J. Phys. Anthropol.* **26**: 149–170.

Riska, B., and Atchley, W. R., 1985, Genetics of growth predict patterns of brain-size evolution, *Science* **229**: 668–671.

Sacher, G., and Staffeldt, M., 1974, Relation of gestation time to brain weight for placental mammals: Implications for the theory of vertebrate growth, *Am. Nat.* **108**: 593–615.

Sargis, E. J., 2002, Primate origins nailed, *Science* **298**: 1564–1565.

Shea, B. T., 1983, Phyletic size change and brain:body allometry: A consideration based on the African pongids and other primates, *Int. J. Primatol.* **4**: 33–62.

Shea, B. T., 1987, Reproductive strategies, body size, and encephalization in primate evolution, *Int. J. Primatol.* **8**: 139–156.

Shea, B. T., 1988, Heterochrony in primates, in: *Heterochrony in Evolution: A Multidisciplinary Approach,* M. L. McKinney, ed., Plenum Press, New York, pp. 237–266.

Shea, B. T., 1990, Dynamic morphology: Growth, life history, and ecology in primate evolution, in *Primate Life History and Evolution,* C. J. DeRousseau, ed., Wiley-Liss, New York, pp. 325–352.

Shea, B. T., 1992a, A developmental perspective on size change and allometry in evolution, *Evol. Anthropol.* **1**: 125–134.

Shea, B. T., 1992b, Ontogenetic scaling of skeletal proportions in the talapoin monkey, *J. Hum. Evol.* **23**: 283–307.

Shea, B. T., 2005, Brain/body allometry: using extant apes to establish appropriate scaling baselines, *Am. J. Phys. Anthro.* **126**: S40: 189.

Silcox, M. T., 2002, Paleoprimatology at the Society of Vertebrate Paleontology, *Evol. Anthropol.* **11**: 1–3.

Soligo, C., and Muller, A. E., 1999, Nails and claws in primate evolution, *J. Hum. Evol.* **36**: 97–114.

Stephan, H., 1972, Evolution of primate brains: A comparative anatomical investigation, in: *The Functional and Evolutionary Biology of Primates*, R. H. Tuttle, ed., Aldine, Chicago, pp. 155–174.

Stephan, H., Bauchot, R., and Andy, O., 1970, Data on the size of the brain and of various parts in insectivores and primates, in: *The Primate Brain*, C. Noback and W. Montagna, eds., Appleton-Century-Crofts, New York, pp. 289–297.

Sussman, R. W., 1991, Primate origins and the evolution of angiosperms, *Am. J. Primatol.* **23**: 209–223.

Symonds, M. R. E., 2005, Phylogeny and life histories of the 'Insectivora': controversies and consequences. *Biol. Rev.* **80**: 3–128.

Szalay, F. S., and Dagosto, M., 1988, The evolution of hallucial grasping in primates, *J. Hum. Evol.* **17**: 1–33.

Szalay, F. S., and Drawhorn, G., 1980, Evolution and diversification of the Archonta in an arboreal milieu, in: *Comparative Biology and Evolutionary Relationships of the Tree Shrews*, W. P. Luckett, ed., Plenum Press, New York, pp. 133–170.

van Shaik, C. P., and Deaner, R. O., 2002, Life history and cognitive evolution in primates, in: *Animal Social Complexity*, F. B. M de Waal and P. L. Tyack, eds., Harvard University Press, Cambridge, MA.

Vaughan, T. A., Ryan, J. M., and Czaplewski, N. J., 2000, *Mammalogy*, 4th Ed., Saunders College Publishing, Orlando, FL.

Walker, E. P., Warnick, F., Hamlet, S. E., Lange, K. I., Davis, M. A., Uible, H. E. et al., 1975, *Mammals of the World*, 3rd Ed., Johns Hopkins University Press, Baltimore.

Wharton, C. H., 1950, Notes on the life history of the flying lemur, *J. Mammal.* **31**: 269–273.

White, J. F., and Gould, S. J., 1965, Interpretation of the coefficient in the allometric equation, *Am. Nat.* **99**: 5–18.

Williams, M. F., 2002, Primate encephalization and intelligence, *Med. Hypotheses* **58**: 284–290.

Youlatos, D., and Godinot, M., 2004, Locomotor adaptations of *Plesiadapis* (Mammalia: Plesiadapiformes) as reflected in selected parts of the postcranium, *Folia Primatol.* (suppl. 75): 352 [Abstract].

Zeveloff, S. I., and Boyce, M. S., 1986, Maternal investment in mammals, *Nature* **321**: 537–538.

CHAPTER EIGHTEEN

Evolutionary Specializations of Primate Brain Systems

Todd M. Preuss

INTRODUCTION

Primates are distinguished from other mammals by a number of anatomical features, including convergent, close-set orbits; enlarged eyes; digits tipped with nails rather than claws; opposable hallux; and elongated calcaneus (Cartmill, 1992; Martin, 1990; Szalay et al., 1987). These shared, derived characters (synapomorphies) are generally thought to have arisen as adaptations in ancestral (stem) primates for nocturnal activity in the fine, terminal branches of trees (Martin, 1990). The behaviors that drove the evolution of primate anatomical synapomorphies remain at issue, with proposals including visually guided predation on insects and small vertebrates (Allman, 1977; Cartmill, 1972, 1974), foraging on fruits and flowers, in addition to predation (Rasmussen, 1990; Sussman, 1991), and a hindlimb-dominated "graspleaping" locomotor pattern (Szalay and Dagosto, 1988).

If evolutionary history left its imprint on the primate body, what mark did it leave on the brain? Traditionally, studies of primate brain evolution have focused on changes in brain size and external morphology. Size and external morphology give little indication of evolutionary changes in internal brain organization, however, and, if modern neuroscience teaches us anything, it is

Todd M. Preuss • Yerkes National Primate Research Center, Emory University, Atlanta, GA 30329

that brains are internally structured and compartmentalized to an extraordinary degree. For example, mammalian cerebral cortex is comprised of several large histological domains (iso-, archi-, and paleocortex), each of which consists of multiple smaller divisions (cortical areas), which are composed of smaller, repeating units (modules), which are themselves composed of even smaller, vertically oriented neuronal clusters (minicolumns). Each of these structural compartments is composed of neurons that are connected to other neurons within the same compartment in very particular ways, and which are connected to the neurons of other cortical compartments in very particular ways. Far from being a diffuse neural net, as once was thought, mammalian cerebral cortex exhibits a degree of internal structural complexity unrivaled by any other biological tissue.

Given such a view of the brain, it is easy to imagine how evolution might have modified its organization. One might expect, for example, that particular lineages evolved new compartments (areas or modules), or that ancestral mammalian compartments were reorganized internally in distinctive ways in different taxa, or that new systems of connections between compartments evolved in certain groups. In fact, many changes like these did occur during mammalian evolution (reviewed by Preuss, 2000, 2001).

We face a considerable obstacle, however, in attempting to reconstruct the brain changes that accompanied primate origins—the paucity of comparative information. As a group, neuroscientists are inclined to concentrate their research on a few model animal taxa, under the assumption that the important features of mammalian brain organization are shared among mammals. In keeping with the modern biomedical research paradigm, neuroscientists have come to treat animals as standardized materials for exploring the organization of "the brain"—as though there were only *one* brain—rather than as resources for exploring the diversity of brains (Logan, 1999, 2001; Preuss, 2000; see also Raff, 1996). The result is while the brains of a select few taxa—especially rats and macaque monkeys—have been studied in great detail; much less effort has been devoted to other mammalian taxa. This places great limitations on our ability to reconstruct primate brain evolution.

Fortunately, neuroscience has maintained a small cadre of investigators willing to buck the trend, maintaining an interest in evolution and a curiosity about differences in mammalian brain organization. A central figure in this group is Irving T. Diamond (see, for example, Diamond, 1973; Diamond and Hall, 1969). Diamond was greatly influenced by the writings of Elliot Smith and Le Gros Clark, and devoted himself to developing an evolutionary neuroscience of mammals. Moreover, he conveyed his enthusiasm for evolution to the very talented graduate students he attracted. Diamond, his students, and their associates have been responsible for a remarkable fraction of all the comparative studies of mammalian brain organization that have been published over the last 35 years. It is largely due to the efforts of these individuals that we know anything at all about the brains of strepsirhine primates, tree shrews, bats, and insectivores, and are therefore in a position to say something about the evolution of primate brain systems.

This review is founded upon earlier works focusing on brain evolution at the level of structures and systems, particularly those of Allman (1977), Kaas (1980), Kaas and Preuss (1993), Pettigrew (1986; Pettigrew et al., 1989), Preuss (1993, 1995a), and Preuss and Kaas (1999). However, effort has been made to provide a more comprehensive and synthetic treatment of the subject of primate brain specializations than has previously been attempted. Many readers will, I fear, find the enumeration of neuroanatomical details rather *too* comprehensive for their liking. This degree of specificity is necessary, however, to illustrate the great number and variety of changes in brain organization that accompanied primate origins. Moreover, the very fact that neural changes were so numerous has important implications for students of behavioral evolution.

SPECIALIZATIONS OF THE VISUAL SYSTEM

Overview

Discussion of visual system evolution requires a very brief review of the anatomy of the visual system (for more details, see especially Kaas and Huerta, 1988; Kaas et al., 1978). Visual information reaches the brain by way of projections from the ganglion cells of the retina. Retinal ganglion cells are of several types, with different morphologies and physiological properties. Currently, three main classes of ganglion cells are distinguished, usually called M, P, and K cells in primates. This designation reflects the fact that these cells project to separate *magnocellular*, *parvocellular*, and *koniocellular* layers of the lateral geniculate nucleus. It is likely that nonprimate mammals have cells homologous to the M, P, and K cells (usually termed Y, X, and W cells, respectively, in nonprimates).

M cells have large cell bodies and dendritic fields; they have large visual receptive fields that integrate information from both rods and cones and respond well to light over a broad range of the visual spectrum. M cells have good temporal resolution, which is to say they respond well to flickering or moving stimuli, but have relatively coarse spatial resolution, so they are not well suited for fine visual discrimination. P cells are smaller than M cells, and have smaller receptive fields that integrate inputs from cones; they respond well to light from a more restricted part of the spectrum than M cells (i.e., they are relatively wavelength selective), have poor temporal resolution (they track stimulus onset and offset poorly), but have good spatial resolution. P cells and their efferent targets process color information in diurnal anthropoids, although they must do more than this, because P cells are present in nocturnal primates, and nocturnal primates lack the density of cones and diversity of cone types necessary to support color-opponent processing, which is the basis of fine color discrimination in diurnal anthropoids (Dkhissi-Benyahya et al., 2001). K cells are poorly characterized at present; they seem to be intermediate between M and P cells morphologically and physiologically. Their heterogeneity suggests that the K class is actually a composite of multiple cell classes. Interestingly, the short-wavelength photoreceptors (S cones), which respond maximally to light in the blue part of the spectrum, have recently been indicated (in anthropoids) to have a privileged anatomical relationship to K cells rather than to P cells (Dacey, 2000), which challenges the conventional idea that color processing is carried out exclusively in the P pathway. It is unclear at present whether all S cones are related to K cells and whether all K cells receive S-cone input.

Retinal ganglion cells project to many structures in the brainstem, the strongest projections targeting the lateral geniculate nucleus (LGN) of the thalamus and the superior colliculus (SC) in the midbrain (Figure 1). The LGN contains separate cell types linked to each of the different retinal ganglion cell types, and these cells are segregated into distinct strata or laminae. In primates, there are separate magnocellular and parvocellular layers, which receive inputs from the M and P retinal ganglion cells, respectively. These layers come in pairs, so that separate M and P layers represent each eye. The layers sometimes subdivide further, as in humans, in which it is conventional to recognize six main LGN layers: one pair of magnocellular layers and two pairs of parvocellular layers (Kaas et al., 1978). In strepsirhine primates, the koniocellular layers also form pairs of layers that segregate input from the two

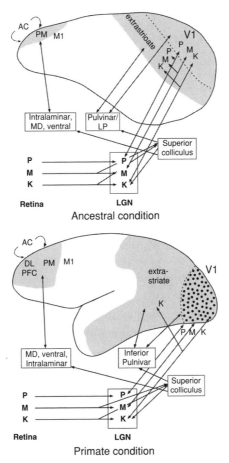

Figure 1. Schematic representation of the distribution of visual information to the thalamus, superior colliculus, and cortex, illustrating ancestral organization and derived characteristics of primates. Visual information originating from the P, M, and K cells of the retina reaches the thalamic lateral geniculate nucleus (LGN) and the superior colliculus. The lateral geniculate nucleus (LGN) relays information from the retina and superior colliculus to the primary visual area (V1) and to extrastriate visual cortex. In primates, the K layers of the LGN target specialized compartments within area V1, the blobs (represented here as an array of dark spots). The LGN also projects to extrastriate cortex, although these projections are weak in primates and appear to arise primarily from the K layers of the LGN. Extrastriate cortex receives additional projections from inferior pulvinar of the thalamus in primates, and from its homologue in nonprimates, known as the pulvinar/lateral posterior (LP) nucleus. The inferior pulvinar receives major inputs from the superficial (visual) layers of the superior

(*Continued*)

retinas; the situation is less clear in anthropoids, because their K layers are not all distinctly separate from the M and P layers (Kaas et al., 1978).

The pattern of LGN lamination varies considerably among mammals (reviewed by Kaas and Preuss, 1993; Kaas et al., 1972; Sanderson, 1986). Cell types that are mixed in primates can be segregated into separate laminae in other taxa. On current evidence, primates are the only mammals in which P cells are known to be completely segregated from M cells. The pattern of lamination in dermopterans and chiropterans is not well understood, however, and they may have a primate-like pattern, as indicated by Pettigrew et al. (1989).

Each geniculate layer receives projections from one hemiretina, which form a map of the contralateral half of the visual scene. So, for example, the left LGN contains a stack of maps of the right visual scene; half of these maps are formed by inputs from the left retina, half from the right retina. Since the right side of the visual scene is projected onto the left side of each retina, the left LGN receives its projections from the left (temporal) side of the left retina and from the left (nasal) side of the right retina. The fibers from the nasal hemiretinas cross the midline on their way to the thalamus, forming the optic chiasm; projections from the temporal hemiretinas are uncrossed.

In most nonprimate mammals, which have eyes set on the side of the head, the left eye sees mainly the left visual field and the right eye the right visual field; there is only a relatively small region of binocular vision. In these animals, the LGN on each side of the brain receives its major input from the contralateral retina; only a small projection arises from the ipsilateral retina, corresponding to the field of binocular overlap. Primates, by contrast, have forward facing eyes and consequently a large region of overlap between the images cast onto the left and right retinas. Correspondingly, the projections of the ipsilateral retina to the LGN are nearly as numerous as the projections from the contralateral LGN.

Figure 1. (*Continued*) colliculus. The deeper layers of the superior colliculus project to thalamic nuclei connected mainly with frontal cortex. These include the mediodorsal nucleus (MD) and the intralaminar and ventral thalamic nuclei, which project primarily to premotor cortex (PM) and dorsolateral prefrontal cortex (DLPFC), located on the lateral surface of the frontal lobe, and to the anterior cingulate (AC) cortex, located on the medial wall of the frontal lobe. In primates, the strongest tectal projections reach MD and the medial division of the ventral anterior nucleus; these nuclei project primarily to DLPFC. In nonprimates, the intralaminar nuclei are the main targets of tectal projections. Additional abbreviations: M1 = primary motor area.

The second major target of retinal projections is the superior colliculus (SC), a layered structure occupying the roof (or "tectum") of the midbrain, posterior to the thalamus (Huerta and Harting, 1984). Only the M and K ganglion cells project to the SC, where they terminate most densely in its superficial layers. Retinal terminations are arranged in an orderly, topographic manner. Inputs from each eye are at least partially segregated, although the patterning of segregation is more complex than the simple laminar pattern of the LGN, and unlike the LGN, some tectal cells have binocular inputs. Moreover, the SC is not solely a visual structure—the middle and deep layers receive inputs from the auditory and somatosensory systems—and the auditory and somatosensory maps are in spatial register with the visual map of the superficial layers.

The deep layers of the SC play an important role in organizing orienting movements, mediating the so-called "visual grasp reflex" by which animals rapidly shift their head and eyes so that the image of potentially significant objects fall onto the central retina, where resolving power is greatest. The role of SC as an eye-movement control center is emphasized especially by researchers who study anthropoid primates, animals that make unusually large-amplitude eye movements. In vertebrate neurobiology more generally, however, the SC is seen as a structure that, among other things, coordinates rapid orienting movements of the head, eyes, pinnae, and perioral face region (Dean et al., 1989). There is, furthermore, evidence for forelimb representation in the colliculi of nonprimate vertebrates, and it is significant that recent studies by Werner and colleagues in macaque monkeys implicate SC in the control of proximal forelimb movements associated with reaching (e.g., Werner et al., 1997).

The SC influences other brain regions by several routes (Huerta and Harting, 1984). Descending projections target the oculomotor nuclei and other motor structures in the brainstem and upper cervical spinal cord. Ascending projections target the thalamus and other diencephalic structures, as illustrated in Figure 1. The upper layers of the colliculus project to thalamic nuclei that provide inputs to visual cortex—the LGN (specifically its K layers) and inferior portions of the pulvinar. Deep layers of the SC project to more anterior structures of the thalamus, particularly the intralaminar, ventral (VA), and mediodorsal (MD) nuclei. The projections to the MD and VA nuclei, which project in turn to frontal cortex, may be especially strong in primates (Huerta and Harting, 1984).

Retinotectal Organization

With this background, we can now consider the best-known specialization of primate brain structure, namely, the distinctive retinotopic organization of the primate superior colliculus (Figure 2). The term *retinotopy* refers to the way retinal information is mapped onto visual structures. In primates, the pattern of retinal projections to SC and the resulting visual field representation in SC are similar to the pattern exhibited by the LGN: each SC contains a complete representation of the contralateral visual field. This representation is supported by major projections from the nasal retina of the contralateral eye and from the temporal retina of the contralateral eye. There is no substantial representation of the ipsilateral visual field in the SC of primates: the representation of the vertical meridian is located at the anterior limit of the colliculus. This pattern of organization has been found in every primate that has been examined with microelectrode mapping methods or by tracing projections from the retina to the SC, a sample that includes numerous New World and Old World anthropoid species, as well as lorisid strepsirhines (see especially Kaas and Huerta, 1988, for a review of the published literature).

Although the primate condition was once thought to be typical of mammals, work in the early 1970s established that the primate condition is unusual. In most nonprimate mammals examined, the visual representation in SC crosses the vertical meridian to include a significant portion of the ipsilateral visual field, in addition to the contralateral visual field. Moreover, rather than receiving nearly equal projections from both retinas, SC projections in nonprimate mammals arise mainly from the contralateral retina, and span a territory that includes both the temporal and nasal hemiretinas, with a relatively small contribution from the ipsilateral retina. This type of organization has been found in a variety of mammalian taxa, including tree shrews, rodents, lagomorphs, artiodactyls, perissodactyls, carnivores, marsupials, and monotremes (see Allman, 1977, Kaas and Huerta, 1988, Pettigrew, 1986, and Rosa and Schmid, 1994, for citations to the extensive primary literature). This is presumably the ancestral mammalian SC organization.

The apparently clear dichotomy between primate SC organization and the ancestral mammalian condition makes SC organization potentially a useful character for sorting out phyletic relationships among the mammalian taxa that have been considered to be particularly closely related to primates (Figure 3). The taxa most commonly touted as close relatives of the Order Primates are

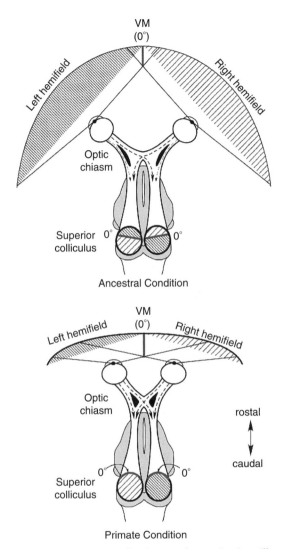

Figure 2. Schematic representation of retinotectal organization, illustrating ancestral organization and derived characteristics of primates. The sections are oriented approximately in the horizontal plane, passing through the eyes, optic chiasm, and superior colliculi. In primates, each colliculus contains a complete representation of the contralateral visual field; the representation of the vertical meridian (VM) lies at the rostral pole of the colliculus, and strong projections reach the colliculus from both eyes. In the ancestral, each colliculus represents the field of view of the contralateral eye; inputs arise mainly from the contralateral eye, with a small contribution from the ipsilateral eye. This schematic is based on Figure 1 of Pettigrew et al. (1989).

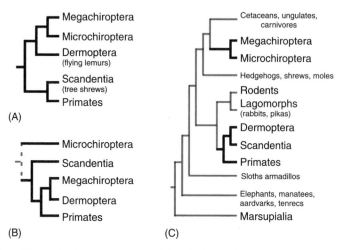

Figure 3. Alternative interpretations of the phyletic relationship of primates to other mammalian orders. Most modern workers accept that primates are closely related to tree shrews (order Scandentia), flying lemurs (order Dermoptera), and bats (order Chiroptera). Bats are comprised of two major subgroups, the Megachiroptera (fruit bats, flying foxes, and related taxa) and the Microchiroptera (echolocating bats). The status of bats is a matter of major disagreement in these interpretations. (A) Many accounts consider bats to be monophyletic and to belong within the Archonta, as in the interpretation of Novacek (1992). (B) In the "flying primate" hypothesis favored by Pettigrew et al. (1989), bats are considered to be diphyletic. Megachiropteran bats are included in Archonta, but microchiropterans are held to be distantly related to both megachiropterans and to primates. (C) The recent molecular results of Murphy et al. (2001) indicate that bats are monophyletic, but distantly related to primates. This analysis, like a number of others, places the superorder Archonta within a larger group, called Euarchontoglires, that includes rodents and lagomorphs.

tree shrews (Scandentia), flying lemurs (Dermoptera), and bats (Chiroptera); collectively, these orders are held to constitute the superorder Archonta (Gregory, 1910; McKenna, 1975). A number of recent studies suggest that rodents and lagomorphs (superorder Glires) are also closely related to primates and tree shrews (e.g., Miyamoto, 1996; Murphy et al., 2001; Shoshani and McKenna, 1998). What is the condition of SC in these groups? Rodents, lagomorphs, and tree shrews all retain the ancestral character state, as noted in an earlier section. Pettigrew (1986), however, presented evidence indicating that some bats—specifically, the megachiropteran bats ("megabats")—possess the primate condition, while microchiropteran bats ("microbats") retain the ancestral SC organization. In addition, Pettigrew et al. (1989) maintained that flying lemurs also show a primate-like condition. Pettigrew and colleagues concluded that a clade comprised of Megachiroptera plus Dermoptera is the sister group of primates. This is the "flying primate" hypothesis. An important corollary of this view is that megachiropterans and microchiropterans are not sister taxa, and therefore Chiroptera is not a natural, monophyletic taxon.

Pettigrew's initial report (Pettigrew, 1986) was based on recordings and tracer injections in six individuals from three species of the megachiropteran genus *Pteropus* (flying foxes) and two individuals of *Macroderma gigas*, a microchiropteran species with a relatively well-developed visual system. In a subsequent monograph, Pettigrew et al. (1989) indicated that the megabat *Rousettus aegyptiacus* and the dermopteran *Cynocephalus variegatus* also have primate-like SCs. They adduced additional anatomical features (mostly features of the visual system) that unite primates, megachiropterans, and dermopterans. Anatomical data were not presented in detail in their monograph, however, and some of the character states they attributed to megachiropterans and dermopterans have been questioned (Kaas and Preuss, 1993).

The claim that megabats have a primate-like SC has also been challenged. Thiele et al. (1991) examined the megabat *Rousettus aegyptiacus*, using tracer injections of the eye and the SC, and recordings from the SC. In contrast to Pettigrew, they concluded that the SC visual representation in *Rousettus* is not restricted to the contralateral visual field, but rather extends at least 25° past the vertical meridian into the ipsilateral field. Furthermore, their tracing studies indicated that major projections to SC arise from both the nasal and temporal portions of the contralateral retina, there being only a small projection from the ipsilateral retina. They concluded, therefore, that *Rousettus* retains the ancestral mammalian condition. The results of Thiele et al. prompted Rosa and Schmid (1994) to reexamine SC organization in the megabat *Pteropus* using microelectrode recording, and they concluded that while visual receptive fields do indeed extend into the ipsilateral visual field, the ipsilateral representation is much less than reported in *Rousettus* by Thiele et al. Rosa and Schmid suggest that the differences between their results and those of Thiele et al. mainly reflect differences in methodology, differences that led Thiele et al. to systematically overestimate the extent of ipsilateral representation. Even if one accepts this, however, megabats would still seem to have more ipsilateral representation than is found in primates. While acknowledging that megabats retain some ancestral features of SC organization (including strong projections from the contralateral eye), Rosa and Schmid affirm the view that the pattern of visuotopic representation in *Pteropus* is primate-like.

At the same time that neuroscientists have been debating the condition of the megabat superior colliculus, phylogenetic studies have reduced its significance as an indicator of phyletic relationships. A growing body of evidence, particularly from comparative molecular investigations, indicates that tree shrews and flying lemurs, rather than bats, are the closest living relatives of primates (e.g., Adkins and Honeycutt, 1991; Ammerman and Hillis, 1992; Bailey et al., 1992; Murphy et al., 2001). Indeed, current evidence suggests that bats are very distantly related to primates and should not be included within Archonta at all (Murphy et al., 2001). In addition, an impressive array of anatomical and molecular data supports bat monophyly (see, for example, Allard et al., 1999; Honeycutt and Adkins, 1993; Murphy et al., 2001; Novacek, 1992; Shoshani and McKenna, 1998; Simmons, 1994). Thus, even if it were to be clearly demonstrated that the megabat SC is primate-like, that similarity would now have to be considered convergent rather than homologous in the context of the full range of comparative information currently available.

Blobs

In primates, and in the nonprimate taxa that have been studied, the SC sends projections to the LGN, which terminate specifically in the koniocellular (K) layers (Huerta and Harting, 1984). The K layers, along with the M and P layers, project to visual areas in the posterior part of cerebral isocortex, where the densest projections of the LGN terminate in the so-called primary visual area (V1). These projections have been studied in great detail in primates (for review, see Casagrande and Kaas, 1994). As shown in Figure 4, the P and M layers project to largely nonoverlapping levels within the middle cortical layer (layer 4) of V1, the P layers terminating in the deep part of layer 4 while the M projections terminate in a band just superficial to the main P projection. In most (but not all) anthropoid primates, an additional, thin band of P-layer projections terminates above the band of M projections. The K projections differ markedly from the M and P projections, terminating primarily in the superficial cortical layers, specifically in layers 1 and 3. Within layer 3, the K terminations are clustered into repeating, regularly spaced territories separated by tissue that lacks K inputs.

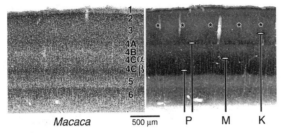

Figure 4. The laminar and compartmental distribution of projections from the LGN to the primary visual cortex (area V1) in macaque monkeys. Left: A section through area V1 stained for cell bodies using a Nissl stain; cortical layers are numbered in conventional fashion. Right: The main targets of projections from the P, M, and K layers of the LGN indicated on a tissue section stained for cytochrome oxidase (CO). The main M and P projections are distributed in horizontal bands at different levels within layer 4. The K projections, however, are distributed mainly to patchy territories of dense CO staining within layer 3 called "blobs"; these are indicated here with asterisks. Additional projections from the M and P cells to layer 6 and from the K cells to layer 1 are not shown.

Conveniently, the punctate zones of K input are marked by high levels of cytochrome oxidase (CO) activity and can be revealed using a histochemical stain for CO. In fact, the discovery of CO-dense zones antedates the discovery of their relationship to the K pathway by Lachica and Casagrande (1992). The patchy distribution of CO staining in layer 3 of primate cortex was initially reported by Horton and Hubel (1981). These patches have come to be called "blobs" or "puffs" (Wong-Riley, 1994). The fact that blobs can be reliably revealed using a relatively simple histochemical stain, provided the tissue to be stained is fresh and not too strongly fixed, has made it possible to study blobs in a much wider array of species than usually receives the attention of neuroscientists. From the start, it has been appreciated that primates generally (and perhaps universally) have blobs, and nonprimates generally (and perhaps universally) do not (Horton and Hubel, 1981). Blobs have been observed in all catarrhine and platyrrhine species examined using suitable tissue, as well as in lorisoids and cheirogaleids (Horton, 1984; Horton and Hedley-Whyte, 1984; Horton and Hubel, 1981; Preuss and Kaas, 1996; Preuss et al., 1993; Wong-Riley, 1988), and evidently in lemurids (Jeo et al., 1997). Definitive evidence for or against the presence of blobs in tarsiers is presently lacking, because it has proven difficult to obtain suitable tissue. By contrast to primates, blobs are absent in tree shrews (Horton, 1984; Jain et al., 1994; Wong-Riley, 1988), murid and sciurid rodents (Horton, 1984; Wong-Riley, 1988), and rabbits (Horton, 1984). Blobs are absent in the one bat genus (*Pteropus*) that has been examined specifically in this regard (Ichida et al., 2000; see also the figures in Rosa, 1999; Rosa et al., 1993, 1994). Blobs appear to be absent in marsupials (L. H. Krubitzer, pers. comm.; see also the published photographs in Kahn et al., 2000; Martinich et al., 1990; Rosa et al., 1999).

Although carnivores were initially reported to lack blobs (Horton, 1984; Wong-Riley, 1988), more recent studies indicate that at least some carnivores—namely, cats and ferrets—possess alternating territories of dark and light CO staining in the upper layers of area V1 (Boyd and Matsubara, 1996; Cresho et al., 1992; Murphy et al., 1995). Like primate blobs, cat blobs appear to receive direct inputs from the K cells of the LGN (Boyd and Matsubara, 1996). Despite their similarities, the presence of blobs in primates and carnivores probably reflects homoplasy rather than homology, because animals considered to be more closely related to primates—tree shrews, rodents, lagomorphs, and bats—lack blobs (Preuss, 2000; Preuss and Kaas, 1996).

What is the function of blobs? Based mainly on microelectrode recordings of the visual properties of area V1 neurons, Livingstone and Hubel (1984, 1988) argued that blobs receive their major inputs from the P pathway and serve as specialized color-processing modules. While this idea has become a fixture of textbooks, it has struck some as problematic (Allman and Zucker, 1990; Casagrande, 1994; Preuss, 2000), if only because well-defined blobs are present in nocturnal primates, which have very limited capacity for color discrimination. Moreover, anatomical studies indicate that blobs receive direct inputs from the K layers of the LGN, as discussed above, but not from the P layers, which are usually thought to carry the color-opponent signals required for acute color discrimination. The blob story has taken a new twist with reports that at least some of the K retinal ganglion cells are specifically related to short-wavelength (S) photoreceptors (Dacey, 2000); these K cells could send information from the S-cone channel to the K layers of the LGN and thence to the blobs in area V1. There is at present, however, no evidence that blobs have a correspondingly strong relationship to the medium- and long-wavelength cones, and so a role for blobs in color vision (which requires the interaction of different cone types) has still not been established. It would also be premature to conclude that there is an exclusive relationship between S cones and blobs. In this regard, it is worth noting that blobs are present in two primate taxa, *Otolemur* and *Aotus,* in which mutations have inactivated the S-pigment gene (Jacobs et al., 1996).

Casagrande (1994) has considered alternatives to the idea that blobs function as color modules, based on the recognition that blobs may have a privileged relationship to the superior colliculus, by virtue of the strong SC projection to the K layers of the LGN. She suggests that the functions of blobs are related to the attentional and eye movement functions of the colliculus. This possibility will be considered in more detail in a subsequent section.

The Critical Role of V1 in Primate Vision

In primates, lesions of the primary visual area have a devastating effect on visual detection and discrimination. This has been demonstrated both in anthropoids and strepsirhines (*Galago*) (Atencio et al., 1975). Humans with V1 lesions report they are blind in the affected parts of the visual field, and indicate they are unaware of the occurrence of stimuli presented therein. In electrophysiological experiments, lesion or deactivation of area V1 (also known as the "striate" area, owing to its possession of a conspicuous, horizontal band of myelinated fibers) results in marked suppression of stimulus-driven activity in regions of higher-order, "extrastriate" visual areas that represent the lesioned part of the visual field (reviewed by Rodman and Moore, 1997). Collectively, these results suggest that much of the visual information that reaches higher-order cortical centers traverses area V1. Remarkably, however, when patients are instructed to guess the location of stimuli presented in the lesioned visual field, or to identify the characteristics of those stimuli, they do better than chance, indicating that some visual processing capacity is retained in the lesioned part of the visual field representation—even though subjects insist they are unaware that stimuli have been presented (Pöppel et al., 1973). This phenomenon, known as "blindsight," can be demonstrated in nonhuman primates as well as in humans (Cowey and Stoerig, 1995; Weiskrantz, 1996).

To understand why lesions of area V1 have such a destructive effect on visual processing in primates, it is necessary to consider the routes through the thalamus by which visual information reaches the cortex. The most numerous visual projections to the cortex arise from the M, P, and K layers of the LGN and terminate in area V1; in turn, area V1 projects to the second visual area (V2) and a variety of other extrastriate visual areas. There are also, however, direct LGN projections to area V2 and other extrastriate areas, although these are very much weaker than the LGN projections to area V1 (Rodman and Moore, 1997). In addition to these geniculocortical projections, the inferior pulvinar nucleus (which receives visual inputs from the superficial layers of the SC) sends projections to V1 as well as to V2 and other extrastriate visual areas. The projections to extrastriate cortex from the LGN and pulvinar presumably provide the anatomical substrates for residual visual capacity following V1 lesions.

Lesions of V1 in nonprimate mammals have much less dramatic effects than in primates. For example, tree shrews with lesions of area V1 retain considerable visual discriminative capacity (Killackey et al., 1971, 1972). In addition, V1 lesions in rats, cats, and bats typically do not produce decrements of visual responsiveness in extrastriate visual areas comparable to those observed in primates (see Funk and Rosa, 1998, and references therein). The reason V1 lesions in nonprimate taxa have relatively modest effects on extrastriate function than in primates may reflect differences in thalamocortical organization. Specifically, the LGN projections to extrastriate visual areas are probably less numerous in primates than in nonprimates, as suggested by the fact that these projections have only been recognized in primates quite recently, while they have long been recognized in nonprimates. Furthermore, the extrastriate projection may arise primarily from K cells in primates (Hendry and Reid, 2000; Rodman et al., 2001), whereas there appear to be more substantial projections from the M and/or P cells in nonprimates, especially to area V2 (e.g., Kawano, 1998). Alternatively, the LGN projections to extrastriate cortex might be less potent physiologically in primates than in nonprimates, and the projections from V1 to extrastriate visual areas more so (in this regard, see also Funk and Rosa, 1998). It is also possible that the strong influence of V1 on extrastriate areas in primates results from an increased potency of V1 projections to the inferior pulvinar, which projects in turn to extrastriate cortex (Cusick, 2002). By whatever mechanism, V1 exerts a much stronger influence on extrastriate areas in primates than in other mammals that have been examined.

Dorsal and Ventral Visual Processing Streams and Their Termini in Higher-Order Parietal and Temporal Cortex

In primates, area V1 projects to multiple extrastriate areas (principally V2, V3, V3A, V4, and MT) which serve in turn as the major sources of visual information to higher-order visual areas (Figure 5). Estimates of the total number of visual areas in anthropoid primates range from at least 15 (Kaas, 1989) to more than 30 (Felleman and Van Essen, 1991). Although strepsirhines have not been investigated as exhaustively as anthropoids, they possess many of the same areas (Collins et al., 2001; Krubitzer and Kaas, 1990; Preuss and Goldman-Rakic, 1991c; Preuss and Kaas, 1996; Preuss et al., 1993; Rosa et al., 1997). By contrast to primates, comparative studies suggest that ancestral eutherians possessed only a few visual areas, which included V1, V2, perhaps two or three additional areas on the lateral surface rostral to V2, and a medial area (Rosa, 1999; Rosa and Krubitzer, 1999). Certain taxa that have large regions of posterior cortex devoted to vision, such as tree shrews and megabats, nonetheless appear not to possess many visual areas in addition to those that were present in ancestral eutherians (Lyon et al., 1998; Rosa, 1999). Thus, the large number of areas present in primates represents a derived condition, and many of the visual areas present in primates must lack homologues in other mammals, and are therefore neomorphic (Allman, 1977; Allman and McGuiness, 1988; Kaas, 1987, 1989; Rosa, 1999).

Studies of the connections between primate visual areas reveal that extrastriate cortex is organized into at least two, partly independent, processing streams, that have been termed the dorsal and ventral pathways (Boussaoud et al., 1990; Felleman and Van Essen, 1991; Livingstone and Hubel, 1988; Ungerleider and Mishkin, 1982; Young, 1992). The dorsal pathway includes the middle temporal area (MT) and areas downstream from MT, which contain neurons that are sensitive to object motion. The ventral pathway includes area V4 and areas downstream from V4; these areas contain neurons sensitive to features of object form and (in diurnal anthropoids, at least) to color. The dorsal and ventral streams terminate in the posterior parietal (PP) cortex and inferior temporal (IT) cortex, respectively, two regions classically regarded as higher-order association cortex. In humans and nonhuman primates, lesions of PP and IT have been found to produce very different kinds of visual deficits. Damage to posterior parietal cortex results in sensory neglect (inattention), errors in the spatial localization of objects, and misdirected reaching. Damage to IT cortex produces an inability to recognize familiar objects (visual agnosia). Individual neurons in portions of IT cortex respond to the sight of specific classes of objects such as faces. In view of these differences, the dorsal/parietal and ventral/temporal pathways have been characterized as

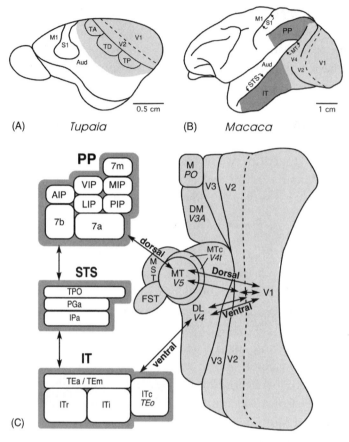

Figure 5. Visual areas of tree shrews and primates. (A) The visual cortical territory of tree shrews (*Tupaia*), denoted by gray shading, as described by Lyon et al. (1998). Tree shrew visual cortex includes a primary visual area (V1), second visual area (V2), and several extrastriate areas (TA, TD, TP). The connections of these visual areas extend forward into the region posterior to the primary somatosensory area (S1), a region that may contain additional visual areas or polysensory cortex. (B) Primates have a very large number of visual areas, which collectively occupy the occipital lobe and large portions of the parietal and temporal lobes. (C) Because visual cortex covers such a large territory, including tissue that in some primates is buried within deep sulci, it is convenient to represent the visual areas as if flattened, as in this schematic drawing. Some primate areas are known by different designations; alternative names are indicated in italics. Included among the primate divisions are three sets of higher-order areas, located in the posterior parietal lobe (PP), superior temporal sulcus (STS), and inferotemporal region (IT), respectively, which are highlighted here with dark shading. Within the visual cortex of

the "where is it" and "what is it" systems, respectively (Ungerleider and Mishkin, 1982).

However, recent evidence suggests there is more to parietal function than spatial localization. Posterior parietal cortex has been found to receive not only visual inputs, but also somatosensory inputs from the forelimb as well as information about eye and head position. The different subdivisions of PP combine these inputs in particular ways to perform very specific types of sensorimotor computations (Andersen et al., 1997; Colby and Duhamel, 1996). For example, some posterior parietal areas have visual receptive fields that shift to accommodate eye movements. This shift allows the receptive fields to be transformed from eye-centered coordinates to head-centered or body-centered coordinates, which presumably are the reference frames in which forelimb movements are programmed. In other posterior parietal areas, neurons respond to tactile stimuli on a particular set of fingertips, say, or on the lips, and also to the sight of an object when it is near the fingertips, or near the lips. Some parietal neurons respond well only to visual stimuli that are within reaching distance and many parietal neurons are responsive during active looking or reaching. For these and other reasons, the functional role of posterior parietal cortex has been characterized as "vision for action" (Goodale and Milner, 1992). (For recent reviews of the structure and function of higher-order parietal and temporal cortex, see Farah, 2000; Jeannerod, 1997; Milner and Goodale, 1995.)

Although not as well known as the posterior parietal and inferotemporal cortex, there is a third region of higher-order cortex located in the depth and upper bank of the superior temporal sulcus (STS). Although this has been dubbed the superior temporal "polysensory" area (Bruce et al., 1981), the main sensory inputs to STS cortex are visual. These inputs arise particularly from the dorsal stream, although some divisions of STS cortex appear to integrate inputs

---

primates, there are two main "streams" of visual processing extending forward from V1: a dorsal stream, which includes area MT and extends into PP, and a ventral stream, which includes area DL and extends into IT. The STS cortex receives inputs from both PP and IT. Patterns of cortical connectivity are represented in a highly simplified fashion here; for more detailed treatment, see Boussaoud et al. (1990) and Felleman and Van Essen (1991). Note that projections between cortical areas tend to be reciprocal, although "forward" and "backward" projections are not functionally equivalent. Additional abbreviations: Aud = auditory cortex; M1 = primary motor cortex.

from the dorsal and ventral streams (Boussaoud et al., 1990; Cusick, 1997). Electrophysiological studies in macaques have revealed that STS neurons are responsive to biological motion: they respond to the sight of animals, or parts of animals (especially faces and hands), moving in particular ways; some neurons are responsive to specific classes of hand-object interactions (e.g., Oram and Perrett, 1994; Perrett et al., 1990). In humans, functional imaging has revealed areas sensitive to biological motion in temporal cortex (Grossman et al., 2000); posterior parietal cortex and ventral premotor cortex exhibit similar properties as well (Buccino et al., 2001).

PP, IT, and STS have been studied most extensively in catarrhine primates (in particular, macaque monkeys and humans), but there is evidence that homologous regions exist in other primates. The posterior parietal region has been little studied in New World monkeys, but strepsirhines (galagos) possess multiple, histologically distinguishable divisions of PP; and the connections of this regions with frontal cortical areas (Preuss and Goldman-Rakic, 1991b) and with the pulvinar nucleus (Glendenning et al., 1975; Preuss and Goldman-Rakic; unpublished observations; Raczkowski and Diamond, 1981), resemble the connections of macaque posterior parietal cortex. As for IT, there is good connectional and architectonic evidence that homologous cortex exists in New World monkeys (Weller and Kaas, 1987). This region has received little attention in strepsirhines, although the histology of the IT region in galagos has been described as resembling that of catarrhines, and what little is known about the connections and functions of this region is consistent with the hypothesis that strepsirhines possess IT cortex (Preuss and Goldman-Rakic, 1991b; Raczkowski and Diamond, 1980). The STS cortex has not been studied in New World monkeys, but a region homologous to STS cortex has been attributed to galagos by Preuss and Goldman-Rakic (1991b, 1991c), based on similarities to macaque STS in histology and frontal lobe connectivity. However, Preuss and Goldman-Rakic also noted that whereas multiple architectonic divisions of the STS cortex could be distinguished in macaques, only a single division could be recognized in galagos. This suggests that the STS region underwent major changes during the evolution of anthropoids or catarrhines.

It is not surprising that a system of partly separate dorsal and ventral visual processing systems like that found in primates is not readily discerned in most other mammals, as most other mammals have relatively few visual areas. It is widely believed that at least some nonprimates possess cortex homologous to the primate parietal cortex. In rodents, for example, there is a very small zone located between primary somatosensory and extrastriate visual cortex that receives inputs from both sensory modalities, that could be homologous to one or more divisions of primate PP (Kolb, 1990; Reep et al., 1994). Others have considered the same region a division of somatosensory cortex, however (e.g., Krubitzer et al., 1986), and the connections of this region do not specifically resemble those of any one of the PP divisions of primates. There is no evidence, moreover, that this region consists of multiple areas in rodents, each with different connectional and functional characteristics, as it does in primates. Furthermore, there is no strong evidence for the existence of homologues of STS or IT cortex in nonprimate mammals (Preuss and Goldman-Rakic, 1991b).

The visual cortex of carnivores provides perhaps the most interesting comparison to primates, because like primates, carnivores possess a large number of visual areas. Payne (1993) has offered a detailed comparison of cat and primate visual cortex, and concludes that most cat areas have homologues in primates. Given that tree shrews—animals generally thought to be more closely related to primates than are cats—have few visual areas, it seems likely that many of the similarities between the extrastriate visual areas of primates and carnivores cited by Payne as evidence of homology are actually the result of convergent evolution. Moreover, cats do not exhibit the connectional segregation of visual areas into dorsal and ventral processing streams found in primates (compare Scannell et al., 1995, and Young, 1992), and even Payne notes that cats lack a strong candidate for homology with primate IT cortex. Interestingly, there is at least one group of mammals in addition to primates— sheep—that have neurons in visual cortex that respond specifically to faces; some of these neurons are responsive to sheep faces, while others are selective for the faces of humans or sheepdogs (Kendrick, 1991). Based on current comparative evidence, this is best interpreted as an instance of convergence rather than homology.

SPECIALIZATIONS OF FRONTAL CORTEX

Dorsolateral Prefrontal Cortex

By convention, neuroscientists recognize two broad divisions of primate frontal cortex, the motor zone, which consists of the primary motor area

(M1) and a variety of "premotor" areas, and the prefrontal zone, which comprises the remainder of the frontal cortex. The prefrontal zone can be divided into a medial region (consisting mainly of anterior cingulate cortex), an orbital region, and a dorsolateral region.

The dorsolateral prefrontal cortex (DLPFC) has long been a region of special interest to neuroscientists because it has been suspected to play an important role in the highest levels of cognitive functioning, as well as in the derangements of cognition that accompany schizophrenia and other neuropsychiatric disorders. Experimental studies of macaque monkeys, supplemented since the late 1970s by functional imaging studies in humans, have confirmed the involvement of DLPFC in several aspects of cognition, among them selective attention (including attention to action), working memory, and the programming of appropriate novel behaviors in nonroutine situations. DLPFC consists of multiple areas, each having a somewhat different pattern of cortical and subcortical connections and presumably making different functional contributions (Figure 6). Most subdivisions of DLPFC have strong connections with the higher-order parietal and temporal regions (PP, IT, STS) discussed in an earlier section. The most intensively studied divisions of DLPFC include the *frontal eyefield* (FEF), which is located along the anterior bank of the arcuate sulcus in macaques, and is involved in controlling the direction of eye movements and attention; and the *principalis cortex*, which occupies the banks of the principal sulcus in macaques, and is involved in working memory for loci in nearby space. (Note that while some workers consider the FEF to be part of prefrontal cortex, others consider it part of premotor cortex. For recent reviews of DLPFC organization and function, see Roberts et al., 1998.)

The dorsolateral prefrontal cortex has also been of great interest from an evolutionary standpoint. In most mammals, the motor region occupies the anterior end of the lateral cortical surface. In primates, by contrast, the motor region is located more posteriorly, near the center of the hemisphere, and the anterior end of the lateral frontal lobe is occupied by the DLPFC (see Figure 1). This suggests that DLPFC is new tissue, added to the anterior end of the frontal lobe during the evolution of stem primates. Furthermore, primate DLPFC is histologically distinctive—it has a band of densely packed, very small cells (granule cells) in layer 4, whereas other parts of the frontal cortex (the motor, cingulate, and orbital regions) are agranular or dysgranular. The granular character of primate DLPFC has been observed in both strepsirhine

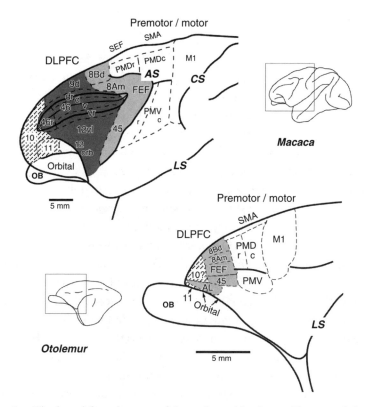

Figure 6. The lateral frontal cortex of the anthropoid primate *Macaca* and the strepsirhine primate *Otolemur*, showing the location of proposed divisions of the dorsolateral prefrontal cortex (DLPFC) and premotor/motor cortex. The DLPFC areas are shaded or stippled here. In *Macaca*, the deep arcuate sulcus (AS) separates the motor region from the DLPC, and the central sulcus (CS) separates motor cortex from somatosensory cortex. *Otolemur* possesses no comparable, deep sulci. Recent studies have identified many new divisions of both cortical regions. *Macaca* and *Otolemur* share a set of areas in DLPFC, which includes the frontal eyefield (FEF), but *Macaca* possess many additional areas, including notably the areas located within and surrounding the principal sulcus (*PS*), which include areas 46, 9, and 12 (and their many subdivisions). These areas are indicated by dark shading. *Otolemur* possesses one prefrontal area (the anterolateral area, AL) that has not been definitely identified in macaques. Motor cortex consists of a primary motor area (M1); dorsal and ventral premotor cortex (PMD, PMV); and a supplementary motor area (SMA), which includes an eye-movement region, the supplementary eye field (SEF). An SEF has not been identified in *Otolemur*. Additional motor areas are located on the medial wall of the frontal lobe and are not shown here. Area PMD consists of rostral and caudal subdivisions (PMDr, PMDc). Rostral and caudal divisions of PMV can be distinguished in macaques, but have not been identified in strepsirhines or in New World monkeys. Based on Preuss and Goldman-Rakic (1991a), Preuss et al. (1996), and Wu et al. (2000). Additional abbreviations: LS = lateral sulcus; OB = olfactory bulb.

and anthropoid primates (Brodmann, 1909; Preuss and Goldman-Rakic, 1991a,c). By contrast, all nonprimates that have been examined lack a histologically granular frontal region in the location of primate DLPFC; the entire frontal mantle in these animals is agranular or dysgranular, like the orbital, cingulate, and motor cortex of primates.

The fact that DLPFC has a distinctive histology in primates led Brodmann (1909) to conclude that this region is really well developed only in primates, and his cortical maps imply that most, if not all, the granular prefrontal areas of primates are unique to primates. Brodmann's views have not been universally acclaimed, for if correct they would mean that a region of great scientific and medical significance is not accessible for study in nonprimate mammals. A major program of research intended to identify nonprimate homologues of DLPFC was begun, which dominated comparative frontal lobe research from the period of the 1940s through the 1980s. This research is reviewed in detail by Preuss (1995a), which provides the basis of the following synopsis.

Investigators took the following tack: perhaps the differences between primates and nonprimates are *merely* histological; in that case, one should be able to use other characteristics—connectional, neurochemical, or functional—to identify the homologue of DLPFC in nonprimates. So, for example, workers noted that primate DLPFC is the recipient of strong projections from the thalamic mediodorsal nucleus (MD) and from dopamine-containing nuclei of the brainstem, and that lesions of primate DLPFC produce deficits on spatial delayed-reaction tasks, which are thought to tap spatial working memory. These characteristics came to be regarded as the defining features of dorsolateral prefrontal cortex.

Given the importance of rats as model animals, it is useful to consider the interpretation of rat–macaque frontal lobe homologies that emerged from this research. In rats, the medial frontal cortex (MFC)—that is, the cortex covering the medial surface of the frontal lobe from about the anterior end of the corpus callosum to the frontal pole—has come to be considered homologous to the dorsolateral prefrontal cortex of primates (Figure 7A, B). The densest projections from nucleus MD and from the dopaminergic nuclei target this medial region and the adjacent orbital cortex (which is presumably homologous to primate orbital cortex, a region that receives a similar complement of projections). In addition, a portion of MFC located in the cingulate region immediately anterior to the genu of the corpus callosum (the prelimbic area, PL), has been proposed as the rat homologue of macaque principalis cortex, because PL appears to be the critical region for spatial delayed-reaction performance in rats. A region along the dorsal rim (or "shoulder") of the medal frontal lobe has come to be regarded as the rat homologue of the FEF, because electrical simulation of this zone elicits eye movements.

While this interpretation of homologies has been widely adopted in rat research, Preuss (1995a) pointed out certain problems with it, and suggested an alternative interpretation (Figure 7C). None of the supposedly defining characteristics of DLPFC is actually diagnostic of that region in primates: MD

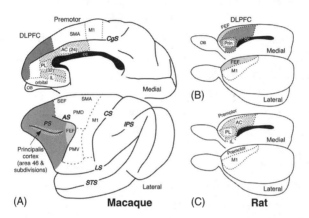

Figure 7. Alternative interpretations of frontal-lobe homologies of macaques (A) and rats (B, C). Traditional interpretations of rat organization (B) identified the cortex occupying the medial wall of the frontal cortex with the dorsolateral prefrontal cortex (DLFPC) of primates. Within the medial frontal region, the cortex located on the dorsal margin has been identified specifically with the FEF and the cortex located anterior to the corpus callosum (CC) has been identified specifically with macaque principalis cortex (Prin). Traditional interpretations of rat frontal cortex either make no attempt to identify homologues of cingulate cortex (*stippled*), or treat the medial cortex as containing a mixture of DLPFC and cingulate characteristics. Preuss's interpretation (C) identifies the medial frontal areas of rats with the medial frontal areas of primates. In particular, the areas in rats situated immediately anterior and dorsal to the corpus callosum are identified with subdivisions of primate cingulate cortex, specifically the anterior cingulate (AC), prelimbic (PL), and infralimbic (IL) areas. The dorsal rim is considered to be premotor cortex. In addition to their positional similarities to the primate medial frontal and premotor areas, these areas closely resemble primate medial frontal and premotor cortex in histology, connectivity, and function. Under this interpretation, rats lack cortex homologous to primate DLPFC. Based on Preuss (1995a).

projections are not limited to DLPFC in primates, but instead reach virtually the entire frontal lobe, including the premotor cortex, orbital cortex, and the cingulate cortex (located on the medial surface of the frontal lobe). Dopaminergic innervation is similarly widespread, and in fact recent research indicates that macaque cingulate cortex receives a denser dopaminergic input than does DLPFC. Also, while lesions of the principalis region of DLPFC do indeed produce deficits on spatial delayed-reaction tasks, lesions of the medial wall of the frontal lobe in primates can also produce such deficits. Likewise, the status of the rat's shoulder cortex as an FEF homologue is compromised by recent findings indicating that eye movements can be elicited by stimulation of dorsomedial premotor cortex (including the so-called supplementary eyefield) in primates, cortical regions that more closely resemble rat shoulder cortex connectionally and histologically than does the DLPFC. For these reasons, Preuss (1995a) suggested that the rat MFC is actually homologous to primate cingulate and premotor cortex, or at least to portions of those regions. This interpretation is supported by many other features of histology, neurochemistry, connectivity, and function.

From an evolutionary standpoint, therefore, there are no good grounds at present for supposing that DLPFC exists in any mammalian group other than primates. The identification of the rat's prelimbic area as a homologue of principalis cortex seems especially problematic: For one thing, the two areas fall within different developmental fields—PL is proisocortical, while principalis cortex is isocortical. For another, the principalis region is not even a universal feature of primate organization: as shown in Figure 6, the DLPFC of *Otolemur*—which in all respects resembles macaque DLPFC more closely than does the rat MFC—lacks a territory having the specific connectional and myeloarchitectonic features of principalis cortex (Preuss and Goldman-Rakic, 1991a,c). Increasingly, therefore, workers are emphasizing the similarities between rat MFC and primate MFC (see, e.g., Öngür and Price, 2000). This reevaluation of cortical homologies is a matter of some biomedical consequence, because MFC may be of even greater clinical importance than DLPFC.

Premotor Cortex

Frontal lobe changes in early primate evolution were not restricted to prefrontal cortex—the motor region was modified as well. Classically, three divisions of motor cortex were recognized in primates: (a) the primary motor area (M1); (b) the premotor area (PM), located on the lateral surface anterior to M1; and (c) the supplementary motor area (SMA or M2), located on the dorsomedial margin of the hemisphere anterior to M1 and medial to PM. Research during the last decade has indicated that the classical M1 and PM areas actually consist of multiple cortical divisions, distinguishable on the basis of differences in histology, cytochemistry, physiological characteristics, and connectivity (Figure 6). In addition, several previously unknown motor areas have been identified on the medial surface anterior and inferior to SMA. No fewer than nine motor divisions are now recognized in the frontal cortex of anthropoids (Preuss et al., 1996) and strepsirhines (Wu et al., 2000). We currently lack critical studies comparing motor cortex organization in primates and a variety of nonprimate mammals. Modern studies of rodents and carnivores, however, suggest that these mammals possess many fewer motor divisions—perhaps two or three in rodents, and perhaps three or four in carnivores (Wise, 1996). Area M1 is the only motor area that can be identified with confidence across eutherian orders (Nudo and Masterton, 1990; Wise, 1996). It thus seems likely that at least some of the premotor areas of primates are unique to primates.

On current evidence, the strongest candidate for a neomorphic motor area is the ventral premotor area (PMV), an area that has been characterized in catarrhine, platyrrhine, and strepsirhine primates (Preuss et al., 1996; Wu et al., 2000). In catarrhines, PMV can be divided into separate caudal and rostral zones, based on differences in connectivity and function. PMV has distinctive characteristics that are unlike any motor area known in nonprimate mammals (see especially the comparative study of Nudo and Masterton, 1990, and the review of Preuss et al., 1993). One particularly noteworthy feature is its somatotopy—whereas most motor areas represent the face, forelimb, and hindlimb, PMV has sizeable forelimb and orofacial representations, but little or no representation of the body below the neck.

In primates, the lateral premotor areas, PMV and PMD, play important roles in visually guided reaching and grasping (Jackson and Husain, 1996; Jeannerod et al., 1995; Wise et al., 1997). Both areas receive visual and visuomotor inputs from the dorsal visual stream via connections with posterior parietal cortex. Area PMD appears to be relatively specialized for the control of orienting and reaching movements. Area PMV is involved in prehensive movements of the mouth and hands: neurons respond selectively when the individual performs a particular movement or observes the movement being performed; neurons that respond in both conditions have been dubbed "mirror neurons" (Gallese et al., 1996). Individual neurons respond in relation to specific prehensive actions, including precision grip, whole-hand grips, tearing movements, hand-to-mouth movements, and so forth (Gentilucci and Rizzolatti, 1990). Rizzolatti and colleagues suggest that PMV contains a "vocabulary" of motor acts (Rizzolatti et al., 1988).

SPECIALIZATIONS OF LIMBIC CORTEX

The limbic cortex, located along the margin (*limbus*) of the cortical mantle, is a functionally heterogeneous region that includes structures involved in emotion and memory. Limbic cortex forms a continuous band that extends from the cingulate region, which lies immediately anterior and superior to the corpus callosum, through the retrosplenial region, which is wrapped around the posterior end (splenium) of the corpus callosum, and into the parahippocampal region of the medial temporal lobe. Much of the cortex of the limbic region has a simplified laminar structure, and is considered allocortex or periallocortex rather than isocortex, but some portions have a well-developed laminar organization and are classified as proisocortex or isocortex. The allocortex (hippocampus and primary olfactory cortex) bears some resemblance to the three-layered cortex of reptiles, which makes it tempting to view the limbic region as being frozen in the evolutionary past. This is not the case: even the early cortical cartographers noted important variations among mammals in the areal subdivisions of the limbic region (Brodmann, 1909). There have been few modern comparative studies of the organization of limbic cortex, but there are reasons to think that stem primates evolved specializations within this territory.

One region of interest is cingulate cortex. In primates, cingulate cortex is usually divided into an anterior region (comprised of Brodmann's areas 24, 25, and 32) and a posterior region (areas 23 and 31) (Brodmann, 1909; Vogt et al., 1987). The cingulate cortex is adjoined posteriorly by retrosplenial cortex. Homologues of the anterior and retrosplenial territories are recognizable in most eutherians that have been examined, and older studies identified a posterior cingulate region in a variety of taxa. Modern studies of rats and rabbits, however, depict anterior cingulate area 24 as adjoining the retrosplenial cortex directly, without an intervening posterior cingulate territory (Vogt, 1985; Vogt et al., 1986). This raises the possibility that primate posterior cingulate cortex is neomorphic. The posterior cingulate cortex underwent further modification later in primate evolution, in stem haplorhines or anthropoids (Armstrong et al., 1986; Preuss and Goldman-Rakic, 1991b; Zilles et al., 1986). The posterior cingulate region is connected with cortical areas known to have visuospatial functions (including PP and DLPFC) and functional studies place it in a network of structures involved in spatial attention and memory, eye movements, and mental navigation (e.g., Maguire, 1997; Vogt et al., 1992).

Another possible locus of evolutionary change is the posterior parahippocampal cortex. The parahippocampal region comprises a set of areas present in a wide variety of eutherians; these include the entorhinal, perirhinal, and prorhinal areas, which are located in the temporal lobe adjacent to the hippocampus. Primates possess an additional pair of areas, termed TH and TF, which lie posterior to the entorhinal–perirhinal region. These areas have been identified in both strepsirhine and anthropoid primates (Preuss and Goldman-Rakic, 1991b) but are not usually attributed to nonprimate mammals. It has been suggested that a territory in rats termed the postrhinal area is homologous to primate posterior parahippocampal cortex (Burwell et al., 1995). It is possible that postrhinal cortex is homologous to area TH—both share a fairly simple laminar organization—but it seems unlikely that postrhinal cortex also includes a homologue of the thick, well-laminated area TF. The parahippocampal areas are important way stations in pathways that link isocortical areas to the hippocampus, a structure involved in the formation of long-term memories (Squire and Zola, 1996), and the likelihood that primates possess one or more new divisions of parahippocampal cortex suggests they may have evolved new memory-related systems.

DORSAL PULVINAR AND RELATED CORTICAL NETWORKS

The largest component of the thalamus in anthropoid primates is called the pulvinar, a structure composed of several major divisions, usually termed the inferior, lateral, medial, and oral (anterior) pulvinar nuclei (Gutierrez et al., 2000). This terminology is generally not used in nonprimate mammals, although it is clear that at least some components of the primate pulvinar have homologues in other mammals. The anterior pulvinar of primates is related to the somatosensory system, and may be homologous to a portion of the posterior thalamic nucleus of rodents. The inferior pulvinar (including the inferior part of the lateral pulvinar nucleus) is the major target of projections from visual structures (superior colliculus, primary visual cortex) in primates. This inferior region is probably homologous to the largest part of the structure designated as the "pulvinar" or "pulvinar-lateral posterior complex" in other mammals, which is characterized by similar visual connections in other mammals that have been studied, including tree shrews, rodents, rabbits, insectivores, and carnivores, at least (for reviews, see Diamond, 1973; Harting et al., 1972; Huerta and Harting, 1984).

Collectively, the anterior somatosensory and inferior visual regions account for all or virtually all the pulvinar of nonprimates, but in primates there remains a large dorsal (or superior) pulvinar region, consisting of the traditional medial pulvinar nucleus and the dorsal part of the traditional lateral pulvinar nucleus (Gutierrez et al., 2000). In contrast to the inferior pulvinar, the dorsal pulvinar is notable for its *lack* of strong inputs from the superior colliculus. Studies in anthropoid primates have shown the dorsal pulvinar to be connected with an astonishing array of cortical areas, spanning the mantle. Its strongest connections are probably with dorsolateral prefrontal, posterior parietal, STS, and IT cortex, although it is also connected with limbic cortex (including posterior cingulate, parahippocampal, and insular cortex) and with higher-order visual, auditory, and somatosensory areas (see especially Goldman-Rakic, 1988; Gutierrez et al., 2000; Romanski et al., 1997; Selemon and Goldman-Rakic, 1988). A number of cortical areas that are connected with the dorsal pulvinar are themselves interconnected: for example, there are strong connections between dorsolateral prefrontal, posterior parietal, and STS areas (Figure 8).

Prosimians have a suite of pulvinar subdivisions similar to anthropoids (Glendenning et al., 1975; Simmons, 1988; Weller and Kaas, 1982). There are few published data on dorsal pulvinar connections, but the studies of Raczkowski and Diamond (1980, 1981) demonstrate connections with posterior parietal and temporal association cortex. Also, tracer injections of prefrontal, posterior parietal, and STS cortex made by Preuss and Goldman-Rakic (unpublished observations) yielded strong labeling of the dorsal pulvinar. These injections also demonstrated strong interconnections among the higher-order cortical areas (Preuss and Goldman-Rakic, 1991c).

In nonprimate mammals that have been examined, there is no thalamic structure in a location comparable to that of the dorsal pulvinar of primates (immediately dorsal and anterior to the visual pulvinar-lateral posterior complex) that has its distinctive connectional characteristics (lack of strong retinal

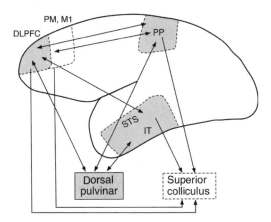

Figure 8. Primates are distinguished among mammals by the presence of strong connections linking higher-order frontal (DLPFC), parietal (PP), and temporal (STS, IT) cortical regions. These regions are connected with a common thalamic structure, the dorsal pulvinar, which appears to be unique to primates. These higher-order cortical regions, along with forelimb representations in premotor (PM) and primary motor (M1) cortex, also send strong projections to the superior colliculus; comparable projections do not exist in nonprimate mammals.

and collicular inputs, connections with far-flung, often mutually interconnected, cortical areas). It is very likely that both the dorsal pulvinar and the transcortical networks of which it is a component are unique to primates. Although expansion of the pulvinar has been cited as a hallmark of primate evolution (Harting et al., 1972; Le Gros Clark, 1959), what is really distinctive about primates is the *dorsal* pulvinar.

The remarkably extensive cortical connections of the dorsal pulvinar place it in a favorable position to influence, and possibly to coordinate and integrate, activity over a very large fraction of the cortical mantle. Despite this, experimenters have devoted little effort to understanding the functions of the dorsal pulvinar. There is some evidence that it plays a role in selective attention, especially in the spatial domain (e.g., LaBerge and Buchsbaum, 1990; Petersen et al., 1987). Connectional studies have prompted similar conclusions, because the prefrontal and parietal areas that have such prominent pulvinar connections are also believed to have attention-related functions (Gutierrez et al., 2000; Romanski et al., 1997; Selemon and Goldman-Rakic, 1988).

SPECIALIZATIONS OF CORTICOTECTAL ORGANIZATION

As noted in an earlier section, the superior colliculus influences activity of cerebral cortex by means of its projections to several thalamic nuclei that innervate the cortex. The cortex, in turn, sends projections back to the colliculus. In most mammals that have been studied, the densest projections to the colliculus arise from visual cortex and other sensory areas of the posterior cerebrum; projections from frontal cortex are present, but are relatively scant (e.g., opossums: Martinich et al., 2000; insectivores: Künzle, 1995; cats: Tortelly et al., 1980; Harting et al., 1992; rabbits: Buchanan et al., 1994; rats: Beckstead, 1979; Leichnetz and Gonzalo-Ruiz, 1987; Reep et al., 1987; tree shrews: Casseday et al., 1979). In primates, by contrast, dense corticotectal projections originate from frontal cortex and from higher-order parietal and temporal areas (Figure 8), as well as from visual areas (Fries, 1984). In Old World and New World monkeys, frontotectal projections arise mainly from portions of dorsolateral prefrontal cortex, including the frontal eyefield (FEF), principalis cortex, the cortex dorsal to the principal sulcus, and motor areas (the supplementary motor area, PMD, and M1) (Fries, 1984; Leichnetz et al., 1981). The frontal territories from which the densest projections arise represent eye and head movements (FEF, PMDr), forelimb movements (the forelimb representations of PMDc and M1), and territories involved in spatial representation (principalis cortex). The density of corticotectal projection neurons in FEF and PMD equals or exceeds that of visual cortex (Fries, 1984). In addition, the frontal projections of anthropoid primates provide substantial innervation of the superficial collicular layers, which are the major targets of retinotectal projections and the major source of projections to the LGN, whereas in the nonprimates that have been examined, frontal projections appear to innervate the middle and deep layers of the SC almost exclusively (Segal et al., 1983). The corticocollicular projections of strepsirhine primates have not been examined in detail, although existing evidence suggests they are similar to those of anthropoids. Kaas and Huerta (1988) review published evidence of projections arising from visual areas in *Galago*. In *Otolemur*, Preuss and Goldman-Rakic (unpublished observations) noted dense collicular labeling resulting from the tracer injections of frontal, parietal, and temporal cortex and described in their study of intracortical connectivity (Preuss and Goldman-Rakic, 1991c); injections that involved the FEF produced strong labeling of the superficial layers as well as deeper layers.

PRIMATE SUPERIOR COLLICULUS IN ATTENTION
AND ACTION

With the cortical projections to the superior colliculus, this account of primate brain specializations comes full circle. The SC, an ancient structure involved in the control of eye, head, and forelimb movements through its descending projections, is also the linchpin of the highly derived forebrain systems of primates involved in visuospatial attention and vision for action. As we have seen, the SC is in a critical position to influence a wide array of cortical areas by means of its ascending projections to the multiple thalamic nuclei, and to be influenced in turn by the dense projections that arise from cerebral cortex.

What influences might the superior colliculus have upon cortical activity? There could be several different effects, given that separate projection systems to the thalamus arise from the superficial and the intermediate/deep layers of the SC. One possibility, as Casagrande (1994) suggests with reference to V1, is that projections to the blobs arising from the superficial layers of the SC (relayed by the K layers of the LGN), serve to modulate the activity of V1 neurons over a restricted portion of visual space. This would constitute a local attentional "spotlight." Another important role of the SC might be to provide information to the cortex about the current direction of the eyes and head or about impending changes in gaze direction. It has long been recognized that the visual receptive fields of some cortical neurons exhibit "gaze effects": that is, the responses of neurons to visual stimulation reflect not only where the stimulus falls on the retina but also the position of the eyes in the orbits (i.e., the direction of gaze). Gaze effects have been most extensively investigated in posterior parietal cortex (Andersen and Mountcastle, 1983; Andersen et al., 1993). However, it is now clear they are present in many additional areas, including V1, multiple extrastriate visual areas, premotor areas PMD and PMV, and FEF (Boussaoud and Bremmer, 1999). In studies of premotor neurons, Jouffrais and Boussaoud (1999) have demonstrated interactions between retinotopic location of visual stimulation, direction of gaze, and the direction of an instructed forelimb movement. Gaze effects are believed to reflect the physiological processes by which the original retinotopic mapping are recoded into head-centered and body-centered coordinate systems, the latter being considered more suitable for programming forelimb movements. The intermediate and deep layers of the SC, where neurons that respond in relation to eye and head movements are concentrated, would likely be the source of gaze-related information to the thalamus, and thence to the cortex.

How might cortical activity influence the superior colliculus? Again, there could be several different influences, depending on which collicular layers are targeted by particular cortical areas. Projections from certain low-order visual areas terminate in the superficial layers of the SC, which could influence the activity of the area V1 blobs through collicular projections to the K layers of the LGN. It may be significant, too, that primates possess a strong projection from FEF and posterior parietal cortex onto the superficial SC layers, which could conceivably modulate the activity of neurons in V1 blobs. Projections to the intermediate and deep layers of the SC, which arise from a much larger portion of the cortex, could influence the activity of collicular neurons that project to brainstem and spinal centers controlling eye, head, and forelimb movements.

In this context, we also need to consider how the visual field specializations of the primate superior colliculus relate to function. Recall that primates are distinctive among mammals in having a collicular representation restricted to the contralateral hemifield, but comprised of nearly equal inputs from both eyes. This suggests that binocular interactions are particularly important in primate SC functioning, perhaps in relation to tracking target distance or to monitoring the location of the hand as it moves toward a target. Reaching accuracy is known to depend critically on visual feedback about the location of the hand as it moves in depth from the periphery toward a foveated stimulus (Jeannerod, 1988).

CONCLUSIONS

Comparative analysis indicates that brain organization was extensively modified early in primate evolution, subsequent to the divergence of primates from their closest relatives and prior to the radiation of living primate taxa. At minimum, these modifications included *changes in the internal organization of existing structures*, as exemplified by the altered retinotopy of the superior colliculus; *changes in the extrinsic connections of existing structures*, such as the advent of terminations within the superficial layers of the superior colliculus arising from prefrontal cortex; the *addition of new structural units*, including the blobs of area V1, many cortical areas, and the dorsal part of the pulvinar nucleus; and the *emergence of new networks of connectivity*, including the dorsal and ventral processing pathways, the strong interconnections between higher-order association areas, and the extensive interconnections between association areas and the dorsal pulvinar.

This analysis suggests that the evolutionary transformations were especially concentrated in the visual system. It should be said, however, that the prominence of the visual system in this analysis probably reflects to some extent the concentration of effort by neuroscientists on the visual system relative to other brain systems. There is every reason to suppose that other regions—olfactory and auditory structures, for example, or the amygdala, hypothalamus, and hippocampus—were modified during the emergence of primates. At present, however, we lack the comparative studies necessary to identify these changes. Similarly, the studies reviewed here emphasize particular levels or dimensions of brain organization, mainly the nuclei and areas of the thalamus and cortex, and the connections between these structures. There are many other dimensions of neurobiological organization that could have undergone important changes during primate origins—such as, for example, the distribution of neurotransmitters and receptor molecules, and the biochemical and morphological phenotypes of neurons. We know that mammals vary along these dimensions of organization (Preuss, 2001), but, once again, we lack the comparative studies required to specify the changes that took place during primate origins.

The great extent of the evolutionary changes that occurred at early stages of primate history challenges conventional ideas about brain evolution held by both neuroscientists and primatologists. Although the neuroscientific literature provides many examples of phyletic variation in brain organization, these variations are rarely treated in a systematic fashion. This has made it possible for neuroscientists to persist in believing that brain organization is highly conserved across mammalian taxa, and therefore that results obtained studying rats, or other convenient laboratory taxa, can be extended with confidence to other mammals. This uniformitarian stance is clearly belied by the evidence. Primate biologists, accepting the uniformitarian premise promulgated by neuroscientists, have focused on the one dimension of brain organization that obviously does vary greatly across taxa—size. Nevertheless, it is clear that evolution can affect brain organization in many ways, and the relationship between changes in internal organization and brain size is by no means very direct. Tree shrews and strepsirhine primates are encephalized to similar degrees (Stephan et al., 1988), but the ways their brains are compartmentalized are very different, as are the patterns of connections between brain structures. In all these respects, strepsirhines resemble tree shrews much less than they do anthropoid primates—animals that are twice as encephalized on average as strepsirhines. Clearly, if one wants to understand how evolution modified the brain structures and systems that subserve cognition and behavior, it is necessary to employ the kinds of neuroscientific techniques that allow one to examine these structures and systems directly, and to employ these techniques within a comparative framework.

If the adoption of comparative neuroscientific approaches is crucial for advancing our understanding of brain evolution, it is no less important for advancing our understanding of behavioral evolution. Ideas about behavioral organization derived from neuroscientific studies can lead to insights about behavioral evolution, insights that would be very difficult to obtain by studying behavior alone.

Consider the results of the present survey. Probably no one will be surprised to learn that brain structures related to vision were modified in stem primates, including systems specifically related to looking and visually guided reaching. This is just what we would expect, based on the results of comparative morphological and behavioral studies carried out by primate origins researchers. The neuroscientific findings indicating modifications of the superior colliculus and dorsal visual pathway seem congenial to the visual predation model, in particular. Primate visual system modifications are not limited to the dorsal pathway, however: primates also have a ventral tier of cortical visual areas, specialized for object discrimination and recognition. This, too, is perhaps not entirely unexpected, since some interpretations of primate origins emphasize the exploitation of fruit and other plant parts, and thus it makes sense that primates evolved specializations of the brain regions that are engaged when animals choose objects based on their visual appearance. We might even regard the evolutionary parcellation of primate visual cortex into distinct dorsal and ventral processing systems as an elegant solution to the problem of how to accommodate the demands of both visually guided predation and plant foraging.

This sort of *post hoc* analysis ignores some issues of real importance for students of behavior, however. Consider that visually guided looking and reaching is an activity pattern exhibited by many groups of mammals, not just primates, yet primates evolved many nervous system specializations related to looking and reaching. Furthermore, these modifications extend to high functional levels of the cerebral cortex—including posterior parietal, premotor, and dorsolateral prefrontal cortex—rather than simply being limited to the sensory and motor systems that control eye and hand movements. At the level of behavior, looking–reaching activity may be homologous across taxa, but at the level of neural mechanisms, the looking and reaching of primates is in several respects quite different from that of other animals. Similar considerations apply to the specialized neural mechanisms of the primate ventral/temporal pathway and their relationship to visual object discrimination—presumably many mammals use vision to scrutinize potential food items, yet only primates evolved a ventral visual pathway.

David Hume, in his *Treatise on Human Nature* (Hume, 1739–1740/1911), argued that if another species exhibits behavior like our own, we can confidently conclude that the internal mechanisms responsible for that behavior in us are present in the other species as well. This "argument by analogy"—shared behaviors imply shared neuropsychological mechanisms—has been a bedrock of comparative psychology (Povinelli et al., 2000). Yet the validity of this rule has been called into question by recent results in comparative cognition (Povinelli et al., 2000). The neurobiological evidence adds to the case against Hume, demonstrating that the mechanisms underlying a particular behavior or set of behaviors can undergo considerable divergence without producing obvious differences in overt behavior.

The key word here, however, is *obvious*. If two taxa possess different behavioral mechanisms, we should strongly suspect they have different behavioral capabilities. Their behavior may be similar in some ways—and perhaps in ways that are especially conspicuous to human observers—but we can reasonably expect to find differences if we probe their behavior deeply enough.

How could the neuropsychology of such a simple activity as looking and reaching vary significantly across taxa? Perhaps the act of looking and reaching is really not so simple. Students of motor control in macaques and humans have decomposed looking and reaching into many component processes (see, e.g., Jeannerod, 1988, 1997; Jeannerod et al., 1995). These include: (1) locating the target in space, (2) discriminating target size and orientation; (3) directing attention to the target location; (4) foveating the target by moving the head and eyes; (5) programming and implementing the initial, ballistic movement of the forelimb toward the target; (6) adjusting the trajectory of the limb as it nears the target, using visual feedback; (7) preshaping the hand to an aperture and orientation appropriate for the target to be grasped; and (8) adjusting grip forces to accommodate the actual weight, texture, and compressibility of the target, using tactile and proprioceptive feedback. When looking and reaching are parsed in this way, one can begin to understand how species might vary, with evolution modifying the component processes that contribute to the action. So, the hand preshaping process might be absent, or function quite differently, in animals that make a whole-hand grasp, such as strepsirhines. There could also be differences in the way components of the looking–reaching system interact with the rest of the cognitive system: expectations about where targets are likely to appear, stored in working memory, could direct attention, eye movements, and reaching prior to the appearance of a target, but only if working memory has access to the relevant attentional and motor mechanisms. Movements could then be made in anticipation of the appearance of the target, or movements could be "primed," so as to be produced more rapidly when the target appears. In other animals, looking and reaching might be more controlled mainly by the immediate sensory characteristics of the external stimuli and less influenced by internal states. A difference of this sort could very well distinguish primates from nonprimates, because in primates, the superior colliculus receives strong inputs from parietal and frontal areas implicated in spatial attention and working memory.

If the foregoing discussion seems unduly hypothetical, one must bear in mind that there is no significant body of experimental research that compares looking and reaching in different mammalian taxa. Evidently, looking and reaching have not seemed very interesting to comparative psychologists. Moreover, as long as a particular behavior is regarded as a unitary thing, it is difficult to imagine how evolution can work on it, except to change its efficiency or magnitude—one taxon might reach more accurately or more quickly than another, but that is about it. The same principle applies to monolithic psychological categories such as "learning" and "intelligence." Once we understand the component processes that generate a particular behavior, or that comprise a particular psychological process, the scope of evolutionary changes one can conceive of (and thus explore empirically) is greatly enlarged. Neuroscientific studies are critical for uncovering these component processes. It is not necessary that all behavioral studies should include direct investigations of brain function, but our thinking about behavior organization and evolution does need to be neuroscientifically informed.

If neuroscientific analysis can reveal unexpected differences in the mechanisms underlying similar behaviors, it can also reveal unexpected commonalities between seemingly disparate cognitive and behavioral functions. Functional imaging studies of semantic representation in the human brain indicate that objects are represented in distributed systems that include not only low-order visual areas, but also certain portions of posterior parietal and ventral premotor cortex (Chao and Martin, 2000; Martin and Chao, 2001). So, for example, ventral premotor cortex is differentially activated when subjects merely view tools as opposed to other kinds of objects, and both ventral premotor cortex and posterior parietal cortex are more strongly activated when subjects name tools than when they name other objects (Chao and Martin, 2000). This is a striking result because the activated zones in ventral premotor and posterior parietal areas either correspond to, or are located very close to, the territories involved in grasping and manipulation. Based on these findings, and on clinical observations of persons with lesions involving the relevant areas, it has been argued that semantic representations of objects comprise both their visual characteristics and the motor schemas appropriate for grasping and utilizing them (Chao and Martin, 2000; Martin and Chao, 2001). As Martin (1998) notes, results like these are consistent with the idea that human knowledge is "embodied," that is, that conceptual systems are directly grounded in structures related to perception and action (Clark, 1999). Remarkably, the frontal and parietal areas involved in grasping in humans are also either identical to, or located very close to, areas that are involved in language (Kimura, 1993; Preuss, 1995b; Rizzolatti and Arbib, 1998), the ventral premotor area corresponding to part of Broca's area and the posterior parietal area to the language cortex of the supramarginal gyrus (Brodmann's area 40). It appears that in the evolution of language, the network of cortical areas that represents objects and object-related prehension were recruited in the service of language. (Whether precisely the same tissue territories are used in language and grasping, or whether there is some functional segregation of these territories, is not clear on current evidence, but the latter result would not necessarily compromise the claim.) This evolutionary analysis, by pointing out a link between forelimb-related and language-related cortical systems, also suggests why it is that language can be conveyed as fully and naturally by manual signing as with speech (Preuss, 1995b).

Primate biologists need to adopt neuroscientific concepts and methods in order to enrich their account of behavioral evolution. But if students of primate evolution need neuroscience, neuroscientists need evolution as well. It is fashionable for neuroscientists to disavow any particular interest in evolution. (One

Nobel laureate in neuroscience is reputed to have stated, and evidently without irony, "I'm not interested in evolution—I just want to understand the monkey.") Many neuroscientists seem to regard evolution as an abstract academic concern, irrelevant to the serious business of understanding how the brain works and how it breaks down in disease. Yet our brains did not arise *de novo* to suit the particular needs of human beings—evolution produced our brains by modifying those of our ancestors. The intimate relationship between language, object representation, and grasping is a particularly dramatic illustration of how evolution co-opts and modifies existing structures to serve new functions. Any attempt to understand the organization of the human brain that does not take account of its evolutionary origins and history is likely destined for extinction.

ACKNOWLEDGMENTS

The author is indebted to Drs. Mary Anne Case, Jon Kaas, Daniel Povinelli, and John Redmond for commenting on earlier versions of the manuscript. The author's research is supported by the James S. McDonnell Foundation (JSMF 20002029).

REFERENCES

Adkins, R. M., and Honeycutt, R. L., 1991, Molecular phylogeny of the superorder Archonta, *Proc. Natl. Acad. Sci. USA* **88**: 10317–10321.

Allard, M. W., Honeycutt, R. L., and Novacek, M. J., 1999, Advances in higher level mammalian relationships, *Cladistics* **15**: 213–219.

Allman, J., and Mcguiness, E., 1988, Visual cortex in primates, in: *Comparative Primate Biology, Volume 4: Neurosciences*, H. D. Steklis and J. Erwin, eds., Alan R. Liss, New York, pp. 279–326.

Allman, J., and Zucker, S., 1990, Cytochrome oxidase and functional coding in primate striate cortex: A hypothesis, *Cold Spring Harbor Symp. Quant. Biol.* **55**: 979–982.

Allman, J. M., 1977, Evolution of the visual system in the early primates, in: *Progress in Psychology and Physiological Psychology*, J. M. Sprague and A. N. Epstein, eds., Academic, New York, pp. 1–53.

Ammerman, L. K., and Hillis, D. M., 1992, A molecular test of bat relationships: Monophyly or diphyly? *Syst. Biol.* **41**: 222–232.

Andersen, R. A., and Mountcastle, V. B., 1983, The influence of the angle of gaze upon the excitability of the light-sensitive neurons of the posterior parietal cortex, *J. Neurosci.* **3**: 532–548.

Andersen, R. A., Snyder, L. H., Bradley, D. C., and Xing, J., 1997, Multimodal representation of space in the posterior parietal cortex and its use in planning movements, *Annu. Rev. Neurosci.* **20**: 303–330.

Andersen, R. A., Snyder, L. H., Li, C. S., and Stricanne, B., 1993, Coordinate transformations in the representation of spatial information, *Curr. Opin. Neurobiol.* **3**: 171–176.

Armstrong, E., Zilles, K., Schlaug, G., and Schleicher, A., 1986, Comparative aspects of the primate posterior cingulate cortex, *J. Comp. Neurol.* **253**: 539–548.

Atencio, F. W., Diamond, I. T., and Ward, J. P., 1975, Behavioral study of the visual cortex of *Galago senegalensis,* **89**: 1109–1135.

Bailey, W. J., Slightom, J. L., and Goodman, M., 1992, Rejection of the "flying primate" hypothesis by phylogenetic evidence from the e-globin gene, *Science* **256**: 86–89.

Beckstead, R. M., 1979, An autoradiographic examination of corticocortical and subcortical projections of the mediodorsal-projection (prefrontal) cortex in the rat, *J. Comp. Neurol.* **184**: 43–62.

Boussaoud, D., and Bremmer, F., 1999, Gaze effects in the cerebral cortex: Reference frames for space coding and action, *Exp. Brain Res.* **128**: 170–180.

Boussaoud, D., Ungerleider, L. G., and Desimone, R., 1990, Pathways for motion analysis: Cortical connections of the medial superior temporal and fundus of the superior temporal visual areas in the macaque, *J. Comp. Neurol.* **296**: 462–495.

Boyd, J. D., and Matsubara, J. A., 1996, Laminar and columnar patterns of geniculocortical projections in the cat: Relationship to cytochrome oxidase, *J. Comp. Neurol.* **365**: 659–682.

Brodmann, K., 1909, *Vergleichende Lokalisationslehre der Grosshirnrhinde*, Barth, Leipzig (reprinted as *Brodmann's 'Localisation in the Cerebral Cortex,'* translated and edited by L. J. Garey, London: Smith-Gordon, 1994).

Bruce, C., Desimone, R., and Gross, C. G., 1981, Visual properties of neurons in a polysensory area in superior temporal sulcus of the macaque, *J. Neurophysiol.* **46**: 369–384.

Buccino, G., Binkofski, F., Fink, G. R., Fadiga, L., Fogassi, L., Gallese, V., et al., 2001, Action observation activates premotor and parietal areas in a somatotopic manner: An fMRI study, *Eur. J. Neurosci.* **13**: 400–404.

Buchanan, S. L., Thompson, R. H., Maxwell, B. L., and Powell, D. A., 1994, Efferent connections of the medial prefrontal cortex in the rabbit, *Exp. Brain Res.* **100**: 469–483.

Burwell, R. D., Witter, M. P., and Amaral, D. G., 1995, Perirhinal and postrhinal cortices of the rat: A review of the neuroanatomical literature and comparison with findings from the monkey brain, *Hippocampus* **5**: 390–408.

Cartmill, M., 1972, Arboreal adaptations and the origin of the order Primates, in: *The Functional and Evolutionary Biology of Primates*, R. Tuttle, ed., Aldine, Chicago, pp. 97–122.

Cartmill, M., 1974, Rethinking primate origins, *Science* **184**: 436–443.

Cartmill, M., 1992, New views on primate origins, *Evol. Anthropol.* **1**: 105–111.

Casagrande, V. A., 1994, A third parallel visual pathway to primate V1, *Trends Neurosci.* **17**: 305–310.

Casagrande, V. A., and Kaas, J. H., 1994, The afferent, intrinsic, and efferent connections of primary visual cortex in primates, in: *Cerebral Cortex, Vol. 10, Primary Visual Cortex in Primates*, A. Peters, and K. Rockland, eds., Plenum, New York, pp. 201–259.

Casseday, J. H., Jones, D. R., and Diamond, I. T., 1979, Projections from cortex to tectum in the tree shrew, *Tupaia glis. J. Comp. Neurol.* **185**: 253–291.

Chao, L. L., and Martin, A., 2000, Representation of manipulable man-made objects in the dorsal stream, *Neuroimage* **12**: 478–484.

Clark, A., 1999, An embodied cognitive science? *Trends Cogn. Sci.* **3**: 345–351.

Colby, C. L., and Duhamel, J. R., 1996, Spatial representations for action in parietal cortex, *Brain Res. Cogn. Brain Res.* **5**: 105–115.

Collins, C. E., Stepniewska, I., and Kaas, J. H., 2001, Topographic patterns of V2 cortical connections in a prosimian primate (*Galago garnetti*), *J. Comp. Neurol.* **431**: 155–167.

Cowey, A., and Stoerig, P., 1995, Blindsight in monkeys, *Nature* **373**: 247–249.

Cresho, H. S., Rasco, L. M., Rose, G. H., and Condo, G. J., 1992, Blob-like pattern of cytochrome oxidase staining in ferret visual cortex, *Soc. Neurosci. Abstr.* **18**: 298.

Cusick, C., 2002, Thalamic systems and the diversity of cortical areas, in: *Cortical Areas: Unity and Diversity*, R. Miller and A. Schuez, eds., Harwood Academic, London, pp. 155–178.

Cusick, C. G., 1997, The superior temporal polysensory region in monkeys, in: *Cerebral Cortex, Vol. 12: Extrastriate Cortex in Primates*, K. S. Rockland, and E. G. Jones, eds., Plenum, New York, pp. 435–468.

Dacey, D. M., 2000, Parallel pathways for spectral coding in primate retina, *Annu. Rev. Neurosci.* **23**: 743–775.

Dean, P., Redgrave, P., and Westby, G. W., 1989, Event or emergency? Two response systems in the mammalian superior colliculus, *Trends Neurosci.* **12**: 137–147.

Diamond, I. T., 1973, The evolution of the tectal-pulvinar system in mammals: Structural and behavioural studies of the visual system, *Symp. Zool. Soc. Lond.* **33**: 205–233.

Diamond, I. T., and Hall, W. C., 1969, Evolution of neocortex, *Science* **164**: 251–262.

Dkhissi-Benyahya, O., Szel, A., Degrip, W. J., and Cooper, H. M., 2001, Short and mid-wavelength cone distribution in a nocturnal strepsirrhine primate (*Microcebus murinus*), *J. Comp. Neurol.* **438**: 490–504.

Farah, M. J., 2000, *The Cognitive Neuroscience of Vision*, Blackwell, Oxford.

Felleman, D. J., and Van Essen, D. C., 1991, Distributed hierarchical processing in the primate cerebral cortex, *Cereb. Cortex* **1**: 1–47.

Fries, W., 1984, Cortical projections to the superior colliculus in the macaque monkey: A retrograde study using horseradish peroxidase, *J. Comp. Neurol.* **230**: 55–76.

Funk, A. P., and Rosa, M. G. P., 1998, Visual responses of neurones in the second visual area of flying foxes (*Pteropus poliocephalus*) after lesions of striate cortex, *J. Physiol., (Lond.)* **513**: 507–519.

Gallese, V., Fadiga, L., Fogassi, L., and Rizzolatti, G., 1996, Action recognition in the premotor cortex, *Brain* **119**: 593–609.

Gentilucci, M., and Rizzolatti, G., 1990, Cortical motor control of arm and hand movements, in: *Vision in Action: The Control of Grasping*, M. A. Goodale, ed., Ablex, Norwood, NJ, pp. 147–162.

Glendenning, K. K., Hall, J. A., Diamond, I. T., and Hall, W. C., 1975, The pulvinar nucleus of *Galago senegalensis*, *J. Comp. Neurol.* **161**: 419–458.

Goldman-Rakic, P. S., 1988, Topography of cognition: Parallel distributed networks in primate association cortex, *Annu. Rev. Neurosci.* **11**: 137–156.

Goodale, M. A., and Milner, A. D., 1992, Separate visual pathways for perception and action, *Trends Neurosci.* **15**: 20–25.

Gregory, W. K., 1910, The orders of mammals, *Bull. Am. Mus. Nat. Hist.* **27**: 1–524.

Grossman, E., Donnelly, M., Price, R., Pickens, D., Morgan, V., Neighbor, G., et al., 2000, Brain areas involved in perception of biological motion, *J. Cogn. Neurosci.* **12**: 711–720.

Gutierrez, C., Cola, M. G., Seltzer, B., and Cusick, C., 2000, Neurochemical and connectional organization of the dorsal pulvinar complex in monkeys, *J. Comp. Neurol.* **419**: 61–86.

Harting, J. K., Hall, W. C., and Diamond, I. T., 1972, Evolution of the pulvinar, *Brain. Behav. Evolut.* **6**: 424–452.

Harting, J. K., Updyke, B. V., and Van Lieshout, D. P., 1992, Corticotectal projections in the cat: Anterograde transport studies of twenty-five cortical areas, *J. Comp. Neurol.* **324**: 379–414.

Hendry, S. H. C., and Reid, R. C., 2000, The koniocellular pathway in primate vision, *Annu. Rev. Neurosci.* **2000**: 127–153.

Honeycutt, R. L., and Adkins, R. M., 1993, Higher-level systematics of eutherian mammals: An assessment of molecular characters and phylogenetic hypotheses, *Annu. Rev. Ecol. Syst.* **24**: 279–305.

Horton, J. C., 1984, Cytochrome oxidase patches: A new cytoarchitectonic feature of monkey visual cortex, *Philos. Trans. R. Soc. Lond. B Biol. Sci.* **304**: 199–253.

Horton, J. C., and Hedley-Whyte, E. T., 1984, Mapping of cytochrome oxidase patches and ocular dominance columns in human visual cortex, *Philos. Trans. R. Soc. Lond. B Biol. Sci.* **304**: 255–272.

Horton, J. C., and Hubel, D. H., 1981, Regular patchy distribution of cytochrome oxidase staining in primary visual cortex of macaque monkey, *Nature* **292**: 762–764.

Huerta, M. F., and Harting, J. K., 1984, The mammalian superior colliculus: Studies of its morphology and connections, in: *Comparative Neurology of the Optic Tectum*, H. Vanegas, ed., Plenum, New York, pp. 687–773.

Hume, D., 1739–1740/1911, *A Treatise on Human Nature*, 2 volumes, A. D. Lindsay, ed., Dent, London.

Ichida, J. M., Rosa, M. G., and Casagrande, V. A., 2000, Does the visual system of the flying fox resemble that of primates? The distribution of calcium-binding proteins in the primary visual pathway of *Pteropus poliocephalus, J. Comp. Neurol.* **417**: 73–87.

Jackson, S. R., and Husain, M., 1996, Visuomotor functions of the lateral pre-motor cortex, *Curr. Opin. Neurobiol.* **6**: 788–795.

Jacobs, G. H., Neitz, M., and Neitz, J., 1996, Mutations in S-cone pigment genes and the absence of colour vision in two species of nocturnal primate, *Proc. R. Soc. Lond. B Biol. Sci.* **263**: 705–710.

Jain, N., Preuss, T. M., and Kaas, J. H., 1994, Subdivisions of the visual system labeled with the Cat-301 antibody in tree shrews, *Vis. Neurosci.* **11**: 731–741.

Jeannerod, M., 1988, *The Neural and Behavioural Organization of Goal-Directed Movements*, Oxford University Press, Oxford.

Jeannerod, M., 1997, *The Cognitive Neuroscience of Action*, Blackwell, Oxford.

Jeannerod, M., Arbib, M. A., Rizzolatti, G., and Sakata, H., 1995, Grasping objects: The cortical mechanisms of visuomotor transformation, *Trends Neurosci.* **18**: 314–320.

Jeo, R., Sereno, M. I., and Allman, J. M., 1997, Cytochrome oxidase blobs in primary visual cortex of *Eulemur fulvus* and *Microcebus murinus, Soc. Neurosci. Abstr.* **23**: 2058.

Jouffrais, C., and Boussaoud, D., 1999, Neuronal activity related to eye-hand coordination in the primate premotor cortex, *Exp. Brain Res.* **128**: 205–209.

Kaas, J. H., 1980, A comparative survey of visual cortex organization in mammals, in: *Comparative Neurology of the Telencephalon*, in: S. O. E. Ebbeson, ed., Plenum Press, New York, pp. 483–502.

Kaas, J. H., 1987, The organization and evolution of neocortex, in: *Higher Brain Function: Recent Explorations of the Brain's Emergent Properties*, S. P. Wise, ed., John Wiley, New York, pp. 347–378.

Kaas, J. H., 1989, Why does the brain have so many visual areas? *J. Cog. Neurosci.* **1**: 121–135.

Kaas, J. H., Guillery, R. W., and Allman, J. M., 1972, Some principles of organization in the dorsal lateral geniculate nucleus, *Brain Behav. Evol.* **6**: 253–299.

Kaas, J. H., and Huerta, M. F., 1988, The subcortical visual system of primates, in: *Comparative Primate Biology, Volume 4: Neurosciences*, H. D. Steklis and J. Erwin, eds., Alan R. Liss, New York, pp. 327–391.

Kaas, J. H., Huerta, M. F., Weber, J. T., and Harting, J. K., 1978, Patterns of retinal terminations and laminar organization of the lateral geniculate nucleus of primates, *J. Comp. Neurol.* **182**: 517–553.

Kaas, J. H., and Preuss, T. M., 1993, Archontan affinities as reflected in the visual system, in: *Mammal Phylogeny: Placentals*, F. S. Szalay, M. J. Novacek, and M. C. McKenna, eds., Springer Verlag, New York, pp. 115–128.

Kahn, D. M., Huffman, K. J., and Krubitzer, L., 2000, Organization and connections of V1 in *Monodelphis domestica*, *J. Comp. Neurol.* **428**: 337–354.

Kawano, J., 1998, Cortical projections of the parvocellular laminae C of the dorsal lateral geniculate nucleus in the cat: An anterograde wheat germ agglutinin conjugated to horseradish peroxidase study, *J. Comp. Neurol.* **392**: 439–457.

Kendrick, K. M., 1991, How the sheep's brain controls the visual recognition of animals and humans, *J. Anim. Sci* **69**: 5008–5016.

Killackey, H., Snyder, M., and Diamond, I. T., 1971, Function of striate and temporal cortex in the tree shrew, *J. Comp. Physiol. Psychol.* **74**(Suppl. 2):1–29.

Killackey, H., Wilson, M., and Diamond, I. T., 1972, Further studies of the striate and extrastriate visual cortex in the tree shrew, *J. Comp. Physiol. Psychol.* **81**: 45–63.

Kimura, D., 1993, *Neuromotor Mechanisms in Human Communication*, Oxford University Press, Oxford.

Kolb, B., 1990, Posterior parietal and temporal association cortex, in: *The Cerebral Cortex of the Rat*, B. Kolb and R. C. Tees, eds., MIT Press, Cambridge, MA, pp. 459–471.

Krubitzer, L. A., and Kaas, J. H., 1990, Cortical connections of MT in four species of primates: Areal, modular, and retinotopic patterns, *Vis. Neurosci.* **5**: 165–204.

Krubitzer, L. A., Sesma, M. A., and Kaas, J. H., 1986, Microelectrode maps, myeloarchitecture, and cortical connections of three somatotopically organized representations of the body surface in the parietal cortex of squirrels, *J. Comp. Neurol.* **250**: 403–430.

Künzle, H., 1995, Regional and laminar distribution of cortical neurons projecting to either superior or inferior colliculus in the hedgehog tenrec, *Cereb. Cortex* **5**: 338–352.

Laberge, D., and Buchsbaum, M. S., 1990, Positron emission tomographic measurements of pulvinar activity during an attention task, *J. Neurosci.* **10**: 613–619.

Lachica, E. A., and Casagrande, V. A., 1992, Direct W-like geniculate projections to the cytochrome oxidase (CO) blobs in primate visual cortex: Axon morphology, *J. Comp. Neurol.* **319**: 141–158.

Le Gros Clark, W. E., 1959, *The Antecedents of Man*, Edinburgh University Press, Edinburgh.

Leichnetz, G. R., and Gonzalo-Ruiz, A., 1987, Collateralization of frontal eye field (medial precentral/anterior cingulate) neurons projecting to the paraoculomotor region, superior colliculus, and medial pontine reticular formation in the rat: A fluorescent double-labeling study, *Exp. Brain Res.* **68**: 355–364.

Leichnetz, G. R., Spencer, R. F., Hardy, S. G., and Astruc, J., 1981, The prefrontal corticotectal projection in the monkey: an anterograde and retrograde horseradish peroxidase study, *Neuroscience* **6**: 1023–1041.

Livingstone, M. S., and Hubel, D. H., 1984, Anatomy and physiology of a color system in the primate visual cortex, *J. Neurosci.* **4**: 309–356.

Livingstone, M. S., and Hubel, D. H., 1988, Segregation of form, color, movement, and depth: Anatomy, physiology, and perception, *Science* **240**: 740–749.

Logan, C. A., 1999, The altered rationale for the choice of a standard animal in experimental psychology: Henry H. Donaldson, Adolf Meyer, and "the" albino rat, *Hist. Psychol.* **2**: 3–24.

Logan, C. A., 2001, "[A]re Norway rats...things?": Diversity versus generality in the use of albino rats in experiments on development and sexuality, *J. Hist. Biol.* **34**: 287–314.

Lyon, D. C., Jain, N., and Kaas, J. H., 1998, Cortical connections of striate and extrastriate visual areas in tree shrews, *J. Comp. Neurol.* **401**: 109–128.

Maguire, E. A., 1997, Hippocampal involvement in human topographical memory: Evidence from functional imaging, *Philos. Trans. R. Soc. Lond. B Biol. Sci.* **352**: 1475–1480.

Martin, A., 1998, Organization of semantic knowledge and the origin of words in the brain, in: *The Origin and Diversification of Language*, N. G. Jablonski and L. C. Aiello, eds., University of California Press, San Francisco, CA, pp. 69–88.

Martin, A., and Chao, L. L., 2001, Semantic memory and the brain: Structure and processes, *Curr. Opin. Neurobiol.* **11**: 194–201.

Martin, R. D., 1990, *Primate Origins and Evolution*, Princeton University Press, Princeton.

Martinich, S., Pontes, M. N., and Rocha-Miranda, C. E., 2000, Patterns of corticocortical, corticotectal, and commissural connections in the opossum visual cortex, *J. Comp. Neurol.* **416**: 224–244.

Martinich, S., Rosa, M. G., and Rocha-Miranda, C. E., 1990, Patterns of cytochrome oxidase activity in the visual cortex of a South American opossum (*Didelphis marsupialis aurita*), *Braz. J. Med. Biol. Res.* **23**: 883–887.

Mckenna, M. C., 1975, Toward a phylogenetic classification of the Mammalia, in: *Phylogeny of the Primates*, W. P. Luckett and F. S. Szalay, eds., Plenum Press, New York, pp. 21–46.

Milner, A. D., and Goodale, M. A., 1995, *The Visual Brain in Action*, Oxford University Press, Oxford.

Miyamoto, M. M., 1996, A congruence study of molecular and morphological data for eutherian mammals, *Mol. Phylogenet. Evol.* **6:** 373–390.

Murphy, K. M., Jones, D. G., and Van Sluyters, R. C., 1995, Cytochrome-oxidase blobs in cat primary visual cortex, *J. Neurosci.* **15:** 4196–4208.

Murphy, W. J., Eizirik, E., O'brien, S. J., Madsen, O., Scally, M., Douady, C. J., et al., 2001, Resolution of the early placental mammal radiation using Bayesian phylogenetics, *Science* **294:** 2348–2351.

Novacek, M. J., 1992, Mammalian phylogeny: Shaking the tree, *Nature* **356:** 121–125.

Nudo, R. J., and Masterton, R. B., 1990, Descending pathways to the spinal cord, III: Sites of origin of the corticospinal tract, *J. Comp. Neurol.* **296:** 559–583.

Öngür, D., and Price, J. L., 2000, The organization of networks within the orbital and medial prefrontal cortex of rats, monkeys and humans, *Cereb. Cortex* **10:** 206–219.

Oram, M. W., and Perrett, D. I., 1994, Responses of anterior superior temporal polysensory (STPa) neurons to "biological motion" stimuli, *J. Cogn. Neurosci.* **6:** 99–116.

Payne, B. R., 1993, Evidence for visual cortical area homologs in cat and macaque monkey, *Cereb. Cortex* **3:** 1–25.

Perrett, D. I., Mistlin, A. J., Harries, M. H., and Chitty, A. J., 1990, Understanding the visual appearance and consequence of hand actions, in: *Vision and Action: The Control of Grasping*, M. A. Goodale, ed., Ablex Publishing, Norwood, NJ, pp. 163–180.

Petersen, S. E., Robinson, D. L., and Morris, J. D., 1987, Contributions of the pulvinar to visual spatial attention, *Neuropsychologia* **25:** 97–105.

Pettigrew, J. D., Jamieson, B. G. M., Robson, S. K., Hall, L. S., Mcanally, K. I., and Cooper, H. M., 1989, Phylogenetic relations between microbats, megabats and primates (Mammalia: Chiroptera and Primates, *Phil. Trans. R. Soc. Lond. B, Biol. Sci.* **325:** 489–559.

Pettigrew, J. P., 1986, Flying primates? Megabats have the advanced pathway from eye to midbrain, *Science* **231:** 1304–1306.

Pöppel, E., Held, R., and Frost, D., 1973, Residual visual function after brain wounds involving central visual pathways in man, *Nature* **243:** 295–296.

Povinelli, D. J., Bering, J. M., and Giambrone, S., 2000, Toward a science of other minds: Escaping the argument by analogy, *Cog. Sci.* **24:** 509–541.

Preuss, T. M., 1993, The role of the neurosciences in primate evolutionary biology: Historical commentary and prospectus, in: *Primates and their Relatives in Phylogenetic Perspective*, R. D. E. MacPhee, ed., Plenum Press, New York, pp. 333–362.

Preuss, T. M., 1995a, Do rats have prefrontal cortex? The Rose-Woolsey-Akert program reconsidered, *J. Cogn. Neurosci.* **7:** 1–24.

Preuss, T. M., 1995b, The argument from animals to humans in cognitive neuro-science, in: *The Cognitive Neurosciences*, M. S. Gazzaniga, ed., MIT Press, Cambridge, MA, pp. 1227–1241.

Preuss, T. M., 2000, Taking the measure of diversity: Comparative alternatives to the model-animal paradigm in cortical neuroscience, *Brain Behav. Evol.* **55**: 287–299.

Preuss, T. M., 2001, The discovery of cerebral diversity: An unwelcome scientific revolution, in: *Evolutionary Anatomy of the Primate Cerebral Cortex*, D. Falk and K. Gibson, eds., Cambridge University Press, Cambridge, pp. 138–164.

Preuss, T. M., Beck, P. D., and Kaas, J. H., 1993, Areal, modular, and connectional organization of visual cortex in a prosimian primate, the slow loris (*Nycticebus coucang*), *Brain Behav. Evol.* **42**: 237–251.

Preuss, T. M., and Goldman-Rakic, P. S., 1991a, Myelo- and cytoarchitecture of the granular frontal cortex and surrounding regions in the strepsirhine primate *Galago* and the anthropoid primate *Macaca*, *J. Comp. Neurol.* **310**: 429–474.

Preuss, T. M., and Goldman-Rakic, P. S., 1991b, Architectonics of the parietal and temporal association cortex in the strepsirhine primate *Galago* compared to the anthropoid primate *Macaca*, *J. Comp. Neurol.* **310**: 475–506.

Preuss, T. M., and Goldman-Rakic, P. S., 1991c, Ipsilateral cortical connections of granular frontal cortex in the strepsirhine primate *Galago*, with comparative comments on anthropoid primates, *J. Comp. Neurol.* **310**: 507–549.

Preuss, T. M., and Kaas, J. H., 1996, Cytochrome oxidase "blobs" and other characteristics of primary visual cortex in a lemuroid primate, *Cheirogaleus medius*, *Brain Behav. Evol.* **47**: 103–112.

Preuss, T. M., and Kaas, J. H., 1999, Human brain evolution, in: *Fundamental Neuroscience*, F. E. Bloom, S. C. Landis, J. L. Robert, L. R. Squire, and M. J. Zigmond, eds., Academic Press, San Diego, pp. 1283–1311.

Preuss, T. M., Stepniewska, I., and Kaas, J. H., 1996, Movement representation in the dorsal and ventral premotor areas of owl monkeys: A microstimulation study, *J. Comp. Neurol.* **371**: 649–676.

Raczkowski, D., and Diamond, I. T., 1980, Cortical connections of the pulvinar nucleus in *Galago*, *J. Comp. Neurol.* **193**: 1–40.

Raczkowski, D., and Diamond, I. T., 1981, Projections from the superior colliculus and the neocortex to the pulvinar nucleus in *Galago*, *J. Comp. Neurol.* **200**: 231–254.

Raff, R. A., 1996, *The Shape of Life: Genes, Development, and the Evolution of Animal Form*, University of Chicago Press, Chicago.

Rasmussen, D. T., 1990, Primate origins: Lessons from a Neotropical marsupial, *Am. J. Primatol.* **22**: 263–277.

Reep, R. L., Chandler, H. C., King, V., and Corwin, J. V., 1994, Rat posterior parietal cortex: Topography of corticocortical and thalamic connections, *Exp. Brain Res.* **100**: 67–84.

Reep, R. L., Corwin, J. V., Hashimoto, A., and Watson, R. T., 1987, Efferent connections of the rostral portion of medial agranular cortex in rats, *Brain Res. Bull.* **19**: 203–221.

Rizzolatti, G., and Arbib, M. A., 1998, Language within our grasp, *Trends Neurosci.* **21**: 188–194.

Rizzolatti, G., Camarda, R., Fogassi, L., Gentilucci, M., Luppino, G., and Matelli, M., 1988, Functional organization of inferior area 6 in the macaque monkey, II. Area F5 and the control of distal movements, *Exp. Brain Res.* **71**: 491–507.

Roberts, A. C., Robbins, T. W., and Weiskrantz, L., 1998, *The Prefrontal Cortex: Executive and Cognitive Functions*, Oxford University Press, Oxford.

Rodman, H. R., and Moore, T., 1997, Development and plasticity of extrastriate visual cortex in monkeys, in: *Cerebral Cortex, Vol. 12: Extrastriate Cortex in Primates*, K. S. Rockland, J. H. Kaas, and A. Peters, eds., Plenum Press, New York, pp. 639–672.

Rodman, H. R., Sorenson, K. M., Shim, A. J., and Hexter, D. P., 2001, Calbindin immunoreactivity in the geniculo-extrastriate system of the macaque: Implications for heterogeneity in the koniocellular pathway and recovery from cortical damage, *J. Comp. Neurol.* **431**: 168–181.

Romanski, L. M., Giguere, M., Bates, J. F., and Goldman-Rakic, P. S., 1997, Topographic organization of medial pulvinar connections with the prefrontal cortex in the rhesus monkey, *J. Comp. Neurol.* **379**: 313–332.

Rosa, M. G., 1999, Topographic organisation of extrastriate areas in the flying fox: Implications for the evolution of mammalian visual cortex, *J. Comp. Neurol.* **411**: 503–523.

Rosa, M. G., Casagrande, V. A., Preuss, T., and Kaas, J. H., 1997, Visual field representation in striate and prestriate cortices of a prosimian primate (*Galago garnetti*), *J. Neurophysiol.* **77**: 3193–3217.

Rosa, M. G., and Krubitzer, L. A., 1999, The evolution of visual cortex: Where is V2? *Trends Neurosci.* **22**: 242–248.

Rosa, M. G., Krubitzer, L. A., Molnar, Z., and Nelson, J. E., 1999, Organization of visual cortex in the northern quoll, *Dasyurus hallucatus*: Evidence for a homologue of the second visual area in marsupials, *Eur. J. Neurosci.* **11**: 907–915.

Rosa, M. G., and Schmid, L. M., 1994, Topography and extent of visual-field representation in the superior colliculus of the megachiropteran *Pteropus*, *Vis. Neurosci.* **11**: 1037–1057.

Rosa, M. G., Schmid, L. M., Krubitzer, L. A., and Pettigrew, J. D., 1993, Retinotopic organization of the primary visual cortex of flying foxes (*Pteropus poliocephalus* and *Pteropus scapulatus*), *J. Comp. Neurol.* **335**: 55–72.

Rosa, M. G., Schmid, L. M., and Pettigrew, J. D., 1994, Organization of the second visual area in the megachiropteran bat *Pteropus*, *Cereb. Cortex* **4**: 52–68.

Sanderson, K. J., 1986, Evolution of the lateral geniculate nucleus, in: *Visual Neuroscience*, J. D. Pettigrew, K. J. Sanderson, and W. R. Levick, eds., Cambridge University Press, Cambridge, pp. 183–195.

Scannell, J. W., Blakemore, C., and Young, M. P., 1995, Analysis of connectivity in the cat cerebral cortex, *J. Neurosci.* **15:** 1463–1483.

Segal, R. L., Beckstead, R. M., Kersey, K., and Edwards, S. B., 1983, The prefrontal corticotectal projection in the cat, *Exp. Brain Res.* **51:** 423–432.

Selemon, L. D., and Goldman-Rakic, P. S., 1988, Common cortical and subcortical targets of the dorsolateral prefrontal and posterior parietal cortices in the rhesus monkey: Evidence for a distributed neural network subserving spatially guided behavior, *J. Neurosci.* **8:** 4049–4068.

Shoshani, J., and Mckenna, M. C., 1998, Higher taxonomic relationships among extant mammals based on morphology, with selected comparisons of results from molecular data, *Mol. Phylogenet. Evol.* **9:** 572–584.

Simmons, N. B., 1994, The case for chiropteran monophyly, *Am. Mus. Novitates*, No. 3103, pp. 1–54.

Simmons, R. M. T., 1988, Comparative morphology of the primate diencephalon, in: *Comparative Primate Biology, Vol. 4: Neurosciences*, H. D. Steklis and J. Erwin, eds., Alan R. Liss, New York, pp. 155–201.

Squire, L. R., and Zola, S. M., 1996, Structure and function of declarative and non-declarative memory systems, *Proc. Natl. Acad. Sci. USA* **93:** 13515–13522.

Stephan, H., Baron, G., and Frahm, H. D., 1988, Comparative size of brains and brain components, in: *Comparative Primate Biology, Vol. 4: Neurosciences*, H. D. Steklis and J. Erwin, eds., Liss, New York, pp. 1–38.

Sussman, R. W., 1991, Primate origins and the evolution of angiosperms, *Am. J. Primatol.* **23:** 209–223.

Szalay, F. S., and Dagosto, M., 1988, Evolution of hallucial grasping in the primates, *J. Human Evol.* **17:** 1–33.

Szalay, F. S., Rosenberger, A. L., and Dagosto, M., 1987, Diagnosis and differentiation of the order Primates, *Yearb. Phys. Anthropol.* **30:** 75–105.

Thiele, A., Vogelsang, M., and Hoffmann, K. P., 1991, Pattern of retinotectal projection in the megachiropteran bat *Rousettus aegyptiacus*, *J. Comp. Neurol.* **314:** 671–683.

Tortelly, A., Reinoso-Suarez, F., and Llamas, A., 1980, Projections from non-visual cortical areas to the superior colliculus demonstrated by retrograde transport of HRP in the cat, *Brain Res* **188:** 543–549.

Ungerleider, L. G., and Mishkin, M., 1982, Two cortical visual systems, in: *Analysis of Visual Behavior*, D. G. Ingle, M. A. Goodale, and R. J. Q. Mansfield, eds., MIT Press, Cambridge, MA, pp. 549–586.

Vogt, B. A., 1985, Cingulate cortex, in: *Cerebral Cortex, Vol. 4: Association and Auditory Cortices*, A. Peters and E. G. Jones, eds., Plenum, New York, pp. 88–149.

Vogt, B. A., Finch, D. M., and Olson, C. R., 1992, Functional heterogeneity in cingulate cortex: The anterior executive and posterior evaluative regions, *Cereb. Cortex* **2**: 435–443.

Vogt, B. A., Pandya, D. N., and Rosene, D. L., 1987, Cingulate cortex of the rhesus monkey: I. Cytoarchitecture and thalamic afferents, *J. Comp. Neurol.* **262**: 256–270.

Vogt, B. A., Sikes, R. W., Swadlow, H. A., and Weyand, T. G., 1986, Rabbit cingulate cortex: Cytoarchitecture, physiological border with visual cortex, and afferent cortical connections of visual, motor, postsubicular, and intracingulate origin, *J. Comp. Neurol.* **248**: 74–94.

Weiskrantz, L., 1996, Blindsight revisited, *Curr. Opin. Neurobiol.* **6**: 215–220.

Weller, R. E., and Kaas, J. H., 1982, The organization of the visual system in *Galago*: Comparisons with monkeys, in: *The Lesser Bushbaby (Galago) as an Animal Model: Selected Topics*, D. E. Haines, ed., CRC Press, Boca Raton, FL, pp. 107–136.

Weller, R. E., and Kaas, J. H., 1987, Subdivisions and connections of inferior temporal cortex in owl monkeys, *J. Comp. Neurol.* **256**: 137–172.

Werner, W., Dannenberg, S., and Hoffmann, K. P., 1997, Arm-movement-related neurons in the primate superior colliculus and underlying reticular formation: Comparison of neuronal activity with EMGs of muscles of the shoulder, arm and trunk during reaching, *Exp. Brain Res.* **115**: 191–205.

Wise, S. P., 1996, Evolutionary and comparative neurobiology of the supplementary sensorimotor cortex, in: *The Supplementary Sensorimotor Cortex (Advances in Neurology, Vol. 70)*, H. Lüders, ed., Raven Press, New York, pp. 71–83.

Wise, S. P., Boussaoud, D., Johnson, P. B., and Caminiti, R., 1997, Premotor and parietal cortex: Corticocortical connectivity and combinatorial computations, *Annu. Rev. Neurosci.* **20**: 25–42.

Wong-Riley, M. T. T., 1988, Comparative study of the mammalian primary visual cortex with cytochrome oxidase histochemistry, in: *Vision: Structure and Function*, D. T. Yew, K. F. So, and D. S. C. Tsang, eds., World Scientific Press, NJ, pp. 450–486.

Wong-Riley, M. T. T., 1994, Primary visual cortex: Dynamic metabolic organization and plasticity revealed by cytochrome oxidase, in: *Cerebral Cortex, Volume 10: Primary Visual Cortex in Primates*, A. Peters and K. Rockland, eds., Plenum Press, New York, pp. 141–200.

Wu, C. W., Bichot, N. P., and Kaas, J. H., 2000, Converging evidence from microstimulation, architecture, and connections for multiple motor areas in the frontal and cingulate cortex of prosimian primates, *J. Comp. Neurol.* **423**: 140–177.

Young, M. P., 1992, Objective analysis of the topological organization of the primate cortical visual system, *Nature* **358**: 152–155.

Zilles, K., Armstrong, E., Schlaug, G., and Schleicher, A., 1986, Quantitative cytoarchitectonics of the posterior cingulate cortex in primates, *J. Comp. Neurol.* **253**: 514–524.

CHAPTER NINETEEN

New Views on the Origin of Primate Social Organization

Alexandra E. Müller, Christophe Soligo, and Urs Thalmann

INTRODUCTION

The origin of primate social organization has attracted considerable interest from primatologists. A so-called solitary pattern, believed to be present in most nocturnal prosimians, has generally been considered to be the most primitive of primate social systems (Crook and Gartlan, 1966; Eisenberg et al., 1972). There have been extensive discussions regarding possible explanations of why some primates forage in groups while others do not. Several factors have been inferred to be responsible for a "solitary" lifestyle. These include nocturnality, small body size, insectivory, and predation pressure (Bearder, 1987; Clutton-Brock and Harvey, 1977; Dunbar, 1988; Kappeler, 1997a; van Schaik, 1983; van Schaik and van Hooff, 1983). However, none of these explanations are satisfactory (Kappeler, 1997a), and it has been suggested that a combination of several factors may be involved in determining social organization in primates (Müller and Thalmann, 2000).

Alexandra E. Müller and Urs Thalmann • Anthropological Institute and Museum, University of Zürich, Winterthurerstrasse 190, 8057 Zürich, Switzerland **Christophe Soligo** • Human Origins Group, Department of Palaeontology, The Natural History Museum, Cromwell Road, London SW7 5BD, UK

With the exception of woolly lemurs (*Avahi* sp.) that live in cohesive groups (Thalmann, 2001, 2002), nocturnal prosimians are most often encountered alone during their nightly activities and are therefore termed "solitary foragers" (Bearder, 1987). This term accurately describes the difference between the lifestyles of nocturnal prosimians and group-living diurnal and cathemeral species. It does not describe the social life of nocturnal prosimians as these are not solitary in the sense of being nonsocial. Although they forage solitarily, they live in social networks that are distinguished through the formation of sleeping groups and the occurrence of regular interactions (Charles-Dominique, 1977, 1978). It is nevertheless still common in the literature to describe the social organization of nocturnal prosimians as solitary and thus to imply that patterns of social organization are the same in all nocturnal prosimians. This is clearly not the case. Instead, patterns of social organization are as diverse in nocturnal prosimians as they are in group-living species (Müller and Thalmann, 2000). Although we may accept that the ancestral primates were solitary foragers, the ancestral pattern of primate social organization cannot simply be described as "solitary," but needs a more detailed assessment.

In attempting to trace the ancestral pattern of primate social organization, the mouse and dwarf lemurs (Cheirogaleidae) as well as the bushbabies and lorises (Lorisiformes) are of special interest because they are believed to approach the ancestral condition most closely (e.g., Charles-Dominique and Martin, 1970; Martin, 1972). Charles-Dominique and Martin (1970) stressed the similarities between gray mouse lemurs (*Microcebus murinus*) and Demidoff's bushbabies (*Galagoides demidoff*) concerning size, activity cycle, habitat preference, diet, vocalization behavior and social organization pattern, and concluded that parameters that are typical for these species also represent those of ancestral primates (Charles-Dominique and Martin, 1970; Martin, 1972). Those parameters also included the pattern of social organization, which was thought to represent a harem system with the range of one adult male overlapping the ranges of several adult females (Charles-Dominique, 1972, 1977, 1995; Martin, 1972, 1981, 1995). However, new results from field studies, especially regarding the social organization of cheirogaleids, have failed to support the hypothesis of a harem system being the ancestral pattern of primate social organization (Müller and Thalmann, 2000).

The suggestion that all solitary foragers live in social networks raises further questions. Sociality is an important attribute of the entire order Primates

(Sussman, 1999), and it has been suggested that social networks were already present in the last common ancestor of living primates (Müller and Thalmann, 2000). The significance of social networks per se, however, has only rarely been treated as an important issue and the question why all primates are social, whereas this is not always the case in other mammals, has never been investigated. Because being social is not necessarily equivalent to living in groups, living in cohesive groups on the one hand and the presence of social networks in solitary foragers on the other hand suggest the presence of different selective factors. Benefits derived from living and foraging in cohesive groups, such as decreased predation pressure and improved food exploitation (e.g., Alexander, 1974; Bertram, 1978; Chapman and Chapman, 2000; Cheney and Wrangham, 1987; Jolly, 1985; Lee, 1994; van Schaik, 1983; van Schaik and van Hooff, 1983) do not apply for solitary foragers and can, therefore, not be regarded as advantages derived from the presence of sociality. Obviously, other determinants must have favored the evolution of a social network in the ancestral primate.

Within this chapter we discuss: (1) a new theory regarding the origin and evolution of primate social organization, (2) the ancestral pattern of primate social organization, and (3) a new hypothesis that aims to explain the presence of social networks in all species of primates.

DEFINITION OF SOCIAL ORGANIZATION AND SOCIALITY

Social organization consists of three different systems: the spatial system, the social system, and the mating system (Sterling, 1993). Of these, the mating system is rarely known. For that reason, we will here only consider the spatial and social systems to describe patterns of social organization. We follow the definitions of Müller and Thalmann (2000):

> The spatial system describes the spatiotemporal distribution of individuals. Spatial relations are similar throughout mammals, where four patterns are generally recognized: (1) male and female home range coincide (monogamy), (2) the home range of a male overlaps the home ranges of several females and *vice versa* (polygynandry or multimale/multifemale system), (3) the home range of one male overlaps those of several females (polygyny or harem), and (4) the home range of one female overlaps those of several males (polyandry).

In contrast, there are major differences between social systems. Social systems describe the behavior and social relations between individuals (mainly males and females). In elephant shrews (Macroscelidea) for example the home ranges of a male and a female coincide and both sexes defend their ranges against conspecifics of the same sex. Males and females, however, neither share nests nor do they exhibit social contacts outside the breeding season (FitzGibbon, 1997; Rathbun, 1979; Sauer, 1973). In fat-tailed dwarf lemurs (*Cheirogaleus medius*), the male and the female also share a home range, but contrary to the elephant shrews, *C. medius* pairs share sleeping sites and have social contacts year-round (Fietz, 1999b; Fietz et al., 2000; Müller, 1998, 1999a,b). Woolly lemurs (*Avahi* sp.) forage in cohesive groups that consist of an adult pair and their offspring (Harcourt, 1991; Thalmann, 2001, 2002; Warren and Crompton, 1997). All of these species must be labeled monogamous even though they show obvious differences in their social relations. We, therefore, distinguish between these three basic social systems and use different terms: (1) animals forage solitarily and have no social contacts with conspecifics (spatial pattern), (2) animals forage solitarily but social networks are present (dispersed pattern), and (3) animals live in cohesive groups (gregarious pattern). In this sense elephant shrews exhibit spatial monogamy, the social organization of fat-tailed dwarf lemurs is dispersed monogamy and woolly lemurs show gregarious monogamy. The same distinction between patterns of social systems is applied to the other types of spatial system (i.e., to polygynandry, polygyny, and polyandry), but for spatial multimale/multifemale systems (spatial polygynandry) we use the term "promiscuity." Species whose social organization pattern is of the spatial type are described as nonsocial, whereas those with dispersed and gregarious systems are termed social.

ORIGIN AND EVOLUTION OF PRIMATE
SOCIAL ORGANIZATION

As it is impossible to investigate the social life of fossil prosimians (but see Krishtalka et al., 1990; Plavcan et al., 2002), a reconstruction of the ancestral condition of primate social organization has to be based on what is known of the biology of extant species. It is widely accepted that the cheirogaleids and the lorisiforms are likely to represent the first primates most closely (Charles-Dominique and Martin, 1970; Martin, 1972). The social organization patterns of these primates will therefore give the best indication of the ancestral primate condition. In addition, the whole lemur group is of major interest because lemurs exhibit a wide range of size and activity types (e.g., Fleagle, 1999; Sussman, 1999; Tattersall, 1982, 1987). The high diversity of lemur adaptations may provide insights into the evolution of primate social organization, since the evolutionary shift from nocturnal to diurnal activity rhythms is believed to have involved the change from solitary foraging to foraging in cohesive groups (Martin, 1981; van Schaik and van Hooff, 1983). In addition, information on other mammals should be included in any analysis that aims to trace evolutionary patterns among primates (Martin, 1981), because "early 'advances' in primate evolution can only be defined in relation to the hypothetical ancestral condition for placental mammals [...]", and "[...] in order to reach a clear formulation of the evolutionary origins of any single order of placental mammals, it is also necessary to take into account the evolution of monotremes and the marsupials [...]" (Martin, 1990: 141).

Müller and Thalmann (2000) reviewed the social organization of the strepsirrhine primates, as well as that of several orders of mammals in order to investigate the ancestral pattern of social organization in primates and mammals in general. The ordinal relationships between primates and other mammals have yet to be resolved (see Müller and Thalmann, 2000 for references). For that reason, Müller and Thalmann (2000) used those orders that have either primitive morphological characteristic and/or are believed to be related to the primates in their comparative analysis of the evolution of social organization in primates and other mammals. The orders of placental mammals that were investigated include the insectivores (Insectivora), elephant shrews (Macroscelidea), tree shrews (Scandentia), and flying lemurs (Dermoptera). They further investigated those marsupials that are believed to approach the ancestral marsupial condition most closely, such as the didelphids (Didelphimorphia) and dasyurids (Dasyuromorphia) (Springer et al., 1997) as well as the monotremes (Monotremata). For every group, the most frequent pattern of social organization was inferred to be the ancestral condition of the group (Müller and Thalmann, 2000).

Inferences from Strepsirrhine Primates

The hypothesis that ancestral primates lived in dispersed harem systems has recently been rejected on the basis of new data on cheirogaleid and lorisiform social organization (Müller and Thalmann, 2000). If it is assumed that cheirogaleids and lorisiforms approach the ancestral primate condition most closely and that the pattern of social organization that is found most frequently in those two groups is most likely to be representative of the first primates, the hypothesis of a dispersed harem system representing the ancestral primate condition can indeed not be upheld. No cheirogaleid species studied so far exhibits a dispersed harem system. *Mirza coquereli* as well as *Microcebus murinus*, *M. ravelobensis* and presumably *M. rufus*, and *M. berthae* (see Rasoloarison et al., 2000) exhibit dispersed multimale/multifemale systems (Atsalis, 2000; Ehresmann, 2000; Fietz, 1999a; Kappeler, 1997b; Radespiel, 2000; Schwab, 2000), whereas *Cheirogaleus medius* and *Phaner furcifer* live in dispersed monogamous family groups (Charles-Dominique and Petter, 1980; Fietz, 1999b; Fietz et al., 2000; Müller 1998, 1999a,b; Schülke and Kappeler, 2003). There are as yet no data on the social organization of the remaining cheirogaleid species.

A reevaluation and reinterpretation of the patterns of social organization in bushbabies and lorises revealed that, with the possible exception of *Galago alleni*, none of the species exhibit a dispersed harem system. Instead, these species are most likely to exhibit dispersed multimale/multifemale systems (Müller and Thalmann, 2000). Bushbaby males generally have bigger home ranges than females, and the females are associated in matriarchies. The females of one group are usually aggressive toward females of another group and the males are often aggressive toward other males (Bearder and Doyle, 1974; Bearder and Martin, 1980; Bearder, 1987; Charles-Dominique, 1974, 1977). In *Galagoides demidoff* and *Galago moholi*, males have been divided into A and B males. The home range of an A male overlaps the range of one or more female groups, but other males have at least spatial access to those females as well (Bearder and Martin, 1980; Charles-Dominique, 1972, 1977; Pullen et al., 2000). In fact, Pullen et al., (2000) observed that *G. moholi* females copulated with more than one male. In *Otolemur crassicaudatus* and *O. garnettii*, an adult male overlaps only one female association but again, the females have access to more than one male (Bearder, 1987; Clark, 1985; Doyle and Bearder, 1977; Nash and Harcourt, 1986). By contrast, *Galagoides zanzibaricus* is organized in dispersed monogamous groups (Harcourt and Nash, 1986a). Data on lorisid social organization are very limited (Müller and Thalmann, 2000) but the Asian slender loris (*Loris tardigradus*) has been reported to live in dispersed multimale/multifemale systems (Nekaris, 2003).

The larger solitary foragers (*Daubentonia madagascariensis* and *Lepilemur* sp.) and the gregarious lemurs do not live in harems either but in multi-male/multifemale systems or monogamously. Based on all of the above data, it was concluded that dispersed multimale/multifemale systems and dispersed monogamy are the predominant patterns of social organization found in strepsirrhine primates (Müller and Thalmann, 2000) (Figure 1). A dispersed multimale/multifemale system is the most frequent pattern among strepsirhines, but monogamy occurs almost as often, especially in lemurs. Which

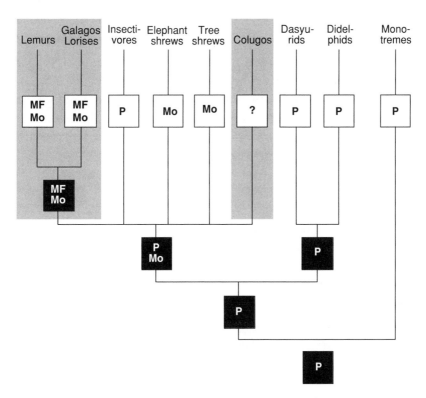

Figure 1. Cladogram indicating the predominant patterns of social organization in strepsirrhine primates, "primitive" placentals (insectivores, elephant shrews, tree shrews, and colugos), "primitive" marsupials (dasyurids and didelphids), and monotremes. P = Promiscuity; MF = Multimale/Multifemale system; Mo = Monogamy; ? = unknown. Data from the literature are in white boxes, possible ancestral patterns are in black boxes. Animal groups that exhibit social networks are underlain with gray. (Adapted from Müller and Thalmann, 2000.)

of these two systems is most likely to represent the ancestral condition for the order Primates, therefore, remains unresolved.

Inferences from "Primitive" Mammals

A review of the patterns of social organization in those groups of placental mammals that are thought to be primitive and/or related to primates also failed to support a harem system as the ancestral pattern for either primates or placental mammals. Very little is known about patterns of social organization in insectivores, but the majority of species exhibit a nonsocial lifestyle (Müller and Thalmann, 2000). Only few species live in social networks. These include the lesser moonrats (*Hylomys suillus*), American least shrews (*Cryptotis parva*), desert shrews (*Notiosorex crawfordi*), southern water shrews (*Neomys anomalus*), some white-toothed shrew species (*Crocidura* sp.), desmans (*Desmana moschata*), American shrew moles (*Neurotrichus gibbsii*), eastern moles (*Scalopus aquaticus*) and star-nosed moles (*Condylura cristata*) (e.g., Cantoni and Vogel, 1989; Harvey, 1976; Krushinka and Rychlik, 1993; Nowak, 1999). The dominant pattern, however, is promiscuity with the ranges of several males overlapping the ranges of several females and an overall lack of social contacts (Figure 1). Males generally compete for estrous females. Only two species are known to be monogamous: the greater white-toothed shrew (*Crocidura russula*) and the Pyrenean desman (*Galemys pyrenaicus*) (Cantoni and Vogel, 1989; Favre et al., 1997; Stone, 1987). To date no species of insectivore has been reported to live in a harem system (Müller and Thalmann, 2000).

By contrast, elephant shrews and tree shrews exhibit monogamy in that the home ranges of an adult male and an adult female coincide (FitzGibbon, 1997; Gould, 1978; Kawamichi and Kawamichi, 1979, 1982; Nowak, 1999; Rathbun, 1979; Sauer, 1973) (Figure 1). In elephant shrews the adult pair has no social contacts (FitzGibbon, 1997; Rathbun, 1979; Sauer, 1973) and their social organization can therefore be referred to as spatial monogamy. In wild tree shrews, the situation remains somewhat unclear. Kawamichi and Kawamichi (1979) report that in common tree shrews (*Tupaia glis*), 18% of encounters were social encounters and that one-third of these social encounters occurred between an adult male and an adult female. The majority of these intersexual encounters occur between partners (Kawamichi and Kawamichi, 1979). It is not known whether partners share their sleeping sites in the wild, but in captivity, the adult pair shares a nest (Martin, 1968). Pen-tailed tree shrews (*Ptilocercus lowii*) also occur in pairs, but Madras tree shrews (*Anathana ellioti*) are mostly seen alone and do not share their nests (Chorazyna and Kurup, 1975; Emmons, 2000; Gould, 1978; Nowak, 1999).

Nothing is known regarding the social organization of flying lemurs, except that individuals forage solitarily and may share sleeping sites (Nowak, 1999). This would indicate that their social organization is of the dispersed type (Figure 1).

Insectivores, tree shrews, and elephant shrews all seem to be relatively close to the ancestral placental stock. It remains unclear, however, which of these orders is most representative of the ancestral placental stock. There are still two possible interpretations regarding the ancestral pattern of social organization in placental mammals: promiscuity and spatial monogamy.

Didelphids and dasyurids generally have a nonsocial lifestyle and the males compete over estrous females (e.g., Bradley, 1997; Buchman and Guiler, 1977; Collins, 1973; Cuttle, 1982; Fleming, 1972; Fox and Whitford, 1982; Jarman and Kruuk, 1996; Lazenby-Cohen and Cockburn, 1988; Morton, 1978; Nowak, 1999; Righetti, 1996; Ryser, 1992; Serena and Soderquist, 1989; Soderquist, 1995; Sunquist et al., 1987; Woolley, 1991). The available data strongly suggest promiscuity to be the dominant pattern of social organization among those marsupials that are believed to approach the ancestral marsupial condition most closely (Müller and Thalmann, 2000) (Figure 1). The available data also suggest promiscuity to be the pattern of social organization present in the monotremes (e.g., Augee et al., 1975; Brattstrom, 1973; Gardner and Serena, 1995; Griffiths, 1978; Gust and Handasyde, 1995; Nowak, 1999; Serena, 1994). It is therefore most likely that promiscuity, being the dominant pattern of social organization among "primitive" mammals, represents the ancestral condition for mammals in general (Müller and Thalmann, 2000) (Figure 1).

Reconstruction of the Ancestral Pattern of Primate Social Organization

The promiscuous systems of the monotremes, didelphids, dasyurids, and insectivores, as well as the spatial monogamy of tree shrews and elephant shrews are very similar to the dispersed multimale/multifemale system and the dispersed monogamy found in cheirogaleids and lorisiforms. The spatial aspects of social organization are in fact the same in promiscuous and dispersed polygynandrous systems on the one hand and in spatial and dispersed monogamy on the other hand. The difference is that in each case social networks are present in the latter (i.e., in cheirogaleids and lorisiforms), but not in the former system (i.e., in all other groups). As a result, Müller and Thalmann (2000) have suggested that the most likely scenario regarding the origin of primate social organization is that a dispersed multimale/multifemale pattern arose from promiscuity. Such a dispersed multimale/multifemale pattern is found in all the species of mouse lemur (*Microcebus* sp.), which have so far been investigated, in *Mirza coquereli*, as well as in the majority of species of bushbaby and in the slender loris.

Many of the species that are organized in dispersed multimale/multifemale groups are characterized by the presence of matriarchies. This is the case in most bushbabies, as well as in mouse lemurs (Bearder, 1987; Charles-Dominique, 1977; Kappeler et al., 2002; Radespiel et al., 2001; Wimmer et al., 2002). This relatedness between the females of multimale/multifemale groups might have been at the origin of the development of social networks as the evolution of sociality is believed to have involved kin selection (West-Eberhard, 1975). We, therefore, argue that the ancestral pattern of primate social organization was a dispersed multimale/multifemale system with the core group being formed by related females, which engaged in social contacts with some males. Thus, in the earliest primates, groups of related females were at the origin of the evolution of social networks from the ancestral system of promiscuity.

CAUSES FOR SOCIALITY IN PRIMATES

The main difference between the promiscuity of primitive placentals and the dispersed multimale/multifemale pattern of some of the nocturnal strepsirhines is the presence of social networks in the latter. Within these social networks, social contacts occur throughout the year and are not restricted to the breeding season. This contrasts with many other "solitary" mammals that have no friendly contacts with conspecifics apart from mother–infant relationships and mating contacts (Müller and Thalmann, 2000). The question why primates have evolved sociality while many other mammals did not has yet to be resolved. Sociality per se can bring several benefits, such as a reduction of ectoparasite-born disease through allogrooming, alloparenting by older offspring, and the sharing of information on food resources and sleeping sites (Clark, 1985; Müller, 1999a). Among small animals, thermoregulatory advantages through nest-sharing are possible (Genoud et al., 1997; Perret, 1998; van Schaik and van Hooff, 1983). It remains unclear, however, why other mammals can do without these suggested advantages, whereas primates, seemingly, cannot.

Since all living primates are social, it is impossible to investigate the natural history of nonsocial primates to infer causes of sociality in primates. For that reason, Müller and Soligo (2002, 2005) used the order Rodentia as a model. Rodents occupy almost all potential ecological niches and exhibit a wide variety of patterns of social organization ranging from intolerant solitary species to eusocial forms. The aim of this approach was to explore possible determinants for the absence or presence of sociality in rodents and to apply the findings to the origin of sociality in primates. The following potential determinants were investigated: substrate utilization (terrestrial *versus* arboreal), activity cycles (diurnal *versus* nocturnal), dietary preferences (faunivore, frugi–omnivore, granivore, frugi–herbivore, or herbivore) and body size (<100 g, 100 g–1 kg, or >1 kg).

Ecological Factors and Sociality in Rodents

Large (>1 kg) arboreal rodent genera that include an important amount of fruits in their diet (i.e., frugi–omnivorous and frugi–herbivorous forms), are more likely to be social than other rodents (Müller and Soligo, 2002, 2005). In contrast, activity cycle did not appear to influence the presence or absence of sociality (Müller and Soligo, 2005), which is in agreement with the fact that all primates exhibit a social lifestyle, whether they are nocturnal or diurnal. A diurnal lifestyle, however, seems to be an important attribute that promotes living in cohesive groups (van Schaik and van Hooff, 1983). The nocturnal group-living species *Avahi* spp. and *Aotus* spp. may not represent valid counterarguments as it has been suggested that the nocturnal lifestyle of these species is a secondary attribute (Eisenberg et al., 1972; Martin, 1990; Müller and Thalmann, 2000; Sussman, 1999, 2000; Thalmann, 2001) and that their gregarious pattern of social organization arose from the gregarious pattern of diurnal primates and not from the dispersed nocturnal type (Müller and Thalmann, 2000).

The fact that large rodents (>1 kg) exhibit a social lifestyle more often than medium-sized and small forms would support the suggestion that small body

size is an important determinant of a "solitary" lifestyle as found among nocturnal prosimians (e.g., Charles-Dominique, 1978; Clutton-Brock and
Harvey, 1977). However, most "solitary" nocturnal prosimians are social, but
weigh less than 1 kg. Large body size in primates might therefore at best be
a determinant for foraging in cohesive groups. Alternatively, sociality may
have evolved in an ancestral primate with a body weight above 1 kg and was
subsequently retained throughout the evolution of the order. As body size
influences many life-history traits and affects social behavior, sociality may in
fact not be directly linked to body size but to size-related factors instead
(Armitage, 1981).

It is puzzling that arboreal rodents are more likely to be social than their
terrestrial counterparts, as this seems to contradict the hypothesis that predation pressure encourages sociality and because terrestrial animals are
assumed to be more susceptible to predation than arboreal species (Dunbar,
1988). However, since the presence of social networks is not equivalent to
living in groups, it does not in itself carry benefits concerning predation
avoidance. Furthermore, at least among primates, terrestrial species do not
necessarily suffer higher predation risks than arboreal forms (Isbell, 1994).
The study on rodents, however, revealed that arboreal and terrestrial genera
differ in their dietary preferences (Figure 2). Whereas the amount of
frugi–omnivorous genera is the same both in the tree and on the ground,
more terrestrial rodents are faunivorous or herbivorous and more arboreal
forms have a granivorous or frugi–herbivorous diet (Müller and Soligo,
2005) (Figure 2). This is not surprising since food availability is not the same

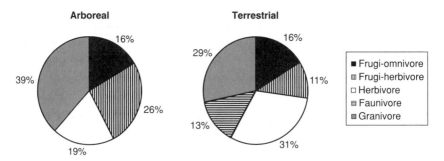

Figure 2. Percentage of dietary categories in arboreal and terrestrial rodent genera
(after Müller and Soligo, 2005).

for arboreal and terrestrial animals: insects occur mainly in the leaf-litter on the ground or in the shrub and underground layer (Cartmill, 1972) and grasses, forbs, roots, and stalks are not found in trees. By contrast, fruits and berries grow in trees and shrubs and their availability to terrestrial species is therefore very limited.

Insects generally occur in small and highly dispersed patches (Oates, 1987). Green vegetation is either uniformly scattered and more-or-less stable (leaves, roots, stems, and stalks) or clumped and unpredictable (grasses, herbs, and forbs) (Clutton-Brock, 1974). By contrast, fruits, berries, seeds, young leaves, and flowers are patchily distributed in space and time and therefore, difficult to find. Hence, the limiting factor for feeding on these items is to find a good site (Krebs and Davies, 1987; Milton, 1981; Oates, 1987). Once a good location is known, the resource can be depended on because fruits, seeds, flowers, and young leaves show a high degree of predictability. Animals can move directly between resources if they know where to find them and when they are fruiting or flowering. This is obviously preferable to wasting time and energy by searching in a random fashion. Because the selective factor enabling animals to locate such resources seems to be increased mental complexity with an emphasis on learning and retention rather than genetic coding, membership in a close social unit can increase the efficiency of food exploitation if information on feeding sites is shared (Cords, 2000; Gautier-Hion, 1988; Milton, 1981). This has also been suggested for nesting colonies and communal roots of birds that may serve as "information centers" where individuals can find out where good feeding sites are (Ward and Zahavi, 1973). The same might be true of nocturnal prosimians and other animals that feed on dispersed but predictable food resources. In a social network, adult animals can teach their offspring or siblings about food locations (Clark, 1985; Müller, 1999a; Müller and Soligo, 2002).

This hypothesis is supported by the fact that the presence of a social lifestyle in rodents varies according to dietary preferences (Figure 3). The largest number of social rodents are found within frugi–omnivorous and frugi–herbivorous genera and the proportion of social rodents within these categories lies well above the rodent average (Müller and Soligo, 2005). Although seeds are believed to be patchily distributed, granivorous rodents are less likely to be social than other rodents. Many granivorous rodents hoard food and limiting pilferage could be more important to those animals than maximizing the efficiency of localizing food (Müller and Soligo, 2005).

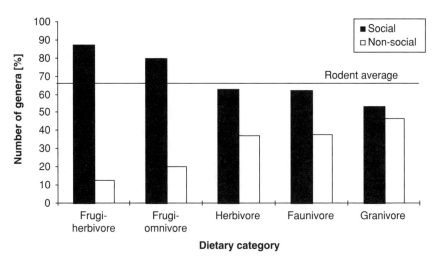

Figure 3. Percentage of dietary categories in social and nonsocial rodent genera. Rodent average (i.e., overall percentage of social genera in rodents) is 66.2% (after Müller and Soligo, 2005).

The fact that large (>1 kg) arboreal genera that feed on fruits and vegetation or on fruits and animal matter are more likely to be social than other rodents supports the notion of a link between a diet that includes an important amount of fruits and the evolution of social networks. The relevant factor between frugivory and the development of social networks lies in the importance of sharing information on food resources. This can be extended to argue that a frugivorous diet lay behind the evolution of social networks in early primates (Müller and Soligo, 2002, 2005). Indeed, social networks are thought to have already been present in the last common ancestor of living primates (Müller and Thalmann, 2000). Based on the ecological correlates of sociality found in rodents, Müller and Soligo (2005) have predicted that social stem-primates were relatively large (>1 kg) and arboreal and had a diet that included a significant amount of fruit. This hypothesis is also supported by data on marsupials where those species that include fruits in their diets exhibit social networks whereas others do not (Müller, 1999b; Rasmussen and Sussman, this volume). It is, however, partially at odds with the current majority view concerning the nature of the first primates, as these are often believed to have been small, insectivorous, nocturnal, and arboreal.

Body Size

The suggestion that the earliest primates were large (>1 kg) contrasts with the current consensus view that the first primates were small with body weights in the range of mouse and dwarf lemurs (e.g., Cartmill, 1974a; Dagosto, 1988; Martin 1972, 1990). Some other evidence, however, has also pointed to the possibility that the first primates were larger than previously thought (Soligo, 2001; Soligo and Müller 1999a,b). The presence of nails as opposed to functional claws on the digits of the foot and hand is thought to be a defining characteristic of the order Primates (see Soligo and Müller, 1999b). Despite this, the issue of why functional claws should have been lost in the lineage that led to the last common ancestor of modern primates is still contentious (e.g., Cartmill, 1974a, 1985, 1992). It has recently been argued that functional claws were reduced to nails as a result of a phyletic increase in body mass in the lineage leading to the last common ancestor of modern primates (Soligo, 2001; Soligo and Müller, 1999a,b). A recent comparison of the size distribution of claw-bearing and nail-bearing species of arboreal mammals has added support to this hypothesis, as it revealed that the nail-bearing species were significantly larger than the claw-bearing species (Soligo, 2001). Based on the details of the size distributions of the two groups, Soligo (2001) argued that a weight of more than 800 g was attained in the lineage that led to the last common ancestor of modern primates. That estimate is in agreement with the findings of Müller and Soligo (2005) discussed earlier.

Diet

Forward-facing eyes and a reduced sense of smell are characteristic features of all modern primates. It is hence likely that these features evolved in the lineage leading to the last common ancestor of modern primates. Based, on the one hand, on the observation that optic convergence is a feature characteristic of many predators, and, on the other hand, on the assumption that most recent strepsirhines are insectivorous, Cartmill (1972, 1974b, 1992) concluded that the first primates were visually oriented predators, which preyed primarily on insects. Diet, however, could not have promoted the presence of social networks in ancestral primates as suggested by Müller and Soligo (2002, 2005), if it consisted mainly of invertebrate prey. By contrast, Sussman (1991, 1999; Sussman and Raven, 1978) sought to explain the typical adaptive traits of primates in the context of the Tertiary radiation of flowering plants (angiosperms) and argued that there are only very few species of primates that are known to include more animal matter than plant material in their diet. In addition, primate gut morphology reflects adaptations for an omnivorous diet (Martin et al., 1985). Sussman (1991, 1999; Sussman and Raven, 1978), therefore, proposed that early primates had an omnivorous diet that included a significant part of plant material and that primates coevolved with the angiosperms. Indeed most of the species that are believed to approach the ancestral primate condition most closely, such as cheirogaleids and lorisiforms, have a frugi–omnivorous diet (Atsalis, 1999; Fietz and Ganzhorn, 1999; Harcourt and Nash, 1986b; Hladik et al., 1980; Nash et al., 1989) and only few species are insectivorous (some of the small bushbabies (*Galagoides* sp.), slender lorises (*Loris tardigradus*) and angwantibos (*Arctocebus* sp.)) (Charles-Dominique, 1977; Harcourt and Nash, 1986b; Nash et al., 1989; Nekaris and Rasmussen, 2003). The suggestion that the earliest primates weighed more than 800 g (Soligo, 2001) adds support to the hypothesis that the first primates lived on a frugi–omnivorous diet. According to Kay (1984), a primate weighing more than 350 g cannot be primarily insectivorous. At more than 800 g, the first primates would, therefore, have had to include a substantial amount of fruit in their diet.

CONCLUSIONS

The ancestral condition of primate social organization is a dispersed multimale/multifemale system (= dispersed polygynandry). Within this system the ranges of adult males overlap the ranges of several adult females and *vice versa*, and males and females exhibit year-round social contacts, i.e., they live in social networks. This contrasts with other mammals that often have the same spatial system but differ from primates through the lack of social networks (= promiscuity). It is hypothesized that promiscuity is the ancestral condition of mammalian social organization and that a dispersed multimale/multifemale system arose from this system through the evolution of social networks in the earliest primates. Social networks are likely to have evolved through a prolongation of mother–infant relations.

The ubiquitous presence of social networks in primates is believed to be due to their primarily frugivorous diet. On the one hand, fruits are patchily distributed in space and time and, therefore, difficult to find, but on the other hand they are predictable and can be depended on once their location is known. Social networks are beneficial as they enable animals to share information about food sites with their offspring or siblings. Social networks are very likely to have been present in the last common ancestor of living primates, in conjunction with a diet that included large amounts of fruit.

ACKNOWLEDGMENTS

We thank Marian Dagosto and Matt Ravosa for inviting us to participate in the conference and this volume. Our special thanks go to our mentor Bob Martin, who gave us any possible support during our studies and who raised our interest in primate evolution.

REFERENCES

Alexander, R. D., 1974, The evolution of social behavior, *Ann. Rev. Ecol. Syst.* **5:** 325–383.

Armitage, K. B., 1981, Sociality as a life-history tactic of ground squirrels, *Oecologia* **48:** 36–49.

Atsalis, S., 1999, Diet of the brown mouse lemur (*Microcebus rufus*) in Ranomafana National Park, Madagascar, *Int. J. Primatol.* **20:** 193–229.

Atsalis, S., 2000, Spatial distribution and population composition of the brown mouse lemur (*Microcebus rufus*) in Ranomafana National Park, Madagascar, and its implications for social organization, *Am. J. Primatol.* **51:** 61–78.

Augee, M. L., Ealey, E. H. M., and Price, I. P., 1975, Movements of echidnas, *Tachyglossus aculeatus*, determined by marking-recapture and radio tracking, *Austr. Wildl. Res.* **2:** 93–101.

Bearder, S. K., 1987, Lorises, bushbabies, and tarsiers: Diverse societies in solitary foragers, in: *Primate Societies*, B. B. Smuts, D. L. Cheney, R. M. Seyfarth, R. W. Wrangham, and T. T. Struhsaker, eds., University of Chicago Press, Chicago, pp. 11–24.

Bearder, S. K., and Doyle, G. A., 1974, Field and laboratory studies of social organization in bushbabies (*Galago senegalensis*), *J. Hum. Evol.* **3:** 37–50.

Bearder, S. K., and Martin, R. D., 1980, The social organisation of a nocturnal primate revealed by radio tracking, in: *A Handbook on Biotelemetry and Radio Tracking*, C. J. Amlaner Jr. and D. W. MacDonald, eds., Pergamon Press, Oxford, pp. 633–648.

Bertram, B. C., 1978, Living in groups: Predators and prey, in: *Behavioural Ecology. An Evolutionary Approach*, 2nd Ed., J. R. Krebs and N. B. Davies, eds., Blackwell Scientific Publications, Oxford, pp. 64–96.

Bradley, A. J., 1997, Reproduction and life history in the red-tailed phascogale, *Phascogale calura* (Marsupialia: Dasyuridae): The adaptive-stress senescence hypothesis, *J. Zool.* **241**: 739–755.

Brattstrom, B. H., 1973, Social and maintenance behavior of the echidna *Tachyglossus aculeatus*, *J. Mammal.* **54**: 50–70.

Buchman, O. L. K., and Guiler, E. R., 1977, Behaviour and ecology of the Tasmanian devil, *Sarcophilus harrisii*, in: *The Biology of Marsupials*, B. Stonehouse and D. Gilmore, eds., The Macmillan Press Ltd, London, pp. 155–168.

Cantoni, D., and Vogel, P., 1989, Social organization and mating system of free-ranging, greater white-toothed shrews, *Crocidura russula*, *Anim. Behav.* **38**: 205–214.

Cartmill, M., 1972, Arboreal adaptations and the origin of the Order Primates, in: *The Functional and Evolutionary Biology of Primates*, R. Tuttle, ed., Aldine-Atherton, Chicago, pp. 97–122.

Cartmill, M., 1974a, Pads and claws in arboreal locomotion, in: *Primate Locomotion*, F. Jenkins, ed., Plenum, New York, pp. 45–83.

Cartmill, M., 1974b, Rethinking primate origins, *Science* **184**: 436–443.

Cartmill, M., 1985, Climbing, in: *Functional Vertebrate Morphology*, M. Hildebrand, ed., Aldine-Atherton, Chicago, pp. 73–88.

Cartmill, M., 1992, New views on primate origins. *Evol. Anthropol.* **1**: 105–111.

Chapman, C. A., and Chapman, L. J., 2000, Determinants of group size in primates: The importance of travel costs, in: *On the Move: How and Why Animals Travel in Groups*, S. Boinski and P. A. Garber, eds., University of Chicago Press, Chicago, pp. 24–42.

Charles-Dominique, M., and Martin, R. D., 1970, Evolution of lorises and lemurs, *Nature* **227**: 257–260.

Charles-Dominique, P., 1972, Ecologie et vie sociale de *Galago demidovii* (Fischer 1808, Prosimii), *Z. Tierpsychol. Beiheft.* **9**: 7–41.

Charles-Dominique, P., 1974, Aggression and territoriality in nocturnal prosimians, in: *Primate Aggression, Territoriality, and Xenophobia. A Comparative Perspective*, R. L. Holloway, ed., Academic Press, New York, pp. 31–48.

Charles-Dominique, P., 1977, *Ecology and Behavior of Nocturnal Primates*, Columbia University Press, New York.

Charles-Dominique, P., 1978, Solitary and gregarious prosimians: Evolution of social structures in primates, in: *Evolution*, D. J. Chivers and K. A. Joysey, eds., Academic Press, New York, pp. 139–149.

Charles-Dominique, P., 1995, Food distribution and reproductive constraints in the evolution of social structure: Nocturnal primates and other mammals, in: *Creatures of the Dark: The Nocturnal Prosimians*, L. Altermann, G. A. Doyle, and M. K. Izard, eds., Plenum Press, New York, pp. 425–438.

Charles-Dominique, P., and Petter, J. J., 1980, Ecology and social life of *Phaner fur-cifer*, in: *Nocturnal Malagasy Primates. Ecology, Physiology, and Behavior*, P. Charles-Dominique, H. M. Cooper, A. Hladik, C. M. Hladik, E. Pagès, G. F. Pariente et al., eds., Academic Press, New York, pp. 75–96.

Cheney, D. L., and Wrangham, R. W., 1987, Predation, in: *Primate Societies*, B. B. Smuts, D. L. Cheney, R. M. Seyfarth, R. W. Wrangham, and T. T. Struhsaker, eds., University of Chicago Press, Chicago, pp. 267–281.

Chorazyna, H., and Kurup, G. U., 1975, Observations on the ecology and behaviour of *Anathana ellioti* in the wild, in: *Contemporary Primatology*, S. Kondo, M. Kawai, and A. Ehara, eds., Karger, Basel, pp. 342–344.

Clark, A. B., 1985, Sociality in a nocturnal "solitary" prosimian: *Galago crassicauda-tus*, *Int. J. Primatol.* **6**: 581–600.

Clutton-Brock, T. H., 1974, Primate social organisation and ecology, *Nature* **250**: 539–542.

Clutton-Brock, T. H., and Harvey, P. H., 1977, Primate ecology and social organiza-tion, *J. Zool.* **183**: 1–39.

Collins, L. R., 1973, *Monotremes and marsupials: A Reference for Zoological Institutions*, Smithsonian Institution Press, Washington.

Cords, M., 2000, Mixed species association and group movement, in: *On the Move: How and Why Animals Travel in Groups*, S. Boinski and P. A. Garber, eds., University of Chicago Press, Chicago, pp. 73–99.

Crook, J. H., and Gartlan, J. S., 1966, Evolution of primate societies, *Nature* **210**: 1200–1203.

Cuttle, P., 1982, Life history of the dasyurid marsupial *Phascogale tapoatafa*, in: *Carnivorous Marsupials*, M. Archer, ed., Mosman: The Royal Zoological Society, New South Wales, pp. 13–22.

Dagosto, M., 1988, Implications of postcranial evidence for the origin of Euprimates, *J. Hum. Evol.* **17**: 35–56.

Doyle, G. A., and Bearder, S. K., 1977, The galagines of South Africa, in: *Primate Conservation*, R. Prince, III and G. H. Bourne, eds., Academic Press, New York, pp. 1–35.

Dunbar, R. I. M., 1988, *Primate Social Systems*, Croom Helm, London.

Ehresmann, P., 2000 *Ökologische Differenzierung von zwei sympatrischen Mausmaki-Arten (Microcebus murinus und M. ravelobensis) im Trockenwald Nordwest-Madagaskars*, Ph.D., University of Hannover.

Eisenberg, J. F., Muckenhirn, N. A., and Rudran, R., 1972, The relation between ecology and social structure in primates, *Science* **176**: 863–874.

Emmons, L. H., 2000, *Tupai: A Field Study of Bornean Treeshrews*, University of California Press, Berkeley.

Favre, L., Balloux, F., Goudet, J., and Perrin, N., 1997, Female-biased dispersal in the monogamous mammal *Crocidura russula*: Evidence from field data and microsatellite patterns, *Proc. R. Soc. Lond. B* **264**: 127–132.

Fietz, J., 1999a, Mating system of *Microcebus murinus*, *Am. J. Primatol.* **48**: 127–133.

Fietz, J., 1999b, Monogamy as a rule rather than exception in nocturnal lemurs: The case of the fat-tailed dwarf lemur (*Cheirogaleus medius*), *Ethology* **105**: 259–272.

Fietz, J., and Ganzhorn, J. U., 1999, Feeding ecology of the hibernating primate *Cheirogaleus medius*: How does it get so fat? *Oecologia* **121**: 157–164.

Fietz, J., Zischler, H., Schwiegk, C., Tomiuk, J., Dausmann, K. H., and Ganzhorn, J. U., 2000, High rates of extra-pair young in the pair-living fat-tailed dwarf lemur, *Cheirogaleus medius*, *Behav. Ecol. Sociobiol.* **49**: 8–17.

FitzGibbon, C. D., 1997, The adaptive significance of monogamy in the golden-rumped elephant-shrew, *J. Zool.* **242**: 167–177.

Fleagle, J., 1999, *Primate Adaptation and Evolution*, Academic Press, New York.

Fleming, T. H., 1972, Aspects of the population dynamics of three species of opossums in the Panama Canal Zone, *J. Mammal.* **53**: 619–623.

Fox, B. J., and Whitford, D., 1982, Polyoestry in a predictable coastal environment: Reproduction, growth and development in *Sminthopsis murina* (Dasyuridae, Marsupialia), in: *Carnivorous Marsupials*, M. Archer, ed., Mosman: The Royal Zoological Society, New South Wales, pp. 39–48.

Gardner, J. L., and Serena, M., 1995, Spatial organisation and movement patterns of adult male platypus, *Ornithorhynchus anatinus* (Monotremata: Ornithorhynchidae), *Austr. J. Zool.* **43**: 91–103.

Gautier-Hion, A., 1988, Polyspecific associations among forest guenons: Ecological, behavioural and evolutionary aspects, in: *A Primate Radiation: Evolutionary Biology of the African Guenons*, A. Gautier-Hion, F. Bourlière, J. P. Gautier, and J. Kingdon, eds., Cambridge University Press, Cambridge, pp. 452–476.

Genoud, M. R., Martin, R. D., and Glaser, D., 1997, Rate of metabolism in the smallest simian primate, the pygmy marmoset (*Cebuella pygmaea*), *Am. J. Primatol.* **41**: 229–245.

Gould, E., 1978, The behavior of the moonrat, *Echinosorex gymnurus* (Erinaceidae) and the pentail shrew, *Ptilocercus lowi* (Tupaiidae) with comments on the behavior of other insectivora, *Z. Tierpsychol. Beiheft* **48**: 1–27.

Griffiths, M., 1978, *The Biology of the Monotremes*, Academic Press., New York.

Gust, N., and Handasyde, K., 1995, Seasonal variation in the ranging behavior of the platypus (*Ornitorhynchus anatinus*) on the Goulburn River, Victoria, *Austr. J. Zool.* **43**: 193–208.

Harcourt, C., 1991, Diet and behaviour of a nocturnal lemur, *Avahi laniger*, in the wild, *J. Zool.* **223**: 667–674.

Harcourt, C. S., and Nash, L. T., 1986a, Social organization of galagos in Kenyan coastal forests: I. *Galago zanzibaricus, Am. J. Primatol.* **10**: 339–355.

Harcourt, C. S., and Nash, L. T., 1986b, Species differences in substrate use and diet between sympatric galagos in two Kenyan coastal forests, *Primates* **27**: 41–52.

Harvey, M. J., 1976, Home range, movements, and diel activity of the eastern mole, *Scalopus aquaticus, Am. Midl. Nat.* **95**: 436–445.

Hladik, C. M., Charles-Dominique, P., and Petter, J. J., 1980, Feeding strategies of five nocturnal prosimians in the dry forest of the West coast of Madagascar, in: *Nocturnal Malagasy Primates*, P. Charles Dominique, H. M. Cooper, A. Hladik, C. M. Hladik, E. Pagès, G. F. Pariente et al., eds., *Ecology, Physiology, and Behavior*, Academic Press, New York, pp. 41–74.

Isbell, L. A., 1994, Predation on primates: Ecological patterns and evolutionary consequences, *Evol. Anthropol.* **3**: 61–71.

Jarman, P. J., and Kruuk, H., 1996, Phylogeny and spatial organisation in mammals, in: *Comparison of Marsupial and Placental Behaviour*, D. B. Croft and U. Ganslosser, eds., Filander Verlag, Fürth, pp. 80–101.

Jolly, A., 1985, *The Evolution of Primate Behavior* (2nd Ed), Macmillan, New York.

Kappeler, P. M., 1997a, Determinants of primate social organization—Comparative evidence and new insights from Malagasy lemurs, *Biol. Rev.* **72**: 111–151.

Kappeler, P. M., 1997b, Intrasexual selection and *Mirza coquereli*: Evidence for scramble competition polygyny in a solitary primate, *Beh. Ecol. Sociobiol.* **45**: 115–127.

Kappeler, P. M., Wimmer, B., Zinner, D., and Tautz, D., 2002, The hidden matrilineal structure of a solitary lemur: Implications for primate social evolution, *Proc. R. Soc. Lond. B* **269**: 1755–1763.

Kawamichi, T., and Kawamichi, M., 1979, Spatial organization and territory of tree shrews (*Tupaia glis*), *Anim. Behav.* **27**: 381–393.

Kawamichi, T., and Kawamichi, M., 1982, Social system and independence of offspring in tree shrews, *Primates* **23**: 189–205.

Kay, R. F., 1984, On the use of anatomical features to infer foraging behavior in extinct primates, in: *Adaptations for Foraging in Nonhuman Primates*, P. S. Rodman and J. G. H. Cant, eds., Columbia University Press, New York, pp. 21–53.

Krebs, J. R., and Davies, N. B., 1987, *An Introduction to Behavioural Ecology* (2nd Ed.), Blackwell Scientific Publications., Oxford.

Krishtalka, L., Stucky, R. K., and Beard, K. C., 1990, The earliest fossil evidence for sexual dimorphism in primates, *Proc. Natl. Acad. Sci.* **87**: 5223–5226.

Krushinka, N. L., and Rychlik, L., 1993, Intra- and interspecific antagonistic behaviour in two sympatric species of water shrews: *Neomys fodiens* and *N. anomalus, J. Ethol.* **11**: 11–21.

Lazenby-Cohen, K. A., and Cockburn, A., 1988, Lek promiscuity in a semelparous mammal, *Antechinus stuartii* (Marsupialia: Dasyuridae)? *Behav. Ecol. Sociobiol.* **22**: 195–202.

Lee, P. C., 1994, Social structure and evolution, in: *Behaviour and Evolution*, P. J. B. Slater and T. R. Halliday, eds., Cambridge University Press, Cambridge, pp. 266–337.

Martin, R. D., 1968, Reproduction and ontogeny in tree-shrews (*Tupaia belangeri*), with reference to their general behaviour and taxonomic relationships, *Z. Säugetierk.* **25**: 409–495, 505–532.

Martin, R. D., 1972, Adaptive radiation and behavior of the Malagasy lemurs, *Phil. Trans. R. Soc. Lond.* **264**: 295–352.

Martin, R. D., 1981, Field studies of primate behaviour, *Symp. Zool. Soc. Lond.* **46**: 287–336.

Martin, R. D., 1990, *Primate Origins and Evolution—A Phylogenetic Reconstruction*, Princeton University Press, Princeton.

Martin, R. D., 1995, Prosimians: From obscurity to extinction? in: *Creatures of the Dark: The Nocturnal Prosimians*, L. Altermann, G. A. Doyle, and M. K. Izard, eds., Plenum Press, New York, pp. 535–563.

Martin, R. D., Chivers, D. J., MacLarnon, A. M., and Hladik, C. M., 1985, Gastrointestinal allometry in primates and other mammals, in: *Size and Scaling in Primate Biology*, W. J. Jungers, ed., Plenum Press, New York, pp. 61–89.

Milton, K., 1981, Distribution patterns of tropical plant foods as an evolutionary stimulus to primate mental development, *Am. Anthropol.* **83**: 534–548.

Morton, D. W., 1978, An ecological study of *Sminthopsis crassicaudata* (Marsupialia: Dasyuridae). II. Behaviour and social organization, *Austr. Wildl. Res.* **5**: 163–182.

Müller, A. E., 1998, A preliminary report on the social organisation of *Cheirogaleus medius* (Cheirogaleidae; Primates) in North-West Madagascar, *Folia Primatol.* **69**: 160–166.

Müller, A. E., 1999a, *The Social Organisation of the Fat-Tailed Dwarf Lemur, Cheirogaleus medius (Lemuriformes; Primates)*, Ph.D., Univesity of Zurich.

Müller, A. E., 1999b, Social organization of the fat-tailed dwarf lemur (*Cheirogaleus medius*) in northwestern Madagascar, in: B. Rakotosamimanana, H. Rasamimanana, J. U. Ganzhorn, and S. M. Goodman, eds., *New Directions in Lemur Studies*, Kluwer Academic/Plenum Publishers, New York, pp. 139–158.

Müller, A. E., and Soligo, C., 2002, Causes for primate sociality: Inferences from other mammals, *Am. J. Phys. Anthropol.* (Suppl 34): 116.

Müller, A. E., and Soligo, C., 2005, Primate sociality in evolutionary context. *Am. J. Phys. Anthropol.* **128**: 399–414

Müller, A. E., and Thalmann, U., 2000 Origin and evolution of primate social organisation: A reconstruction, *Biol. Rev.* **75**: 405–435.

Nash, L. T., Bearder, S. K., and Olson, T. R., 1989, Synopsis of *Galago* species characteristics, *Int. J. Primatol.* **10**: 57–80.

Nash, L. T., and Harcourt, C. S., 1986, Social organization of galagos in Kenyan coastal forests: II: *Galago garnettii*, *Am. J. Primatol.* **10**: 357–369.

Nekaris, K. A. I., 2003, Spacing system of the Mysore Slender Loris (*Loris tardigradus lydekkerianus*), *Am. J. Phys. Anthropol.* **121**: 86–96.

Nekaris, K. A. I., and Rasmussen, D. T., 2003, Diet and feeding behavior of Mysore slender loris, *Int. J. Primatol.* **24**: 33–46.

Nowak, R. M., 1999, *Walker's Mammals of the World* (6th Ed.), The Johns Hopkins University Press, Baltimore.

Oates, J. F., 1987,. Food distribution and foraging behavior, in B. B. Smuts, D. L. Cheney, R. M. Seyfarth, R. W. Wrangham, and T. T. Struhsaker, eds., *Primate Societies*, University of Chicago Press, Chicago, pp. 197–209.

Perret, M., 1998, Energetic advantage of nest-sharing in a solitary primate, the lesser mouse lemur (*Microcebus murinus*), *J. Mammal.* **79**: 1093–1102.

Plavcan, J. M., Kay, R. F., Jungers, W. L., and van Schaik, C. P., 2002, *Reconstructing Behavior in the Primate Fossil Record*, Kluwer Academic/Plenum Publishers, New York.

Pullen, S. L., Bearder, S. K., and Dixson, A. F., 2000, Preliminary observations on sexual behavior and the mating system in free-ranging lesser galagos (*Galago moholi*), *Am. J. Primatol.* **51**: 79–88.

Radespiel, U., 2000, Sociality in the gray mouse lemur (*Microcebus murinus*) in northwestern Madagascar, *Am. J. Primatol.* **51**: 21–40.

Radespiel, U., Sarikaya, Z., Zimmermann, E., and Bruford, M., 2001, Socio-genetic structure in a free-living nocturnal primate population: Sex-specific differences in the grey mouse lemur (*Microcebus murinus*), *Behav. Ecol. Sociobiol.* **50**: 493–502.

Rasmussen D. T., and Sussman R. W., 2006, Parallelisms among possums and primates, in: *Primate Origins: Adaptation and Evolution*, M. Ravosa and M. Dagosto (eds.). Springer, New York, pp. 775–803.

Rasoloarison, R. M., Goodman, S. M., and Ganzhorn, J. U., 2000, Taxonomic revision of mouse lemurs (*Microcebus*) in the western portions of Madagascar, *Int. J. Primatol.* **21**: 963–1019.

Rathbun, G., 1979, The social structure and ecology of elephant-shrews, *Adv. Ethol.* **20**: 1–76.

Righetti, J., 1996, A comparison of the behavioural mechanisms of competition in shrews (Soricidae) and small dasyurid marsupials (Dasyuridae), in *Comparison of Marsupial and Placental Behaviour*, D. B. Croft and U. Ganslosser, eds., Filander Verlag, Fürth, pp. 293–299.

Ryser, J., 1992, The mating system and males mating success of the Virginia opossum *Didelphis virginiana* in Florida, *J. Zool.* **228**: 127–139.

Sauer, E. G. F., 1973, Zum Sozialverhalten der kurzohrigen Elefantenspitzmaus, *Macroscelides proboscidens*, *Z. Säugetierk.* **38**: 65–97.

Schülke, O., and Kappeler, P. M., 2003, So near and yet so far: Territorial pairs but low cohesion between pair partners in a nocturnal lemur, *Phaner furcifer*, *Anim. Behav.* **65**: 331–343.

Schwab, D., 2000, A preliminary study of spatial distribution and mating system of pygmy mouse lemurs (*Microcebus* cf *myoxinus*), *Am. J. Primatol.* **51:** 41–60.

Serena, M., 1994, Use of time and space by platypus (*Ornithorhynchus anatinus*: Monotremata) along a Victorian stream, *J. Zool.* **232:** 117–131.

Serena, M., and Soderquist, T. R., 1989, Spatial organization of a riparian population of the carnivorous marsupial *Dasyurus geoffroyii*, *J. Zool.* **219:** 373–384.

Soderquist, T. R., 1995, Spatial organization of the arboreal carnivorous marsupial *Phascogale tapoatafa*, *J. Zool.* **237:** 385–398.

Soligo, C., 2001, *Adaptions and Ecology of the Earliest Primates*, Ph.D., University of Zurich.

Soligo, C., and Müller, A. E., 1999a, Body weight trajectories in primate evolution: The evidence from nails and claws, *Am. J. Phys. Anthro.* (Suppl 28): 255.

Soligo, C., and Müller, A. E., 1999b, Nails and claws in primate evolution, *J. Hum. Evol.* **36:** 97–114.

Springer, M. S., Kirsch, J. A. W., and Case, J. A., 1997, The chronicle of marsupial evolution, in *Molecular Evolution and Adaptive Radiation*, T. J. Givnish and K. J. Sytsma, eds., Cambridge University Press, Cambridge, pp. 129–161.

Sterling, E. J., 1993, Patterns of range use and social organization in aye-ayes (Daubentonia madagascariensis) on Nosy Mangabe, in *Lemur Social Systems and Their Ecological Basis*, P. M. Kappeler and J. U. Ganzhorn, eds., Plenum Press, New York, pp. 1–10.

Stone, R. D., 1987, The social ecology of the Pyrenean desman (*Galemys pyrenaicus*) (Insectivora: Talpidae), as revealed by radiotelemetry, *J. Zool.* **212:** 117–129.

Sunquist, M. E., Austad, S. N., and Sunquist, F., 1987, Movement patterns and home range in the common opossum (*Didelphis marsupialis*), *J. Mammal.* **68:** 173–176.

Sussman, R. W., 1991, Primate origins and the evolution of angiosperms, *Am. J. Primatol.* **23:** 209–223.

Sussman, R. W., 1999, *Primate Ecology and Social Structure. Vol. 1. Lorises, Lemurs and Tarsiers*, Pearson Custom Publishing, Needham Heights.

Sussman, R. W., 2000, *Primate Ecology and Social Structure. Vol. 2. New World Monkeys*, Pearson Custom Publishing, Needham Heights.

Sussman, R. W., and Raven, P. H., 1978, Pollination by lemurs and marsupials: An archaic coevolutionary system, *Science* **200:** 731–736.

Tattersall, I., 1982, *The Lemurs of Madagascar*, Columbia University Press, New York.

Tattersall, I., 1987, Cathemeral activity in primates: A definition, *Folia Primatol.* **49:** 200–202.

Thalmann, U., 2001 Food resource characteristics in two nocturnal lemurs with different social behavior: *Avahi occidentalis* and *Lepilemur edwardsi*, *Int. J. Primatol.* **22:** 287–324.

Thalmann, U., 2002, Contrasts between two nocturnal leaf-eating lemurs, *Evol. Anthropol.* (Suppl 1): 105–107.

van Schaik, C. P., 1983, Why are diurnal primates living in groups? *Behaviour* **87**: 120–140.

van Schaik, C. P., and van Hooff, J.A.R.A.M., 1983, On the ultimate causes of primate social systems, *Behaviour* **85**: 91–117.

Ward, P., and Zahavi, A., 1973, The importance of certain assemblages of birds ad "information centers" for food finding, *Ibis* **115**: 517–534.

Warren, R. D., and Crompton, R. H., 1997, A comparative study of the ranging behaviour, activity rhythms and sociality of *Lepilemur edwardsi* (Primates, Lepilemuridae) and *Avahi occidentalis* (Primates, Indriidae) at Ampijoroa, *J. Zool.* **243**: 397–415.

West-Eberhard, M. J., 1975, The evolution of social behavior by kin selection, *Q. Rev. Biol.* **50**: 1–33.

Wimmer, B., Tautz, D., and Kappeler, P. M., 2002, The genetic population structure of the gray mouse lemur (*Microcebus murinus*), a basal primate from madagascar, *Behav. Ecol. Sociobiol.* **52**: 166–175.

Woolley, P. A., 1991, Reproduction in *Dasykaluta rosamundae* (Marsupialia, Dasyuridae): Field and laboratory observations, *Austr. J. Zool.* **39**: 549–568.

CHAPTER TWENTY

Primate Bioenergetics: An Evolutionary Perspective

J. Josh Snodgrass, William R. Leonard, and Marcia L. Robertson

INTRODUCTION

Energy dynamics represent an important interface between an organism and its environment. A variety of factors, including body mass, locomotor strategy, and foraging behavior, determine an animal's energy demands. Body mass is the most important determinant in predicting metabolic costs both for resting metabolic rate (RMR; the amount of energy used by an inactive animal under thermoneutral conditions) (Kleiber, 1961) and total daily energy costs (TEE or FMR) (Nagy, 1987; Nagy et al., 1999). The Kleiber (1961) scaling relationship correlates RMR with adult body mass and demonstrates that RMR scales to the three-quarters power of body mass in mammals, from the very small (e.g., mice) to the very large (e.g., elephants). While most mammals have RMRs predicted by body size, certain groups (e.g., marsupials, edentates) deviate significantly from this relationship.

Primates as a group do not significantly differ from the mammalian scaling relationship, though there exists a great deal of variation within the order. For example, strepsirrhines differ from other primates in having depressed RMRs from those predicted for their size based on the Kleiber

J. Josh Snodgrass, • Department of Anthropology, University of Oregon, Eugene, OR 97403. William R. Leonard and Marcia L. Robertson • Department of Anthropology, North Western University, Evanston, IL 60208

scaling relationship. Although a number of explanations have been offered to explain hypometabolism[1] in strepsirrhines, the phenomenon remains enigmatic. At least four hypotheses for strepsirrhine hypometabolism have been proposed: (1) adaptation to arboreal folivory, (2) adaptation to a diet deviant for body size, (3) a thermoregulatory adaptation, and (4) phylogenetic inertia (i.e., hypometabolism is a primitive mammalian trait) (Kurland and Pearson, 1986; Ross, 1992). Since Kurland and Pearson's (1986) review of strepsirrhine[2] hypometabolism, RMR has been measured on numerous additional primate species, doubling the available data for strepsirrhines. Additional data are available on diet and ecology in primate species.

In this chapter, we examine data on resting metabolic rates from a large sample of primate species to investigate variation in RMRs within the primate order. First, we explore the nature of metabolic variation in strepsirrhines and haplorhines, specifically focusing on strepsirrhine hypometabolism. We then consider whether specific ecological factors, such as folivory, arboreality, or activity cycle (i.e., diurnal or nocturnal), can explain strepsirrhine hypometabolism. After evaluating these proximate explanations, we then examine whether strepsirrhine hypometabolism may be a primitive characteristic shared with other closely related mammalian species. Finally, the implications of primate metabolic variation and strepsirrhine hypometabolism for early primate evolution are addressed.

SAMPLE AND METHODS

We obtained information on body mass (kg) and RMR (kcal/day) for 41 primate species, including 17 species of strepsirrhine and 24 species of haplorhine from published sources, from which we calculated a single unweighted average for each species (Table 1). All RMR values are expressed as kilocalories per day (kcal/day) and were converted from other units when necessary.

Data on brain mass (g) and body mass (kg) for 15 strepsirrhine species and 21 haplorhines were obtained from Bauchot and Stephan (1969) and Stephan et al. (1981). For each species, we calculated a single unweighted average for both brain mass and body mass (Table 2). Humans were excluded from the

---

[1] We follow Kurland and Pearson (1986) in defining hypometabolism as having a RMR more than 20% below that predicted for body size by the Kleiber scaling relationship. This conservative definition is used in order to avoid the misclassification of a species as hypometabolic as a result of measurement error, the measurement of an animal during sleep, or due to lack of standardized procedures.

[2] Kurland and Pearson (1986) used the traditional Prosimii–Anthropoidea taxonomic split but, since they did not include *Tarsius* in their analysis, there is no difference between their use of prosimian and our use of strepsirrhine.

Table 1. Sample information for primate metabolic and ecological data

Species	Metabolic data[a]			Ecological data[b]		
	RMR (kcal/day)	Body mass (kg)	Deviation[c]	DQ[d]	Habitat[e]	Activity cycle[f]
Suborder Strepsirrhini						
Arctocebus calabarensis	15.2	0.206	−28.99	327.5	A	N
Cheirogaleus medius	22.7	0.300	−20.00	–	A	N
Eulemur fulvus	42.0	2.397	−68.85	129.0	A	D
Euoticus elegantulus	25.1	0.260	−1.52	230.0	A	N
Galago moholi	13.9	0.155	−19.62	–	A	N
Galago senegalensis	18.1	0.215	−18.11	278.0	A	N
Galagoides demidoff	6.3	0.058	−23.85	305.0	A	N
Lemur catta	45.1	2.678	−69.22	166.0	A	D
Lepilemur ruficaudatus	27.6	0.682	−47.46	149.0	A	N
Loris tardigradus	14.8	0.284	−45.65	327.5	A	N
Microcebus murinus	4.9	0.054	−37.51	–	A	N
Nycticebus coucang	32.4	1.380	−63.65	–	A	N
Otolemur crassicaudatus	47.6	0.950	−29.33	195.0	A	N
Otolemur garnettii	47.8	1.028	−33.13	275.0	A	N
Perodicticus potto	41.3	1.000	−41.00	190.0	A	N
Propithecus verreauxi	86.8	3.080	−46.67	200.0	A	D
Varecia variegata	69.9	3.512	−61.08	–	A	D
Suborder Haplorhini						
Alouatta palliata	231.9	4.670	+4.28	136.0	A	D
Aotus trivirgatus	52.4	1.020	−26.25	177.5	A	N
Callithrix geoffroyi	27.0	0.225	+18.07	235.0	A	D
Callithrix jacchus	22.8	0.356	−29.23	235.0	A	D
Cebuella pygmaea	10.1	0.105	−21.78	249.5	A	D
Cercopithecus mitis	407.7	8.500	+17.00	201.5	T	D
Cercocebus torquatus	196.2	4.000	−0.90	234.0	A	D
Colobus guereza	357.9	10.450	−12.03	126.0	A	D
Erythrocebus patas	186.9	3.000	+17.13	–	T	D
Homo sapiens	1400.0	53.500	+1.10	–	T	D
Hylobates lar	123.4	1.900	+8.93	181.0	A	D
Leontopithecus rosalia	51.1	0.718	−6.41	–	A	D
Macaca fascicularis	400.9	7.100	+31.67	200.0	T	D
Macaca fuscata	485.4	9.580	+27.34	223.0	T	D
Macaca mulatta	231.9	5.380	−6.22	159.0	T	D
Pan troglodytes	581.9	18.300	−6.05	178.0	T	D
Papio anubis	342.9	9.500	−9.47	207.0	T	D
Papio cynocephalus	668.9	14.300	+29.95	184.0	T	D
Papio papio	297.3	6.230	+7.70	–	T	D
Papio ursinus	589.3	16.620	+2.27	189.5	T	D
Pongo pygmaeus	569.1	16.200	+0.68	172.5	A	D
Saguinus geoffroyi	50.5	0.500	+21.43	263.0	A	D
Saimiri sciureus	68.8	0.850	+11.03	323.0	A	D
Tarsius syrichta	8.9	0.113	−34.80	350.0	A	N

[a]McNab and Wright (1987); Leonard and Robertson (1994); Thompson et al. (1994); Kappeler (1996).

[b]Richard (1985); Sailer et al. (1985); Nowak (1991); Napier and Napier (1994); Rowe (1996).

[c]Metabolic deviation from predicted by Kleiber equation.

[d]Dietary quality.

[e]A = primarily arboreal; T = primarily terrestrial.

[f]D = diurnal; N = nocturnal.

Table 2. Sample information for primate brain data

Species	Brain mass (g)[a]	Body mass (kg)[a]
Suborder Strepsirrhini		
Arctocebus calabarensis	7.2	0.323
Cheirogaleus medius	3.1	0.177
Eulemur fulvus	25.2	2.397
Euoticus elegantulus	7.2	0.274
Galago senegalensis	4.8	0.186
Galagoides demidoff	3.4	0.081
Lemur catta	25.6	2.678
Lepilemur ruficaudatus	7.6	0.682
Loris tardigradus	6.6	0.322
Microcebus murinus	1.8	0.054
Nycticebus coucang	12.5	0.800
Otolemur crassicaudatus	10.3	0.850
Perodicticus potto	14.0	1.150
Propithecus verreauxi	26.7	3.480
Varecia variegata	34.2	3.512
Suborder Haplorhini		
Alouatta palliata	51.0	6.400
Aotus trivirgatus	16.0	0.850
Callithrix geoffroyi	7.6	0.280
Callithrix jacchus	7.6	0.280
Cebuella pygmaea	4.5	0.140
Cercopithecus mitis	76.0	6.500
Cercocebus torquatus	104.0	7.900
Colobus guereza	73.0	7.000
Erythrocebus patas	118.0	8.000
Hylobates lar	102.0	6.000
Macaca fascicularis	74.0	5.500
Macaca fuscata	84.0	5.900
Macaca mulatta	110.0	8.000
Pan troglodytes	420.0	46.000
Papio anubis	205.0	26.000
Papio cynocephalus	195.0	19.000
Papio papio	190.0	18.000
Papio ursinus	190.0	18.000
Pongo pygmaeus	370.0	55.000
Saguinus geoffroyi	10.0	0.380
Saimiri sciureus	22.0	0.680

[a]Bauchot and Stephan (1969); Stephan et al. (1981).

analysis because they are outliers for brain size in relation to body size and consequently substantially alter regressions. Because of differences between the body masses of animals used for brain studies and those for metabolic studies, when comparing metabolic rates to brain size, we calculated an adjusted RMR for each species to account for this difference.

Information on dietary quality (DQ) was obtained for 12 strepsirrhine and 20 haplorhine species (Table 1) from Richard (1985), Rowe (1996) and Sailer et al. (1985). Diet quality was assessed using an index, developed by Sailer et al. (1985), which considers the relative energy and nutrient density of dietary items. The DQ index is a weighted average of the proportions of foliage, reproductive plant material, and animal material. The DQ is calculated as:

$$DQ = s + 2(r) + 3.5(a).$$

Here s = percent of diet derived from structural plant parts (e.g., leaves, stems, and bark), r = percent of diet derived from reproductive plant parts (e.g., fruits, flowers, nectar, and resin), and a = percent of diet derived from animal parts (including both vertebrates and invertebrates). The DQ ranges from a minimum of 100 (100% foliage) to a maximum of 350 (100% animal material). Humans were excluded from the dietary analysis because the range of possible diets is larger than any nonhuman primate species, and consequently an all-inclusive DQ for the human species is not possible.

To assess functional consequences of substrate and habitat use, we classified species as arboreal or terrestrial based on primary habitat (Table 1); this determination was derived from relevant literature (Nowak, 1991; Rowe, 1996). While this dichotomy is overly simplified, it is used simply to get a general picture of habitat use. Additionally, we obtained information on activity cycle (i.e., nocturnal or diurnal) from published sources for all 17 species of strepsirrhine and all 24 haplorhine species (Rowe, 1996; Table 1).

To examine the evolutionary context of RMR in primates, we compiled metabolic data for closely related mammalian orders. We obtained information on RMR (kcal/day) and body mass (kg) for bats (order Chiroptera) and tree shrews (order Scandentia) from published sources, from which we calculated a single unweighted average for each species (Table 3). No metabolic data were available for colugos (order Dermoptera). All RMR values are expressed as kilocalories per day (kcal/day) and were converted from other units when necessary.

We compiled data on body mass (kg) estimates for 16 species of subfossil Malagasy lemurs from Godfrey et al. (1997; Table 4). Body mass reconstructions, based on regressions of humeral and femoral midshaft circumferences

Table 3. Sample information for RMR and body mass for selected mammalian species

Species	RMR (kcal/day)[a]	Body mass (kg)[a]
Order Chiroptera		
Anoura caudifer	4.07	0.012
Artibeus lituratus	9.82	0.070
Carollia perspicilla	3.64	0.015
Chalinolobus gouldii	2.92	0.018
Chrotopterus auritus	11.80	0.096
Cynopterus brachyotis	5.45	0.037
Desmodus rotundus	3.06	0.029
Diaemus youngi	3.99	0.037
Diphylla ecaudata	3.96	0.028
Dobsonia minor	12.71	0.087
Eonicterus spelaea	5.61	0.052
Glossophaga longirostris	3.07	0.014
Glossophaga soricina	2.50	0.010
Hipposideros galeritus	1.08	0.009
Histiotus velatus	1.16	0.011
Leptonycteris curasoae	3.95	0.024
Leptonycteris sanborni	5.10	0.022
Macroderma gigas	10.94	0.107
Macroglossus minimus	2.39	0.016
Megaloglossus woermanni	2.52	0.012
Miniopterus schreibersii	3.01	0.011
Molossus molossus	4.61	0.056
Noctilio albiventris	2.75	0.027
Noctilio leporinus	5.44	0.061
Nyctimene albiventer	4.64	0.028
Nyctophilus geoffroyi	1.32	0.008
Nyctophilus major	2.36	0.014
Paranyctimene raptor	3.36	0.021
Phyllostomus discolor	4.06	0.034
Phyllostomus elongatus	4.55	0.036
Phyllostomus hastatus	8.18	0.084
Pteropus poliocephalus	36.74	0.598
Pteropus scapulatus	28.12	0.362
Rhinonycteris aurantius	1.88	0.008
Rhinophylla fisherae	1.88	0.010
Rousettus aegyptiacus	14.22	0.146
Sturnira lilium	4.56	0.022
Syconycteris australis	3.92	0.018
Tonatia bidens	4.48	0.027
Uroderma bilobatum	3.08	0.016
Order Scandentia		
Ptilocercus lowii	5.04	0.058
Tupaia glis	10.84	0.123

[a]Bradley and Hudson (1974); Whittow and Gould (1976); McNab (1988); Arends et al. (1995); Geiser et al. (1996); Hosken (1997); Hosken and Withers (1997, 1999); Bartels et al. (1998); Baudinette et al. (2000).

Table 4. Reconstructed body masses (kg) and cranial capacities (cc) for selected subfossil Malagasy lemur species

Species	Body mass (kg)[a]	Cranial capacity (cc)[b]
Family Archaeolemuridae		
Archaeolemur edwardsi	22.0	104[c,d]
Archaeolemur majori	17.0	
Hadropithecus stenognathus	28.0	
Family Daubentoniidae		
Daubentonia robusta	10.0	
Family Lemuridae		
Pachylemur insignis	10.0	
Pachylemur jullyi	12.5	46[c]
Family Megaladapidae		
Megaladapis edwardsi	75.0	
Megaladapis grandidieri	65.0	
Megaladapis madagascariensis	40.0	118[c]
Family Palaeopropithecidae		
Archaeoindris fontoynontii	200.0	
Babakotia radofilai	15.0	49[d]
Mesopropithecus dolichobrachion	12.0	
Mesopropithecus globiceps	10.0	
Mesopropithecus pithecoides	11.0	
Palaeopropithecus ingens	45.0	
Palaeopropithecus maximus	55.0	99[c]

[a]Godfrey et al. (1997).
[b]Ravosa (unpublished data).
[c]British Museum (Natural History).
[d]Duke University Primate Center.

indicate that the subfossil lemurs were all larger than living strepsirrhine primates. Some species had body masses slightly greater than the largest living strepsirrhines (*Indri indri* and *Propithecus diadema*) (Smith and Jungers, 1997); however, all known species appear to have had body masses of at least 10 kg (Godfrey et al., 1997). Numerous species were considerably larger, including *Archaeoindris fontoynontii*, which is estimated to have reached an adult mass of 200 kg. We additionally present data on cranial capacity (cc) for five species of subfossil Malagasy lemur, which were collected by M. Ravosa (unpublished data) (Table 4).

Allometric relationships were determined using ordinary least squares regressions (OLS)[3] of $\log_{10}$-transformed data. Additionally, allometric relationships were calculated using reduced major axis (RMA); however, RMA values are not reported because they were not significantly different from parameters calculated using OLS. Differences in regression parameters were assessed using Student's t-tests. All analyses were performed using SPSS (Version 8.0), except RMA equations, which were calculated using BIOMstat (Version 3.30a).

RESULTS

Metabolic Variation in Strepsirrhines and Haplorhines

Metabolic rates in the strepsirrhines are significantly lower than those predicted by the Kleiber scaling relationship, averaging $38.6 \pm 4.7\%$ *below* the norm (Table 1). The range of predicted values for strepsirrhines is from -1.52 to -69.22% below those predicted by the Kleiber scaling relationship, and 14 of the 17 strepsirrhine species are hypometabolic by criteria described above (i.e., >20% below predicted by Kleiber scaling relationship). *Euoticus elegantulus*, *Galago moholi*, and *Galago senegalensis* are the three strepsirrhine species not classified as hypometabolic, with metabolic deviations from those predicted by Kleiber scaling relationship of -1.52, -19.62, and -18.11%, respectively. Within strepsirrhines, lorisiforms (lorises, pottos, and bushbabies) ($n = 10$) have RMRs that average 30.5% below predicted, while the lemuriforms (Malagasy lemurs) ($n=7$) average 50.1% below predicted. Bushbabies (Galagonidae) ($n = 6$) are slightly hypometabolic averaging 20.9% below predicted.

The relationship between RMR and body mass in strepsirrhines significantly differs in both scaling coefficient (slope) and y-intercept from the haplorhine regression ($P < 0.001$) (Figure 1). Differences in metabolic rates are also evident from the standardized residuals (z-scores) of the RMR to body mass regression for the pooled sample. Mean z-scores are significantly lower

---

[3] There has been debate in the anthropological literature in recent years regarding the most appropriate line fitting technique for describing allometric equations; some favor the use of ordinary least squares regressions (OLS), some the major axis (MA), and others the reduced major axis (RMA). OLS may underestimate the true slope (when the coefficient of determination [r^2] is low) because it does not consider error in the X variable (Harvey and Pagel, 1991). However, the preferable method for accurate and effective controls (especially with high r^2 values) for the effects of body mass is OLS (Harvey and Pagel, 1991), which we use in this study.

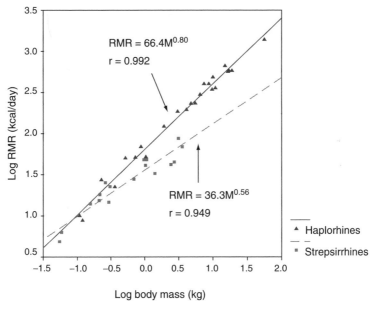

Figure 1. Log$_{10}$ plot of RMR (kcal/day) versus body mass (kg) for strepsirrhine species ($n=17$) and haplorhine species ($n=24$). The strepsirrhine regression has a significantly lower scaling coefficient than predicted by the Kleiber equation (RMR = 70M$^{0.75}$).

in strepsirrhines than haplorhines ($P<0.001$). The strepsirrhine regression substantially deviates from the Kleiber scaling relationship. The scaling relationship in the strepsirrhines is: RMR = 36.3M$^{0.56}$, whereas in haplorhines the relationship is: RMR = 66.4M$^{0.80}$; the latter is almost identical to the Kleiber scaling relationship prediction (i.e., RMR = 70M$^{0.75}$). Haplorhines average 1.9 ± 3.8% *above* predicted values and do not significantly differ from Kleiber scaling relationship predictions. The relationship between body mass and RMR for the combined sample of primates ($n=41$) is RMR = 54.7M$^{0.81}$.

Ecological Correlates of Strepsirrhine Hypometabolism

Hypometabolism is often observed in species consuming a nutrient-poor (low quality) diet (Kurland and Pearson, 1986; McNab, 1986). In the present sample, relatively lower DQ is associated with depressed metabolic rates in both strepsirrhine ($n=12$) and haplorhine primates ($n=20$). However, while variation in DQ helps to explain within-group differences in RMR, the metabolic differences between strepsirrhines and haplorhines cannot be explained by dietary differences.

All strepsirrhines are primarily arboreal, including *Lemur catta*, which spends roughly 25% of its time on the ground (Martin, 1990). Across all primates, arboreal species ($n = 30$) have significantly lower metabolic rates than terrestrial species ($n = 11$) (-24.1% versus 10.2%) ($P < 0.001$). However, even after controlling for habitat use, strepsirrhines ($n = 17$) have significantly lower metabolic rates than arboreal haplorhines ($n = 13$) (-38.6 ± 4.7% versus -5.2 ± 5.1%; $P < 0.001$). Both arboreal haplorhines and terrestrial haplorhines ($n = 11$) significantly differ from strepsirrhines ($P < 0.001$).

The relationship between activity cycle (i.e., nocturnal or diurnal) and body mass to deviation from predicted RMR among species demonstrates that the degree of hypometabolism is significantly greater among the larger species. The diurnal[4] strepsirrhine species ($n = 4$), of which all are Malagasy lemurs, have the largest body sizes and have significantly lower metabolic rates than the nocturnal strepsirrhine species ($n = 13$) (61.5% versus -31.5%; $P < 0.01$).

Metabolic Variation and Body Composition

The relationship between brain mass and body mass in strepsirrhine ($n = 15$) and haplorhine species ($n = 21$) demonstrates that strepsirrhines have relatively smaller brains than haplorhine species (Figure 2). The scaling coefficient for strepsirrhine species is significantly lower than that of the haplorhines (0.75 versus 0.64; $P < 0.05$). Additionally, the y-intercept of strepsirrhines is significantly lower than haplorhines ($P < 0.001$). The relative size difference in brain mass of strepsirrhines and haplorhines is also evident from the z-scores of the brain mass to body mass regression for the pooled sample. Mean z-scores are significantly lower in strepsirrhines than haplorhines ($P < 0.001$).

When the relationship of brain size and RMR in strepsirrhines ($n = 15$) and haplorhines ($n = 21$) is examined, the scaling coefficients for the strepsirrhines and haplorhine regressions are comparable and both scale isometrically (1.02 versus 0.96; n.s.). This suggests that both groups spend similar proportions of RMR on brain metabolism and that species with different body sizes have similar relationships between brain size and RMR. The y-intercepts in strepsirrhines and haplorhines are not significantly different. Additionally, strepsirrhines do

---

[4] One of these species, *Eulemur fulvus*, is more appropriately classified as "cathemeral," which reflects its activity period both during the day and night (Fleagle, 1999); it has been collapsed into the category "diurnal" to allow statistical treatment.

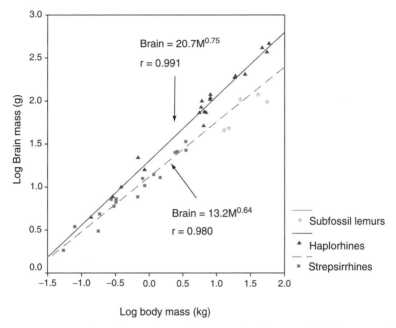

Figure 2. $\log_{10}$ plot of brain mass (kg) versus body mass (kg) for strepsirrhine species ($n = 15$) and haplorhine species ($n = 21$; humans excluded). The scaling coefficient for strepsirrhine species is significantly lower than that of the haplorhines (0.75 versus 0.64; $P < 0.05$). Asterisks represent values for subfossil Malagasy lemurs based on reconstructions (Table 4).

not significantly differ from haplorhines in mean z-scores of the brain mass to RMR regression for the pooled sample.

Phylogenetic Influences on Strepsirrhine Hypometabolism: Comparative Metabolic Data

From our previous analyses, it appears that proximate ecological factors do not provide a full explanation for strepsirrhine hypometabolism. Consequently, we next considered whether the distinctive metabolic pattern of strepsirrhines is a primitive trait that is shared with other closely related mammalian species. To evaluate this explanation, we considered metabolic data for selected nonprimate species (Figure 3).

In the two studies of RMR in tree shrews that were conducted under standardized conditions, both species measured were shown to be hypometabolic by above criteria. *Ptilocercus lowii*, the only nocturnal tree shrew species and

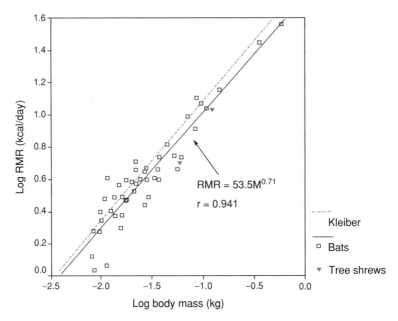

Figure 3. Log_{10} plot of RMR (kcal/day) versus body mass (kg) for selected bat (order Chiroptera; $n=46$) and tree shrew (order Scandentia; $n=2$) species. Also shown is the Kleiber equation ($RMR=70M^{0.75}$).

one of the few that is also predominantly arboreal, has an RMR 39.1% below that predicted by the Kleiber equation (Whittow and Gould, 1976). *Tupaia glis*, which is diurnal and partially terrestrial (Martin, 1990; Nowak, 1991), is also hypometabolic and has an RMR 25.5% below that predicted by the Kleiber equation.[5]

Bats have metabolic rates that average 10% below those predicted by the Kleiber equation, though this masks considerable variation found within the order. Bats are not hypometabolic by previously described criteria. The scaling relationship between RMR and body mass in bats is: RMR $=53.5M^{0.71}$. Microbats ($n=34$) have metabolic rates nearly identical to megabats (family Pteropodidae; $n=12$) and a similar range of variation in body size is seen within the two groups. Microbats on average deviate from that predicted by the Kleiber equation by −10% and megabats deviate by on average −9%.

[5] The body mass for this species is not given in the original publication (Bradley and Hudson, 1974) and was obtained from McNab (1988).

Phylogenetic Influences on Strepsirrhine Hypometabolism: Implications for Subfossil Lemurs

The extant lemurs of Madagascar are diverse in species number, morphological adaptation, and ecology, but there is evidence that this diversity was much larger in the recent past (Godfrey and Jungers, 2002; Mittermeier et al., 1994; Simons, 1997). Known primarily from recent paleontological sites (i.e., "subfossils"), there is evidence for at least 17 species representing at least five families of extinct lemurs: Archaeolemuridae, Daubentoniidae, Lemuridae, Megaladapidae, and Palaeopropithecidae (Godfrey and Jungers, 2002). Only two of these, Daubentoniidae and Lemuridae, have living representatives. Humans initially colonized Madagascar only about 2000 years ago and are implicated in the extinction event of numerous lemur species, as well as three species of hippopotamus, two species of bibymalagasy, a medium-sized carnivore, two genera of flightless birds, and a species of giant tortoise (Garbutt, 1999). Human activities, such as hunting, habitat alteration, introduction of nonnative species (e.g., wild cattle), and possibly, the introduction of nonnative diseases, likely played a major role in the extinction. There is also some evidence that Late Holocene climatic changes might have contributed to the extinctions. The range of dates for subfossil lemurs runs from about 26,000 years BP (*Megaladapis* from Antsiroandoha Cave in northern Madagascar) to about 500 years BP for *Palaeopropithecus* from Manamby Plateau in southwest Madagascar (Simons, 1997). Additionally, there is a historical report that suggests the presence of a large-bodied lemur in Madagascar in the 17th century (Flacourt, 1658). Ethnographic sources also suggest that a large-bodied lemur might have survived in Madagascar into the 20th century (Burney and Ramilisonina, 1999). By all indication, the subfossil and extant lemurs are part of the same contemporary fauna and the former should not be considered as ancestors of the latter (Mittermeier et al., 1994).

All of the subfossil lemurs, with the exception of *Daubentonia robusta* (a relative of the aye-aye), are thought to have been diurnal, based on relative orbit size and body size (Simons, 1997). An enormous range of locomotor diversity is seen in the subfossil lemurs, but it seems likely that, based on body size and postcranial morphology, most of the subfossil lemurs likely spent at least some time on the ground (Godfrey et al., 1997). Certain groups, such as *Hadropithecus* and *Archaeolemur*, may have spent considerable time on the ground (Godfrey et al., 1997). With the possible exception of *D. robusta*, all extinct lemur species likely included some leaves in their diet, supplementing this diet with fruit, seeds, and possibly fauna. Many of the larger species are inferred to be highly folivorous on the basis of both body size and morphological adaptations.

Body mass reconstructions, based on regressions of humeral and femoral midshaft circumferences indicate that the subfossil lemurs were all larger than living strepsirrhine primates (Godfrey et al., 1997; Table 4). Some species had body masses slightly greater than the largest living strepsirrhines (*Indri indri* and *Propithecus diadema*) (Smith and Jungers, 1997); however, all known species appear to have had body masses of at least 10 kg (Godfrey et al., 1997). Numerous species were considerably larger, including *Archaeoindris fontoynontii*, which is estimated to have reached an adult body mass of about 200 kg.

The RMR predictions for 16 species of subfossil lemur are presented in Figure 4 and are based on body masses reconstructed for subfossil taxa (Table 4). For a given body mass, we calculated RMR based on the Kleiber scaling relationship ($70M^{0.75}$) and a strepsirrhine-only regression from this study ($36.3M^{0.56}$), which assumes that the subfossil lemurs were hypometabolic (based on the retention of the primitive condition). Assuming metabolic rates similar to those seen in living strepsirrhines, there would have been considerable energy savings in all species, which would have been amplified at larger body sizes. For example, in the largest of the subfossil lemurs, *A. fontoynontii*, with an estimated body mass of 200 kg, would have had an RMR (using the strepsirrhine-only regression) of only about 20% of that predicted by Kleiber scaling relationship. This energy savings likely would have been further amplified through low-total energy costs, as is likely based on morphological evidence, which indicates a highly folivorous sloth-like creature that probably spent considerable time on the ground (Simons, 1997). One of the consequences of depressed metabolic rates is that they may have had the effect of limiting competition for resources (McNab, 1980). However, there are also reproductive consequences of depressed metabolic rates as there is some indication that mammalian species with depressed metabolic rates also have low-intrinsic rates of population increase (McNab, 1980, 1986).

Data on cranial capacity for five species of subfossil lemurs (Table 4) demonstrated a similar scaling relationship of brain size and body mass as in

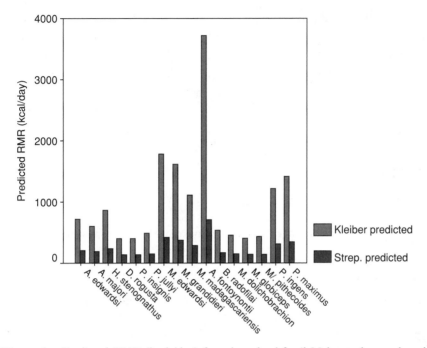

Figure 4. Predicted RMR (kcal/day) for selected subfossil Malagasy lemurs based on reconstructed body weights from Godfrey et al. (1997). Gray bars represent predicted RMRs based on the Kleiber scaling relationship ($70M^{0.75}$) and black bars represent predicted RMRs based on the strepsirrhine regression ($36.3M^{0.56}$).

extant strepsirrhines (Figure 2). *M. madagascariensis* and *P. maximus* have similar body masses (40 kg and 55 kg, respectively) as common chimpanzees (*P. troglodytes*) though their brain sizes are only about one-quarter the size. However, there is variation in the subfossil lemurs in relative brain size; data from *A. edwardsi* indicate that it was relatively encephalized when compared to the larger subfossil lemurs. These results are in general agreement with Jungers (1999), and demonstrate that strepsirrhines have larger brains than similar-sized mammals but considerably smaller brains than haplorhine primates. This may be the result of physiological limitations in supporting brain metabolism (Armstrong, 1983, 1985).

It seems likely that the earliest strepsirrhine colonizers of Madagascar were hypometabolic and small bodied. Purvis (1995) suggests that mouse and dwarf lemurs of the family Cheirogaleidae are the most ancestral of extant lemurs and, on these grounds, it seems likely that the initial colonizers of Madagascar were small-bodied and later diversified in terms of body size. However, this evidence is currently untestable given the dearth of pre-Holocene primate fossils from Madagascar. Depressed metabolic rates and the ability to enter torpor may also have increased chances of survival during a transoceanic rafting to Madagascar from Africa (Kappeler, 2000; Warren and Crompton, 1996). These low-metabolic rates may have had important consequences for the survival and diversification of Malagasy primates.

DISCUSSION

Ecological Correlates of Primate Metabolic Variation

Although primates do not significantly differ from other mammals in the scaling of body mass and metabolic rate, there is considerable variation within the order. Haplorhines, as a group, have a scaling relationship similar to other mammals. In contrast, strepsirrhines are hypometabolic and have a scaling relationship that markedly deviates from that of haplorhines and other mammals. Indeed, all strepsirrhine species have RMRs lower than those predicted for body mass based on the Kleiber scaling relationship.

A number of ecological factors, such as low-quality diet, arboreality, and nocturnality, have been linked with hypometabolism, both in strepsirrhines and other mammalian groups (Kurland and Pearson, 1986; McNab, 1978, 1980, 1986; Ross, 1992). In these explanations, adaptations to particular ecological factors are postulated to have led to depressed metabolic rates.

Hypometabolism, which slows passage rates to allow increased nutrient extraction, has been linked to low-quality diets associated with the inclusion of large amounts of foliage (McNab, 1978, 1980, 1986). This may be particularly important in the smaller species with folivorous diets. It has also been suggested that depressed metabolic rates allow a species to consume greater quantities of toxic insects without experiencing deleterious effects (Charles-Dominique, 1977; McNab, 1980, 1986; Oates, 1984).

Our examination of the role of low-diet quality in hypometabolism produced mixed results. Strepsirrhine species with low-quality diets for body size tended to have depressed metabolic rates, suggesting that hypometabolism in this group is partially influenced by low-diet quality, particularly in the larger species. Depressed metabolic rates also appear to be associated with low-quality diets in haplorhines; species with lower quality diets than predicted for body size have lower metabolic rates than predicted for body size. However, the regressions describing the relationship for each group are parallel, and thus, dietary differences alone cannot explain the metabolic differences between strepsirrhines and haplorhines.

McNab (1978, 1986) also raises the possibility that folivory in the context of an arboreal habitat may differentially depress metabolic rates in certain mammalian groups, including primates. Specifically, the depressed metabolic rates of arboreal folivores may be attributable to a combination of factors, including a low-quality diet, relatively sedentary habits, and the consequent decreases in skeletal muscle mass (McNab, 1978, 1986). Our results suggest that, like diet, habitat does exert a significant influence on metabolic rates, as arboreal species have lower RMRs than terrestrial species. However, among only arboreal species, strepsirrhines have significantly lower metabolic rates than haplorhines. *Lemur catta*, the strepsirrhine that spends the largest proportion of time on the ground, is also the most hypometabolic of the strepsirrhines (−69.22% from predicted), though the confounding effects of body mass and diet make it difficult to separate out habitat preference. While there is some support for a model that considers both low-dietary quality and arboreality in strepsirrhines, it cannot explain metabolic rates in some species, including some haplorhines. For example, *Alouatta palliata*, a folivorous and arboreal haplorhine, has a metabolic rate slightly above that predicted by Kleiber scaling relationship (+4.28%). Additionally, hominoids, such as *Pongo pygmaeus* and *Hylobates lar* have low-quality diets, but have RMRs (+0.68 and +8.93%, respectively) at or slightly above that predicted for body size.

The results presented in the current study indicate that there is a relationship between depressed metabolic rates and lower dietary quality in both strepsirrhines and haplorhines, but the metabolic differences between strepsirrhines and haplorhines cannot be explained by dietary differences alone. In fact, as discussed previously, the regressions that describe the relationships in each of the groups are parallel. These results echo those of Leonard and Robertson (1994), but with an enlarged sample size.

Another ecological variable that has been discussed in reference to hypometabolism is waking cycle, largely because of its importance in the context of thermoregulation. A relatively low RMR has been proposed to be a thermoregulatory adaptation in strepsirrhines (Charles-Dominique, 1974; Müller and Jaksche, 1980) and in certain haplorhine species, such as *Aotus trivirgatus* (Le Maho et al., 1981). It is suggested that this adaptation would be seen in well-insulated animals living in tropical environments with high-daytime temperatures and low-nighttime temperatures. In this view, nocturnal activity increases heat production during the coldest part of the 24-h cycle, while inactivity during the day reduces heat production during the hottest part of the 24-h cycle. This hypothesis is a continuation of a larger literature that suggests that strepsirrhines use behavioral adjustments, such as reduced activity levels and sunning behavior, in order to efficiently thermoregulate (e.g., Morland, 1993). It has been proposed that the cath-emeral behavior of *Eulemur fulvus*, which is active at night during the cool dry season, is an adaptation to minimize cold stress and energy costs (Curtis et al., 1999). Thus, the thermoregulatory hypothesis for strepsirrhine hypometabolism predicts that depressed RMRs in well-insulated, nocturnal primates living in hot environments are the result of thermoregulatory adaptations.

Data from the study demonstrate that among strepsirrhines, nocturnal species have relatively higher RMRs than diurnal species, rather than lower as would be predicted by the thermoregulatory model. This difference, how-ever, may be partially an artifact of body size and diet, as the diurnal species are the four largest of the strepsirrhines and additionally are some of the most folivorous. It should be noted that the two nocturnal haplorhine species, *Aotus trivirgatus* and *Tarsius syrichta*, have RMRs that fall substantially below those predicted by the Kleiber scaling relationship, but both are also relatively small bodied. Interestingly, *Aotus* and *Tarsius* are thought to be secondarily nocturnal (Martin, 1990). While current data do not support the ther-moregulatory hypothesis, a recent colonization of the diurnal niche has been suggested for the diurnal (and cathemeral) strepsirrhines (Ross, 1996; van Schaik and Kappeler, 1996), which could partially explain metabolic rates of these diurnal strepsirrhines.

In summary, none of the ecological arguments entirely explain the level of hypometabolism observed in strepsirrhines. Depressed metabolic rates are exhibited by strepsirrhines of a range of body sizes, with diverse dietary strate-gies, and different activity patterns and waking cycles. While it is clear that adaptations to proximate ecological factors, such as diet, play a role in struc-turing metabolic costs, these factors cannot entirely explain hypometabolism in strepsirrhines.

Phylogenetic Influence on Primate Metabolic Variation

In addition to proximate ecological factors, phylogenetic inertia has been suggested as an explanation for hypometabolism in strepsirrhines and other mammal groups (e.g., Eisentraut, 1961; Elgar and Harvey, 1987; Martin, 1989; Ross, 1992). This hypothesis suggests that hypometabolism is a primitive mammalian trait that has been retained in extant strepsirrhines. While Kurland and Pearson (1986) discuss the possibility that strepsirrhines are hypometabolic because of phylogenetic inertia, they do not test this hypothesis. Ross (1992) lends some support to the role of phylogenetic effects on strepsirrhine hypometabolism, though problems with the methodology[6] preclude acceptance of her results.

The results of the present study provide support for the phylogenetic inertia model, but an understanding of the metabolic rates of closely related species is important to test this hypothesis. The superorder Archonta was originally proposed by Gregory (1910) to contain primates, bats (order Chiroptera), colugos or "flying lemurs" (order Dermoptera), and the tree shrews and elephant shrews (order Menotyphla). McKenna (1975) later removed the elephant shrews leaving primates, colugos, bats, and tree shrews (order Scandentia) in the superorder. The superorder Archonta has been the subject of numerous investigations, using morphological studies of living species, paleontological studies, and molecular investigations (see review in Sargis, 2002). The validity of Archonta has received its greatest support from the result of comparative studies of skeletal characters of the ankle region (e.g., Szalay, 1977). However, testing the integrity of Archonta is problematic because of the dearth of fossil evidence in all but the primates. Additionally, the use of morphological traits that are primitive or convergent (rather than shared derived characters), has led to false support for Archonta (Martin, 1990).

No consensus exists on the integrity of Archonta as a monophyletic unit, but there are data both from morphological and molecular studies that support the close relationship of primates with other mammalian orders.

---

[6] Ross (1992) compared metabolic data for primates with a regression generated by Stahl (1967). The Stahl regression is not based on 349 mammalian species, as claimed by Ross, but 349 data points. The paper does not provide information on which species were used and the number of data points for each and, additionally, does not control for animals in the resting condition. The scaling coefficient is higher and, consequently, Ross' calculations of metabolic deviations are invalid.

However, a number of recent studies have not supported Archonta as a monophyletic group, but instead support close relationships between subsets of the members. In particular, Euarchonta, which includes primates, colugos, and tree shrews (but not bats) has received support from molecular studies (Adkins and Honeycutt, 1991, 1993; Madsen et al., 2001; Murphy et al., 2001a,b; Stanhope et al., 1993), as well as combined morphological and molecular evidence (Liu and Miyamoto, 1999; Liu et al., 2001). Interestingly, some studies have indicated a close evolutionary relationship between Euarchonta and Glires (rodents and lagomorphs), together forming Euarchontoglires (Madsen et al., 2001; Murphy et al., 2001a,b).

Metabolic data for members of the superorder Archonta were available for tree shrews and bats. Unfortunately, no metabolic data are available for colugos. Both tree shrews and bats have, on average, lower metabolic rates than similar-sized mammals according to the Kleiber scaling relationship. In the two studies of RMR in tree shrews that were conducted under standardized conditions, both species measured were shown to be hypometabolic. *Ptilocercus lowii*, a nocturnal and arboreal tree shrew species, has an RMR 39.1% below that predicted by the Kleiber equation. *Tupaia glis*, which is diurnal and partially terrestrial (Martin, 1990; Nowak, 1991), is also hypometabolic and has an RMR 25.5% below that predicted by Kleiber scaling relationship. Both species are omnivorous and include various amounts of insects and fruits as the main items in their diet (Martin, 1990). Tree shrews have often been used as models of early primate morphology and behavior (and were classified by some authorities [e.g., Simpson, 1945] at one time as members of the primate order), largely because of their inferred close phylogenetic relationship and certain morphological similarities shared with primates. However, there is a good deal of morphological and behavioral variation between species of tree shrew, and many of the shared morphological traits may actually be either primitive or convergent (Martin, 1990). That said, there are indications from both molecular and morphological studies that Scandentia is closely related to primates, possibly as a sister group.

Bats have metabolic rates that average 10% below those predicted by the Kleiber equation, though this average masks considerable variation found within the order. While bats have metabolic rates lower than expected, they are not hypometabolic by previously described criteria. The scaling relationship between RMR and body mass in bats is: $RMR = 53.5M^{0.71}$. Microbats

($n = 34$) have metabolic rates nearly identical to megabats (family Pteropodidae; $n = 12$) and a similar range of variation in body size is seen in the two groups. Microbats on average deviate from that predicted by the Kleiber equation by −10%, whereas megabats deviate by on average −9%. Despite these low-metabolic rates, bats have the highest capacity gas exchange system found in living mammals (Szewczak, 1997).

The phylogenetic position of tarsiers among primates makes them an important group to examine in the phylogenetic argument since they possess numerous primitive mammalian traits that were subsequently lost in anthropoids (Martin, 1990). Molecular studies, using protein and DNA sequence evidence (Bonner et al., 1980; DeJong and Goodman, 1988; Dijan and Green, 1991; Koop et al., 1989a,b; Miyamoto and Goodman, 1990; Pollock and Mullin, 1987; Porter et al., 1995; Shoshani et al., 1996; Zietkiewicz et al., 1999), lend support to the classification of tarsiers as a sister clade of the anthropoids, both subsumed within the suborder Haplorhini. Additionally, many morphological studies based on derived features support the grouping of tarsiers as haplorhines (Beard et al., 1991; Martin, 1990; Ross, 1994; Szalay et al., 1987).

The only tarsier species with available metabolic measurements taken under standardized conditions is *Tarsius syrichta*, which has an RMR well below (−34.8%) that predicted by the Kleiber equation. Tarsiers are nocturnal, arboreal, and small-bodied, and the only primates that consume 100% animal material (mostly insects and some vertebrates). The depressed metabolic rates of tarsiers may be the result of the retention of a primitive mammalian trait, as is hypothesized for the strepsirrhines.

Taken as whole, metabolic rates in the closest living relatives of primates provide some evidence for hypometabolism as a primitive trait that has been retained in living strepsirrhines. However, further resolution of primate superordinal relationships, as well as further studies of metabolism in close relatives, are needed.

The phylogenetic explanation is often used as an unenlightening nonexplanation (e.g., Hayssen and Lacy, 1985), but in order to fully understand phylogenetic inertia as an explanation, the reasons for the evolution of hypometabolism must be addressed. Additional questions that must be addressed are why hypometabolism was maintained in descendant lineages and whether there was active selection to maintain it in the extant species, or whether it was retained because there was not active selection against it.

Unfortunately, it is often difficult to unravel the effects of phylogeny from current adaptations, since phylogenetically close animals also tend to have similarities in both ecology and biology (Elgar and Harvey, 1987; McNab, 1986).

Body Composition and Primate Metabolic Variation

Differences in body composition are important in influencing variation in metabolic energy requirements, given marked differences in mass-specific metabolic rates across tissues. Muscle mass, for example, varies from 24 to 61% of total body weight in mammals, with slow-moving arboreal mammals, such as sloths, occupying the low end and terrestrial carnivores, such as lions, occupying the high end (Calder, 1984; Grand, 1977; Muchlinski et al., 2003). McNab (1978) postulates that the depressed RMRs of arboreal mammals are partly the result of low levels of muscularity. Thus, variation in tissue size and concomitant variation in tissue metabolic rates contribute to the structuring of energy costs and provide a mechanism for deviations from predicted metabolic rates.

The relative size of the brain has been linked by a number of researchers to metabolic rate, since both scale to the three-quarters power of body mass (i.e., 0.75) (Armstrong, 1983, 1985; Hoffman, 1983; Martin, 1981, 1996). While some have hypothesized that brain size and its associated high tissue metabolic rates partially structure RMR (e.g., Holliday, 1986), others have taken the opposite approach and hypothesized that metabolic rates influence brain size (e.g., Armstrong, 1983, 1985). The latter relates to a proposed relationship between the size of the brain and the ability of the body to support brain metabolism. Martin's (1981) maternal energy hypothesis is an extension of this reasoning and postulates that brain size is related to maternal metabolic rate. The relatively small brains of strepsirrhines (compared to haplorhines) could be related to depressed metabolic rates in strepsirrhine females and specifically to the transfer of nutrients during pregnancy and lactation. Importantly, the corollary is that the evolution of higher metabolic rates in anthropoids (or possibly in haplorhines, depending on the position of tarsiers) may have allowed these animals to grow and support relatively larger brains. Strepsirrhines invest less in the prenatal development of their offspring than haplorhines, but when controlled for metabolic rate, this difference disappears (Richard and Dewar, 1991; Young et al., 1990). Female feeding priority is also most common in strepsirrhines, especially those with low-metabolic rates, and this has been suggested to help females cope with high maternal energy costs associated with reproduction (Richard and Dewar, 1991).

The study clearly demonstrates that strepsirrhines are less encephalized than haplorhines for any given body size. However, the available primate data also show that the relationship between brain size and RMR is comparable between the two groups. This result suggests that strepsirrhines and haplorhines spend comparable proportions of their RMR on brain metabolism; this supports the conclusions of Armstrong (1985), who used a smaller sample of species. It is possible that the lower levels of encephalization in strepsirrhines relative to haplorhines may be a consequence of metabolic stress (i.e., the low-metabolic rates of strepsirrhines are unable to support relatively large brains). However, this picture is overly simplistic since numerous species deviate from the brain size to RMR relationship. Additionally, as noted by Martin (1996), the range of variation in the relationship between brain size and body mass exceeds that between RMR and body mass. Finally, humans have extraordinarily large brains that account for roughly 20–25% of RMR but do not have elevated RMRs compared to those predicted for body mass (Leonard and Robertson, 1994). How this could have evolved has been the subject of intense debate (e.g., Aiello and Wheeler, 1995; Leonard and Robertson, 1994).

Implications for Models of Primate Origins

The nature and origins of metabolic variation in strepsirrhines and haplorhines have important implications for our understanding of the ecology and evolution of the earliest primates. For any given size, primates and other mammals consume considerably more energy than similar-sized reptiles. Mammals have total daily energy costs that average about 17 times that of comparably sized reptiles (Nagy, 1987).

The earliest true primates appeared in the Late Paleocene and Early Eocene and are defined by a suite of cranial and postcranial features not present in the plesiadapiforms or other mammals. While once considered within the primate order (as archaic primates), plesiadapiforms have recently been removed (Fleagle, 1999; Martin, 1990; Rose, 1995). Some authorities place the plesiadapiforms with colugos within the order Dermoptera (Beard, 1993), while others consider them a separate mammalian order (Fleagle, 1999). In fact, all plesiadapiforms with the exception of *Purgatorius* are too derived dentally to be ancestral to living primates (Rose, 1995). The first true primates (i.e., Euprimates) show a suite of derived cranial characters, such as orbital convergence and frontation, which are associated with increased reliance on vision. They also show derived postcranial features, such as nails instead of claws, and grasping hands and feet, which have been linked to increased manipulative abilities within an arboreal environment. The earliest fossils attributed to primates are highly fragmentary but nonetheless show characters that link them with living primates (Rose, 1995). These species include *Altiatlasius koulchii* from the Late Paleocene of Morocco and *Altanius orlovi* from the Early Eocene of Mongolia (Fleagle, 1999; Rose, 1995). *Altiatlasius* is thought to have had a body size on the order of between 50 and 100 g, while *Altanius* is thought to have had a body size of about 10 g. These fossils appear to be more primitive and generalized than adapoids or omomyoids. Early mammalian forms were similarly small bodied and all appear to have been under 500 g. Fossils of *Hadrocodium wui* from the Early Jurassic of China had an adult body weight estimated to be about 2 g (Luo et al., 2001), while other groups were slightly larger, (e.g., *Morganucodon*, at 27–89 g and *Sinoconodon*, at 13–517 g; Luo et al., 2001).

The Eocene primates are typically divided into two major groups, the adapoids (superfamily Adapoidea) and the omomyoids (superfamily Omomyoidea) (Fleagle, 1999). The former have been compared to living lemurs in certain aspects of craniodental and postcranial morphology, while the latter have been likened to living tarsiers; however, the exact phylogenetic relationship with living primates remains unclear (Martin, 1990). While both groups exhibit considerable diversity, the earliest members of each are similar in many aspects. Some of the earliest genera include *Donrussellia* and *Cantius* of the adapoids and *Teilhardina* and *Steinius* of the omomyoids (Rose, 1995). Reconstructed body size of *D. provincialis* was about 140 g (Rose, 1995), though some of the other species may have been slightly larger (210–730 g; Fleagle, 1999). *Cantius* is thought to have been considerably larger and had a body mass range on the order of 1–3 kg for nine species (Fleagle, 1999). *Teilhardina*, like most other omomyoids, was small bodied, with estimates for the genus (four species) ranging from 60 to 135 g in adult body size (Fleagle, 1999; Rose, 1995). *Steinius* was on the order of about 300–400 g (Fleagle, 1999). Adapoids later diversified and obtained body sizes up to 7–8 kg (Fleagle, 1999). Some exhibited sexual dimorphism, most appear to have been diurnal, and most were likely frugivores or folivores (Rose, 1995).

Omomyoids remained primarily small bodied (<100 g) though a radiation of omomyoids took place in North America and included larger bodied species (exceeding 2 kg) after the extinction of most of the adapoids. Most adapoid and omomyoid species went extinct at the Grande Coupure extinction event, which occurred about 34 MYA at the end of the Eocene and appears to be associated with decreased temperature and humidity in the Northern Hemisphere (Fleagle, 1999; Köhler and Moyà-Solà, 1999).

A number of models have been offered to explain the evolution of primates. Early models of primate origins (e.g., Jones, 1916) explained the suite of distinctive primate characters as adaptations to life in an arboreal environment that favored emphasis on the visual system and grasping hands and feet. However, as pointed out by numerous critics (e.g., Cartmill, 1974), this explanation ignores the fact that most nonprimate arboreal animals possess claws rather than nails, do not have grasping hands and feet, have laterally directed eyes, and rely heavily on olfaction. Thus, a generalized adaptation to an arboreal environment is unlikely to account for the evolution of these derived traits in primates. Additionally, it has become evident that the closest living relatives of primates (the archontans) are all at least partly arboreal; suggesting that the adaptive shift in early primates involved something beyond simply colonization of the trees.

More recent models have sought to explain the origin of primates as result of specific adaptive shifts within the arboreal environment. Cartmill's visual predation model (1974, 1992) explains the evolution of primate characteristics as an adaptive suite of features related to visual prey detection and predation (primarily on insects) on terminal branches and in the forest undergrowth. In contrast, Sussman (1991) has argued that it was not visual predation that led to the evolution of the primate traits but instead they are related to terminal branch feeding on the products of flowering plants (e.g., fruit, nectar, etc.), as well as the insects that pollinate these flowering plants. Terminal branch feeding as the impetus for the evolution of prehensile hands and feet, irrespective of diet, has received support in comparative studies of didelphid marsupials (Lemelin, 1999).

Information on primate bioenergetics has important implications for evaluating alternative models of primate origins. In particular, since body size has important energetic consequences and is critical in shaping dietary patterns, information on the size of early primate ancestors provides an important link to energetics and metabolism.

In general, among primate species there is an inverse relationship between body size and DQ (Leonard and Robertson, 1994; Sailer et al., 1985). This relationship (the "Jarman-Bell" relationship; Bell, 1971; Gaulin, 1979; Jarman, 1974) appears to be a consequence of the Kleiber scaling relationship between mass and metabolic rate. Large primates have high total energy needs, but relatively low mass-specific requirement, and are able to meet their energy demands by feeding on resources that are widely abundant but lower in quality (e.g., leaves, other foliage). In contrast, small primates have low total energy needs, but extremely high requirements per unit mass. These species tend to subsist on food items that are limited in their abundance but rich in energy and nutrients (e.g., insects, small vertebrates, saps, and gums).

Thus, as data on extant species show, body size greatly shapes and constrains the types of foods on which a primate can subsist. For example, insectivorous diets can only be sustained in very small animals and folivorous diets can only be sustained in considerably larger animals (Kay, 1984). Insects are excellent energy and protein sources for small animals, given their high relative energy demands. Conversely, leaves can provide an ample source of energy for larger bodied animals because of relatively lower energy requirements and longer gut passage times that allow for more nutrient extraction. However, animals smaller than about 700 g have a difficult time sustaining themselves energetically on a diet largely based on leaves. Fruits typically provide an ample source of available carbohydrates but are limited in terms of available protein. Frugivorous animals must supplement their diet with other sources of protein such as insects, leaves, or vertebrates.

Fossil and comparative studies of living animals suggests that the earliest primates were small bodied, with body sizes considerably smaller than 500 g and likely under 100 g. These early primates were likely primarily arboreal, nocturnal, inhabited tropical forests, and were adapted for climbing, grasping, and leaping in a fine-branch niche (Rose, 1995; Martin, 1990). As noted by Martin (1990), this ancestor was similar in many respects to living mouse lemurs and dwarf bushbabies, and contrasts markedly with the tree shrews, which are commonly used as early-primate analogs.

Considering the metabolic data on two strepsirrhines under 100 g (*Galagoides demidoff* and *Microcebus murinus*), we find that both are hypometabolic, with deviations from predicted RMR of −23.85 and −37.51%, respectively. While no haplorhines in the sample are below 100 g, the 105 g *Cebuella pygmaea* and the 113 g *Tarsius syrichta* are both also hypometabolic, with deviations of −21.78 and −34.80%, respectively. These species all have relatively high-quality diets and all obtain considerable energy from insects. While tarsiers are the only living primates to subsist on 100% animal prey (primarily insects), the living strepsirrhines under 100 g, *G. demidoff* and members of the genus *Microcebus* (including *M. murinus* and *M. rufus*), consume high-quality diets with varying amounts of insects and vertebrates (Atsalis, 1999; Charles-Dominique, 1977; Mittermeier et al., 1994). *Galagoides demidoff* consumes roughly 70% insects and supplements these primarily with fruit and gums (Charles-Dominique, 1977). *Microcebus murinus* has an omnivorous diet that includes insects, fruits, flowers, small vertebrates, insect secretions, gums, nectars, and other plant products (Corbin and Schmid, 1995; Hladik et al., 1980; Martin, 1973). *Microcebus rufus* appears to be heavily reliant on both fruit and insects, and has been described as a frugivore–faunivore (Atsalis, 1999). Interestingly, while *M. rufus* consumes a variety of plant species, it is heavily reliant upon several varieties of *Bakerella* (a type of mistletoe) known to have a very high fat content. Both *M. murinus* and *M. rufus* show seasonal shifts in diet (Atsalis, 1999; Hladik et al., 1980).

The ancestral primate most likely relied heavily on insects, especially during certain seasons, and supplemented its diet with high-quality plant parts, such as fruits, as well as small vertebrates. As pointed out by Martin (1990) it is in the terminal branches of tropical trees and shrubs that insects and fruit resources would have been most readily available to the earliest primates. Terminal branch feeding and its associated anatomical features in primates may have evolved to exploit changing patterns of insect and fruit availability that resulted from radiation of angiosperms during the Early Cenozoic. Low maintenance and total energy costs may have enhanced survival in early primates, especially in environments with low overall productivity and/or marked seasonality.

While hypometabolism can enhance survival in certain environments, there are important reproductive consequences of hypometabolism. Mammalian species with relatively low-metabolic rates also tend to have low-intrinsic rates of population growth (McNab, 1980, 1986). However, while population growth may be slower in hypometabolic species, there are environments where this would clearly be favored. The depressed metabolic rates of some mammal and bird species from isolated oceanic islands appear to be the result of selection for resource minimization in an environment with limited resources (McNab, 1994). It has been suggested that hypometabolic insectivores are better able to deal with seasonal fluctuations in food abundance (McNab, 1980). There is also evidence from bats that indicates that low-metabolic rates are important for coping with variation in food availability (i.e., avoiding starvation during periods of low-insect availability) (Audet and Thomas, 1997). Additionally, nonseasonal torpor and hibernation can confer considerable energy savings to small-bodied mammals (Wang and Wolowyk, 1988).

Thus, the physiological ecology of extant small-bodied strepsirrhines strongly suggests that the earliest primates were hypometabolic and heavily reliant on insects. The specific explanations for why hypometabolism is so common among small-bodied primates remain unclear; however, the patchy and seasonally variable nature of key food resources for these species may have played an important role. Further, it appears that the low-metabolic rates common among all extant strepsirrhines may have a deep evolutionary history. Such an interpretation implies that increased rates of metabolic turnover (and greater encephalization) occurred with the evolution of larger-bodied primates that were reliant on a different suite of food resources.

ACKNOWLEDGMENTS

We thank M. Muchlinski, M. Ravosa, B. Shea, M. Sorensen, and C. Terranova for discussions of the project. We are grateful to M. Ravosa for access to unpublished data on subfossil lemur cranial capacities. We thank M. Dagosto and M. Ravosa for inviting us to participate in this symposium.

REFERENCES

Adkins, R. M., and Honeycutt, R. L., 1991, Molecular phylogeny of the superorder Archonta, *Proc. Natl. Acad. Sci. USA* **88**: 10317–10321.

Adkins, R. M., and Honeycutt, R. L., 1993, A molecular examination of Archontan and Chiropteran monophyly, in: *Primates and Their Relatives in Phylogenetic Perspective*, R. D. E. MacPhee, ed., Plenum Press, New York, pp. 227–249.

Aiello, L. C., and Wheeler, P., 1995, The expensive-tissue hypothesis: The brain and the digestive system in human and primate evolution, *Curr. Anthropol.* **36**: 199–221.

Arends, A., Bonaccorso, F. J., and Genoud, M., 1995, Basal rates of metabolism of nectarivorous bats (Phyllostomidae) from a semiarid thorn forest in Venezuela, *J. Mammal.* **76**(3): 947–956.

Armstrong, E., 1983, Relative brain size and metabolism in mammals, *Science* **220**: 1302–1304.

Armstrong, E., 1985, Relative brain size in monkeys and prosimians, *Am. J. Phys. Anthropol.* **66**: 263–273.

Atsalis, S., 1999, Diet of the brown mouse lemur (*Microcebus rufus*) in Ranomafana National Park, Madagascar, *Int. J. Primatol.* **20**(2): 193–229.

Audet, D., and Thomas, D. W., 1997, Facultative hypothermia as a thermoregulatory strategy in the phyllostomid bats, *Carollia perspicillata* and *Sturnira lilium*, *J. Comp. Physiol. [B]* **167**: 146–152.

Bartels, W., Law, B. S., and Geiser, F., 1998, Daily torpor and energetics in a tropical mammal, the northern blossom-bat *Macroglossus minimus* (Megachiroptera), *J. Comp. Physiol. [B]* **168**: 233–239.

Bauchot, R., and Stephan, H., 1969, Encephalisation et niveau evolutif chez les simiens, *Mammalia* **33**: 225–275.

Baudinette, R. V., Churchill, S. K., Christian, K. A., Nelson, J. E., and Hudson, P. J., 2000, Energy, water balance and the roost microenvironment in three Australian cave-dwelling bats (Microchiroptera), *J. Comp. Physiol. [B]* **170**: 439–446.

Beard, K. C., Krishtalka, L., and Stucky, R. K., 1991, First skulls of the early Eocene primate *Shoshonius cooperi* and the anthropoid-tarsier dichotomy, *Nature* **349**: 64–66.

Bell, R. H., 1971, A grazing ecosystem in the Serengeti, *Sci. Am.* **225**: 86–93.

Bonner, T. I., Heinemann, R., and Todaro, G. J., 1980, Evolution of DNA sequences has been retarded in Malagasy primates, *Nature* **286**: 470–473.

Bradley, S. R., and Hudson, J. W., 1974, Temperature regulation in the tree shrew *Tupaia glis*, *Comp. Biochem. Physiol. A Mol. Integr. Physiol.* **48**: 55–60.

Burney, D. A., and Ramilisonina, 1999, The Kilopilopitsofy, Kidoky, and Bokyboky: Accounts of strange animals from Belo-sur-mer, Madagascar, and the megafaunal "extinction window," *Am. Anthropol.* **100**(4): 957–966.

Cartmill, M., 1974, Rethinking primate origins, *Science* **184**: 436–443.

Cartmill, M., 1992, New views on primate origins, *Evol. Anthropol.* **1**(3): 105–111.

Calder, W. A., 1984, *Size, Function, and Life History*, Harvard University Press, Cambridge.

Charles-Dominique, P., 1974, Ecology and feeding behavior of five sympatric lorisids in Gabon, in: *Prosimian Biology*, R. D. Martin, G. A. Doyle, and A. C. Walker, eds., Duckworth, London, pp. 131–150.

Charles-Dominique, P., 1977, *Ecology and Behavior of Nocturnal Primates*, Columbia University Press, New York.

Corbin, G. D., and Schmid, J., 1995, Insect secretions determine habitat use pattern by a female lesser mouse lemur (*Microcebus murinus*), *Am. J. Primatol.* **37**: 317–324.

Curtis, D. J., Zaramody, A., and Martin, R. D., 1999, Cathemerality in the mongoose lemur, *Eulemur mongoz*, *Am. J. Primatol.* **47**: 279–298.

DeJong, W. W., and Goodman, M., 1988, Anthropoid affinities of *Tarsius* supported by lens α-crystallin sequences, *J. Hum. Evol.* **17**: 575–582.

Dijan, P., and Green, H., 1991, Involucrin gene at tarsioids and other primates: Alternatives in evolution of segment repeats, *Proc. Natl. Acad. Sci. USA* **88**: 5321–5325.

Eisentraut, M., 1961, Beobachtungen über den Wärmehaushalt bei Halbaffen, *Biol. Zbl.* **80**: 319–325.

Elgar, M. A., and Harvey, P. H., 1987, Basal metabolic rates in mammals: Allometry, phylogeny and ecology, *Funct. Ecol.* **1**: 25–36.

Flacourt, E., de., 1658, *Histoire de la Grande Isle Madagascar*, Paris.

Fleagle, J. G., 1999, *Primate Adaptation and Evolution*, Academic Press, San Diego.

Garbutt, N., 1999, *Mammals of Madagascar*, Yale University Press, New Haven.

Gaulin, S. J. C., 1979, A Jarman/Bell model of primate feeding niches, *Hum. Ecol.* **7**: 1–20.

Geiser, F., 1996, Thermoregulation, energy metabolism, and torpor in blossom-bats, *Syconycteris australis* (Megachiroptera), *J. Zool.* **239**: 583–590.

Godfrey, L. R., and Jungers, W. L., 2002, Quaternary fossil lemurs, in: *The Primate Fossil Record*, W. Hartwig, ed., Cambridge University Press, New York, pp. 97–121.

Godfrey, L. R., Jungers, W. L., Reed, K. E., Simons, E. L., and Chatrath, P. S., 1997, Subfossil lemurs: Inferences about past and present primate communities in Madagascar, in: *Natural Change and Human Impact in Madagascar*, S. M. Goodman and B. D. Patterson, eds., Smithsonian Institution Press, Washington, DC, pp. 218–256.

Grand, T. I., 1977, Body weight: Its relation to tissue composition, segment distribution, and motor function, *Am. J. Phys. Anthropol.* **47**(2): 211–239.

Gregory, W. K., 1910, The orders of mammals, *Bull. Am. Mus. Nat. Hist.* **27**: 1–524.

Harvey, P. H., and Pagel, M. D., 1991, *The Comparative Method in Evolutionary Biology*, Oxford University Press, Oxford.

Hayssen, V., and Lacy, R. C., 1985, Basal metabolic rates in mammals: Taxonomic differences in the allometry of BMR and body mass, *Comp. Biochem. Physiol. A Mol. Integr. Physiol.* **81**: 741–754.

Hladik, C. M., Charles-Dominique, P., and Petter, J. J., 1980, Feeding strategies of five nocturnal prosimians in the dry forest of the west coast of Madagascar, in: *Nocturnal Malagasy Primates: Ecology, Physiology, and Behavior*, P. Dominique, H. M. Cooper, A. Hladik, C. M. Hladik, E. Pages, P. G. Pariente et al., eds., Academic Press, New York, pp. 41–73.

Hoffman, M. A., 1983, Energy metabolism, brain size and longevity in mammals, *Q. Rev. Biol.* **58**: 495–512.

Holliday, M. A., 1986, Body composition and energy needs during growth, in: *Human Growth: A Comprehensive Treatise*, F. Falkner and J. M. Tanner, eds., vol. 2, pp. 101–107, Plenum Press, New York.

Hosken, D. J., 1997, Thermal biology and metabolism of the greater long-eared bat, *Nyctophilus major* (Chiroptera: Vespertilionidae), *Aust. J. Zool.* **45**: 145–156.

Hosken, D. J., and Withers, P. C., 1997, Temperature regulation and metabolism of an Australian bat, *Chalinolobus gouldii* (Chiroptera: Vespertilionidae), *J. Comp. Physiol. [B]* **167**: 71–80.

Hosken, D. J., and Withers, P. C., 1999, Metabolic physiology of euthermic and torpid lesser long-eared bats, *Nyctophilus geoffroyi* (Chiroptera: Vespertilionidae), *J. Mammal.* **80**(1): 42–52.

Jarman, P. J., 1974, The social organization of antelope in relation to their ecology, *Behaviour* **58**: 215–267.

Jones, F. W., 1916, *Arboreal Man*, Arnold, London.

Jungers, W. L., 1999, Brain size and body size in subfossil Malagasy lemurs, *Am. J. Phys. Anthropol. (Supp)* **28**: 163.

Kappeler, P. M., 1996, Causes and consequences of life-history variation among strepsirhine primates, *Am. Nat.* **148**(5): 868–891.

Kappeler, P. M., 2000, Lemur origins: Rafting by groups of hibernators? *Folia Primatol.* **71**: 422–425.

Kay, R. F., 1984, On the use of anatomical features to infer foraging behavior in extinct primates, in: *Adaptations for Foraging in Nonhuman Primates*, P. S. Rodman and J. G. H. Cant, eds., Columbia University Press, New York, pp. 21–53.

Kleiber, M., 1961, *The Fire of Life*, Wiley, New York.

Köhler, M., and Moyà-Solà, S., 1999, A finding of Oligocene primates on the European continent, *Proc. Nat. Acad. Sci. USA* **96**(25): 14664–14667.

Koop, B. F., Siemieniak, D., Slightom, J. L., Goodman, M., Dunbar, J., Wright, P. C., and Simons, E. L., 1989a, *Tarsius* δ- and β-globin genes: Conversions, evolution, and systematic implications, *J. Biol. Chem.* **264**: 68–79.

Koop, B. F., Tagle, D. A., Goodman, M., and Slightom, J. L., 1989b, A molecular view of primate phylogeny and important systematic and evolutionary questions, *Mol. Biol. Evol.* **6**: 580–612.

Kurland, J. A., and Pearson, J. D., 1986, Ecological significance of hypometabolism in nonhuman primates: Allometry, adaptation, and deviant diets, *Am. J. Phys. Anthropol.* **71**: 445–457.

Le Maho, Y., Goffart, M., Rochas, A., Felbabel, H., and Chatonnet, J., 1981, Thermoregulation in the only nocturnal simian: The night monkey *Aotus trivirgatus*, *Am. J. Physiol.* **240**: R156–165.

Lemelin, P., 1999, Morphological correlates of substrate use in didelphid marsupials: Implications for primate origins, *J. Zool.* **247**: 165–175.

Leonard, W. R., and Robertson, M. L., 1994, Evolutionary perspectives on human nutrition: The influence of brain and body size on diet and metabolism, *Am. J. Hum. Biol.* **6**: 77–88.

Liu, F. G. R., and Miyamoto, M. M., 1999, Phylogenetic assessment of molecular and morphological data for Eutherian mammals, *Syst. Biol.* **48**: 54–64.

Liu, F. G. R., Miyamoto, M. M., Freire, N. P., Ong, P. Q., Tennant, M. R., Young, T. S. et al., 2001, Molecular and morphological supertrees for eutherian (placental) mammals, *Science* **291**: 1786–1789.

Luo, Z. H., Crompton, A. W., and Sun, A. L., 2001, A new mammaliaform from the Early Jurassic and evolution of mammalian characteristics, *Science* **292**: 1535–1540.

Madsen, O., Scally, M., Douady, C. J., Kao, D. J., DeBry, R. W., Adkins, R. et al., 2001, Parallel adaptive radiations in two major clades of placental mammals, *Nature* **409**: 610–614.

Martin, R. D., 1973, A review of the behaviour and ecology of the lesser mouse lemur (*Microcebus murinus* J. F. Miller 1777), in: *Comparative Ecology and Behaviour of Primates*, R. P. Michael and J. H. Crook, eds., Academic Press, London, pp. 1–68.

Martin, R. D., 1981, Relative brain size and metabolic rate in terrestrial vertebrates, *Nature* **293**: 57–60.

Martin, R. D., 1989, Size, shape and evolution, in: *Evolutionary Studies: A Centenary Celebration of the Life of Julian Huxley*, M. Keynes and G. A. Harrison, eds., Macmillan, Hampshire, pp. 96–141.

Martin, R. D., 1990, *Primate Origins and Evolution: A Phylogenetic Reconstruction*, Princeton University Press, Princeton.

Martin, R. D., 1996, Scaling of the mammalian brain: The maternal energy hypothesis, *News Physiol. Sci.* **11**: 149–156.

McKenna, M. C., 1975, Toward a phylogenetic classification of the mammalia, in: *Phylogeny of the Primates*, W. P. Luckett and R. S. Szalay, eds., Plenum Press, New York, pp. 21–46.

McNab, B. K., 1978, Energetics of arboreal folivores: Physiological problems and ecological consequences of feeding on an ubiquitous food supply, in: *The Ecology of Arboreal Folivores*, G. G. Montgomery, ed., Smithsonian Institution Press, Washington, DC, pp. 153–162.

McNab, B. K., 1980, Food habits, energetics, and the population biology of mammals, *Am. Nat.* **116**(1): 106–124.

McNab, B. K., 1986, The influence of food habits on the energetics of eutherian mammals, *Ecol. Monogr.* **56**(1): 1–9.

McNab, B. K., 1988, Complications inherent in scaling the basal rate of metabolism in mammals, *Q. Rev. Biol.* **63**(1): 25–54.

McNab, B. K., 1994, Resource use and the survival of land and freshwater vertebrates on oceanic islands, *Am. Nat.* **144**(4): 643–660.

McNab, B. K., and Wright, P. C., 1987, Temperature regulation and oxygen consumption in the Philippine tarsier *Tarsius syrichta*, *Physiol. Zool.* **60**(5): 596–600.

Mittermeier, R. A., Tattersall, I., Konstant, W. R., Meyers, D. M., and Mast, R. B., 1994, *Lemurs of Madagascar*, Conservation International, Washington, DC.

Miyamoto, M. M., and Goodman, M., 1990, DNA systematics and evolution of primates, *Annu. Rev. Ecol. Syst.* **2**: 197–220.

Morland, H. S., 1993, Seasonal behavioral variation and its relationship to thermoregulation in ruffed lemurs (*Varecia variegata variegata*), in: *Lemur Social Systems and Their Ecological Basis*, P. M. Kappeler and J. U. Ganzhorn, eds., Plenum Press, New York, pp. 193–204.

Muchlinski, M. N., Snodgrass, J. J., and Terranova, C. J., 2003, Scaling of muscle mass in primates, *Am. J. Phys. Anthropol. (Supp.)* **36**: 155.

Müller, E. F., and Jaksche, H., 1980, Thermoregulation, oxygen consumption and evaporative water loss in the thick-tailed bushbaby (*Galago crassicaudatus* Geoffroy 1812), *Z. Säugetierkd.* **45**: 269–278.

Murphy, W. J., Eizirik, E., Johnson, W. E., Zhang, Y. P., Ryder, O. A., and O'Brien, S. J., 2001a, Molecular phylogenetics and the origins of placental mammals. *Nature* **409**: 614–618.

Murphy, W. J., Eizirik, E., O'Brien, S. J., Madsen, O., Scally, M., Douady, C. J. et al., 2001b, Resolution of the early placental mammal radiation using Bayesian phylogenetics, *Science* **294**: 2348–2351.

Nagy, K. A., 1987, Field metabolic rate and food requirement scaling in mammals and birds, *Ecol. Monogr.* **57**: 111–128.

Nagy, K. A., Girard, I. A., and Brown, T. K., 1999, Energetics of free-ranging mammals, reptiles, and birds, *Annu. Rev. Nutr.* **19**: 247–277.

Napier, J. R., and Napier, P. H., 1994, *The Natural History of the Primates*, MIT Press, Cambridge.

Nowak, R. M., 1991, *Walker's Mammals of the World*, Johns Hopkins University Press, Baltimore.

Oates, J. F., 1984, The niche of the potto, *Perodicticus potto*, *Int. J. Primatol.* **5**: 51–61.

Pollock, J. I., and Mullin, R. J., 1987, Vitamin C biosynthesis in prosimians: Evidence for the anthropoid affinity of *Tarsius*, *Am. J. Phys. Anthropol.* **73**: 65–70.

Porter, C. A., Sampaio, I., Schneider, H., Schneider, M. P. C., Czelusniak, J., and Goodman, M., 1995, Evidence on primate phylogeny from ε-globin gene sequences and flanking regions, *J. Mol. Evol.* **40**: 30–55.

Purvis, A., 1995, A composite estimate of primate phylogeny, *Philos. Trans. R. Soc. Lond. B Biol. Sci.* **346**: 405–421.

Richard, A. F., 1985, *Primates in Nature*, W. H. Freeman, New York.

Richard, A. F., and Dewar, R. E., 1991, Lemur ecology, *Annu. Rev. Ecol. Syst.* **22**: 145–175.

Rose, K. D., 1995, The earliest primates, *Evol. Anthropol.* **3**(5): 159–173.

Ross, C., 1992, Basal metabolic rate, body weight and diet in primates: An evaluation of the evidence, *Folia Primatol.* **58**: 7–23.

Ross, C., 1994, The craniofacial evidence for anthropoid and tarsier relationships, in: *Anthropoid Origins*, J. Fleagle and R. E. Kay, eds., Plenum Press, New York, pp. 469–547.

Ross, C., 1996, An adaptive explanation for the origin of the Anthropoidea (Primates), *Am. J. Primatol.* **40**: 205–230.

Rowe, N., 1996, *The Pictorial Guide to Living Primates*, Pogonias Press, New York.

Sailer, L. D., Gaulin, S. J. C., Boster, J. S., and Kurland, J. A., 1985, Measuring the relationship between dietary quality and body size in primates, *Primates* **26**:14–27.

Sargis, E. J., 2002, The postcranial morphology of *Ptilocercus lowii* (Scandentia, Tupaiidae): An analysis of primatomorphan and volitantian characters, *J. Mammal. Evol.* **9**: 137–160.

Shoshani, J., Groves, C. P., Simons, E. L., and Gunnell, G. F., 1996, Primate phylogeny: Morphological vs. molecular results, *Mol. Phylogenet. Evol.* **5**: 102–154.

Simons, E. L., 1997, Lemurs: Old and new, in: *Natural Change and Human Impact in Madagascar*, S. M. Goodman and B. D. Patterson, eds., Smithsonian Institution Press, Washington, DC, pp. 142–166.

Simpson, G. G., 1945, The principles of classification and a classification of the mammals, *Bull. Amer. Mus. Nat. Hist.* **85**: 1–350.

Smith, R. J., and Jungers, W. L., 1997, Body mass in comparative primatology, *J. Hum. Evol.* **32**: 523–559.

Stahl, W. R., 1967, Scaling of respiratory variables in mammals, *J. Appl. Physiol.* **22**: 453–460.

Stanhope, M. J., Bailey, W. J., Czelusniak, J., Goodman, M., Si, J.-S., Nickerman, J. et al., 1993, A molecular view of primate supraordinal relationships from the analysis of both nucleotide and amino acid sequences, in: *Primates and Their Relatives in Phylogenetic Perspective*, R. D. E. MacPhee, ed., Plenum Press, New York, pp. 251–292.

Stephan, H., Frahm, H., and Baron, G., 1981, New and revised data on volumes of brain structures in insectivores and primates, *Folia Primatol.* **35**:1–29.

Sussman, R. W., 1991, Primate origins and the evolution of angiosperms, *Am. J. Primatol.* **23**: 209–223.

Szalay, F. S., 1977, Phylogenetic relationships and a classification of the eutherian Mammalia, in: *Major Patterns in Vertebrate Evolution*, M. K. Hecht, P. C. Goody, and B. M. Hecht, eds., Plenum Press, New York, pp. 315–374.

Szalay, F. S., Rosenberger, A. L., and Dagosto, M., 1987, Diagnosis and differentiation of the order Primates, *Yearb. Phys. Anthropol.* **30**: 75–105.

Szewczak, J. M., 1997, Matching gas exchange in the bat from flight to torpor, *Am. Zool.* **37**: 92–100.

Thompson, S. D., Power, M. L., Rutledge, C. E., Kleiman, D. G., 1994, Energy metabolism and thermoregulation in the golden lion tamarin (*Leontopithecus rosalia*), *Folia Primatol.* **63**: 131–143.

van Schaik, C. P., and Kappeler, P. M., 1996, The social systems of gregarious lemurs: Lack of convergence with anthropoids due to evolutionary disequilibrium? *Ethology* **102**: 915–941.

Wang, L. C., and Wolowyk, M. W., 1988, Torpor in mammals and birds, *Can. J. Zool.* **16**: 133–137.

Warren, R. D., and Crompton, R. H., 1996, Lazy leapers: Energetics. Phylogenetic inertia and the locomotor differentiation of the Malagasy primates, in: *Biogeographie de Madagascar*, W. R. Lourenco, ed., Editions ds l'ORSTOM, Paris, pp. 259–266.

Whittow, G. C., and Gould, E., 1976, Body temperature and oxygen consumption of the pentail shrew (*Ptilocercus lowii*), *J. Mammal.* **57**: 754–756.

Young, A. L., Richard, A. F., and Aiello, L. C., 1990, Female dominance and maternal investment in strepsirhine primates, *Am. Nat.* **135**: 473–488.

Zietkiewicz, E., Richer, C., and Labuda, D., 1999, Phylogenetic affinities of tarsier in the context of primate Alu repeats, *Mol. Phylogenet. Evol.* **11**(1): 77–83

CHAPTER TWENTY-ONE

Episodic Molecular Evolution of Some Protein Hormones in Primates and Its Implications for Primate Adaptation

Soojin Yi and Wen-Hsiung Li

INTRODUCTION

The order Primates is one of the most speciose placental orders. According to the tabulation of the living mammalian species by Wilson and Reeder (1993), there are 233 primate species. Only four other orders: Rodentia, Chiroptera, Carnivora, and Eulipotyphla consist of more species than primates. There are also roughly twice as many fossil species of primates. Therefore, a large number of speciation events occurred during the course of primate evolution since the initial radiation of plesiadapiforms in the Paleocene (Fleagle, 1999). Primates exhibit a diverse array of evolutionary tempos. For instance, the genera *Aotus*, *Tarsius*, and *Macaca* seem to have stayed morphologically the same for tens of millions of years, whereas some genera show remarkable rates of evolution within a relatively short time period (Fleagle, 1999). A good

Soojin Yi and Wen-Hsiung Li • Department of Ecology and Evolution, University of Chicago, Chicago, IL 60637

example of the latter situation is *Homo*—the very genus that includes our own species. Currently, primates occupy many different types of habitats and show a great diversity in their adaptive traits such as behavior, diet, and locomotion (Fleagle, 1999). Understanding the bases of such adaptive evolution is one of the ultimate goals of the study of evolution.

Protein hormones might have played an important role in primate adaptation because of their essential role in physiology. In the last several decades, much data has accumulated on the structure and function of protein hormones. In parallel to this, the molecular evolutionary patterns of some protein hormones have been investigated in depth, especially in primates, due to their implications for medicine. This chapter focuses on several cases of rapid evolution in protein hormones that might have caused significant physiological changes in some primate lineages.

The study of physiological mechanisms that are controlled by protein hormones has a long history. Early development of this field depended heavily on animal models. For example, in the 1920s, the discovery of insulin was largely based on studies with dogs. From the 1930s to the present, all hormonal substitution therapies have benefited from pharmacological trials on a variety of mammalian species. For some hormone therapies, purified animal products, such as porcine insulin, have been the choice before the advent of genetic engineering (for references, see De Pablo, 1993). In view of the fact that nonprimate hormones usually worked on humans, it was not surprising to find that the amino acid sequences of hormone proteins have been well conserved among species (see Li, 1997). However, when the molecular evolutionary features of some protein hormones were investigated in detail, many cases of "episodic" evolution were found. Episodic evolution refers to the situation in which the rate of evolution of a biomolecule changes dramatically in a short time period (Li, 1997; Wallis, 1994, 1996). It has been shown, and is described in a later section, that such dramatic acceleration is usually confined to a few lineages, while the "basal" rate of evolution in the majority of lineages remains approximately constant. Episodic evolution is often considered to be the signature of adaptive evolution. However, it may also occur by relaxation of functional constraints; so determining the cause of an episodic event can be difficult. Interestingly, to date some of the best-characterized examples of episodic evolution occurred in primate lineages. In this chapter, we describe some of the best-studied cases.

To determine whether some molecular changes in evolution are due to positive selection requires some statistical methods. In this chapter, we first explain some current statistical tools for this purpose. Then, we use the example of the molecular evolution of lysozyme in some primate lineages to demonstrate one such method. The reason for using lysozyme is twofold. First, even though lysozyme itself is not a protein hormone, it is also a secretive protein. Second, some molecular changes of this protein have been linked to the adaptive evolution of a physiological trait in some primate lineages (Stewart and Wilson, 1987). This example will help readers understand how positive selection of amino acid substitutions can be investigated using statistical tools. We then describe in detail the molecular evolution of growth hormone (GH) and growth hormone receptor (GHR) in primates. The GH was the first protein hormone noticed to exhibit an episodic mode of molecular evolution (Wallis, 1994). This was possible mostly because of the abundant data on this protein, reflecting the long interests of both evolutionary biologists and biochemists. Together GH and GHR provide a rare opportunity to investigate the functional basis of molecular evolutionary changes underlying the coevolution of two proteins.

The GH in higher primates demonstrates another means of adaptive evolution, namely, gene duplication. A gene duplication initially increases the protein production, but later, the two genes may diverge in tissue expression and become specialized in different tissues. As will be described, the GH gene has been duplicated to multiple copies in higher primates. While one copy is still expressed only in the pituitary, the other copies are expressed in a new tissue, the placenta.

Next, we describe the evolution of Chorionic Gonadotropin (CG) in primates. The CG is a member of a tightly regulated network of reproductive hormones that establish and maintain pregnancy. The CG hormone arose from a gene duplication but has acquired the specialized role of keeping the pregnancy immediately after fertilization. Therefore, it is another example for the evolutionary significance of gene duplication. The usage of CG for this purpose, as well as the presence of this hormone itself, is confined to some lineages of primates. We will examine the evolution of CG in these lineages. This example shows two essential steps in the evolution of a new protein hormone following a gene duplication: first, a new expression pattern is established; and second, a novel protein coding sequence evolves by adaptive evolution.

Finally, we describe the episodic modes of molecular evolution of several protein hormones in other mammalian species. This section provides a glimpse of the extent of the phenomenon of episodic molecular evolution of hormones in mammals.

MOLECULAR EVOLUTION OF LYSOZYME: AN EXAMPLE OF ADAPTIVE EVOLUTION

Evolution of Lysozymes in Colobine Monkeys

Lysozyme is a ubiquitous bacteriolytic enzyme found in virtually all animals. Its function is to cleave the $\beta(14)$ glycosidic bonds between N-acetyl glucosamine and C-acetyl muramic acid in the cell walls of bacteria. As it is present in body fluids, such as saliva, serum, tears, etc., it is often the first line of defense against foreign bacteria. In foregut fermenters, which are animals whose anterior part of the stomach functions as a chamber for bacterial fermentation of ingested plant matters, lysozyme is secreted in the posterior parts of the digestive system so that it can be used to free nutrients from within the bacterial cell. This type of digestion has independently arisen twice in the evolution of placental mammals: once in the ruminants and once in leaf-eating colobine monkeys. In both cases, lysozyme has been recruited to degrade the cell walls of bacteria, which carries on fermentation in the foregut.

Therefore, the usage of lysozyme in the digestive system is a derived trait that has evolved to suit eating leaves as their main source of nutrition. Another trait evolved to suit this life history of colobine monkeys is the evolution of an enlarged stomach with numerous sections, similar to but much less elaborate than that of cows (Fleagle, 1999). Stewart and Wilson (1987) noticed that there are five uniquely shared amino acids between the lysozyme sequences from cows and langurs compared to only one amino acid uniquely shared by those from cows and horses. Since cows and langurs diverged much earlier than the separation of the cow and horse lineages, the uniquely shared amino acids in these two species are likely to be the results of a series of adaptive parallel substitutions that occurred independently in both lineages (i.e., an example of convergent evolution at the molecular level). The adaptive nature of these substitutions is such that some of them contribute to a better performance of lysozyme at low pH values (see Li, 1997).

For the above reason, the molecular evolution of lysozyme has been a favorite example of adaptive evolution, and it often serves as a model example to assess the performance of statistical methods for detecting selection from DNA sequence data. In the next section, one such method is described.

Statistical Analyses to Detect Positive Selection
in Lysozyme Sequences

Protein coding DNA sequences can be divided into two types of sites. First, substitutions at some sites can change the encoded amino acids. These are called nonsynonymous substitutions. Substitutions that do not cause any amino acid changes, due to the degeneracy of the genetic code, are synonymous (Li, 1997). Nonsynonymous mutations have direct phenotypic consequences (changes in the protein product) and, therefore, may be subject to natural selection. Synonymous mutations are not subject to selection at the protein level, although selection may operate at the RNA or translation level.

These differences in the effects of selection on the two types of mutation in the protein coding regions form the basis of inferring the underlying forces on DNA sequence evolution. The rate of synonymous substitutions (d_S) is considered to reflect the rate of mutation in that region, while the rate of nonsynonymous substitutions (d_N) is shaped by specific types of selection for that region. Therefore, a d_N/d_S ratio smaller than 1 means that nonsynonymous mutations have been fixed more slowly than the mutation rate or the neutral rate. This can be explained by selection to preserve the existing protein sequences, often called, negative or purifying selection. In fact, most protein sequences are assumed to evolve according to this fashion because most of the changes in protein sequences are likely to be deleterious in effect. A d_N/d_S ratio equal to 1 (statistically) suggests that mutations on the sequences are all equal in fitness, regardless of the consequences. This is often referred as a neutral mode of evolution. On the other hand, a d_N/d_S ratio significantly greater than 1 means that more nonsynonymous substitutions occurred than did synonymous mutations. As the mutation rate within the same gene is likely to be similar, this strongly suggests that many nonsynonymous mutations were selectively fixed (i.e., positive selection had driven the fixation of such mutations).

Yang (1998) developed a maximum likelihood approach to estimate the d_N/d_S ratio along each lineage in the phylogenetic tree of the species under study. This method takes into account the transition/transversion rate bias and nonuniform codon usage; it is often not straightforward to accommodate these factors by approximate pairwise methods. His method can accommodate the uniform-ratio model, with a single d_N/d_S ratio over all lineages of interest, as well as a free-ratio model at the other extreme, which assumes different underlying d_N/d_S ratios for different lineages. Intermediate models are also available to implement. Then a likelihood ratio test can be performed to compare the performances of different models.

Yang (1998) used this method to test whether the presence of the presumed positive selection can be detected from the DNA sequences of lysozyme. The lysozyme gene sequences of 24 primate species were analyzed. The result from an analysis utilizing a subset of seven sequences is shown in Figure 1.

The free-ratio model, which assumes different d_N/d_S ratios for different branches, performed significantly better than the one-ratio model, which assumes a single d_N/d_S ratio for all the branches. The branch leading to colobine monkeys (branch c) and the branch leading to hominoids (branch h) are long (i.e., they have accumulated many changes) and have very high d_N/d_S ratios. The d_N/d_S ratios along the c and h branches were significantly greater than the background ratios. The d_N/d_S ratio along the h branch was significantly greater than 1, indicating that positive selection had operated during the lysozyme evolution along this lineage. This is in agreement with

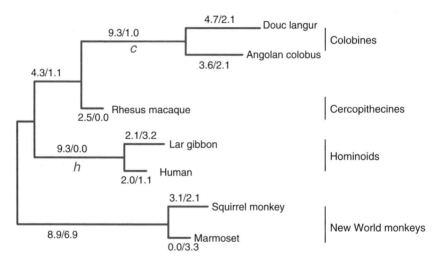

Figure 1. The maximum-likelihood estimates of the numbers of nonsynonymous and synonymous substitutions in each branch for the entire lysozyme coding sequences in seven primate species. The "free-ratio" model is used, which assumes different d_N/d_S ratios for different branches. Branches are proportional to the total numbers of substitutions.

a previous analysis (Messier and Stewart, 1997). The d_N/d_S ratio along the c branch was not significantly greater than 1. However, the hypothesis that the d_N/d_S ratio along this branch was greater than 1 was never rejected. Therefore, this result is compatible with both relaxed selective constraints and operation of positive selection along the c lineage. Since lysozyme did not lose function along branch c, but acquired a new function, the hypothesis of positive selection appears more plausible than reduced selective constraints.

There are a variety of likelihood methods developed to detect natural selection on the nucleotide level (for a review, see Yang and Bielawski, 2000). This is to account for more realistic evolutionary models. For example, the method described above assumes that all amino acid sites are under the same selective pressure, with the same d_N/d_S ratio. The analysis effectively averages the d_N/d_S ratios across all sites and positive selection is detected only if that average is significantly greater than 1. This assumption is very conservative; it is more realistic to imagine positive selection operating only on a few amino acid sites, while most of the amino acid sites are under strong purifying selection due to functional constraint. To address this possibility, Nielsen and Yang (1998) implemented a likelihood-ratio test to account for several classes of sites with different intensities of selective pressure. This method is more realistic and may provide an *a priori* hypothesis that certain structural and functional domains of the protein are under positive selection.

However, if adaptive evolution occurs only in a short time interval and affects only a few crucial amino acids, then this method is not likely to be powerful because this approach can detect positive selection only if the d_N/d_S ratio averaged over all lineages is greater than 1. Yang and Nielsen (2002) subsequently extended their model so that it allows the d_N/d_S ratio to vary both among sites and among lineages in a likelihood framework. These models may be useful for identifying positive selection along prespecified branches that affects only a few sites in the protein.

In reality, however, some models require unnecessarily large numbers of evolutionary parameters. Also, comparisons between different submodels are often biologically meaningless. In addition, implementing a model with a large number of parameters requires long amino acid coding sequences and large sample sizes; otherwise, the power of the tests are usually low. Particularly, to test whether some particular branches were under positive selection requires additional information. Nevertheless, in the case of lysozyme, the branch leading to the ancestor of the colobine monkeys has consistently been shown to be under positive selection (Yang, 1998; Yang and Nielsen, 2002).

EVOLUTION OF GROWTH HORMONE
AND ITS RECEPTOR

The method described earlier is an example of how to detect selection from sequence data. To infer the functional significance of specific substitutions, however, requires detailed biochemical and structural data. The GH and GHR proteins provide an example with such desired data. They have been studied extensively for their biochemical properties, as well as evolutionary features. Using current technology of molecular biology and biochemistry, it may be even possible to experimentally determine whether a specific amino acid substitution has important functional consequences.

GH and GHR as a Model System of Coevolution

Three characteristics of GH and GHR are worth noting from the evolutionary perspective. First, they provide a good system to study the coevolution of two proteins. As the functional pathway of GH begins with its binding to GHR, the two proteins are constrained to evolve together. Understanding the evolutionary trajectories of these two proteins will therefore help us understand how two proteins coevolve. Second, unlike nonprimate mammals, which possess only one GH gene, higher primates possess multiple copies of GH and GH-related genes, indicating the importance of GH gene duplication during the evolution of higher primates. Finally, GH shows a conspicuous pattern of "episodic" molecular evolution at the protein and DNA sequence level. For these reasons, GH and GHR provide an interesting case for the study of molecular evolution.

Another advantage of this system is that one can measure the effect of each amino acid substitution at the functional level (i.e., the effect on binding interactions between GH and GHR). Using *in vitro* binding assays, the contribution of each substitution on its phenotype (binding affinity) can be measured in a controlled environment. An example of this approach will be described.

Biology of GH and GHR and their Interactions

Mammalian GH plays the role of a central endocrine regulator, controlling many different biochemical pathways (e.g., the metabolism of proteins, carbohydrates, and lipids). In humans, GH is also known to be involved in diabetes and to play a major role in carcinogenesis. Abnormal levels of GH directly induce specific phenotypes: dwarfism when hyposecreted and gigantism when hypersecreted, before puberty. In adults, hypersecretion of GH caused by pituitary adenomas leads to a condition, known as acromegaly, distinguished by large fingers, hands, and feet (see Okada and Kopchick, 2001 for a recent review).

The biochemical pathways induced by GH begin with the formation of a biologically active ternary complex comprised of one GH bound to two GHR molecules. First, one GH molecule binds to one GHR molecule through the high-affinity site of GH, called site1. Second, the resulting 1:1 complex then attracts a second GHR molecule to bind through the low-affinity site2 to form a 1:2 structure (Cunningham et al., 1989; Wells, 1996). This ternary complex is then able to elicit subsequent signal transduction pathways and participate in myriads of biochemical functions (Kossiakoff, 1995; Wells, 1994, 1996).

The study of GH and GHR interaction is greatly facilitated by a special characteristic of the GHR molecule. GHR is composed of three domains: extracellular, transmembrane, and intracellular. The part that participates in binding with the GH is the extracellular domain. This domain of GHR is found freely circulating in the bloodstream, independent of the other regions. When purified, they exhibit the same activity as the full-length counterpart (Fuh et al., 1990). Hence this domain is also called as the growth hormone binding protein (GHBP). Experiments to elucidate the structural and biochemical aspects of GH and GHR interactions can be performed using GHBPs, making experiments much more manageable.

The interfaces between molecules in the hGH-(hGHR)$_2$ complex have been resolved in detail: structurally, by means of a high resolution X-ray crystallography (De Vos et al., 1992) and functionally, through mutational analyses (Cunningham and Wells, 1989; Cunningham et al., 1989; Clackson and Wells, 1995; Clackson et al., 1998). These two approaches generally agreed on the importance of specific residues. That is, the residues shown to reside structurally in the interfaces between the GH and GHR molecules had significant consequences in the stability of the GH–GHR complex when replaced with alanine—a relatively inert amino acid residue. This is because most of the residues in the binding interfaces form salt bridges and hydrogen bonds, which stabilize the intermolecular contact areas (De Vos et al., 1992; Kossiakoff, 1995). Recent studies of GH–GHR interactions suggest additional indirect contribution of some residues located relatively distant from the molecular interfaces (Behncken et al., 1997; Clackson et al., 1998).

Gene Duplications Leading
to Multiple GH-Related Loci in Primates

Another characteristic of primate GH that may have significant evolutionary consequences is the presence of multiple GH-related genes. In human and rhesus monkey, there are five copies of GH and GH-related genes, while there is only a single GH-related gene in nonprimate mammals. In human, the five genes that comprise the GH cluster from 5′ to 3′ are: hGH-N, hPL-1, hPL-2, hGH-V, and hPL-3 (hPL stands for human placental lactogen: Figure 2). These genes show the same transcriptional orientation, but their expressions differ widely from one another in terms of both the tissues expressed and the level of transcription (Figure 2). Only hGH-N is produced in the pituitary and is referred as the GH, while all the other genes are expressed in the placenta. The locus hPL-1 carries a mutation in the 5′ splice site so that only incompletely processed forms are produced and may not be functional (Barrera-Saldaña, 1998).

The questions of when and how the duplications occurred are parts of ongoing investigation. Adkins et al. (2001) showed that there is a single GH gene in the genome of the bushbaby—a prosimian. According to Wallis et al. (2001), there are multiple GH-related genes in the marmoset—a New World monkey. These studies indicate that the amplification of the GH gene cluster occurred after the separation of the haplorhine lineage from the strepsirhine lineage. However, it is unclear whether all the loci in human and rhesus monkey are the products of the same duplications as in the marmoset, or whether some duplications are unique to the human or rhesus monkey lineage. It is often difficult to infer the evolutionary relationships of duplicated genes because of the possibility of extensive gene conversion among loci (Wallis, 1996).

The fact that the GH gene has been duplicated to multiple copies in higher primates suggests selective advantage for the retention of duplicate GH genes.

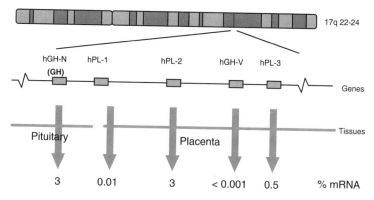

Figure 2. Genomic organizations of the five GH and GH-related genes on human chromosome 17q22-24. The tissue in which each gene is expressed and the mRNA level are shown. (Adapted from Figure 1 of Barrera-Saldaña, 1998.)

Also, the divergence of the duplicate genes into the pituitary-expressed GH gene and the placenta-expressed GH-like genes supports the view that gene duplication allows the opportunity of tissue specialization (Force et al., 1999; Lynch et al., 2001).

We note that an episodic mode of molecular evolution has occurred following in the duplicated GH loci. Was it due to adaptive evolution? Ohta (1993) analyzed the ratio of nonsynonymous to synonymous substitutions among species and among duplicated copies of GH and GHR. She found that the ratio was significantly greater in the comparisons of GH-related genes within species than that between species, implying that there were more non-synonymous substitutions between the duplicated GH-related genes than between the same genes in different species. However, it is not clear whether the accelerated amino acid substitution was due to relaxation of functional constraints or positive Darwinian selection. In the next section, we describe the pattern of molecular evolution of the GH-N locus, which is expressed in the pituitary and has retained the same function of mammalian GH.

Episodic Molecular Evolution of GH in Mammals

It was noticed early on that there is a considerable rate variation in the evolution of GH among mammalian lineages (Gillespie, 1991; Wallis, 1981). Wallis (1994) extended the investigation to include 16 mammalian species belonging to 7 eutherian orders. He inferred the "ancestral" mammalian sequence (which is identical to the pig GH) and showed that human and rhesus monkey GH differ dramatically from the ancestral sequence, by 62 and 64 residues, respectively. This is in a stark contrast to the observation that the GH sequences from other mammalian species differ from the ancestral sequence by only a few amino acids (Wallis, 1994). The accelerated amino acid substitution observed in the evolution of GH in the hominoid lineage is among the fastest rate of molecular evolution of known mammalian proteins so far (see the compilation in Li, 1997). Wallis (1994) also noted that following a brief period of rapid "burst" of substitutions, the rate of evolution soon returned to the slow "basal" rate.

Subsequently, the DNA sequences of GHs from other primate species were obtained to pinpoint the time where the rapid evolution of GH began (Adkins et al., 2001; Liu et al., 2001; Wallis et al., 2001). The GHs from two prosimian species, the bushbaby (Adkins et al., 2001) and the slow loris (Wallis et al., 2001), show a slow rate of evolution comparable to that in other mammalian species, whereas three New World monkey species studied (Liu et al., 2001; Wallis et al., 2001) have evolved at a rate close to that in human GH. These observations concluded that the rapid "burst" of evolution of primate GHs occurred before the divergence of the New and Old World monkey lineages.

What is the basis of this rapid evolution is a puzzling question. A selectionist view is that it is due to positive Darwinian selection (Ohta, 1993; Wallis, 1994, 2001). If so, what is the functional basis of this phenomenon? The other extreme is a neutralist interpretation: for some reason, possibly related to the functional redundancy conferred by the gene duplication(s), the selective constraints upon the coding sequences of GH were lifted briefly before the split of Old World and New World monkey lineages, allowing rapid amino acid substitutions to occur.

Test of Positive Selection in DNA Sequences of Primate GHs and GHRs

Liu et al. (2001) investigated the above two hypotheses using DNA sequence data from the bushbaby, tarsier, squirrel monkey, rhesus monkey, and human. They conducted a maximum-likelihood analysis of the nonsynonymous rate/synonymous rate ratio (the d_N/d_S ratio; see an earlier section). We also performed a similar analysis, adding the sequence data from the slow loris and marmoset, and obtained virtually the same result (Figure 3A). The free-ratio model, which assumes an independent d_N/d_S ratio for each branch, fitted the data significantly better than the one-ratio model, which assumed the same d_N/d_S ratio for all the species considered. Rapid evolution started after the divergence between the tarsier and simian lineages. The highest rate occurred in branch s, which connects the common ancestor of the simian and tarsier lineages and the common ancestor of simians. However, the estimated d_N/d_S ratio along this branch was still less than 1; so this analysis did not decisively detect positive selection.

Liu et al. (2001) also examined a subset of amino acid sites: the "function-ally important" sites, which include sites that form salt bridges and hydrogen bonds with GHRs and sites that are involved in the interaction with the pro-lactin receptor (Cunningham and Wells, 1989, 1991; Somers et al., 1994). They showed that significantly more amino acid changes have accumulated at functionally important sites than the other sites in simian GHs. This suggests the presence of positive selection in the evolution of these sites. The GHR sequences from primate species were also analyzed using two mammalian out-groups: the pig and the rabbit (Liu et al., 2001). Again, the d_N/d_S ratios for the primate lineage and outgroups were significantly different by the likeli-hood ratio test, best described as episodic molecular evolution (see Figure3B). The ancestral branch of higher primates showed a significantly accelerated nonsynonymous substitution rate. The d_N/d_S ratio of GHR along the human lineage was even greater than 1.

As seen from these analyses, statistical analyses done without functional and structural information often do not produce clear-cut results. This is because positive selection may not continue for a very long time. For exam-ple, it is easy to imagine a period of rapid evolution driven by adaptive natural selection for some time in a specific lineage. If this period was short compared to the total length of the branches used in the analysis, then the conventional analyses will not have enough statistical power to detect such a period, unless there is an *a priori* assignment of branches to be considered. If the back-ground rate of evolution was very low, due to strong purifying selection, then even in the presence of positive selection for the majority of sites, some sites will still be governed by the negative selective force and, therefore, reduce the overall d_N/d_S ratio. Even though several statistical models implementing more realistic likelihood assumptions have been devised, the power of such

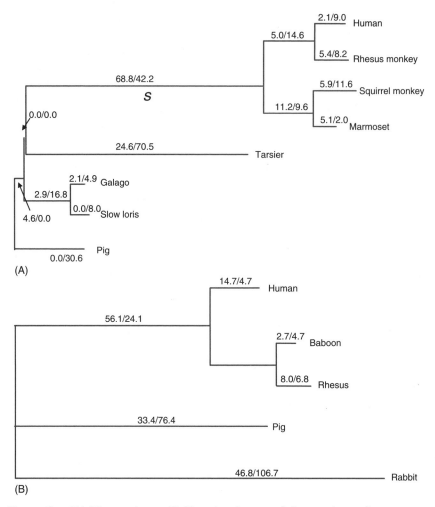

Figure 3. (A) The maximum-likelihood estimates of the numbers of nonsynonymous and synonymous substitutions in the GH genes under the "free-ratio" model.
S indicates the branch of the simian common ancestor. Branches are proportional to
the total numbers of substitutions. (B) The maximum-likelihood estimates of the
numbers of nonsynonymous and synonymous substitutions in the GHR genes under
the "free-ratio" model. (Adapted from Figure 2B of Liu et al., 2001.)

tests is usually low (see above). Therefore, an informative and complementary approach will be to directly investigate the functional consequences of specific substitutions that have occurred.

A Case Study of Functional Evolution—The Emergence of Species Specificity

In this section, we describe a series of studies in an effort to elucidate the underlying mechanism of specific amino acid substitutions that are responsible for the evolution of a functional trait. One of the major differences between primate GH and GHRs, from those of other mammalian species, is as follows: GHs from humans and rhesus monkey can bind and activate nonprimate GHRs, as well as primate GHRs. In contrast, nonprimate GHs have extremely low affinities for human GHR and, therefore, cannot stimulate growth in rhesus monkey (Carr and Freisen, 1976). This phenomenon is referred to as the "species specificity" of primate GHR (Carr and Freisen, 1976; Peterson and Brooks, 2000).

A simple hypothesis of the evolution of species specificity of primate GHR entails that an amino acid substitution specific for the primate lineage is responsible for the emergence of species specificity. With this idea, Souza et al. (1995) examined the GHR sequences from various species to find residues at the GH–GHR interfaces that differentiate primate GHRs from nonprimate GHRs. The interaction between Asp171 of GH and Arg43 of GHR caught their attention. Arg43 of GHR forms two hydrogen bonds with Asp171 and Thr175 of hGH (De Vos et al., 1992). While the Thr175 residue is conserved between primates and nonprimates, Asp171 is not; nonprimates have His instead of Asp at the site equivalent to 171. As for GHR, nonprimates have Leu instead of Arg at position 43. These two are the only pair of complementary residues at the site1 interface that differ between primates and nonprimates. Based on this observation, Souza et al. (1995) proposed that the incompatibility between GH171 and GHR43 is a major determinant of species specificity between primate GHRs and nonprimate GHs.

Biochemists have tested the validity of this proposal directly by engineering changes in GH and GHR and performing binding experiments (Behncken et al., 1997; Gobius et al., 1992; Laird et al., 1991; Souza et al., 1995). For example, Souza et al. (1995) showed that a change from Leu43 to Arg43 in bovine or rat GHR greatly reduced their site1 affinity for bovine GH and rat GH but had little effect on their affinity to hGH, whereas, a change from Arg to Leu in hGHR enabled it to bind bGH (also see Laird et al., 1991). The results from these experiments showed that residues GH171His and GHR43Arg are mainly responsible for the incompatibility of these two proteins in distantly related species.

Some interesting questions arise. First, "When did the critical substitution that caused the species specificity occur during primate evolution?" Liu et al. (2001) and Wallis et al. (2001) showed that the change from His to Asp at site 171 of GH occurred in New and Old World monkeys. Liu et al. (2001) also found that the Arg substitution at site 43 of GHR occurred only in the Old World monkeys; the New World monkey species still have Leu—the "nonprimate" residue. This implies that the His171Asp change in GH preceded the Leu43Arg change in GHR, supporting the hypothesis based on stereochemical inferences that the His $\rightarrow$ Asp mutation in the hormone must have preceded the Leu $\rightarrow$ Arg mutation in the receptor (Behncken et al., 1997).

Second, since the GH of the New World monkeys has the "primate" form but the GHR still has the "nonprimate" form, can the GHR of New World monkeys bind to both primate and nonprimate GH? Recently, *in vitro* assays showed that the GHBP of squirrel monkey can indeed bind to both hGH and rat GH (Yi et al., 2002). Thus, the GHR of New World monkeys represent a transitional phase in the emergence of the species specificity of GHR in Old World primates.

Third, was the emergence of species specificity driven by positive selection? If the GHR43Leu $\rightarrow$ Arg substitution had a positive effect on the binding affinity between GH and GHR, then this should be measurable from the binding interactions of GH and a mutant encoding this new mutation. Yi et al. (2002) therefore engineered the GHR of the squirrel monkey to encode the Leu43Arg mutation and determined the binding affinity of this GHR toward the GH of the squirrel monkey, in comparison with that of the wild-type squirrel monkey GHR. The mutant GHR performed no better than the wild type GHR; in fact, the binding affinity was about twofold lower than that of the wild-type receptor. Therefore, the new mutation may have no selective advantage over the "nonprimate" residue Leu, at least for the squirrel monkey GHR. This suggests that the emergence of species specificity was due to random drift. A second line of evidence supporting this view is the distribution of the intermediary stage (GH171Asp – GHR43Leu). Liu et al. (2001) reported that the intermediary phase persists in both the squirrel monkey and the spider monkey—the two most diverged New World monkey species. These two species are estimated to have diverged about 25 MYA (Goodman et al., 1998). If the emergence of species specificity had a selective advantage, the intermediary state should have been short lived. From these two lines of evidence, they proposed that the emergence of species specificity, which is an example of dramatic functional difference between two variants of the same protein, may not be due to an adaptive evolution (Yi et al., 2002).

EVOLUTION OF CHORIONIC GONADOTROPIN IN PRIMATES

In this section, we describe the molecular evolution of a new protein hormone, the Chrionic Gonadotropin (CG) that emerged in higher primates. As CG is an essential hormone in the regulation of female reproduction in higher primates, we shall first describe the female reproductive cycle and the role of CG within this cycle.

Hormonal Regulation of Primate Reproduction and Role of CG

The human menstrual cycle is divided into two phases: the "follicular phase" before the ovulation and the "luteal phase" after the ovulation. During the follicular phase, Luteinizing Hormone (LH) and Follicle-Stimulating Hormone (FSH) from the anterior pituitary gland induce the growth of a cluster of follicles in the ovary. At the midpoint of the cycle, an LH surge stimulates ovulation, the release of a mature oocyte from the follicle into the oviduct. When this occurs, the remaining cells of the follicle are converted into the corpus luteum, which functions as a temporary endocrine gland. As the luteal cycle begins, the corpus luteum produces progesterone, which signals the uterine endometrium to prepare for a potential pregnancy. This preparation involves significant growth of the uterine endometrial lining, marked by increased vasculature and prolific secretory activity. The luteal phase typically lasts 11–15 days in humans. If the oocyte is not fertilized during the luteal phase of the cycle, the corpus luteum atrophies. This will lead to the sharp decline of progesterone production and as a result both the oocyte and the specialized uterine lining are shed, in the period called menstruation. Then the next cycle begins. The menstrual cycles in other mammals are similar to that in humans, though the timing and duration of each part of the menstrual cycle vary among species. The hormones involved are the same and the cycle is generally subject to the same regulatory control as well, except for the role of CG (see in a later section).

When fertilization occurs, however, the menstrual cycle must be interrupted and adopted for the proliferation of the new offspring. Since the developing fetus depends entirely on mother's body for all the necessary resources, it has to signal the maternal body its existence as soon as possible. This is to prohibit the turnover of the menstrual cycle in order to keep the pregnancy. Different mammalian taxa have developed different mechanisms for this signal to establish pregnancy (for a review, see Niswender et al., 2000).

In humans and other higher primates, the establishment and maintenance of pregnancy is mainly achieved by CG. CG is detected as early as 1 week after fertilization. CG functions by binding to the LH receptors in the corpus luteum so that LH cannot initiate the follicular phase. As a result, the corpus luteum continues to produce progesterone and the pregnancy can be maintained. CG is the major agent to block the menstrual cycle up to end of the first trimester (13 weeks). By then the fully developed human placenta becomes capable of producing progesterone on its own. At this time the corpus luteum is no longer needed. Thus, CG provides a bridge for the developing fetus to get to the point where it can support the pregnancy on its own. Besides this direct interaction with the LH receptor, CG is also known to function in an indirect way to establish pregnancy by blocking the action of another protein hormone, prostaglandin. This function of CG is relatively little understood and not discussed in detail here.

Molecular Structure and Origin of CG

CG is a member of the glycoprotein family, which includes CG, LH, FSH, and thyroid stimulating hormone (TSH). These four hormones are all made up of two peptide subunits: the α and β subunits (Albanese et al., 1996). All four hormones share the same α subunit, encoded by the single-copy glycoprotein hormone α-subunit gene. It is the unique β subunit of each glycoprotein hormone that confers biological specificity to each hormone (Albanese et al., 1996).

The β subunit of CG is most similar to the β subunit of LH. It was proposed that the β subunit of CG evolved from a duplication of the ancestral β subunit of the LH gene (Fiddes and Goodman, 1980; Talmadge et al., 1984). One observation supporting this is that the CGβ-subunit and LHβ-subunit genes are both located on chromosome 19q13.32 in human, as would have been the result of a tandem duplication. Compared to this, the FSHβ-subunit gene is located on chromosome 11p13 and TSHβ-subunit on chromosome 1p13.

There are two major differences between the CGβ-subunit gene and the LHβ-subunit gene. First, they are expressed in different tissues. While LH (also FSH and TSH) is produced by the pituitary gland, CG is expressed in the placenta. This means that both the α and β subunits of CG are expressed in the placenta. The α subunit of CG is common to the other glycoprotein hormones, implying that the α subunit should be expressed not only in the pituitary gland but also in the placenta. This is not the case for the α-subunit gene in other mammalian species, where CG is not present. The second difference is that the CGβ subunit is 24 amino acids longer than the LHβ subunit (see later). In the following subsections, the molecular biology and evolution of the CG subunits will be described in detail.

Evolution of Placental Expression of the CGα-Subunit Gene in Primates

How did the α subunit of the glycoprotein hormones acquire the ability to express in the placenta in some primates? Analysis of the sequence of the CG α-subunit promoter has shown that two types of promoter elements control the placental expression in humans. The first is the trophoblast-specific element (TSE) and the second is a pair of enhancer elements, called the cyclic-AMP response elements (CREs) (Nilson et al., 1991). The exact sequence recognition site in TSE is not yet clearly resolved. CRE is an eight-base palindromic sequence (TGACGTCA) to which the transcription enhancing proteins bind (Figure 4A).

Experimental evidence suggests that TSE is not sufficient for placental expression but requires at least one CRE with it (Nilson et al., 1991; Roberts and Anthony, 1994). Therefore, the presence of CRE is critical for placental expression. Alignment of promoter sequences of various mammalian species demonstrated that only the human and the gorilla possess two CRE elements

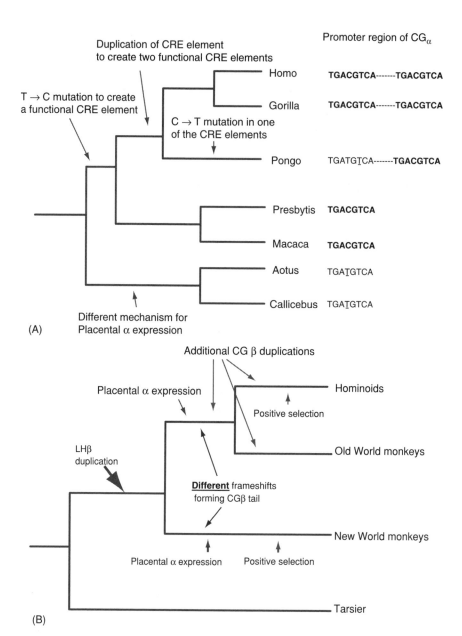

Figure 4. (A) The proposed sequential molecular evolutionary events in the promoter region of the CGα region leading to the observed numbers of functional CRE elements from several primate species (see text). Partial nucleotide sequences of the CRE elements in the promoter region of the CGα gene in each species are also shown. The functional CRE palindrome sequence is shown in bold. The nonfunctional copy is shown with an underlined T base at which a T to C mutation is required for it to become functional. (B) A proposed molecular evolutionary history of chorionic gonadotropin in primates. (Adapted from Maston, 2001.)

(Maston and Ruvolo, 2002; the chimpanzee has not been studied). The orangutan also has two copies of CRE, but one copy differs from the perfect palindrome by one base pair (TGA*T*GTCA instead of TGA*C*GTCA). *In vitro* expression studies showed that this one base change reduces the expression of a reporter gene by more than 200 times compared to the promoter containing the correct palindrome CRE (Bokar et al., 1989). Therefore, the orangutan probably possesses only one effectively functional CRE element (Maston and Ruvolo, 2002). Maston and Ruvolo (2002) also showed that Old World monkeys have only one copy of CRE, and New World monkeys and nonprimates possess a single copy of the imperfect palindrome sequence (TGA*T*-GTCA). No species outside of the primates is known to express the α subunit in the placenta.

It is therefore parsimonious to assume that the ancestral CRE element (not functional) is TGA*T*GTCA as found in nonprimate mammals. A T→C transition resulted in the functional CRE element in the common ancestor of Old World primates. A plausible scenario is that the duplication to create the second CRE occurred in the ancestor of the hominoid lineage, to be shared only by the human, the gorilla, and the orangutan. Since *Aotus* and *Callicebus* have a single CRE sequence that is ancestral, they do not have a functional CRE element. In other words, these species must have evolved a different mechanism to enable the placental expression of the α-subunit gene (Maston and Ruvolo, 2002).

Number of Gene Copies and Episodic Molecular Evolution of CGβ-Subunit Genes in Primates

In contrast to the α-subunit gene, which exists as a single copy, the CGβ-subunit gene is present in multiple copies in some higher primates. Maston and Ruvolo (2002) showed that Strepsirhine primates possess only the LHβ-subunit gene but not the CGβ-subunit gene. *Callicebus* and *Aotus*—the two New World monkeys studied—each has one LHβ and one CGβ gene; *Aotus* may have an additional CGβ gene but the data were not conclusive. Rhesus monkey was found to have four LHβ/CGβ genes (presumably three CGβ genes), while *Colobus* and *Presbytis* have six (five CGβ genes). Orangutans have five copies (four CGβ genes), and humans have seven (six CGβ genes). Based upon this observation, Maston and Ruvolo (2002) proposed a series of gene duplication events (Figure 4B). The first duplication event to create the

CGβ-subunit gene must have occurred in the common ancestor of the New World monkeys. Two additional CGβ genes arose in the catarrhine common ancestor and then also in the common ancestor of *Colobus* and *Presbytis*. In addition, one aditional CGβ gene arose in the common ancestor of hominoids and two additional CGβ genes in the human lineage.

The evolution of the CGβ-subunit following the duplication of the LHβ-subunit gene should entail at least two steps. First, as mentioned earlier, CG is expressed exclusively in the placenta, while LH is expressed exclusively in the pituitary gland. This means that the promoter region of CGβ had to evolve a new sequence to enable the unique placental expression of CGβ. *In vitro* expression studies have shown that there are specific regions in the CGβ-subunit gene promoter that are required for placental expression (Hollenberg et al., 1994). However, the exact sequence changes that were needed to establish the placental expression have not been determined yet.

The second step is as follows. The coding regions of the β subunits of human CG and LH are similar up to a frameshift mutation in the third exon. This frameshift was caused by one nucleotide deletion, which is 22 nucleotides upstream of the LHβ-subunit stop codon. This deletion destroyed the original stop codon and added 72 nucleotides to the coding region of the CGβ-subunit. Thus, the β subunit of CG has a unique carboxyl tail of 32 amino acids with no homology to the coding region of the β subunit of LH (Fiddes and Goodman, 1980; Talmadge et al., 1984). Interestingly, there are four sugar chains that are attached to this part of CG.

Maston (2001) found that the CG sequences from human, orangutan, *Presbytis*, *Colobus*, and *Macaca* have the same frameshift mutation. This strongly suggests that there was a single deletion event in the common ancestor of those species (the catarrhine common ancestor) that generated the new carboxyl tail of CGβ-subunit. However, this was not the case for the CGβ-subunit genes from the New World monkey species. They also possess a single base deletion to create the tail, but the most likely position of this deletion is inferred to be six nucleotides upstream from the frameshift found in catarrhines. However, it is difficult to decide whether this was truly a different deletion event because hypothesizing the same deletion event as in catarrhines requires only two or three extra mutations.

The newly created carboxyl end of the CGβ-subunit gene shows an episodic mode of molecular evolution (Maston, 2001; Maston and Ruvolo, 2002). Likelihood test analyses indicated that the d_N/d_S ratio was unusually high (significantly greater than 1) in four branches: the branch leading to the hominoid common ancestor, the terminal branch leading to humans, the branch leading to the platyrrhine common ancestor, and the terminal branch leading to *Callicebus*. The ratio was also high in three other branches, though the branches were too short for the ratio to be statistically significant. Therefore, it was suggested that there have been periods of positive selection acting on this portion of the CGβ-subunit gene during the evolution of higher primates.

Summary of Molecular Evolutionary Events in the Evolution of CG in Primates

Figure 4B summarizes some of the molecular evolutionary events during the evolution of CG in higher primates. According to Maston (2001), the new hormone CG achieved the function of pregnancy establishment twice in primate evolution: once in the catarrhine common ancestor and once in the platyrrhine common ancestor.

After the duplication of the LHβ-subunit gene in the common ancestor, one of the two duplicate genes appears to have evolved independently in the two groups of anthropoids: the platyrrhines and the catarrhines, supported by several lines of evidence (Maston, 2001). First, as mentioned earlier, there were possibly different single-base pair frameshift mutations within the coding sequence. Second, the promoter region of the α subunit shows different CRE binding sites in the two groups. In addition, there seem to be different underlying forces for the evolution of the β-subunit genes in the two groups. The hominoid genes show a steady increase of serine residues, which are used as binding sites for glycosylation, while there is no such pattern in the amino acid changes in the platyrrhine sequences.

The evolution of CG shows how a new duplicate protein hormone gene in primates acquired a new expression profile. Also, the DNA sequences of the β-subunit specific for CG adopted a frameshift mutation to encode a novel carboxyl tail. This shows two aspects of the evolution of proteins (hormones); evolution of a novel expression pattern in tissue or developmental time by modulating regulatory sequences, and protein sequence evolution to acquire new or changed (improved) function.

EPISODIC EVOLUTION OF OTHER
PROTEIN HORMONES

By now we hope to have conveyed a message that the molecular evolution of a protein hormone is not always constant, but may occasionally show an episodic mode. In some cases, the "burst" of substitutions often coincides with changes at the functional or genetic level, such as the emergence of a new function or new expression profile, or such as a gene duplication event. In this section, we present a compilation of cases of episodic evolution of protein hormones in mammals, not restricted to the primates. Lack of functional information and the paucity of sequence data from many species make it difficult to perform extensive analyses as was done earlier.

The compilation is mainly from Wallis (1996, 2001), who investigated the mode of molecular evolution of protein hormones in mammals under the condition that at least one noneutherian tetrapod outgroup sequence was available. Six of the eight hormones he investigated (including the GH) showed the episodic mode of evolution in some mammalian lineages (Figure 5, except for the GH, which was already shown above).

Pituitary prolactin shows bursts of amino acid substitutions in at least four branches during eutherian evolution: primates, artiodactyls, rodents, and elephants (Figure 5A). This rate increase was shown to be characteristic of the mature protein coding sequence but not apparent in the sequences for signal peptides (Wallis, 2000, 2001). This implies the role of selection on the protein product in shaping the episodic evolutionary pattern.

The second case is insulin (Figure 5B). As mentioned in the introduction, insulin was considered to be a good example of the conservatism of protein evolution. However, when the sequences from different mammalian species were compared, there was a burst of evolutionary change in hystricomorph rodents and in New World monkeys. In contrast, the rate of evolution in other mammals has been slow. What is more, the changes seen in the rodent and New World monkey insulins are known to be associated with a substantial loss of activity in standard assays (Beintema and Campagne, 1987; Seino et al., 1987). This observation fits the view that rapid evolution of DNA sequences is often associated with functional changes.

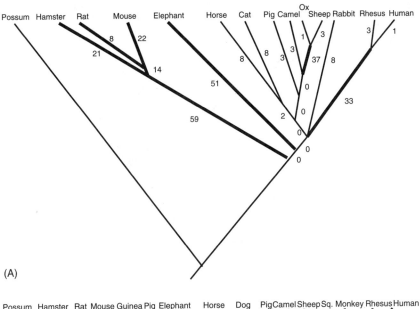

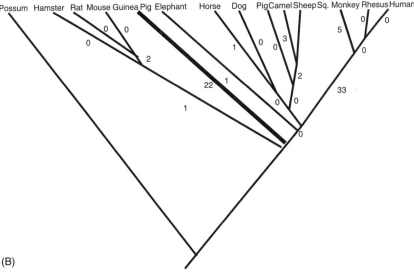

Figure 5. Phylogenetic trees illustrating the molecular evolution of: (A) prolactin, (B) insulin, (C) parathyroid hormone, (D) glycoprotein hormone α subunit, (E) luteinising hormone β subunit, (F) follicle stimulating hormone β subunit, and (G) thyroid stimulating hormone β subunit. Periods of rapid molecular evolution are indicated with thick lines. (Adapted from Figures 1 and 2 of Wallis, 2001.)

(Continued)

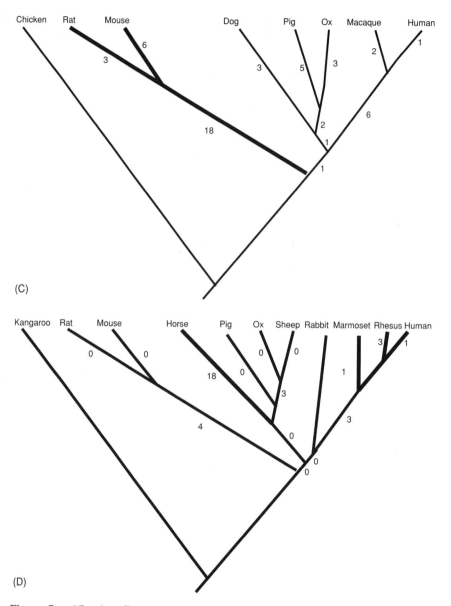

(C)

(D)

Figure 5. *(Continued)*

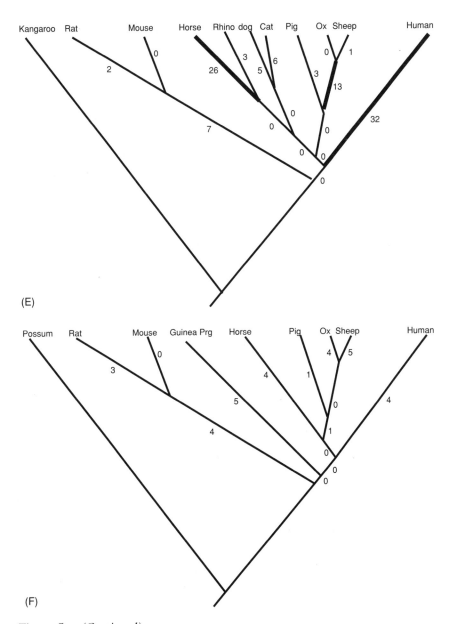

Figure 5. *(Continued)*

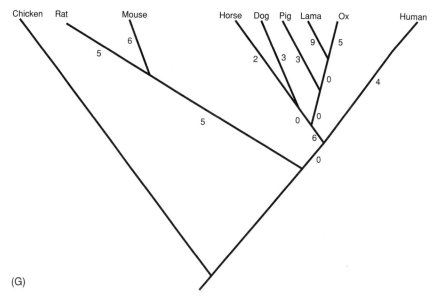

(G)

Figure 5. *(Continued)*

Parathyroid hormone (PTH) sequences in rodents appear to have gone through a phase of rapid evolution (Figure5C). However, additional sequence information is necessary to define the pattern more precisely. Wallis (2001) also looked at the molecular evolution of the α subunit of glycoprotein hormones. This is a common subunit of the four glycoprotein hormones: CG, FSH, TSH, and LH (see an earlier section). This shows an episodic mode of evolution in primates and perissodactyls (Figure 5D). We noted above that the regulatory region of the α subunit has gone through dramatic changes to accommodate the expression of CG in primates. The analyses of Maston (2001) and Maston and Ruvolo (2002) also showed the episodic mode of molecular evolution for the CGα in higher primates.

The molecular evolution of the β subunit of luteinizing hormone (LH) also exhibits an episodic mode of evolution in perissodactyls and primates (Figure 5E). Maston and Ruvolo (2002) also confirmed this pattern through d_N/d_S analyses. The rapid evolution was also seen in Artiodactyla. In comparison, the β subunit of two other members of the glycoprotein family, FSH and TSH, do not show such a pattern. The evolutionary rates estimated from the available sequences from various mammalian species including rodents, artiodactyls and perissodactyls, and hominoids, show a rather constant slow rate of evolution (Figures 5F and G).

CONCLUSIONS

A prevailing view of molecular evolution of protein hormones was that they evolve at a constant rate and the functions are conserved among species. In this chapter, we describe cases where this view does not hold. When investigated in detail, many protein hormones reveal an episodic mode of evolution in some mammalian lineages, that is, while evolving at an approximately constant rate, most of the times the evolutionary rate was dramatically accelerated in some specific lineages.

One possible explanation for this phenomenon of episodic evolution is that positive evolution was driving the spread of some amino acid substitutions. This might have been caused by specific needs for adaptation. Currently, a test widely used to detect the presence of positive selection is the maximum likelihood analysis that compares the rates of nonsynonymous substitution and of synonymous substitution (the d_N/d_S ratio test) in each lineage to detect any perturbation of the background rates of evolution in the lineages studied. In this chapter, some of the cases where positive evolution was detected from the DNA sequences by this method are described.

A pattern that has repeatedly emerged in genes that showed the episodic mode of molecular evolution is the expansion of gene copies by gene duplication. This is shown for two examples in this chapter: the case of GH and the case of CG. In both cases, one copy maintains the original expression in the pituitary, while the other copies are expressed in a new organ, the placenta. The expression of CG in the placenta has given rise to a new mechanism of maintaining pregnancy. The expression of GH-like genes in the placenta is also likely to have evolved by natural selection (for a possible scenario, see Wallis, 1997); otherwise, they would have accumulated degenerative mutations or become lost (Force et al., 1999; Lynch et al., 2001). As mentioned earlier, an analysis of the GH and GH-related gene sequences suggested that these genes have evolved at a fast rate, possibly reflecting the influence of positive selection.

Considerable insight can be gained if one can study the functional and structural consequences of an observed amino acid substitution. With such information, it may be possible to infer the selective pressure associated with the specific amino acid substitution. In this respect the GH and GHR in primates provide an ideal model system. Some studies on the structural and functional consequences of amino acid changes have been described in an earlier section.

Lastly, when a protein hormone evolves a new function or undergoes positive selection, a new expression pattern may need to be established. In the case of the evolution of CG, it involved establishing the expression of the CGα subunit in a new tissue, the placenta. This was achieved through a series of novel molecular events. Whether the case of CG represents a general mechanism of achieving a new expression pattern is a topic to be pursued in the future.

In conclusion, the molecular evolution of protein hormones provides a paradigm to understand adaptive evolution at the nucleotide or amino acid level. They are often involved in the evolution of adaptive traits because protein hormones play an essential role in vertebrate physiology. The abundance of structural and functional data on protein hormones from primates, mainly due to their importance in medicine, is another advantage for pursuing research in the evolution of protein hormones. With the rapid development in molecular biology tools, the molecular evolution of protein hormones can be studied in detail. Thus, this topic holds a great potential for future research.

REFERENCES

Adkins, R. M., Nekrutenko, A., and Li, W-H., 2001, Bushbaby growth hormone is much more similar to nonprimate growth hormones than to Rhesus monkey and human growth hormone, *Mol. Biol. Evol.* **18**: 55–60.

Albanese, C., Colin, I. M., Crowley, W. F., Ito, M., Pestell, R. G., Weiss, J., et al., 1996, The gonadotropin genes: Evolution of distinct mechanisms for hormonal control, *Recent Prog. Hor. Res.* **51**: 23–61.

Barrera-Saldaña, H. A., 1998, Growth hormone and placental lactogen: Biology, medicine and biotechnology, *Gene* **211**: 11–18.

Behncken, S. N., Rowlinson, S. W., Rowland, J. E., Conway-Campbell, B. L., Monks, T. A., and Waters, M. J., 1997, Aspartate 171 is the major primate-specific determinant of human growth hormone, *J. Biol. Chem.* **272**: 27077–27083.

Beintema, J. J., and Campagne, R. N., 1987, Molecular evolution of rodent insulins, *Mol. Biol. Evol.* **4**: 10–18.

Bokar, J. A., Keri, R. A., Farmerie, T. A., Fenstermaker, R. A., Anderson, B., Hamernik, D. L., et al., 1989, Expression of glycoprotein hormone alpha-subunit gene in the placenta requires a functional cyclic AMP response element, whereas a different *cis*-acting element mediates pituitary-specific expression, *Mol. Cell. Biol.* **9**: 5113–5122.

Carr, D., and Friesen, H. G., 1976, Growth hormone and insulin binding to human liver, *J. Clin. Endocrinol. Metab.* **42**: 484–493.

Chen, E. Y., Liao, Y. C., Smith, D. H., Barrera-Saldana, H. A., Gelinas, R. E., and Seeburg, P. H., 1989, The human growth hormone locus: Nucleotide sequence, biology, and evolution, *Genomics* **4**: 479–497.

Clackson, T., and Wells, J. A., 1995, A hot spot of binding energy in a hormone-receptor interface, *Science* **267**: 383–386.

Clackson, T., Ultsch, M. H., Wells, J. A., and De Vos, A. M., 1998, Structural and functional analysis of the 1:1 growth hormone:receptor complex reveals the molecular basis for receptor affinity, *J. Mol. Biol.* **277**: 1111–1128.

Cunningham, B. C., and Wells, J. A., 1989, High-resolution epitope mapping of hGH-receptor interaction by alanine-scanning mutagenesis, *Science* **244**: 1081–1085.

Cunningham, B. C., and Wells, J. A., 1991, Rational design of receptor-specific variants of human growth hormone, *Proc. Natl. Acad. Sci. USA* **88**: 3407–3411.

Cunningham, B. C., and Wells, J. A., 1993, Comparison of a structural and a functional epitope, *J. Mol. Biol.* **234**: 554–563.

Cunningham, B. C., Jhurani, P., Ng, P., and Wells, J. A., 1989, Receptor and antibody epitopes in human growth hormone identified by homolog-scanning mutagenesis, *Science* **243**: 1330–1336.

Cunningham, B. C., Ultsch, M., De Vos, A. M., Mulkerrin, M. G., Clauser, K. R., and Wells, J. A., 1991, Dimerization of the extracellular domain of the human growth hormone receptor by a single hormone molecule, *Science* **254**: 821–825

De Pablo, F., 1993, Introduction, in: *The Endocrinology of Growth, Deveopment, and Metabolism in Vertebrates*, M. P. Schreibman, C. G. Scanes, and P. K. T. Pang, eds., Academic Press, San Diego, pp. 3–12.

De Vos, A. M., Ultsch, M., and Kossiakoff, A., 1992, Human growth hormone and extracellular domain of its receptor: Crystal structure of the complex, *Science* **255**: 306–312.

Douady, C. J., Chatelier, P. I., Madsen, O., de Jong, W. W., Catzeflis, F., Springer, M. S., et al., 2002, Molecular phylogenetic evidence confirming the Eulipotyphla concept and in support of hedgehogs as the sister group to shrews, *Mol. Phyl. Evol.* **25**: 200–209.

Fiddes, J. C., and Goodman, H. M., 1980, The cDNA for the β-subunit of human chorionic gonadotropin suggests evolution of a gene by readthrough into the 3′-untranslated region, *Nature* **286**: 684–687.

Fiddes, J. C., and Talmadge, K., 1984, Structure, expression, and evolution of the genes for the human glycoprotein hormones, *Recent Prog. Hor. Res.* **40**: 43–78.

Fleagle, J. G., 1999 *Primate Adaptation and Evolution*, Academic Press, San Diego.

Force, A., Lynch, M., Pickett, F. B., Amores, A., Yan, Y. L., and Postlethwait, J., 1999, Preservation of duplicate genes by complementary, degenerative mutations, *Genetics* **151**: 1531–1545.

Fuh, G., Mulkerrin, M. G., Bass, S., McFarland, N., Brochier, M., Bourell, J. H., et al., 1990, The human growth hormone receptor, *J. Biol. Chem.* **265**: 3111–3115.

Gillespie, J. H., 1991, *The Causes of Molecular Evolution*, Oxford University Press, New York.

Gobius, K. S., Rowlinson, S. W., Barnard, R., Mattick, J. S., and Waters, M. J., 1992, The first disulphide loop of the rabbit growth hormone receptor is required for binding to the hormone, *J. Mol. Endo.* **9**: 213–220.

Goldman, N., and Yang, Z., 1994, A codon-based model of nucleotide substitution for protein-coding DNA sequences, *Mol. Biol. Evol.* **11**: 725–736.

Goodman, M., Porter, C. A., Czelusniak, J., Page, S. L., Schneider, H., Shoshani, J., et al., 1998, Toward a phylogenetic classification of primates based on DNA evidence complemented by fossil evidence, *Mol. Phyl. Evol.* **9**: 585–598.

Hollenberg, A. M., Pestel, A. G., Albanese, C., Boers, M. E., and Jameson, J. L., 1994, Multiple promoter elements in the human chorionic gonadotropin β subunit genes distinguish their expression from luteinizing hormone β gene, *Mol. Cell. Endo.* **106**: 111–119.

Johnson, M. H., and Everitt, B. J., 1995, *Essential Reproduction*, Blackwell Scientific, Oxford.

Kawauchi, H., and Yasuda, A., 1989, Evolutionary aspects of growth hormones from nonmammalian species, in: *Advances in Growth Hormone and Growth Factor Research*, E. E. Muller, D. Cocchi, and V. Locatelli, eds., Springer, Berlin, pp. 51–68.

Kossiakoff, A., 1995, Structure of the growth hormone-receptor complex and mechanism of receptor signalling, *J. Nuc. Med.* **36**: 14S–16S.

Laird, D. M., Creely, D. P., Hauser, S. D., and Krivi, G. G., 1991, Analysis of the ligand binding characteristics of chimeric human growth hormone/bovine growth hormone binding proteins, *Proc. U.S. Endocrine Soc.* **73**: 1545.

Levi-Montalcini, R., 1966, The nerve growth factor: Its mode of action on sensory and sympathetic nerve cells, *Harvey Lect. Ser.* **60**: 217–259.

Li, W-H., 1997, *Molecular Evolution*, Sinauer Associates, Sunderland.

Liu, J.-C., Makova, K. D., Adkins, R. M., Gibson, S., and Li, W-H., 2001, Episodic evolution of growth hormone in Primates and emergence of the species specificity of human growth hormone receptor, *Mol. Biol. Evol.* **18**: 945–953.

Lynch, M., O'Hely, M., Walsh, B., and Force, A., 2001, The probability of preservation of a newly arisen gene duplicate, *Genetics* **159**: 1789–1804.

Maston, G., 2001, Ph.D. Thesis, Harvard University.

Maston, G., and Ruvolo, M., 2002, Chorionic gonadtropin has a recent origin within primates and an evolutionary history of selection, *Mol. Biol. Evol.* **19**: 320–335.

Messier, W., and Stewart, C-B., 1997, Episodic adaptive evolution of primate lysozymes, *Nature* **385**: 151–154.

Nielsen, R., and Yang, Z., 1998, Likelihood models for detecting positively selected amino acid sites and applications to the HIV-1 envelope gene, *Genetics* **148**: 929–936.

Nilson, J. H., Bokar, J. A., Clay, C. M., Farmerie, T. A., Fenstermaker, R. A., Jamernik, D. L., et al., 1991, Different combinations of regulatory elements may explain why placenta-specific expression of the glycoprotein hormone α-subunit gene occus only in primates and horses, *Biol. Repro.* **44**: 231–237.

Niswender, G. D., Juengel, J. L., Silva, P. J., Rollyson, M. K., and McIntush, E. W., 2000, Mechanisms controlling the function and life span of the corpus luteum, *Physiol. Rev.* **80**: 1–29.

Nowak, R., 1999, *Walker's Guide to the Mammals*, Johns Hopkins Press, Baltimore.

Ohta, T., 1993, Pattern of nucleotide substitutions in growth hormone-prolactin gene family: A paradigm for evolution by gene duplication, *Genetics* **134**: 1271–1276.

Okada, S., and Kopchick, J. J., 2001, Biological effects of growth hormone and its antagonist, *Trends. Mol. Med.* **7**: 126–132.

Otani, T., Otani, F., Krych, M., Chaplin, D. D., and Boime, I., 1988, Identification of a promoter region in the CGβ gene cluster, *J. Biol.Chem.* **263**: 7322–7329.

Patton, P. E., and Stouffer, R. L., 1991, Current understanding of the corpus luteum in women and nonhuman primates, *Cli. Obst. Gyn.* **34**: 127–143.

Peterson, F. C., and Brooks, C. L., 2000, The species specificity of growth hormone requires the cooperative interaction of two motifs, *FEBS Lett.* **472**: 276–282.

Roberts, R. M., and Anthony, R. V., 1994, Molecular biology of trophectoderm and placental hormones, in: *Molecular Biology of the Female Reproductive System*, J. K. Findlay, ed., Academic Press, San Diego, pp. 395–440.

Seino, S., Steiner, S. F., and Bell, G. I., 1987, Sequence of a New World primate insulin having low biological potency and immunoreactivity, *Proc. Natl. Acad. Sci. USA* **84**: 7423–7427.

Somers, W., Ultsch, M., De Vos, A. M., and Kossiakoff, A. A., 1994, The X-ray structure of a growth hormone-prolactin receptor complex, *Nature* **372**: 478–481.

Souza, S. C., Peter Frick, G., Wang, X., Kopchick, J. J., Lobo, R. B., and Goodman, H. M., 1995, A single arginine residue determines species specificity of the human growth hormone receptor, *Proc. Natl. Acad. Sci. USA* **92**: 959–963.

Stanhope, M. J., Waddell, V. G., Madsen, O., De Jong, W., Hedges, S. B., Cleven, G. C., et al., 1998, Molecular evidence for multiple origins of Insectivora and for a new order of endemic African insectivore mammals, *Proc. Natl. Acad. Sci. USA* **95:** 9967–9972.

Stewart, C. B., and Wilson, A. C., 1987, Sequence convergence and functional adaptation of stomach lysozymes from foregut fermenters, *Cold Spring Harbor Symp. Quant. Biol.* **52:** 891–899.

Talmadge, K., Vamvakopoulos, N. C., and Fiddes, J. C., 1984, Evolution of the genes for the β-subunit of human chorionic gonadotropin and luteinizing hormone, *Nature* **307:** 37–40.

Waddell, P. J., Okaka, N., and Hasegawa, M., 1999, Towards resolving the interordinal relationships of placental mammals, *Syst. Biol.* **48:** 1–5.

Wallis, M., 1981, The molecular evolution of pituitary growh hormone, prolactin and placental lactogen: A protein family showing variable rates of evolution, *J. Mol. Evol.* **17:** 10–18.

Wallis, M., 1994, Variable evolutionary rates in the molecular evolution of mammalian growth hormones, *J. Mol. Evol.* **38:** 619–627.

Wallis, M., 1996, The molecular evolution of vertebrate growth hormone: A pattern of near-stasis interrupted by sustained bursts of rapid change, *J. Mol. Evol.* **43:** 93–100.

Wallis, M., 1997, Function switching as a basis for bursts of rapid change during the evolution of pituitary growth hormone, *J. Mol. Evol.* **44:** 348–350.

Wallis, M., 2000, Episodic evolution of protein hormones: Molecular evolution of pituitary prolactin, *J. Mol. Evol.* **50:** 465–473.

Wallis, M., 2001, Episodic evolution of protein hormones in mammals, *J. Mol. Evol.* **53:** 10–18.

Wallis, O. C., Zhang, Y.-P., and Wallis, M., 2001, Molecular evolution of GH in primates: Characterization of the GH genes from slow loris and marmoset defines an episode of rapid evolutionary change, *J. Mol. Endocrinol.* **26:** 249–258.

Wells, J. A., 1994, Structural and functional basis for hormone binding and receptor oligomerization, *Curr. Biol.* **6:** 163–173.

Wells, J. A., 1996, Binding in the growth hormone receptor complex, *Proc. Natl. Acad. Sci. USA* **93:** 1–6.

Wilson, D. E., and Reeder, D. E., 1993, *Mammal Species of the World: A Taxonomic and Geographic Reference*, Smithsonian Institution Press, Washington DC.

Yang, Z., 1998, Likelihood ratio tests for detecting positive selection and application to primate lysozyme evolution, *Mol. Biol. Evol.* **15:** 568–573.

Yang, Z., and Bielawski, J. P., 2000, Statistical methods for detecting molecular adaptation, *Trends Ecol. Evol.* **15:** 496–503.

Yang, Z., and Nielsen, R., 2002, Codon-substitution models for detecting molecular adaptation at individual sites along specific lineages, *Mol. Biol. Evol.* **19**: 908–917.

Yi, S., Bernat, B., Pál, G., Kossiakoff, A., and Li, W-H., 2002, Functional promiscuity of squirrel monkey growth hormone receptor toward both primate and nonprimate growth hormones, *Mol. Biol. Evol.* **19**: 1083–1092.

CHAPTER TWENTY-TWO

Parallelisms Among Primates and Possums

D. Tab Rasmussen and Robert W. Sussman

INTRODUCTION

Primates originated once, long ago, in an unknown place, without leaving a fossil record of the event. At face value, learning anything about primate origins seems implausible. Can our curiosity about such a singular, unobservable, historical event be investigated within the realm of science?

The evolutionary history of any biological lineage is like a single experimental lab trial in one important way—both are unique events unfolding along a particular sequence of causes and effects. Once the sequence is over, it is history. Much has been made of the problem that unique historical events are inexplicable to science (e.g., Cartmill, 1990; Popper, 1957). This is why lab scientists never rely on a single experimental trial—they make several parallel runs and analyze the various outcomes comparatively. Lab scientists have learned to control as many variables as possible before launching a trial, so that the diverse outcomes of a set of trials can be interpreted more easily as a consequence of one theoretically interesting factor that is allowed to vary. The lab scientist is not attempting to explain a single experimental run *per se*, but rather to explain which variables influence the outcome of a set of runs.

The evolution of a single lineage is like a single lab trial in this way. Despite the philosophical similarities between a single experimental run and an evolutionary

D. Tab Rasmussen and Robert W. Sussman • Department of Anthropology, Washington University, St. Louis, MO 63130-4899

run, there are fundamental practical differences between the two. Evolutionary time exceeds the lifetime of the scientist, evolutionary lineages have much more complex cause-and-effect sequences, and parallel evolutionary runs were not designed with controls to highlight the effect of a single interesting variable. The evolutionist is in the position of a chemist who walks into a lab after several poorly controlled trials are over, with no lab notes, and maybe without even knowing what the question was.

But these problems are not insurmountable. First, an evolutionary biologist interested in processes of adaptation has the advantage of knowing what the questions are. Evolutionary theory is robust enough that many meaningful questions can be generated about the variable outcomes of a set of evolutionary runs, including questions about how organisms are morphologically adapted to their ways of life. Second, the evolutionist has no shortage of evolutionary runs—there are literally millions of them. Even within mammalogy there are thousands, and within primatology there are over 200 trials still living and many more extinct (but represented in the fossil record). The biggest problem confronting the biologist is that the evolutionary runs cannot be controlled in advance. Instead, the challenge is to identify after the fact which outcomes are relevant to a given question and then to control for confounding variables through proper comparisons.

To understand the singular origin of the order Primates, one must look for other animal lineages with parallel outcomes. By definition, no other evolutionary run yielded a primate, so the questions really center around the origin of key primate-like attributes that occur in other animals, either in combination or dismantled piece by piece. Which experimental runs on Earth have yielded grasping hands and feet? Which have yielded large brains? Primate-like visual systems? This kind of approach to primate origins is epitomized by Cartmill's (1972, 1974a) landmark studies on the mammalian visual system and the grasping extremities, which highlighted how the comparative study of completed evolutionary experiments could be used to test hypotheses about primate adaptations.

This comparative approach to primate origins requires that we find and examine as many parallel independent evolutionary runs as possible. The phalangeroid marsupials of Australia and New Guinea are one such mammalian group that shows parallel development of primate-like traits. Smith (1984a) wrote that the phalangeroid diversification "has led to some remarkable convergences of form, function and behavior with the arboreal lemurs, bush babies, monkeys and squirrels of other continents." While New World marsupials have received some attention regarding questions of primate origins (Cartmill, 1972, 1974a, 1992; Hamrick, 2001; Larson et al., 2000; Lemelin, 1996, 1999; Rasmussen, 1990), few studies have drawn on information about the Australasian marsupial radiations (Cartmill, 1972, 1974a; Larson et al., 2000). In a recent analysis of the primate gait, the marsupial *Phascolarctos* (the koala) was also examined and found to be quite primate-like (Larson et al., 2000), but the study included no phalangeroids, several species of which could be expected to be even closer to primates in their adaptations. Other studies have compared phalangeroid and prosimian radiations from an ecological point of view, but not with a focus on the issue of primate origins (Smith and Ganzhorn, 1996; Winter, 1996). When morphological parallels were first identified between the grasping extremities and the convergent orbits of some phalangeroids and those of prosimian primates, there was inadequate ecological and behavioral data to correctly interpret what these features meant. With growth of knowledge about free-ranging phalangeroid behavior we know now, for example, that the one phalangeroid model (*Cercartetus*) held up to epitomize the visual predation hypothesis of primate origins (Cartmill, 1974a) is, in reality, a flower specialist (Lee and Cockburn, 1985; Turner, 1984).

In this paper we review what is known ecologically and behaviorally about the phalangeroid marsupials of Australia and New Guinea. The purpose of this paper is to highlight which phalangeroid species and behaviors may offer promise in researching primate origins. We believe that some of the hypotheses generated in this symposium can be examined further by investigating phalangeroid marsupials. Of course, the converse is also true, that the wealth of studies on the living primates may prove valuable in testing hypotheses about the origin and radiation of phalangeroids.

OVERVIEW OF PHALANGEROID PHYLOGENY

Until recently all marsupials were usually classified in the indisputably monophyletic order Marsupialia, but scientific certainty being what it is to taxonomists, this tidy order has now been cleaved into a controversial array of ordinal level classifications—some of which utilize more than a half dozen orders (Aplin and Archer, 1987; Archer, 1984; Ride, 1964; Szalay, 1993, 1994). A currently popular ordinal designation for the particular marsupials under consideration in this paper is Diprotodontia—the kangaroos, wombats, koalas, and phalangeroids—which share several derived specializations including reduction of the lower incisor series to a single procumbent pair (diprotodonty), and partial fusion of the second and third toes of the hind-foot. The phalangeroids, in turn, are a diverse subgroup of diprotodonts that encompass "many unique or relict species and genera that are united only by a common adaptation to life in forested environments" (Smith, 1984b), a phrase that nicely echoes the old arboreal theory of primate origins. Phalangeroids typically have claws on all five digits of each extremity except the first digit of the hindfoot, which is clawless and opposable. The hands of most phalangeroids are capable of grasping, some with the first digit opposed to the others, others with digits 1–2 opposed to digits 3–5.

Among phalangeroids, several natural clades are easily recognizable, while other groupings are less certain (Baverstock, 1984; Figure 1). Most researchers agree that one phylogenetic outlier is the honey possum, *Tarsipes*

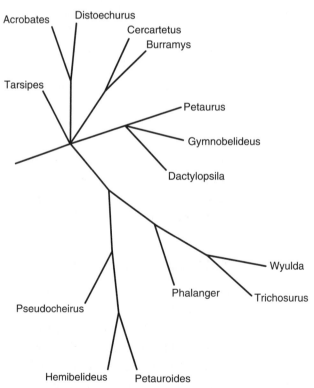

Figure 1.   Unrooted phylogenetic tree showing hypothesized relationships among phalangeroids.

rostratus, classified in its own family, Tarsipedidae. Several miniature species of feathertail possums (*Acrobates* and *Distoechurus*) form a distinct clade, and they are now usually placed in the family Acrobatidae. Another confident clade contains the pygmy-possums of the family Burramyidae (*Cercartetus* and *Burramys*). The nocturnal, arboreal *Cercartetus* is particularly relevant as a parallel to primitive primates (Cartmill, 1974a). Together, the living tarsi-pedids, acrobatids, and burramyids comprise an ecologically definable guild containing tiny (5–70 g) pollen and nectar feeders that will be discussed below under the category "Miniature Flower Specialists."

The family Petauridae is a natural grouping of arboreal species containing the familiar sugar gliders and their gliding congeners (*Petaurus* spp.), along with a few nongliding species of the genus *Dactylopsila* (including its sub-genus *Dactylonax*), and the nongliding Leadbeater's possum, *Gymnobelideus leadbeateri*. The petaurids range in size from about 100 to 700 g and rely on diets containing significant amounts of insects, tree exudates, fruit, and flower products. In size, arboreality, and diet, this radiation is of obvious interest to students of primate evolution. This clade will be discussed below in the section on "Small-bodied Omnivores."

The final grouping of phalangeroids contains the typical possums and cuscuses of Australasia. Recently, it has been popular to recognize two fami-lies, Phalangeridae and Pseudocheiridae, which nevertheless are more closely related to each other than to any of the lineages outlined above. These largest of the phalangeroids (0.7–5 kg) are typically nocturnal, arboreal mammals specialized to a folivorous diet, particularly to leaves of *Eucalyptus*. The fam-ily Phalangeridae includes the brushtail possums (*Trichosurus*), the best stud-ied of all the phalangeroid marsupials; the monotypic scaly tailed possum (*Wyulda*); and the slow-climbing cuscuses, traditionally put in a single genus *Phalanger*, but now inevitably split into several, following recognition that the group is paraphyletic (= monophyletic with some weird descendents) with respect to the clade of *Trichosurus* and *Wyulda*. Genera now often used within the phalanger group are *Spilocuscus, Strigocuscus,* and *Ailurops.* The family Pseudocheiridae contains the largest of the gliding possums (*Petauroides volans*), and several nongliding taxa, including the lemuroid ringtail possum (*Hemibelideus lemuroides*) and a diversity of other ringtail possums (so named because their prehensile tails coil into a tight spiral). The ringtails were tradi-tionally classified in one genus, *Pseudocheirus*, but they are now recognized as being paraphyletic with respect to *Hemibelideus* and *Petauroides* (and are therefore undergoing taxonomic revisions, usually including the use of the genera *Pseudochirops* and *Petropseudes*). The phalangerid–pseudocheirid radiation is discussed below under "Larger-bodied Folivores."

MINIATURE FLOWER SPECIALISTS

Sussman (1991, 1995, 1999) suggested that the key ecological factor leading to the origin of primates was adaptation to the utilization of angiosperm products—fruit and flowers—at the tips of terminal branches. There is still some question as to whether one particular food item was crucial in the evolution of the earliest primates, or whether primate adaptations are better seen as related to a general exploitation of terminal branch resources. The presence in Australasia of several independent lineages of small mammals specially adapted to the use of flowers has obvious relevance to the angiosperm exploitation theory. Are flower specialists among the phalangeroids particularly primate-like in their hand and foot structure, visual systems, brain sizes, or reproductive patterns?

The most widespread flower specialists in Australia are the pygmy-possums of the genus *Cercartetus*, distributed in the rainforests of New Guinea and Queensland (*C. caudatus*), rainforest to dry forest of southeastern Australia (*C. nanus*), a variety of drier forests of southern Australia and Tasmania (*C. lepidus*), and dry forest, heath, and scrub of southwestern Australia (*C. concinnus*). These are tiny mammals, with the smallest, *C. lepidus*, being only 6–9 g, while the largest, *C. caudatus*, is 25–70 g. In other words, the largest *Cercartetus* overlaps in size with the smallest living primate, *Microcebus*, several species of which fall within the range of 26–77 g (Atsalis et al., 1996). Several early Tertiary primates are as small as *Cercartetus*, including the earliest certain euprimate from the very base of the Eocene, *Teilhardina*, and a community of tiny primates from the middle Eocene of China, which includes species possibly as small as 12 g (Gebo et al., 2000). Key morphological attributes of *Cercartetus* include relatively short snout, large eyes, moderate degrees of orbital convergence, prehensile tail, and grasping hands and feet with reduced claws and expanded lobes on the distal pad. Several authors have repeated the observation that in proportions the forefoot looks like a human hand. These animals show a close physical resemblance to *Microcebus* in their overall shape, in the short soft fur, and in the grayish to reddish brown coloration often with dark spectacles around the eyes (Figure 2). Marsupial biologists have noted the close resemblance

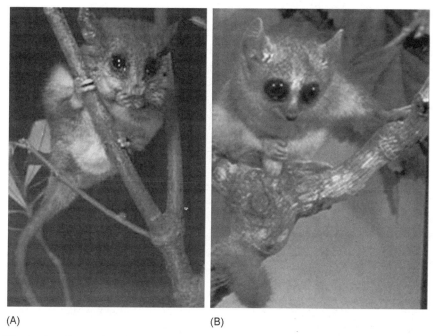

(A) (B)

Figure 2. Comparison between the possum *Cercartetus* (A) and the primate *Microcebus* (B). Photograph (A) by R. Whitford, reproduced with permission from Nature Focus © Australian Museum; photograph (B) by R. Mittermeier, reproduced with his permission.

between pygmy-possums, the didelphid *Marmosa*, and the cheirogaleid *Microcebus* (Lee and Cockburn, 1985). Cartmill (1974a) emphasized the relevance of *Cercartetus* to primate origins because of their reduced claws and convergent orbits, writing "the adaptations of *Cercartetus* ... represent plausible structural antecedents for the traits that distinguish the extant primates." He pointed out that the orbital marginal orientation (a measure of convergence) of *Cercartetus caudatus* is almost identical to that of *Galago*, *Avahi*, and *Tarsius*.

Species of *Cercartetus* are arboreal and nocturnal. They have been noted to be agile scramblers and leapers (Atherton and Haffenden, 1983). *Cercartetus* has been characterized as a flower specialist, with the diet usually including a significant proportion of nectar and pollen (Lee and Cockburn, 1985; Turner, 1984). They also eat blossoms, fruit, and arthropods (Smith, 1986). Individuals apparently forage solitarily at night, but groups of up to four individuals have been found sleeping together in nests, another similarity to some cheirogaleids (Collins, 1973). Physiologically, they also show some interesting parallels with prosimians—towards the end of the summer, the base of the tail in *Cercartetus* expands as a result of fat storage, as in the primate *Cheirogaleus medius*, and they undergo seasonal periods of torpor, also as in cheirogaleids (Hladik et al., 1980; Wright and Martin, 1995). *Cercartetus* diverges from cheirogaleids in its higher rates of reproductive output—females have four nipples but may occasionally produce five or six offspring (Ward, 1992). From an ecomorphological perspective, *Cercartetus* appears to offer much promise in helping to understand parallel evolution of primate-like attributes in a small, terminal-branch feeding mammal.

The case of the burramyids highlights how important it is to obtain precise ecological and behavioral data, even in the presence of solid morphological data, in order to test ideas about primate origins. Cartmill's (1974a) generalization that *Cercartetus* and *Burramys* "forage for fruit and insects in the shrub layer of Australian forests and heaths" is no longer a fair characterization of their habits in light of what has been learned recently about these animals in the wild. *Burramys*, the one member of the group that falls well outside the characterization of "flower specialist," is a divergent taxon adapted to alpine and subalpine habitats, where it lives terrestrially in boulder screes, eats fruit, and stores winter supplies of seeds (Lee and Cockburn, 1985). *Burramys* shares plagiaulacoid lower premolars (high serrated blade-like teeth) with a very few other mammals, including the extinct, early Tertiary plesiadapiform *Carpolestes* (Rose, 1975; Simpson, 1933). The interpretation by the paleontologist Rose (1975) that the carpolestid diet may have consisted of tough herbage, fruits, and seeds, along with insects, is buttressed now by initial data on the actual diet of the dentally similar *Burramys*, which consists of grasses, fruits, seeds and insects (Calaby, 1983; Lee and Cockburn, 1985). The functional use of the plagiaulacoid p4 is illuminated by observations of *Burramys*: "hard-shelled seeds and insects with hard cuticles are held at the sides of the mouth and broken or cut up with the premolars. Efficient use of these teeth requires that the animal also have highly manipulative forepaws" (Calaby, 1983). In addition, *Burramys* uses the specialized premolars to slice grasses for storage in nests (Calaby, 1983). Although insects are fed upon by *Burramys*, we believe that characterization of this genus as a model of a primitive arboreal visual predator (Cartmill, 1974a) was premature given their dental specializations and what is now known of their diet. But as outlined below, its relevance for other models of primate origins may be more relevant given the recent postcranial discoveries about *Carpolestes* (Bloch and Boyer, 2002).

The adaptations of *Cercartetus* also differ significantly from Cartmill's (1974a) characterization. The species that has been studied best (*C. nanus*) is a flower specialist, particularly dependent on pollen and nectar, rather than eating the blossoms themselves; the species also feeds on insects associated with their food flowers (Turner, 1984; Figure 3). Indeed, most of the insects eaten by *C. nanus* are suspected to have been obtained directly from the flowers of *Banksia* (Lee and Cockburn, 1985). Less is known of the diet in *C. caudatus*, the one specific item of the natural diet listed by Atherton and Haffenden (1983) is nectar from a species of eucalypt (note this is the possum classified in the genus *Eudromicia* by Cartmill, 1972). The western species, *C. concinnus*, lives in a variety of drier forests and heath habitats, and is known to be partly terrestrial in its activity. The tiny species *C. lepidus* is reported to be primarily insectivorous, in contrast to its congeners (Green, 1983), but it is also poorly studied, with little information available about where and how it finds insects, and how this relates to other food resources. All species of the genus probably eat arthropods, and interestingly, captive *C. nanus* have been

(A)   (B)

Figure 3. Flower feeding by the possum *Cercartetus* (A) and the primate *Cheirogaleus,* photograph (A) © Kathie Atkinson; photograph (B) by David Baum, reproduced with his permission.

seen capturing flying moths by snatching them out of the air with two hands
(Atherton and Haffenden, 1983). While the observed behavior associated
with insect capture fits well with expectations from Cartmill's visual predation
hypothesis, it occurs in an ecological context that matches Sussman's
angiosperm exploitation hypothesis. Based on current data, the best-known
species of *Cercartetus* may be characterized as a flower specialist that captures
insects opportunistically from the flowers themselves. This is very similar to
the ecological pattern observed in the primate-like didelphid (*Caluromys*),
which captures insects flushed while foraging among fruit and flowers at the
tips of branches (Rasmussen, 1990; Steiner, 1981). We suggest that both
cases offer support for the idea that it is the combined windfall of insects
associated with fruit and flowers that provides an ecological stage for the
development of primate-like adaptations (Crompton, 1995; Martin, 1986,
1990; Rasmussen, 1990, 2001; Sussman and Raven, 1978; Sussman, 1999).

The acrobatids closely resemble *Cercartetus* in many respects, and were traditionally classified with that genus in the family Burramyidae based on close
physical similarity, but molecular studies appear to have eliminated a close phylogenetic relationship between acrobatids and burramyids (Baverstock, 1984).
The most salient physical difference between acrobatids and *Cercartetus* relates
to the tail—both acrobatid genera (*Acrobates* of Australia, *Distoechurus* of
New Guinea) have a pair of stiff fringes of hair extending laterally from the
tail, resembling a feather. In addition, *Acrobates* possesses a gliding membrane
stretching from elbow to knee, and is adept at sailing. *Acrobates* also has finely
serrated distal pads on their hands and feet that allow them to gain purchase on
smooth surfaces, including vertical panes of glass (Russell, 1983). *Distoechurus*
has small, sharp, curved claws and no expansion of the distal digital pads. Both
genera are tiny, agile, nocturnal animals that apparently rely on flower products
and associated insects in a variety of forest types (Goldingay and Kavanagh,
1995; Lee and Cockburn, 1985). Up to forty individuals of *Acrobates* have been
spotted together in a single flowering tree crown. The two acrobatids parallel
primates but diverge from *Cercartetus* in having small litters; the 1-4 offspring
produced by them is less than would be predicted from their tiny body masses
(Ward, 1990). Like *Cercartetus* and some cheirogaleids, *Acrobates* undergoes
periods of torpor (Jones and Geiser, 1992). This group offers another evolutionary view of a small-bodied mammalian flower specialist, this time with the
trick of gliding added (in *Acrobates*) as a way to get around among the terminal
branches of a tropical flowering tree (Figure 4).

(A)   (B)

Figure 4.   Two genera of miniature flower specialists, *Acrobates* (A) and *Distoechurus* (B). Photograph (A) reproduced with permission from Nature Focus © Australian Museum; Photograph (B) Kathie, Atkinson, reproduced with permission from Nature Focus © Australian Museum.

The final miniature flower specialist is the honey possum, *Tarsipes rostratus*, the genus name being derived from its tarsier-like foot. While *Tarsipes* does indeed have primate-like hands and feet with reduced claws, capable of fine grasping (Cartmill, 1992; Renfree et al., 1984), the honey possum is much less primate-like in other respects than are *Cercartetus* and the acrobatids. Compared to all other phalangeroids, it has an elongated snout (even described as "beaklike" by Szalay, 1994), pronounced development of facial vibrissae, and eyes lacking much convergence. *Tarsipes* has a specialized, bristle-tipped tongue—which can be extended 25 mm beyond the nose—for harvesting nectar and pollen, and the cheek teeth are reduced to peg-like rudiments. Insects are taken infrequently, if ever (Lee and Cockburn, 1985; Renfree et al., 1984; Wiens et al., 1979; but see Wooller et al., 1984). A bristled tongue used in procuring pollen and nectar is also found in the primate *Eulemur rubriventer* (Overdorff, 1992; Sussman, 1999). The scansorial *Tarsipes* lives on sandplain heaths in southwestern Australia, where it is often on the ground; e.g., it is found in pitfalls such as fence postholes. The elongated snout is supposed to be an adaptation for probing flowers, but in proportions, it is reminiscent of rostra seen in other small-bodied terrestrial mammals such as tenrecs, shrews and elephant shrews. *Tarsipes* can be held up as the ultimate nectar specialist among mammals. Unlike *Cercartetus*, *Tarsipes* is not highly arboreal nor does it seem to feed on insects attracted to flowers or the fruit that are generated from flowers. *Tarsipes* therefore represents an excellent evolutionary trial to help tease apart factors that contribute to primate-like hands and feet, but not primate-like faces. In this case, the obvious inference is that prehensile hands and feet are used for fine-branch grasping, not for insect capture.

SMALL-BODIED OMNIVORES

In contrast to the miniature flower specialists, petaurids are larger in size and apparently less dependent on nectar and pollen. Perhaps the most primate-like member of the family is the Leadbeater's possum, *Gymnobelideus leadbeateri*, which is restricted to moist eucalypt forests in mountains of southern Australia. *Gymnobelideus* was believed to be extinct for many decades, until it was rediscovered in the wild in the 1960s (Smith, 1984a). *Gymnobelideus* weighs between 100 and 170 g, so it is within the size range of the ecologically similar cheirogaleids; it looks superficially like a South Australian version of Madagascar's *Phaner* (Figure 5). It is quite prosimian-like in appearance, with a short face, large and moderately convergent orbits, black facial markings and a dorsal stripe, with grasping hands and feet. Unlike many phalangeroids, it lacks both a gliding membrane and a prehensile tail. The species is noctur-

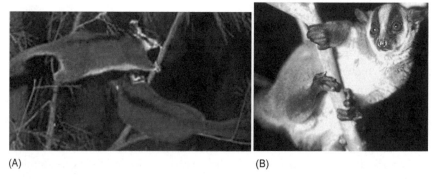

(A)   (B)

Figure 5. Comparison between the possum *Gymnobelideus* (A) and the primate *Phaner* (B). Photograph (A) by A. Smith; photograph (B) by R. Mittermeier reproduced with his permission.

nal, and is an agile arborealist, capable of making leaps of a meter or more in distance. The diet is very similar to those of cheirogaleids, consisting of insects, tree exudates, flower products, and even secretions of homopteran insects, as also observed in *Phaner* and *Mirza* (Hladik et al., 1980; Pages, 1980; Smith and Ganzhorn, 1996; Smith, 1984c, 1984d; Sussman, 1999; Wright and Martin, 1995). *Gymnobelideus* resembles the primates *Callithrix*, *Cebuella*, and *Phaner* in gouging or scraping tree trunks to generate the flow of exudates (Charles-Dominique and Petter, 1980). The arthropod prey of *Gymnobelideus* consists of tree crickets, beetles, moths, and spiders (Lee and Cockburn, 1985). In addition, these marsupials feed on manna (a carbohydrate exudate from eucaplypt leaves). Among phalangeroids, *Gymnobelideus* is one of the best overall matches ecomorphologically to the primitive primate *Mirza*, the exudate specialist *Phaner*, and the primate-like didelphid *Caluromys*.

The other nongliding genus in Petauridae is *Dactylopsila*, which has already gained fame in comparison to primates by sharing a specialized feeding adaptation with the Malagasy aye-aye, *Daubentonia* (Cartmill, 1974b; Rand, 1937). Both taxa have enlarged incisors for gouging wood, which they utilize to open the tunnels of wood-boring larvae, and an elongated finger on the hand tipped with a hooked nail, which they use to ream out beetle larvae. Both tap the wood with their forefeet, apparently using auditory clues to detect subsurface features (Erickson, 1995). The small-bodied *Dactylopsila* is even more similar in size and shape to the extinct apatemyids than it is to the much larger *Daubentonia*; apatemyids were a very successful group of arboreal Eocene mammals widespread in North America and Europe (Koenigswald, 1987). It has been pointed out that Madagascar, Australasia, and the Eocene all lack woodpeckers (Picidae), birds that specialize on wood-boring larvae on most landmasses today (Cartmill, 1974b; Koenigswald, 1987). *Dactylopsila* is reported to be a frenetic, extremely quick and active arborealist, quite a contrast to the more deliberate, heavy *Daubentonia*. In addition to its wood-boring activities, *Dactylopsila* breaks into the nests of eusocial insects (Smith, 1982b), and also feeds on crickets, spiders, and other arthropods (Lee and Cockburn, 1985). *Dactylopsila* shows greater orbital convergence than phalangerids proper (Cartmill, 1972). In the future, research should investigate the extent to which the adaptations of *Dactylopsila* match primate attributes aside from the aye-aye parallels.

The gliding petaurids (*Petaurus*), such as the sugar glider (*P. breviceps*), have been noted by many zoologists to be reminiscent of primates. The larger-bodied and more folivorous species, such as *P. australis* (up to 700 g), are much more heavily clawed than is the small species, *P. breviceps* (100–160 g), which has more primate-like hands. This difference presumably indicates that the larger forms are utilizing thicker, supportive branches and trunks that exceed the grasping diameter of their hands. *P. breviceps* also differs from its larger congeners in having a flatter, primate-like face (reflected in the species name), and it apparently exhibits more orbital convergence than the larger forms (Figure 6). *P. breviceps* resembles some cheirogaleid and galagine primates in having a diet consisting of insects, fruit, tree exudates, and homopteran secretions (Henry and Suckling, 1984; Smith, 1982a). In a fine display of visually directed predation, *P. breviceps* has been observed to leap at and catch moths in flight (Nowak and Paradiso, 1983). Members of the genus

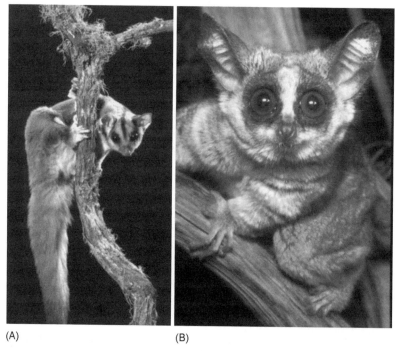

(A)   (B)

Figure 6.   The possum *Petaurus norfolcensis* (A), one of the larger species of gliding petaurids, compared to *Galago moholi* (B), a small galagine. Photograph (A) by E. Beaton, www.estherbeaton.com; photograph (B) T. Rasmussen.

communicate chemically using well-developed scent glands, and vocally by using a diversity of calls (Goldingay, 1992). *P. breviceps* nests gregariously in apparent kin groups of up to about seven individuals. Its life history is also similar to that of small-bodied prosimians, producing one litter per year of one or two offspring, which grow up fairly slowly. The young are independent at three months and disperse by about 10 months of age.

Gymnobelideus, Dactylopsila, and *Petaurus* have the highest encephalization quotients of any marsupials, a final important parallelism with primates (Lee and Cockburn, 1985; Nelson and Stephan, 1982).

LARGER-BODIED FOLIVORES

The phalangeroids that weigh from about 1 kg in body mass and higher (up to a maximum of about 5 kg) make their living eating leaves, particularly leaves of *Eucalyptus*. Microbial fermentation of cellulose is accomplished in an enlarged cecum (Smith, 1984a). Several of these larger phalangeroids present interesting comparisons to primates based on obvious external physical characteristics, as is evident from the name of *Hemibelideus lemuroides*. Compelling comparisons have been drawn between the slow-climbing cuscuses (*Phalanger* and relatives) and the Asian slow lorises (*Nycticebus*), which differ from each other in the presence of a long prehensile tail in the former and only a remnant stub in the latter (Figure 7). The cuscuses have orbits more convergent than those of other phalangerids (Cartmill, 1972) but less so than in *Nycticebus*, snouts that are relatively truncated, and short lorisine ears. They utilize their hands and feet—which have reduced claws compared to other phalangerids and pseudocheirids—to travel like lorises via cautious slow climbing. They prefer to maintain a grip on branches with three extremities at a time, and only attempt leaps when moving along open ground. Interestingly, the cuscuses are the least folivorous of these larger phalangeroids, relying on a diet of leaves, fruit, seeds, and animal prey (Hume et al., 1997). They are dietary generalists but locomotor specialists.

In contrast, the closely related brushtails (*Trichosurus*) are both locomotor generalists and dietary generalists, and have proven to be successful colonizers following their introduction to New Zealand. They usually eat a high proportion of leaves in addition to smaller amounts of fruit, shoots, animal prey, and other items (Kerle, 1984; Proctor-Gray, 1984; Statham, 1984).

(A)   (B)

Figure 7. The possum *Phalanger* (A) and the primate *Nycticebus* (B), both slow-climbing arboreal mammals. Photograph (A) Reproduced with permission from Nature Focus © Australian Museum; photograph (B) by D.T.Rasmussen.

Brushtails have more of a galago cast than do the cuscuses, with sharp, projecting snouts, large erect ears, usually bushy tails, and sternal scent glands, but postcranially, they are more closely comparable to *Didelphis* (the familiar New World opossum). *Trichosurus* moves in the trees and on the ground, and occurs in a wide variety of habitats. A third member of this clade, *Wyulda*, is restricted in distribution to southern Australia where it lives in terrestrial rock piles but forages up into trees (Humphreys et al., 1984; Muncie, 1999). *Wyulda* shows considerably fewer primate-like attributes to casual inspection than do its relatives—if anything, it might make an interesting analogy to the tree shrew *Anathana*, which also lives in rock piles and forages in trees.

Pseudocheirids, or ringtail possums, eat a highly folivorous diet, mainly consisting of eucalypt leaves (Proctor-Gray, 1984; Thompson and Owen, 1964). They are distinguished from phalangerids by having higher, more elaborate shearing crests on their molars for processing leaves. While most ringtails are arboreal, one species, the rock ringtail possum (*Petropseudes dahli*) lives on rock outcrops. Apparently in correlation with its terrestrial habits, it shows a "longer snout, shorter tail, shorter legs, and shorter claws" than do its arboreal relatives (Nelson and Kerle, 1983). Typically, members of *Pseudocheirus* and *Pseudochirops* are slow, deliberate climbers, resembling the cuscuses of the family Phalangeridae. Indeed, the similarities between certain members of the paraphyletic ringtails and the paraphyletic cuscuses suggest that their common ancestor (the ancestor of the combined clade of Phalangeridae and Pseudocheiridae) was a deliberate quadruped with a generalist to folivorous diet; the species *Phalanger orientalis* and *Pseudocheirus herbertensis* are noted to be particularly similar to each other despite their family level separation. In contrast, the lemuroid ringtail possum (*H. lemuroides*) is a highly arboreal rainforest possum with a flatter face and longer limbs than those of *Pseudocheirus dahli*; this species is expert at leaping through the canopy (Figure 8). Remarkably, *Hemibelideus* has slight folds of skin less than 25 mm in width along the sides of the trunk (Johnson-Murray, 1987), and it spreads the limbs wide during a leap like a gliding possum—whether these flaps represent the origin of a gliding membrane from an ancestor like *Pseudocheirus* or the remnants of one from an ancestor like *Petauroides* (see below) is a fascinating question, with profound implications either way. The prehensile tail of *Hemibelideus* shows an interesting parallel with ateline platyrrhines in being heavily furred over most of its length except the ventral surface of the grasping tip (also characteristic of some other phalangeroid folivores). *Hemibelideus* is an animal with a lemur-like countenance, the tail of a spider monkey, and incipient or vestigial gliding membranes. Although nocturnal, the lemuroid ringtail is reported to be gregarious, with

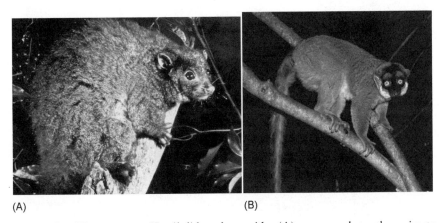

(A)   (B)

Figure 8. The possum *Hemibelideus lemuroides* (A) compared to the primate *Eulemur fulvus* (B). Photograph (A) reproduced with the permission of the Environmental Protection Agency, Queensland, © State of Queensland; Photograph (B) by D.T.Rasmussen.

groups of two or three being common (Winter and Atherton, 1984)—one group of eight animals was reported in a single tree (Winter, 1983).

Another large, heavily clawed pseudocheirid is the greater glider (*Petauroides volans*), which resembles the greater galago (*Galago crassicaudatus*) in body size, shape, and general facial proportions, including the large ears (Figure 9). The similarity even extends to having gray and black color phases (Ride, 1970). In terms of its diet and social system, the greater glider has been specifically compared to the prosimian *Lepilemur* (Henry, 1984). The greater glider produces one young at a time which takes two years to mature, a remarkably slow reproductive rate for a mammal of its size. Ecologically, the glider differs from the galago in its folivorous habits and, of course, in its exceptional ability to glide (several dozens of meters per flight). The parallel development of gliding in several lineages of phalangeroids and in primate-like dermopterans raises the question of why there are no gliding primates.

(A)   (B)

Figure 9. The possum *Petauroides volans* (A) compared to the primate *Galago crassicaudatus* (B). Photograph (A) by A. Smith; photograph (B) by D.T.Rasmussen.

CRITICAL PRIMATE ADAPTATIONS

Neglecting the petrosal bulla and a few other pedantries of uncertain adaptive significance, there is widespread agreement that some of the most important shared primate attributes reflecting something about the adaptive context of primate origins include the following: (1) grasping hands and feet with nails rather than claws, and locomotion utilizing diagonal footfall patterns, (2) notable development of functional stereoscopy through orbital convergence, (3) large brain relative to body size, and (4) relatively small litters and reduced developmental rates. There is no particular reason to believe that primate origins involved a near-simultaneous acquisition of all these traits in one grand evolutionary event. Rather, they must have been acquired one after the other, probably each one by subtle degrees, over the course of significant geological time, and involving several nested adaptive radiations. The fossil record of anthropoid origins—a particularly well-known "origins" event—have demonstrated how the discovery of more and more fossil intermediates between extant-grade prosimians and extant-grade anthropoids breaks down the assemblage of shared specializations initially believed to characterize the dichotomy (Simons and Rasmussen, 1995, 1996). The earliest undoubted anthropoids do have postorbital closure, but not fusion of the mandibular symphysis, reduction of the molar trigonids, brain enlargement, or other formerly diagnostic characters. Similarly, fossils demonstrate that the origin of hominids involved acquisition of bipedalism first, followed much later by significant expansion of the brain and extensive use of stone tools. (Remember that one century ago this simple fact was unknown and was an issue of debate; this example also serves to show how we can acquire confident knowledge about unique historical events through the fossil record, in contrast to Cartmill, 1990). Undoubtedly, the same pattern will hold true for primate origins—the entire assemblage of "primate traits" will break down into a chronological sequence of adaptive shifts if the fossil record ever becomes sufficiently dense.

The recent description of postcranial adaptations in the archaic primate *Carpolestes* provides exactly this kind of insight into the sequence of acquisition of primate attributes (Bloch and Boyer, 2002, this volume; Sargis, 2002). A beautifully preserved skeleton of this genus shows well-developed grasping extremities, but the absence of stereoscopic vision, or other primate specializations of the face. The postcranial remains provide compelling evidence that

Carpolestes can be placed as a close relative of the euprimate clade, and that it was a climber in a small-branch milieu, but not a stereoscopic specialist. Bloch and Boyer (2002) concluded:

> It has been suggested that "If the first euprimates had grasping feet and blunt teeth adapted for eating fruit, but retained small, divergent orbits like those of *Plesiadapis. . ."* [Cartmill, 1992], the terminal branch feeding hypothesis for primate origins [Sussman's hypothesis] would be supported. The fossil find presented here is consistent with the hypothesis that early euprimates evolved grasping first and convergent orbits later . . . and inconsistent with the visual predation hypothesis."

Carpolestes contributes an important paleontological sample to our pool of experimental runs showing an incomplete assemblage of "primate-like" traits, and phalangeroids provide additional living samples. Primate-like traits can be found in just about every phalangeroid, but developed to varying degrees (or in other words, phalangeroid-like traits occur in most prosimian primates). One advantage of looking at the phalangeroid radiation as a whole, or at ecological communities of phalangeroids (Lindenmayer, 1997; Smith and Ganzhorn, 1996), rather than at a single "primate-like species" is that the radiation represents the complex, diverse outcomes of many evolutionary trials, any one of which is potentially relevant in one or another way to the adaptive features believed to be important for primate origins—orbital convergence, eye size, color vision, grasping hands, diagonal footfall patterns, large brains, slow development, and low metabolic and reproductive rates. Many parallel evolutionary runs are more useful than just one or a few. A complex radiation of organisms alive today offers great potential to understand the diverse, nested hierarchies of sequential adaptive radiations through time that must have occurred in the origin of primates. Indeed, at a broader taxonomic scope, the phalangeroids make fascinating comparisons to plesiadapiforms, dermopterans, and scandentians. At an extreme, reductionist, cladistic level, the origin of primates could be traced (theoretically) to a single speciation event, but from the point of view of the origin of primate adaptations, such a view is simplistic and insufficient.

The greatest convergence of primate-like traits occurs in the small-bodied omnivores, *Gymnobelideus*, *Dactylopsila*, and *Petaurus* and the miniature flower specialist *Cercartetus*. These observations clearly offer broad support for the idea that primates, too, may have diversified as terminal-branch-foraging omnivores making significant use of flowers, fruits and the associated insects (Rasmussen, 2001; Sussman, 1999). More specifically, our current knowledge of the phalangeroid radiations suggest that primate-like hands with great prehension, reduced claws, and expanded tactile pads, evolved in flower specialists and other small-bodied forms that utilize small-diameter supports in terminal branches or fine undergrowth. This observation is consistent with recent research on the evolution of the primate hand (Hamrick, 1998, 2001; Lemelin, 1996, 1999). Visual convergence is found among animals that move actively in arboreal, terminal branches harvesting angiosperm products and catching arthropod prey found there. The dasyurid marsupials, which were not discussed in this paper, are small-bodied predators but not arboreal angiosperm users; they may provide a good test in comparison to phalangeroids that might untangle how insect predation on the ground differs from insect capture at fruit and flowers in terminal branches (Righetti, 1996).

We believe that some of the hypotheses being considered in the symposium can be examined further by research on living marsupials. For example, do any other marsupials besides *Caluromys* have diagonal-sequence walking gaits and, if so, how does this relate to their use of supports? How are the adaptations for grasp-leaping distributed among phalangeroids and does this represent one or a number of distinct ecomorphological locomotor modes? What is the distribution of traits such as hypometabolism, or various forms of color vision? How do primates and phalangeroids that are similar ecologically differ from each other? Morphological analyses of phalangeroids in an ecological context have lagged behind the growth of comparable information on primates, a problem lamented by Szalay (1994, p. 248): "The osteological aspects of the living phalangeriforms...have not been adequately analyzed from an ecomorphological perspective.... A great deal of ecological and behavioral information, to be obtained from focused observations, is still missing."

PARALLELISM AND PRIMITIVENESS

After parallelisms have been identified, a final critical point that must be addressed is to distinguish specializations from the primitive condition for a group. The close resemblance between food procurement in *Daubentonia* and *Dactylopsila* does not imply that ancestral primates or phalangeroids were specialists on wood-boring larvae, because these are convergent specializations generated well after the origins of primates and phalangeroids. The many ecomorph parallelisms suggested by this literature review (*Lepilemur* and *Petauroides*, *Phaner* and *Petaurus*, *Mirza* and *Gymnobelideus*, *Microcebus* and *Cercartetus*, *Daubentonia* and *Dactylopsila*, *Nycticebus* and *Phalanger*) cannot all reflect similarities that were critical in the sequence of events leading to primate origins. If Cartmill (1972) erred in his visual predation hypothesis, it may have been because he relied too heavily on those cases of animals that showed specialized, extreme orbital convergence—such as lorises, owls, and cats—rather than on animals that have moderate convergence and are more likely to reflect the primitive primate condition (Crompton, 1995; Sussman, 1991, 1995). Among prosimians, the ones that behave like Cartmill predicted for an early primate are indeed those with the greatest orbital convergence: *Tarsius* and *Loris* are both nearly completely faunivorous and both rely partly (but not completely) on visually directed predation and prey capture with the hands (Crompton, 1995; MacKinnon and MacKinnon, 1980; Nekaris and Rasmussen, 2003; Niemitz, 1984). But these unusual specialists are unlikely to reflect critical events at the basal radiations of primates. Both *Tarsius* and *Loris* are not only specialized in their facial structure and diet, but also in their divergent postcranial adaptations relative to the primitive primate condition (Crompton, 1995).

The question of which came first—visually oriented predation or agile movements in fine branches of angiosperms—appears to be a chicken-and-egg type of proposition. However, it seems to us that the ability to navigate in the precarious, three-dimensional world of the terminal branches would be a prerequisite to any ability for visual hunting of fast-moving insects. When a mammal is maneuvering through fine branches the entire habitat becomes dynamic, as the animal plunges and the fine branches bearing small fruits and flowers bob and sway (Figure 10). It is not clear why fine branches, fruits and flowers moving quickly across an early primate's visual field should require optical specializations fundamentally different from those suggested by Cartmill to be associated specifically with insect prey moving across the visual field. We cannot identify a cogent ecological context that would compel early primates to adapt to visual predation of insects outside of the terminal branch milieu—after all, there are many insect-eating mammals that do not have primate-like traits. The complexity of the terminal branch environment is further highlighted by the fact that, unlike terrestrial mammals on a stable surface, a primate or possum in branches much smaller than its own body size must

Figure 10. *Cercartetus* foraging among terminal angiosperm products, a possible good model of ancestral primate adaptations. Photograph is reproduced with the permission of the Environmental Protection Agency, Queensland; the copyright in the reproduced material belongs to the State of Queensland.

perceive the three-dimensional depth below, thereby encountering problems perhaps similar to those faced by birds and fish (T. M. Preuss, personal communication). This interpretation of the chicken-and-egg problem now seems to be solidly supported by the paleontological data on *Carpolestes* (Bloch and Boyer, 2002).

Marsupials such as *Caluromys, Marmosa, Cercartetus, Gymnobelideus, Petaurus,* and others, offer compelling similarities to primitive primates, and represent a virtually untapped resource for work on primate origins. Because the phalangeroids represent a diverse, complex radiation, they also offer a realistic view of how subtle divergences among closely related forms in an ecological context may relate to the origin of higher-level taxa.

REFERENCES

Aplin, K. P., and Archer, M., 1987, Recent advances in marsupial systematics with a new syncretic classification, in: *Possums and Opossums: Studies in Evolution*, vol. I, M. Archer, ed., Surrey Beatty and Sons Pty Limited, Chipping Norton, Australia, pp. xv–lxxii.

Archer, M., 1984, The Australian marsupial radiation, in: *Vertebrate Zoogeography and Evolution in Australasia-Animals in Space and Time*, M. Archer and G. Clayton, eds., Herperian Press, Carlisle, Australia, pp. 585–625.

Atherton, R. G., and Haffenden, A. T., 1983, Long-tailed pygmy-possums, in: *Complete Book of Australian Mammals*, R. Strahan, ed., Angus and Robertson Publishers, Sydney, pp. 166–167.

Atsalis, S., Schmid, J., and Kappeler, P., 1996, Metrical comparisons of three species of mouse lemur, *J. Hum. Evol.* **31:** 61–68.

Baverstock, P. R., 1984, The molecular relationships of Australasian possums and gliders, in: *Possums and Gliders*, A. Smith and I. Hume, eds., Surrey Beatty & Sons Pty Limited, Chipping Norton, Australia, pp. 1–8.

Bloch, J. I., and Boyer, D. M., 2002, Grasping primate origins, *Science* **298:** 1606–1610.

Calaby, J. H., 1983, Mountain pygmy-possum, *Burramys parvus*, in: *The Australian Museum Complete Book of Australian Mammals*, R. Strahan, ed., Angus and Robertson, Sidney, pp. 168–169.

Cartmill, M., 1972, Arboreal adaptations and the origin of the order Primates, in: *The Functional and Evolutionary Biology of Primates*, R. Tuttle, ed., Aldine Atherton, Chicago, pp. 97–122.

Cartmill, M., 1974a, Rethinking primate origins, *Science* **184:** 436–443.

Cartmill, M., 1974b, *Daubentonia, Dactylopsila*, woodpeckers and klinorhynchy, in: *Prosimian Biology*, R. D. Martin, G. A. Doyle, and A. C. Walker, eds., Duckworth, London, pp. 655–670.

Cartmill, M., 1990, Human uniqueness and theoretical content in paleoanthropology, *Int. J. Primatol.* **11:** 173–192.

Cartmill, M., 1992, New views on primate origins, *Evol. Anthropol.* **1:** 105–111.

Charles-Dominique, P., and Petter, J. J., 1980, Ecology and social life of *Phaner furcifer*, in: *Nocturnal Malagasy Primates: Ecology, Physiology, and Behavior*, P. Charles-Dominique, H. M. Cooper, A. Hladik, C. M. Hladik, E. Pages, G. E. Pariente, A. Petter-Rousseaux, J. J. Petter, and A. Schilling, eds., Academic Press, New York, pp. 75–95.

Collins, L., 1973, *Monotremes and Marsupials*, Smithsonian Institution Press, Washington, D.C., 323 p.

Crompton, R. H., 1995, Visual predation, habitat structure, and the ancestral primate niche, in: *Creatures of the Dark: The Nocturnal Prosimians*, L. Alterman, G. A. Doyle, and M. K. Izard, eds., Plenum Press, New York, pp. 11–30.

Erickson, C. J., 1995, Perspectives on percussive foraging in the aye-aye (*Daubentonia madagascariensis*), in: *Creatures of the Dark: The Nocturnal Prosimians*, L. Alterman, G. A. Doyle, and M. K. Izard, eds., Plenum Press, New York, pp. 251–260.

Popper, K. R., 1957, *The Poverty of Historicism,* Routledge and Kegan Paul, London.

Proctor-Gray, E., 1984, Dietary ecology of the coppery brushtail possum, green ringtail possum and Lumholtz's tree-kangaroo in North Queensland, in: *Possums and Gliders,* A. Smith and I. Hume, eds., Surrey Beatty & Sons Pty Limited, Chipping Norton, Australia, pp. 129–136.

Rand, A. S., 1937, Some original observations on the habits of *Dactylopsila trivirgata* Gray, *Am. Mus. Novit.* **957:** 1–7.

Rasmussen, D. T., 1990, Primate origins: Lessons from a Neotropical marsupial, *Am. J. Primatol.* **22:** 263–277.

Rasmussen, D. T., 2001, Primate origins, in: *The Primate Fossil Record,* W. C. Hartwig, ed., Cambridge University Press, London, pp. 5–10.

Renfree, M. B., Russell, E. M., and Wooller, R. D., 1984, Reproduction and life history of the honey possum, *Tarsipes rostratus,* in: *Possums and Gliders,* A. Smith and I. Hume, eds., Surrey Beatty & Sons Pty Limited, Chipping Norton, Australia, pp. 359–373.

Righetti, J., 1996, A comparison of the behavioural mechanisms of competition in shrews (Soricidae) and small dasyurid marsupials (Dasyuridae), in: *Comparison of Marsupial and Placental Behavior,* D. B. Croft, ed., Filander Verlag, Fürth.

Ride, W. D. L., 1964, A review of Australian fossil marsupials, *J. R. Soc. Wes. Aus.* **47:** 97–131.

Ride, W. D. L., 1970, *A Guide to the Native Mammals of Australia,* Oxford University Press, Melbourne.

Rose, K. D., 1975, The Carpolestidae: Early Tertiary primates from North America, *Bull. Museum Comp. Zool.* **147:** 1–74.

Russell, R., 1983, Feathertail glider, in: *Complete Book of Australian Mammals,* R. Strahan, ed., Angus and Robertson Publishers, Sydney, pp. 170–171.

Sargis, E. J., 2002, Primate origins nailed, *Science* **298:** 1564–1565.

Simons, E. L., and Rasmussen, D. T., 1995, A whole new world of ancestors: Eocene anthropoideans from Africa, *Evol. Anthropol.* **3:** 128–139.

Simons, E. L., and Rasmussen, D. T., 1996, The skull of *Catopithecus browni,* an early Tertiary catarrhine, *Am. J. Phys. Anthropol.* **100:** 261–292.

Simpson, G. G., 1933, The "plagiaulacoid" type of mammalian dentition, *J. Mammal.* **14:** 97–107.

Smith, A. P., 1982a, Diet and feeding strategies of the marsupial sugar glider in temperate Australia, *J. Animal Ecol.* **51:** 149–166.

Smith, A. P., 1982b, Is the striped possum an arboreal anteater? *Australian Mammalogy* **5:** 229–235.

Smith, A. P., 1984a, Ringtails, pygmy possums, gliders, in: *The Encyclopedia of Mammals,* D. D. MacDonald, ed., Equinox Ltd., Oxford, England, pp. 856–861.

Smith, A. P., 1984b, The species of living possums and gliders, in: *Possums and Gliders*, A. Smith and I. Hume, eds., Surrey Beatty & Sons Pty Limited, Chipping Norton, Australia, pp. xiii–xv.

Smith, A. P., 1984c, Demographic consequences of reproduction, dispersal and social interaction in a population of Leadbeater's possum (*Gymnobelideus leadbeateri*), in: *Possums and Gliders*, A. Smith and I. Hume, eds., Surrey Beatty & Sons Pty Limited, Chipping Norton, Australia, pp. 359–373.

Smith, A. P., 1984d, Diet of Leadbeater's possum, *Aust. Wildlife Res.* **11**: 265–273.

Smith, A. P., 1986, Stomach contents of the long-tailed pygmy-possum *Cercartetus caudatus* (Marsupialia: Burramyidae), *Aus. Mammal.* **9**: 135–137

Smith, A. P., and Ganzhorn, J. U., 1996, Convergence in community structure and dietary adaptation in Australian possums and gliders and Malagasy lemurs, *Aust. J. Ecol.* **21**: 31–46.

Statham, H. L., 1984, The diet of *Trichosurus vulpecula* (Kerr) in four Tasmanian forest locations, in: *Possums and Gliders*, A. Smith and I. Hume, eds., Surrey Beatty & Sons Pty Limited, Chipping Norton, Australia, pp. 213–219.

Steiner, K. E., 1981, Nectarivory and potential pollination by a Neotropical marsupial, *Ann. Mo. Bot. Gard.* **68**: 505–513.

Sussman, R. W., 1991, Primate origins and the evolution of angiosperms, *Am. J. Primatol.* **23**: 209–223.

Sussman, R. W., 1995, How primates invented the rainforest and vice versa, in: *Creatures of the Dark: The Nocturnal Prosimians*, L. Alterman, G. A. Doyle, and M. K. Izard, eds., Plenum Press, New York, pp. 1–10.

Sussman, R. W., 1999, *Primate Ecology and Social Structure, vol. 1: Lorises, Lemurs and Tarsiers*, Pearson Custom Publishing, Needham Heights, MA, 284 p.

Sussman, R. W., and Raven, P. H., 1978, Pollination by lemurs and marsupials: An archaic coevolutionary system, *Science* **200**: 731–736.

Szalay, F. S., 1993, Metatherian taxon phylogeny: Evidence and interpretation from the cranioskeletal system, in: *Mammal Phylogeny: Mesozoic Differentiation, Multituberculates, Monotremes, Early Therians, and Marsupials*, F. S. Szalay, M. J. Novacek, and M. C. McKenna, eds., Springer-Verlag, New York, pp. 216–242.

Szalay, F. S., 1994, *Evolutionary History of the Marsupials and an Analysis of Osteological Characters*, Cambridge University Press, Cambridge, England, 481 pp.

Thompson, J. A., and Owen, W. H., 1964, A field study of the Australian ringtail possum *Pseudocheirus peregrinus* (Marsupialia: Phalangeridae), *Ecol. Monogr.* **34**: 27–52.

Turner, V. B., 1984, *Banksia* pollen as a source of protein in the diet of two Australian marsupials, *Cercarcetus nanus* and *Tarsipes rostratus*, *Oikos* **43**: 53–61.

Ward, S. J., 1990, Life history of the feathertail glider, *Acrobates pygmaeus* (Acrobatidae: Marsupialia) in south-eastern Australia, *Aust. J. Zool.* **38:** 503–17.

Ward, S. J., 1992, Life history of the little pygmy-possum, *Cercartetus lepidus*, in the Big Desert, Victoria, *Aust. J. Zool.* **40:** 43–55.

Wiens, D., Renfree, M., and Wooller, R. O., 1979, Pollen loads of honey possums (*Tarsipes spencerae*) and nonflying mammal pollination in southwestern Australia, *Ann. Mo. Bot. Gard.* **66:** 830–838.

Winter, J. W., 1983, Lemuroid ringtail possum, in: *Complete Book of Australian Mammals*, R. Strahan, ed., Angus and Robertson Publishers, Sydney, p. 133.

Winter, J. W., 1996, Australian possums and Madagascan lemurs: Behavioural comparison of ecological equivalents, in: *Comparison of Marsupial and Placental Behavior*, D. B. Croft, ed., Filander Verlag, Fürth.

Winter, J. W., and Atherton, R. G., 1984, Social group size in North Queensland ringtail possums of the genera *Pseudocheirus* and *Hemibelideus*, in: *Possums and Gliders*, A. Smith and I. Hume, eds., Surrey Beatty & Sons Pty Limited, Chipping Norton, Australia, pp. 311–319.

Wooller, R. D., Russell, E. M., and Renfree, M. B., 1984, Honey possums and their food plants, in: *Possums and Gliders*, A. Smith and I. Hume, eds., Surrey Beatty & Sons Pty Limited, Chipping Norton, Australia, pp. 439–443.

Wright, P. C., and Martin, L. B., 1995, Predation, pollination and torpor in two nocturnal prosimians: *Cheirogaleus major* and *Microcebus rufus* in the rain forest of Madagascar, in: *Creatures of the Dark: The Nocturnal Prosimians*, L. Alterman, G. A. Doyle, and M. K. Izard, eds., Plenum Press, New York, pp. 45–60.

CHAPTER TWENTY-THREE

Perspectives on Primate Color Vision

Peter W. Lucas, Nathaniel J. Dominy,
Daniel Osorio, Wanda Peterson-Pereira,
Pablo Riba-Hernandez,
Silvia Solis-Madrigal, Kathryn E. Stoner,
and Nayuta Yamashita

INTRODUCTION

This chapter is an attempt to sketch the evolution of the primate lineage in the broadest possible ecological terms. Inevitably, it is based partly on the ideas of others. At minimum, any novelty that we can offer probably derives from insights gleaned from fieldwork on living primates. From that field backdrop, we extrapolate to other epochs. At maximum, this account is likely to be thought to be completely beyond the bounds of acceptable speculation. For this excess, however, we are unrepentant and hope that at least some of

Peter W. Lucas • Department of Anthropology, George Washington University, 2110 G Street NW, Washington DC 20037; **Nathaniel J. Dominy** • Department of Anthropology, University of California, 1156 High Street, Santa Cruz, CA 95064; **Daniel Osorio** • School of Biological Sciences, University of Sussex, Brighton, BN1 9QG. UK; **Wanda Peterson-Pereira, Pablo Riba-Hernandez, and Silvia Solis-Madrigal** • Escuela de Biologia, Universidad de Costa Rica, San Pedro, San José, Costa Rica; **Kathryn E. Stoner** • Centro de Investigaciones en Ecosistemas, Universidad Nacional Autónoma de México, Apartado Postal 27-3 (Xangari), Morelia, Michoacán 58089, México; **Nayuta Yamashita** • Department of Cell and Neurobiology, BMT 408, Keck School of Medicine, University of Southern California, 1333 San Pablo Street, Los Angeles, CA 90089-9112

the ideas possess credibility. Our general focus is on correlating the evolution of primates with that of angiosperms, interpreting this association via diet. We take our lead here from various sources, but the influences of Cartmill (1972) and Sussman (1991) have been very strong in this regard, as will be obvious. Our explanations are simplistic and based on a likely sequence of sensory cues that animals use to find foods, starting from color vision, which acts most efficiently at longer distances, 5–25 m or so (Janson and Di Bitetti, 1997), to those acting in or around the mouth—like texture and taste (Dominy et al., 2001).

VISION

In Color

Color is an important physical characteristic that can be used as a cue by primates to procure and ingest foods. Undoubtedly, the major importance of color vision in primates is in relation to angiosperms because these plants use color extensively in a reproductive context to attract pollinators and seed dispersers. Lemurs are effective pollinators (Birkinshaw and Colquhoun, 1998; Kress et al., 1994; Nilsson et al., 1993), which Sussman and Raven (1978) have suggested might have great significance for early primate evolution. However, color does not seem to play an important role in the lemur pollination syndromes that these authors describe. Some primates are also effective seed dispersers (Chapman, 1995), which connects to the proposal that the main reason for good color vision in primates is to find fruits (Allen, 1879; Mollon, 1989; Polyak, 1957), even though seed dispersal was not originally associated with this idea. However, recent evidence suggests that color vision that allows the discrimination of red hues largely evolved in primates in response to a search for young new red leaves (Lucas et al. 2003a; Stoner et al. 2005).

Analysis of the colors of fruits in primate diets indicates that yellow, orange-brown, and red fruits are common (Dominy, 2004a). As described later, the discrimination of these hues from a monotonous green background of mature foliage requires a type of color vision found, at least partially, in many primate species, called trichromacy. The wide variety of colors displayed by flowers and ripe fruit in tropical regions has not yet been fully explained in terms of the color preferences of presumed target pollinators or seed dispersers. In contrast, at least from the perspective of a reader living in temperate zones, foliage appears a dark uniform green and holds little interest for the color vision debate. However, depending on the time of year, a visitor to tropical rainforests often sees isolated flashes of red produced, not by flowers and fruits, but by young leaves. In fact, between 20% and 60% of forest plants in the tropics flush leaves with reddish tints (Dominy et al., 2002). The reasons are uncertain, but a major one appears to be crypsis (camouflage) from potential herbivores. Most herbivores appear to lack receptors for detecting the long wavelengths that dominate the reflectance spectra (which determines the color) of these leaves. If fruits are first in importance in the diets of most primates, then leaves run a close second for many larger species and dominate in some. Lucas et al. (1998, 2003a) have suggested that a reliance on leaves, at least as a fallback when fruits are unavailable, influenced the evolution of color vision in primates.

Color vision allows animals to discriminate light spectra by comparing responses of two or more photoreceptors with differing spectral sensitivities (Jacobs, 1993; Surridge et al., 2003). Spectral sensitivity is defined here as the relative probability of a photon of a given wavelength incident on the eye producing a response in the photoreceptor cell. The number of spectral receptors limits the dimensionality of color vision: two spectral types allow dichromacy with two primaries needed to match any spectrum; three spectral types allow trichromacy with three primaries being required, and so forth. As well as number of receptors, the ability to discriminate light spectra is affected by neural mechanisms of vision, and by spectral tuning of the photo receptors.

Cones are active only in daylight and the number of spectral types varies between taxonomic groups. In birds, for example, there are four types giving relatively even spectral coverage of 350–600 nm, and good color discrimination. By comparison most diurnal mammals have only two types of cone: a short-wavelength cone containing a pigment (opsin) with a peak absorbance in primates of about 420–440 nm and a long-wavelength cone with a variable peak absorbance of about 520–560 nm (Jacobs, 1993). Two groups of anthropoid primate—the catarrhines and the howler monkeys (genus *Alouatta*)—possess three cone types. The third cone has been produced by duplication of the gene coding for the opsin pigment in the long-wavelength cone. This occurred independently in *Alouatta* and in catarrhines, but the three cone types in both have remarkably consistent peak absorbances of around 430, 530, and 560 nm, respectively (Hunt et al., 1998).

The brain of a mammal with two cone types can compare the relative strengths of their outputs to give a "chromatic" signal. Color vision that is based on comparison of just two receptor types is termed dichromatic. Following human color psychology, these chromatic signals lie on what we call the "blue-yellow" perceptual axis (e.g., Regan et al., 2001). A dichromatic mammal will confuse many spectra that we can discriminate, especially those we recognize as reds, yellows, and greens. This is because, with three cone types, we can make two independent sets of comparisons between three outputs to give two sets of chromatic signals, and hence trichromatic color vision.

Many primate species with only two opsin gene loci have polymorphisms of the long-wavelength opsin (of still generally unknown gene frequencies) that, lying on the X chromosome, confer trichromacy on some females. These species include all New World monkey genera to be tested except *Alouatta*, which is fully trichromatic (Jacobs et al., 1996), and *Aotus*, which has only one cone type and is colorblind (Jacobs et al., 1993). Recently, the polymorphism has also been found among several strepsirrhines, including *Propithecus verreauxi* and *Varecia variegata* (Jacobs et al., 2002; Tan and Li, 1999).

Measurement of color is now beginning in primate field studies (Smith et al., 2003 a,b), and its analysis is developing rapidly (Kelber et al., 2003; Osorio et al., 2004), indicating its potential value to primates. Importantly for this chapter, mature leaves vary in color almost entirely in yellow-blueness, not in red-greenness (Dominy and Lucas, 2001; Osorio and Vorobyev, 1996; Regan et al., 2001; Sumner and Mollon, 2000) while both very young leaves and ripening fruits can show great differences in their red-green signal from mature foliage (Dominy and Lucas, 2001; Sumner and Mollon, 2000). In fruits though, this varies greatly depending on the final hue. Compared to mature foliage, some fruits stay green when ripe (and primates eat a surprisingly large amount of green fruit—Dominy, 2004a), while others ripen with change in the blue-yellow signal, the red-green signal, or with both.

In Black and White

Nocturnal vision utilizes rod photoreceptors that are sensitive to low-light intensities. In all mammals, these cells have only one spectral tuning (Ahnelt and Kolb, 2000). Since cones do not operate at low-light levels, no comparators are available with the result that rod vision is colorblind. Achromatic vision provided by rods or summed (rather than compared) red and green cone signals is used for many visual tasks including perception of motion, stereo-depth, and (for cones) fine spatial detail, which are all colorless (Livingstone and Hubel, 1988). Like most mammals, primates have both rods and cones in their retina, which implies activity in both night and day. Some primates are in fact capable of varying their activity patterns, day and night, to environmental conditions (e.g., the mongoose lemur, *Eulemur mongoz*—Curtis et al., 1999). A cone-rod balance is not universal. Some mammals have cone-dominated retinas, like squirrels and tupaiid tree shrews (Jacobs, 1993), while others like the pen-tailed tree shrew (*Ptilocercus lowii*) possess a retina that is almost entirely filled with rods (Emmons, 2000). The importance of light levels to all uses of color vision has been recognized in some work now being reported (Yamashita et al., 2005).

TEXTURE

Texture is another physical characteristic that primates can use to identify food items for ingestion. Texture is a term that refers to the physical hand or mouth "feel" of objects and provides important information about the potential edibility of plant parts. Most of the food textures that primates encounter are actually mechanical defenses erected by plants to deter herbivores. These defenses are of two types (Lucas, 2004: Lucas et al., 2000), aiming either to prevent cracks in structures from starting (crack initiation) or to prevent small cracks from growing (crack propagation). A primate attempting to detach a plant part could face one of two possible limits preventing its success: either it could not generate sufficient stress—a limit that relates to problems with crack initiation or, it could run out of displacement—a criterion relevant to crack propagation. Defenses that rely on these limits to protect the plant are called stress-limited and displacement-limited defenses, respectively. These limits are valuable concepts for mechanical analyses in any context, devised by Ashby (1999) to help engineers select materials for specific applications. In a biological context, we can recognize them as forming the basis of plant defenses against herbivore damage.

Lucas et al. (2000) gave an example of these two defenses that is appropriate in the context of this chapter and so we repeat it here. Consider attempting to fracture a horizontal branch on a small tree. Grasping it with your arm, you may bend it downwards with the intention of fracturing and detaching it purely with this downward motion. However, there are two possible limits that could prevent you from succeeding. Firstly, you could bend it through 90°, right against the trunk without success. This is *displacement limitation*: there was sufficient stress but you ran out of displacement. Alternatively, the branch may resist with little deflection, but not crack because you cannot generate sufficient stress. This is *stress limitation*: the displacement was sufficient displacement, but you lacked the force to generate the stress.

Three major material properties of plants largely determine the effectiveness of these defenses. *Young's modulus* is the ratio of stress to strain in which stress is the concentration of a force (obtained by dividing a force by the surface area over which it acts) and strain is the intensity of the force's effect on dimensional change (measured as the ratio of linear displacement in the direction of the force to the size of the structure). Young's modulus is an elastic measure; it holds only for small elastic strains and when there is a proportionate change in strain with stress. *Toughness* is the ability to resist crack growth and defined as the energy consumed in growing a crack of given area. The boundary between distortion or deformation and fracture (i.e., crack initiation) is not marked by any particular property and depends on the size of object that is breaking and the manner by which it is loaded. The *yield stress* is the boundary between elastic (recoverable) and plastic (permanent) strain in any material that exhibits it.

As shown elsewhere (Ashby, 1999, Lucas et al., 2000), stress-limited defenses are typified either by a high-yield stress or by a high value of K_{IC}, which is roughly equal to the square root of the product of toughness to Young's modulus. These "hard" defenses contrast with "tough" displacement-limited defenses, which are controlled by the square root of the ratio of toughness to Young's modulus.

The structure of plants designed in these two ways look very different. Stress-limited defenses, which try to prevent cracks from initiating, tend to involve hard outer surfaces. Dense outer layers raise Young's modulus, particularly in bending, which is the most likely way in which high stresses are realized during ingestion. Hard, sharp features also act to deter herbivores due to their high-yield strengths. Familiar contemporary examples are spines, thorns, and stiff hairs—the latter being common in plant parts that primates feed on. These structures often have amorphous silica incorporated in them because this is harder than plant cell wall material. Some hard tissues are, however, defended by their very thick cell walls—many seed shells are like this. In contrast to this, displacement-limited defenses, that try to prevent cracks growing, are typified by a woody structure. Elongated woody cells have a cellular arrangement whereby the cell wall buckles plastically into the cellular lumen absorbing large amounts of energy, thus impeding crack growth. This greatly increases the toughness of tissue containing such cells while lowering the Young's modulus slightly. Rather than be disposed at the plant surface, wood is disposed throughout the plant interior wherever possible.

These two defenses are actually incompatible in plants, the simplest example being the mechanism of toughening (Lucas et al., 2000). The toughness of woody cells depends on the arrangement of the cell wall, but also on the presence of a cell lumen into which the wall must buckle. In contrast, the exact arrangement of the cell wall is immaterial to its hardness, which is maximized by infilling virtually the whole cell with cell wall, as in seed shells. Very dense seed shells can have cell walls organized exactly as in wood, but their toughness is very much lower.

Stress-limited defenses are actually antiingestion defenses that will evolve whenever major herbivores are large enough to threaten the survival of a plant by a single bite. Just one fracture is then life threatening for the plant. Savannah grasses are a familiar contemporary example. These small plants have long thin parallel-veined leaves reinforced by thick-walled sclerenchyma barely hidden under superficial ridges. Resistance to the tensile element of the bite of large ungulates is optimized by the sclerenchyma being aligned parallel to the direction of the bite force, maximizing the stresses at which the leaves start to fracture. This stress-limited design appears to explain the parallel venation of the leaves of many plants, including most monocotyledonous angiosperms. They often also contain large amounts of superficially located silica, consistent with the same defensive strategy. Such grasses must have been subject to predation by relatively large herbivores for much of their evolutionary history. Recent extinctions can, however, confuse the contemporary picture because plants evolve slower than animals. The survival of many stress-limited defenses in the neotropics, such as spines, thorns, and hard seeds, has been associated with megafauna, like the gomphotheres of the neotropics, which died out about 10,000 years ago (Janzen and Martin, 1982).

The spacing of thorns and spines seen on many stress-limited plants, such as the acacias of the African savanna, is probably linked to herbivore size—the larger the spacing, the larger the predator. The small siliceous hairs found on some angiosperm leaves can be interpreted as miniature spines that probably represent a stress-limited defense against some very small herbivores, probably cell suckers, which are often significant causes of leaf damage (Leigh, 1999). In contrast, displacement-limited defenses are antifragmentation (antimastication) defenses that typify plants attacked by very small "chewing" herbivores. It is extremely difficult to stop all damage at small scale. In particular, leaves are very difficult to protect because photosynthesis is compromised by any structures positioned in the light path leading to chloroplasts. The major predators of most angiosperms are "chewing" invertebrates (Leigh, 1999). In consequence, displacement-limited defenses predominate in most angiosperms.

TASTE

The vast array of chemical and mechanical defenses in plants shows evidence of economy when viewed at a fine enough scale (Choong, 1996), meaning tissues with mechanical defenses (e.g., thick cell walls) tend not to have chemical defenses and vice versa. As described in an earlier section, "hard" stress-limited defenses are designed to stop cracks initiating. If cracks do not start, then cells are not opened. Accordingly, plants or plant parts designed this way would be predicted to have few chemical defenses. Many monocotyledonous plants are an example of this. Even though silica and cell wall materials are relatively cheap components for a plant, "hard" defenses have a major drawback: they are (generally) dead. On the other hand, "tough" displacement-limited defenses rely not on resisting crack initiation, but on opposing crack propagation. So cells will inevitably be opened, suggesting immediately that chemical defenses have a role as herbivore deterrents. The major problem with displacement-limited defense is that they are slow to develop, leaving young tissue vulnerable. The greatest concentration of chemicals is found in young tissues, an example being tannins and phenolic compounds in young leaves (Coley, 1983).

The amount of cell wall in a food is called by nutritionists its fiber content. Fiber is a key element of food quality for a primate because cell walls either have to be fractured (by the teeth or by gut peristalsis) to expose cell contents or else microorganisms need to be housed in the gut so as to achieve this enzymatically. The efficiency of digestion depends on fiber. Unfortunately, cell walls are largely colorless, odorless, and tasteless, so the major means of sensing them must be via their texture. As shown by Lucas et al. (2000), the volume fraction of a tissue that is occupied by cell wall is not proportional to its toughness and this must displace foraging away from the optimum.

EARLY PRIMATE EVOLUTION

Primate origins stretch back to the Cretaceous, perhaps over 80 MYA (million years ago) (Martin, 1993, 2000; Tavaré et al., 2002). The date is just a prediction, but one that happens to coincide, very roughly, with the appearance of a significant number of angiosperms in the fossil record. It is not very clear what early flowering plants looked like. However, evidence from exceptionally well-preserved sites suggest that early dicotyledonous angiosperms (dicots for short) were predominantly herbaceous, clumped together with large (monocotyledonous) palms in windless humid patches, perhaps around streams, and interspersed between large areas dominated by ferns, various gymnosperms, and cycads (Wing et al., 1993). Evaporation from the streams might have been responsible for the high humidity. It is increasingly likely that angiosperm evolution was linked to an aquatic or semiaquatic habitat and such a relatively still and damp environment was surely required in order that insect pollination (which typified even the very earliest dicots—Crane et al., 1995) would be favored over the accidental effect of wind.

Throughout most of the Mesozoic era, herbivorous dinosaurs were the major consumers of plants. We believe that, faced with such megaherbivore pressure, stress-limited defenses were developed. Being small relative to these herbivores, food plants of dinosaurs needed extensive external mechanical protection because a single bite represented potentially life-threatening loss. In order to deter these herbivores, many of these plants probably had widely spaced sharp thorns and spines and any "woody" tissue that they possessed would be dense and positioned as externally as possible in the plant in order to prevent penetration. There would, in any case, not have been much point in attempting to stop a dinosaur feeding by chemical means because the residence time in the mouth of, largely unbroken, plant parts was too short. This assumes, though, that plant chemical defenses are not a furtive attempt to poison animals without their sensing it, but are intended to provide clear clues via taste receptors that poisoning would result if feeding continued. We suspect that, as in living birds, dinosaurs had few taste buds. Poisonous chemicals with aversive tastes would have been no deterrent to dinosaurs because most dinosaurs did not open cells by chewing.

We suggest that the ancestor of primates was a small dichromatic predator that lived in angiosperm patches immune from attack by dinosaurs because of the small size of the plants. They hunted for moving insects at night, viewed against smooth tree trunks. Even if insects were hunted against this background in the day, color vision would have no particular value because there would not be sufficient chromatic contrast. However, we think that they may have also searched foliage for static insects in daylight—an activity that could well have been aided by their blue-yellow color vision. This suggested cathemeral activity pattern is simply a merger of the foraging behavior of nocturnal and diurnal tree shrews, as reported by Emmons (2000). Later, as angiosperms diversified in the Later Cretaceous, early primates moved onto larger plants. Herbivorous dinosaurs of the Later Cretaceous were smaller, down even to 30 kg (Norman and Weishampel, 1991) and probably consumed such angiosperms (Crane et al., 1995; Norman and Weishampel, 1991). These newly targeted angiosperms defended themselves against these dinosaurs by developing tightly spaced thorns and spines. Very small primates developed tactile pads (backed by nails) to sense and avoid these sharp obstructions, coupling this with leaping behavior to jump between them. This is the most plausible explanation for leaping behavior in early primates (Dagosto, this volume; but see Lemelin and Schmitt, this volume for an alternative view).

Our scenario is congruent in other senses with the suggestion that primates moved on fine branches (Cartmill, 1972; Sussman, 1991). Movement of primates on plants that were stress-limited means that the Young's moduli of their branches would be likely to be very high and able, therefore, to support their weight much better than angiosperms that were not under attack from such large predators. This behavior would further assist early primates in avoiding predation by allowing them to escape from potential predators.

Angiosperms were probably not very common even towards the end of the Cretaceous. After that point in time, however, they started to diversify rapidly. A switch to a warmer, wetter climate in the Early Tertiary may have triggered this (Eriksson et al., 2000). The loss of megaherbivores, replaced by invertebrates, triggered a switch to "woody" displacement-limited defenses that better resist chewing invertebrates by obstructing crack growth. Since cells were inevitably opened, there would have been a rapid diversification of chemical defenses. As angiosperms began to dominate the land early in the Tertiary, multispecies, multilayered, forest canopies began to appear, becoming most diverse at the end of the Paleocene (Eriksson et al., 2000). Specialized seed dispersal syndromes probably date back to this period. In fact, dispersal by larger vertebrates may produce such clumping that tree species adapt so that seedlings can prosper under the crowns of their parents (Dominy and Duncan, 2005). Howe (1989) has argued that the dispersal activities of vertebrates were essential to the evolution of angiosperm diversity. Dispersal by the latter produced clumped tree species, so much so that seedlings could even prosper under the crowns of their parent trees (Dominy, 2001; Dominy and Duncan, 2005). This Late Paleocene world offered expanded habitats for primate species, which may have expanded their fruit-eating activities, initially stealing fruits from targeted dispersers during the day. Since larger vertebrates tend to eat larger fruits with larger seeds (Howe, 1989), the heavier anthropoid primates that evolved through the Later Eocene and Oligocene were probably feeding on highly clumped resources. This may have put pressure on these primates to aggregate into social groups, simply because their target plant species were themselves patchily distributed. This tendency may have favored opsin polymorphism in anthropoids (and, at some as yet undetermined point, in some larger prosimians) with trichromatic females being favored as group members for identifying fruit patches, as has been suggested by Mollon (1989).

We believe that routine trichromacy in catarrhines probably only dates back to the start of the Miocene and considerably later in the ancestral howler monkey. Dominy and Lucas (2001) and Lucas et al. (2003a) have argued that these independent gene duplications are likely to have been associated with the need for primates to rely on leaves as a fallback food when fruits were unavailable.

FUTURE RESEARCH

The consequences for any primate foraging with or without the red-green signal depend very much on the relationship of color to the actual quality of their preferred food items. The integration of research in color vision in primates with sensory modalities, such as smell, taste, and texture is clearly vital, some of which is now proceeding (e.g., Dominy, 2004b; Riba-Hernandez et al., 2004). We hope that field technology (Lucas et al.,

2003b,c) will soon improve to the degree that it will be routinely possible to obtain quantified estimates of what primates sense and what they use those senses for.

REFERENCES

Ahnelt, P. K., and Kolb, H., 2000, The mammalian photoreceptor mosaic-adaptive design, *Prog. Retin. Eye Res.* **19**: 711–777.

Allen, G., 1879, *The Colour-Sense: Its Origin and Development.* Trübner & Co., London.

Ashby, M. F., 1999, *Materials Selection in Mechanical Design*, 2nd edn., Butterworth-Heinemann, Oxford.

Birkinshaw, C. R., and Colquhoun, I. C., 1998, Pollination of *Ravenala madagascariensis* and *Parkia madagascariensis* by *Eulemur macaco* in Madagascar, *Folia Primatol.* **69**: 252–259.

Cartmill, M., 1972, Arboreal adaptations and the origin of the order Primates, in: *The Functional and Evolutionary Biology of Primates*, R. Tuttle, ed., Aldine, Chicago, pp. 97–122.

Chapman, C. A., 1995, Primate seed dispersal: Coevolution and conservation implications, *Evol. Anthropol.* **4**: 74–82.

Choong, M. F., 1996, What makes a leaf tough and how this affects the pattern of *Castanopsis fissa* leaf consumption by caterpillars, *Funct. Ecol.* **10**: 668–674.

Coley, P. D., 1983, Herbivory and defensive characteristics of tree species in a lowland tropical rainforest, *Ecol. Monogr.* **53**: 209–233.

Corlett, R. T., and Lucas, P. W., 1990, Alternative seed-handling strategies in primates: Seed-spitting by long-tailed macaques, *Oecologia* **82**: 166–171.

Crane, P. R., Friis, E. M., and Pederson, K. R., 1995, The origin and early diversification of angiosperms, *Nature* **363**: 342–344.

Curtis, D. J., Zaramody, A., and Martin, R. D., 1999, Cathemerality in the mongoose lemur, *Eulemur mongoz*, *Am. J. Primatol.* **47**: 279–298.

Dominy, N. J., 2001, *Trichromacy and the Ecology of Food Selection in Four West African Primates*, Ph.D. Thesis, University of Hong Kong.

Dominy, N. J., 2004a, Color as an indicator of food quality to anthropoid primates: Ecological evidence and an evolutionary scenario, in: *Anthropoid Origins: New Visions*, C. Ross and R. F. Kay, eds., Kluwer, New York, pp. 615–644.

Dominy, N. J., 2004b, Fruits, fingers, and fermentation: The sensory cues available to foraging primates, *Int. Comp. Biol.* **44**: 295–303.

Dominy, N. J., and Duncan, B. W., 2005, Seed-spitting primates and the conservation and dispersion of large-seeded trees, *Int. J. Primatol.* **26**: 631–649.

Dominy, N. J., and Lucas, P. W., 2001, The ecological value of trichromatic vision to primates, *Nature* **410**: 363–366.

Dominy, N. J., Lucas, P. W., Osorio, D., and Yamashita, N., 2001, The sensory ecology of primate food perception, *Evol. Anthropol.* **10**: 171–186.

Dominy, N. J., Lucas, P. W., Ramsden, L., Riba-Hernandez, P., Stoner, K. E., and Turner, I. M., 2002, Why are young leaves red? *Oikos* **98**: 163–176.

Emmons, L. H., 2000, *Tupai: A Field Study of Bornean Treeshrews*, University of California Press, Berkeley.

Eriksson, O., Friis, E. M., and Löfgren P., 2000, Seed size, fruit size, and dispersal systems in angiosperms from the early Cretaceous to the late Tertiary, *Am. Nat.* **156**: 47–58.

Howe, H. F., 1989, Scatter- and clump dispersal and seedling demography: Hypothesis and implications, *Oecologia* **79**: 417–426.

Hunt, D. M., Dulai, K. S., Cowing, J. A., Juillot, C., Mollon, J. D., Bowmaker, J. K., et al., 1998, Molecular evolution of trichromacy in primates, *Vision Res.* **38**: 3299–3306.

Jacobs, G. H., 1993, The distribution and nature of colour vision among the mammals, *Biol. Rev.* **68**: 413–471.

Jacobs, G. H., Deegan, J. F., Neitz, J., Crognale, M. A., and Neitz, M., 1993, Photopigments and color vision in the nocturnal monkey, *Aotus. Vis. Res.* **33**: 1773–1783.

Jacobs, G. H., Neitz, M., Deegan, J. F., and Neitz, J., 1996, Trichromatic color vision in New World monkeys, *Nature* **382**: 156–158.

Jacobs, G. H., Deegan II, J. F., Tan, Y., and Li, W.-H., 2002, Opsin gene and photopigment polymorhism in a prosimian primate, *Vis. Res.* **42**: 11–18.

Janson, C. H., and Di Bitetti, M. S., 1997, Experimental analysis of food detection in capuchin monkeys: Effects of distance, travel speed, and resource size, *Behav. Ecol. Sociobiol.* **41**: 17–24.

Janzen, D. H., and Martin, P. S., 1982, Neotropical anachronisms: The fruits the Gomphotheres ate, *Science* **215**: 19–27.

Kelber, A., Vorobyev, M., and Osorio, D., 2003, Animal colour vision—Behavioural tests and physiological concepts, *Biol. Rev.* **78**: 81–118.

Kress, W. J., Schatz, G. E., Andrianifahanana, M., and Morland, H. S., 1994, Pollination of *Ravenala madagascariensis* (Strelitziaceae) by lemurs in Madagascar: Evidence for an archaic coevolutionary system? *Am. J. Bot.* **81**: 542–551.

Kursar, T., and Coley, P. D., 1992, Delayed greening in tropical leaves: An antiherbivore defense? *Biotropica* **24**: 256–262.

Leigh, E. G., Jr., 1999, *Tropical Forest Ecology*, Oxford University Press, New York.

Livingstone, M., and Hubel, D., 1988, Segregation of form, color, movement, and depth: Anatomy, physiology, and perception, *Science* **240**: 740–749.

Lucas, P. W., 2004, *Dental Functional Morphology*, Cambridge University Press, Cambridge.

Lucas, P. W., Darvell, B. W., Lee, P. K. D., Yuen, T. D. B., and Choong, M. F., 1998, Colour cues for leaf food selection by long-tailed macaques (*Macaca fascicularis*) with a new suggestion for the evolution of trichromatic colour vision, *Folia Primatol.* **69:** 139–152.

Lucas, P. W., Turner, I. M., Dominy, N. J., and Yamashita, N., 2000, Mechanical defences to herbivory, *Ann. Bot.* **86:** 913–920.

Lucas, P. W., Beta, T., Darvell, B. W., Dominy, N. J., Essackjee, H. C., Lee, P. K. D., et al., 2001, Field kit to characterize physical, chemical and spatial aspects of potential foods of primates, *Folia Primatol.* **72:** 11–15.

Lucas, P. W., Dominy, N. J., Riba-Hernandez, P., Stoner, K. E., Yamashita, N., Loría-Calderón, L., et al., 2003a, Adaptive function of primate trichromatic vision, *Evolution* **57:** 2636–2643.

Lucas, P. W., Osorio, D., Yamashita, N., Prinz, J. F., Dominy, N. J., and Darvell, B. W., 2003b, Dietary analysis I: Physics, in: *Field and Laboratory Methods in Primatology*, J. Setchell and D. Curtis, (eds., Cambridge University Press, Cambridge, pp. 184–198.

Lucas, P. W., Corlett, R. T., Dominy, N. J., Essackjee, H. C., Riba-Hernandez, P., Stoner, K. E. and Yamashita, N., 2003c, Dietary analysis II: Chemistry, in: *Field and Laboratory Methods in Primatology*, J. Setchell and D. Curtis, eds., Cambridge University Press, Cambridge, pp. 199–213.

Martin, R. D., 1993, Primate origins: Plugging the gaps, *Nature* **363:** 222–334.

Martin, R. D., 2000, Origin, diversity and relationships of lemurs, *Int. J. Primatol.* **21:** 1021–1049.

Mollon, J. D., 1989, "Tho' she kneel'd in that place where they grew..." The uses and origins of primate colour vision, *J. Exp. Biol.* **146:** 21–38.

Nilsson, L. A., Rabakonandrianina, E., Pettersson, B., and Gruenmeier, R., 1993, Lemur pollination in the Malagasy rainforest liana *Strongylodon craveniae* (Leguminosae), *Evol. Trend. Plant.* **7:** 49–56.

Norman, D. B., and Weishampel, D. B., 1991, Feeding mechanisms in some small herbivorous dinosaurs: Processes and patterns, in: *Biomechanics in Evolution*, J. M. V. Rayner and R. J. Wootton, eds., Cambridge University Press, Cambridge, pp. 161–181.

Osorio, D., and Vorobyev, M., 1996, Colour-vision as an adaptation to frugivory in primates, *P. Roy. Soc. Lond.* **B263:** 593–599.

Osorio, D., Smith, A. C., Vorobyev, M., and Buchanan-Smith, H. M., 2004, Detection of fruit and the selection of primate visual pigments for color vision, *Am. Nat.* **164:** 696–708.

Polyak, S., 1957, *The Vertebrate Visual System*, University of Chicago Press, Chicago.

Regan, B. C., Julliot, C., Simmen, B., Vienot, F., Charles-Dominique, P., and Mollon, J. D., 2001, Fruits, foliage and the evolution of primate colour vision, *Philos. T. Roy. Soc. Lond.* **B356**: 229–283.

Riba-Hernandez, P., Stoner, K. E., and Osorio, D., 2004, Effect of polymorphic colour vision for fruit detection in the spider monkey *Ateles geoffroyi*, and its implications for the maintenance of polymorphic colour vision in platyrrhine monkeys, *J. Exp. Biol.* **207**: 2465–2470.

Smith, A. C., Buchanan-Smith, H. M., Surridge, A. K., Osorio, D., and Mundy, N. I., 2003, The effect of colour vision status on the detection and selection of fruits by tamarins (*Saguinus* spp.), *J. Exp. Biol.* **206**: 3159–3165.

Smith, A. C., Buchanan-Smith, H. M., Surridge, A. K., and Mundy, N. I., 2003b, The effect of colour vision status and sex on leadership within wild mixed-species groups of saddleback (*Saguinus fuscicollis*) and moustached tamarins (*S. mystax*), *Am. J. Primatol.* **61**: 145–157.

Stoner, K. E., Riba-Hernández, P. & Lucas, P., 2005, Comparative use of color vision for frugivory by sympatric species of platyrrhines, *Am. J. Primatol.* **67**: 399–409.

Sumner, P., and Mollon, J. D., 2000, Catarrhine photopigments are optimized for detecting targets against a foliage background, *J. Exp. Biol.* **203**: 1987–2000.

Surridge, A. K., Osorio, D., and Mundy, N. I., 2003, Evolution and selection of trichromatic vision in primates, *Trends Ecol. Evol.* **51**: 198–205.

Sussman, R. W., and Raven, P. H., 1978, Pollination by lemurs and marsupials: An archaic coevolutionary system, *Science* **200**: 731–736.

Sussman, R. W., 1991, Primate origins and the evolution of angiosperms, *Am. J. Primatol.* **23**: 209–223.

Tan, Y., and Li, W.-H., 1999, Trichromatic vision in prosimians, *Nature* **402**: 36.

Tavaré, S., Marshall, C. R., Will, O., Soligo, C., and Martin, R. D., 2002, Using the fossil record to estimate the age of the last common ancestor of extant primates, *Nature* **416**: 726–729.

Vorobyev, M., and Osorio, D., 1998, Receptor noise as a determinant of colour thresholds, *Proc. R. Soc. Lond.* **B265**: 351–358.

Wing, S. L., Hickey, L. J., and Swisher, C. C., 1993, Implications of an exceptional fossil flora for Late Cretaceous vegetation, *Nature* **363**: 342–344.

Yamashita, N., Stoner, K. E., Riba-Hernandez, P., Dominy, N. J., & Lucas, P. W., 2005, Light levels used during feeding by primate species with different color vision phenotypes, *Behav. Ecol. Sociobiol.* **58**: 628–629.

TAXON INDEX